Membranes: Specialized Functions in Plants

Membranes: Specialized Functions in Plants

M. Smallwood
The Plant Laboratory, Department of Biology, University of York, PO Box 373, York YO1 5YW, UK

J.P. Knox
Centre for Plant Biochemisty and Biotechnology, University of Leeds, Leeds LS2 9JT, UK

D.J. Bowles
The Plant Laboratory, Department of Biology, University of York, PO Box 373, York YO1 5YW, UK

βIOS
SCIENTIFIC
PUBLISHERS

© BIOS Scientific Publishers Limited, 1996

First published 1996

A CIP catalogue record for this book is available from the British Library.

ISBN 1 85996 200 9

BIOS Scientific Publishers Ltd
9 Newtec Place, Magdalen Road, Oxford OX4 1RE, UK
Tel. +44 (0) 1865 726286. Fax +44 (0) 1865 246823
World Wide Web home page: http://www.Bookshop.co.uk/BIOS/

DISTRIBUTORS

Australia and New Zealand
  DA Information Services
  648 Whitehorse Road, Mitcham
  Victoria 3132

India
  Viva Books Private Limited
  4325/3 Ansari Road
  Daryaganj
  New Delhi 110002

Singapore and South East Asia
  Toppan Company (S) PTE Ltd
  38 Liu Fang Road, Jurong
  Singapore 2262

USA and Canada
  BIOS Scientific Publishers
  PO Box 605, Herndon
  VA 20172-0605

Typeset by Chandos Electronic Publishing, Stanton Harcourt, UK.
Printed by Biddles Ltd, Guildford, UK.

Cover illustration: The inset on the front cover shows the cellular distribution of the endoplasmic reticulum (ER) in two daughter cells just after division of a tobacco cotyledon cell. The ER is visualized by the specific immunological detection of the ER resident protein calreticulin in combination with green fluorescent secondary antibodies. The DNA of the two daughter nuclei is visualized in orange by a subsequent propidium iodide staining. Note the accumulation of ER in the fragmoplast and the nuclear envelopes. The inset on the back cover shows a cell prior to cell division in the anaphase/metaphase, stained with the same procedure. The ER (green) is associated with the spindle figure on either side of the chromosomes (orange). The staining patterns suggest that, during cell division in plant cells, the ER moves along the microtubules in the opposite direction to the chromosomes and accumulates where the new cell plate is being formed. Original photographs courtesy of Dr Anna-Stina Höglund, Uppsala Genetic Center, Sweden.

Contents

Membrane Lipid Metabolism

Regulation of Membrane Permeability

Membrane Compartments within the Cell

Plasmodesmata

Modified Membranes in Symbiotic Relationships

Contributors

Askerlund, P. Department of Plant Biochemistry, Lund University, P.O. Box 117, S-221 00 Lund, Sweden

Baker, A. Centre for Plant Biochemistry and Biotechnology, University of Leeds, Leeds LS2 9JT, UK

Blanton, R.L. Department of Biological Sciences, Texas Tech University, Box 43131, Lubbock, TX 79409-3131, USA

Block, M.A. Laboratoire de Physiologie Cellulaire Végétale, URA CNRS 567, Département de Biologie Moléculaire et Structurale, Centre d'Etudes Nucléaires de Grenoble et Université Joseph Fourier, F-38054 Grenoble Cedex 9, France

Boutry, M. Unité de Biochimie Physiologique, Université Catholique de Louvain, Place Croix du Sud 2-20, B-1348 Louvain-la-Neuve, Belgium

Brewin, N.J. Department of Genetics, John Innes Centre, Colney, Norwich NR4 7UH, UK

Callow, J.A. School of Biological Sciences, The University of Birmingham, Birmingham B15 2TT, UK

Cooke, T.J. Department of Plant Biology, University of Maryland, College Park, MD 20742, USA

Crawford, J.W. Scottish Crop Research Institute, Invergowrie, Dundee DD2 5DA, UK

Denecke, J. Department of Biology, University of York, Heslington, York YO1 5DD, UK

Ding, B. Section of Plant Biology, University of California, Davis, California CA 95616, USA. Present address: Department of Botany, Oklahoma State University, Stillwater, OK 74078, USA

Douce, R. Laboratoire de Physiologie Cellulaire Végétale, URA CNRS 567, Département de Biologie Moléculaire et Structurale, Centre d'Etudes Nucléaires de Grenoble et Université Joseph Fourier, F-38054 Grenoble Cedex 9, France

Dove, S.K. Department of Cell Biology, John Innes Centre, Colney, Norwich NR4 7UH, UK

Drøbak, B.K. Department of Cell Biology, John Innes Centre, Colney, Norwich NR4 7UH, UK

Faye, L. LTI-CNRS URA 203, Université de Rouen, 76821 Mont Saint Aignan, France

Findlay, G.P. School of Biological Sciences, Flinders University, GPO Box 2100, Adelaide, South Australia, Australia 5001

Frommer, W.B. Institut für Genbiologische Forschung, Ihnestrasse 63, D-14195 Berlin, Germany

Garrill, A. School of Biological Sciences, Flinders University, GPO Box 2100, Adelaide, South Australia, Australia 5001. Present address: Molecular Genetics Group, Department of Zoology, University of Western Ontario, London, Ontario, Canada

Gray, J. Department of Plant Sciences, University of Cambridge, Downing Street, Cambridge CB2 3EA, UK

Green, J.R. School of Biological Sciences, The University of Birmingham, Birmingham B15 2TT, UK

Grosbois, M. Laboratoire de Physiologie Cellulaire et Moléculaire des Plantes, Université Pierre et Marie Curie, CNRS URA 1180 Tour 53 - Case 154, 4 place Jussieu, F-75252 Paris Cedex 05 France

Guerbette, F. Laboratoire de Physiologie Cellulaire et Moléculaire des Plantes, Université Pierre et Marie Curie, CNRS URA 1180 Tour 53 - Case 154, 4 place Jussieu, F-75252 Paris Cedex 05 France

Hahn, M.G. The University of Georgia, Complex Carbohydrate Research Center and Department of Botany, 220 Riverbend Road, Athens, GA 30602-4712, USA

Haigler, C.H. Department of Biological Sciences, Texas Tech University, Box 43131, Lubbock, TX 79409-3131, USA

Harmon, A.C. Department of Botany and the Graduate Program in Plant Molecular and Cellular Biology, University of Florida, P.O. Box 118526, Gainesville, FL 32611-8526, USA

Harms, K. Institut für Genbiologische Forschung, Ihnestrasse 63, D-14195 Berlin, Germany

Hawes, C. School of Biological and Molecular Sciences, Oxford Brookes University, Gipsy Lane Campus, Oxford OX3 0BP, UK

Hepler, P.K. Department of Biology, University of Massachusetts, Amherst, MA 01003-5180, USA

Hirner, B. Institut für Genbiologische Forschung, Ihnestrasse 63, D-14195 Berlin, Germany

Jauh, G.Y. Department of Botany and Plant Sciences, University of California, Riverside, CA 92521-0124, USA

Jolliot, A. Laboratoire de Physiologie Cellulaire et Moléculaire des Plantes, Université Pierre et Marie Curie, CNRS URA 1180, Tour 53 - Case 154, 4 place Jussieu, F-75252 Paris Cedex 05, France

Joyard, J. Laboratoire de Physiologie Cellulaire Végétale, URA CNRS 567, Département de Biologie Moléculaire et Structurale, Centre d'Etudes Nucléaires de Grenoble et Université Joseph Fourier, F-38054 Grenoble Cedex 9, France

Kader, J.-C. Laboratoire de Physiologie Cellulaire et Moléculaire des Plantes, Université Pierre et Marie Curie, CNRS URA 1180 Tour 53 - Case 154, 4 place Jussieu, F-75252 Paris Cedex 05, France

Kauss, H. FB Biologie der Universität, Postfach 3049, D-67653 Kaiserslautern, Germany

Knox, J.P. Centre for Plant Biochemistry and Biotechnology, University of Leeds, Leeds LS2 9JT, UK

Kühn, C. Institut für Genbiologische Forschung, Ihnestrasse 63, D-14195 Berlin, Germany

Lee, J.-Y. Department of Botany and the Graduate Program in Plant Molecular and Cellular Biology, University of Florida, P.O. Box 118526, Gainesville, FL 32611-8526, USA

Lichtscheidl, I.K. Institute of Plant Physiology, University of Vienna, Althanstrasse 14, A-1091 Vienna, Austria

Lloyd, C.W. Department of Cell Biology, John Innes Centre, Colney, Norwich NR4 7UH, UK

Loake, G.J. Institute of Cell and Molecuar Biology, Daniel Rutherford Building, King's Buildings', Mayfield Road, Edinburgh EH9 3JH, UK

Lord, E.M. Department of Botany and Plant Sciences, University of California, Riverside, CA 92521-0124, USA

Lucas, W.J. Section of Plant Biology, University of California, Davis, CA 95616, USA

Marechal, E. Laboratoire de Physiologie Cellulaire Végétale, URA CNRS 567, Département de Biologie Moléculaire et Structurale, Centre d'Etudes Nucléaires de Grenoble et Université Joseph Fourier, F-38054 Grenoble Cedex 9, France

Martin, T. Institut für Genbiologische Forschung, Ihnestrasse 63, D-14195 Berlin, Germany

Michelet, B. Unité de Biochimie Physiologique, Université Catholique de Louvain, Place Croix du Sud 2-20, B-1348 Louvain-la-Neuve, Belgium. Present address: Laboratory of Plant Molecular Biology, The Rockefeller University, 1230 York Avenue, New York, NY 10021-6399, USA

Oparka, K.J. Scottish Crop Research Institute, Invergowrie, Dundee DD2 5DA, UK

Oursel, A. Laboratoire de Physiologie Cellulaire et Moléculaire des Plantes, Université Pierre et Marie Curie, CNRS URA 1180 Tour 53 - Case 154, 4 place Jussieu, F-75252 Paris Cedex 05, France

Palme, K. Max Plancke Institut für Züchtungsforschung, Carl-von-Linné-Weg 10, Koln, D-50829 Köln, Germany

Prior, D.A.M. Scottish Crop Research Institute, Invergowrie, Dundee DD2 5DA, UK

Raikhel, N.V. Department of Energy-Plant Research Laboratory, Michigan State University, East Lansing, MI 48823, USA

Reynolds, T.L. Monsanto Company Mail Zone AA3I, 700 Chesterfield Parkway North, St. Louis, MO 63198, USA

Riesmeier, J.W. Institut für Genbiologische Forschung, Ihnestrasse 63, D-14195 Berlin, Germany

Robinson, C. Department of Biological Sciences, University of Warwick, Coventry CV4 7AL, UK

Satiat-Jeunemaitre, B. CNRS, UPR 40, Institute des Sciences Végétales, 91198 Gif-sur-Yvette Cedex, France

Schulz, B. Institut für Genbiologische Forschung, Ihnestrasse 63, Berlin, D-14195 Germany. Present address: Institut für Botanik, Universität zu Köln, Gyrhofstrasse 15, D-50931 Köln, Germany

Shao, J. Department of Botany, University of Florida, P.O. Box 118526, Gainesville, FL 32611-8526, USA

Smith, S.E. Department of Soil Science and The Cooperative Research Centre for Soil and Land Mangement, Waite Campus, The University of Adelaide, South Australia, Australia 5064

Smith, F.A. Department of Botany, The University of Adelaide, South Australia, Australia 5005

Sommarin, M. Department of Plant Biochemistry, Lund University, P.O. Box 117, S-221 00 Lund, Sweden

Staiger, C.J. Department of Cell Biology, John Innes Centre, Colney, Norwich NR4 7UH, UK. Present address: Department of Biological Sciences, Purdue University, West Lafayette, IN 47907-1392, USA

Tester, M. Department of Plant Sciences, University of Cambridge, Downing Street, Cambridge CB2 3EA, UK

Thain, J.F. School of Biological Sciences, University of East Anglia, Norwich NR4 7TJ, UK

Tilney, L.G. Department of Botany, University of Pennsylvania, Philadelphia, PA 19104, USA

Tilney, M.S. Department of Botany, University of Pennsylvania, Philadelphia, PA 19104, USA

Tyerman, S.D. School of Biological Sciences, Flinders University, GPO Box 2100, Adelaide, South Australia, Australia 5001

Walling, L.L. Department of Botany and Plant Sciences, University of California, Riverside, CA 92521-0124, USA

Wildon, D.C. School of Biological Sciences, University of East Anglia, Norwich NR4 7TJ, UK

Willmitzer, L. Institut für Genbiologische Forschung, Ihnestrasse 63, D-14195 Berlin, Germany

Yoo, B.-C. Department of Botany and the Graduate Program in Plant Molecular Cellular Biology, University of Florida, P.O. Box 118526, Gainesville, FL 32611-8526, USA

Abbreviations

ABA	abscisic acid
ABP	actin-binding protein
ACC	aminocyclopropane-1-carboxylic acid
ACP	acyl carrier protein
ADF	actin depolymerizing protein
AGP	arabinogalactan-protein
ALP	alkaline phosphatase
AO	active oxygen
ARF	ADP-ribosylation factor
BFA	brefeldin A
BiP	binding protein
BL	barley lectin
BrHMA	bromohexamethylene amiloride
CaM	calmodulin
CAM	cell adhesion molecule
CaMV	cauliflower mosaic virus
CCV	clathrin-coated vesicle
CDPK	calmodulin-like domain kinase/calcium-dependent protein kinase
CHAPS	3-[(3-cholamidopropyl)dimethylammonio]-1-propanesulphonate
CHS	chalcone synthase
CLSM	confocal laser scanning microscopy
COP	coat protein
CPMV	cowpea mosaic virus
CPY	carboxypeptidase Y
CTPP	carboxyl-terminal propeptide
DAPI	4,6-diamidino-2-phenylindole
DCB	2, 6-dichlorobenzonitrile
DCCD	N, N'-dicyclohexylcarbodiimide
DCMU	3-(3,4-dichlorophenyl)-1,1-dimethylurea
DEAE	diethylaminoethyl
DEPC	diethylpyrocarbonate
DES	diethylstilbesterol
DG	diacylglycerol
DGDG	digalactosyldiacylglycerol
DHFR	dihydrofolate reductase
DMSO	dimethylsulphoxide
DP	degree of polymerization
DPAP	dipeptidyl aminopeptidase
DTT	dithiothreitol
ECM	extracellular matrix
EDTA	ethylenediamine tetra-acetic acid
EF	extracytoplasmic fracture
EGF	epidermal growth factor
EGTA	ethylene glycol-bis(β-aminoethyl) ether N, N, N', N'-tetraacetic acid
EHM	extrahaustorial membrane
ELISA	enzyme-linked immunosorbent assay
EM	electron microscopy
ER	endoplasmic reticulum

ES	extracytoplasmic surface
EST	expressed sequence tag
F-actin	filamentous actin
FBPase	fructose-1, 6-bisphosphatase
FC	fusicoccin
FCR	ferricyanide reductase
FITC	fluorescein isothiocyanate
FN	fibronectin
FNR	ferredoxin:NADPH oxidoreductase
FPLC	fast protein liquid chromatography
βGlcY	β-glucosyl Yariv reagent
G-actin	globular actin
GAP	GTPase-activating protein
GC	generative cell
GC-MS	gas chromatography–mass spectroscopy
gMDH	glyoxysomal malate dehydrogenase
GNRP	guanine nucleotide release protein
GnT-I	N-acetylglucosaminlytransferase
GO	glycolate oxidase
GS I	β-glucan synthase I
GS II	β-glucan synthase II
GT	glucose transporter
GUS	β-glucuronidase
HAO	haemagglutinin
HATS	high-affinity transport system
HC	haustaurial complex
HPLC	high-pressure liquid chromatography
HPR	hydroxypyruvate reductase
HR	hypersensitive response
HRGP	hydroxyproline-rich glycoprotein
Hsc	heat shock cognate
Hsp	heat shock protein
HST	host-specific toxin
Hyp	hydroxyproline
IAA	indole-3-acetic acid
IC	intermediate compartment
ICL	isocitrate lyase
IF	intermediate filament
IMP	intramembrane particle
Ins(1,4,5)P$_3$	inositol(1, 4, 5)triphosphate
IPA	inositol phosphate
JA	jasmonic acid
JIP	jasmonate-induced protein
LATS	low-affinity transport system
LHCP	light-harvesting chlorophyll-binding protein
LIMP	lysosomal integral membrane protein
LOX	lipoxygenase
LPS	lipopolysaccharide
LRR	leucine-rich repeat
LTP	lipid transfer protein
LYCH	Lucifer Yellow CH
M6P	mannose 6-phosphate
mAb	monoclonal antibody
MAP	microtubule-associated protein
Me-JA	methyl jasmonate

MF	microfilament
MFS	major facilitator superfamily
MGDG	male germ unit
MI	major intrinsic protein
MP	movement protein
MS	mechanosensitive
MT	microtubule
NAA	naphthyl-1-acetic acid
NBD-PC	nitrobenzoxadiazol-phosphatidylcholine
NEM	N-ethylmaleimide
NMR	nuclear magnetic resonance
NPTII	neomycin phosphotransferase II
NSF	N-ethylmaleimide-sensitive fusion protein
nsL-TP	non-specific lipid transfer protein
NTPP	amino-terminal propeptide
12-oxo-PDA	12-oxo-phytodienoic acid
P_o	channel open probability
PAL	phenylalanine ammonia-lyase
PAM	periarbuscular membrane
PAT	phosphoinothrian acetyl transferase
PBM	peribacteroid membrane
PC	phosphatidylcholine
PC-TP	proteins specifically transferring PC
PCMBS	p-chloromercuribenzyl sulphonic acid
PCR	partially coated reticulum
p.d.	potential difference
PDGF	platelet-derived growth factor
PE	phosphatidylethanolamine
PEP	phosphoenolpyruvate
PF	protoplasmic fracture
PG	phosphatidylglycerol
3-PGA	3-phosphoglycerate
PGA	polygalacturonic acid
PGA/RG-I	polygalacturonan/rhamnogalacturonan I
PHA	phytohaemagglutinin
PhyA	active phytochrome
PI	phosphatidylinositol
PI 3-kinase	phosphatidylinositol 3-kinase
PI-TP	proteins specifically transferring PI
PIC	phosphoinositidase C
PKC	protein kinase C
PLC	phospholipase C
PLD	phospholipase D
PM	plasma membrane
PPB	preprophase band
PP_i	pyrophosphate
PP_iase	pyrophosphatase
PPM	peripheral plasma membrane
PrA	proteinase A
PRK1	receptor-like kinase
PS	protoplasmic surface
PtdIns	phosphatidylinositol
PtdIns(4)P	phosphatidylinositol(4)phosphate
PtdIns(4,5)P_2	phosphatidylinositol(4,5)bisphosphate
PTS	class-1 peroxisomal targeting signal

PUFA	polyunsaturated fatty acid
rER	rough endoplasmic reticulum
Rubisco	ribulose 1,5-bisphosphate carboxylase/oxygenase
SA	salicylic acid
SAM	substrate adhesion molecule
SAR	systemic acquired resistance
SBP	sucrose binding protein
SDS–PAGE	sodium dodecyl sulphate–polyacrylamide gel electrophoresis
SEL	size exclusion limit
SGAT	serine:glyoxylate amino transferase
SLG	S-locus glycoprotein
SNAP	soluble NSF attachment protein
SPP	stromal processing peptidase
SQDG	sulphoquinovosyldiacylglycerol
SRK	S-locus receptor kinase
SRP	signal recognition particle
SSU	small subunit
SUT	sucrose transporter
t-SNARE	target membrane SNAP receptor
TC	terminal complex
TEM	transmission electron microscope
Tes	N-[tris(hydroxymethyl)methyl]-2-aminomethanesulfonic acid
TGN	trans-Golgi network
TIP	tonoplast intrinsic protein
TLC	thin-layer chromatography
TMV	tobacco mosaic virus
TNF	tumour necrosis factor
TPP	thylakoidal processing peptidase
TPT	triose-phosphate translocator
triose-P	triose phosphate
UV	ultraviolet
v-SNARE	vesicle-specific SNAP receptor
VA	vesicular–arbuscular
VAMP	vesicle-associated membrane protein
VN	vitronectin
VSP	vegetative storage protein
XET	xyloglucan endotransglycosylase
XG	xyloglucan

Preface

When we were invited to design a book on 'membranes' we decided to highlight features of membrane biology that are specific to plants. We chose this strategy because we wanted to develop a reference book and survey of membrane biology *in relation* to plant biology. First and foremost we are plant biologists and the contents of this book reflect our enthusiasm and excitement about plants fuelled by the ever-increasing understanding of plant form and function at the molecular level. In a sense, we used the opportunity as editors to invite chapters on all of the areas of plant membranes that we find fascinating. This may have produced a rather personal viewpoint of the range of topics included, but we hope that there is also some intellectual foundation to our choice!

The Contents List begins with a section entitled "Membranes and the Cell Surface". The plasma membrane surrounds the protoplast and represent the junction and gateway linking it to the outer world of the symplast. It is increasingly recognized that the cell wall, plasma membrane and cytoskeleton interact to relay information across the hydrophobic barrier. The chapters in this section discuss the specialized functions of the plant plasma membrane within its cellular context: as a key site of synthesis of cell wall polysaccharides, as the major site of molecular recognition events, and as the location for the start of signalling cascades that link external stimuli to end-effect.

The second section, "Membrane Lipid Metabolism", addresses specialized lipids of plant cell membranes. It covers the structural role(s) of the galactolipids and sterols, the putative role of inositol-containing lipids in cell signalling, and the lipid-transfer proteins that are now known to exist in both the symplast and the apoplast and play important roles trafficking lipids around and between the two hydrophilic compartments.

In Section 3, "Regulation of Membrane Permeability", the authors address ion transport at the cell surface and vacuolar membranes. Movement of ions and the regulation of ion transport processes are fundamental to most, if not all, plant cell functions. We are now beginning to understand these events at the molecular level, with the aid of patch-clamp analyses, and this in turn is demonstrating the complexity of regulation both at the plasma membrane and, significantly for plants, at the tonoplast. Given the importance of turgor in the biology of plants, it is perhaps not surprising that the number of mechanosensitive channels identified is increasing. Ion fluxes have been linked causally as the stimulus–response coupling mechanism for some cellular responses. Interestingly, electrical activity is also now recognized as being of importance in long-distance systemic signalling although we are still some way from understanding the underlying molecular events leading to the variation potentials and action potentials that can be recorded.

The fourth section, "Membrane Compartments within the Cell", addresses the endomembranes, and again highlights specialized functions in plants. Authors discuss the problem of protein targeting to the vacuole, how plant peroxisomes are formed and the action of the endoplasmic reticulum, Golgi apparatus and secretory vesicles in assembling, sorting and transporting newly synthesized proteins, glycoproteins and polysaccharides. The cortical endoplasmic reticulum is also a feature of all plant cells, yet is rarely highlighted in the literature on membranes. In particular, the chloroplast is discussed with respect to events that control its division and thereby its biogenesis as well as events that lead to the import of nuclear-encoded gene products into the organelle.

Plasmodesmata, the topic of the fifth section, may seem an odd choice for a book on membranes, given that they are generally considered in texts discussing cell–cell junctions and cell signalling. Nevertheless, we chose to include plasmodesmata in this volume because of their significance to the understanding of the role of membranes in long-range communication throughout the plant. Given that membrane continuity through plasmodesmatal junctions is proven, fluidity of the lipid bilayer and mobility of protein and lipid components through plasmodesmata must be considered with the same emphasis as solute transport through the pore.

The book ends with a section on changes to membranes when plant cells are in intimate association with another organism; whether beneficial, in the context of symbiotic relationships, or detrimental, as when the cells of the plant are interacting with pathogens and pests. As yet, molecular information on these changes is limited. However, we wished to highlight this topic both for the light shed on fundamental processes in membrane biology and for the insights to be gained that may lead to novel crop protection applications.

We hope you as the reader will enjoy the selection of topics, and the very excellent reviews provided by the authors who have kindly contributed their time and enthusiasm to this book.

Margaret Smallwood
Paul Knox
Dianna Bowles

Chapter 1

Interactions between the plasma membrane and the cytoskeleton in plants

Clive W. Lloyd, Bjorn K. Drøbak, Stephen K. Dove and Christopher J. Staiger

1. Introduction

The cortical cytoskeleton is dynamic

A review of the interactions between the cytoskeleton and the plasma membrane is largely restricted to the period outside mitosis and cytokinesis when cortical microtubules (MTs) construct first the interphase MT array and then the preprophase band (PPB) before depolymerizing around prometaphase. Such interactions between the plasma membrane and cortical MTs are mainly involved with shape control. Interphase MTs are usually wound transversely around elongating cells and encourage the formation of corresponding patterns of cellulose deposition (see Giddings and Staehelin, 1991, for review) and it is still a major objective to determine exactly how the microtubular pattern is transmitted across the plasma membrane. Currently, the indirect model is favoured, in which cellulose synthase complexes embedded in the lipid bilayer, propelled across the face of the membrane by the act of polymerizing a linear crystallite, are confined to channels between adjacent membrane-linked MTs. However, our knowledge of transmembrane linkages in plants is not sufficiently reliable for us to be able to rule out a direct link between synthase and microtubule. Either way, the patterns that MTs are capable of forming and the links between them and the plasma membrane have a direct bearing on the texture of the wall. Not only do MTs influence the extracellular matrix, but, in turn, the wall seems to confer stability and order on the MTs. These cortical MTs are not permanently fixed on the plasma membrane, since they adopt different configurations according to the presence of particular growth factors (see Shibaoka, 1994, for review). Ethylene, for example, can cause MTs to align parallel to the long axis of a cell to inhibit cell elongation, whereas gibberellic acid is capable of stimulating elongation by increasing the transverse alignment of MTs. One of the surprising findings of recent microinjection studies has been that cortical MTs are highly dynamic, and this has now become an important part of discussions of cortical behaviour. In cells microinjected with rhodamine–tubulin, the cortical MT array labels to steady state within about 20 min, indicating a rapid turnover of subunits (Wasteneys *et al.*, 1993). After photobleaching an area of the labelled cortical array, the fluorescence recovers rapidly (Yuan *et al.,* 1994), and quantification of fluorescence recovery rates reveals interphase and preprophase band MTs of *Tradescantia* stamen hair cells to be several times more dynamic than the MTs of some fibroblasts (Hush *et al.*, 1994). One explanation for this behaviour is that the plant MTs are displaying dynamic instability (Mitchison and Kirschner, 1984); this is the process whereby growing and shrinking MTs coexist and stochastically convert between the two phases. Dynamic microtubules would therefore seem to account for changes or shifts in MT organization, but the discovery of motor microtubule-associated proteins (MAPs) (see below) in plants could mean that MTs might also slide into new positions. The reorientation of cortical MTs, from transverse to longitudinal, has been observed in living pea epidermal cells microinjected with rhodamine-tubulin (Yuan *et al.*, 1994). This is probably a wound-induced effect, but the reverse reorientation, from longitudinal to transverse, has been observed in microinjected cells treated with gibberellic acid (Lloyd *et al.*, 1996). In such cells, MTs in one alignment grow whilst MTs in the pre-existing alignment shrink, and so apart from the dynamic behaviour itself it becomes important to consider what causes the differential stability of sets of MTs. Some MTs may be linked to the plasma membrane to a greater or lesser extent than others, and some may be more extensively cross-linked to neighbouring MTs than others. In addition, three-dimensional reconstruction of the cortical array of pea epidermal cells microinjected with rhodamine–tubulin shows that MTs at the outer epidermal wall can be more steeply pitched than MTs along the adjoining radial walls of the same cell (Yuan *et al.*, 1995). This suggests that MT reorientation could occur largely (or at least initially) at the outer tangential cell cortex, which indicates that there is regional differentiation of the cytoskeleton, and it may be no coincidence that it is the environmentally sensitive outer surface which behaves differently.

During G2 the cortical MTs begin to vacate the ends of cells and to concentrate in an increasingly tight preprophase band of MTs which anticipates where the cell plate will join the mother wall. Not only do MTs therefore change their position and/or attachment to the plasma membrane whilst shifting the direction of cell expansion during interphase, but they do so in G2 as they form at the perimeter of the future division plane. Hush *et al.* (1994) found that the half-time for recovery of fluorescence after photobleaching is similar for both interphase and PPB microtubules in *Tradescantia* stamen hair cells, so the reorganization does not at first sight appear to be due to major changes in MT dynamics.

Actin filaments are also a component of the cortical cytoskeleton, and they remain in the cytoplasm of mitotic cells when cortical MTs are absent. They occur in close proximity to interphase MTs, but also extend deeper into the cytoplasm and, within transvacuolar strands, support cytoplasmic streaming. The way in which this pervasive actin network interacts with MTs and plasma membrane is unclear, but, like the MTs, it is probably susceptible to various agonists that impinge upon the plasma membrane. It is at this interface between the plasma membrane and the cytoskeleton that external signals are likely to be perceived and transduced into effects on cell shape and division.

2. Microtubule/plasma membrane interactions

The wall influences MT stability and alignment

Seen against the new background of a highly dynamic cytoskeleton, previous studies on the attachment of MTs to the plasma membrane become especially important and such attachments are likely to be the target of modulatory signalling chains. By attaching protoplasts to poly-L-lysine-coated substrates and then bursting them in various (usually hypotonic) buffers, the cytoplasmic face of the adherent membrane disc is exposed, revealing cortical elements, particularly of the cytoskeleton. This method of preparing 'footprints' offers a direct functional assay of the link between microtubules and the plasma membrane. Marchant (1978) burst *Mougeotia* protoplasts on polylysine-treated electron microscope (EM) grids to show that cortical MTs were still attached. These could be removed if the protoplasts were treated at low temperatures or with colchicine before bursting. In this system, protoplasts appear to need some time to recover since MTs become attached to the plasma membrane over a 3-h period. Marchant (1979) exploited this in order to add inhibitors of either MT polymerization or cellulose biosynthesis. He found that in the presence of anti-MT drugs the protoplasts had no membrane-associated MTs yet still regenerated a wall. However, the fibrils were presumably random, since the cell remained spherical. Conversely, the inhibitor of cellulose biosynthesis, coumarin, inhibited the formation of a proper wall and the protoplast remained spherical even though cortical MTs were present and could be assayed as being membrane-associated by bursting. Marchant concluded that MTs had no direct cytoskeletal role in determining asymmetrical shape.

In 1980, Lloyd *et al.* used varying lengths of cellulase treatments to prepare cells of varying shapes for anti-tubulin staining. Elongated carrot suspension cells had transversely arranged cortical MTs, as did wider, ovoid cells in the process of conversion to full protoplasts. Spherical protoplasts, on the other hand, contained disorganized, non-parallel MTs. At all stages the cortical MTs conformed to the shape of the cell during this transformation, indicating two important points. First, microtubules must be capable of some dynamic behaviour that allows them to move relative to one another as the widening cell becomes a sphere. Despite a threefold increase in the circumference, MTs remain attached to the plasma membrane with a tenacity which prevents their subsequent removal from footprints with 10% Triton-X-100. However, bearing in mind the dynamic MT behaviour revealed by microinjection studies, we should consider whether the stable membrane-associated MTs are a subset, or whether they might have become converted to a more stable form by the act of attaching the protoplast to the poly-L-lysine-coated coverslip. Secondly, MTs do not by themselves appear to maintain the asymmetrical shape of the cell. This was confirmed by Melan (1990), who showed that protoplasts prepared from pea epicotyls pretreated with the MT-stabilizing drug, taxol, retained parallel MT arrays, unlike control protoplasts in which MTs were random. However, because both groups of protoplasts were spherical, it would appear that ordered MT arrays are not by themselves sufficient to retain asymmetrical cell shape. These experiments suggest that some components in the wall either directly or indirectly stabilize the MTs and allow them to retain parallelism. These lateral and radial interactions between molecules of the cortical region are evidently crucial to morphology, and suggest a form of transmembrane signalling not too dissimilar to the types of interactions that occur between cytoplasmic actin, plasma membrane integrins and extracellular matrix in animal cells, except that here the major emphasis is on cortical MTs. In tip-growing filamentous cells of *Adiantum* (Kagawa *et al.,* 1992), there is a ring of MTs and actin microfilaments at the apical hemisphere. Upon plasmolysis, plasma membrane is detached from the wall except where this circular array remains to attach the membrane to the wall. Colchicine does not affect the microfilaments, but it removes MTs as well as the subapical junction between plasma membrane and wall, indicating that transmembranous links between wall and MTs, rather than wall and actin microfilaments, represent the stabilizing axis.

Further indications that components of the wall impinge upon the stability of cortical MTs have been provided by Shibaoka's group (Akashi and Shibaoka, 1991; Akashi *et al.,* 1990). These workers showed that MTs were not isolated on the plasma membrane ghosts if tobacco BY-2 protoplasts were pretreated with proteases. It was inferred from this that the extracellular portion of a transmembrane glycoprotein, involved in MT stability or membrane attachment, had been proteolytically cleaved. In untreated protoplasts the MTs became increasingly stable to cold when extensin or poly-L-lysine was added to protoplasts, but not if they had been previously treated with proteases. When the wall was allowed to regenerate, cortical MTs became parallel to one another, but this did not occur if the protoplasts were treated with trypsin. One possible interpretation is that cortical MTs are immobilized by being attached to transmembrane proteins, the external portions of which are cross-linked by extensin.

Taken together, the results of these various experiments suggest that a detergent-insoluble link spans the plasma membrane to unite MTs and some component of the wall.

Calcium and phosphorylation as modulators of MT behaviour

Transmembrane signalling mechanisms are likely to be involved in agonist-induced switches in MT alignment, and there is an urgent need to identify the components of the signalling chains. One possibility (Hepler *et al.,* 1990) is that the endoplasmic reticulum, which is closely associated with the plasma membrane in the form of a tubular network, is involved in releasing Ca^{2+} in response to extracellular agonists. Several lines of experimentation emphasize a link between Ca^{2+} and MT stability. For instance, using footprints derived from *Mougeotia* protoplasts, Kakimoto and Shibaoka (1986) found that about half the MTs were stable to 10 µM Ca^{2+} over a 10-min period. Neither calcium nor Triton-X-100 alone depolymerized MTs, but when given successively these treatments caused removal of MTs from the footprints. Regular cross-bridges were seen between some neighbouring MTs. Lancelle *et al.* (1986) have also observed fine cross-links between adjacent cortical MTs in *Tradescantia* stamen hairs, and they have reported bridges between the plasma membrane and the MTs. It is such radial links with the plasma membrane that presumably resist detergent extraction of exposed MTs, whilst the lateral bridges between MTs probably have different solubilization properties. Throwing some light on the MT–membrane linkage, Sonobe and Takahashi (1994) found that MTs could reassemble on membrane ghosts from which pre-existing MTs had been stripped with Ca^{2+}, but that they could not reassemble on membranes stripped by Na_2CO_3 or KCl, the implication being that the latter treatments removed a MAP responsible for binding MTs to the plasma membrane.

Other experiments on footprints derived from maize suspension cells confirm that Triton-X-100 increases the sensitivity of MTs to Ca^{2+} (Wang *et al.,* 1989), although there was little effect on sensitivity to Mg^{2+}. Using tobacco BY-2 ghosts, Sonobe (1990) found that incubation with 3 mM ATP and 5 mM $MgCl_2$ caused partial disappearance of the MTs, whereas 1 mM $CaCl_2$ had no effect. Likewise, a 30–50% ammonium sulphate fraction of BY-2 cells had no effect on cortical MTs but, when added in the presence of Mg/ATP, the extract completely removed the MTs. Occasionally, PPB ghosts were seen on the coverslips, but they appeared to be no more stable to the depolymerizing conditions than interphase MTs. The addition of taxol prevented the disappearance of MTs. In contrast, MTs of the spindle and phragmoplast were stable under these conditions. The activity in the ammonium sulphate fraction remains uncharacterized, but the experimental results hint at the possibility that phosphorylation of cortical MT protein may be associated with stability. In a subsequent study, Katsuta and Shibaoka (1992) prepared footprints from synchronized BY-2 tobacco cells at a time when the PPB is known to be present. ATP, and not its non-hydrolysable analogue AMP-PNP, caused the disappearance of cortical MTs, but this effect could be inhibited by the kinase inhibitor staurosporine. It is possible that a kinase is responsible for controlling the association of cortical MTs with the membrane.

One kinase recently shown to be present in the PPB is the 34-kDa protein coded by the cell division cycle gene, *cdc2*. Mineyuki *et al.* (1991) found that an antibody to the PSTAIR motif found in p34[cdc2] kinase (and also in other kinases) stained the PPB. With a more specific antibody raised against the 17 amino acids of the C-terminus of the functional homologue of maize cdc2, Colasanti *et al.* (1993) have also located this kinase to the PPB. The interphase MT band in the developing stomatal complex of maize is also a dense accumulation of MTs, but since, unlike the PPB, this does not stain with the anti-cdc2 antibody, it would appear that the cortical location of the kinase is under cell-cycle control. The substrate for the kinase is unknown, but clarification of this issue will have an enormous impact on our understanding of the cortical division site, and of the formation and disappearance of the PPB.

In a series of experiments that linked Ca^{2+} sensitivity and MAPs, Cyr (1991) demonstrated that 1 mM free Ca^{2+} in the protoplast lysis buffer destabilized the cortical MTs of carrot cells. However, if the calcium chelator EGTA was included in the lysis buffer and then removed, subsequent addition of 1 mM free Ca^{2+} was no longer able to destabilize the MTs. On the assumption that EGTA might be removing Ca/calmodulin, Cyr added exogenous calmodulin together with free Ca^{2+} to show that Ca-sensitivity could be restored to carrot cortical MTs prepared in the presence of EGTA. Since calmodulin without Ca^{2+} did not destabilize MTs, it was concluded that Ca^{2+}/calmodulin is the active complex which interacts with MTs. However, washing MTs with salt caused a decrease in the ability of the complex to destabilize MTs, implying that Ca^{2+}/calmodulin exerts its effect via MT-associated protein(s). Later, Fisher and Cyr (1993) used a monoclonal antibody to plant calmodulin to show that this protein decorates cortical MTs in detergent-extracted tobacco and carrot footprints. This insoluble form of calmodulin is likely to be important for the behaviour of cortical MTs, and again emphasizes the potential role of Ca^{2+} in cytoskeletal signalling events.

Microtubule-associated proteins (MAPs)

Using a calmodulin affinity column, Cyr (1991) found that two proteins (76 and 129 kDa) previously reported to bind to MTs (Cyr and Palevitz, 1989) were amongst those from a carrot extract that bound to the column. The intriguing possibility is that calcium-mediated processes affect the stability of cortical MTs, perhaps by influencing the MAPs.

There are likely to be two broad categories of MAPs at the cortex: those that link MTs to the plasma membrane, and those that link MTs to each other. It is beyond the scope of this review to discuss the latter in detail but they, too, can be divided into two categories: the structural, filamentous MAPs which stabilize MTs and perhaps cross-link them, and the motor MAPs which allow MTs and organelles to move relative to one another. Several research groups have identified proteins that would appear to belong to the structural class (Cyr and Palevitz, 1989; Jiang and Sonobe, 1993; Maekawa *et al.,* 1990; Schellenbaum *et al.,* 1993). Antibodies to the 65-kDa tobacco MAP decorate all four MT arrays throughout the cell cycle, so there is no indication, as yet, of cycle-specific modifications that might be expected to influence the interaction of cortical MTs with each other and with the plasma membrane (Jiang and Sonobe, 1993). As for motor MAPs that support MT-based motility, there is biochemical (Cai *et al.,* 1993), genetic (Mitsui *et al.,* 1993) and immunolgical (Tiezzi *et al.,* 1992) evidence for the existence of kinesins or kinesin-like proteins in plants. Moscatelli *et al.* (1995) reported two 400-kDa proteins from tobacco pollen tubes which resemble dynein biochemically, and Asada and Shibaoka (1994) have identified 120- and 125-kDa nucleotide-sensitive proteins from phragmoplasts which support MT gliding. The role of such motor proteins is not yet established, but could involve the movement of particles, and of organelles such as the endoplasmic reticulum, relative to the MTs, as well as the movement of MTs relative to one another.

Tubulin itself has been reported to behave as an integral membrane protein, raising the possibility that MT polymerization could occur directly upon the plasma membrane (Laporte *et al.,* 1993). Antibodies to γ-tubulin have shown that this member of the tubulin superfamily, found in other eukaryotes at the minus ends of MTs (i.e. at MT organizing centres), can be detected at the cell periphery (Liu *et al.,* 1993). There is no indication that this is a membrane protein, but this finding, too, raises the possibility that MT nucleation may occur at the cortex.

Other components of the cortical cytoskeleton

In focusing on the cytoskeleton, we should not lose sight of the fact that other organelles also segregate with membrane disks torn out of protoplasts, indicating a wider range of interactions. Traas (1984) critical-point-dried footprints to show that the MTs were surrounded by a meshwork of filaments of diameter 5–10 μm. These 'microfilaments' were seen to be associated with coated pits and vesicles. In addition, Hepler *et al.* (1990) have shown that polygonal arrays of tubular endoplasmic reticulum segregate with plasma membrane disks. The identity of the cortical meshwork observed by some techniques is unclear. Fine filaments that parallel the MTs are likely to be actin filaments seen in rhodamine–phalloidin studies (see Section 3). One candidate for other, more anastomosing material has been intermediate filaments (IFs). Several antibodies to IFs have been found to label cortical MTs (for review, see Shaw *et al.,* 1991), but it is not clear that the conditions which preserve the antigenicity of these proteins also allow them to be seen in a filamentous form. In carrot footprints, anti-IF antibodies decorate electron-dense material which co-distributes with the MTs; such material could be a collapsed IF network, but this remains unproven (Goodbody *et al.,* 1989). In one intriguing communication, Su *et al.* (1990) used colloidal gold-tagged anti-keratin antibodies to label an extensive 'keratin intermediate filament-like system' in detergent-extracted maize and tobacco footprints. However, following similar extraction conditions, the broadly cross-reactive antibody against intermediate filaments, ME101, was found in immunogold experiments to decorate cortical MTs but not the fine, filamentous, cortical meshwork (Fairbairn *et al.,* 1994). The idea that this meshwork is composed of intermediate filaments is an attractive one, but despite the fact that several lines of evidence favour the existence of IF antigens as MT-decorating proteins and as non-MT-associated fibrillar bundles, there is little evidence that these antigens constitute a separate cortical meshwork. It would seem wise, therefore, not to name these antigens 'intermediate filaments' (even in fibrillar bundles where they actually do form insoluble filaments) until the full extent of sequence similarity with animal IF proteins is known.

3. Microfilaments

Microfilament/plasma membrane interactions

In contrast to animal cells, plant cells appear to have less specialized, or at least less obvious, cortical actin cytoskeletons. Several observations might help to explain these differences. Plant cells are not motile and therefore do not require an active cortical cytoskeleton to send out membrane protrusions or to fix attachments to the substrate. A rigid cell wall has partially or wholly supplanted the function of the cytoskeleton in maintaining shape. Furthermore, the highly developed cortical microtubule system observed in most plant cells can probably accomplish many of the functions attributed to the actin membrane skeleton in animal cells. Nevertheless, actin does coexist with the cortical microtubule array, and the normal functions of both arrays are certain to be highly interdependent. In the next section, we shall review what is currently known about the actin membrane skeleton.

Using phallotoxins or antibodies which bind to filamentous actin, fluorescence microscopy studies have expanded our knowledge of the distribution of actin arrays in the plant cell cortex, but only recently have we obtained hints as to their potential function. Most interphase plant cells contain a complex network of actin filaments and actin bundles which surround the nucleus, extend throughout the cytoplasm and divide into a fine meshwork in the cell periphery, and which in places have been seen to parallel the cortical MTs (Lloyd, 1989; Seagull, 1989; Traas, 1990). Improved methods for actin preservation and staining at the ultrastructural level have led to better images of this cortical actin network: small bundles of actin are found coaligned with cortical MTs (Ding *et al.,* 1991a; Lancelle and Hepler, 1991; Lancelle *et al.,* 1987) and associated with cortical endoplasmic reticulum (Lichtscheidl *et al.,* 1990; see also Chapter 22 by Lichtscheidl and Hepler). In addition to actin bundles, individual microfilaments, whose identity as actin has not been confirmed by antibody labelling, are coaligned with the cortical microtubules (Ding *et al.,* 1991a; Lancelle and Hepler, 1991) and also appear to be closely associated with the plasma membrane (Ding *et al.,* 1991a).

Unlike the microtubule cytoskeleton, a network of actin microfilaments remains in the cortical cytoplasm throughout cell division (for reviews see Liu and Palevitz, 1992; Lloyd, 1989). Soon after mitosis has been initiated, the cortex is reorganized and shifts in the distribution of microtubules and microfilaments occur. The preprophase band, a marker for the division plane and the site where the new cell plate fuses with the mother wall, consists of both microtubules and microfilaments (Kakimoto and Shibaoka, 1987; Palevitz, 1987; Traas *et al.*, 1987). At the ultrastructural level, predominantly single microfilaments are observed either alone or in close association with microtubules (Ding *et al.*, 1991b). A functional interaction between the microtubules and microfilaments of the PPB is indicated by the fact that cytochalasin D treatment leads to the formation of broader PPBs (Mineyuki and Palevitz, 1990). During karyokinesis, microfilaments connect the spindle poles to the cortex and may be involved in maintaining or determining spindle placement (Lloyd and Traas, 1988; Traas *et al.*, 1987). The edges of the cytokinetic phragmoplast are also connected to the cortex by actin filaments in vacuolate cells (Goodbody and Lloyd, 1990; Lloyd and Traas, 1988), leading to a model in which these microfilaments guide the leading edge of the phragmoplast to the division site (Lloyd and Traas, 1988; Lloyd, 1989). Other cell types, however, appear to exclude cortical actin from the division plane following PPB disappearance, and in this context F-actin is a negative marker for the division plane (Cleary *et al.*, 1992; Cleary, 1995; Liu and Palevitz, 1992). Further studies are required to resolve whether or not F-actin guides the phragmoplast to the division site.

In addition to structural data, further evidence for a connection between actin filaments and the plasma membrane comes from experimental studies on protoplasts and living walled cells. Typically, vacuolated cells have strands of cytoplasm, exhibiting vigorous cytoplasmic streaming, which criss-cross the vacuole; these suspend the nucleus in a central position in preparation for division and, in a few cases, have been shown to contain actin filaments (e.g. Kakimoto and Shibaoka, 1987; Traas *et al.*, 1987; Zhang *et al.*, 1993). Hahne and Hoffman (1984) observed that callus protoplasts of *Hibiscus* had small invaginations on their surface at sites corresponding to cytoplasmic strands. By ablating a single strand with a laser microbeam or treating the cells with cytochalasin B, the strands and the surface irregularities were eliminated. Invaginations reappeared concurrent with the resumption of streaming and re-formation of strands. Similar tension generation has been noted in the transvacuolar strands which suspend the premitotic nucleus in *Nautilocalyx* epidermal cells; laser ablation of a strand caused the severed ends to recoil and the position of the nucleus to be altered (Goodbody *et al.*, 1991). More recently, Schindler and co-workers (Grabski *et al.*, 1994) have developed a quantitative assay for measuring the elasticity of plant cytoplasm. The technique, called a cell optical displacement assay, uses a laser beam to capture and displace transvacuolar or cortical cytoplasmic strands and to move these for a defined distance through the cell. Using a range of external stimuli, it was demonstrated that cytoskeletal tension could be regulated by pH and calcium as well as by fatty acids and diacylglycerol. These experiments demonstrate that the cytoskeleton, through force generation, is able to develop tension which is somehow translated to the plasma membrane. The nature of the attachment of cytoskeletal components to the plasma membrane is unknown, but several models can be proposed. The connection could be direct, via linkages between actin and integral or peripheral membrane proteins, or indirect, via connections to macromolecular complexes or other organelles. Some compelling structural evidence supports an indirect attachment of actin to the plasma membrane through connections with the endoplasmic reticulum (ER) (Lichtscheidl *et al.*, 1990) and with microtubules (Lancelle *et al.*, 1987; Lancelle and Hepler, 1991). In addition, the existence of a transmembrane linkage is suggested by the sensitivity of cytoplasmic streaming and actin organization to proteases applied to walled mesophyll cells (Masuda *et al.*, 1991). By plasmolysing onion epidermal cells, Pont-Lezica *et al.*

(1993) found that the plasma membrane remained attached to the wall by drawn-out Hechtian strands which still exhibited streaming and exerted tension on the membrane. These authors argued, from indirect evidence, that the strands represent transmembrane points of attachment of the ER/cytoskeleton complex to foci in the wall where hydroxyproline-rich glycoproteins cluster. Since the strands can attach to the outer epidermal wall, which has no neighbouring structures, such attachment points cannot be solely intercellular plasmodesmata.

In animal cells (Luna and Hitt, 1992), microfilaments (and intermediate filaments) are focused on common intercellular junctions which bond cells together at 'spot welds'. These mutual *adherens*-type junctions are important for tissue stability, but one-sided hemi-junctions with a substratum allow the cell to exert tension and to 'crawl'. The transmembrane linkage, from specific components of the extracellular matrix to transmembrane receptors to the cytoskeleton, is complex and contains a respectable list of components. The biology of non-motile plant cells would seem to argue against the kind of arrangement where the cytoskeleton is organized in response to the matrix, but occasional reports indicate that cytoplasmic strands coalign between adjoining cells. By staining *Tradescantia* leaf epidermal cells with rhodamine–phalloidin, Goodbody and Lloyd (1990) observed that actin 'cables in' neighbouring cells focused at a common point on the separating wall, giving the impression that the actin passes from cell to cell. One possible explanation for this intriguing staining pattern was suggested by a microscopic examination of plasmodesmata (White *et al.*, 1994) – the plasma membrane-lined cytoplasmic channels that interconnect most plant cells. Actin was detected at the ultrastructural level by immunogold staining, as well as by confocal microscopy following rhodamine–phalloidin labelling. The significance of this interaction is revealed by the striking morphological change observed when microfilament organization is perturbed by cytochalasin: the majority of channels are open in treated root cells, whereas most of the channels in untreated roots are closed. Information about the presence of transmembrane receptors that could link the adhesion molecules in the extracellular matrix to the cytoskeleton is limited (Sanders *et al.*, 1992, also see Chapter 2 by Lord and Walling), but tentative identification of integrin-like molecules has been achieved by immunoblotting extracts of *Fucus*, maize (Quatrano *et al.*, 1991) and soybean (Schindler *et al.*, 1989). The integrin class of transmembrane receptors binds to a specific amino acid sequence (RGD) found in many of the extracellular substrate adhesion molecules. Synthetic RGD peptides can inhibit adhesion and locomotion by competitively binding to integrins on the surface of animal cells. When applied to plant cells, the RGD peptide inhibits cell–cell adhesion between salt-adapted tobacco protoplasts (Zhu *et al.*, 1993). However, the role of the cytoskeleton in this process has not been examined. Cytoplasmic streaming in *Chara* is sensitive to gravitational fields and differs in two directions. Wayne and co-workers (1992) found that the RGD peptide inhibited graviperception and differential streaming rates. Convincing evidence, in the form of sequence data, biochemical purification and *in-vitro* activity, will be required before we can state unequivocally that integrin receptors exist in plants.

Control of actin organization: actin-binding proteins (ABPs)

Several classes of actin-associated protein control the morphology of actin arrays and regulate the interaction of actin with membranous structures (Hartwig and Kwiatkowski, 1991; Luna and Condeelis, 1990). These include capping proteins, such as cap 32/34, that bind to the ends of microfilaments. Severing proteins, such as gelsolin, villin and severin, probably represent a specialized subset of capping proteins that cut filaments in the presence of micromolar calcium and remain bound to the newly created free end. Proteins that bind to the sides of actin filaments are able to bundle microfilaments, or to cross-link them into a network, depending on the binding stoichiometry and the length of the molecule. Motor molecules, or myosins, cause the sliding of actin filaments relative to each other, attach filaments to membranes, or translocate small, membrane-bound cargo along filaments. Another class of actin-binding proteins associates with membranes. In platelets, for example, ABP-280 or filamin dimers link actin bound at its N-terminus to membranes via a C-terminal site. Another protein, ponticulin, is a 17-kDa integral membrane glycoprotein with the capacity to bind to the side of microfilaments (Wuestehube and Luna, 1987), and it is required for actin nucleation activity on isolated *Dictyostelium* plasma membranes (Luna *et al.*, 1990). No such proteins have been identified in plants, although we can point to myosin, several monomer-binding proteins and spectrin as potential linkers of the actin cytoskeleton to the plasma membrane.

In erythrocytes, actin (as part of a two-dimensional actin–spectrin lattice) is anchored by peripheral membrane proteins to integral membrane proteins, such as glycophorin. This network is thought to stabilize the cell membrane and to help to determine cell shape. Immunological evidence is beginning to accumulate for spectrin-related proteins in plants. The first such study was that by Michaud *et al.* (1991), who used an antibody to human erythrocyte β spectrin to immunoblot a 220-kDa homologue in tomato. By immunofluorescence, the antigen appeared to be located on the plasma membrane. De Ruijter and Emons (1993) have also detected a 220-kDa polypeptide in plants with anti-spectrin antibodies, although the labelling of a more heavily cross-reactive 85-kDa band suggested the presence of breakdown products. Furthermore, Faraday and Spanswick (1993) used an anti-spectrin antibody to detect a 230-kDa polypeptide in high-purity plasma membranes prepared from rice roots by aqueous-polymer two-phase partitioning. Spectrin homologues therefore appear to exist in plants, although nothing is known about potential partners in a submembrane skeleton, or about the function of such complexes in cells whose shape appears to be firmly controlled by interactions between the wall and the microtubular cytoskeleton. However, the erythrocyte model for the involvement of spectrin in shape control need not necessarily be extended to other cell types. It is possible that the primary function of the membrane skeleton in plant cells is to organize membrane signalling and transport complexes, and that it has little or nothing to do with the determination of cell shape.

Myosins are actin-stimulated mechano-chemical enzymes, capable of supporting movement, which were originally classified as conventional or unconventional on the basis of the potential of the heavy chains to form dimers (for reviews see Cheney and Mooseker, 1992; Titus, 1993). All myosins share a well conserved head domain containing both the actin- and the ATP-binding domains. Different myosin molecules are most readily distinguished by the structure and sequence of the tail domain, which probably specifies the 'cargo' that the molecule can carry and thereby defines its function. The tails of unconventional myosins vary considerably: several contain domains rich in basic amino acids and are capable of binding to acidic phospholipids (Adams and Pollard, 1989; Hayden *et al.,* 1990). Moreover, immunofluorescence studies demonstrate that these myosins associate with cellular membranes, including the plasma membrane (Baines and Korn, 1990). Such myosins are therefore implicated in dynamic interactions between actin and the plasma membrane in motile cells and in specialized structures such as microvilli. Most organisms, including plants, contain multigene families of myosins. The genes for six *Arabidopsis thaliana* myosin-like molecules have been cloned and sequenced (Kinkema and Schiefelbein, 1992; Kinkema *et al.,* 1994; Knight and Kendrick-Jones, 1993), and these fall into two major classes that could reflect distinct cytoplasmic functions. The motility and membrane-binding properties of the cloned plant myosins have not yet been determined. However, it is likely that these myosins will play a fundamental role not only in cytoplasmic streaming and vesicle movement, but also in cytoskeletal tension generation and dynamic interactions between microtubules and microfilaments and between microfilaments and the endoplasmic reticulum.

Actin-binding proteins belonging to the monomer-binding class have recently been identified in plants (Staiger *et al.,* 1993; Valenta *et al.,* 1991). Profilin is the best characterized of these molecules. It binds to G-actin with relatively high affinity (dissociation constants are in the low micromolar range), and it is known to have interactions with three types of ligand: linear tracts of proline residues, actin and polyphosphoinositide lipids. Simple models predict that sequestering proteins control the amount of polymerized actin by regulating the size of the free monomer pool. Profilin was first discovered in plants, quite fortuitously, as a pollen allergen. Valenta and co-workers screened a birch pollen cDNA library with sera from patients who were allergic to tree pollen, and among the clones characterized was one with significant sequence homology to profilins (Valenta *et al.,* 1991). Subsequently, additional profilin genes have been characterized from maize (Staiger *et al.,* 1993), timothy grass (Valenta *et al.,* 1993, 1994), tobacco (Mittermann *et al.,* 1995) and bean (Vidali *et al.,* 1995). At the amino acid sequence level, plant profilins are only distantly related to other eukaryotic profilins, as only a few residues are completely conserved in all profilins (Staiger *et al.,* 1993). Restricted comparisons, especially with lower eukaryotic profilins, reveal functionally conserved motifs. The greatest similarity occurs at the N- and C-termini of the profilin molecule. This is significant with respect to conservation of function, as a set of highly conserved hydrophobic residues contributed by the two terminal α-helices and the underlying β-sheet is known to be necessary for binding to poly-L-proline (Metzler *et al.,* 1994). Recombinant and native plant profilins all possess the ability to bind to poly-L-proline, and recombinant birch profilin is able to find a proline-rich focal adhesion protein, VASP, identified in vertebrate cells (Reinhard *et al.,* 1995). This ability to bind proline-rich sequences may reflect the evolutionary conservation of a mechanism for spatial regulation of profilin within cells. Its significance for plant cell behaviour has yet to be determined. Both recombinant and native profilin can bind to actin, demonstrating that the plant molecule is a functional homologue (Giehl *et al.,* 1994; Ruhlandt *et al.,* 1994; Valenta *et al.,* 1993). Profilin–actin complexes can be purified from pollen and other plant tissues (Staiger *et al.,* 1993; Valenta *et al.,* 1993), and the complex can be reconstituted *in vitro* with purified and bacterially expressed constituents (Valenta *et al.,* 1993). Moreover, profilin can act as a sequestering protein in living cells. When microinjected into *Tradescantia* stamen hair cells, bacterially expressed profilin has a rapid and dose-dependent effect on cytoplasmic streaming and on the organization of cytoplasmic strands; cytoplasmic strands break and streaming stops within 2 to 3 min (Staiger *et al.,* 1994). Fluorescein–phalloidin staining confirms that this is due to depolymerization of F-actin. These effects are mimicked by purified pig brain profilin and another G-actin-binding protein, DNAase I. This suggests that the integrity of cytoplasmic strands depends on a dynamic equilibrium between G- and F-actin. Profilin has the potential, therefore, to be a powerful regulator of actin organization in living plant cells.

Recent data from the animal literature indicate that straightforward sequestration of G-actin by profilin may be too simple a model. It is argued that the amount of profilin and its affinity for actin are unlikely to account for the size of the unpolymerized actin pool known to exist in certain cells (Machesky *et al.,* 1990). Additional data demonstrate that the association of profilin with actin has a unique consequence: it stimulates the exchange of adenine nucleotides on G-actin (Goldschmidt-Clermont *et al.,* 1992; Mockrin and Korn, 1980). This means that profilin will catalyse the conversion of ADP–actin to ATP–actin, a form which polymerizes more rapidly and has slower dissociation kinetics, thus leading to net assembly of new polymer (Goldschmidt-Clermont *et al.,* 1992; Pollard and Cooper, 1986). Moreover, by a careful evaluation of the effects of profilin on actin polymerization, it was determined that profilin could actually lead to net polymer assembly under certain conditions (Pantaloni and Carlier, 1993). In pollen grains, profilin represents 2–6% of total protein and has a concentration of 20–40 µM (Staiger, unpublished data). In theory, therefore, profilin appears to be abundant enough to sequester a large proportion of the soluble actin in particular plant cells. Bacterially expressed profilin, unlike pig brain profilin, is unable to stimulate nucleotide exchange on purified rabbit skeletal G-actin (Staiger, unpublished data). This may simply represent a decreased affinity of plant profilin for animal actin, as suggested by blot-overlay assays (Valenta *et al.,* 1993), or it may indicate the loss of this activity through evolutionary divergence of the profilin molecule. Further work will extend our understanding of the function of profilin in plants, but the evidence to date points to a role in controlling microfilament dynamics through binding to G-actin. Members of another family of actin-binding proteins, known as actin-depolymerizing factors (ADF), have now been identified in plants. Several research groups have identified cDNAs from lily, *Brassica* (Kim *et al.,* 1993), maize (Rozycka *et al.,* 1995) and *Arabidopsis* (Staiger and Ashworth, 1995; Ashworth and Staiger, unpublished data). The predicted polypeptides share 40–60% amino acid sequence similarity with ADF members from lower eukaryotes and vertebrates. Several of the non-plant ADFs, including cofilin and ADF from vertebrates, and actophorin from *Acanthamoeba,* have been well characterized biochemically. They share the ability to cause actin depolymerization at steady state through the combined ability to bind monomeric actin and to sever actin filaments, thereby creating many new ends for disassembly. Cofilin is also able to inhibit nucleotide exchange on actin, a property that would serve to slow down filament re-formation by keeping the actin subunits in the ADP-bound form. ADF interactions with actin are apparently insensitive to the concentration of actin, but can be regulated by pH, phosphorylation and polyphosphoinositides. From preliminary studies it is clear that recombinant plant ADF (*Arabidopsis* ADF1) retains the ability to sever and destroy actin filaments *in vitro* and in living cells (Staiger and Ashworth, 1995; Yuan, Ashworth and Staiger, unpublished data; Hussey *et al.,* personal communication). Future studies will concentrate on the involvement of members of the ADF family during processes that involve large-scale destruction and reorganization of the actin cytoskeleton, such as mitosis and pollen germination.

Control of actin organization: polyphosphoinositides

Unable to move as entire organisms away from adverse conditions, plants are especially adept at reorganizing their cellular contents and cytoskeleton in response to extracellular signals (for reviews see Seagull, 1989; Staiger and Lloyd, 1991). In particular, the actin cytoskeleton changes in response to light, during photo-orientation of chloroplasts (Kadota and Wada, 1992), in response to wounding (Goodbody and Lloyd, 1990) and in response to moisture (e.g. during pollen germination; Heslop-Harrison and Heslop-Harrison, 1989). In animals and lower eukaryotes, actin organization is modulated by actin-associated proteins, many of which are themselves regulated by intracellular second messengers, including calcium and poly-phosphoinositides.

An exciting development in this field is the finding that molecules of the phosphoinositide signalling pathway associate with the cytoskeleton. Both the nuclear and the cytoskeletal compartments of mammalian cells have been shown to possess distinct phosphoinositide systems with potential physiological roles that may differ from those of the plasma membrane system. Little is known about the exact functions of either of the non-plasma membrane phosphoinositide systems, but in a study of epidermal growth factor (EGF) stimulation of animal tissue culture cells it was found that the activity of the cytoskeletal phosphoinositide kinases was dramatically upregulated after EGF stimulation, and that the kinase was associated with the actin cytoskeleton (Payrastre *et al.,* 1991). An investigation into the possible roles of the nuclear phosphoinositide system has been carried out by Divecha *et al.* (1991), who demonstrated that cell stimulation of 3T3 cells by insulin-like growth factor 1 led to a rapid and transient decrease in nuclear polyphosphoinositides, and a concurrent increase in diacylglycerol. Evidence for the apparent existence of distinct nuclear and cytoskeletal phosphoinositide systems in plant cells has emerged from several studies. Hendrix *et al.* (1989) reported that both PtdIns(4)P and PtdIns(4,5)P2 can be synthesized by nuclei isolated from carrot protoplasts. These findings are doubly significant. First, they demonstrate that the biosynthetic enzymes are associated with this organelle, and secondly, they indicate the presence of the necessary lipid substrates. However, these experiments do not clarify whether the kinases and their substrates are genuinely associated with the nucleus itself or with other components that segregate with the nucleus, such as the cytoskeleton. Using cytoskeletons prepared from carrot protoplasts, Xu *et al.* (1992) have shown that a significant proportion of the total cellular PtdIns 4-kinase activity is associated with the detergent-resistant cytoskeleton as well as with the nucleus which segregates with this fraction. In a subsequent study, it was also shown that PtdIns 3-kinase activity is associated with detergent-resistant cytoskeletons (Dove *et al.*, 1994). Tan and Boss (1992) demonstrated that the cytoskeletal PtdIns 4-kinase was associated with actin isolated from a plasma membrane fraction which also contained some DG (diacylglycerol)-kinase and PtdIns(4)P 5-kinase activity. The association of the latter two enzymes with the cytoskeleton may depend on preparative conditions. One possibility highlighted by these two studies is that PtdIns 4-kinase may exist in two forms, membrane and cytoskeletal, although it is also possible that the 'membrane' form of the enzyme is actually due to actin attached to the plasma membrane.

The physiological roles of the cytoskeletal (and nuclear) PtdIns 3- and 4-kinases are far from clear, and more information about the precise cellular localization of this enzyme would aid the making of educated guesses. An intriguing possibility is that the PtdIns 4-kinase product (i.e. PtdIns(4)P) itself has regulatory functions. As this possibility has been discussed in detail elsewhere (Drøbak, 1993; Pike, 1992), only one example will be given here. One of the few documented functions of PtdIns(4)P in plants is its apparent ability to modulate the *in vitro* activity of a vanadate-sensitive plasma membrane ATPase (Memon and Boss, 1990). As parts of the peripheral cytoskeleton and the plasma membrane can be in close proximity in plants, the possibility exists that the cytoskeletal PtdIns 4-kinase might directly phosphorylate plasma membrane-associated lipids. If this were the case, one could envisage a possible link between cytosolic events, activation of the cytoskeletal (or plasma membrane) PtdIns 4-kinase(s), and transmembrane ion fluxes.

Many ABPs are concentrated near the plasma membrane, where they can perceive signals and transduce them into reorganization of the cytoskeleton. The recent identification of actin- and phospholipid-binding proteins in plants represents a major step forward in understanding the link between signalling and cell structure. Actin is not the only cellular constituent known to associate with profilin, which also has a strong affinity for the polyphosphoinositides, PtdIns(4)P and PtdIns(4,5)P2. In fact, the affinity of profilin for PtdIns(4,5)P2 is at least 10-fold greater than its affinity for actin (Goldschmidt-Clermont *et al.*, 1990), so that the binding of PtdIns(4,5)P2 to profilin causes dissociation of profilactin complexes (Lassing and Lindberg, 1985). At intracellular PtdIns(4,5)P2 and actin concentrations, the prediction is that most profilin will be bound to phospholipids (Goldschmidt-Clermont *et al.*, 1990; Goldschmidt-Clermont and Janmey, 1991). In support of this model, profilin can be localized near the plasma membrane (Hartwig *et al.*, 1989). Interestingly, the yeast profilin gene has been identified as a suppressor of a mutation in a gene (*CAP*) involved in the ras/adenyl-cyclase signalling pathway (Vjotek *et al.*, 1991). It has recently been demonstrated that plant profilin can bind to polyphosphoinositides (Drøbak *et al.*, 1994) and it is clearly possible that such interactions may be involved in the control of cytoskeletal dynamics (see also Chapter 12 by Drøbak).

Other actin-associated proteins such as cofilin, villin, gelsolin, cap 32/34 and cap 100 could also play a key role in translating transmembrane signals into cytoskeletal reorganization (for reviews see Hartwig and Kwiatkowski, 1991; Luna and Hitt, 1992). The interaction of each of these proteins with actin is modulated by binding to polyphosphoinositides. Cofilin-like genes have been identified in higher plants (Kim *et al.*, 1993), and the putative lipid and actin-binding domains are well conserved. Experiments similar to those described for plant profilin should enable us to determine whether these genes encode functional homologues.

Control of actin organization: calcium-dependent protein kinase (CDPK)

Another signalling molecule with the potential to regulate the actin cytoskeleton is calcium-dependent protein kinase (for review see McCurdy and Williamson, 1991; see also Chapter 8 by Harmon *et al.*). CDPKs are encoded by a small family of genes, and each contains a catalytic kinase domain and a calmodulin-like Ca^{2+}-binding domain (Harper *et al.*, 1991). At least one isoform of CDPK interacts with the cytoskeleton and has been found to co-localize with F-actin arrays in onion root cells, *Tradescantia* pollen tubes (Putnam-Evans *et al.*, 1989) and *Chara* internodal cells (McCurdy and Harmon, 1992a). It is unlikely that CDPK binds to actin directly, because the purified components do not interact *in vitro*. It is more probable that CDPK interacts with an actin-associated protein which could also be its substrate *in vivo*. Several lines of evidence implicate myosin as a substrate. Purified CDPK can phosphorylate gizzard myosin light chain (Putnam-Evans *et al.*, 1990). In *Chara,* CDPK is found on actin bundles, on the surface of small organelles and on other motile membrane components, a pattern virtually indistinguishable from myosin heavy chain (McCurdy and Harmon, 1992a). Streaming in plants is inhibited by increased levels of cellular Ca^{2+}, and probably involves phosphorylation (Tominaga *et al.*, 1987). CDPK phosphorylates a putative myosin light chain homologue from *Chara* in a Ca^{2+}-dependent manner (McCurdy and Harmon, 1992b). This evidence strongly suggests that CDPK plays a key role in the Ca^{2+}-dependent inhibition of cytoplasmic streaming in plant cells (for review see Williamson, 1993). Not only does CDPK associate with the cytoskeleton, but an oat CDPK has been shown to bind to purified plasma membranes (Schaller *et al.*, 1992). While it is not yet clear whether isoforms have different intracellular locations and functions, a membrane-associated CDPK is in a prime position to transduce extracellular signals rapidly and effectively into cessation of streaming or reorganization of the cortical cytoskeleton.

GTP-binding proteins and cytoskeletal dynamics

More than 40 members of the ras superfamily of low-molecular-weight (*c.* 20–30 kDa) GTP-binding proteins have been identified in eukaryotes. This family of proteins can be divided into four main subfamilies, based on amino acid sequence similarities and functional characteristics: ras-like, rho-like, rab(ypt)-like and ARF-like. The biological activity of low-molecular-weight GTP-binding proteins is regulated by cycling between GTP- and GDP-bound forms ('GTP/GDP switches'). In their GTP-bound forms these proteins are activated, and hydrolysis of GTP to GDP by a GTPase activity intrinsic to the proteins then leads to inactivation. Members of the rho family have attracted particular attention since several members appear to play very specific roles both in modulation of cytoskeletal dynamics and in transducing signals between the plasma membrane and the cytoskeleton (for review see Hall *et al.*, 1993).

A large number of low-molecular-weight GTP-binding proteins have been shown to be present in plant cells, but little is known about their function. The initial identification of these proteins was based either on their ability to bind GTP (e.g. Drøbak *et al.*, 1988), or on cross-reactions with antibodies raised to known mammalian GTP-binding proteins (Clarkson *et al.*, 1991). More recently, a number of cDNAs and genes which encode low-molecular-weight proteins in plants have been identified. Of the plant genes sequenced so far, most show highest homology to the rab(ypt) family which, in mammalian cells and yeast, appear to be involved in intracellular transport and secretion (for overview see Terryn *et al.*, 1993). However, a cDNA was recently cloned from pea which encodes a rho equivalent (Yang and Watson, 1993). The deduced pea rho protein (Rho1Ps) contains 197 amino acids and shows 45–64% sequence similarity to other members of the eukaryotic rho family. The function of this protein is not known, but in view of the ability of rho proteins to promote actin-associated plasma membrane processes in fibroblasts (Ridley and Hall, 1992; Ridley *et al.*, 1992), it would be prudent to look for effects on plasma membrane/cytoskeletal interactions in plants.

In addition to their direct (or indirect) effects on cytoskeletal activity, low-molecular-weight GTP-binding proteins may also become directly associated with the cytoskeletal matrix. One example is a member of the ras-like subfamily, rap2B, which has been shown to associate directly with the cytoskeleton during platelet aggregation, in contrast to the situation in resting platelets, where the rap2B protein is completely soluble in Triton-X-100 (Torti *et al.*, 1993). It has been suggested that the translocation and interaction of rap2B with the cytoskeleton causes a functional link between the cytoskeleton and trafficking of secretory granules in platelets. Recent evidence (Drøbak *et al.*, unpublished data) indicates that low-molecular-weight GTP-binding proteins can also become specifically associated with the plant cytoskeleton, and we are trying to identify them.

4. Prospects

In summary, it is clear that this interface between the plasma membrane and its associated cytoskeleton is an important one, for which we are beginning to see hints of controlling regulatory processes. Progress in this area will tell us more about the control of cell shape, and, although very little has been said here about plant growth regulators, it is at this interface that shape-controlling agonists such as light, gravity, hormones and herbicides are likely to exert their effects. Plant physiology cannot really be said to have absorbed the cytoskeleton into its thinking, despite the fact that microtubules have been known for over 30 years, but it seems a reasonably safe prediction that forthcoming research on the control of cytoskeleton/membrane interactions will profoundly affect the way in which we think about plant growth. Microtubules in plants are now known to be dynamic, not only in the sense that they reorientate, but also in that their rapid incorporation of labelled, microinjected tubulin indicates that they are in dynamic equilibrium with the soluble tubulin pool (Zhang *et al.*, 1990; also Hepler *et al.*, 1993; Hush *et al.*, 1994; Yuan *et al.*, 1994, 1995). Microtubules may appear to be stationary, but in reality they are dancing on the spot. Another reasonably safe prediction is that microtubule dynamics will be found to be modulated by the microtubule-associated proteins, kinases and other regulatory enzymes that are now being described, and the next stage will be to determine how such processes respond to external signals.

It is often said that the line between the protoplast and the cell wall is an arbitrary one, and several papers reviewed in this and other chapters in the present volume point to the links between the cell wall and the cytoplasm. An underlying but only loosely outlined theme of this review is that there are transmembranous links between certain wall components and filamentous proteins of the cytoskeleton. These are beginning to be studied and will eventually enhance our understanding of shape control in interphase cells, but since the cortical cytoskeleton is normally dismantled prior to the formation of the mitotic spindle, it is possible that some of the controls of MT stability will also be on the pathway for division control. Perhaps it is at the interface between plasma membrane and cytoskeleton that cell biology and plant physiology can find a common language.

References

Adams RJ, Pollard TD. (1989) Binding of myosin I to membrane lipids. *Nature* **340:** 565–568.

Aderem A. (1992) Signal transduction and the actin cytoskeleton: the roles of MARCKS and profilin. *Trends Biol. Sci.* **17:** 438–443.

Akashi T, Shibaoka H. (1991) Involvement of transmembrane proteins in the association of cortical microtubules with the plasma membrane in tobacco BY-2 cells. *J. Cell Sci.* **98:** 169–174.

Akashi T, Kawasaki S, Shibaoka H. (1990) Stabilization of cortical microtubules by the cell wall in cultured tobacco cells. *Planta* **182:** 363–369.

Asada T, Shibaoka H. (1994) Isolation of polypeptides with microtubule-translocating activity from phragmoplasts of tobacco BY-2 cells. *J. Cell Sci.* **107:** 2249–2257.

Baines EC, Korn ED. (1990) Localization of myosin IC and myosin II in *Acanthamoeba castellanii* by indirect immunofluorescence and immunogold electron microscopy. *J. Cell Biol.* **111:** 1895–1904.

Cai G, Bartalesi A, Del Casino C, Moscatelli A, Tiezzi A, Cresti M. (1993) The kinesin-immunoreactive homologue from *Nicotiana tabacum* pollen tubes: biochemical properties and subcellular localization. *Planta* **191:** 496–506.

Cheney RE, Mooseker MS. (1992) Unconventional myosins. *Curr. Opin. Cell. Biol.* **4:** 27–35.

Cleary AL. (1995) F-actin redistributions at the division site on living *Tradescantia* stomatal complexes as revealed by microinjection of rhodamine–phalloidin. *Protoplasma* **185:** 152–165.

Cleary AL, Gunning BES, Wasteneys GO, Hepler PK. (1992) Microtubule and F-actin dynamics at the division site in living *Tradescantia* stamen hair cells. *J. Cell Sci.* **103:** 977–988.

Colasanti J, Cho S-O, Wick S, Sundaresan V. (1993) Localization of the functional p34^{cdc2} homolog of maize in root tip and stomatal complex cells: association with predicted division sites. *Plant Cell* **5:** 1101–1111.

Cyr RJ. (1991) Calcium/calmodulin affects microtubule stability in lysed protoplasts. *J. Cell Sci.* **100:** 311–317.

Cyr RJ, Palevitz BA. (1989) Microtubule-binding proteins from carrot. I. Initial characterization and microtubule bundling. *Planta* **177:** 245–260.

Ding B, Turgeon R, Parthasarathy MV. (1991a) Microfilament organization and distribution in freeze substituted tobacco plant tissues. *Protoplasma* **165:** 96–105.

Ding B, Turgeon R, Parthasarathy MV. (1991b) Microfilaments in the preprophase band of freeze substituted tobacco root cells. *Protoplasma* **165:** 209–211.

Diveccha N, Banfic H, Irvine RF. (1991) The polyphosphoinositide cycle exists in the nuclei of Swiss 3T3 cells under the control of a receptor (for IGF-1) in the plasma membrane, and the stimulation of the cycle increases nuclear diacylglycerol and apparently induces translocation of protein kinase C to the nucleus. *EMBO J.* **10:** 3207–3214.

Dove SK, Lloyd CW, Drøbak BK. (1994) Identification of a phosphatidylinositol 3-kinase in plant cells: association with the cytoskeleton. *Biochem. J.* **303:** 347–350.

Drøbak BK. (1993) Plant phosphoinositide and intracellular signalling. *Plant Physiol.* **102:** 705–709.

Drøbak BK, Allan EF, Comerford JG, Roberts K, Dawson AP. (1988) Presence of guanine-nucleotide binding proteins in a plant hypocotyl microsomal fraction. *Biochem. Biophys. Res. Commun.* **150:** 899–903.

Drøbak BK, Watkins PAC, Valenta R, Dove SK, Lloyd CW, Staiger CJ. (1994) Inhibition of plant plasma membrane phosphoinositide phospholipase C by the actin-binding protein, profilin. *Plant J.* **6:** 389–400.

Fairbairn DJ, Goodbody KC, Lloyd CW. (1994) Simultaneous labelling of microtubules and fibrillar bundles in tobacco BY-2 cells by the anti-intermediate filament antibody, ME101. *Protoplasma* **182:** 160–169.

Faraday CD, Spanswick RM. (1993) Evidence for a membrane skeleton in higher plants. *FEBS Lett.* **318:** 313–316.

Fisher DF, Cyr RJ. (1993) Calcium levels affect the ability to immunolocalize calmodulin to cortical microtubules. *Plant Physiol.* **103:** 543–551.

Giddings TH, Staehelin LA. (1991) Microtubule-mediated control of microfibril deposition: a re-examination of the hypothesis. In: *The Cytoskeletal Basis of Plant Growth and Form* (Lloyd CW, ed.). London: Academic Press, pp. 85–99.

Giehl K, Valenta R, Rothkegel M, Ronsiek M, Mannherz H-G, Jockusch BM. (1994) Interaction of plant profilin with mammalian actin. *Eur. J. Biochem.* **226:** 681–689.

Goldschmidt-Clermont PJ, Janmey PA. (1991) Profilin, a weak CAP for actin and RAS. *Cell* **66:** 419–421.

Goldschmidt-Clermont PJ, Machesky LM, Baldassare JJ, Pollard TD. (1990) The actin-binding protein profilin binds to PIP2 and inhibits its hydrolysis by phospholipase C. *Science* **247:** 1575–1578.

Goldschmidt-Clermont PJ, Furman MI, Wachsstock D, Safer D, Nachmias V, Pollard TD. (1992) The control of actin nucleotide exchange by thymosin B4 and profilin. A potential regulatory mechanism for actin polymerization in cells. *Mol. Biol. Cell* **3:** 1015–1024.

Goodbody KC, Lloyd CW. (1990) Actin filaments line up across *Tradescantia* epidermal cells, anticipating wound-induced division planes. *Protoplasma* **157:** 92–101.

Goodbody KC, Venverloo, CJ, Lloyd CW. (1991) Laser microsurgery demonstrates that cytoplasmic strands anchoring the nucleus across the vacuole of premitotic plant cells are under tension. Implications for division plane alignment. *Development* **113:** 931–939.

Goodbody KC, Hargreaves AJ, Lloyd CW. (1989) On the distribution of microtubule-associated intermediate filament antigens in plant suspension cells. *J Cell Sci.* **93:** 427–438.

Grabski S, Xie XG, Holland JF, Schindler M. (1994) Lipids trigger changes in the elasticity of the cytoskeleton in plants: a cell optical displacement assay for live cell measurements. *J. Cell Biol.* **126:** 713–726.

Hahne G, Hoffman F. (1984) The effect of laser microsurgery on cytoplasmic strands and cytoplasmic streaming in isolated plant protoplasts. *Eur. J. Cell Biol.* **33:** 175–179.

Hall A, Paterson HF, Adamson P, Ridley AJ. (1993) Cellular responses regulated by rho-related small GTP-binding proteins. *Phil. Trans. R. Soc. Lond. B* **340:** 267–271.

Harper JF, Sussman MR, Schaller GE, Putnam-Evans C, Charbonneau H, Harmon AC. (1991) A calcium-dependent protein kinase with a regulatory domain similar to calmodulin. *Science* **252:** 951–954.

Hartwig JH, Kwiatkowski DJ. (1991) Actin-binding proteins. *Curr. Opin. Cell Biol.* **3:** 87–97.

Hartwig JH, Chambers KA, Hopcia KL, Kwiatkowski DJ. (1989) Association of profilin with filament-free regions of human leukocyte and platelet membranes and reversible membrane binding during platelet activation. *J. Cell Biol.* **109:** 1571–1579.

Hayden SM, Wolenski JS, Mooseker MS. (1990) Binding of brush border myosin I to phospholipid vesicles. *J. Cell Biol.* **111:** 443–451.

Hendrix KW, Qasefa HA, Boss WF. (1989) The polyphosphoinositides, phosphatidylinositol monophosphate and phosphatidylinositol bisphosphate are present in nuclei isolated from carrot protoplasts. *Protoplasma* **151:** 62–72.

Hepler PK, Palevitz BA, Lancelle SA, McCanley MM, Lichtscheidl I. (1990) Cortical endoplasmic reticulum in plants. *J. Cell Sci.* **96:** 355–373.

Hepler PK, Cleary AL, Gunning BES, Wadsworth P, Wasteneys GO, Zhang DH. (1993) Cytoskeletal dynamics in living plant cells. *Cell Biol. Int.* **17:** 127–142.

Heslop-Harrison J, Heslop-Harrison Y. (1989) Conformation and movement of the vegetative nucleus of the angiosperm pollen tube: association with the actin cytoskeleton. *J. Cell Sci.* **93:** 299–308.

Hush JH, Wadsworth P, Callaham DA, Hepler PK. (1994) Quantification of microtubule dynamics in living cells using fluorescence redistribution after photobleaching. *J. Cell Sci.* **107:** 775–784.

Jiang, C-J, Sonobe S. (1993) Identification and preliminary characterization of a 65-kDa higher plant microtubule-associated protein. *J. Cell Sci.* **105:** 891–901.

Kadota A, Wada M. (1992) Photoinduction of formation of circular structures by microfilaments on chloroplasts during intracellular orientation in protonemal cells of the fern *Adiantum capillus-veneris*. *Protoplasma* **167:** 97–107.

Kagawa T, Kadota A, Wada M. (1992) The junction between the plasma membrane and the cell wall in fern protonemal cells, as visualized after plasmolysis, and its dependence on arrays of cortical microtubules. *Protoplasma* **170:** 186–190.

Kakimoto T, Shibaoka H. (1987) Actin filaments and microtubules in the preprophase band and phragmoplast of tobacco cells. *Protoplasma* **140:** 151–156.

Kakimoto T, Shibaoka H. (1986) Calcium sensitivity of cortical microtubules in the green alga *Mougeotia*. *Plant Cell Physiol.* **27:** 91–101.

Katsuta T, Shibaoka H. (1992) Inhibition by kinase inhibitors of the development and the disappearance of the preprophase band of microtubules in tobacco BY-2 cells. *J. Cell Sci.* **103:** 397–405.

Kim S, Kim Y, An G. (1993) Molecular cloning and characterization of anther-preferential cDNA encoding a putative actin-depolymerizing factor. *Plant Mol. Biol.* **21:** 39–45.

Kinkema MD, Schiefelbein JW. (1992) A myosin from *Arabidopsis thaliana*. *Mol. Biol. Cell* **3:** 46a.

Kinkema M, Wang H, Schiefelbein J. (1994) Molecular analysis of the myosin gene family in *Arabidopsis thaliana. Plant Mol. Biol.* **26:** 1139–1153.

Knight AE, Kendrick-Jones J. (1993) A myosin-like protein from a higher plant. *J. Mol. Biol.* **231:** 154–158.

Lancelle SA, Hepler PK. (1991) Association of actin with cortical microtubules revealed by immunogold localization in *Nicotiana* pollen tubes. *Protoplasma* **165:** 167–172.

Lancelle SA, Callaham DA, Hepler PK. (1986) A method for rapid freeze fixation of plant cells. *Protoplasma* **131:** 153–165.

Lancelle SA, Cresti M, Hepler PK. (1987) Ultrastructure of the cytoskeleton in freeze-substituted pollen tubes of *Nicotiana alata. Protoplasma* **140:** 141–150.

Laporte K, Rossignol M, Traas JA. (1993) Interaction of tubulin with the plasma membrane: tubulin is present in purified plasmalemma and behaves as an integral membrane protein. *Planta* **191:** 413–416.

Lassing I, Lindberg U. (1985) Specific interaction between phosphatidylinositol 4,5-bisphosphate and profilactin. *Nature* **314:** 472–474.

Lichtscheidl IK, Lancelle SA, Hepler PK. (1990) Actin–endoplasmic reticulum complexes in *Drosera.* Their structural relationship with the plasmalemma, nucleus, and organelles in cells prepared by high pressure freezing. *Protoplasma* **155:** 116–126.

Liu B, Palevitz BA. (1992) Organization of cortical microfilaments in dividing root cells. *Cell Motil. Cytoskel.* **23:** 252–264.

Liu B, Marc J, Joshi HC, Palevitz BA. (1993) A γ-tubulin-related protein associated with the microtubule arrays of higher plants in a cell cycle-dependent manner. *J. Cell Sci.* **104:** 1217–1228.

Lloyd CW. (1989) The plant cytoskeleton. *Curr. Opin. Cell Biol.* **1:** 30–35.

Lloyd CW, Traas JA. (1988) The role of F-actin in determining the division plane of carrot suspension cells. Drug studies. *Development* **102:** 211–221.

Lloyd CW, Slabas AR, Powell AJ, Lowe SB. (1980) Microtubules, protoplasts and plant cell shape. *Planta* **147:** 500–506.

Lloyd CW, Shaw PJ, Warn RM, Yuan M. (1996) Gibberellic acid-induced reorientation of cortical microtubules in living plant cells. J. *Microscopy* **181:** (in press).

Luna EJ, Condeelis JS. (1990) Actin-associated proteins in *Dictyostelium discoideum. Dev. Genetics* **11:** 328–337.

Luna EJ, Hitt AL. (1992) Cytoskeleton–plasma membrane interactions. *Science* **258:** 955–964.

Luna EJ, Wuestehuebe LJ, Chia CP, Shariff A, Hitt AL, Ingalls HM. (1990) Ponticulin, a developmentally regulated plasma membrane glycoprotein, mediates actin binding and nucleation. *Dev. Genetics* **11:** 354–361.

McCurdy DW, Williamson RE. (1991) Actin and actin-associated proteins. In: *The Cytoskeletal Basis of Plant Growth and Form* (Lloyd CW, ed.). London: Academic Press, pp. 3–14.

McCurdy DW, Harmon AC. (1992a) Calcium-dependent protein kinase in the green alga *Chara. Planta* **188:** 54–61.

McCurdy DW, Harmon AC. (1992b) Phosphorylation of a putative myosin light chain in *Chara* by calcium-dependent protein kinase. *Protoplasma* **171:** 85–88.

Machesky LM, Goldschmidt-Clermont PJ, Pollard TD. (1990) The affinities of human platelet and *Acanthamoeba* profilin isoforms for polyphosphoinositides account for their relative abilities to inhibit phospholipase C. *Cell Regul.* **1:** 937–950.

Maekawa T, Ogihara S, Murofushi H, Nagai R. (1990) Green algal microtubule-associated protein with a molecular weight of 90 kDa which bundles microtubules. *Protoplasma* **158:** 10–18.

Marchant HM. (1978) Microtubules associated with the plasma membrane isolated from protoplasts of the green alga *Mougeotia. Exp. Cell Res.* **115:** 25–30.

Marchant HM. (1979) Microtubules, cell wall deposition and the determination of plant cell shape. *Nature* **278:** 167–168.

Masuda Y, Takagi S, Nagai R. (1991) Protease-sensitive anchoring of microfilament bundles provides tracks for cytoplasmic streaming in *Vallisneria. Protoplasma* **162:** 151–159.

Melan MA. (1990) Taxol maintains organized microtubule patterns in protoplasts which lead to the resynthesis of organized cell wall microfibrils. *Protoplasma* **153:** 169–177.

Memon AR, Boss WF. (1990) Rapid light-induced changes in phosphoinositide kinases and H$^+$-ATPase in plasma membrane of sunflower hypocotyls. *J. Biol. Chem.* **265:** 14817–14821.

Metzler WJ, Bell AJ, Ernst E, Lavoie TB, Mueller L. (1994) Identification of poly-L-proline binding site on human profilin. *J. Biol. Chem.* **269:** 4620–4625.

Michaud D, Guillet G, Rogers PA, Charest PH. (1991) Identification of a 200kD membrane-associated plant cell protein immunologically related to β-spectrin. *FEBS Lett.* **294:** 77–80.

Mineyuki Y, Palevitz BA. (1990) Relationship between preprophase band organization, F-actin and the division site in *Allium*: fluorescence and morphometric studies on cytochalasin-treated cells. *J. Cell Sci.* **97**: 283–295.

Mineyuki Y, Yamashita M, Nagahama Y. (1991). p34^{cdc2} kinase homologue in the preprophase band. *Protoplasma* **162**: 182–186.

Mitchison TM, Kirschner M. (1984) Dynamic instability of microtubule growth. *Nature* **312**: 237–242.

Mitsui H, Yamaguchi-Shinozaki K, Shinozaki K, Nishikawa N, Takahashi H. (1993) Identification of a gene family (kat) encoding kinesin-like proteins in *Arabidopsis thaliana* and the characterization of secondary structure of Kat A. *Mol. Gen. Genet.* **238**: 362–368.

Mitterman I, Swoboda I, Pierson E, Eller N, Kraft D, Valenta R, Heberle-Bors E. (1995) Molecular cloning and characterization of profilin from tobacco (*Nicotiana tabacum*): increased expression during pollen maturation. *Plant Mol. Biol.* **27**: 137–146.

Mockrin S, Korn ED. (1980) *Acanthamoeba* profilin interacts with G-actin to increase the rate of exchange of actin-bound adenosine 5'-triphosphate. *Biochemistry* **19**: 5359–5362.

Moscatelli A, del Casino C, Lozzi L, Cai G, Scali M, Tiezzi A, Cresti M. (1995) High molecular weight polypeptides related to dynein heavy chains in *Nicotiana tabacum* pollen tubes. *J. Cell Sci.* **108**: 1117–1125.

Palevitz BA. (1987) Actin in the preprophase band of *Allium cepa*. *J. Cell Biol.* **104**: 1515–1519.

Pantaloni D, Carlier MF. (1993) How profilin promotes actin filament assembly in the presence of thymosin β4. *Cell* **75**: 1007–1014.

Payrastre B, Van Bergen en Henegouwen PMP, Breton M, Den Hartigh JC, Plantavid M, Verkleij AJ, Boonstra J. (1991) Phosphoinositide kinase, diacylglycerol kinase and phospholipase C activities associated with the cytoskeleton: effect of epidermal growth factor. *J. Cell Biol.* **115**: 121–128.

Pike LJ. (1992) Phosphatidyl 4-kinases and the role of polyphosphoinositides in cellular regulation. *Endocr. Rev.* **13**: 692–706.

Pollard TD, Cooper JA. (1986) Actin and actin-binding proteins. A critical evaluation of mechanisms and functions. *Annu. Rev. Biochem.* **55**: 987–1035.

Pont-Lezica RF, McNally JG, Pickard BG. (1993) Wall-to-membrane linkers in onion epidermis: some hypotheses. *Plant Cell Environ.* **16**: 111–123.

Putnam-Evans C, Harmon AC, Palevitz BA, Fechheimer M, Cormier MJ. (1989) Calcium-dependent protein kinase is localized with F-actin in plant cells. *Cell Motil. Cytoskel.* **12**: 12–22.

Putnam-Evans C, Harmon AC, Cormier MJ. (1990) Purification and characterization of a novel calcium-dependent protein kinase from soybean. *Biochemistry* **29**: 2488–2495.

Quatrano RS, Brian L, Aldridge J, Schultz T. (1991) Polar axis fixation in *Fucus* zygotes: components of the cytoskeleton and extracellular matrix. *Development* Suppl. 1: 11–16.

Reinhard M, Giehl K, Abel K, Haffner C, Jarchau T, Hoppe V, Jockusch BM, Walter U. (1995) The proline-rich focal adhesion and microfilament protein VASP is a ligand for profilins. *EMBO J.* **14**: 1583–1589.

Ridley AJ, Hall A. (1992) The small GTP-binding protein rho regulates the assembly of focal adhesions and actin stress fibers in response to growth factors. *Cell* **70**: 389–399.

Ridley AJ, Paterson HF, Johnston CL, Diekmann D, Hall A. (1992) The small GTP-binding protein rac regulates growth factor-induced membrane ruffling. *Cell* **70**: 401–410.

Rozycka M, Khan S, Lopez I, Greenland AJ, Hussey PJ. (1995) A *Zea mays* pollen cDNA encoding a putative actin-depolymerizing factor. *Plant Physiol.* **107**: 1011–1012.

Ruhlandt G, Lange U, Grolig F. (1994) Profilins purified from higher plants bind to actin from cardiac muscle and to actin from green alga. *Plant Cell Physiol.* **35**: 849–854.

de Ruijter N, Emons A-M. (1993) Immunodetection of spectrin antigens in plant cells. *Cell Biol. Int.* **17**: 169–182.

Sanders LA, Wang CS, Walling LL, Lord EM. (1991) A homolog of the substrate adhesion molecule vitronectin occurs in four species of flowering plants. *Plant Cell* **3**: 629–635.

Schaller GE, Harmon AC, Sussman MR. (1992) Characterization of a calcium-dependent and lipid-dependent protein kinase associated with the plasma membrane of oat. *Biochemistry* **31**: 1721–1727.

Schellenbaum P, Vantard M, Peter C, Fellows A, Lambert A-M. (1993) Co-assembly properties of higher plant microtubule-associated proteins with purified brain and plant tubulins. *Plant J.* **3**: 253–260.

Schindler M, Meiners S, Cheresh DA. (1989) RGD-dependent linkage between plant cell wall and plasma membrane: consequences of growth. *J. Cell Biol.* **108**: 1955–1965.

Seagull RW. (1989) The plant cytoskeleton. *Crit. Rev. Plant Sci.* **8**: 131–167.

Shaw PJ, Fairbairn DJ, Lloyd CW. (1991) Cytoplasmic and nuclear intermediate filament antigens in higher plants. In: *The Cytoskeletal Basis of Plant Growth and Form* (Lloyd CW, ed.). London: Academic Press, pp. 69–81.

Shibaoka H. (1994) Plant hormone-induced changes in the orientation of cortical microtubules: alterations in the cross-linking between microtubules and the plasma membrane. *Ann. Rev. Plant Physiol. Plant Mol. Biol.* **45**: 527–544.

Sonobe S. (1990) ATP-dependent depolymerization of cortical microtubules by an extract in tobacco BY-2 cells. *Plant Cell Physiol.* **31**: 1147–1153.

Sonobe S, Takahashi S. (1994) Association of microtubules with the plasma membrane of tobacco BY-2 cells *in vitro. Plant Cell Physiol.* **35**: 451–460.

Staiger CJ, Ashworth A. (1995) Functional studies of actin depolymerizing factors from *Arabidopsis thaliana. J. Cell. Biochem.* **19A**: 141.

Staiger CJ, Lloyd CW. (1991) The plant cytoskeleton. *Curr. Biol.* **3**: 33–42.

Staiger CJ, Goodbody KC, Hussey PJ, Valenta R, Drøbak BK, Lloyd CW. (1993) The profilin multigene family of maize: differential expression of three isoforms. *Plant J.* **4**: 631–641.

Staiger CJ, Yuan M, Valenta R, Shaw PJ, Warn RM, Lloyd CW. (1994) Microinjected profilin affects cytoplasmic streaming by rapidly depolymerizing actin microfilaments. *Curr. Biol.* **4**: 215–219.

Su F, Gu W, Zhai Z. (1990) The keratin intermediate filament-like system in the plant mesophyll cells. *Science in China (Series B)* **33**: 1084–1091.

Tan Z, Boss WF. (1992) Association of phosphatidylinositol kinase, phosphatidylinositol monophosphate kinase, and diacylglycerol kinase with the cytoskeleton and F-actin fractions of carrot (*Daucus carota*) cells grown in suspension culture. *Plant Physiol.* **100**: 2116–2120.

Terryn N, Van Montagu M, Inze D. (1993) GTP-binding proteins in plants. *Plant Mol. Biol.* **22**: 143–152.

Tiezzi A, Moscatelli A, Cai G, Bartalesi A, Cresti M. (1992) An immunoreactive homolog of mammalian kinesin in *Nicotiana tabacum* pollen tubes. *Cell Motil. Cytoskel.* **21**: 132–137.

Titus MA. (1993) Myosins. *Curr. Opin. Cell Biol.* **5**: 77–81.

Tominaga Y, Wayne R, Tung HYL, Tazawa M. (1987) Phosphorylation–dephosphorylation is involved in Ca^{2+}-controlled cytoplasmic streaming of characean cells. *Protoplasma* **136**: 161–169.

Torti M, Ramaschi G, Sinigaglia F, Lapetina EC, Balduini C. (1993) Association of the low molecular weight GTP-binding protein rap2B with the cytoskeleton during platelet aggregation. *Proc. Natl Acad. Sci. USA* **90**: 7553–7557.

Traas JA. (1984) Visualization of the membrane bound cytoskeleton and coated pits by means of dry-cleaving. *Protoplasma* **119**: 212–218.

Traas JA. (1990) The plasma membrane-associated cytoskeleton. In: *The Plant Plasma Membrane: Structure, Function and Molecular Biology* (Larsson C, Moller IM, ed.). Berlin: Springer-Verlag, pp. 269–292.

Traas JA, Doonan JH, Rawlins DJ, Shaw PJ, Watts J, Lloyd CW. (1987) An actin network is present in the cytoplasm throughout the cell cycle of carrot cells and associates with the dividing nucleus. *J. Cell Biol.* **105**: 387–395.

Valenta R, Duchene M, Pettenburger K, *et al.* (1991) Identification of profilin as a novel pollen allergen; IgE autoreactivity in sensitized individuals. *Science* **253**: 557–560.

Valenta R, Ferreira F, Grotes M, *et al.* (1993) Identification of profilin as an actin-binding protein in higher plants. *J. Biol. Chem.* **268**: 22777–22781.

Valenta R, Ball T, Vrtala S, Duchene M, Kraft D, Scheiner O. (1994) cDNA cloning and expression of timothy grass (*Phleum pratense*) pollen profilin in *Escherichia coli. Biochem. Biophys. Res. Commun.* **199**: 106–118.

Vidali L, Perez HE, Lopez VV, Noguez R, Zamudio F, Sanchez F. (1995) Purification, characterization and cDNA cloning of profilin from *Phaseolus vulgaris. Plant Physiol.* **108**: 115–123.

Vjotek A, Haarer B, Field J, Gerst J, Pollard TD, Brown S, Wigler M. (1991) Evidence for a functional link between profilin and CAP in the yeast *S. cerevisiae. Cell* **66**: 497–505.

Wang H, Cutler AJ, Saleem M, Fowke LC. (1989) Microtubules in maize protoplasts derived from cell suspension cultures: effect of calcium and magnesium ions. *Eur. J. Cell Biol.* **49**: 80–86.

Wasteneys GO, Gunning BES, Hepler PK. (1993) Microinjection of fluorescent brain tubulin reveals dynamic properties of cortical microtubules in living plant cells. *Cell Motil. Cytoskel.* **24**: 205–213.

Wayne R, Staves MP, Leopold AC. (1992) The contribution of the extracellular matrix to gravisensing in characean cells. *J. Cell Sci.* **101**: 611–623.

White RG, Badelt K, Overall RL, Vesk M. (1994) Actin associated with plasmodesmata. *Protoplasma* **180**: 169–184.

Williamson RE. (1993) Organelle movements. *Annu. Rev. Plant Physiol. Plant Mol. Biol.* **44:** 181–202.

Wuestehube LJ, Luna EJ. (1987) F-actin binds to the cytoplasmic surface of ponticulin, a 17-kDa integral glycoprotein from *Dictyostelium discoideum. J. Cell Biol.* **105:** 1741–1751.

Xu P, Lloyd CW, Staiger CJ, Drøbak BK. (1992) Association of phosphatidylinositol 4-kinase with the plant cytoskeleton. *Plant Cell* **4:** 941–951.

Yang Z, Watson JC. (1993) Molecular cloning and characterization of rho, a ras-related small GTP-binding protein from the garden pea. *Proc. Natl Acad. Sci. USA* **90:** 9732–9736.

Yuan M, Shaw PJ, Warn RM, Lloyd CW. (1994) Dynamic reorientation of cortical microtubules, from transverse to longitudinal, in living plant cells. *Proc. Natl Acad. Sci. USA* **91:** 6050–6053.

Yuan M, Warn RM, Shaw, PJ, Lloyd CW. (1995) Dynamic microtubules under the radial and outer tangential walls of microinjected pea epidermal cells observed by computer reconstruction. *Plant J.* **7:** 17–23.

Zhang D, Wadsworth P, Hepler PK. (1990) Microtubule dynamics in living dividing plant cells: confocal imaging of microinjected fluorescent brain tubulin. *Proc. Natl Acad. Sci. USA* **87:** 8820–8824.

Zhang D, Wadsworth P, Hepler PK. (1993) Dynamics of microfilaments are similar, but distinct from microtubules during cytokinesis in living, dividing plant cells. *Cell Motil. Cytoskel.* **24:** 151–155.

Zhu JK, Shi J, Singh U, Wyatt SE, Bressan RA, Hasegawa PM, Carpita NC. (1993) Enrichment of vitronectin- and fibronectin-like proteins in NaCl-adapted plant cells and evidence for their involvement in plasma membrane–cell wall adhesion. *Plant J.* **3:** 637–646.

Chapter 2

Cell adhesion in plants and its role in pollination

Elizabeth M. Lord, Linda L. Walling and Guang Yuh Jauh

1. Introduction

Animal development is characterized by cell adhesion and movement events that are not thought to occur in plants as a rule. The molecules involved are typically extracellular, but they bind to transmembrane proteins that interact with cytoskeletal elements inside the cell, providing direct links between the exterior and interior of the cell (Adair and Mecham, 1990). The significance of these physical and chemical links between the extracellular matrix (ECM) and the interior of the cell is only now being appreciated as pivotal in development and differentiation in animals (see *Current Opinion in Cell Biology*, Vol. 7, No. 5, 1995, which is dedicated to the topic). The mouse mammary gland (Boudreau *et al.*, 1995) is the model system for regulation by the ECM of tissue-specific gene expression in animals. Response elements have been discovered in promoters of genes dependent on adhesion to a specific ECM. The adhesion event is purported to regulate the expression of transcription factors that bind to these elements in order to activate transcription of the gene. The structural links from the ECM to the nucleus are speculative, but we now have a better understanding of the presence of a nuclear matrix (Penman, 1995), and how such links could be the transducers of information from the exterior to the interior of the cell via adhesion events (Ingber, 1993).

Investigations focusing on the structure, function and composition of the plant cell wall, plasma membrane and cytoskeleton have demonstrated that, although the plant cell wall is structurally distinct from the animal extracellular matrix, similar communication channels may exist in plants. Studies investigating adhesion events and the localization of glycoproteins and proteoglycans during plant development and in response to the environment have changed the view that the plant cell wall or ECM is an inert structure primarily involved in support (Bolwell, 1993; Lord and Sanders, 1992; Reuzeau and Pont-Lezica, 1995; Showalter, 1993; Ye *et al.*, 1991).

Types of adhesion in plants

In plants, adhesion can occur between walls (ECMs) of two cells, between plasma membranes of two cells, or between the plasma membrane and ECM of one cell. Adhesion between plant cells is most often due to a wall-to-wall linkage via the middle lamella. This is initiated during every cell division at the newly formed cell plate, and leaves living connections between cells via plasmodesmata. In this manner, a three-dimensional tissue system is formed in higher plants whereby every cell adheres to its neighbor initially, and during organ expansion separation may occur to form intercellular spaces. Knox (1992) has reviewed this aspect of plant cell adhesion, which is the basis for tissue formation, and has noted that we still do not know which factors in the middle lamella are involved in adhesion. Recent experimental work on somatic embryogenesis documents the critical nature of cell–cell adherence for the complex aspects of morphogenesis as well (Van Engelen and De Vries, 1992).

There are a number of examples in plants where cells produce highly modified walls and are capable of communicating from cell to cell across their walls without the benefit of cytoplasmic connections such as plasmodesmata. The best-known case is in pollen–stigma interactions, where the pollen grain first adheres to the stigma, hydrates and germinates, and then the pollen tube interacts with the transmitting tract cells of the stigma and style and either a compatible or an incompatible reaction ensues (Knox, 1984; Preuss *et al.,* 1993). Presumably, these wall-to-wall adhesion events precipitate signaling across the walls that involves the plasma membranes of the two cells. The best-described system of this type is sporophytic self-incompatibility, where pollination events are arrested on the stigma if the cross is incompatible. In *Brassica,* two related genes at the S-locus appear to be involved in this signaling event, which results in self-incompatibility. One of these genes encodes the S-locus glycoprotein, which is an ECM molecule in the stigmatic papillae involved in the specificity of the recognition event, and the other encodes a membrane-anchored protein that is a receptor kinase (Nasrallah and Nasrallah, 1993; Nasrallah *et al.,* 1994). These two molecules, one in the ECM and the other spanning the plasma membrane, appear to function in a signaling system very similar to those which involve adhesion events in animal cells. We have proposed that a similar adhesion event occurs between the pollen tube and the stigma/stylar transmitting tract during compatible pollinations where the pollen tube moves toward the ovary unimpeded (Sanders and Lord, 1989). Here, the specialized wall of the pollen tube may adhere to the secretory matrix of the transmitting tract cells, and these adhesion sites may facilitate the movement of the tube cell and sperm cells from the stigma to the ovary. In our model, both cell–cell adhesion and cell movement phenomena are mediated by wall–plasma membrane and cytoskeletal connections (Sanders and Lord, 1992). We shall describe the details of this system in the following sections.

Another wall–wall interaction that triggers significant changes in the behavior of the cells involved is that of carpel fusion (Verbeke and Walker, 1986). This ubiquitous event in flowering plants is an example of adhesion between previously free epidermi of two carpels. An increase in secretion into the ECM occurs when the two walls meet and deposition of unknown materials apparently causes the walls to fuse (Walker, 1975). This is interesting in view of the pollen tube–transmitting tract interaction mentioned above, since the event of contact and fusion of the two carpels results in production of the transmitting tract of the style, the central core of specialized cells that secrete matrices which may facilitate pollen tube extension. Lolle *et al.* (1992) have isolated a mutant in *Arabidopsis* in which the fusion of epidermi is enhanced. Instead of fusion being limited to developing carpels, epidermi in diverse locations on the plant fuse. Fusion results in the production of a secretion that is similar to a transmitting tract ECM and which supports pollen tube extension between fused leaves. This mutation alters the organ specificity of adhesion events; the floral genetic program that directs carpel adhesion and dedifferentiation to produce the transmitting tract is expressed in the vegetative organs. Another example of developmentally controlled fusion events occurs in a red alga. In this analogous system, rhizoids fuse with each other and with a substrate. The glycoprotein responsible for triggering this fusion event has been isolated, but is as yet unidentified (Watson and Waaland, 1983).

We know little about what triggers the production of adhesive compounds in plants or about the composition of the compounds themselves, except perhaps in algal systems such as *Chlamydomonas.* This unicellular green alga provides the best-described system of cell–cell adhesion to date. When gametes fuse to form the zygote, an early event is adhesion of the flagellar tips of the two cell types. This adhesion is mediated by an agglutinin, a hydroxyproline-rich glycoprotein (HRGP) molecule localized in the ECM. The HRGP has a free globular end and a hook domain that may be attached to the plasma membrane via a transmembrane protein (Van den Ende, 1992). The polysaccharide moiety of this glycoprotein probably provides the specificity for the gamete–gamete adhesion event, because deglycosylation prevents mating. A cytoskeletal connection is also implicated, since functional microtubules are necessary for adhesion. Gamete adhesion initiates a series of events that includes signaling by cyclic AMP (cAMP) and the production of enzymes for cell wall release to allow for protoplast fusion, and so represents a clear example of signal transduction in the algae. Yeasts also have agglutinins which are involved in the initial stages of mating (Gooday, 1992). Adhesion via ECM glycoproteins may be a common event in single-celled organisms. This view is further supported by adhesion studies of *Fucus*. In this alga, adhesion to the substrate occurs after fertilization. Vreeland *et al.* (1993a, b) have provided evidence that a nonspecific adhesive is produced by the zygote and that the secretion of a peroxidase catalyzes phenolic cross-links of ECM carbohydrates (alginate and sulfated fucans), resulting in adhesive site formation. The investigators were able to mimic this adhesion event *in vitro* by using inert microspheres as an assay system for stickiness. These data are consistent with Fry's (1986) conclusions that 'cross-links between walls are responsible for cell adherence'. Wall-bound peroxidases can cross-link glycoproteins and pectins in the wall, and may be involved in adhesion and subsequent prevention of cell expansion. Peroxidases are heme-glycoproteins found throughout the plant in many forms, but their functions are largely unknown (Gaspar *et al.,* 1982). A recent study has implicated an ECM peroxidase in triggering somatic embryogenesis in cell cultures (Van Engelen and De Vries, 1992), an event that is correlated with cell adhesion and a lack of cell expansion.

In the pathogenic fungi, wall–wall adhesion plays a role in contact sensing during infection (Gooday, 1992; Kwon and Epstein, 1993; Nicholson and Epstein, 1991; Read *et al.,* 1992). The rusts show close adherence to the host, usually growing along its anticlinal cell walls. When actin microfilaments are visualized in these hyphae, areas of concentration very like focal adhesions in animal cells are seen in the adhesion zone. When grown on inert surfaces, the hyphae produce an adhesive ECM in the zone of contact. Certainly adhesion plays a role in pathogenesis and symbiotic associations between plants and other organisms, but as yet we know little about the agents involved.

The second class of adhesion events in plants involves plasma membrane interactions. In animals, the plasma membranes of cells can interact directly via cell adhesion molecules, or CAMs. These transmembrane proteins attract like to like in the formation of homogeneous tissues from collections of migrating cells (Öbrink, 1991). In plants, such interactions are rare because of the presence of a substantial ECM around every cell. One exception occurs during fertilization, when the sperm cells enter the synergid and come into contact with the egg plasma membrane. There is little if any wall surrounding the egg or sperm cells. It is therefore reasonable to propose that CAM analogs may be involved in the process of fertilization in plants. The observations that sperm are dimorphic in many species, and that only one type of sperm cell fuses with the egg cell, provide more evidence for a cell-specific membrane recognition event in fertilization that may involve CAM-like molecules (Russell, 1992).

A third type of adhesion event occurs between the plasma membrane and the wall; this class of interactions is well documented in animal cells. The molecules involved are substrate adhesion molecules (SAMs) in the ECM, SAM receptors (integrins) in the plasma membrane, and the cytoskeletal proteins within the cell that are bound by the integrins (Geiger, 1989; Hynes, 1992; Roberts, 1990) (*Figure 1*). When a cell adheres to a substrate molecule, it forms discrete regions, called focal adhesions, where SAMs, integrins and the cytoskeleton are localized (Burridge *et al.,*

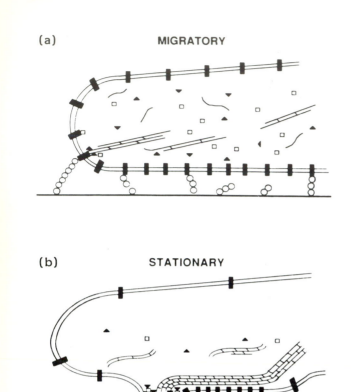

(a) MIGRATORY

(b) STATIONARY

Figure 1. Speculative models for modes of cell–substratum adhesion of migratory and stationary cells. (a) Migratory cells have a disorganized cytoskeleton; the fibronectin (FN) receptors are diffuse in the plane of the membrane and may be mobile. A number of receptors are not occupied by FN and may constitute a reserve pool for new adhesion. (b) In stationary cells, the cytoskeleton is well organized and the receptors are localized mostly in clusters close to focal contact sites. These receptors have restricted lateral mobility in the membrane. They are linked directly both to the cytoskeleton and to FN fibers. In focal contacts, FN receptors interact indirectly with actin microfilaments through talin and vinculin. ■, FN receptors; ○, FN; □, talin; ▲, vinculin; α-actinin is indicated by cross-links between actin microfilaments. Adapted from Dufour *et al.* (1988) with permission from Elsevier Trends Journals.

1988). In *Chlamydomonas,* agglutinins in the ECM attach to membrane-spanning molecules and represent an analogous adherence event. There is good evidence that links between the wall and the plasma membrane occur in plants, and although arabinogalactan-proteins (AGPs) (Roberts, 1989) and other HRGPs such as extensin (Pont-Lezica *et al.,* 1993) have been suggested as participating in this linkage, we have little knowledge of the factors involved. The evidence for such links is threefold. First, circumstantial evidence indicates that the production of an ECM is necessary for basic developmental events such as cell division (Meyer and Abel, 1974), establishment of polarity (Kropf, 1992), and cell–cell recognition events (Callow and Green, 1992). Secondly, visual evidence has accumulated which indicates that, upon plasmolysis, the plasma membrane maintains close contacts with the wall in regions which resemble focal adhesions and have been referred to as Hechtian strands (Pont-Lezica *et al.,* 1993). These are especially prominent in cells forming secondary walls (Roberts and Haigler, 1989). There are numerous examples of induction of wall–membrane adhesion by stress such as high salt levels (Zhu *et al.,* 1993a), freezing (Johnson-Flanagan and Singh, 1986) and pathogen infection (Lee-Stadelmann *et al.,* 1984). Finally, there is immunological evidence. Cross-reactivity of antibodies to the animal SAM molecules, vitronectin and fibronectin, and to an integrin has been observed in a number of instances in plants, algae and fungi (Gens *et al.,* 1993; Kaminskyj and Heath, 1995; Mityazaki *et al.,* 1992; Pont-Lezica *et al.,* 1993; Quatrano, 1991; Sanders *et al.,* 1991; Schindler *et al.,* 1989; Wagner *et al.,* 1992; Wagner and Matthysse, 1992; Wang *et al.,* 1994; Watson and Waaland, 1983; Zhu *et al.,* 1993a, 1994).

Animal SAMs such as fibronectin and vitronectin are extracellular glycoproteins that mediate cell adhesion and cell spreading and bind to integral membrane receptors (integrins). The binding of SAMs to integrins is mediated through an RGD peptide located in the N-terminal domain of SAMs. Cross-reaction of antibodies with an integrin subunit of a protein in fungi suggests that this transmembrane receptor is present and perhaps analogous to those in animal systems (Marcantonio and Hynes, 1988). Over the past 3 years, data have accumulated from a number of laboratories which suggest the presence of an integrin-like receptor in plants, fungi and algae (Gens *et al.*, 1993; Kaminskyj and Heath, 1995; Quatrano *et al.*, 1991; Schindler *et al.*, 1989; Wagner and Matthysse, 1992; Watson and Waaland, 1983; Wayne *et al.*, 1992), and substrate adhesion molecule analogs in higher plants (Sanders *et al.*, 1991; Wagner and Matthysse, 1992; Wang *et al.*, 1994; Zhu *et al.*, 1993a, 1994), a brown alga (Wagner *et al.*, 1992) and an acellular slime mold (Mityazaki *et al.*, 1992). Detection of the presence of substrate adhesion analogs in plants is primarily based on the ability to detect plant proteins that cross-react with polyclonal antisera to human vitronectin (Preissner, 1991); there are two reports of a protein antigenically related to animal fibronectin (Gens *et al.*, 1993; Zhu *et al.*, 1993a). Collectively, these investigations form an intriguing body of data which suggests that proteins antigenically related to animal SAMs exist in higher plants, algae and slime molds. At present, no genes encoding a plant vitronectin-like molecule have been isolated.

There is additional evidence from the cytoskeletal literature which supports the view that adhesion zones in plant cells contain microfilaments and microtubules. In fern protodermal cells which show tip growth, the plasma membrane in the subapical region remains attached to the wall under plasmolysis. Only colchicine can disrupt this tight connection, implicating microtubules in adherence (Kagawa *et al.*, 1992). The alignments of cortical microtubules under zones of wall–membrane adherence are also seen in developing xylem cells in the regions where secondary wall materials are being deposited in the ECM (Picket-Heaps, 1968). Colchicine can be used in cell suspension cultures to induce cell expansion and separation of the aggregates that form in culture (Hayashi and Yoshida, 1988). This circumstantial evidence implicates the cortical microtubules in adhesion events but, to date, the mechanisms involved are unknown. In *Fucus,* microfilaments occur in cytoplasmic strands that extend from the wall during plasmolysis; once broken, the strands leave a footprint of actin on the wall (D. Kropf, personal communication). It is known that a wall is necessary for axis fixation in the fertilized *Fucus* egg (Kropf *et al.*, 1988). The model of Quatrano *et al.* (1991) for axis fixation in *Fucus* incorporates the basic elements for interconnections between the cytoskeleton, plasma membrane and wall, including a proposed vitronectin-like molecule. Heath's work on the fungus *Saprolegnia* (Heath, 1990; Jackson and Heath, 1993; Kaminskyj and Heath, 1995) reveals a complex interaction between the F-actin microfilaments, spectrin and integrin homologs in this tip-growing organism. His group has provided good evidence for cytoplasm–wall adhesion via integrins and co-localization of actin to these same sites. There are immunological data supporting the presence of integrins in plant cells (Schindler *et al.*, 1989), but further work has not led to the isolation of such molecules. However, there is substantial circumstantial evidence that cytoskeleton, plasma membrane and ECM links occur in plants (Reuzeau and Pont-Lezica, 1995; Wyatt and Carpita, 1993). In pollen tubes, extensive microfilament networks occur around the vegetative nucleus and its associated sperm cells, and this network extends close to the plasma membrane (Tiwari and Polito, 1988). When lily pollen tubes are grown *in vivo*, clusters of microfilaments form that resemble focal adhesions in animal cells (Jauh and Lord, 1995; Pierson *et al.*, 1986) (*Figure 2*). If the pollen tube is grown *in vitro*, no such formations occur, indicating the importance of the stylar matrix in their induction.

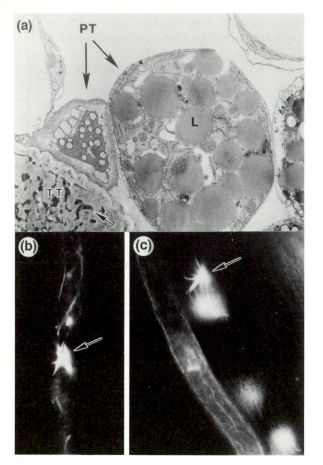

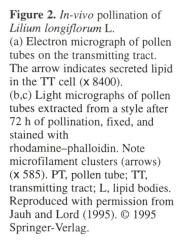

Figure 2. *In-vivo* pollination of *Lilium longiflorum* L.
(a) Electron micrograph of pollen tubes on the transmitting tract. The arrow indicates secreted lipid in the TT cell (x 8400).
(b,c) Light micrographs of pollen tubes extracted from a style after 72 h of pollination, fixed, and stained with rhodamine–phalloidin. Note microfilament clusters (arrows) (x 585). PT, pollen tube; TT, transmitting tract; L, lipid bodies. Reproduced with permission from Jauh and Lord (1995). © 1995 Springer-Verlag.

Antibodies raised against nonplant proteins have played a major role in deciphering the structure and function of the plant cytoskeleton (Lloyd, 1987). A similar approach is now being used with antibodies to various nonplant ECM, cytoskeletal and plasma membrane proteins (Reuzeau and Pont-Lezica, 1995; Roberts, 1990; Wyatt and Carpita, 1993). The stage has been set for a re-examination of adhesion events in plants using the models designed for animal cell movement and attachment during embryogenesis (*Figure 1*). Although little progress has been made in identifying the component molecules in these models, a new paradigm has emerged for plant cell biologists in which the wall or ECM is no longer seen as inert and separate from the living cell.

Evidence for adhesion molecules in pollination

In our previous work we proposed that adhesion events during pollination were responsible for the faster, directed movement of the tube cell in the style compared to its behavior *in vitro* (Sanders and Lord, 1992). We postulated the existence of SAMs (substrate adhesion molecules) or CAMs (cell adhesion molecules) (*Figure 1*) in the transmitting tract ECM of the style and in the pollen tube ECM that functioned in both self-recognition and facilitation of tube cell movement in the style. The only experimental evidence for a movement molecule in the style was provided by the study showing transport of inert particles in the transmitting tracts of three species (Sanders and Lord, 1989).

To test the ability of the tube cell to move in the style, independent of the pollen grain and spent tube, we excised portions of the stigma and style that the tube cell had traversed during a lily pollination, and recorded the ability of the remaining pollen tube tip, containing the tube cell cytoplasm and the two sperm cells, to effect fertilization (Jauh and Lord, 1995). This positive result demonstrated the independence of the tube cell once it is sequestered at the tip of the pollen tube by a callose wall or plug. Recent work by O'Driscoll *et al.* (1993) showed that this callose wall at the rear of the tube cell was impenetrable and that the tube cell was capable of endocytosis as well as exocytosis at the tip.

There is a sizeable literature on the ultrastructure and biochemistry of the lily style (Kroh and Knuiman, 1982; Labarca and Loewus, 1972; Rosen, 1971), but no mention of the adhesive nature of the transmitting tract or its secretions. When fixation methods are used to preserve adhesives (Chaubal *et al.,* 1991), a fibrillar wall component which appears to fuse the tube cell to the stylar transmitting tract in lily can be seen (*Figure 2a*). This wall layer is less prominent, but also evident, between pollen tube cells *in vivo*. When *in-vivo*-grown pollen tubes are removed from the style and processed with rhodamine–phalloidin to observe the F-actin microfilaments, star-shaped clusters of F-actin, similar to focal adhesions in moving animal cells, are observed (*Figure 2b, c*). These configurations were first described in lily by Pierson *et al.* (1986), and were not detected in pollen tubes grown *in vitro*. The adhesion events, the 10-fold faster growth rates of pollen tubes, and the F-actin configurations seen only in *in-vivo*-grown lily pollen tubes are all indicative of an ECM influence in pollination in lily.

We used a variety of monoclonal antibodies (mAbs) to known ECM molecules, pectins and AGPs in order to characterize the adhesive components and obtain structural evidence that these compounds are candidates for adhesion molecules in lily (Jauh and Lord, 1996). Previous work by Li *et al.* (1992, 1994) using such mAbs demonstrated the presence of these molecules in pollen tubes of a number of species, and showed that they occurred in fascinating banded arrays in the wall of tobacco pollen tubes. Our interest is in their role as possible adhesion molecules; we have focused on the *in-vivo* condition in lily. As expected, the mAbs to unesterified pectins localized to the entire length of the lily pollen tube and those to esterified pectins localized to the tip region only (*Figure 3a, b*). When the pollen tubes are extracted from the lily style, they adhere tightly to one another and can be separated by a brief pectinase treatment. The stylar transmitting tract showed extensive presence of unesterified pectins on the surface of the epidermis that conducts the pollen tubes (*Figure 3d*). Immunogold localizations show the esterified pectins localized to the contents of the vesicles secreted at the tip of the pollen tube, with little presence in the wall itself (*Figure 3f*).

The AGP mAbs (JIM 13–16, LM2 and Mac207) all localized to the tip of the pollen tube alone (*Figure 3c*), unless the tubes were pretreated with pectinase, when the whole tube showed localization to the wall and/or plasma membrane. The stylar transmitting tract showed specific localization patterns when AGP mAb JIM13 was used (*Figure 3e*), but no localization when mAb LM2 was used. At the transmission electron microscope (TEM) level, immunogold-labelled mAb JIM13 bound to the plasma membrane of the vesicles that were being secreted into the tube tip, and some localization occurred in the wall (*Figure 3g*).

The presence of AGPs secreted into the pollen tube tip has not been reported before, and indeed tobacco pollen tubes do not show such a pattern of localization (Li *et al.,* 1992). We have obtained evidence for AGP secretion at the tip of pollen tubes (unpublished data) of corn and cherimoya (a primitive dicot). To demonstrate further the ability of lily pollen tubes to secrete AGPs, we used the diagnostic Yariv reagent to detect their presence. Yariv reagent [(β-D-Glc)$_3$] binds AGPs and is red in color, so it can be visualized easily (Jermyn, 1978). Yariv reagent added to lily pollen tubes grown *in vitro* bound predominantly to the tip region and effectively arrested

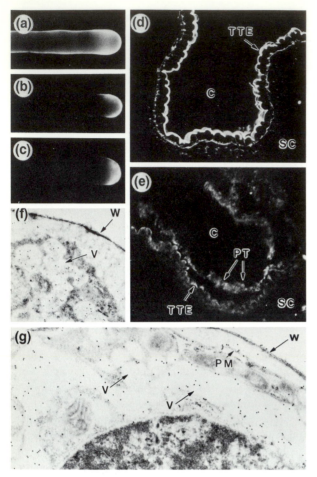

Figure 3. AGPs and pectins localized in pollen tubes and styles of lily by mAbs to unesterified pectins (JIM5), esterified pectins (JIM7) and AGPs (JIM13, LM2) conjugated with fluorescein isothiocyanate (FITC) secondary antibody. Labeling pattern of (a) unesterified pectins, (b) esterified pectins, and (c) AGPs on *in-vitro*-grown pollen tubes (x 430). Unesterified pectins are mainly restricted to TTE cells in the style (d), and one AGP probe (JIM13) localizes to the TTE cells alone (e) (note that the pollen tubes are also labeled in the style, but they have pulled away from the TTE cells) (x 80). Ultrastructure and immunolocalization of AGPs and esterified pectins on *in-vivo*-grown lily pollen tubes in LR White resin. Esterified pectins are localized to vesicles in the pollen tube cell primarily, but can also be seen in the wall (f) (x 17 220). AGPs localized to vesicle membranes (V) in the pollen tube and to the wall (g) (x 20 571). C, hollow stylar canal; PT, pollen tubes; SC, stylar cortex; TTE, transmitting tract epidermis; V, vesicle; W, pollen tube wall; PM, plasma membrane. Reproduced with permission from Jauh and Lord (1996). © 1996 Springer-Verlag.

their growth (*Figure 4*). When such Yariv-treated tubes are transferred to fresh growth medium in the absence of Yariv reagent, the pollen tubes resume growth, with new tube tips emerging from the flanking region of the 'fixed' tube tip. Recent work by Serpe and Nothnagel (1994) showed a similar inhibitory effect of Yariv reagent on the growth of cell suspensions. Collectively, these data suggest the presence of both esterified pectins and AGPs in the vesicles which fuse to the growing pollen tube tip in lily, providing both ECM materials (pectin) and plasma-membrane-bound components (AGPs) that may function in adhesion of the plasma

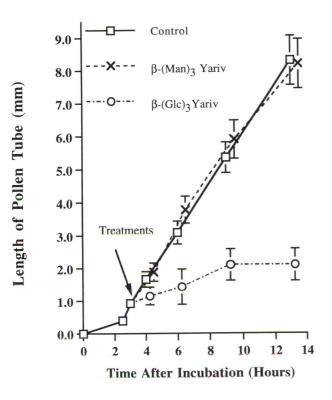

Figure 4. The effects of 30 µM (β-D-Glc)$_3$ and (β-D-Man)$_3$ Yariv reagent on the growth of *in-vitro*-grown lily pollen tubes. Treatments began after 3 h of growth in normal medium. Effects were measurable after 15 min. Reproduced with permission from Jauh and Lord (1996). © 1996 Springer-Verlag.

membrane to the ECM and of the tube cell to the transmitting tract. The transmitting tract surface is also covered by secreted AGPs as well as by pectins, and both may be acting as SAMs in the style. Our model (*Figure 5*) now includes these ECM components as candidates in the adhesion and cell movement events in lily pollination. No doubt there are other components involved in this interaction, such as lipids known to be involved in adhesion events on the stigma (Preuss *et al.,* 1993) and found in great abundance as secreted molecules in lily styles (*Figure 2a*). It is now necessary to isolate the AGPs, pectins and lipids from the style and to devise *in-vitro* assays that can test the ability of these compounds to cause adhesion, faster tube cell movements and the F-actin configurations that occur *in vivo*.

Recently, other laboratories have isolated molecules from pollen tubes and styles that either have adhesive properties (Cheung *et al.,* 1995) or show homology to known adhesive compounds (Rubinstein *et al.,* 1995). In the lily system, we can easily extract *in-vivo*-grown pollen tubes and carry out subtractive library preparations as well as differential mRNA display techniques (Oh *et al.,* 1995) in order to compare the genes expressed in pollen tubes grown *in vitro* with those grown *in vivo*. In this way, we hope to characterize genes from the tube cell induced by contact with the style and perhaps by the adhesion event itself.

2. Models for cell movement: is there a similar mechanism operating in plant cell expansion, especially in tip-growing cells?

Actin-based motility models for animal cells

Cell movement is a common phenomenon in animal development, and locomotion of simple organisms such as amoebae or *Dictyostelium* has been the focus of study for many years (Cooper,

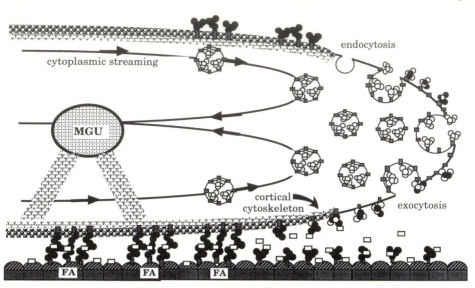

Stylar Transmitting Tract Epidermal Cells

Figure 5. A model for the role of SAMs in cell movement during pollination. The pollen tube tip is illustrated as it progresses through the stylar ECM. The cell wall of the pollen tube is thin at the tip and thicker and more complex back from the tip. The male germ unit (MGU) is composed of the tube cell nucleus and the generative cell (or two sperm). FA, focal adhesion; ■, membrane-bound AGP; □, stylar secreted component; ●, unesterified pectins; ○, esterified pectins.

1991; Stossel, 1993). The assumption has been that the mechanisms involved were universal and that an actin–myosin model as a motive force was probably applicable in most cases. Recently, however, mutants of *Dictyostelium* that lack myosin II and a major isoform of myosin I, but which are still able to move, have been isolated (Spudich, 1989). New proposals have emerged suggesting that actin polymerization and cross-linking can account for the motive force underlying cell movement (Cooper, 1991). This hypothesis is strongly supported by the fact that it is possible to isolate the moving fronts of some cells, which are organelle-free and full of F-actin, and to show that they retain rudimentary locomotion. This suggests that all the equipment for such motion resides at these sites (Euteneuer and Schliwa, 1984). One model for actin-based motility in *Dictyostelium* suggests that there are three steps in this process (Condeelis, 1992). The first step is focal polymerization of actin into microfilaments in response to a signal induced by adhesion of the cell to a matrix molecule, or by perception of a chemoattractant. The second step involves cross-linking of actin microfilaments to form bundles, and the third step involves expansion of the volume of the gel created by the cross-linked actin filaments, which serves to push the membrane out at the moving front. The gel swells by imbibing water, so the motive force would be a combination of actin polymerization at the moving front and osmotic swelling.

Pollination: a case of cell movement in plants?

We have proposed that pollen tube extension in the style is a special case of cell movement in plants (Sanders and Lord, 1989). Our proposal was based on evidence that stylar matrices could transport inert particles in the same way as was seen in animal embryos (Bronner-Fraser, 1982), and that the isolated pollen tube cell was capable of delivering sperm cells to ovules (Jauh and Lord, 1995). One can consider the vegetative cell and two sperm cells that reside at the tip of the tube as it traverses the style as a moving group of cells that leave a trail of cell wall (the spent pollen tube) behind. Our original model was based on those current in the animal literature (*Figure 1*), implicating ECM molecules and their transmembrane receptors as the means of adhesion and creation of foci for actin filament elongation at the moving tip of the tube. Both actin and myosin have been localized in the pollen tube tip (Condeelis, 1974; Tang *et al.,* 1989), so either the actin–myosin model could apply in tube cell extension, or the force for this movement could be solely actin-based, similar to that proposed for *Dictyostelium* (Condeelis, 1992).

This latter model proposes that an adhesion event is the first step in the process of pollination, and that it could be mediated *in vivo* by the interactions of transmembrane receptors in the pollen tube tip that directly or indirectly adhere to a stylar ECM molecule. This binding could initiate the focal points for actin filament polymerization. Alternatively, the adhesion event could be less direct and involve wall–wall interactions. It is possible that glycoproteins in the wall of the pollen tube and in the stylar matrix interact, thereby providing traction for the extending tip. Pollen tubes readily extend *in vitro*, but never as fast or for as long a distance as they do *in vivo*. The fact that they 'move' at all *in vitro* means that their own ECM or wall may be acting as the means of adhesion of the vegetative cell plasma membrane, allowing for the foci of actin polymerization and hence extension (Steer, 1990). In this context, pollen tubes may be making a matrix (the wall) and crawling along it, much as has been proposed recently for the fungi (Heath, 1990). Heath's group has established that turgor pressure is necessary but not sufficient to explain apical extension rates in the fungus *Saprolegnia* (Kaminskyj *et al.,* 1992). They have proposed that Ca^{2+} regulates actin–myosin interactions which promote cytoplasmic migration inside the fungal wall or ECM, and indeed they have recorded contractions in the cytoplasm that may be involved in such movement (Jackson and Heath, 1992). Measurements of turgor pressure in growing and nongrowing pollen tubes are necessary to establish whether this force is sufficient to explain tube extension (Steer and Steer, 1989).

Our original proposal for the mode of cell movement (Sanders and Lord, 1992) relied on data which suggested that F-actin was prominent in the tube tip (Steer and Steer, 1989). Hepler's work using *in-vitro* pollen tubes and antibodies to plant actin suggests that this is not the case; only small amounts of F-actin are present at the tip (P. Hepler, personal communication). In the light of this recent finding, we propose that the sites of polymerization of F-actin for both cytoplasmic streaming and the cortical cytoskeleton are in the region back from the clear tip zone where streaming reverses itself, at the apex of the forward streaming actin where adhesion of the pollen tube to the style is evident, and at the site where the secondary callose wall is beginning to form (*Figure 5*). This we also suggest is the site of movement of the tube cell cytoplasm against the tube wall. The force for movement could be the fast streaming of cytoplasm in the forward direction fueled by the F-actin internal cytoskeleton. This pattern of fast forward streaming on the periphery of the cell and a slow return down the center (called reverse fountain by its discoverer; Iwanami, 1956) is only evident in pollen tubes that are growing. If the tubes cease to grow, the streaming reverts to a circular pattern typical of immobile plant cells in tissues. Continual renewal of the cytoskeletal links between the male germ unit (MGU) and the cortical cytoskeleton are necessary to keep this vital unit in the tip of the pollen tube. Microtubules have been implicated in these linkages in tobacco (Joos *et al.,* 1994), and F-actin microfilaments in other species (Tiwari and Polito, 1988). The MGU does not stream with the cytoplasm, but progresses forward in a manner that resembles walking along the cortical cytoskeleton towards the tip, always remaining at a discrete distance.

The presence of a clear zone of vesicles at the tip and the absence of a callose wall are also correlated with this pattern of reverse fountain streaming. The production of a pectin matrix at the tip of the tube cell, a primary cell wall in the absence of a secondary callose wall, may be the result of forward movement of the tube cell cytoplasm behind the tip, keeping the ECM of the tip in a perpetually young state much like that of a cell plate. If the tube cell ceases to move forward, the secondary wall of callose 'catches up' and deposits a rigid wall at the tip, preventing further movement. The steep calcium gradient at the tip is also correlated with movement of the tube cell and secretion of vesicles at the tip (Battey and Blackbourn, 1993; Miller *et al.*, 1992). If tube growth ceases, streaming enters the tip, the clear zone disappears, the calcium gradient is abolished, the secondary callose wall forms around the tip and the reverse fountain pattern is replaced by the circular streaming of stationary cells. Continued movement of the tube cell prevents all of the above from happening. When Yariv reagent is added to the pollen growth medium, it binds to the exposed AGPs at the tip, preventing them from binding to pectins and immobilizing them in the membrane and/or wall (Jauh and Lord, 1996) (*Figure 4*). An immediate effect is the production of callose in this region, suggesting that such immobilization of the AGPs induces the enzyme callose synthase to be active on the plasma membrane. The normal callose wall back from the tip is initially separated from this newly formed callose wall, leaving flanking regions capable of renewing the growth center for the tube cell. When this reorganization of secretion occurs, Yariv reagent can again act to bind up this center and callose forms at these sites as well. If the Yariv reagent-treated tube is transferred to fresh growth medium before these flanking growth centers are also impeded by callose (≤ 3 h of Yariv reagent treatment), regeneration of pollen tube growth can occur. This suggests that, normally, exocytosis is restricted to the extreme tube tip and forward of the secondary wall, with the endocytosis that apparently accompanies it occurring on the flanks (*Figure 5*). The fact that the tip can be immobilized by Yariv reagent and regeneration of movement can be restored on the flanking regions suggests the inherent capacity of the tube cell to move along its own ECM. The ability of the clear zone at the tube tip to initiate new growth centers when the one nearest the tip is blocked shows a flexibility in this 'growth' zone that is lost if polarization of the tube cell is lost. Its polarized nature can be maintained temporarily even when Yariv reagent has bound up the tip and growth has ceased (Jauh and Lord, 1996). In this 'idling' mode it begins to reorganize another center of growth on the flanks, and can regenerate the tube if no further 'capping' occurs with Yariv reagent. Perhaps the Yariv reagent binding to AGPs in the plasma membrane prevents their normal ease of dispersal in the membrane (lipid-bilayer fluidity). The 'capping' itself may provide a signal that no further vesicle fusion can occur here and that secondary wall formation (callose synthase activity) may now occur. A picture is emerging of pollination as a moving tube cell system that makes an ECM and crawls on it, using the microfilaments of actin to power this movement via a surging forward of cytoplasm against the wall, with some links between the cytoskeleton in the cortex of the cell and the plasma membrane and wall providing for traction. When the tube cell is in the style, *in vivo*, the added traction provided by adhesion to the transmitting tract and other pollen tubes speeds up the process. Whether the tube cell cytoplasm moves with or without the cortical cytoskeleton and plasma membrane complex is unclear. Remnants of both are found in the spent pollen tube, as well as some cytoplasm that gets trapped during formation of the callose walls (plugs) that sequester the tube cell at the tip. Moving animal cells also leave behind such remnants. The major difference between them and the tube cell is the considerable amount of ECM left as a train behind the moving tube cell system.

To discriminate between actin- or actin–myosin-mediated cell movement, we need myosin mutants in plants, as we have in *Dictyostelium,* to allow examination of the effect on pollination. In addition, the model must be supported by identification of the molecules in the stylar ECM and the tube cell wall and plasma membrane which are responsible for the haptotactic extension of pollen tubes. These ECM molecules should be capable of inducing focal adhesions in pollen tubes, and, once isolated, shown to facilitate the rapid extension of tubes *in vitro* in the same way as occurs *in vivo*.

During growth, are plant cells attaching to and spreading on their ECMs?

The view that pollen tubes can be considered as moving cell systems rather than as a growing cell is supported by the fact that the vegetative cell cytoplasm and the two sperm cells are sequestered in the tip of the pollen tube by continual production of walls (callose plugs) behind the vacuole of the vegetative cell. Therefore, the tube grows in volume and length, but the cells do not do so. Plant cells expand as they leave the vicinity of the apical meristems. This expansion may not be due to an increase in cytoplasmic volume, but may be caused by vacuolar expansion which acts to push the cytoplasm close to the periphery of the cell where the wall is expanding (Cosgrove, 1993). There must be continual production of plasma membrane and wall to account for this increase in size. If one compares this process to tip growth, the parallels are striking.

The actin-facilitated movement proposed in pollen tube extension could very well be occurring around the whole cell during expansion growth. The 'movement' in cell expansion growth would be in place, analogous to the attachment and spreading of animal cells on ECMs, and not along a track, as occurs in pollen tube extension or locomoting animal cells. Evidence for this comes from data on membrane–wall adhesion in plant cells, the presence of actin filaments associated with the adhesion zones on the cytoplasmic face, and numerous studies demonstrating that cell wall matrix materials are continuously deposited during expansion (Kutschera, 1991). Extensin or other HRGPs like the arabinogalactin-proteins (AGPs) could be the wall molecules involved in this process (Zhu *et al.,* 1993b). Many of the AGPs are known to be anchored in the plasma membrane (Komalavilas *et al.,* 1992; Pennell, 1992; Roberts, 1989) and may be important for cell division (Serpe and Nothnagel, 1994).

A recent study of the role of actin in cell elongation in *Avena* cells by Thimann *et al.* (1992) provides some support for the notion that actin polymerization itself is causally related to cell expansion. These workers were able to show that three treatments which destroyed actin microfilaments (cytochalasin D, the divalent ions calcium and magnesium, and iodoacetate) inhibited cell expansion, whereas the inhibition of growth by a change in osmoticum did not lead to actin microfilament breakdown. Thimann *et al.* proposed that polymerization of actin alone could provide the motive force to stretch the polysaccharide wall; this could lead to induction of enzymatic changes that break cross-links in the wall, and hence allow for expansion. We would add that adhesion molecules in the plasma membrane and wall are a necessary component of this model, as well as being needed to provide the anchor for the cytoskeleton. This proposal for cell expansion is very similar to the model for tip growth in pollen tubes, which is based on the models for cell attachment and spreading in animal systems. The prevailing notion is that parameters which control water uptake into cells determine the rate of cell growth, and that molecules which cause relaxation of cross-linking in the wall, thereby reducing turgor pressure, are the control points behind cell expansion (Cosgrove, 1993). Osmotic swelling is also a component of the actin-based model. Could a combination of these two mechanisms be operating in cell expansion during growth? Since many initial indicators of differentiation in cells which have left the meristems occur in their walls or ECMs, it is intriguing to speculate that links between the ECM, plasma membrane, cytoskeleton and nucleus at this time are the transducers of positional information in plant development.

Adhesions between the cytoskeleton, plasma membrane and wall are critical components of the models for pollination and plant development, but we know little about the molecules involved in plant systems, other than F-actin. Certainly, a number of known glycoproteins and proteoglycans in the plant cell wall could be involved in these interactions, including the HRGP extensin (Cassab and Varner, 1988; Miller and Fry, 1993; Showalter, 1993), the arabinogalactan-proteins (Roberts, 1989), glycine-rich glycoproteins (Wyatt and Carpita, 1993), or even pectins (Knox, 1992). The data so far are circumstantial but intriguing, and hopefully sufficient to stimulate further investigation.

References

Adair WS, Mecham RP. (1990) *Organization and Assembly of Plant and Animal Extracellular Matrix.* New York: Academic Press.

Battey NN, Blackbourn HD. (1993) Tansley Review No. 57. The control of exocytosis in plant cells. *New Phytol.* **125:** 307–338.

Bolwell GP. (1993) Dynamic aspects of the plant extracellular matrix. *Int. Rev. Cytol.* **146:** 261–323.

Boudreau N, Myers C, Bissell MJ. (1995) From laminin to lamin: regulation of tissue-specific gene expression by the ECM. *Trends Cell Biol.* **5:** 1–4.

Bronner-Fraser M. (1982) Distribution of latex beads and retinal pigment epithelial cells along the ventral neural crest pathway. *Dev. Biol.* **91:** 50–63.

Burridge K, Fath K, Kelly T, Nuckolls G, Turner C. (1988) Focal adhesion: transmembrane junctions between the extracellular matrix and the cytoskeleton. *Annu. Rev. Cell Biol.* **4:** 487–525.

Callow JA, Green JR (ed.). (1992) *Perspectives in Plant Cell Recognition.* Society for Experimental Biology Seminar Series 48. Cambridge: Cambridge University Press.

Cassab GI, Varner JE. (1988) Cell wall proteins. *Annu. Rev. Plant Physiol. Plant Mol. Biol.* **39:** 321–353.

Chaubal R, Wilmot VA, Willard KW. (1991) Visualization, adhesiveness, and cytochemistry of the extracellular matrix produced by urediniospore germ tubes of *Puccinia sorghi. Can. J. Bot.* **69:** 2044–2054.

Cheung AY, Hong Wang, Hen-ming Wu. (1995) A floral transmitting tissue-specific glycoprotein attracts pollen tubes and stimulates their growth. *Cell* **82:** 383–393.

Condeelis JS. (1974) The identification of F-actin in the pollen tube and protoplast of *Amaryllis belladona. Exp. Cell. Res.* **88:** 435–439.

Condeelis JS. (1992) Are all pseudopods created equal? *Cell Motil. Cytoskel.* **22:** 1–6.

Cooper JA. (1991) The role of actin polymerization in cell motility. *Annu. Rev. Physiol.* **53:** 585–605.

Cosgrove DJ. (1993) Water uptake by growing cells: an assessment of the controlling roles of wall relaxation, solute uptake, and hydraulic conductance. *Int. J. Plant Sci.* **154:** 10–21.

Dufour S, Duband J-L, Kornblihtt AR, Thiery JP. (1988) The role of fibronectins in embryonic cell migrations. *Trends Genet.* **4:** 198–203.

Euteneuer U, Schliwa M. (1984) Persistent, directional motility of cells and cytoplasmic fragments in the absence of microtubules. *Nature* **310:** 58–61.

Fry SC. (1986) Cross-linking of matrix polymers in the growing cell walls of angiosperms. *Annu. Rev. Plant Physiol.* **37:** 165–186.

Gaspar T, Penel CL, Thorpe T, Greppin H. (1982) *Peroxidases: a Survey of Their Biochemical and Physiological Roles in Higher Plants.* Geneva: University of Geneva Press.

Geiger B. (1989) Cytoskeleton-associated cell contacts. *Curr. Opin. Cell Biol.* **1:** 103–109.

Gens JS, McNally JG, Pickard BG. (1993) Resolution of binding sites for antibodies to integrin, vitronectin and fibronectin on onion epidermis protoplasts and depectinated walls. *Am. Soc. Grav. Space Biol.* (in press).

Gooday GW. (1992) The fungal surface and its role in sexual interactions. In: *Perspectives in Plant Cell Recognition* (Callow JA, Green JR, ed.). Society for Experimental Biology Seminar Series 48. Cambridge: Cambridge University Press, pp. 33–58.

Hayashi T, Yoshida K. (1988) Cell expansion and single-cell separation induced by colchicine in suspension-cultured soybean cells. *Proc. Natl Acad. Sci. USA* **85:** 2618–2622.

Heath IB. (1990) The roles of actin in tip growth of fungi. *Int. Rev. Cytol.* **123:** 95–127.

Hynes RO. (1992) Integrins: versatility, modulation, and signaling in cell adhesion. *Cell* **69:** 11–25.

Ingber DE. (1993) Cellular tensegrity: defining new rules of biological design that govern the cytoskeleton. *J. Cell Sci.* **104:** 613–627.

Iwanami Y. (1956) Protoplasmic movement in pollen grains and tubes. *Phytomorphology* **6:** 288–295.

Jackson SL, Heath IB. (1992) UV microirradiations elicit Ca^{2+}-dependent apex-directed cytoplasmic contractions in hyphae. *Protoplasma* **170:** 46–52.

Jackson SL, Heath IB. (1993) The dynamic behavior of cytoplasmic F-actin in growing hyphae. *Protoplasma* **173:** 23–34.

Jauh GY, Lord EM. (1995) Movement of the tube cell in the lily style in the absence of the pollen grain and the spent pollen tube. *Sex Plant Reprod.* **8:** 168–172.

Jauh GY, Lord EM. (1996) Localization of pectins and arabinogalactan-proteins in lily (*Lilium longiflorum* L.) pollen tube and style and their possible roles in pollination. *Planta* (in press).

Jermyn MA. (1978) Comparative specificity of concanavalin A and the β-lectins. *Aust. J. Plant Physiol.* **5:** 687–696.

Johnson-Flanagan AM, Singh J. (1986) Membrane deletion during plasmolysis in hardened and non-hardened plant cells. *Plant Cell Environ.* **9:** 299–305.

Joos U, van Aken J, Kristen U. (1994) Microtubules are involved in maintaining the cellular polarity in pollen tubes of *Nicotiana sylvestris. Protoplasma* **179:** 5–15.

Kagawa T, Kadota A, Wada M. (1992) The junction between the plasma membrane and the cell wall in fern protonemal cells, as visualized after plasmolysis, and its dependence on arrays of cortical microtubules. *Protoplasma* **170:** 186–190.

Kaminskyj SGW, Heath IB. (1995) Integrin and spectrin homologues, and cytoplasm–wall adhesion in tip growth. *J. Cell Sci.* **108:** 849–856.

Kaminskyj SGW, Garrill A, Heath IB. (1992) The relation between turgor and tip growth in *Saprolegnia ferax:* turgor is necessary, but not sufficient to explain apical extension rates. *Exp. Mycol.* **16:** 64–75.

Knox JP. (1992) Cell adhesion, cell separation and plant morphogenesis. *Plant J.* **1:** 175–183.

Knox RB. (1984) Pollen pistil interactions. *Annu. Rev. Plant Physiol.* **17:** 508–608.

Komalavilas P, Zhu J-K, Nothnagel EA. (1992) Arabinogalactan-proteins from the suspension culture medium and plasma membrane of rose cells. *J. Biol. Chem.* **266:** 15956–15965.

Kroh M, Knuiman B. (1982) Ultrastructure of cell wall and plug of tobacco pollen tubes after chemical extraction of polysaccharides. *Planta* **154:** 241–250.

Kropf DL. (1992) Establishment and expression of cellular polarity in fucoid zygotes. *Microbiol. Rev.* **56:** 316–339.

Kropf DL, Kloarge B, Quatrano RS. (1988) Cell wall is required for fixation of the embryonic axis in *Fucus* zygotes. *Science* **239:** 187–190.

Kutschera U. (1991) Regulation of cell expansion. In: *The Cytoskeletal Basis of Plant Growth and Form* (Lloyd C, ed.). London: Academic Press, pp. 149–158.

Kwon YH, Epstein L. (1993) A 90-kDa glycoprotein associated with adhesion of *Nectria haematococca* macroconidia to substrata. *Mol. Plant-Microbe Interact.* **6:** 481–487.

Labarca C, Loewus F. (1972) The nutritional role of pistil exudate in pollen tube wall formation in *Lilium longiflorum. Plant Physiol.* **50:** 7–14.

Lee-Stadelmann OY, Bushnell WR, Stadelmann EJ. (1984) Changes of plasmolysis form in epidermal cells of *Hordeum vulgare* infected by *Erysiphe graminis:* evidence for increased membrane–wall adhesion. *Can. J. Bot.* **62:** 1714–1723.

Li Y-Q, Bruun L, Pierson E, Cresti M. (1992) Periodic deposition of arabinogalactan epitopes in the cell wall of pollen tubes of *Nicotiana tabacum* L. *Planta* **188:** 532–538.

Li Y-Q, Chen F, Linskens HF, Cresti M. (1994) Distribution of unesterified and esterified pectins in cell walls of pollen tubes of flowering plants. *Sex Plant Reprod.* **7:** 145–152.

Lloyd CW. (1987) The plant cytoskeleton: the impact of fluorescence microscopy. *Annu. Rev. Plant Physiol.* **38:** 119–139.

Lolle SJ, Cheung AY, Sussex IM. (1992) *Fiddlehead:* an *Arabidopsis* mutant constitutively expressing an organ fusion program that involves interactions between epidermal cells. *Dev. Biol.* **152:** 383–392.

Lord EM, Sanders LC. (1992) Roles for the extracellular matrix in plant development and pollination: a special case of cell movement in plants. *Dev. Biol.* **153:** 16–28.

Marcantonio EE, Hynes RD. (1988) Antibodies to the conserved cytoplasmic domain of the integrin β-subunit react with proteins in vertebrates, invertebrates, and fungi. *J. Cell Biol.* **106:** 1765–1772.

Meyer Y, Abel WD. (1974) Importance of the cell wall for cell division and in the activity of the cytoplasm in cultured tobacco protoplasts. *Planta* **123:** 33–40.

Miller DB, Callaham DA, Gross DJ, Hepler PK. (1992) Free Ca^{2+} gradient in growing pollen tubes of *Lilium. J. Cell Sci.* **101:** 7–12.

Miller JG, Fry SC. (1993) Spinach extensin exhibits characteristics of an adhesive polymer. *Acta. Bot. Neerl.* **42:** 221–231.

Mityazaki K, Hamano T, Hayashi M. (1992) *Physarum* vitronectin-like protein: an Arg-Gly-Asp-dependent cell-spreading protein with a distinct NH_2-terminal sequence. *Exp. Cell. Res.* **199:** 106–110.

Nasrallah JB, Nasrallah ME. (1993) Pollen-stigma signalling in the sporophytic self-incompatibility response of *Brassica. Plant Cell* **5:** 1325–1335.

Nasrallah JB, Stein JC, Kandasamy MK, Nasrallah ME. (1994) Signaling the arrest of pollen tube development in self-incompatible plants. *Science* **226:** 1505–1508.

Nicholson RL, Epstein L. (1991) Adhesion of fungi to the plant surface: prerequisite for pathogenesis. In: *The Fungal Spore and Disease Initiation in Plants and Animals* (Cole GT, Hoch HC, ed.). New York: Plenum Press, pp. 3–23.

Öbrink B. (1991) C-CAM (Cell-CAM 105) — a member of the growing immunoglobulin superfamily of cell adhesion proteins. *BioEssays* **13:** 227–234.

O'Driscoll D, Hann C, Read SM, Steer MW. (1993) Endocytotic uptake of fluorescent dextrans by pollen tubes grown *in vitro. Protoplasma* **175:** 126–130.

Oh BJ, Balint DE, Giovannoni JJ. (1995) A modified procedure for PCR-based differential display and demonstration of use in plants for isolation of genes related to fruit ripening. *Plant Mol. Biol. Rep.* **13:** 70–81.

Penman S. (1995) Rethinking cell structure. *Proc. Natl Acad. Sci. USA* **92:** 5251–5257.

Pennell RI. (1992) Cell surface arabinogalactan proteins, arabinogalactans and plant development. In: *Perspectives in Plant Cell Recognition* (Callow JA, Green JR, ed.). Society for Experimental Biology Seminar Series 48. Cambridge: Cambridge University Press, pp. 105–121.

Picket-Heaps JD. (1968) Xylem wall deposition: radioautographic investigation using lignin precursors. *Protoplasma* **65:** 181–205.

Pierson ES, Derksen J, Traas JA. (1986) Organization of microfilaments and microtubules in pollen tubes grown *in vitro* or *in vivo* in various angiosperms. *Eur. J. Cell Biol.* **41:** 14–18.

Pont-Lezica RF, McNally JG, Pickard BG. (1993) Wall-to-membrane linkers in onion epidermis: some hypotheses. *Plant Cell Environ.* **16:** 111–123.

Preissner KT. (1991) Structure and biological role of vitronectin as multifunctional regulator in the homostatic and immune systems. *Blut* **59:** 419–431.

Preuss D, Lemieux B, Yen G, Davis RW. (1993) A conditional sterile mutation eliminates surface components from *Arabidopsis* pollen and disrupts cell signaling during fertilization. *Genes Dev.* **7:** 974–985.

Quatrano RS, Brian L, Aldridge J, Schultz T. (1991) Polar axis fixation in *Fucus* zygotes: components of the cytoskeleton and extracellular matrix. *Development* Suppl. 1: 11–16.

Read ND, Kellock LJ, Knight H, Trewavas AJ. (1992) Contact sensing during infection by fungal pathogens. In: *Perspectives in Plant Cell Recognition* (Callow JA, Green JR, ed.). Society for Experimental Biology Seminar Series 48, Cambridge: Cambridge University Press, pp. 137–172.

Reuzeau G, Pont-Lezica RF. (1995) Comparing plant and animal extracellular matrix-cytoskeleton connections — are they alike? *Protoplasma* **186:** 113–121.

Roberts AW, Haigler CH. (1989) Rise in chlorotetracycline fluorescence accompanies tracheary element differentiation in suspension cultures of *Zinnia. Protoplasma* **152:** 37–45.

Roberts K. (1989) The plant extracellular matrix. *Curr. Opin. Cell Biol.* **1:** 1020–1027.

Roberts K. (1990) Structures at the plant cell surface. *Curr. Opin. Cell Biol.* **2:** 920–928.

Rosen WG. (1971) Pistil–pollen interactions in *Lilium.* In: *Pollen: Development and Physiology* (Heslop-Harrison J, ed.). London: Butterworth, pp. 239–254.

Rubinstein AL, Broadwater AH, Lowrey KB, Bedinger PA. (1995) *Pex1,* a pollen-specific gene with an extension-like domain. *Plant Biol.* **92:** 3086–3090.

Russell SD. (1992) Double fertilization in sexual reproduction in flowering plants. In: *Sexual Reproduction in Flowering Plants* (Russell SD, Dumas C, ed.). *Int. Rev. Cytol.* **140:** 357–388.

Sanders LC, Lord EM. (1989) Directed movement of latex particles in the gynoecia of three species of flowering plants. *Science* **243:** 1606–1608.

Sanders LC, Lord EM. (1992) A dynamic role for the stylar matrix in pollen tube extension. In: *Sexual Reproduction in Flowering Plants* (Russell SD, Dumas C, ed.). *Int. Rev. Cytol.* **140:** 297–318.

Sanders LC, Wang C-S, Walling LL, Lord EM. (1991) A homolog of the substrate adhesion molecule, vitronectin, occurs in four species of flowering plants. *Plant Cell* **3:** 629–635.

Schindler M, Meiners S, Cheresh DA. (1989) RGD-dependent linkage between plant cell wall and plasma membrane: consequences for growth. *J. Cell Biol.* **108:** 1955–1965.

Serpe MD, Nothnagel EA. (1994) Effects of Yariv phenylglycosides on *Rosa* cell suspensions: evidence for the involvement of arabinogalactan-proteins in cell proliferation. *Planta* **1193:** 542–550.

Showalter AM. (1993) Structure and function of plant cell wall proteins. *Plant Cell* **4:** 9–23.

Spudich JA. (1989) In pursuit of myosin function. *Cell Regul.* **1:** 1–11.

Steer MW. (1990) Role of actin in tip growth. In: *Plant and Fungal Cells* (Heath IB, ed.). London: Academic Press, pp. 119–145.

Steer MW, Steer JM. (1989) Tansley Review No. 16. Pollen tube tip growth. *New Phytol.* **111**: 323–358.

Stossel TP. (1993) On the crawling of animal cells. *Science* **260**: 1086–1094.

Tang X, Hepler PK, Scordilis SP. (1989) Immunochemical and immunocytochemical identification of a myosin heavy chain polypeptide in *Nicotiana* pollen tubes. *J. Cell Sci.* **92**: 569–574.

Thimann KV, Reese K, Nachmias VT. (1992) Actin and the elongation of plant cells. *Protoplasma* **171**: 153–166.

Tiwari SC, Polito VS. (1988) Organization of the cytoskeleton in pollen tubes of *Pyrus communis*: a study employing conventional and freeze-substitution electron microscopy, immunofluorescence and rhodamine–phalloidin. *Protoplasma* **174**: 100–112.

Van den Ende H. (1992) Sexual signalling in *Chlamydomonas*. In: *Perspectives in Plant Cell Recognition* (Callow JA, Green JR, ed.). Society for Experimental Biology Seminar Series 48. Cambridge: Cambridge University Press, pp. 1–18.

Van Engelen FA, De Vries SC. (1992) Extracellular proteins in plant embryogenesis. *Theor. Appl. Genet.* **8**: 66–70.

Verbeke JA, Walker DB. (1986) Morphogenetic factors controlling differentiation and dedifferentiation of epidermal cells in the gynoecium of *Catharanthus roseus*. *Planta* **168**: 43–49.

Vreeland V, Grotkopp E, Espinosa S, Quiroz D, Laetsch WM, West J. (1993a) The pattern of cell wall adhesive formation by *Fucus* zygotes. *Hydrobiologia* **260/261**: 485–491.

Vreeland V, Parungo C, West JA. (1993b) Vanadate peroxidase in the development of wall adhesive in *Fucus* zygotes. *Mol. Biol. Cell* **4**: 147a (abstract).

Wagner VT, Matthysse AG. (1992) Involvement of a vitronectin-like protein in attachment of *Agrobacterium tumefaciens* to carrot suspension culture cells. *J. Bacteriol.* **174**: 5999–6003.

Wagner VT, Brian L, Quatrano RS. (1992) Role of a vitronectin-like molecule in embryo adhesion of the brown alga *Fucus*. *Proc. Natl Acad. Sci. USA* **89**: 3644–3648.

Walker DB. (1975) Postgenital carpel fusion in *Catharanthus roseus*. III. Fine structure of the epidermis during and after fusion. *Protoplasma* **86**: 43–63.

Wang CS, Walling LL, Gu YQ, Ware CF, Lord EM. (1994) Two classes of proteins and mRNAs in *Lilium longiflorum* identified by human vitronectin probes. *Plant Physiol.* **104**: 711–717.

Watson BA, Waaland SD. (1983) Partial purification and characterization of a glycoprotein cell fusion hormone from *Griffithsia pacifica*, a red alga. *Plant Physiol.* **71**: 327–332.

Wayne R, Staves MP, Leopold AC. (1992) The contribution of the extracellular matrix to gravisensing in characean cells. *J. Cell Sci.* **101**: 611–623.

Wyatt SE, Carpita NC. (1993) The plant cytoskeleton–exocellular matrix continuum. *Trends Cell Biol.* **3**: 413–417.

Ye Z-H, Song Y-R, Marcus A, Varner JE. (1991) Comparative localization of three classes of cell wall proteins. *Plant J.* **1**: 175–183.

Zhu J-K, Shi J, Singh U, Wyatt SE, Bressan RA, Hasegawa PM, Carpita NC. (1993a) Enrichment of vitronectin- and fibronectin-like proteins in NaCl-adapted plant cells and evidence for their involvement in plasma membrane–cell wall adhesion. *Plant J.* **3**: 637–646.

Zhu J-K, Bressan RA, Hasegawa PM. (1993b) Loss of arabinogalactan-proteins from the plasma membrane of NaCl-adapted tobacco cells. *Planta* **190**: 221–226.

Zhu J-K, Damsz B, Kononowicz AK, Bressan RA, Hasegawa PM. (1994) A higher plant extracellular vitronectin-like adhesion protein is related to the translational elongation factor-1α. *Plant Cell* **6**: 393–404.

Chapter 3

Membrane conservation during plasmolysis

Karl J. Oparka, Denton A.M. Prior and John W. Crawford

1. Introduction

Plants and bacteria possess the unique ability to contract their protoplasts in response to osmotic shock, allowing them to withstand extremes of osmotic imbalance. This phenomenon, termed plasmolysis, was defined over a century ago by De Vries (1877) as the separation of the living protoplasmic envelope from the cell wall, caused by the action of an external water-withdrawing solution. Various definitions of plasmolysis can be found in the botanical literature, but the simple generic definition given above will be adopted here.

Despite the extensive uses of plasmolysis as a tool in plant cell biology (see Oparka, 1994), considerable ignorance surrounds our knowledge of the fundamental membrane changes which occur during this process. It is now clear that plasmolysis is accompanied by a complex series of substructural changes which allow the plant cell to conserve membrane area during a plasmolysis/deplasmolysis cycle. How such drastic changes are achieved, while still allowing the cell to continue functioning, will be explored in this chapter. Emphasis will be placed on several recent studies in which new insights have been gained into the behaviour of membranes during plasmolysis. This chapter will thus impinge on other related topics covered in this book. We have been selective in our choice of subject material and have focused our attention on potential strategies for membrane conservation during plasmolysis. For a more complete treatise on the structural changes which accompany plasmolysis, the reader is referred to the recent review by Oparka (1994), and for information on the changes in cell–water relations which occur during plasmolysis, the comprehensive reviews of Lee-Stadelmann and Stadelmann (1989) and Stadelmann (1956, 1966) should be consulted.

2. Alterations of the plasma membrane (PM) during plasmolysis

One of the first visible, discernible effects of plasmolysis is the separation of the PM from the cell wall. Several published works often refer to 'rounding up' of the protoplast, and the final form of plasmolysis which is achieved is often indicative of the nature of the plasmolyticum surrounding the cell (Lee-Stadelmann and Stadelmann, 1989; Oparka, 1994; Stadelmann, 1956, 1966). During plasmolysis, protoplast volume may decrease to as little as 15% of the original volume without loss of membrane semi-permeability (Palta and Lee-Stadelmann, 1983). Such massive changes in volume clearly have important implications for the PM and tonoplast and yet, surprisingly, the membrane changes which accompany plasmolysis have received little attention.

The concept that the protoplast 'shrinks' during plasmolysis is in many respects a misleading one. What occurs, in effect, is a massive rearrangement of the protoplast's surface area-to-volume ratio. Initially, separation of the PM may occur at discrete locations within the cell. With progressive plasmolysis (as the protoplast 'rounds up'), the impression, from conventional light microscopy, is that the PM maintains contact with the cell wall over only a small part of its surface. As will be seen, this is a misleading image of the plasmolysed plant cell.

One of the major obstacles encountered by the plant cell during a plasmolysis/deplasmolysis cycle is the inelasticity of the PM. This membrane can withstand a maximum elastic stretching of only about 2%, larger expansions necessitating the incorporation of new material into the membrane (Wolfe *et al.*, 1985). Such membrane inelasticity underlies the need for mechanisms which allow for conservation of PM surface area during the rigours of a plasmolysis/deplasmolysis cycle.

3. Hechtian strands

Many early microscopists reported the presence of thread-like strands connecting the protoplasts of plasmolysed cells with the cell wall. Hecht (1912) is generally credited with the initial observation of these delicate structures (for an historical overview see Oparka, 1994), which subsequently became referred to as 'Hechtian strands' (Drake and Carr, 1978; Oparka *et al.*, 1994; Pont-Lezica *et al.*, 1993). Because of their narrow diameter, only small portions of Hechtian strands have been observed under the electron microscope (Drake and Carr, 1978; Oparka *et al.*, 1994), although they are considerably more conspicuous under the scanning electron microscope (Attree and Sheffield, 1985). Recently, structures resembling Hechtian strands have been reported to occur during wound-induced contraction of the protoplast of the giant-celled green alga *Ernodesmis* (Goddard and La Claire, 1993).

Hechtian strands are barely visible by conventional light microscopy. However, Oparka *et al.* (1994) showed recently that, in plasmolysed onion epidermal cells, Hechtian strands stain intensely with the fluorescent membrane-potential probe $DiOC_6$ (see *Figures 1, 2* and *3*). The use of this probe highlights the extent to which the PM is conserved as fine strands outside the main protoplast body, and also emphasizes the extremely high number of wall–membrane attachment sites which form (or are revealed) during plasmolysis.

Since a large proportion of the PM surface area may potentially be accommodated external to the contracting protoplast, Oparka *et al.* (1994) considered the possibility that extensive Hechtian strand networks (such as those shown in *Figure 1c*) might conserve PM surface area during a plasmolysis/deplasmolysis cycle. Since the surface area of a cylindrical strand increases linearly with its length, while the area of the contracting plasma membrane decreases with the square of its radius, the strands cannot conserve the surface area of the PM unless (a) some become detached during contraction, (b) the radius decreases as the PM contracts, or (c) the network becomes branched. In fact, branching of Hechtian strands (option c) was observed routinely, and as the protoplasts contracted the strands remained under tension, frequently branching close to the surface of the wall (*Figure 1c*). By approximating the branched network by a fractal, with a time-dependent fractal dimension, it was possible to demonstrate that Hechtian strands can potentially accommodate the entire surface area of contraction during plasmolysis (Oparka *et al.*, 1994).

4. Hechtian attachment sites

Although it is generally agreed that Hechtian strands are formed from the PM during plasmolysis, there has been considerable debate as to their site of origin at the cell wall. A popular concept is that Hechtian strands arise from individual plasmodesmata (Drake *et al.*, 1978; Strasburger *et al.*, 1983). Some Hechtian strands may in fact be drawn out from sites of plasmodesmata, maintaining PM continuity between cells (Oparka *et al.*,1994; Strasburger *et al.*,

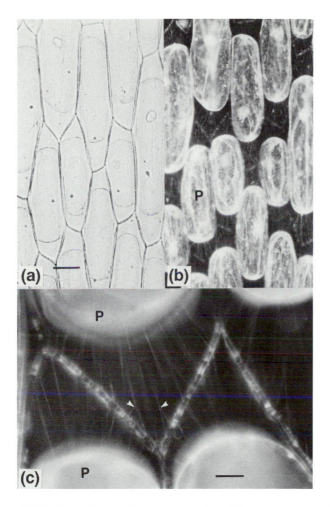

Figure 1. Convex plasmolysis of onion epidermal cells. The same cells are seen in both (a) bright field and (b) fluorescence after staining with the membrane-potential probe $DiOC_6$. Scale bar = 100 μm. Extensive Hechtian strand networks connect the protoplasts to the cell walls. (c) The same cells shown at higher magnification, which reveals branching of the Hechtian strands (darts) and the presence of $DiOC_6$-positive membranes attached to the wall. Scale bar = 20 μm. P, protoplast. Reproduced from Oparka *et al.* (1994) with permission from Blackwell Science Ltd.

1983). Recently, we have used confocal laser scanning microscopy (CLSM) to examine the attachment sites of Hechtian strands. As can be seen from *Figures 2* and *3,* several of the strands arise at or close to the entrances of plasmodesmata. However, it is now clear that the strands may originate from wall sites other than those containing plasmodesmata. Recently, Pont-Lezica *et al.* (1993) found that, in onion epidermal cells, in addition to forming on anticlinal walls which abut other epidermal cells, the Hechtian strands were just as abundant on periclinal walls which lack plasmodesmata. They referred to these sites as 'Hechtian attachment sites', a terminology which will be adopted here. In leaf trichomes of *Nicotiana clevelandii*, abundant Hechtian strands can be seen to attach the contracted protoplast to lateral walls lacking plasmodesmata, as well as to end walls containing abundant plasmodesmata (Oparka and Prior, unpublished data). Thus, it is important to establish that, although Hechtian strands may associate closely with plasmodesmata, they are not dependent on these structures for anchorage.

5. Wall–membrane interactions during plasmolysis

The means by which the PM remains strongly anchored to the wall at specific sites during plasmolysis has not yet been determined, although it has been speculated that transmembrane linkages, similar to those found in animal systems, may be responsible for anchorage. In certain

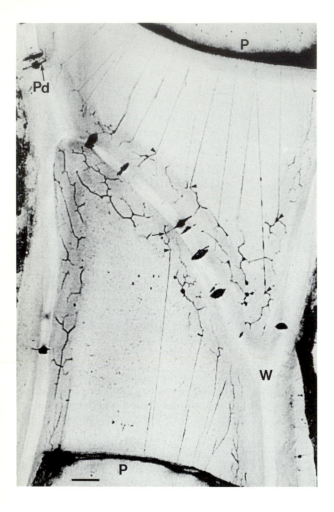

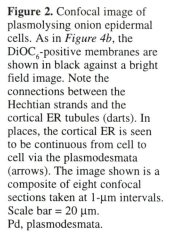

Figure 2. Confocal image of plasmolysing onion epidermal cells. As in *Figure 4b*, the DiOC$_6$-positive membranes are shown in black against a bright field image. Note the connections between the Hechtian strands and the cortical ER tubules (darts). In places, the cortical ER is seen to be continuous from cell to cell via the plasmodesmata (arrows). The image shown is a composite of eight confocal sections taken at 1-µm intervals. Scale bar = 20 µm. Pd, plasmodesmata.

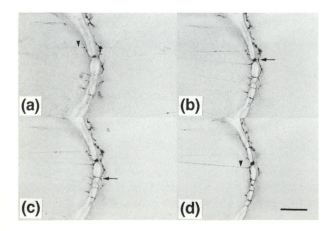

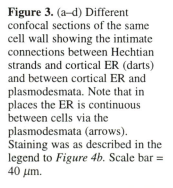

Figure 3. (a–d) Different confocal sections of the same cell wall showing the intimate connections between Hechtian strands and cortical ER (darts) and between cortical ER and plasmodesmata. Note that in places the ER is continuous between cells via the plasmodesmata (arrows). Staining was as described in the legend to *Figure 4b*. Scale bar = 40 µm.

animal cells, specific PM receptors (integrins) recognize a family of extracellular adhesive glycoproteins via the amino acid sequence Arg-Gly-Asp (RGD; Burridge *et al.*, 1988; Ruoslahti and Pierschbacher, 1987). Some of these proteins bind to the RGD sequence of a single adhesion protein only, whereas others recognize groups of them. On the cytoplasmic side of the PM, the receptors connect the extracellular matrix to the cytoskeleton.

Growing evidence has accumulated for the presence of RGD-recognition systems in plants (see Chapter 2 by Lord *et al.*). Schindler *et al.* (1989) introduced synthetic peptides containing the RGD sequence into growing suspension cultures of soybean, and found that these induced dramatic morphological and proliferative changes, apparently mediated by a 70–72 kDa polypeptide localized to the PM–wall interface. In the presence of RGD-containing peptides, the PM separated readily from the cell wall under non-plasmolysing conditions. Subsequent work has shown that the RGD peptide has significant effects on the gravity-dependent polarity of cytoplasmic streaming in internodal cells of *Chara* (Wayne *et al.*, 1992). In this case, tight adhesion of the PM to the cell wall after plasmolysis was observed at the ends of cells, and these appeared to be the sites for sensing the effects of gravity.

Do Hechtian attachment sites reveal the location of RGD-binding proteins in plant cells? Pont-Lezica *et al.* (1993) have argued strongly that specific wall-bound entities are loosely bound to the mural face of the PM, providing physical continuity between wall and PM, and thus performing a function comparable to that of integrins and other matrix receptor complexes. They have also suggested that, as in animal cells, cytoskeletal elements can bind to the cytosolic face of the membrane of these complexes. Pont-Lezica *et al.* (1993) examined the distribution of hydroxyproline-rich glycoprotein (HRGP) as a possible candidate for a wall-to-membrane linker in onion epidermal cells using fluoresence staining with antibodies. They speculated that the antigen was responsible for adhesion of PM to wall at the same Hechtian attachment sites that were visualized by plasmolysis. Arabinogalactan proteins have also been implicated as sites of PM–wall attachment, due to their ability to bind to β-glycan (Pennell *et al.*, 1989). It must be stressed, however, that to date no direct evidence has been forthcoming to suggest that the Hechtian attachment sites do indeed represent the foci for RGD-binding proteins.

6. Hechtian strands arise during freezing of cold-hardened plant cells

Plasmolysis is a common feature in higher plant cells during a freeze–thaw cycle. As the temperature of the cell is lowered, ice grows outside the cell by withdrawal of water from inside, and the cell is subjected to a freeze-induced dehydrative stress (Singh, 1979). Freezing injury arises through a complex series of events, although the PM is generally considered to be the primary site of freezing injury (Steponkus, 1990).

Although much attention has been paid to the behaviour of the PM during freezing (Palta and Li, 1978; Yoshida, 1982), particularly in isolated protoplasts (Steponkus, 1990), the relationship between the cell wall and the PM has been studied by relatively few authors. In an early study of cold hardiness, Scarth (1941) noticed that the ability to form a Hechtian strand network differed in cold-hardened and non-hardened cells of *Hydrangea*. In the non-hardened cells the Hechtian strands were observed to 'snap and crumple' at high external solute concentrations, while those in the hardened cells remained intact. This important observation was recently explored in more detail by Johnson-Flanagan and Singh (1986), who examined the plasmolytic behaviour of cell suspensions of three species, with and without prior cold-hardening at low temperature. The non-hardened cells plasmolysed intensely in external salt solutions, and underwent expansion-induced lysis upon deplasmolysis. During plasmolysis the PM was deleted as relatively few Hechtian strands stretched between the cell wall and the protoplast. During deplasmolysis the initial rapid circularization of the protoplast resulted in the strands breaking and vesiculating.

These vesicles remained free and could not be reincorporated into the expanding PM. During plasmolysis of the hardened cells, the volume-to-surface area ratio was uncoupled through the formation of many Hechtian strands, and these remained continuous with the receding protoplast. Deplasmolysis of the hardened cells proceeded in such a way that the rate of expansion never exceeded the rate of membrane reincorporation. Consequently, the PM was successfully retrieved from the strands as the protoplast expanded.

Johnson-Flanagan and Singh (1986) suggested that the Hechtian strands might confer survival through one of two possible mechanisms. First, by uncoupling the volume-to-surface area ratio, the need to delete membrane to maintain this ratio is circumvented. Secondly, the availability of membrane material in hardened cells may reduce the tension exerted on the strands to a level that can be maintained, whereas a lack of membrane material in the non-hardened cells causes the tension to become too great, and so the strands are released.

The collapse of the cell wall with the PM is a generally accepted phenomenon during extracellular freezing (Palta and Li, 1978), and this event may reduce the need for the PM to lose surface area during freeze-induced dehydration (Singh and Miller, 1985). Palta and Li (1978) have suggested that plasmolysis may not in fact occur in most plants and plant organs during freezing, and that shrinkage and collapse of the cell wall together with the PM are the norm. However, plasmolysis appears to be a general feature that occurs following the freezing of suspension-cultured cells.

7. PM vesiculation

Hechtian strand formation is only one of a number of strategies for conserving PM surface area during plasmolysis. While debate still continues about the evidence for fluid-phase and receptor-mediated endocytosis in turgid plant cells (Oparka *et al.*, 1993; Robinson and Hedrich, 1991; Robinson and Hillmer, 1990), several studies have now clearly demonstrated that endocytic vesiculation of the PM may accompany osmotic contraction of the plant-cell protoplast. While this phenomenon cannot be considered as endocytosis in the strictest sense (Robinson and Hillmer, 1990), the ability of the PM to become deleted as endocytic vesicles is a feature which has been overlooked in the numerous investigations which have utilized plasmolysis as a tool in plant-cell biology. Indeed, recent evidence would suggest that both Hechtian strand formation and endocytic vesiculation of the PM may occur concurrently during plasmolysis (see Oparka, 1994).

Robinson *et al.* (1991) have proposed that the term 'osmocytosis' should be used for osmotic deletion of the PM in order to distinguish this phenomenon from other types of endocytosis in which vesicle internalization is followed by coordinated fusion with other subcellular compartments. Cocking and co-workers (see Cocking, 1972) were instrumental in drawing attention to the occurrence of PM vesiculation in their initial studies of the isolation of plant protoplasts. Such vesiculation was at first a problem, causing extensive uptake of crude enzyme solutions into the protoplasts, and for this reason it became advantageous to pre-plasmolyse the cells before the addition of cell-wall-degrading enzymes (Withers and Cocking, 1972). Cocking's research group was also able to demonstrate the uptake of electron-opaque probes such as ferritin (Power and Cocking, 1970) and thorium dioxide (Withers and Cocking, 1972) into PM vesicles during protoplast preparation. Particles as large as bacteria could be incorporated into the protoplasts provided that they were added during plasmolysis (Davey and Cocking, 1972).

Oparka *et al.* (1990) plasmolysed onion epidermal cells in the presence of Lucifer Yellow CH (LYCH), and demonstrated that during the onset of plasmolysis a discrete population of PM-derived vesicles became labelled with the probe. Immediately following deplasmolysis, the vesicles adhered to the PM, but with time they became mobile and underwent vigorous streaming in the cytosol. By plasmolysing the cells a second time in the presence of the fluorescent probe Cascade Blue, a second distinct population of vesicles became labelled. The vesicle populations remained separate and did not fuse with one another, or with any other internal compartment, presumably because they did not possess the membrane signals that would allow them to do so (Oparka *et al.*, 1991). Interestingly, the membrane incorporated into PM vesicles was not used to replace the PM during deplasmolysis.

In a recent study, Deikmann *et al.* (1993) examined the osmotic behaviour of guard-cell protoplasts compared with intact guard cells. During osmotic contraction, guard-cell protoplasts produced numerous PM vesicles, while intact guard cells underwent volume changes without the production of vesicles. Thus it appears that intact guard cells must undergo some other as yet unidentified means of PM conservation during osmotic contraction and expansion. The PM deleted into vesicles during osmotic contraction is smooth (Gordon-Kamm and Steponkus, 1984; Oparka *et al.*, 1994) and, with time, wall materials may be deposited within such vesicles (Pearce *et al.*, 1974), clearly indicating their derivation from the PM.

8. Deletion of the tonoplast during plasmolysis

Despite the massive changes in vacuolar surface area which accompany plasmolysis, there have been extremely few observations of the fate of the vacuole during osmotic contraction. In non-hardened plant cells, Johnson-Flanagan and Singh (1986) observed that the tonoplast vesiculated into the vacuole as 'sac-like, rod-like or doughnut-shaped structures'. In guard-cell protoplasts, osmotic contraction results in the fragmentation of the vacuole into many small sub-vacuoles or 'mini-vacuoles'. Under the electron microscope, such vacuoles are apparently indistinguishable from those formed from the PM by endocytic vesiculation (Deikmann *et al.*, 1993). Since Hechtian strand formation is clearly not an option open to the tonoplast, the massive surface area changes incurred by the tonoplast imply vesiculation of this membrane on a massive scale, and, furthermore, rapid reinsertion of this membrane at a pace that matches that of the PM during deplasmolysis. As yet we know little about the mechanisms whereby these membrane alterations are achieved and coordinated.

9. PM vesiculation during freezing

Vesiculation of the PM forms an integral part of the freezing resistance mechanism of isolated protoplasts, a phenomenon studied extensively by Steponkus and co-workers (see Steponkus, 1990). When the tension in the PM is reduced to zero in non-hardened rye leaf protoplasts, deletion of membrane occurs via endocytic vesiculation of the PM. Dowgert and Steponkus (1984) used computer-enhanced video microscopy to observe this phenomenon during osmotic contraction. The vesicles were incorporated into the cytoplasm and had a diameter of 0.1–1.5 µm. Vesicle formation was not injurious, but sufficiently large area reductions were irreversible, causing the non-hardened protoplasts to undergo expansion-induced lysis. This phenomenon occurs when a critical tension in the PM is attained during volumetric expansion. Since the intrinsic elastic expansion of the PM is only 2–3% (Wolfe *et al.*, 1985), the larger changes in area (30–50%) that occurred during osmotic expansion required new membrane to be incorporated into the PM. However, as in the case of deplasmolysing onion epidermal cells (Oparka *et al.*, 1990, 1994), the material deleted during osmotic contraction of isolated protoplasts was not incorporated into the plane of the membrane during expansion (Gordon-Kamm and Steponkus, 1984), and the source of the membrane material used for re-expansion is at present unknown (Steponkus, 1990).

10. Hechtian strands versus PM vesicles

From the foregoing evidence, it would appear that both Hechtian strand formation and endocytic vesiculation of the PM may occur concurrently during plasmolysis. By comparing predicted and observed network characteristics, we have shown previously (Oparka et $al.$, 1993) that Hechtian strands are capable of accommodating the surface area of the PM during contraction. It is also possible to calculate whether the observed numbers of vesicles which are produced during plasmolysis are sufficient to account for the necessary reduction in surface area. If we represent the PM as a cylinder of radius a and length b, with hemispherical caps on either end, then the number, n, of vesicles formed during contraction is

$$n = \left(\frac{a_1}{r}\right)^2 \left[1 - \left(\frac{a_0}{a_1}\right)^2\right]\left(\frac{1}{2}\frac{b}{a}+1\right)$$

where a_1, a_0 are the cylinder radii before and after contraction, respectively, and the ratio b/a is assumed to remain constant during contraction. Taking $a_1 = 5$ μm and $a_0 = 4$ μm (implying a 36% reduction in surface area), $r = 1$ μm and $b/a = 8$, we find that $n = 45$. This is of the same order of magnitude as the observed numbers of vesicles formed during plasmolysis.

Since endocytic PM vesicles must be accommodated within the contracting membrane, the accumulated volume of vesicles formed must clearly be less than the volume available for packing. If v is the total volume of vesicles formed, and V_0 is the enclosed volume of the contracted PM, simple geometrical arguments can be used to show that

$$\frac{v}{V_0} = \frac{r}{2a_1}\left(\frac{a_1}{a_0}\right)\left[\frac{1-\left(a_0a_1\right)^2}{1+3/4\left(b/a\right)}\right]$$

If ε is the fraction of V_0 available for packing by vesicles (having a maximum value of 74% for close-packed spherical vesicles in an otherwise empty PM), there is a lower limit on the value of a_0/a_1, obtained from the solution of

$$\varepsilon - \frac{r}{2a_1}\left(\frac{a_1}{a_0}\right)^3\left[\frac{1-\left(a_0a_1\right)^2}{1+3/4\left(b/a\right)}\right]$$

Taking r, a_1 and b/a as before, we find that $a_0/a_1 > 0.35$ for $\varepsilon = 30\%$, and $a_0/a_1 > 0.27$ for $\varepsilon = 70\%$. Thus the limiting value lies around 0.3, corresponding to a maximum possible reduction in surface area by endocytic vesiculation of 90%. Since this is far larger than any observed reduction during plasmolysis, it is clear that packing constraints imposed on the production of endocytic vesicles would not limit PM contraction during plasmolysis, and, as such, the latter is a plausible mechanism (in addition to Hechtian strand formation) for conserving surface area.

However, one of the key questions which has seldom been addressed concerns the determination of precisely which factors regulate the balance between the formation of Hechtian strands and endocytic vesicles. Clearly, the use of protoplasts underestimates the contribution of Hechtian strands to PM conservation during osmotic contraction of intact plant cells during plasmolysis.

Johnson-Flanagan and Singh (1986) performed one of several studies which have examined the relationship between Hechtian strands and PM vesicles. They found that, while cold-hardened cells formed an abundance of Hechtian strands, the non-hardened cells produced few strands and deleted the PM as endocytic vesicles. These vesicles could not be reincorporated into the PM during osmotic expansion, causing the protoplasts to undergo lysis. During plasmolysis of the non-hardened cells, the PM withdrew from the cell wall in an irregular manner. Since endocytic vesiculation occurs at zero PM tension (Steponkus, 1990), they suggested that localized areas of zero tension must still have existed in order to allow vesiculation to proceed during plasmolysis. This would explain how both endocytic vesicles and Hechtian strands may appear in plasmolysed onion epidermal cells (Oparka *et al.*, 1990, 1994).

Another, perhaps simpler, explanation arises from consideration of the physical properties of the membrane. It seems possible (Oster and Moore, 1989) that conditions near to, or within, the surface of the membrane promote non-linear feedback mechanisms which are important for positive reinforcement of small deformations in the membrane to produce vesicles. The resulting high bending moment in the membrane, perhaps in conjunction with some local biochemical event, results in the vesicle breaking free of the membrane. Similar processes taking place in the vicinity of the points of attachment of the Hechtian strand network with the cell wall could conceivably weaken the strand and result in dissociation from the cell wall. Thus the processes responsible for promoting vesicle production in the PM could also result in destabilization of the Hechtian strand network.

11. PM–endoplasmic reticulum interactions during plasmolysis

The endomembrane system of higher plant cells includes an extensive system of tubular and cisternal endoplasmic reticulum (ER) elements. In addition to mobile ER elements, higher plant cells are characterized by an extensive cortical ER system which is located close to the cell wall (for details see Chapter 22 by Lichtscheidl and Hepler). Current evidence suggests that the cortical ER is tightly bound to the PM (Lichtscheidel and Url, 1990; Quader *et al.*, 1987).

It is commonly assumed that the ER and other cell organelles are incorporated into the protoplast as it contracts during plasmolysis (Lee-Stadelmann and Stadelmann, 1989). However, a number of recent observations suggest that this may not be the case. Using Hoffman optics, Pont-Lezica *et al.* (1993) described a 'membranous webbing' underlying the wall of plasmolysed onion epidermal cells, which they termed the 'Hechtian reticulation' because of its close association with the vertices of Hechtian strands, and they drew a tentative comparison between it and the cortical ER system. Other studies have also hinted at the presence of an additional membrane system outside the protoplast of plasmolysed cells (Hecht, 1912; Johnson-Flanagan and Singh, 1986; Price, 1914).

Recently, Oparka *et al.* (1994) showed that, in plasmolysed onion epidermal cells, the cortical ER network stains intensely with the fluorescent probe $DiOC_6$, confirming that part of this system remains strongly attached to the wall during plasmolysis (*Figure 4a*). By contrast, mobile ER elements were readily incorporated into the contracting protoplast. The Hechtian strands, drawn out from their attachment sites, were intimately connected to the stationary cortical ER network, the latter retaining its characteristic polygonal shape at the cell periphery. Such attachments are seen clearly in *Figures 2* and *3,* in which CLSM was used to section the cell periphery optically where the Hechtian strands and cortical ER network coincide. Electron microscopy, utilizing the PM stain PTA, confirmed that the ER profiles became tightly bounded by the PM as it retracted from the cell wall. In effect, it appeared that the strongly anchored cortical ER system became encased by the PM as it retracted into the centre of the cell, the Hechtian strands being continuous from the cortical ER to the surface of the protoplast (see also *Figure 2*).

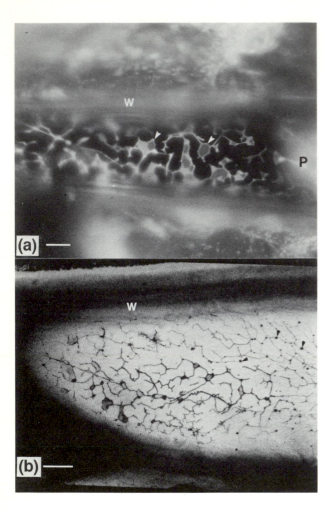

Figure 4. The cortical ER system in onion epidermal cells, composed of a polygonal network of sheets and tubules (the latter predominating), remains firmly attached to the wall during plasmolysis. The tissue was stained with DiOC$_6$. (a) Conventional fluorescence image, in which some of the cisternal ER appears to be completely surrounded by tubules (darts). Scale bar = 40 μm. Reproduced from Oparka *et al.* (1994) with permission from Blackwell Science Ltd. (b) Confocal series in which the fluorescence of DiOC$_6$ is shown in black and superimposed on a bright field image of the cell. The axial depth covered by the series was 9 μm. Note the extensive branching of the tubular network. The dark circles (darts) appear to be attachment points at which the Hechtian strands join the cortical ER. Scale bar = 40 μm. W, cell wall.

12. The cytoskeleton during plasmolysis

Oparka *et al.* (1994) have suggested that the same membrane-spanning linkages which attach the PM to the wall on its external face might also anchor the ER on its internal face (see *Figure 5*). They also speculated that the rigid adhesion of the cortical ER network might provide a scaffolding for the rapid reaffixture or realignment of cytoskeletal elements during deplasmolysis, allowing the protoplast to return to fixed 'reference points' during deplasmolysis.

Actin has been shown to be closely associated with specific foci on the cortical ER (Allen and Brown, 1988; Hepler *et al.*, 1990; Lichtscheidl and Url, 1990), and indirect evidence that actin might be a component of Hechtian strands was presented by Pont-Lezica *et al.* (1993). Microtubules have been found in Hechtian strands within plasmolysed cells of the moss *Funaria* (Schnepf *et al.*, 1986). Cytoskeletal elements such as microfilaments function as tracks for cytoplasmic streaming (Masuda *et al.*, 1991), and plasmolysis is known to induce unusual patterns of cytoplasmic streaming (Küster, 1910; Masuda *et al.*, 1991). In fern protonemal cells, Kagawa *et al.* (1992) found that plasmolysis caused detachment of the PM around the entire cell, with the exception of the sub-apex. Here the protoplast remained firmly attached to the wall by a ring-like band of PM. The location of this junction coincided with that of a circular array of microtubules and microfilaments in the cell cortex. These workers concluded that the junction

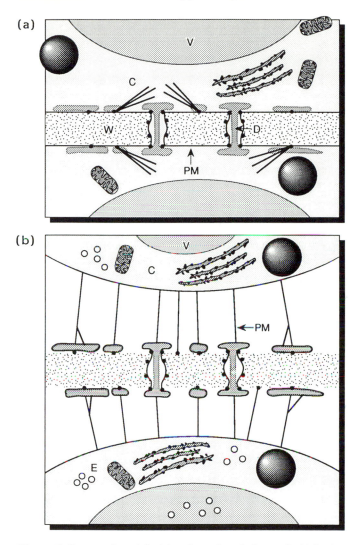

Figure 5. Structural model of the plasmolysed plant cell. (a) In the unplasmolysed cell the plasma membrane (PM) lines the wall and is continuous from one cell to the next via the plasmodesmata. The cortical ER is anchored to the PM which is in turn anchored by putative linkages (•) to the cell wall (W). Elements of the cytoskeleton may be associated with these sites. The cortical ER is also continuous between cells via the central desmotubules of the plasmodesmata. (b) In the plasmolysed cell the anchored cortical ER network is encased by the PM as it retracts from the cell wall. Membrane surface area is conserved by Hechtian strands which connect the surface of the main protoplast body with anchoring points (Hechtian attachment sites) at the cell wall. When the PM reaches zero tension, membrane is deleted in the form of endocytic vesicles (E). Tonoplast vesicles are also deleted internally into the vacuole. Cortical ER continuity is maintained between cells via the central desmotubules of the plasmodesmata. The PM is continuous between the surfaces of the contracted protoplasts and is strongly anchored along the lining of the plasmodesmatal pores. During deplasmolysis, the Hechtian strands (but not endocytic vesicles) are reincorporated into the main body of the expanding protoplast and the PM returns to its original position lining the wall. The cortical ER remains firmly anchored during the entire plasmolysis/deplasmolysis cycle. C, cytoplasm; V, vacuole. Reproduced from Oparka (1994) with permission from The New Phytologist.

between the PM and the cell wall was sustained by a cortical array of microtubules, in some way linked to the presence of a specific and localized transmembrane structure. Thus, elements of the cytoskeleton appear to remain closely associated with PM–cell wall attachment sites during plasmolysis.

13. Plasmodesmata

Plasmolysis has long been used as an effective means of disrupting intercellular transport in order to investigate the role of plasmodesmata in the transport of solutes (Barclay *et al.*, 1982; Hayes *et al.*, 1985; Jarvis and House, 1970; Oparka and Prior, 1987; Weiner *et al.*, 1988), hormones (Drake and Carr, 1978; Drake *et al.*, 1978; Kwiatkowska, 1991) and viruses (Coutts, 1978). However, few of these studies have addressed the effects of a plasmolysis/deplasmolysis cycle on the appearance (and function) of plasmodesmata. The intimate connection between adjacent cells, via both the PM (which is continuous through the plasmodesmatal pore) and the cortical ER (which is continuous between cells via the desmotubule (Hepler, 1982)) clearly poses problems for neighbouring cells that are about to part company during plasmolysis (for more complete descriptions of plasmodesmatal structure see Chapters 27 by Cooke *et al.* and 28 by Ding and Lucas).

   Views differ with regard to the extent to which plasmodesmata are disrupted during plasmolysis (for a discussion of this issue see Oparka, 1994). The pioneering work of Erwee and Goodwin (1984), in which fluorescent probes of differing molecular weights were injected directly into the cytosol of *Egeria* cells, showed that plasmolysis completely prevented the intercellular movement of fluorescent probes. However, movement was restored approximately 10 min following deplasmolysis, after which time the size exclusion limit of the plasmodesmata had increased from 670 Da to 1678 Da. Similarly, Drake *et al.* (1978) found that intercellular electrical coupling was restored about 6 h after plasmolysis of oat coleoptile cells, considerably longer than the time scale of recovery reported by Erwee and Goodwin (1984). It would appear that virus transmission also can be restored following a plasmolysis/deplasmolysis cycle (Coutts, 1978).

14. Structure of plasmodesmata following plasmolysis

Most published micrographs show clear evidence of a continued association of the PM with the lining of the plasmodesmatal pore, even when the cells are severely plasmolysed (Burgess, 1971; Drake *et al.*, 1978; Gigot *et al.*, 1972; Oparka *et al.*, 1994). Tilney *et al.* (1991) examined the effects of a range of treatments, including plasmolysis, on plasmodesmata in fern gametophytes. They found that the plasmodesmata remained embedded in the wall, although rarely they appeared to be pulled out completely, leaving a hole in the wall. Several of the plasmodesmata had large membrane blebs attached to one or both ends of the plasmodesmatal canal, and these often connected several adjacent plasmodesmata (see also Burgess, 1971; Drake *et al.*, 1978). They interpreted these structures as being portions of PM that had been ripped away as the protoplast contracted.

   The above studies indicate that the PM is strongly bound to the cell wall within the plasmodesmatal pore. Whether or not this occurs by the same putative wall-to-membrane linkers that anchor the PM to the wall at Hechtian attachment sites has yet to be demonstrated. *Figure 5a* shows the preponderance of such putative linkages within the plasmodesmatal pore as dark circles spanning the PM. The strong PM–cell wall anchorage within plasmodesmata may explain the ability to isolate plasmodesmata, or at least portions of them, from isolated plant cell-wall fractions (Kotlizky *et al.*, 1992).

Recently, Turner *et al.* (unpublished data) used plasmolysis as the starting point for the isolation of plasmodesmatal proteins from maize root tips, a tissue which contains an abundance of primary plasmodesmata. Following plasmolysis, the tissue was frozen in liquid nitrogen and then ground. After passage through a French press at 2000 psi, the wall fragments were washed and rotary shadowed with platinum/palladium. Electron microscopy of the resulting replicas revealed an extensive retention of membrane blebs within primary plasmodesmata (*Figure 6a*). Subsequent treatment of the wall preparations with Triton-TX-100 removed the central membranous portion of the plasmodesmata, leaving behind an array of doughnut-shaped structures surrounding the plasmodesmatal pores (*Figure 6b*). Such ring structures appeared to be located in the wall around the plasmodesma, rather than in the pore itself, and may correspond to the wall-located 'sphincters' reported by other authors (for review see Robards and Lucas, 1990).

15. Retention of desmotubules

It appears also that the desmotubule remains within the plasmodesmatal pore following plasmolysis, although it is often extended beyond the pore aperture (Burgess, 1971; Gigot *et al.*,

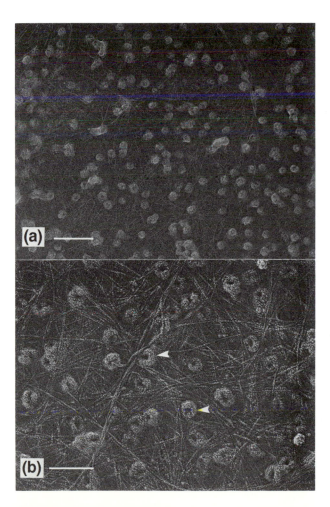

Figure 6. Appearance of plasmodesmata after plasmolysis of maize root tips in 3 M sorbitol. Following plasmolysis, the tissue was frozen in liquid nitrogen, ground and passed through a French press at 2000 psi. After repeated washing, the resulting wall fragments were rotary shadowed with platinum/palladium prior to examination by electron microsopy. (a) Extensive membrane blebs retained within the plasmodesmata. Scale bar = 500 nm. (b) Wall fragments incubated with agitation in 2% Triton-TX-100 for 2 h at 10°C. Note that this treatment removes the membranous portion from the plasmodesmatal pore, leaving behind an array of doughnut-shaped structures located in the wall (darts). Scale bar = 250 nm. Micrographs courtesy of Dr Adrian Turner (John Innes Institute, Norwich, UK).

1972; Oparka *et al.*, 1994; Tilney *et al.*, 1991; see also *Figure 7*). Thus both the PM and the desmotubule appear to be strongly anchored within plasmodesmata. Tilney *et al.* (1991) have suggested that the desmotubule may function as a cytoskeletal element, stabilizing the plasmodesma and giving rise to plasmodesmata of fixed pore sizes.

16. ER–plasmodesmata interactions

While the PM and desmotubule appear to be strongly anchored within the plasmodesmatal pore, the extent to which the ER remains associated with plasmodesmata may vary depending on the severity of plasmolysis. For example, Burgess (1971) described a clear association of the cortical ER with the stretched desmotubules of plasmodesmata in plasmolysed oat coleoptile cells, and Oparka *et al.* (1994) showed that, in onion epidermal cells, regions of wall corresponding to pit fields stained strongly with $DiOC_6$, indicating a preponderance of ER at these sites (*Figure 1c*). More recently, using CLSM, we have shown that the cortical ER network retains apparent continuity between cells via the plasmodesmata, even when the cells are severely plasmolysed (*Figures 2 and 3*). Under the electron microscope, the stretched desmotubules of plasmodesmata were shown to be connected to inflated blebs of ER on either side of the wall (*Figure 7*). By contrast, Tilney *et al.* (1991) found that the desmotubule readily breaks its connection with the ER, and they suggested that the connection of the desmotubule to the ER may be weak by comparison with the connection between the desmotubule and the PM lining the pore. In the final analysis, the extent to which physical membrane continuity is retained between cells might well reflect the severity of plasmolysis, and for this reason it is difficult to draw definitive conclusions from published studies. For example, Burgess (1971) plasmolysed oat coleoptile cells with 0.5 M mannitol and found that, although the protoplasts showed noticeable shrinkage, in most cases the plasmodesmata remained intact. In fact, the plasmodesmata appeared to exert a stabilizing influence on the protoplast, the separation between the PM and the cell wall being less in regions of wall containing plasmodesmata (see also Gigot *et al.*, 1972). However, plasmolysis for 1 h in 0.8 M mannitol resulted in the majority of plasmodesmata being broken by the extreme contraction of the protoplast.

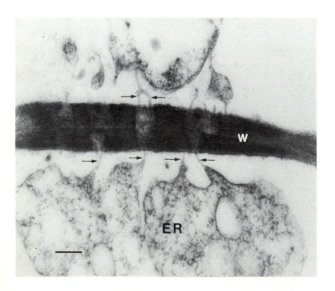

Figure 7. Electron micrograph of plasmodesmata connecting plasmolysed onion epidermal cells. Desmotubules (arrows) remain within the plasmodesmatal pores and are stretched beyond the pore apertures due to the tension created by contraction of the protoplast. The desmotubules are continuous with the ER adjacent to the pit field. Note that in the upper cell, the desmotubule branches near the pore aperture. Scale bar = 30 nm. Reproduced from Oparka *et al.* (1994) with permission from Blackwell Science Ltd.

17. A structural model of the plasmolysed plant cell

Using a combination of fluorescence microscopy and electron microscopy (and incorporating the results of several other published reports), we have proposed a structural model of the plasmolysed plant cell which attempts to interpret at least some of the complex membrane interactions that occur during a plasmolysis/deplasmolysis cycle (*Figure 5*). This model incorporates both Hechtian strand formation and PM vesiculation as major factors contributing to PM conservation, although, as stated earlier, the exact relationship between these two forms of PM reorganization is by no means clear. While Hechtian strands may be reincorporated into the expanding protoplast, the fate of endocytic PM vesicles has yet to be established. The available evidence suggests that this membrane is not readily reincorporated into the PM (Oparka *et al.*, 1990; Steponkus, 1990). Given the inelasticity of the PM, does this mean that synthesis/reinsertion of new membrane is required during deplasmolysis? The PM and tonoplast alterations which occur during plasmolysis remain a remarkable example of how to change the surface area-to-volume ratio of membranes while still maintaining cell function, but the mechanism whereby this is achieved has seldom been addressed. In particular, how are the PM and tonoplast surface areas coordinated during a plasmolysis/deplasmolysis cycle?

According to the above model, the initial closure of plasmodesmata is brought about by close appression of the PM and ER at the neck region of the plasmodesmatal pore, rather than by breakage of plasmodesmata *per se*. Hechtian strands are envisaged as maintaining physical continuity, but not necessarily functional continuity, between adjacent cells.

The cortical ER, retained near the wall during plasmolysis, is also envisaged as being continuous from one cell to the next via the desmotubules of the plasmodesmata. The reason why the cortical ER remains so intimately connected to the cell periphery during plasmolysis, rather than being incorporated into the contracting protoplast, is unclear. Do the same linkages which bind the PM strongly to the wall at specific locations also bind the PM to the ER at the same sites, or is some other form of transmembrane attachment responsible for anchoring the ER?

In many ways plasmolysis remains an enigma to plant cell biologists. The above model is an attempt to interpret some of the complex membrane interactions which occur during plasmolysis, an area of plant cell biology which has been sadly neglected. It is not meant to represent a definitive statement on the subject, but rather has been proposed as a potential starting point for further studies on the behaviour of membranes during plasmolysis.

Acknowledgements

We are grateful to Dr Adrian Turner and Professor Keith Roberts (John Innes Institute, Norwich, UK) for helpful discussions and for allowing us to reproduce *Figure 6*. This research was funded by the Scottish Office Agriculture and Fisheries Department.

References

Allen NA, Brown DT. (1988) Dynamics of the endoplasmic reticulum in living onion epidermal cells in relation to microtubules, microfilaments, and intracellular particle movement. *Cell Motil. Cytoskel.* **10**: 153–163.

Attree SM, Sheffield E. (1985) Plasmolysis of *Pteridium* protoplasts. A study using light and scanning electron microscopy. *Planta* **165**: 151–157.

Barclay GF, Peterson CA, Tyree MT. (1982) Transport of fluorescein in trichomes of *Lycopersicon esculentum*. *Can. J. Bot.* **60**: 397–402.

Burgess J. (1971) Observations on structure and differentiation in plasmodesmata. *Protoplasma* **73**: 83–95.

Burridge K, Fath K, Kelly T, Nuckolls G, Turner C. (1988) Focal adhesions: transmembrane junctions between the extracellular matrix and the cytoskeleton. *Annu. Rev. Cell Biol.* **4**: 487–525.

Cocking EC. (1972) Plant cell protoplasts. Isolation and development. *Annu. Rev. Plant Physiol.* **23**: 29–50.

Coutts RHA. (1978) Suppression of virus induced local lesions in plasmolyzed leaf tissue. *Plant Sci. Lett.* **12**: 77–86.

Davey MR, Cocking EC. (1972) Uptake of bacteria by isolated higher plant protoplasts. *Nature* **239**: 455–456.

Deikmann W, Hedrich R, Raschke K, Robinson DG. (1993) Osmocytosis and vacuolar fragmentation in guard cell protoplasts: their relevance to osmotically induced volume changes in guard cells. *J. Exp. Bot.* **44**: 1569–1577.

De Vries H. (1877) *Untersuchungen über die Mechanischen Ursachen der Zellstrechung Ausgehend von der Einwirung von Salzlösungen auf den Turgor Wachsender Pflanzenzellen.* Leipzig: Engelmann.

Dowgert MF, Steponkus PL. (1984) Behaviour of the plasma membrane of isolated protoplasts during a freeze-thaw cycle. *Plant Physiol.* **75**: 1139–1151.

Drake G, Carr DJ. (1978) Plasmodesmata, tropisms and auxin transport. *J. Exp. Bot.* **29**: 1309–1318.

Drake G, Carr DJ, Anderson WP. (1978) Plasmolysis, plasmodesmata and the electrical coupling of oat coleoptile cells. *J. Exp. Bot.* **29**: 1205–1214.

Erwee MG, Goodwin PB. (1984) Characterization of the *Egeria densa* leaf symplast. Response to plasmolysis, deplasmolysis and to aromatic amino acids. *Protoplasma* **122**: 162–168.

Gigot C, Schmitt C, Hirth L. (1972) Ultrastructural modifications observed during the preparation of protoplasts from tobacco tissue cultures. *J. Ultrastr. Res.* **41**: 418–432.

Goddard RH, La Claire JW. (1993) Novel changes in the plasma membrane and cortical cytoplasm during wound-induced contraction in a giant-celled green alga. *Protoplasma* **176**: 75–83.

Gordon-Kamm WJ, Steponkus PL. (1984) The behaviour of the plasma membrane following osmotic contraction of isolated protoplasts: implications for freezing injury. *Protoplasma* **123**: 83–94.

Hayes PM, Offler CE, Patrick JW. (1985) Cellular structures, plasma membrane surface area and plasmodesmatal frequencies in the stem of *Phaseolus vulgaris* L. in relation to photosynthate transfer. *Ann. Bot.* **56**: 125–138.

Hecht K. (1912) Studien über den Vorgang der Plasmolyse. *Beitr. Biol. Pflan.* **11**: 133–145.

Hepler PK. (1982) Endoplasmic reticulum in the formation of the cell plate and plasmodesmata. *Protoplasma* **111**: 121–133.

Hepler PK, Palevitz BA, Lancelle SA, McCauley MM, Lichtscheidl I. (1990) Cortical endoplasmic reticulum in plants. *J. Cell Sci.* **96**: 355–373.

Jarvis P, House CR. (1970) Evidence for symplasmic ion transport in maize roots. *J. Exp. Bot.* **21**: 83–90.

Johnson-Flanagan AM, Singh J. (1986) Membrane deletion during plasmolysis in hardened and non-hardened plant cells. *Plant Cell Environ.* **9**: 299–306.

Kagawa T, Kadota A, Wada M. (1992) The junction between the plasma membrane and the cell wall in fern protonemal cells, as visualised after plasmolysis and its dependence on arrays of cortical microtubules. *Protoplasma* **170**: 186–190.

Kotlizky G, Schurtz S, Yahalom A, Malik Z, Traub O, Epel BL. (1992) An improved method for the isolation of plasmodesmata embedded in clean cell walls. *Plant J.* **2**: 623–630.

Küster E. (1910) Uber Inhaltsverlagerungen in plasmolysierten Zellen. *Flora* **100**: 267–287.

Kwiatkowska M. (1991) Autoradiographic studies on the role of plasmodesmata in the transport of gibberellin. *Planta* **183**: 294–299.

Lee-Stadelmann OY, Stadelmann EJ. (1989) Plasmolysis and deplasmolysis. *Meth. Enzymol.* **174**: 225–247.

Lichtscheidl I, Url WG. (1990) Organization and dynamics of cortical endoplasmic reticulum in inner epidermal cells of onion bulb scales. *Protoplasma* **157**: 203–215.

Masuda Y, Takagi S, Nagai R. (1991) Protease-sensitive anchoring of microfilament bundles provides tracks for cytoplasmic streaming in *Vallisneria*. *Protoplasma* **162**: 151–159.

Oparka KJ. (1994) Plasmolysis: new insights into an old process. *New Phytol.* **126**: 571–591.

Oparka KJ, Prior DAM. (1987) Carbon-14 sucrose efflux from the perimedulla of growing potato tubers. *Plant Cell Environ.* **10**: 667–676.

Oparka KJ, Prior DAM, Harris N. (1990) Osmotic induction of fluid-phase endocytosis in onion epidermal cells. *Planta* **180**: 555–561.

Oparka KJ, Cole L, Wright KM, Coleman JOD, Hawes CR, Evans DE. (1991) Fluid-phase endocytosis and intracellular compartmentation of fluorescent probes. In: *Endocytosis, Exocytosis and Vesicle Traffic in Plants*. SEB Seminar Series. Cambridge: Cambridge University Press, pp. 81–102.

Oparka KJ, Wright KM, Murant EA, Allan EJ. (1993) Fluid-phase endocytosis: do plants need it? *J. Exp. Bot.* **44**: 247–255.

Oparka KJ, Prior DAM, Crawford JW. (1994) Behaviour of plasma membrane, cortical ER and plasmodesmata during plasmolysis of onion epidermal cells. *Plant Cell Environ.* **17:** 163–171.

Oster GF, Moore H-P. (1989) The budding of membranes. In: *Cell to Cell Signalling: From Experiments to Theoretical Models* (Goldbeter A, ed.). London: Academic Press.

Palta JP, Li PH. (1978) Cell membrane properties in relation to freezing injury. In: *Plant Cold Hardiness and Freezing Stress. Mechanisms and Crop Implications* (Li PH, Sakai A, ed.). New York: Academic Press, pp. 93–115.

Palta JP, Lee-Stadelmann OY. (1983) Vacuolated plant cells as ideal osmometers. Reversibility and limits of plasmolysis and estimation of protoplasm volume in control and water stress tolerant cells. *Plant Cell Environ.* **6:** 601–610.

Pearce RS, Withers LA, Willison JHM. (1974) Bodies of wall-like material ('wall bodies') produced intracellularly by cultured isolated protoplasts and plasmolyzed cells of higher plants. *Protoplasma* **82:** 223–236.

Pennell RI, Knox JP, Scofield GN, Selvendran RR, Roberts K. (1989) A family of abundant plasma membrane-associated glycoproteins related to the arabinogalactan proteins is unique to flowering plants. *J. Cell Biol.* **108:** 1967–1977.

Pont-Lezica RF, McNally JG, Pickard BG. (1993) Wall-to-membrane linkers in onion epidermis; some hypotheses. *Plant Cell Environ.* **16:** 111–123.

Power JB, Cocking EC. (1970) Isolation of leaf protoplasts: macromolecule uptake and growth substance response. *J. Exp. Bot.* **21:** 64–70.

Price SR. (1914) Some studies on the structure of the plant cell by the method of dark-ground illumination. *Ann. Bot.* **28:** 601–632.

Quader H, Hofmann A, Schnepf E. (1987) Shape and movement of the endoplasmic reticulum in onion bulb epidermis cells: possible involvement of actin. *Eur. J. Cell Biol.* **44:** 17–26.

Robards HW, Lucas WJ. (1990) Plasmodesmata. *Annu. Rev. Plant Physiol. Plant Mol. Biol.* **41:** 369–419.

Robinson DG, Hillmer S. (1990) Endocytosis in plants. *Physiol. Plant.* **79:** 96–104.

Robinson DG, Hedrich R. (1991) Vacuolar Lucifer Yellow uptake in plants: endocytosis or anion transport; a critical opinion. *Bot. Acta* **104:** 257–264.

Robinson DG, Hedrich R, Herkt B, Diekmann W, Robert-Nicoud M. (1991) Endocytosis in plants: problems and perspectives. In: *Endocytosis.* NATO ASI Series H62. Heidelberg: Springer-Verlag, pp. 459–466.

Ruoslahti E, Pierschbacher MD. (1987) New perspectives in cell adhesion: RGD and integrins. *Science* **238:** 491–496.

Scarth GW. (1941) Dehydration injury and resistance. *Plant Physiol.* **16:** 171–179.

Schindler M, Meiners S, Cheresh DA. (1989). RGD-dependent linkage between plant cell wall and plasma membrane: consequences for growth. *J. Cell Biol.* **8:** 1955–1965.

Schnepf E, Deichgraber G, Bopp M. (1986) Growth, cell wall formation and differentiation in the protonema of the moss *Funaria hygrometrica*. Effects of plasmolysis on the developmental program and its expression. *Protoplasma* **133:** 50–65.

Singh J. (1979) Ultrastructural alterations in cells of hardened and non-hardened winter rye during hyperosmotic and extracellular freezing stresses. *Protoplasma* **98:** 329–341.

Singh J, Miller RW. (1985) Biophysical and ultrastructural studies of membrane alterations in plant cells during extracellular freezing: molecular mechanisms of membrane injury. In: *Cryopreservation of Plant Cells and Organs* (Kartha K, ed.). Boca Raton, FL: CRC Press, pp. 61–74.

Stadelmann E. (1966) Evaluation of turgidity, plasmolysis and deplasmolysis of plant cells. In: *Methods in Cell Physiology, Vol. II* (Prescott DM, ed.). New York: Academic Press, pp. 143–216.

Stadelmann EJ. (1956) Plasmolyse und Deplasmolyse. In: *Encyclopedia of Plant Physiology, Vol. 2* (Ruhland W, ed.). Berlin: Springer-Verlag, pp. 71–115.

Steponkus PL. (1990) Cold acclimation and freezing injury from a perspective of the plasma membrane. In: *Environmental Injury to Plants* (Katterman F, ed.). New York: Academic Press, pp. 1–16.

Strasburger E, Noll F, Schenck H, Schimper AFW. (1983) *Lehrbuch der Botanik*, 32nd edn (von Denfer D, Ehrendorfer F, Dresinsky A, ed.). Stuttgart: G. Fischer, p. 84.

Tilney LG, Cooke TJ, Connelly PS, Tilney MS. (1991) The structure of plasmodesmata as revealed by plasmolysis, detergent extraction and protease digestion. *J. Cell Biol.* **112:** 739–747.

Wayne R, Staves MP, Leopold AC. (1992) The contribution of the extracellular matrix to gravisensing in characean cells. *J. Cell Sci.* **101:** 612–623.

Weiner H, Burnell JN, Woodrow IE, Heldt HW. (1988) Metabolite diffusion into bundle sheath cells from C-4 plants in relation to C-4 photosynthesis and plasmodesmatal function. *Plant Physiol.* **88:** 815–822.

Withers LA, Cocking EC. (1972) Fine-structural studies on spontaneous and induced fusion of higher plant protoplasts. *J. Cell Sci.* **11**: 59–75.

Wolfe J, Dowgert MF, Steponkus PL. (1985) Dynamics of membrane exchange of the plasma membrane and the lysis of isolated protoplasts during rapid expansions in area. *J. Memb. Biol.* **86**: 127–138.

Yoshida S. (1982) Phospholipid degradation and its control during freezing of plant cells. In: *Plant Cold Hardiness and Freezing Stress. Mechanisms and Crop Implications* (Li PH, Sakai A, ed.). New York: Academic Press, pp. 117–139.

Chapter 4

Cellulose biogenesis

Richard L. Blanton and Candace H. Haigler

1. Introduction

Cellulose biogenesis remains one of the great unsolved problems of plant biochemistry, as has been reviewed frequently (a partial list includes: Bolwell, 1988, 1993; Delmer, 1987, 1990, 1991; Delmer and Amor, 1995; Delmer and Stone, 1988; Delmer *et al.*, 1993a; Haigler, 1985; Iyama *et al.*, 1993; Mullins, 1990; Read and Delmer, 1991; Ross *et al.*, 1991; Seitz and Emmerling, 1990). Membrane fractions or solubilized extracts of plants have not yet been shown to synthesize cellulose selectively and at high rates *in vitro*, and no genes relevant to cellulose polymerization and crystallization have been characterized from plants. This paucity of knowledge tends to astonish those unfamiliar with the field, especially in view of the molecular simplicity of cellulose as a homopolymer of $(1\rightarrow4)$-β-D-glucose, the importance of cellulose for plant function and economic use (wood and fiber), recent increases in knowledge of the biochemical and molecular regulation of bacterial cellulose synthesis, and the advanced biochemical and molecular knowledge of the biosynthesis of other related polysaccharides, such as chitin (a homopolymer of $(1\rightarrow4)$-β-N-acetyl-glucosamine; reviewed by Bulawa, 1993).

However, it is appropriate that the subject of cellulose synthesis be discussed in a book on plant membranes, since all the available evidence suggests that cellulose biogenesis is plasma membrane-associated in plants, as it is in bacteria (see below). Ordered proteins in the plasma membrane are closely associated with cellulose microfibril deposition in diverse species (Emons, 1991; Quader, 1991). In addition, transmembrane potentials and pH gradients appear to be required for cellulose synthesis in plants and bacteria (Bacic and Delmer, 1981; Delmer *et al.*, 1982), and membranes damaged by freeze-thawing have a diminished ability to synthesize cellulose (Amor *et al.*, 1995; Jacob and Northcote, 1985). Therefore, it may well be some aspect of this membrane association that makes reconstitution of a selective synthesizing system with a high rate *in vitro* so difficult. The discussion below will summarize the little that is known about plant cellulose synthesis, and will describe approaches that hold promise for future progress.

2. The plasma membrane specializations associated with cellulose synthesis

There are several lines of evidence which suggest that a structural template under precise genetic control is part of cellulose microfibril biogenesis, which involves both biochemical polymerization and biophysical crystallization. First, the size of the cellulose microfibrils is regulated in an orderly manner between species and at different stages of development (Haigler, 1985; Kuga and Brown, 1991). The smallest microfibrils are 1 to 2 nm wide (in quince seed slime and primary cell walls; Chanzy *et al.*, 1979; Franke and Ermen, 1969), and the largest are about 30 nm wide (in the siphonocladalean alga *Boergesenia forbesii*; Itoh *et al.*, 1984). In cell walls of higher plants, many microfibrils are about 4 nm wide, which may well correlate with the hypothesis that a 3.5-nm microfibril with about 36 glucan chains is the smallest aggregate that can form a true crystal of cellulose I (Chanzy *et al.*, 1978), the most common natural crystalline polymorph (Blackwell, 1982). Smaller aggregates have too many glucan chains in the disorderly surface environment, and often exist as cellulose IV, which has good longitudinal but poor lateral chain order (Chanzy *et al.*, 1978). Secondly, cell walls are often highly organized, with cellulose microfibrils oriented in precise patterns relative to each other and to the cell axis, and the microfibrils often change orientation between wall lamellae (Haigler, 1985). Molecular self-assembly between cellulose and other wall polysaccharides and proteins may help to mediate this organization, particularly in the case of cell walls with arced helicoidal layers of polymers (Vian and Reis, 1991), but deposition of microfibrils in particular patterns also occurs in many cases (Haigler, 1985). Thirdly, cellulose I is not formed when solubilized cellulose molecules crystallize acellularly (Blackwell, 1982), suggesting control of this particular type of crystallization by the living cell.

The probable existence of ordered membrane-associated synthetic complexes for cellulose microfibril formation was predicted (Preston, 1964) before morphological evidence for such complexes was obtained in freeze-fracture replicas of the cellulose-synthesizing chlorococcalean alga, *Oocystis apiculata* (Brown and Montezinos, 1976). Fractured plasma membranes revealed rectangular arrays of intramembrane particles (IMPs) (three rows of 30 particles each) at the ends of cellulose microfibril impressions in the extracytoplasmic fracture (EF) face. (According to established freeze-fracture terminology (Branton *et al.*, 1975), the plasma membrane is split by freeze fracture to reveal four faces, which are designated protoplasmic surface (PS), protoplasmic fracture (PF), extracytoplasmic fracture (EF) and extracytoplasmic surface (ES) in order of progression from inside to outside the cell.) The IMP aggregates were called terminal complexes (TCs) to reflect their presence at the end of cellulose microfibril impressions. Subsequently, similar rectangular arrays of IMPs of diameter 5–17 nm have been observed in numerous species of chlorophycean, ulvophycean and other algae (Hotchkiss, 1989; Mizuta and Brown, 1992b; Okuda and Brown, 1992; Okuda *et al.*, 1994; Quader, 1991). The linear TCs of different species do not always associate with the same face of the membrane after freeze fracture (Hotchkiss, 1989), probably indicating species differences in membrane composition or the proteins of the TC and their relationship to cytoskeletal elements or microfibrils. It should be noted that, in the alga *Pleurochrysis*, the cellulosic scales of which the wall is composed are synthesized in the Golgi apparatus and then secreted (Brown *et al.*, 1969). *Pleurochrysis* is therefore an exception to the general rule that cellulose synthesis occurs on the plasma membrane (and its Golgi apparatus is a good example of the cisternal progression rather than the vesicle-mediated model for Golgi function; Becker *et al.*, 1995).

The existence of organized IMP aggregates associated with cellulose microfibril biogenesis was demonstrated for higher plants by the observation of rosettes of six IMPs in the PF face of *Zea mays* plasma membranes that were replicated after freeze fracture (Mueller and Brown, 1980). *Figure 1* shows rosettes in developing tracheary elements of *Zinnia*. Simultaneously, rosettes were observed in the plasma membrane PF face of a zygnemetalean alga with a cellulosic cell wall, *Micrasterias denticulata* (Giddings *et al.*, 1980). On the basis of numerous observations, each IMP is about 8 nm in diameter, with six forming a rosette of overall diameter 25 ± 5 nm (Emons, 1991). Since IMPs can be pulled out of a membrane face by the fracture event toward the side where they are more firmly anchored (Sleytr and Robards, 1982), rosette IMPs have frequently been assumed to be transmembrane proteins. This has been confirmed by clear visualization of a corresponding rosette of pits in the complementary EF face in double replicas (Grimson and Haigler, unpublished data). Rosettes were often correlated with globular impressions in the opposite face of the membrane (EF face), which have been called terminal globules. The true existence or not of terminal globules and their possible significance are still

Figure 1. Rosettes in developing tracheary elements of *Zinnia elegans* revealed by freeze fracture. The main micrograph shows a cell from the perspective of 'outside looking in'. The PF face of the membrane contains the rosettes (most of which are indicated by small arrowheads). Part of the cell wall appears in the upper left area of the micrograph. Note that the rosettes are confined to a particular region of the membrane, which corresponds to a site of secondary wall thickening of the tracheary element. The inset (upper right) shows three rosettes at higher magnification. The large arrow indicates the direction of platinum/carbon shadowing. Scale bar represents 1 μm on the main figure and 100 nm on the inset.

debated; they could be impressions of other parts of the cellulose synthase complex, artefacts of freezing, membrane deformations resulting from the pulling out of rosette IMPs from the EF face, or impressions of uncrystallized cellulose accumulated on the extracytoplasmic surface of the plasma membrane (Emons, 1991; Hotchkiss, 1989).

Much correlative evidence suggests that these plasma membrane TCs are involved in cellulose microfibril deposition, although there is no direct evidence that they are cellulose synthases *per se* (Emons, 1991). Briefly, the density of rosettes corresponds closely to the rapidity of cellulose synthesis in each cell, with the maximum observed density (191 μm^{-2}) occurring in tracheary elements (Herth, 1985). Particularly for rosettes and less specifically for linear algal TCs, TC size and aggregation pattern correspond to microfibril size, shape and crystallinity (Hotchkiss, 1989). The most dramatic example occurs during secondary wall synthesis in *M. denticulata*, when 6 to 175 rosettes are organized into elaborate hexagonal arrays, and the number of rosettes in each row corresponds to the size of the aligned microfibril within a banded array of microfibrils *rosettes 6 €and 2€ L exocytosis*

(Giddings *et al.*, 1980). Furthermore, on the basis of limited correlative studies conducted so far, the type of TC seems to correspond to the proportions of Iα or Iβ allomorphs that are mixed within the cellulose I of a particular sample. Some linear TCs are associated with majority synthesis of triclinic, metastable Iα, perhaps due to the strain of assembly of such a large microfibril, and smaller rosettes are associated with majority synthesis of monoclinic, stable Iβ (Atalla and VanderHart, 1989; Sugiyama *et al.*, 1991). Finally, the appearance, behavior and apparent movement of linear TCs in algae can be disturbed by the addition of inhibitors of cellulose polymerization or crystallization (Herth, 1989; Mizuta and Brown, 1992a; Quader, 1983; Quader *et al.*, 1983).

Visualization of the rosette type of TC within Golgi vesicles of *M. denticulata* (Giddings *et al.*, 1980) and higher plants (Haigler and Brown, 1986) suggests that they are assembled in the endomembrane system and inserted by exocytosis into the plasma membrane. The possibility cannot be excluded, however, that such observations only indicate a retrieval mechanism for obsolete rosettes. There has been one report of rosettes in the endoplasmic reticulum (Rudolph, 1987). In contrast, there is evidence that at least some of the rectangular algal TCs are assembled within the plasma membrane (Mizuta, 1985; Mizuta and Brown, 1992b), which is perhaps related to their much larger size. Morphometric calculations suggest that individual rosettes persist for only 10–20 min in the plasma membrane, a turnover that could be regulated specifically or that could occur merely as a consequence of rapidly recycling plasma membrane (for review see Delmer, 1990). It has been speculated that the persistence time of individual rosettes in the plasma membrane could be a mechanism for controlling the degree of polymerization (DP) of cellulose (Delmer, 1990), which varies between primary and secondary cell walls.

The current model for cellulose microfibril formation suggests that TCs move through the plasma membrane, driven by the force of crystallization itself (Herth, 1980; Quader, 1983; Staehelin and Giddings, 1982). The controlled orientation of their movement, perhaps restricted by plasma membrane fluidity domains under cytoskeletal control, then becomes one mechanism whereby cellulose microfibrils could become oriented in cell walls (Giddings and Staehelin, 1991). The existence and nature of such fluidity domains is a completely unexplored area.

The aggregation of synthetic sites in the plasma membrane is necessary to ensure that sufficient glucan chains co-crystallize to form cellulose I (Haigler and Benziman, 1982), a concept that has been supported by the abiotic assembly of cellulose I by a cellulase-catalyzed polymerization if catalytic sites are aggregated (Lee *et al.*, 1994). Furthermore, because cellulose polymerization and crystallization are separated by a short temporal gap (Haigler *et al.*, 1980; Herth, 1980), control of IMP organization can help to regulate the size, shape, crystallinity and orientation in the wall of the cellulose microfibrils typical of a particular species or developmental stage (Okuda *et al.*, 1994; Sugiyama *et al.*, 1994). An additional point of control, however, is the presence and abundance of cellulose-binding cell wall matrix molecules in the apoplastic space (e.g. xyloglucan and xylan) that can intercalate with cellulose microfibrils before crystallization (Baba *et al.*, 1994; Haigler, 1991; Uhlin *et al.*, 1995). Crystallization of biological fibrils at least several monomer units away from the polymerization site can be predicted on the basis of non-equilibrium thermodynamics (Blackwell, 1982), and it appears that plant cells have exploited this physical principle to control microfibril characteristics through the organization of plasma membrane IMPs.

The above discussion cannot extend beyond correlation because no proteins involved in higher-plant cellulose polymerization or crystallization have been identified conclusively. The organized IMPs could contain the cellulose synthases and/or represent membrane channels for the passage of regulatory molecules or substrate to a peripheral cellulose synthase (*Figure 2*; Emons, 1991). It has also been suggested that channels are required to pass small groups of pre-

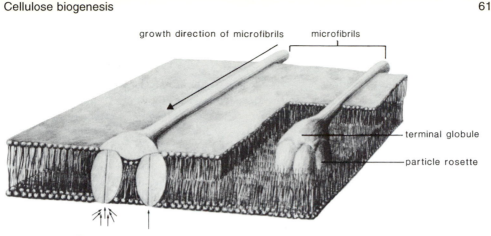

growth direction of microfibrils   microfibrils

terminal globule

particle rosette

UDP-glucose   channel in particle

Figure 2. Proposed schematic diagram of higher plant microfibril TCs embedded in the plasma membrane lipid bilayer. This model proposes that the rosette functions both as a channel and as a synthetic complex, whereas the terminal globule represents an impression of uncrystallized cellulose in the plasma membrane. Other speculative models of how rosettes and terminal globules function in microfibril biogenesis have been proposed (see text and cited reviews). Reprinted from Emons (1991) by courtesy of Marcel Dekker, Inc and the author.

polymerized glucan chains to the apoplastic space, a hypothesis based on the topology of callose synthase in the plasma membrane (Wu and Wasserman, 1993) and on co-purification with callose synthase activity of a plasma membrane protein with high homology to the major intrinsic protein (MIP) family of membrane channels (Qi *et al.*, 1995). A central depression suggestive of a pore has been visualized in each rosette IMP (Rudolph *et al.*, 1989), and this observation was extended by the use of improved freeze-etching and shadowing methods to reveal a deep, hydrophilic central opening within a multi-subunit IMP (Grimson and Haigler, unpublished data). Therefore, rosettes would differ from the structurally similar connexons of gap junctions not only by being larger, but perhaps also by having water channels within each of the six IMPs (Emons, 1991). In the connexon, the six monomeric connexin subunits surround a common water channel (Evans, 1994).

What kinds of proteins would be necessary in a cellulose-synthesizing complex that might, therefore, be represented in particles revealed by freeze-fracture preparations? Northcote (1991) indicated a minimum of three types of protein: (1) a transporter protein to bring the UDP-Glc to the polymerization site; (2) a glucosyltransferase; and (3) a binding (acceptor) protein that would hold the donor and acceptor compounds in the proper orientation for linkage formation. Additional proteins could be transmembrane linkers to the cytoskeleton, extrusion pores or channels, and subsidiary proteins that would serve to organize the complex or to guide crystallization (Delmer and Amor, 1995; Northcote, 1991). Only when particular proteins associated with cellulose biogenesis are purified will it be possible to attempt reconstitution experiments and/or to label particular proteins with specific antibodies in order to clarify further the function of microfibril-associated plasma membrane IMPs.

3. Biochemistry of cellulose polymerization in higher plants

The regulation of cellulose polymerization in plants has been a historically difficult problem due to the inability to produce a high-rate *in-vitro* synthesizing system generating a well-characterized $(1 \rightarrow 4)$-β-glucan product, or to identify related genes. Based on theoretical considerations and the most promising experimental results, many previous attempts to achieve this goal relied on the addition of [^{14}C]UDP-Glc to particulate membrane fractions. However, a frequent problem was that synthesis of (1→3)-β-glucan (callose, which is also synthesized as a wound response *in vivo*) overwhelmed the low levels of (1→4)-β-glucan synthesis (about 5% of total glucan). These results contrast with those obtained for bacterial and protistan preparations, which do use UDP-Glc efficiently as a cellulose synthesis substrate (see below). The higher plant callose synthesis activity (often referred to as glucan synthase II, GS II) is localized at the plasma membrane and stimulated by high levels of UDP-Glc (>500 μM), Ca^{2+} and β-glucoside, with the latter two effectors perhaps becoming elevated upon wounding *in vivo* (see also Chapter 5 by Kauss; also Hayashi *et al.*, 1987; Kauss, 1987; Ohana *et al.*, 1992). In contrast, the (1→4)-β-glucan activity was most commonly optimized by the addition of Mg^{2+} and a decrease in substrate concentration (< 50 μM) (Delmer, 1987; Delmer *et al.*, 1993a). The small amount of (1→4)-β-glucan often synthesized from exogenous UDP-Glc might have been cellulose synthesized by plasma membrane vesicles, or it could have resulted from glucan synthase I (GS I) activity, which is a Golgi-localized xyloglucan glucosyltransferase that produces a γ(1→4)-β-glucan backbone to which xylose side-chains are added (White *et al.*, 1993). As indicated by recent research reports and commentaries therein (Brown *et al.*, 1994; Delmer *et al.*, 1993a; Kudlicka *et al.*, 1995; Li and Brown, 1993; Li *et al.*, 1993; Okuda *et al.*, 1993), the synthesis of cellulose *in vitro* from particulate or solubilized preparations remains a controversial research area. However, especially pertinent to the relationship between cellulose synthesis and the plasma membrane is the observation that mild plasma membrane solubilization conditions (0.05% digitonin in buffer with 0.25 M sucrose) increased the crystalline cellulose synthesized in the presence of added UDP-Glc to 32% of the total product (Kudlicka *et al.*, 1995).

There is evidence that UDP-Glc required for plant cell-wall glucan synthesis might be generated directly from the cleavage of sucrose by sucrose synthase to yield UDP-Glc plus fructose (Chourey *et al.*, 1991; Rollit and Maclachlan, 1974). This would be an energetically conservative mechanism compared to the generation of UDP-Glc by UDP-pyrophosphorylase, and abundant sucrose (the major transport sugar) is available in plants. Recently, a plasma membrane-bound isoform of sucrose synthase was identified from cotton fibers (Amor *et al.*, 1995). Semi-permeabilized fibers used [^{14}C] sucrose much more efficiently than UDP-Glc for the high-rate synthesis of well-characterized cellulose and callose. The rate of cellulose synthesis from sucrose was as high as 0.78 nmol glucose incorporated min^{-1} mg^{-1} dry weight of fiber, whereas the calculated rate of *in-vivo* synthesis was 0.7–1.5 nmol glucose incorporated min^{-1} mg^{-1} dry weight of fiber. Consistent with previous data on the regulation of (1→4)-β- compared to (1→3)-β-glucan synthesis summarized above, the synthesis of cellulose predominated (55–65% of product) when a Ca^{2+} chelator was added to the system, whereas synthesis of callose predominated if Ca^{2+} and cellobiose were added. Immunolocalization showed that sucrose synthase was present at the cell wall/plasma membrane interface in different fibers in patterns consistent with the synthesis of both polymers. These data gave rise to the hypothesis that sucrose synthase is closely coupled to cellulase synthase, so that the pool of cytosolic UDP-Glc is not used for cellulose synthesis; UDP-Glc is instead channeled directly to the growing polymer, with recycling of UDP. However, high-rate cellulose synthesis from sucrose in the semi-permeabilized fibers was not always reproducible, nor was it sustainable for more than 10 min or obtainable in membrane preparations (Amor *et al.*, 1995). Until these difficulties are overcome, a significant barrier to the study of the regulation of plant cellulose synthesis will remain. However, these problems do suggest that cellular organization at the interface of the cortex (with cytoskeletal elements), plasma membrane and cell wall is critical for sustained cellulose biogenesis in plants.

Because it is easy to produce callose and difficult to produce cellulose *in vitro*, it has been proposed that the same or a related enzyme complex under differential regulation is responsible for the synthesis of both of these glucan polymers (Delmer, 1977, 1987; Jacob and Northcote, 1985). Since the potential for wound callose synthesis may remain constitutively available in the plasma membrane of many cells, it would be efficient to use at least parts of the same enzyme complex to generate cellulose microfibrils. Observations supporting this possibility include the findings that (a) membrane-bound sucrose synthase can transmit UDP-Glc to both synthases (Amor *et al.*, 1995) and (b) an *in-vitro* preparation from *Lolium multiflorum* synthesized (1→3)-β, (1→4)-β and (1→3,1→4)-β glucans in varying ratios depending upon the substrate concentration and specific cofactors, yet only a single 31-kDa polypeptide photolabeled with a UDP-Glc photoaffinity analog (Meikle *et al.*, 1991). Furthermore, a model for callose synthase based on biochemical data is conceptually and diagrammatically consistent with the rosettes of IMPs observed in cellulose-synthesizing cells of higher plants (Wu and Wasserman, 1993). However, until the essential polypeptides for callose and cellulose synthesis are conclusively identified, this common-subunit hypothesis will remain only a heuristic possibility.

An additional challenge to the possibility of demonstrating successful *in-vitro* cellulose synthesis is the need to document that the product formed in the reaction is indeed cellulose. Incorporation of glucose from UDP-Glc into an alkali-insoluble polymer can result from the synthesis of cellulose, callose, xyloglucan, mannoglucan and mixed-link glucans. There are three generally accepted criteria for the demonstration of *in-vitro* cellulose synthesis (Blaschek *et al.*, 1983).

(1) The formation of homopolymeric glucan chains with exclusively (1→4)-β linkages must be demonstrated. Total hydrolysis followed by high-pressure liquid chromatography (HPLC) and thin-layer or paper chromatography represent methods for determining sugar composition (Fry, 1988). Methylation analysis is required for linkage determination (Delmer, 1987; Gibeaut and Carpita, 1993; York *et al.*, 1985). The preparation of a radioactively labeled polymer increases the sensitivity of the methods and also allows a distinction to be made between pre-existing polymers and those synthesized *in vitro*. It should be noted that hydrolytic enzymes are generally not suitable for the initial analysis of a new *in-vitro* system because of the potential presence of multiple hydrolytic activities in many available enzyme preparations, and the questionable specificity of others (Delmer, 1987; Gibeaut and Carpita, 1993). The use of enzymes is particularly problematic if several different polymers are synthesized. However, the use of carefully characterized enzymes in a well-defined synthase system (where enzyme results have previously been confirmed by methylation analysis) can be a convenient method of analyzing *in-vitro* products (Meikle *et al.*, 1991).

(2) The (1→4)-β-glucan chains must have a DP characteristic of cellulose (at least several hundred residues). This can be established by nitration of the polymer (Alexander and Mitchell, 1949), or by dissolving it in a solvent such as dimethylacetamide-lithium chloride (McCormick *et al.*, 1985) and then determining the average chain length by non-aqueous gel permeation chromatography. Not only are these technically difficult and hazardous procedures, but they can be subject to error because of differences in physical/chemical behavior of the tested unknown polymer and the polystyrene standards used for generating the molecular-weight standard curve (Timpa, 1991). The latter difficulty can be overcome by the use of universal calibration methods (Timpa, 1991). However, the methods do not distinguish between a polymer synthesized entirely *de novo* and one that results from the addition of a few glucose residues to a pre-existing polymer or other high-molecular-weight acceptor (Henry and Stone, 1985). A more accurate method is methylation analysis, which

allows the average chain length to be estimated from the ratio of internal chain to reducing terminal residues (Henry and Stone, 1985).

(3) The $(1\rightarrow4)$-β-glucan chains of high DP should form crystalline regions characteristic of cellulose. One method for demonstrating crystalline cellulose is insolubility in acetic–nitric reagent (Updegraff, 1969). This is perhaps too strict a criterion for a polymer synthesized *in vitro*; for instance, the acetic–nitric reagent will solubilize up to 50% of the cellulose synthesized *in vitro* by the *Acetobacter* enzyme (a product well characterized as cellulose by other methods) (Delmer *et al.*, 1993a). Nor has the reagent been rigorously demonstrated to be diagnostic for cellulose; for example, crystalline chitin is also insoluble in acetic–nitric reagent (Taylor and Haigler, unpublished data). Selected-area electron diffraction used to generate and record a cellulose crystal pattern is an excellent method, allowing the use of small quantities of material and visualization of the object from which the diffraction pattern is obtained (Chanzy *et al.*, 1978). X-ray diffraction is also used, although the amount of sample required for this method can be difficult to obtain from an *in-vitro* system, and the results can be confounded by the presence of pre-existing polymers. Since polymerization and crystallization of cellulose are separate processes (Haigler *et al.*, 1980; Herth, 1980), the polymers formed *in vitro* would not necessarily have the same crystallinity as those found *in vivo*. Furthermore, because there is little information about the crystal patterns of some glucan polymers, the possibility exists that similar diffraction patterns may be produced by different polymers.

4. Non-plant models for cellulose synthesis

Faced with the difficulties of plant cellulose synthesis, biochemists have looked to organisms other than plants or to enzymes other than plant cellulose synthase to answer questions about cellulose and glucan synthases. The hope is that experimental approaches or nucleic acid or antibody probes developed from studies on these other organisms and enzymes will lead to increased understanding of cellulose synthesis in plants.

Bacterial cellulose synthesis

Most of our knowledge concerning the biochemistry and molecular biology of cellulose synthesis derives from studies of the gram-negative bacterium *Acetobacter xylinum*, a cellulose-synthesizing bacterium that has long been of interest because it produces an extracellular mat of cellulose in fermenting wine or vinegar. It continues to attract industrial interest today on account of its potential to produce industrially useful cellulose of controlled crystallinity. The mat of cellulose is the product of the bacterial population; each cell synthesizes a single, composite fibril of cellulose (referred to as a ribbon) in association with a row of pores on the surface of the outer lipopolysaccharide (LPS) layer (Brown *et al.*, 1976; Zaar, 1979). Freeze fracture splits the LPS to reveal a row of particles (Brown *et al.*, 1976), each of which is probably a manifestation of the deformed pore complex. *Acetobacter* differs from plants in that the cellulose is not deposited in a wall associated with an individual cell, and therefore it has been postulated that the synthetic sites are fixed in the plasma membrane below the extrusion pores in the LPS, in contrast to the presumed mobile sites of synthesis in higher plants and algae (Brown *et al.*, 1978; Brown, 1983).

Progress in understanding the biochemistry of cellulose synthesis in *Acetobacter* has been well reviewed (Ross *et al.*, 1991). Crude membrane and solubilized membrane preparations from *Acetobacter* are highly active (Aloni *et al.*, 1982, 1983) during *in-vitro* synthesis of a product that was well characterized as cellulose (Bureau and Brown, 1987; Lin *et al.*, 1985). A novel activator of the cellulose synthase has been characterized (Ross *et al.*, 1987) and its role in the regulation of cellulose synthesis elucidated (Benziman *et al.*, 1991; Ross *et al.*, 1991). The activator, bis-

(3',5')-cyclic diguanylic acid (c-di-GMP), stimulates the activity of the bacterial enzyme by 200-fold (Ross *et al.*, 1987). The cellulose synthase activity has been shown to be localized in the cytoplasmic membrane (Bureau and Brown, 1987).

Two independent efforts led to the identification of the genes for polypeptides involved in bacterial cellulose synthesis. In one approach, a four-gene bacterial cellulose synthase *(bcs)* operon was identified by complementation of mutants deficient in cellulose synthesis (Wong *et al.*, 1990). The second approach employed photoaffinity labeling with [β-^{32}P]5'-N_3-UDP-Glc to identify the UDP-Glc-binding polypeptide in product-entrapped, purified preparations of the cellulose synthase (Saxena *et al.*, 1990). The N-terminus was sequenced, oligonucleotides synthesized, and a genomic clone isolated. The gene with a product corresponding to the UDP-Glc-binding polypeptide was sequenced and named *acsA* (for *Acetobacter* cellulose synthase; Saxena *et al.*, 1990); *acsA* and *bcsA* showed high sequence identity. The genomic clone also contained a second gene in the operon (called the *acsB* gene) (Saxena *et al.*, 1991), which showed high sequence identity to *bcsB*. Subsequent analysis of the *acs* operon indicated that the *acsA* and *acsB* genes are in fact one gene (the *acsAB* gene) that codes for a 168-kDa polypeptide (Saxena *et al.*, 1994). Presumably, this polypeptide is post-translationally processed to yield the 93-kDa and 83-kDa polypeptides found in purified preparations from *Acetobacter*. In addition, a second cellulose synthase gene *(acsAII)*, coding for a 175-kDa polypeptide, has also been reported (Saxena *et al.*, 1994).

The specific functions of the products of the three (or four) genes in the *acs/bcs* operon have been investigated using genetic complementation of cellulose synthesis-deficient mutants (Wong *et al.*, 1990), photoaffinity labeling of proteins with substrate analogs and c-di-GMP (Mayer *et al.*, 1991; Saxena *et al.*, 1990), and insertional mutagenesis (Saxena *et al.*, 1994). The *bcsA* gene (or the N-terminal part of the *acsAB* gene) has been identified as coding for the substrate-binding polypeptide (Saxena *et al.*, 1990, 1994; Wong *et al.*, 1991). There is evidence that the product of the *bcsB* gene (or the C-terminal part of the *acsAB* gene product) is the activator-binding polypeptide (Mayer *et al.*, 1991; Saxena *et al.*, 1991; Wong *et al.*, 1991). These polypeptides were the ones that co-purified in product-entrapped preparations, suggesting their tight association *in vivo* (Saxena *et al.*, 1991). Sequence analysis suggested that both polypeptides are integral membrane proteins (Saxena *et al.*, 1990, 1991, 1994; Wong *et al.*, 1991).

The *bcsC* and *bcsD* gene products could be involved either in transport and crystallization of the glucan chains or in regulation of the cellulose synthase complex (Wong *et al.*, 1990). The large BcsC polypeptide has a signal peptide and was assumed to be a membrane protein (perhaps in the outer membrane) (Wong *et al.*, 1990). The BcsC and BcsD polypeptides did not co-purify with the BcsA and BcsB polypeptides, suggesting that they are not tightly associated with them *in vivo*. Their expression was required for maximal cellulose synthesis, but their presence is not an absolute requirement for cellulose synthesis to occur (Wong *et al.*, 1990). Mutants in the *acsD* gene produced reduced amounts of cellulose *in vivo*, but similar levels to the wild type *in vitro* (Saxena *et al.*, 1994). The mutant cells synthesized a mixture of cellulose I and II (wild-type cells synthesize exclusively cellulose I), suggesting a role for the *acsD* gene product in cellulose crystallization (Saxena *et al.*, 1994).

Clearly much work is needed to clarify the function of the four gene products of the *bcs* operon and to determine the subunit structure of the large cellulose synthase complex, which was shown by gel filtration to be a multimeric complex in the range of 360–490 kDa (Lin and Brown, 1989). These exciting results obtained from *Acetobacter* have not yet led to identification of the plant cellulose synthase. Application of the preparative procedures used with *Acetobacter* to plant membranes did not yield active plant membrane preparations. Although c-di-GMP does not substantially or specifically activate plant cellulose synthesis activity (Delmer, 1990), two plant polypeptides were labeled when [^{32}P]c-di-GMP was used in direct photolabeling (Amor *et al.*,

1991). Given the unique nature of the activator, it is reasonable to assume that a c-di-GMP-binding protein in plants would be associated with cellulose synthesis. A short region (11 amino acids) of homology (45% identical; 55% similar) was found between the N-terminus of one of the plant polypeptides and an internal sequence in the BcsB polypeptide sequence (Amor *et al.*, 1991). However, this work has not yet led to the isolation of plant genes, and the research has been technically thwarted (Delmer and Amor, 1995).

There has been no reported success in the use of probes prepared from the *Acetobacter bcs* operon sequence to identify homologous sequences in plants. This is perhaps not surprising, given the 28% nucleotide sequence divergence between the two *Acetobacter* sequences. More cellulose synthase sequences are required to enable the identification of regions of conservation that might be useful as the source of heterologous probes. An antiserum prepared against synthetic peptides derived from the BcsB sequence labeled the BcsB polypeptide in *Acetobacter* and identified polypeptides in extracts from a number of plants (Mayer *et al.*, 1991). However, there have been no reports of this approach leading to the identification of plant genes involved in cellulose synthesis.

Agrobacterium tumefaciens also synthesizes cellulose, which enables it to associate with the host plant. Complementation of cellulose-minus mutants of *A. tumefaciens* resulted in the identification of two operons required for cellulose synthesis: *celABC* and *celDE*. The *celA* gene had high sequence similarity to *acsA*, *bcsA* and *acsAII*, suggesting that it, too, encodes the catalytic subunit of the cellulose synthase (Matthysse *et al.*, 1995b). The *celC* gene showed sequence similarity to endoglucanases. The other genes had no database matches, but biochemical studies of the cellulose-minus mutants suggested that *A. tumefaciens* uses a novel pathway of cellulose synthesis involving lipid-linked intermediates (Matthysse *et al.*, 1995a). The products of the *celABC* operon were active in membrane fractions, while the products of the *celDE* operon were active in soluble fractions. The biochemical studies suggested that the *celB*, *celD* and *celE* gene products were involved in the formation and transfer of the lipid-linked intermediates (Matthysse *et al.*, 1995a). Since there is no evidence of lipid-linked intermediates in *Acetobacter* cellulose synthesis (Ross *et al.*, 1991), these intriguing results obtained from *A. tumefaciens* reveal the existence of a significantly different cellulose synthesis mechanism, although the catalytic subunits show sequence homology (Matthysse *et al.*, 1995b).

Protistan cellulose synthesis

Hyphae of the water mold, *Saprolegnia monoica*, synthesize both $(1\rightarrow3)$-β-glucan and $(1\rightarrow4)$-β-glucan. For both $(1\rightarrow3)$-β-glucan and $(1\rightarrow4)$-β-glucan synthesis, *in-vitro* activity has been obtained (Bulone *et al.*, 1990), the activities have been localized to the plasma membrane (Girard and Fèvre, 1984) and solubilized (Girard and Fèvre, 1991), the enzymes have been partially purified by product entrapment procedures (Bulone *et al.*, 1990), and monoclonal antibodies to the enzymes have been prepared (Nodet *et al.*, 1988). Assay conditions can distinguish between the two glucan synthase activities; preparations incubated at high substrate concentrations and in the absence of Mg^{2+} synthesized predominantly $(1\rightarrow3)$-β-glucan. In contrast, $(1\rightarrow4)$-β-glucan synthase activity was observed when low substrate concentrations were used and Mg^{2+} was present (Fèvre and Rougier, 1981). The *Saprolegnia* $(1\rightarrow4)$-β-glucan synthase is stimulated by c-di-GMP; however, much higher levels are required to obtain a much lower stimulation than is observed in *Acetobacter* (Girard *et al.*, 1991). No genes for polypeptides involved in cellulose synthesis in *Saprolegnia* have yet been isolated.

The cellular slime mold *Dictyostelium discoideum* is attractive as a model organism for the study of cellulose synthesis because of its widely demonstrated amenability to experimental and molecular genetic approaches. Cellulose synthesis occurs in most if not all cells during some stage of multicellular development, but not in the feeding and dividing amoebae; cellulose synthesis is, therefore, inducible at high rates. Specifically, cellulose is found in slime trails, stalk tubes, stalk cell walls and spore walls, none of which are known to contain other glucan polymers, but more thorough analyses would be appropriate.

Membrane preparations from multicellular stages engaged in cellulose synthesis (slugs and beyond) incorporate glucose from UDP-Glc into an insoluble polymer that has been shown to be a $(1 \rightarrow 4)$-β-glucan by various methods, including methylation analysis (Blanton and Northcote, 1990). Interestingly, especially given the apparent lack of $(1 \rightarrow 3)$-β-glucans *in vivo*, some $(1 \rightarrow 3)$-β linkages were detected in the *in-vitro* polymer by methylation analysis (Blanton and Northcote, 1990).

There are several noteworthy points regarding *Dictyostelium* cellulose synthesis assayed *in vitro*: (1) the enzyme(s) is(are) highly active, with *in-vitro* rates approaching the *in-vivo* rates of cellulose synthesis (Blanton and Northcote, 1990); (2) the enzyme activity is completely dependent on the presence of Mg^{2+} and is stimulated by the presence of cellobiose (Blanton and Northcote, 1990); and (3) c-di-GMP has no effect on enzyme activity (Blanton, unpublished data). In contrast to *Acetobacter*, the cellulose synthase activity of crude membranes cannot be depleted by washing with chelators (Blanton, unpublished data). It is possible that the enzyme is activated by a covalent modification rather than by the non-covalent association of an activator such as c-di-GMP. Cellobiose is not required by the *Acetobacter* enzyme; however, it is required by the *Saprolegnia* enzymes (Fèvre and Dumas, 1977) and by callose synthase. The role of the cellobiose is unknown. As is the case in other systems, it does not appear to act as a primer, but it may mimic the role of a native activator or regulator (Callaghan *et al.*, 1988a,b). Unfortunately, efforts to solubilize the enzyme activity have not yet been successful, hampering further biochemical progress. However, *Dictyostelium* is potentially promising for the isolation of mutants affected in various aspects of cellulose biogenesis, and for the identification of relevant genes.

5. New approaches to plant cellulose biogenesis

Investigation of cellulose biogenesis by the use of model organisms or the study of potentially related enzymes has not yet led to the isolation of any genes for plant cellulose synthesis. In recent years there have been renewed efforts to approach the plant cellulose synthase directly, by photoaffinity labeling, by isolating binding proteins for cellulose synthesis inhibitors, and by mutational analysis. In addition, proteins that may interact with and regulate the functional cellulose synthase are being identified.

UDP-Glc photoaffinity probes

Affinity labeling with UDP-Glc substrate analogs allows the identification of UDP-Glc-binding polypeptides, although care must be taken in the interpretation of photolabeling data, and the experiments are best performed using highly purified preparations. Probes that have been used include [^{32}P]UDP-Glc (e.g. Delmer *et al.*, 1991; Dhugga and Ray, 1994; Mayer *et al.*, 1991), UDP-pyridoxal (e.g. Mason *et al.*, 1990; Read and Delmer, 1987), [β-^{32}P]5'-N_3-UDP-Glc (e.g. Drake *et al.*, 1989; Frost *et al.*, 1990; Lin *et al.*, 1990), and [^{125}I]5-[3-*p*-azidosalicylamide]allyl-UDP-Glc (Meikle *et al.*, 1991). These probes have led to the identification of UDP-Glc-binding polypeptides in preparations of callose synthase (Dhugga and Ray, 1994; Frost *et al.*, 1990; Lawson *et al.*, 1989; Meikle *et al.*, 1991; Slay *et al.*, 1992), bacterial cellulose synthase (Lin *et al.*, 1990; Mayer *et al.*, 1991), sucrose-phosphate synthase (Salvucci *et al.*, 1990; Salvucci and Klein, 1993) and sucrose synthase (Amor *et al.*, 1995; Delmer *et al.*, 1991). Li *et al.* (1993) have used photoaffinity labeling with [β-^{32}P]5'-N_3-UDP-Glc to identify, in membrane preparations from cotton fiber, candidate UDP-Glc-binding polypeptides for the glucan synthases. The identification of these polypeptides by photoaffinity labeling could lead to the identification of the cognate gene(s), a good example being the *Acetobacter* cellulose synthase (Saxena *et al.*, 1990). However, it should also be noted that, despite a number of successful photolabeling experiments, no genes have yet been isolated for the catalytic subunit of the plant callose synthase.

Cellulose synthesis inhibitors

Herbicides that inhibit cellulose synthesis selectively are 2,6-dichlorobenzonitrile (DCB) (Delmer *et al.*, 1987), isoxaben (Heim *et al.*, 1990b) and phthoxazolin (Omura *et al.*, 1990), with DCB and isoxaben used in most studies. Isoxaben is effective at substantially lower concentrations than DCB. In all higher plant studies so far, these herbicides are only about 80% inhibitory (Heim *et al.*, 1990b; Shedletzky *et al.*, 1992; Taylor *et al.*, 1992). Based on this evidence and the *rsw1* and *tbr* mutants described below (which exhibit a partial reduction in cellulose synthesis), it has been suggested (Delmer, personal communication) that herbicide-sensitive and herbicide-insensitive pathways of cellulose synthesis might exist.

The mode of action of these inhibitors is not understood, but if their effect is directly on cellulose synthesis, then these compounds could be useful for the identification of polypeptides involved in cellulose synthesis. A photoreactive analog of DCB, namely 2,6-dichlorophenyl-azide, was used as a photoaffinity probe to identify an 18-kDa DCB-binding polypeptide (Delmer *et al.*, 1987). However, it did not pellet with the membrane fraction of cell lysates (as does the glucan synthase and presumably cellulose synthase activity). The precise role of this polypeptide is unknown; it is perhaps a regulatory subunit of the cellulose synthase complex (Delmer, 1987). Isoxaben-resistant mutants have been isolated (Heim *et al.*, 1989, 1990a) and are being used to characterize the mode of action of the herbicide (Heim *et al.*, 1990b). Phthoxazolin may prove to be of particular interest, since it alone among these herbicides affects *Acetobacter* cellulose synthesis both *in vivo* and *in vitro* (Omura *et al.*, 1990).

Genetic approaches to cellulose synthesis

Mutants have great potential for eventually elucidating the molecular control of cellulose synthesis, particularly because the first forays into this type of research have indicated the possibility of multiple controls of cellulose synthesis in different cell types and tissues. Although suspension-cultured cells can adapt to greatly reduced levels of cellulose synthesis (Shedletzky *et al.*, 1992), the complete elimination of this process in whole plants would almost certainly be a lethal mutation, given the central role of cellulose in plant development and function. Mutants with partially impaired cellulose synthesis include (a) three independent-gene brittle culm barley mutants with cellulose in the stems reduced by about 80% (Kokubo *et al.*, 1989, 1991), possibly due to lesions in secondary wall synthesis (Yeo *et al.*, 1995), (b) temperature-sensitive mutants of *Arabidopsis thaliana* with altered root morphology (*rsw*), some of which exhibit a reduction in cellulose synthesis at high temperatures (Arioli *et al.*, 1995; Baskin *et al.*, 1992; Williamson *et al.*, 1992), and (c) a constitutive homozygous recessive mutant of *A. thaliana* (*tbr*) that is unable to synthesize the cellulosic secondary walls in some cells, including stem and leaf trichomes. The *tbr* mutant also shows some reduction in the cellulose content of leaf xylem secondary walls, exhibiting reductions in trichome and overall leaf cellulose content of 82% and 30%, respectively (Potikha and Delmer, 1995). Furthermore, its growth is completely inhibited by DCB, whereas the growth of the wild type is only partially inhibited by this compound (Delmer, personal communication). In contrast to specific impairment of secondary wall synthesis in *tbr*, the *rsw1* mutation is lethal at high temperatures, suggesting that it may have impaired primary-wall cellulose synthesis, without which plant growth cannot occur (Williamson *et al.*, 1992).

Sequence analysis

Four sequences exist for genes for the catalytic subunit of cellulose synthase (*bcsA*, *acsA*, *acsAII*, and *celA*). Interestingly, the *Escherichia coli* genome-sequencing project yielded sequences homologous to the *Acetobacter bcsA*, *bcsB* and *bcsC* genes (Delmer and Amor, 1995; Matthysse *et al.*, 1995b; Sofia *et al.*, 1994); *E. coli* is not known to synthesize cellulose, so the functions of these genes are unknown. Comparisons of the cellulose synthase sequences with those of other proteins that bind to UDP-Glc and UDP-*N*-acetyl glucosamine provide the opportunity to search for conserved regions that could serve as the basis for potential probes for cellulose synthase genes in other organisms, but the motifs identified to date are small (Delmer and Amor, 1995; Saxena *et al.*, 1994). Detection of regions of structural homology by hydrophobic cluster analysis of β-glycosyltransferase sequences led to their classification as a single protein family with a similar modular architecture of multiple domains, one of which was characteristic of the polymerizing enzymes in the family (including cellulose synthase). The analysis led to the proposal of a model explaining the mechanism of the polymerization reaction involved in the formation of β-linkages (Saxena *et al.*, 1995).

Related proteins

Another promising area for future research is the identification of accessory proteins that interact with the cellulose synthase complex to modulate its function. As reviewed recently, candidate proteins that exhibit some relationship to cellulose or callose synthase activity include two Ca^{2+}-binding proteins, annexin (Andrawis *et al.*, 1993) and calnexin (Delmer *et al.*, 1993b), and the small GTP-binding protein Rac, which may help to regulate changes in actin organization, or even activate glucan synthases (Delmer *et al.*, 1995).

6. Concluding remarks

From the foregoing discussion, it should be obvious that the field of plant cellulose synthesis is wide open to and in need of new perspectives and new approaches. We have discussed the difficulties encountered in this field not to discourage new researchers, but rather to inform them, since most scientists who are not familiar with cell walls do not appreciate that so little is known about them, and need to know why the problem has been so difficult to resolve. It seems to us that the genetic approach (either involving mutational analysis or identification of genes) is the most promising one, even if someone is able to discover how to repress callose synthesis and stimulate cellulose synthesis in membrane preparations. It is obvious that there will be interesting questions to study for many years, given the complexity of the process of cellulose biogenesis in higher plants and the number of polypeptides that are likely to be involved in polymerization, crystallization and cellular control of deposition. Since cellulose biogenesis is so closely associated with plant plasma membranes, this increasing knowledge should greatly add to our general understanding of plasma membrane structure and function. We can also expect that further progress will illuminate how the cell cortex, plasma membrane and cell wall are integrated into a functional unit, which is an emerging and significant area in plant biology (Wyatt and Carpita, 1993).

Acknowledgements

R.L. Blanton's research on *Dictyostelium* cellulose synthase is supported by the Office of Basic Energy Sciences, US Department of Energy (DE-FG05-90ER20006; DE-FG03-95ER20172).

References

Alexander WJ, Mitchell RL. (1949) Rapid measurement of cellulose viscosity by the nitration method. *Anal. Chem.* **21:** 1497–1500.

Aloni Y, Delmer DP, Benziman M. (1982) Achievement of high rates of *in vitro* synthesis of 1,4-β-D-glucan: activation by cooperative interaction of the *Acetobacter xylinum* enzyme system with GTP, polyethylene glycol, and a protein factor. *Proc. Natl Acad. Sci. USA* **79:** 6448–6452.

Aloni Y, Cohen R, Benziman M, Delmer DP. (1983) Solubilization of the UDP-glucose:1,4-β-D glucosyltransferase (cellulose synthase) from *Acetobacter xylinum. J. Biol. Chem.* **258:** 4419–4423.

Amor Y, Mayer R, Benziman M, Delmer D. (1991) Evidence for a cyclic diguanylic acid-dependent cellulose synthase in plants. *Plant Cell* **3:** 989–995.

Amor Y, Haigler CH, Johnson S, Wainscott M, Delmer DP. (1995) A membrane-associated form of sucrose synthase and its potential role in synthesis of cellulose and callose in plants. *Proc. Natl Acad. Sci. USA* **92:** 9353–9357.

Andrawis A, Solomon M, Delmer DP. (1993) Cotton fiber annexins: a potential role in the regulation of callose synthase. *Plant J.* **3:** 763–772.

Arioli T, Betzner A, Peng L, *et al.* (1995) Radial swelling mutants deficient in cellulose biosynthesis. In: *Seventh Cell Wall Meeting Abstracts* (Zarra I, Revilla G, ed.). Santiago: Universidad de Santiago de Compostela, p. 182.

Atalla RH, VanderHart DL. (1989) Studies on the structure of cellulose using Raman spectroscopy and solid state ^{13}C NMR. In: *Cellulose and Wood; Chemistry and Technology* (Schuerch C, ed.). New York: Wiley-Interscience, pp. 169–188.

Baba K, Sone Y, Misaki A, Hayashi T. (1994) Localization of xyloglucan in the macromolecular complex composed of xyloglucan and cellulose in pea stems. *Plant Cell Physiol.* **35:** 439–444.

Bacic A, Delmer DP. (1981) Stimulation of membrane-associated polysaccharide synthetases by a membrane potential in developing cotton fibers. *Planta* **152:** 346–351.

Baskin TI, Berzner AA, Hoggart R, Cork A, Williamson RE. (1992) Root morphology mutants in *Arabidopsis thaliana. Aust. J. Plant Physiol.* **19:** 427–437.

Becker B, Bölinger B, Melkonian M. (1995) Anterograde transport of algal scales through the Golgi complex is not mediated by vesicles. *Trends Cell Biol.* **5:** 305–307.

Benziman M, Mayer R, Weinhouse H, *et al.* (1991) The c-di-GMP regulatory system of bacterial cellulose synthesis. In: *Cellulose '91* (Chum HL, ed.). Washington, DC: American Chemical Society, Abstract #256.

Blackwell J. (1982) The macromolecular organization of cellulose and chitin. In: *Cellulose and Other Natural Polymer Systems* (Brown RM Jr, ed.). New York: Plenum, pp. 403–428.

Blanton RL, Northcote DH. (1990) A 1,4-β-D-glucan synthase system from *Dictyostelium discoideum. Planta* **180:** 324–332.

Blaschek W, Haass D, Koehler H, Semler U, Franz G. (1983) Demonstration of a β-1,4-primer in cellulose-like glucan synthesized *in vitro. Z. Pflanzenphysiol.* **111:** 357–364.

Bolwell GP. (1988) Synthesis of cell wall components: aspects of control. *Phytochemistry* **27:** 1235–1253.

Bolwell GP. (1993) Dynamic aspects of the plant extracellular matrix. *Int. Rev. Cytol.* **146:** 261–324.

Branton D, Bullivant S, Gilula NB, *et al.* (1975) Freeze-etching nomenclature. *Science* **190:** 54–56.

Brown RM Jr. (1978) Biogenesis of natural polymer systems, with special reference to cellulose assembly and deposition. In: *Structure and Biochemistry of Natural Biological Systems: Proceedings of the Third Philip Morris Science Symposium* (Walk EM, ed.). Richmond, Virginia: Philip Morris, Inc., pp. 51–144.

Brown RM Jr, Montezinos D. (1976) Cellulose microfibrils: visualization of biosynthetic and orienting complexes in association with the plasma membrane. *Proc. Natl Acad. Sci. USA* **84:** 6985–6989.

Brown RM Jr, Franke WW, Kleinig H, Falk H, Sitte P. (1969) Cellulosic wall component produced by the Golgi apparatus of *Pleurochrysis scherffelii. Science* **166:** 894–897.

Brown RM Jr, Willison JHM, Richardson CL. (1976) Cellulose biosynthesis in *Acetobacter xylinum*: visualization of the site of synthesis and direct measurement of the *in vivo* process. *Proc. Natl Acad. Sci. USA* **73:** 4565–4569.

Brown RM, Haigler CH, Suttie J, White AR, Roberts E, Smith C, Itoh T, Cooper K. (1983) The biosynthesis and degradation of cellulose. *J. Appl. Polym. Sci. Appl. Polym. Symp.* **37:** 33–78.

Brown RM Jr, Li L, Okuda K, Kuga S, Kudlicka K, Drake R, Santos R, Clement S. (1994) *In vitro* cellulose synthesis in plants. *Plant Physiol.* **105**: 1–2.

Bulawa CE. (1993) Genetics and molecular biology of chitin synthesis in fungi. *Annu. Rev. Microbiol.* **47**: 505–534.

Bulone V, Girard V, Fèvre M. (1990) Separation and partial purification of 1,3-β-glucan and 1,4-β-glucan synthases from *Saprolegnia. Plant Physiol.* **94**: 1748–1755.

Bureau TE, Brown RM. (1987) *In vitro* synthesis of cellulose II from a cytoplasmic membrane fraction of *Acetobacter xylinum. Proc. Natl Acad. Sci. USA* **84**: 6985–6989.

Callaghan T, Ross P, Weinberger-Ohana P, Garden G, Benziman M. (1988a) β-glucoside activators of mung bean UDP-glucose:β-glucan synthase. I. Identification of an endogenous β-linked glucolipid activator. *Plant Physiol.* **86**: 1099–1103.

Callaghan T, Ross P, Weinberger-Ohana P, Garden G, Benziman M. (1988b) β-glucoside activators of mung bean UDP-glucose:β-glucan synthase. II. Comparison of effects of an endogenous β-linked glucolipid with synthetic n-alkyl β-D-monoglucopyranosides. *Plant Physiol.* **86**: 1104–1107.

Chanzy H, Imada K, Vuong R. (1978) Electron diffraction from the primary wall of cotton fibers. *Protoplasma* **94**: 299–306.

Chanzy H, Imada K, Mollard A, Vuong R, Barnoud F. (1979) Crystallographic aspects of sub-elementary cellulose fibrils occuring in the wall of rose cells cultured *in vitro. Protoplasma* **100**: 317–322.

Chourey PS, Chen YC, Miller ME. (1991) Early cell degeneration in developing endosperm is unique to the *shrunken* mutation in maize. *Maydica* **36**: 141–146.

Delmer DP. (1977) Biosynthesis of cellulose and other plant cell wall polysaccharides. *Recent Adv. Phytochem.* **11**: 45–77.

Delmer DP. (1987) Cellulose biosynthesis. *Annu. Rev. Plant Physiol.* **38**: 259–290.

Delmer DP. (1990) Role of the plasma membrane in cellulose synthesis. In: *The Plant Plasma Membrane* (Larsson C, Møller IM, ed.). Berlin: Springer-Verlag, pp. 256–268.

Delmer DP. (1991) The biochemistry of cellulose synthesis. In: *The Cytoskeletal Basis of Plant Growth and Form* (Lloyd C, ed.). New York: Academic Press, pp. 102–107.

Delmer DP, Stone BA. (1988) Biosynthesis of plant cell walls. In: *The Biochemistry of Plants* (Stumpf PK, Conn EE, ed.). San Diego: Academic Press, pp. 373–420.

Delmer DP, Amor Y. (1995) Cellulose biosynthesis. *Plant Cell* **7**: 987–1000.

Delmer DP, Benziman M, Padan E. (1982) Requirement for a membrane potential for cellulose synthesis in intact cells of *Acetobacter xylinum. Proc. Natl Acad. Sci. USA* **79**: 5282–5286.

Delmer DP, Read SM, Cooper G. (1987) Identification of a receptor protein in cotton fibers for the herbicide 2,6-dichlorobenzonitrile. *Plant Physiol.* **84**: 415–420.

Delmer DP, Solomon M, Read SM. (1991) Direct photolabeling with [^{32}P]UDP-glucose for identification of a subunit of cotton fiber callose synthase. *Plant Physiol.* **95**: 556–563.

Delmer DP, Ohana P, Gonen L, Benziman M. (1993a) *In vitro* synthesis of cellulose in plants: still a long way to go! *Plant Physiol.* **103**: 307–308.

Delmer DP, Volokita M, Solomon M, Fritz U, Delphendahl W, Herth W. (1993b) A monoclonal antibody recognizes a 65 kDa higher plant membrane polypeptide which undergoes cation-dependent association with callose synthase *in vitro* and co-localizes with sites of high callose deposition. *Protoplasma* **176**: 33–42.

Delmer DP, Pear JR, Andrawis A, Stalker DM. (1995) Genes encoding small GTP-binding proteins analogous to mammalian Rac are preferentially expressed in developing cotton fibers. *Mol. Gen. Genet.* **248**: 43–51.

Dhugga KS, Ray PM. (1994) Purification of a 1,3-β-D-glucan synthase activity from pea tissue – two polypeptides of 55 kDa and 70 kDa copurify with enzyme activity. *Eur. J. Biochem.* **220**: 943–953.

Drake RR Jr, Evans RK, Wolf MJ, Haley BE. (1989) Synthesis and properties of 5-azido-UDP-glucose. Development of photoaffinity probes for nucleotide diphosphate sugar binding sites. *J. Biol. Chem.* **264**: 11928–11933.

Emons AMC. (1991) Role of particle rosettes and terminal globules in cellulose synthesis. In: *Biosynthesis and Biodegradation of Cellulose* (Haigler CH, Weimer PJ, ed.). New York: Marcel Dekker, pp. 71–98.

Evans WH. (1994) Hiroshima welcomes gap junction communicants. *Trends Cell Biol.* **4**: 26–29.

Fèvre M, Dumas C. (1977) β-glucan synthases from *Saprolegnia monoica. J. Gen. Microbiol.* **103**: 297–306.

Fèvre M, Rougier M. (1981) β-1-3- and β-1,4-glucan synthesis by membrane fractions from the fungus *Saprolegnia. Planta* **151**: 232–241.

Franke WW, Ermen B. (1969) Negative staining of plant slime cellulose: an examination of the elementary fibril concept. *Z. Naturforsch.* **24b:** 917–927.

Frost DJ, Read SM, Drake RR, Haley BE, Wasserman BP. (1990) Identification of the UDP-glucose-binding polypeptide of callose synthase from *Beta vulgaris* L. by photoaffinity labeling with 5-azido-UDP-glucose. *J. Biol. Chem.* **265:** 2162–2167.

Fry SC. (1988) *The Growing Plant Cell Wall: Chemical and Metabolic Analysis.* Harlow: Longman Scientific and Technical.

Gibeaut DM, Carpita NC. (1993) Synthesis of (1→3), (1→4)-β-D-glucan in the Golgi apparatus of maize coleoptiles. *Proc. Natl Acad. Sci. USA* **90:** 3850–3854.

Giddings TH, Brower DL, Staehelin LA. (1980) Visualization of particle complexes in the plasma membrane of *Micrasterias denticulata*, associated with the formation of cellulose fibrils in primary and secondary cell walls. *J. Cell Biol.* **84:** 327–339.

Giddings TH Jr, Staehelin LA. (1991) Microtubule-mediated control of microfibril deposition: a re-examination of the hypothesis. In: *The Cytoskeletal Basis of Plant Growth and Form* (Lloyd CW, ed.). London: Academic Press, pp. 85–99.

Girard V, Fèvre M. (1984) β-1-4- and β-1-3-glucan synthases are associated with the plasma membrane of the fungus *Saprolegnia. Planta* **160:** 400–406.

Girard V, Fèvre M. (1991) Solubilization of a membrane-bound stimulator of 1,3-β-glucan synthase from *Saprolegnia. Plant Sci.* **76:** 193–200.

Girard V, Fèvre M, Mayer R, Benziman M. (1991) Cyclic diguanylic acid stimulates 1,4-β-glucan synthase from *Saprolegnia monoica. FEMS Microbiol. Lett.* **82:** 293–296.

Haigler CH. (1985) The functions and biogenesis of native cellulose. In: *Cellulose Chemistry and its Applications* (Nevell TP, Zeronian SH, ed.). Chichester: Ellis Horwood, pp. 30–83.

Haigler CH. (1991) Relationship between polymerization and crystallization in microfibril biogenesis. In: *Biosynthesis and Biodegradation of Cellulose* (Haigler CH, Weimer PJ, ed.). New York: Marcel Dekker, pp. 99–124.

Haigler CH, Benziman M. (1982) Biogenesis of cellulose I microfibrils occurs by cell-directed self-assembly in *Acetobacter xylinum*. In: *Cellulose and Other Natural Polymer Systems. Biogenesis, Structure, and Degradation* (Brown RM Jr, ed.). New York: Plenum Press, pp. 273–297.

Haigler CH, Brown RM Jr. (1986) Transport of rosettes from the Golgi apparatus to plasma membrane in isolated mesophyll cells of *Zinnia elegans* during differentiation to tracheary elements in suspension culture. *Protoplasma* **134:** 111–120.

Haigler CH, Brown RM Jr, Benziman M. (1980) Calcofluor White ST alters cellulose synthesis in *Acetobacter xylinum. Science* **210:** 903–906.

Hayashi T, Read SM, Bussell J, Thelen M, Lin F-C, Brown RM, Delmer DP. (1987) UDP-glucose: (1-3)-β-glucan synthases from mung bean and cotton. Differential effects of Ca^{2+} and Mg^{2+} on enzyme properties and on macromolecular structure of the glucan product. *Plant Physiol.* **83:** 1054–1062.

Heim DR, Roberts JL, Pike PD, Larrinua IM. (1989) Mutation of a locus of *Arabidopsis thaliana* confers resistance to the herbicide isoxaben. *Plant Physiol.* **92:** 858–861.

Heim DR, Roberts JL, Pike PD, Larrinua IM. (1990a) A second locus *lxrB1* in *Arabidopsis thaliana*, that confers resistance to the herbicide isoxaben. *Plant Physiol.* **92:** 858–861.

Heim DR, Skomp JR, Tschabold EE, Larrinua IM. (1990b) Isoxaben inhibits the synthesis of acid insoluble cell wall materials in *Arabidopsis thaliana. Plant Physiol.* **93:** 695–700.

Henry RJ, Stone BA. (1985) Extent of β-glucan chain elongation by ryegrass (*Lolium muliflorum*) enzymes. *Carbohydr. Polym.* **5:** 1–12.

Herth W. (1980) Calcofluor white and congo red inhibit microfibril assembly of *Poterioochromonas*: evidence for a gap between polymerization and microfibril polymerization. *J. Cell Biol.* **87:** 442–450.

Herth W. (1985) Plasma membrane rosettes involved in localized wall thickening during xylem vessel formation of *Lepidium sativum. Planta* **164:** 12–21.

Herth W. (1989) Inhibitor effects on putative cellulose synthetase complexes of vascular plants. In: *Cellulose and Wood; Chemistry and Technology* (Schuerch C, ed.). New York: Wiley-Interscience, pp. 795–810.

Hotchkiss A. (1989) Cellulose biosynthesis: the terminal complex hypothesis and its relationship to other contemporary research topics. In: *Plant Cell Wall Polymers; Biogenesis and Biodegradation* (Lewis NG, Paice MG, ed.). Washington, DC: American Chemical Society, pp. 232–247.

Itoh T, O'Neil RM, Brown RM Jr. (1984) Interference of cell wall regeneration of *Boergesenia forbesii* protoplasts by Tinopal LPW, a fluorescent brightening agent. *Protoplasma* **123:** 174–183.

Iyama K, Lam TBT, Meikle PJ, Ng K, Rhodes DI, Stone BA. (1993) Cell wall biosynthesis and its regulation. In: *Forage Cell Wall Structure and Digestibility* (Jung HG, Buxton DR, Hatfield RD, Ralph

J, ed.). Madison, Wisconsin: American Society of Agronomy, Crop Science Society of America, and Soil Science Society of America, pp. 621–683.

Jacob SR, Northcote DH. (1985) *In vitro* glucan synthesis by membranes of celery petioles: the role of the membrane in determining the type of linkage formed. *J. Cell Sci. Suppl.* **2**: 1–11.

Kauss H. (1987) Some aspects of calcium-dependent regulation in plant metabolism. *Annu. Rev. Plant Physiol.* **38**: 47–72.

Kokubo A, Kuraishi S, Sakurai N. (1989) Culm strength of barley; correlation among maximum bending stress, cell wall dimensions, and cellulose content. *Plant Physiol.* **91**: 876–882.

Kokubo A, Sakurai N, Kuraishi S, Takeda K. (1991) Culm brittleness of barley (*Hordeum vulgare* L.) mutants is caused by smaller number of cellulose molecules in cell wall. *Plant Physiol.* **97**: 509–514.

Kudlicka K, Brown RM Jr, Li L, Lee JH, Shin H, Kuga S. (1995) β-glucan synthesis in the cotton fiber. IV. *In vitro* assembly of the cellulose I allomorph. *Plant Physiol.* **107**: 111–123.

Kuga S, Brown RM Jr. (1991) Physical structure of cellulose microfibrils: implications for biogenesis. In: *Biosynthesis and Biodegradation of Cellulose* (Haigler CH, Weimer PJ, ed.). New York: Marcel Dekker, pp. 125–142.

Lawson SG, Mason TL, Sabin RD, Sloan ME, Drake RR, Haley BE, Wasserman BP. (1989) UDP-glucose:(1,3)-β-glucan synthase from *Daucus carota* L. *Plant Physiol.* **90**: 101–108.

Lee JH, Brown RM Jr, Kuga S, Shoda S, Kobayashi S. (1994) Assembly of synthetic cellulose I. *Proc. Natl Acad. Sci. USA* **91**: 7425–7429.

Li L, Brown RM Jr. (1993) β-Glucan synthesis in the cotton fiber. II. Regulation and kinetic properties of β-glucan synthases. *Plant Physiol.* **101**: 1143–1148.

Li L, Drake RJ, Clement S, Brown RM Jr. (1993) β-Glucan synthesis in the cotton fiber. III. Identification of UDP-glucose-binding subunits of β-glucan synthases by photoaffinity labeling with [β-^{32}P]5'-N$_3$-UDP-glucose. *Plant Physiol.* **101**: 1149–1156.

Lin F-C, Brown RM Jr. (1989) Purification of cellulose synthase from *Acetobacter xylinum*. In: *Cellulose and Wood – Chemistry and Technology* (Schuerch C, ed.). New York: Wiley-Interscience, pp. 473–492.

Lin F-C, Brown RM Jr, Cooper JB, Delmer DP. (1985) Synthesis of fibrils *in vitro* by a solubilized cellulose synthase from *Acetobacter xylinum*. *Science* **230**: 822–825.

Lin F-C, Brown RM Jr, Drake RR Jr, Haley BE. (1990) Identification of the uridine 5'-diphosphoglucose (UDP-Glc) binding subunit of cellulose synthase in *Acetobacter xylinum* using the photoaffinity probe 5-azido-UDP-Glc. *J. Biol. Chem.* **265**: 4782–4784.

McCormick CL, Callais PA, Hutchinson RH. (1985) Solution studies of cellulose in lithium chloride and N,N-dimethylacetamide. *Macromolecules* **18**: 2394–2401.

Mason TL, Read SM, Frost DJ, Wasserman BP. (1990) Inhibition and labeling of red beet uridine 5'diphospho-glucose:(1,3)-β-glucan synthase by chemical modification with formaldehyde and uridine 5'diphospho-pyridoxal. *Physiol. Plant.* **79**: 439–447.

Matthysse AG, Thomas DL, White AR. (1995a) Mechanism of cellulose synthesis in *Agrobacterium tumefaciens*. *J. Bacteriol.* **177**: 1076–1081.

Matthysse AG, White S, Lightfoot R. (1995b) Genes required for cellulose synthesis in *Agrobacterium tumefaciens*. *J. Bacteriol.* **177**: 1069–1075.

Mayer R, Ross P, Weinhouse H, *et al.* (1991) Polypeptide composition of bacterial cyclic diguanylic acid-dependent cellulose synthase and the occurrence of immunologically cross-reacting proteins in higher plants. *Proc. Natl Acad. Sci. USA* **88**: 5472–5476.

Meikle PJ, Ng KF, Johnson E, Hoogenraad NJ, Stone BA. (1991) The β-glucan synthase from *Lolium multiflorum*. Detergent solubilization, purification using monoclonal antibodies, and photoaffinity labeling with a novel photoreactive pyrimidine analogue of uridine 5'-diphosphoglucose. *J. Biol. Chem.* **266**: 22569–22581.

Mizuta S. (1985) Assembly of cellulose synthesizing complexes on the plasma membrane of *Boodlea coacta*. *Plant Cell Physiol.* **26**: 1443–1453.

Mizuta S, Brown RM Jr. (1992a) Effects of 2,6-dichlorobenzonitrile and Tinopal LPW on the structure of the cellulose synthesizing complexes of *Vaucheria hamata*. *Protoplasma* **166**: 200–207.

Mizuta S, Brown RM Jr. (1992b) High resolution analysis of the formation of cellulose synthesizing complexes in *Vaucheria hamata*. *Protoplasma* **166**: 187–199.

Mueller SC, Brown RM Jr. (1980) Evidence for an intramembrane component associated with a cellulose microfibril synthesizing complex in higher plants. *J. Cell Biol.* **84**: 315–326.

Mullins JT. (1990) Regulatory mechanisms of β-glucan synthases in bacteria, fungi, and plants. *Physiol. Plant.* **78**: 309–314.

Nodet P, Grange J, Fèvre M. (1988) Dot-blot assays and their use as a direct antigen-binding method to screen monoclonal antibodies to 1,4-β- and 1,3-β-glucan synthases. *Anal. Biochem.* **174**: 662–665.

Northcote DH. (1991) Site of cellulose synthesis. In: *Biosynthesis and Biodegradation of Cellulose* (Haigler CH, Weimer PJ, ed.). New York: Marcel Dekker, pp. 165–176.

Ohana P, Delmer DP, Volman G, Steffens JC, Matthews DE, Benziman M. (1992) Furfuryl-β-glucoside: an endogenous activator of higher plant UDP-glucose:(1-3)-β-glucan synthase. Biological activity, distribution, and *in vitro* synthesis. *Plant Physiol.* **98:** 708–715.

Okuda K, Brown RM Jr. (1992) A new putative cellulose-synthesizing complex of *Coleochaete scutata. Protoplasma* **168:** 51–63.

Okuda K, Li L, Kudlicka K, Kuga S, Brown RM Jr. (1993) β-Glucan synthesis in the cotton fiber. I. Identification of β-1,4- and β-1,3-glucans synthesized *in vitro. Plant Physiol.* **101:** 1131–1142.

Okuda K, Tsekos I, Brown RM Jr. (1994) Cellulose microfibril assembly in *Erythrocladia subintegra* Rosenv.: an ideal system for understanding the relationship between synthesizing complexes (TCs) and microfibril crystallization. *Protoplasma* **180:** 49–58.

Omura S, Tanaka Y, Kanaya I, Shinose M, Takahashi Y. (1990) Phthoxazolin, a specific inhibitor of cellulose biosynthesis, produced by a strain of *Streptomyces* sp. *J. Antibiotics* **43:** 1034–1036.

Potikha T, Delmer DP. (1995) A mutant of *Arabidopsis thaliana* displaying altered patterns of cellulose deposition. *Plant J.* **7:** 453–460.

Preston RD. (1964) Structural and mechanical aspects of plant cell walls with particular evidence of synthesis and growth. In: *The Formation of Wood in Forest Trees* (Zimmermann MH, ed.). New York: Academic Press, pp. 169–201.

Qi X, Tai C-Y, Wasserman BP. (1995) Plasma membrane intrinsic proteins of *Beta vulgaris* L. *Plant Physiol.* **108:** 387–392.

Quader H. (1983) Morphology and movement of cellulose synthesizing (terminal) complexes in *Oocystis solitaria*: evidence that microfibril assembly is the motive force. *Eur. J. Cell Biol.* **32:** 174–177.

Quader H. (1991) Role of linear terminal complexes in cellulose synthesis. In: *Biosynthesis and Biodegradation of Cellulose* (Haigler CH, Weimer PJ, ed.). New York: Marcel Dekker, pp. 51–69.

Quader H, Robinson DG, van Kempen R. (1983) Cell wall development in *Oocystis solitaria* in the presence of polysaccharide dyes. *Planta* **157:** 317–323.

Read SM, Delmer DP. (1987) Inhibition of mung bean UDP-glucose:(1-3)-β-glucan synthase by UDP-pyridoxal. *Plant Physiol.* **85:** 1008–1015.

Read SM, Delmer DP. (1991) Biochemistry and regulation of cellulose synthesis in higher plants. In: *Biosynthesis and Biodegradation of Cellulose* (Haigler CH, Weimer PJ, ed.). New York: Marcel Dekker, pp. 177–200.

Rollit J, Maclachlan GA. (1974) Synthesis of wall glucan from sucrose by enzyme preparations from *Pisum sativum. Phytochemistry* **13:** 367–374.

Ross P, Mayer R, Benziman M. (1991) Cellulose biosynthesis and function in bacteria. *Microbiol. Rev.* **55:** 35–58.

Ross P, Weinhouse H, Aloni Y, *et al.* (1987) Regulation of cellulose synthesis in *Acetobacter xylinum* by cyclic diguanylic acid. *Nature* **325:** 279–281.

Rudolph U. (1987) Occurrence of rosettes in the ER membrane of young *Funaria hygrometrica* protonemata. *Naturwissenschaften* **74:** 439.

Rudolph U, Gross H, Schnepf E. (1989) Investigations of the turnover of the putative cellulose-synthesizing particle 'rosettes' within the plasma membrane of *Funaria hygrometrica* protonema cells. II. Rosette structure and the effects of cycloheximide, actinomycin D, 2,6-dichlorobenzonitrile, biofluor, heat shock, and plasmolysis. *Protoplasma* **148:** 57–69.

Salvucci ME, Klein RE. (1993) Identification of the uridine-binding domain of sucrose-phosphate synthase. *Plant Physiol.* **102:** 529–536.

Salvucci ME, Drake RR, Haley BE. (1990) Purification and photoaffinity labeling of sucrose phosphate synthase from spinach leaves. *Arch. Biochem. Biophys.* **281:** 212–218.

Saxena IM, Lin FC, Brown RM Jr. (1990) Cloning and sequencing of the cellulose synthase catalytic subunit gene of *Acetobacter xylinum. Plant Mol. Biol.* **15:** 673–683.

Saxena IM, Lin FC, Brown RM Jr. (1991) Identification of a new gene in an operon for cellulose biosynthesis in *Acetobacter xylinum. Plant Mol. Biol.* **16:** 947–954.

Saxena IM, Kudlicka K, Okuda K, Brown RM Jr. (1994) Characterization of genes in the cellulose-synthesizing operon (*acs* operon) of *Acetobacter xylinum*: implications for cellulose crystallization. *J Bacteriol.* **176:** 5735–5752.

Saxena IM, Brown RM Jr, Fèvre M, Geremia RA, Henrissat B. (1995) Multidomain architecture of β-glycosyl transferases: implications for mechanism of action. *J. Bacteriol.* **177:** 1419–1424.

Seitz HU, Emmerling M. (1990) The cellulose synthase problem. *Bot. Acta* **103:** 7–8.

Shedletzky E, Shmuel M, Trainin T, Kalman S, Delmer D. (1992) Cell wall structure in cells adapted to growth on the cellulose-synthesis inhibitor 2,6-dichlorobenzonitrile. A comparison between two dicotyledonous plants and a graminaceous monocot. *Plant Physiol.* **100**: 120–130.

Slay RM, Watada AE, Frost DJ, Wasserman BP. (1992) Characterization of the UDP-glucose:(1,3)-β-glucan (callose) synthase from plasma membranes of celery: polypeptide profiles and photolabeling patterns of enriched fractions suggest callose synthase complexes from various sources share a common structure. *Plant Sci.* **86**: 125–136.

Sleytr UB, Robards AW. (1982) Understanding the artefact problem in freeze-fracture replication: a review. *J. Microscopy* **126**: 101–122.

Sofia HJ, Burland V, Daniels DL, Plunkett G III, Blattner FR. (1994) Analysis of the *Escherichia coli* genome. V. DNA sequence of the region from 76.0 to 81.5 minutes. *Nucleic Acids Res.* **22**: 2576–2586.

Staehelin LA, Giddings TH Jr. (1982) Membrane-mediated control of cell wall microfibrillar order. In: *Developmental Order: Its Origins and Regulation* (Subtelny S, Green PB, ed.). New York: Alan R. Liss, pp. 133–147.

Sugiyama J, Persson J, Chanzy H. (1991) Combined infrared and electron diffraction study of the polymorphism of native celluloses. *Macromolecules* **24**: 2461–2466.

Sugiyama J, Chanzy H, Revol JF. (1994) On the polarity of cellulose in the cell wall of *Valonia*. *Planta* **193**: 260–265.

Taylor JT, Owen TP Jr, Koonce LT, Haigler CH. (1992) Dispersed lignin in tracheary elements treated with cellulose synthesis inhibitors provides evidence that molecules of the secondary cell wall mediate wall patterning. *Plant J.* **2**: 959–970.

Timpa JD. (1991) Application of universal calibration in gel permeation chromatography for molecular weight determinations of plant cell wall polymers: cotton fiber. *J. Agric. Food Chem.* **39**: 270–275.

Uhlin KI, Atalla RH, Thompson NS. (1995) Influence of hemicelluloses on the aggregation patterns of bacterial cellulose. *Cellulose* **2**: 129–144.

Updegraff DM. (1969) Semimicro determination of cellulose in biological materials. *Anal. Biochem.* **32**: 420–424.

Vian B, Reis D. (1991) Relationship of cellulose and other cell wall components: supramolecular organization. In: *Biosynthesis and Biodegradation of Cellulose* (Haigler CH, Weimer PJ, ed.). New York: Marcel Dekker, pp. 25–50.

White AR, Xin Y, Pezeshk V. (1993) Xyloglucan glucosyltransferase in Golgi membranes from *Pisum sativum* (pea). *Biochem. J.* **294**: 231–238.

Williamson RE, Baskin TI, Cork A, Birch R. (1992) Genetic analysis of the microtubule- and microfibril-dependent mechanisms that determine cell and root shape in *Arabidopsis thaliana*. In: *Sixth Cell Wall Meeting Abstracts* (Sassen MMA, Derksen JWM, Emons AMC, Wolters-Arts AMC, ed.). Nijmegen: University Press, p. 128.

Wong HC, Fear AL, Calhoon RD, *et al.* (1990) Genetic organization of the cellulose synthase operon in *Acetobacter xylinum*. *Proc. Natl Acad. Sci. USA* **87**: 8130–8134.

Wong HC, Fear AL, Calhoon RD, Eichinger GH, Mayer R, Ross R, Benziman M, Ben-Bassat A, Tal R. (1991) Genetic organization and regulation of the cellulose synthase operon in *Acetobacter xylinum*. In: *Cellulose '91* (Chum HL, ed.). Washington, DC: American Chemical Society, Abstract #12.

Wu A, Wasserman BP. (1993) Limited proteolysis of (1,3)-β-glucan (callose) synthase from *Beta vulgaris* L: topology of protease-sensitive sites and polypeptide identification using Pronase E. *Plant J.* **4**: 683–695.

Wyatt SE, Carpita NC. (1993) The plant cytoskeleton–cell wall continuum. *Trends Cell Biol.* **3**: 413–417.

Yeo U-D, Soh W-Y, Tasaka H, Sakurai N, Kuraishi S, Takeda K. (1995) Cell wall polysaccharides of callus and suspension-cultured cells from three cellulose-less mutants of barley (*Hordeum vulgare* L.). *Plant Cell Physiol.* **36**: 931–936.

York WS, Darvill AG, McNeil M, Stevenson TT, Albersheim P. (1985) Isolation and characterization of plant cell walls and cell wall components. *Meth. Enzymol.* **118**: 3–40.

Zaar K. (1979) Visualization of pores (export sites) correlated with cellulose production in the envelope of the gram-negative bacterium *Acetobacter xylinum*. *J. Cell Biol.* **80**: 773–777.

Chapter 5

Callose synthesis

Heinrich Kauss

1. Occurrence and functions of callose

Cytological observations dating back to the end of the last century provide many examples of cases where the 'regular' plant cell wall appears to be altered by localized appositions termed 'callosities'. Based on differential staining properties, their chemical structure was assumed to differ from that of the cell wall as a whole, and thus the material concerned was termed 'callose'. Subsequently, callose was identified as a linear 1,3-β-glucan. Following lengthy and often spirited discussion of the specificity of stains, resorcin blue became widely accepted as an indicator of callose in light microscopy (Eschrich, 1956). These staining procedures have given way to a more easily applicable technique using decolorized, water-soluble aniline blue, which provides a bright yellow–white fluorescence with callose (e.g. Kauss, 1989). The fluorophore active in this procedure is present as an impurity in commercial aniline blue and is called 'Sirofluor' (Evans *et al.*, 1984). Aniline blue fluorescence may also be used to quantify callose after extraction with alkali or dimethylsulphoxide (Kauss, 1989). At the transmission electron microscope (TEM) level, callose remains fully translucent with the usual contrasting techniques, but can be identified with special chemical procedures (see *Figure 1*). Alternatively, callose in TEM sections can be immunogold-labelled using 1,3-β-glucanase (Benhamou, 1992) or monoclonal and polyclonal antibodies (Xu and Mendgen, 1994) as specific reagents. The latter technique even allows differentiation between 1,3-β-glucans of plant and fungal origin.

The function of callose is, in many cases, indicated by the anatomical and physiological context in which it occurs. Numerous earlier reports have been reviewed by Eschrich (1956), while an encyclopaedic work summarizes more recent advances in the chemistry and biology of callose (Stone and Clarke, 1992). A few examples will be described here to document the widespread distribution of callose in the plant kingdom and its multipurpose function as a building material for the transient reconstruction and rapid repair of plant cell walls.

2. Callose in higher plants

Developmentally regulated callose deposition

Callose is often found where the existing cell wall has been altered during developmental changes in the cell shape. The transiently observed callose in the cell plate of dividing cells (Kakimoto and Shibaoka, 1992) or the appearence of callose in cotton fibres during the early stages of secondary wall synthesis without obvious exogenous induction (Maltby *et al.*, 1979) are examples of this type of deposition.

Another example of transient callose deposition occurs in the pollen mother cell of angiosperms, which is lined by a massive callose layer that also surrounds the tetrad of microspores. Deposition of this layer starts shortly before and continues during meiosis, and the callose disappears once the pollen grain exine is complete (see Worrall *et al.*, 1992). It has been

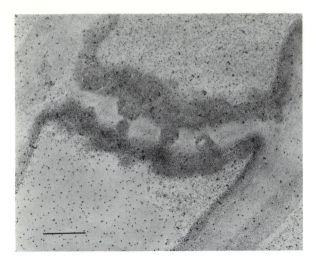

Figure 1. Callose deposited in the sieve pores and on the sieve plate in the phloem of a stem from *Abutilon*. With the special contrasting technique used, callose is rendered more electron-dense (contrasted) than the general cell wall, whereas with the usual TEM techniques callose remains completely translucent. Staining of callose was performed with silver methenamine after previous oxidation by permanganate followed by periodate (Brander and Wattendorff, 1987). Scale bar represents 1 μm. Micrograph reproduced courtesy of J. Wattendorff (Fribourg, Switzerland).

suggested that callose degradation by 'callase' is related to the nutrition of the developing pollen grains (Eschrich, 1956), but it is more likely that it is simply a prerequisite for the final liberation of the mature pollen. Malfunctions in the timing of callose degradation can be the cause of some genetically analysed male sterility, evidence for which is now available from transgenic tobacco plants (Worrall *et al.*, 1992).

The growing pollen tube also deposits callose in a developmentally regulated fashion. This callose obviously fulfils two quite different functions. The tube wall consists of an outer layer composed of an arabinan, beneath which lie some cellulose fibres, followed by an inner layer consisting of callose which constitutes about 80% of the wall material and represents a regular part of the pollen tube wall (Meikle *et al.*, 1991b; Schlüpmann *et al.*, 1993). In addition, callose plugs are commonly formed behind the protoplast at intervals during elongation of the tube. The plugs presumably maintain the two nuclei in the proximity of the tip as it grows towards the mother cell (nicely illustrated by Haring *et al.*, 1990). Thus the pollen tube cytosol is maintained within a reasonably small volume. This coordinated process is disrupted if self-pollination occurs. In this case, the plugs may form in front of the nuclei and prevent fertilization (Haring *et al.*, 1990).

Still another example of callose deposition under developmental control is the formation of lenticels (Eschrich, 1956), which are organs that enable gas exchange to occur in the otherwise suberized surface of stems or petioles. The cells that loosely fill the lenticels arise from a cambial layer and exhibit callose during the process of cell migration towards an aperture in the ruptured periderm. The callose disappears again in the outer cluster of older cells, which become hydrophobic. Whether the callose, in this case, is degraded or simply impregnated by hydrophobic material, and therefore becomes inaccessible to stains, is apparently unknown.

Callose induced by stress

A special role of callose appears to be the sealing of connections that unite the protoplasts of contiguous cells, such as plasmodesmata or sieve pores. In fresh and even fixed plant material, these gaps in the cell wall are always lined by callose, a characteristic that readily allows visualization under the light microscope. To what extent this callose was already present in the functional phloem or was rapidly deposited as a consequence of cutting remains unclear. Physiological observations suggest it is likely that the sieve pores remain open as long as the phloem is functional. In some cases, however, sieve pores can be entirely filled and the surrounding sieve plate covered with callose (*Figure 1*), especially in autumn when the phloem has presumably ceased to function. In plants where the phloem strands function for more than one vegetative season, the callose covering the sieve pores disappears in the spring, and this disappearance is correlated with an increase in 1,3-β-glucanase activity, an enzyme capable of degrading callose (Krabel *et al.*, 1993). It is generally believed, therefore, that callose can rapidly and sometimes transiently seal the sieve pores in order to prevent leakage of the phloem sap under various stress conditions such as puncture by aphids, senescence, mechanical wounding or even cutting of tissue for rapid fixation prior to microscopical examination.

Callose also represents a universal building material for the reinforcement or repair of cell walls during the course of pathogenesis. As a result of hypersensitive reactions, callose can become deposited on all walls of single epidermal cells penetrated by fungal hyphae, or only on that part of the epidermal cell wall which is shared with a dead cell (e.g. at the rim of local virus-induced lesions; Kauss, 1987). Such cell walls may subsequently become suberized to facilitate lesion closure.

Still more localized is the callose contained in 'papillae', cell wall appositions located immediately below the sites of attempted invasion of host cells by fungi. Although papillae often also contain other materials, the inhibition of callose synthesis in barley coleoptiles can decrease the resistance of epidermal cells to powdery mildew, indicating that callose deposition is indeed one of the many factors responsible for the resistance of plant cells to pathogens (Bayles *et al.*, 1990). In the same biotrophic plant/pathogen interaction, at a later stage of infection, fully developed fungal haustoria may cease to function and then become encased by callose, similar to calcium oxalate crystal idioblasts (Brander and Wattendorff, 1987; Eschrich, 1956). Successful formation of papillae is also associated with systemically acquired resistance in cucumber plants. In the leaves rendered resistant by previous infection of other leaves, the callose synthase (GS II; see Section 6) in purified plasma membrane vesicles exhibits a 2- to 4-fold higher specific activity (Schmele and Kauss, 1990). This provides further evidence that the ability of plant cells to produce callose under stress conditions is under long-term regulation, in addition to short-term regulation by effectors as described in Sections 6 and 7.

An agriculturally important example of callose induction is the deposition of callose in the root tip of soybean during Al^{3+} toxicity (Wissemeier *et al.*, 1987). This phenomenon may be used as a convenient indicator for early signs of aluminium toxicity and for screening of aluminium-tolerant cultivars (Zhang and Taylor, 1993).

3. Callose in lower plants

The typical staining properties of callose have also been reported for certain cell wall components of various lower plants. Fungi will not be considered here, although 1,3-β-glucans are widely distributed in their cell walls, irrespective of the phylogenetic location (Eschrich, 1956; Ruiz-Herrera, 1991; Stone and Clarke, 1992). Two examples from diatoms and mosses will be mentioned here in order to illustrate in more detail the universal occurrence of callose.

In large freshwater diatoms of the genus *Pinnularia*, a continuous callose strip always surrounds the rim of the siliceous hypotheca, presumably forming a plastic joint with the epitheca (Waterkeyn and Bienfait, 1987). This joint persists throughout cell division, but disappears once the daughter cells have formed their own joint. In addition, callose is permanently present at other specific sites in the complex diatom cell wall, and it is also transiently deposited shortly after cell division in the fenestrated alveoli of the newly formed hypotheca of each daughter cell. It has been suggested (Waterkeyn and Bienfait, 1987) that, in *Pinnularia,* callose is always deposited at sites where the plasma membrane would otherwise be directly exposed to external water. However, it should also be noted that the solutes in the protoplast result in a turgor pressure of up to 8 bars in freshwater diatoms (A. Schmid, Salzburg, personal communication). The callose joint might therefore also function as a removable, strain-resistant organic 'glue' between the two siliceous valves.

An example of the occurrence of callose in mosses is the unistratose thallus of the water moss *Riella* (Grotha, 1986). As in higher plants, cell-plate callose as well as callose induced by physical stress (see Section 7) can be observed at every part of the surface. In addition, in the centre of both outer surfaces in the adult cells there is a distinct region characterized by intense Golgi vesicle-mediated exocytosis where deposition of callose can be chemically induced by incubation with, for example, tetracyclines (Grotha, 1986). In this latter case, the function of callose remains obscure but may simply represent stress callose sealing cell parts. Grotha (1986) also employed aniline blue fluorescence to make a direct quantitative estimation of callose *in situ* using the exposure-meter of the microscope.

4. Chemical nature of callose

The main limitation on research on the chemical structure of callose was and still is its very localized occurrence in close association with large amounts of other cell wall polysaccharides. Although it was initially suggested (Eschrich, 1956) that callose might consist of D-glucose, in 1958 it was shown that the glucose molecules in callose are mainly 1,3-β-linked (Kessler, 1958). In order to demonstrate this, callose was purified from dormant grape-vine phloem by combined mechanical disintegration and sedimentation, followed by solubilization of the remaining cellulose in ammoniacal copper hydroxide (Cuoxam), in which callose is insoluble (Kessler, 1958). Purity was evaluated at each step by light microscopy. The resulting preparation was apparently free of cellulose, soluble in dilute alkali and shown by permethylation to consist of linear 1,3-β-glucan chains. The average degree of polymerization (DP) was estimated from the rather low periodate consumption to be at least 90 (Kessler, 1958). However, the majority of callose produced *in vivo* appears to have a longer chain length. Using gel permeation (e.g. for a callose preparation from cotton fibres), a DP of between 300 and 1200 has been estimated (Hayashi *et al.*, 1987), although the solubilization in alkali and the precipitation procedure used to prepare the material are likely to have caused a loss of shorter callose molecules and breaks of longer ones. It should be noted that complete solubility of callose in dilute alkali is more the exception than the rule. Part of the callose produced *in vivo* and *in vitro* (see Section 6) is not soluble under these conditions even at high temperatures, and the proportion of alkali-insoluble material appears to depend on the plant and tissue type. Autoclaving in dimethylsulphoxide (Kauss, 1989) or enzymatic digestion with 1,3-β-glucanases may help to avoid the problems caused by the fact that a portion of the callose remains insoluble even after heating in alkali. The method of callose preparation from grape-vine phloem (Kessler, 1958) has been described above in some detail because it represents a rare example of callose in presumably pure form. The procedure appears to be worthy of use in conjunction with modern methods of carbohydrate chemistry to establish in more detail the properties of sieve-tube callose. For example, the eventual minor degree of 1,6-branching points as suggested for the callose found constitutively in cotton fibres (Maltby *et al.*, 1979) and pollen tubes (Meikle *et al.*, 1991b), or the possible deposition of short-chain callose *in vivo*, are questions that still need to be resolved.

Preparations of callose from other sources appear to have been less pure than sieve-plate callose. For instance, the callose plugs produced in the germinating pollen grain and pollen tube during self-pollination of rye have also been isolated (Vithanage *et al.*, 1980). In addition to callose, they also appeared to contain some cellulose and a considerable amount of mixed 1,3/1,4-β-glucan, which is a typical hemicellulose of grasses and does not bind to the fluorophore active in aniline blue used for callose identification (Kauss, 1989). These results appear to reflect the fact that, under natural conditions, callose often occurs in an intimate mixture with other materials. For instance, in papillae from various plants, histochemical evidence suggests that callose can be associated with phenolics, proteins or silica, although these substances may occur in the papillae as distinct layers. Papillae from potato tubers infected by *Phytophthora infestans* also retained their form after complete removal of callose with 1,3-β-glucanase, whereas a combination of this enzyme and cellulase was able to remove the papillae completely (Hächler and Hohl, 1982). These findings, together with negative results from histochemical tests for other materials, suggest that in this case papillae are constructed from a close association of callose and cellulose.

5. Which membranes synthesize callose?

Plant cell wall polysaccharides, in general, can be synthesized by two different pathways. Cellulose fibres emerge directly from plasma membrane-localized enzyme complexes (see Chapter 4 by Blanton and Haigler), while matrix polysaccharides (hemicelluloses and pectins) are thought to be synthesized in the Golgi apparatus and to reach the wall after fusion of transport vesicles with the plasma membrane (Driouich *et al.*, 1993).

There are some suggestions in the literature that in certain cases callose may be synthesized in the Golgi apparatus, although the only doubtful evidence supporting this view might be the demonstration of the so-called type I β-glucan synthase activity in Golgi-enriched membrane fractions (see Section 6). In pollen tubes, the presence of electron-translucent vesicles near the plasma membrane was also taken as evidence for the transport of callose via the endomembrane pathway (Cresti and van Went, 1976). However, without special contrasting procedures, the TEM appearance provides no information about chemical composition. It appears more likely, although as yet unproven, that a steady deposition of callose during pollen-tube growth occurs as a result of the activity of the plasma membrane-localized callose synthase. Interestingly, the atypical callose synthase from pollen tubes has considerable *in-vitro* activity in the absence of Ca^{2+}, indicating regulatory properties that differ from those of the enzyme in other tissues (Schlüpmann *et al.*, 1993; see Sections 7 and 8).

Cell-plate callose might at first glance appear to be another example of Golgi-mediated callose synthesis. However, using immunocytological methods, it has been shown that the Golgi apparatus does not contain callose under conditions where callose was demonstrated in the cell plate and at the plasmodesmata of the primary wall (Northcote *et al.*, 1989). In addition, isolated phragmoplasts contain a β-glucan synthase which exhibits low substrate affinity and Ca^{2+} dependence (Kakimoto and Shibaoka, 1992), properties typical of the plasma membrane-localized type II β-glucan synthase (see Section 6). These results suggest that the phragmoplasts can synthesize callose directly, and that this synthetic capacity might therefore represent an early manifestation of the developing plasma membrane.

More convincing arguments are available for the direct production of callose from the plasma membrane, at least in all cases where callose deposition is related to obvious external stimuli. This evidence arises from several sources and can best be understood if some *in-vitro* properties of the callose synthase are first described, followed by a discussion of the induction of callose synthesis *in vivo*.

6. Properties of β-glucan synthases *in vitro*

β-Glucan synthase I (GS I)

In the course of early experiments on the fractionation of plant membranes, two types of β-glucan synthase activity were observed (Ray, 1979). Type I (GS I) was characterized by a high affinity for the substrate UDP-glucose, and was found to be associated with membrane vesicles derived from the Golgi apparatus. For diagnostic purposes, the enzyme activity is assayed at 5 to 10 μM UDP-glucose and 10 mM Mg^{2+}. GS I typically exhibits rather low activity, is unstable, and its products are rarely identified, although in some cases they appear to include 1,3-β-glucan (e.g. in Golgi-enriched membrane fractions from maize coleoptiles) (Gibeaut and Carpita, 1993). One should bear in mind, however, that GS I activity in gradient fractions from soybean suspension cells is increased by about twofold in the presence of 70 μM Ca^{2+} (Fink et al., 1987). This suggests that part of the observed enzyme activity was due to the presence of contaminating GS II activity, detectable even at the low UDP-glucose concentrations used for GS I assays which, due to low activity, require a high specific radioactivity of substrate. The function of GS I in vivo, which might consist of several different enzyme activities, is not clear. A latent precursor of GS II or participation in the synthesis of xyloglucans exhibiting a 1,4-β-glucan backbone are possibilities (Dhugga et al., 1991). In this latter case, the GS I products also include a glycoprotein which is reversibly glucosylated from UDP-glucose and may function as an intermediate in xyloglucan synthesis.

β-Glucan synthase II (GS II, callose synthase)

The second type of β-glucan synthase was demonstrated by fractionation of plant membranes, and was found to be localized on the plasma membrane (Ray, 1979). It still represents the most reliable marker for plasma membranes and is present with high activity in all plant tissues so far investigated. The products of the reaction are linear chains of 1,3-β-glucan with a degree of polymerization estimated by gel permeation chromatography to range between 300 and 1200 (Hayashi et al., 1987). The callose synthesized in vitro appears to lack the few branching points described for callose deposited in vivo (Hayashi et al., 1987; Schlüpmann et al., 1993), suggesting that the branching enzyme might not reside on the plasma membrane but possibly in the plant cell wall, as suggested in the case of the yeast Candida (Hartland et al., 1991).

Activators. Due to the low affinity of GS II for UDP-glucose, the enzyme is routinely assayed at a substrate concentration of 0.5–1 mM. It is further activated by certain low-molecular-weight sugars, a fact that was originally obscured by the relatively high concentrations of sucrose contained in the gradients used for membrane fractionation. Various β-glucosides are at present used as activators of GS II, with cellobiose being the most active commercially available compound. However, cellobiose is not incorporated into the reaction product (Hayashi et al., 1987). More recently, it has been shown that various β-glucosides with hydrophobic aglycones are even better activators than cellobiose, and a possibly related endogenous activator of GS II was identified as β-furfuryl-β-glucoside (Ohana et al., 1992).

The action of some of the in-vitro activators for callose synthase is shown in Table 1. The significant activation of the native enzyme by micromolar concentrations of Ca^{2+} (Hayashi et al., 1987; Kauss et al., 1983; Kauss and Jeblick, 1985) appears to be of physiological importance. The Ca^{2+} concentration used in the experiment shown in Table 1 is saturating; half-activation is attained at approximately 0.5 μM Ca^{2+} (Kauss et al., 1983; Kauss and Jeblick, 1985, 1986a). Ca^{2+} activation only becomes evident when trace Ca^{2+}, present in all chemicals and distilled water, is chelated by EGTA. Mg^{2+} at comparatively high concentrations apparently also causes some activation (Table 1). However, it has been shown that Mg^{2+} also results in a higher molecular weight of the enzyme complex and the synthesis of a more aggregated and less soluble form of callose (Hayashi et al., 1987). These observations may in part explain the apparent activation of the enzyme by Mg^{2+}, since GS II assay procedures always involve washing steps for the reaction products which may lead to the loss of shorter callose chains.

Table 1. Activation of callose synthase *in vitro* in microsomes from dark-grown mung bean hypocotyls[a]

Additions to the assay	Enzyme activity[b]
Native microsomes	
EGTA (Ca^{2+} <10^{-8} M)	0.2
Ca^{2+} (about 70 µM)	8.1
EGTA + 4 mM $MgCl_2$	2.5
Ca^{2+} + $MgCl_2$	14.4
EGTA + 0.2 mM spermine	4.2
Ca^{2+} + spermine	16.4
Ca^{2+} + spermine + $MgCl_2$	17.2
EGTA + 25 µg poly-L-ornithine[c]	15.7
Ca^{2+} + 25 µg poly-L-ornithine	24.6
Trypsinized microsomes	
EGTA + $MgCl_2$	12.0

[a] Washed microsomes (Kauss *et al.,* 1983) from 5 g of tissue were suspended in 8 ml of Tes/NaOH, pH 7.0, containing 1 mM DTT. For trypsinization, a portion of the microsomes was supplemented with 0.02% (w/v) digitonin and 25 µg ml^{-1} trypsin for 1 min, followed by 100 µg ml^{-1} soybean trypsin inhibitor. The standard enzyme assay (Kauss *et al.,* 1983) contained 10 mM cellobiose, 8% (w/v) glycerol, 0.02% (w/v) digitonin and 4 mM EGTA to decrease the Ca^{2+} concentration to <10^{-8} M. The addition of 4 mM $CaCl_2$ brought the free Ca^{2+} concentration to about 70 µM. The assay was started by the addition of 0.8 mM labelled UDP-glucose, and the callose was determined as described previously (Kauss *et al.,* 1983). Unpublished results.
[b] x 10^{-3} cpm/50 µl microsomes/10 min.
[c] See Kauss and Jeblick (1985) for properties of the compound.

On limited proteolysis with trypsin, the callose synthase from soybean suspension cells was activated and rendered Ca^{2+} independent (Kauss *et al.,* 1983; Kauss and Jeblick, 1985). Although proteolytic activation was reported not to occur with cotton and mung bean membranes (Hayashi *et al.,* 1987), data will be presented here from an experiment on mung bean microsomes in order to demonstrate the possibility of proteolytic activation in this species as well (*Table 1*). To observe activation by trypsin it is essential that trypsinization is performed in the presence of digitonin, terminated at the optimal time with trypsin inhibitor, and that the assay mixture contains 4 mM Mg^{2+}. It remains, however, to be clarified whether the proteolysis affects the callose synthase directly in a zymogen-like fashion, or whether it alters or removes one of the regulatory peptides that are presumably associated with it (see below).

Other *in-vitro* activators of callose synthase, recognized first for soybean suspension cells (Kauss and Jeblick, 1985), are polyamines (spermine, spermidine) and polycations (poly-L-ornithine, poly-L-lysine, chitosan, ruthenium red). *Table 1* shows that these substances are also effective for mung bean microsomes. Polyamines and polycations appear to bind to sites on the enzyme different from Ca^{2+}, since they act in the presence of EGTA (*Table 1*), with trypsinized enzyme and in the presence of La^{3+} (Kauss and Jeblick, 1985). At 1 to 2 µM Ca^{2+}, spermine and poly-L-ornithine act synergistically with Ca^{2+} (Kauss and Jeblick, 1986a). The callose synthase exhibits sigmoidal kinetics of an allosteric type with respect to substrate concentration (e.g. Hayashi *et al.,* 1987, Kauss and Jeblick, 1986a), and therefore an increase in apparent affinity for its substrate is observed in the presence of the above-mentioned activators. *In vivo* this might lead to an especially pronounced activation under conditions of low substrate concentration.

A large but variable proportion of the plasma membrane vesicles present in microsomes as well as plasma membrane vesicles isolated by phase partitioning are predominantly 'outside-out' and impermeable to ions. Thus detergents must be included in GS II assay mixtures in order to allow access of the substrate to the enzyme. The most suitable detergent for this purpose is digitonin. The obligatory presence of detergents under assay conditions probably obscures the fact that the plasma membrane-localized callose synthase is also activated by certain lipids. Evidence for activation by lipids is derived from reconstitution experiments (Sloan *et al.*, 1987) and from the observation that, in the presence of saturating concentrations of digitonin, further stimulation of callose synthase activity can be effected by lysophosphatidylcholine, platelet-activating factor or acylcarnitine (Kauss and Jeblick, 1986b). With these lipids, the stimulation is only observed at a certain critical concentration, whereas higher concentrations are inhibitory. Unsaturated free fatty acids are inhibitory at all concentrations (Kauss and Jeblick, 1986b). Thus digitonin, when added to allow access of substrate and ionic effectors to the cytoplasmic side of plasma membrane vesicles, might also partly function as an *in-vitro* activator, making investigation of the effect of endogenous lipophilic activators difficult. The physiological significance of these results may be that such substances can arise, for example, in the course of pathogenesis and thereby modulate the activity of callose synthase *in vivo*.

Solubilization and subunits of the GS II enzyme complex. In recent years there have been many attempts to solubilize, purify and reconstitute callose synthase. As these studies are still in progress, they will only be briefly summarized here. Starting from enriched plasma membranes, the most pure soluble preparations of callose synthase still contain several peptides (Delmer *et al.*, 1991; Dhugga and Ray, 1991; Fink *et al.*, 1990; Frederikson *et al.*, 1991; Li *et al.*, 1993; Meikle *et al.*, 1991a; Wu and Wasserman, 1993). Immunological and chromatographic techniques and product entrapment have been used with advantage in the purification. Substrate analogues were also used as photoaffinity labels in the presence or absence of Ca^{2+}, which is considered to aid the recognition of the substrate-binding peptide. In so far as a generalization between plants, researchers and techniques is yet possible, the most promising peptide constituents of callose synthase have a molecular mass range in the 31–37 and 52–57 kDa classes. Peptides of molecular mass 31 kDa and 48 kDa have been purified to apparent homogeneity from rice and peanut plasma membranes, respectively, and were assumed to exhibit GS II activity without the presence of additional peptides (Kamat *et al.*, 1992; Kuribayashi *et al.*, 1993). The main unanswered question is which and how many of the peptides found in most preparations represent regulatory peptides associated with the catalytic subunit of GS II, and which of the observed peptides are impurities. For instance, an 84-kDa peptide able to bind to [^{32}P]UDP-glucose in a Ca^{2+}-dependent manner is abundant in membrane preparations from cotton (Delmer *et al.*, 1991). This peptide was sequenced and identified as sucrose synthase (Amor *et al.*, 1995), an enzyme which is normally soluble. When present in a membrane-bound form it may possibly have a role in the locally concentrated production of UDP-glucose, the substrate for callose synthesis. Complications probably also arise when the enzyme purification starts with microsomes rather than with purified plasma membranes (e.g. Meikle *et al.*, 1991a). In this case, the UDP-glucose-binding peptides found might possibly be related to GS I activity, or, in grasses, to the synthesis of the hemicellulosic mixed 1,3/1,4-β-glucans, which are Golgi-derived. In contrast, a 65-kDa peptide which can associate with the enzyme in a cation-dependent manner may be of regulatory significance for callose synthase (Delmer *et al.*, 1993b). Similarly, at least three annexins were characterized from cotton fibres and shown to inhibit callose synthase (Andrawis *et al.*, 1993; Clark and Roux, 1995). These 34-kDa peptides interact in a cation-dependent fashion with membranes, serve *in vitro* as substrates for endogenous protein kinases and might, therefore, be involved in the regulation of callose synthase.

Relationship between callose synthase and cellulose synthase. In conjunction with problems related to the purification of peptides involved in callose synthesis, the question of a possible relationship between this enzyme and the cellulose synthase complex (Delmer, 1987) should be addressed. Cellulose synthase is treated in detail in Chapter 4 by Blanton and Haigler. The universal occurrence of callose synthase and its properties have for decades obscured the results of researchers working on cellulose synthesis *in vitro*. Callose tends to be partially insoluble even in concentrated alkali, especially when Mg^{2+} is present in the assay mixture (Hayashi *et al.,* 1987). *In-vitro* cellulose synthase assays with UDP-glucose as a substrate are always hampered, therefore, by the excess amounts of callose produced. The optimal conditions under which membrane preparations synthesize a small proportion of 1,4-β-glucan (cellulose?) still greatly favour callose synthesis (Li *et al.,* 1993; Okuda *et al.,* 1993). The anticipated relationship between cellulose and callose synthesis has been discussed more than once (Amor *et al.,* 1995; Brown *et al.,* 1994; Delmer *et al.,* 1993a).

7. Induction of callose synthesis *in vivo*

Cases where Golgi complexes are not present at the site of callose formation represent clear evidence of deposition of callose from the plasma membrane (e.g. at the sieve pores in mature phloem). Similarly, plasmolysis of onion scale epidermis induces callose, which becomes evident over wide areas of the plasma membrane (Eschrich, 1957), suggesting that it is formed directly by membrane-localized enzymes. During our experiments on callose induction in suspension cells (see below), we also observed that callose was layered on to the inner side of the cell wall as long as the plasma membrane was in contact with the cell wall, but that it could also be detected directly over wide areas at the plasma membrane once it separated from the wall due to plasmolysis caused by the extensive leakage of solutes. All these examples suggest that the plasma membrane is the origin of callose induced by stress.

The relationship between local perturbation of the plasma membrane and callose deposition can also be demonstrated by pressing a leaf surface with a sharp edge for a few seconds, without causing obvious damage. Within the next 30 min, the deposition of callose patches extends along the indented parts of the surface of several cells (for example see Grotha, 1986). Based on these and similar examples, it appeared that stress-induced callose deposition might be the result of local perturbation of plasma membrane, an assumption that would also explain the often very localized occurrence of callose. In this case, physical stretching or compression of the plasma membrane would be the stress causing the induction (*Figure 2*).

For the purposes of physiological and biochemical studies related to callose induction, we have used suspension-cultured plant cells and taken the elicitor approach, which involves searching for chemicals that cause deposition of callose. For a rapid determination of callose, we developed a quantitative version of the aniline blue fluorescence assay (Kauss, 1989; Köhle *et al.,* 1985). In recent years many chemically diverse compounds have been recognized as callose elicitors (*Figure 2*). The saponins and polyene antibiotics are known to interact with the plasma membrane by combining with sterols, although effects on sterol-free artificial membranes are also known (Bolard, 1986). Direct interaction with the membrane is also likely for certain detergents and for acylated peptides that presumably intercalate with the lipophilic phase of the plasma membrane. In contrast, polycations appear to exert their effects in a different way. Chitosan, one of the best callose elicitors found to date, increased in activity with an increase in the degree of polymerization up to 5000, equivalent to an average molecular weight of 1000 kDa and an average chain length of 2.5 µm (Kauss *et al.,* 1989). At a comparable degree of polymerization, chitosan fragments exhibiting 23% *N*-acetylation were less effective than fully deacetylated fragments, both in protoplasts and in suspension cells from *Catharanthus* (Kauss *et al.,* 1989). Chitosan therefore appears to interact at the plasma membrane surface at many sites and over long distances, presumably binding simultaneously in a co-operative manner to the

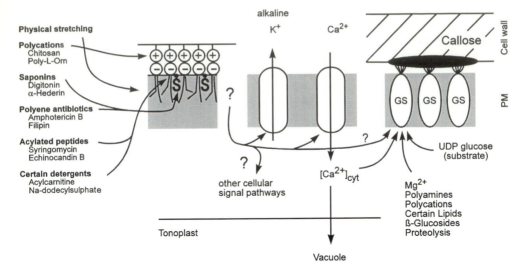

Figure 2. Parameters likely to be involved in stress-induced callose deposition. Diverse callose elicitors as well as physical stretching of the plasma membrane (left-hand module) are suggested to disturb the membrane integrity, creating a common but unidentified signal that is able to activate ion-transport processes (centre module), leading to an increase in cytoplasmic Ca^{2+} concentration. In addition, the unknown signal must be able either to influence the GS II directly or to activate other cellular signal pathways, causing the mobilization of an additional enzyme activator which is possibly identical to one of those known from *in-vitro* experiments (shown in right-hand module). The three modules should be envisaged as operating in combination in the same membrane area, so explaining the often localized callose deposition at the stressed site. For further details see text. GS, 1,3-β-glucan synthase II; S, sterols; PM, plasma membrane.

polar heads of many membrane lipid molecules. From a physiological point of view, the low-molecular-weight callose elicitors mentioned above may represent models for microbial compounds that are able to diffuse through the plant cell wall. In contrast, partially acetylated chitosan of high molecular weight is an insoluble component of many fungal cell walls, and thus may be regarded as a model for charged microbial surfaces which, *in situ,* are likely to have an elicitor action when they come into direct contact with the plant plasma membrane surface (Kauss, 1990). Interestingly, this situation appears to be prevented in biotrophic interactions by the 'extracellular matrix', which presumably insulates the fungal wall from the plant plasma membrane, allowing transient coexistence of the fungal and plant cells (Kauss, 1990).

Physical stretching of plasma membranes as well as the lipophilic or polycationic elicitors (*Figure 2*) all appear to create a similar signal at the plasma membrane by regulating ion-transport processes, as indicated by a limited leakage of cellular ions, predominantly K^+ (Kauss *et al.,* 1989, 1990; Köhle *et al.,* 1985; Waldmann *et al.,* 1988; *Table 2*). Whether the observed concomitant external alkalinization indicates an additional change in H^+-transport processes, or is merely the consequence of an imbalance between the leakage of cations and anions, is not yet clear. However, we have consistently observed an increase in net Ca^{2+} uptake, determined either as a decrease in external Ca^{2+} concentration (*Table 2*) or as uptake of $^{45}Ca^{2+}$ (Kauss and Jeblick, 1991; Kauss *et al.,* 1991). When the induced Ca^{2+} influx is prevented by complexing the external free Ca^{2+} with EGTA, callose synthesis stops, suggesting that the influx of Ca^{2+} is a prerequisite for callose synthesis *in vivo* (Kauss *et al.,* 1991; Köhle *et al.,* 1985). Inhibition of the elicitor-induced net Ca^{2+} influx by the Ca^{2+}-channel blocker nifedipine (Waldmann *et al.,* 1988) suggested that the Ca^{2+} influx might be due to opening of controlled transport sites rather than to a non-specific change in general permeability. The protein phosphatase inhibitor okadaic acid also causes a decrease in induced Ca^{2+} uptake and concomitant callose synthesis (Kauss and Jeblick, 1991), supporting the hypothesis that Ca^{2+} enters at gated sites, and possibly also indicating that the respective transport proteins are controlled by phosphoryl-ation/dephosphorylation. The results of studies involving metabolic inhibitors must be interpreted cautiously, however, since the possibility of unexpected side-reactions cannot be excluded. In this context it is of interest that stress-activated Ca^{2+} and K^+ channels also occur in plants (Blatt and Thiel, 1993), and this might in part explain the induction of callose by physical stretching of plasma membranes (*Figure 2*).

The massive Ca^{2+} influx into the cytoplasm certainly causes some increase in Ca^{2+} concentration on the cytosolic side of the plasma membrane prior to the ion being sequestered in the vacuole, the main Ca^{2+} store in plant cells. Unfortunately, to date we have not been able to measure the cytoplasmic Ca^{2+} concentration directly under the conditions necessary for the determination of callose induction. The main reason for this appears to be that the various methods available are only reliable when the cytoplasmic Ca^{2+} concentration is determined in the same individual cell or protoplast in which a correlated physiological event can be quantitatively monitored. This is not the case for the determination of induced callose in a population of suspension cells or protoplasts, in which individual cells or protoplasts do not react to the same extent.

Inconsistent with the above hypothesis of Ca^{2+}-regulated induction of callose deposition is the observation that the Ca^{2+} ionophores, ionomycin and A23187, induce little callose formation compared to the callose elicitor digitonin (Waldmann *et al.*, 1988; *Table 2*). A large increase in net Ca^{2+} uptake caused by A23187 resulted in only a small increase in callose formation, whereas a slight increase in Ca^{2+} uptake caused by increasing the digitonin concentration to 10 µM results in a dramatic increase in callose deposition (*Table 2*). Similarly, the degree of net Ca^{2+} uptake, ion leakage and callose synthesis induced by chemically unrelated elicitors showed no correlation. For instance, the bacterial phytotoxin syringomycin caused maximal callose induction at 0.6 µg ml^{-1}, but higher concentrations greatly increased K^+ release and net Ca^{2+} uptake without further increasing callose induction (Kauss *et al.*, 1991). In contrast, the polyene antibiotic amphotericin B at a concentration of 2 µM induced half-maximal net Ca^{2+} uptake and K^+ release with little callose induction, and amphotericin B concentrations of up to 25 µM were

Table 2. Effect of the Ca^{2+} ionophore A23187 on digitonin-induced electrolyte leakage, decrease in external Ca^{2+} concentration and callose synthesis by suspension-cultured cells of *Catharanthus roseus*[a]

Substances added	Leakage of electrolytes (µS)	Decrease in external Ca^{2+} (µM)	Callose synthesis (µg g^{-1} fresh weight)
A23187	49	35	9
Digitonin (5 µM)	24	7	67
Digitonin (5 µM) + A23187	48	34	73
Digitonin (10 µM)	47	14	183
Digitonin (10 µM) + A23187	74	35	183

[a] The experiment was performed over a period of 3 h in a diluted growth medium containing about 70 µM free Ca^{2+}. A23187 was added in 0.5% (v/v) DMSO at a final concentration of 10 µM, and the other samples had the same final solvent concentration. Values from untreated control samples are subtracted and mean values from four parallel samples are given. Data are from Waldmann (1992); for experimental details see Waldmann *et al.* (1988).

required for maximal callose induction (Kauss *et al.,* 1990). These results indicate that induced Ca^{2+} uptake and the presumably associated increase in cytoplasmic Ca^{2+} are prerequisites for callose induction but are not sufficient for the overall regulation of this process. It is also interesting in this context that under the conditions used for the experiment shown in *Table 2,* fusicoccin (0.5 μM) enhances the digitonin-induced callose synthesis, whereas this toxin alone does not elicit callose synthesis (Waldmann, 1992). This indicates that an increase in electric potential and pH gradient at the plasma membrane favours callose synthesis once induced by other means, and that a decrease in membrane potential may not be an initial step in callose elicitation.

In addition to Ca^{2+}, several other *in-vitro* activators of callose synthase are summarized in *Figure 2.* It seems quite possible that these substances may also play a role *in vivo.* It has been shown in barley suspension cells that the cytoplasmic concentration of β-furfuryl-β-glucoside increases under conditions that favour callose synthesis, presumably by release of the activating substance from vacuoles (Ohana *et al.,* 1993). The authors suggested that this event might help to activate the callose synthase *in vivo* synergistically with Ca^{2+}. A similar possibility has previously been suggested for polyamines (Kauss and Jeblick, 1985; Kauss, 1990), but in this case experimental data are still lacking. A possible activation of callose synthase *in vivo* by degradation products from membrane phospholipids (e.g. lysophosphatidylcholine) known to activate the enzyme *in vitro* (Kauss and Jeblick, 1986b) has also been suggested. Interestingly, the β-glucoside of salicyl alcohol is among the alkyl-β-glucosides found to be good activators of callose synthase *in vitro* (Ohana *et al.,* 1992). It appeared possible, therefore, that the β-glucoside of salicylic acid which accumulates in plant tissues under conditions of systemic acquired resistance (Raskin, 1992) might play a role *in vivo* as a callose synthase activator. It has been shown (H. Kauss and E. Niederberger, unpublished results), however, that this substance is completely inactive as an activator of callose synthase *in vitro.* Instead, long-term regulation of GS II in systemically resistant cucumber leaves appears to occur by an increase in enzyme activity (Schmele and Kauss, 1990).

Phosphorylation/dephosphorylation might represent yet another regulatory mechanism involved in callose induction. The phosphatase inhibitor okadaic acid inhibits callose synthesis induced by syringomycin, digitonin and amphotericin B, with a concomitant decrease in ^{45}Ca^{2+} uptake (Kauss and Jeblick, 1991). These results indirectly indicate that a decrease in the phosphorylation status of unknown regulatory membrane proteins might represent part of the callose induction process. However, all attempts to show a direct role for protein phosphorylation in *in-vitro* callose synthesis have been inconclusive up until now.

8. Some unanswered questions

The as yet unresolved problems relating to the purification of callose synthase and to possible regulatory peptides associated with this enzyme, as well as the other major difficulties associated with the mode of callose induction under stress, have been mentioned above. In the case of stress-induced callose, it has been suggested that callose is deposited at sites on the cell surface where the stress is localized (*Figure 2*). In contrast, no obvious correlation with the stressed part of the cell exists for the deposition of callose in plasmodesmata or sieve pores (*Figure 1*). This process can be induced, for example, by cutting a stem or petiole some distance from the site of callose formation. In this case, one must assume that the callose synthase or a regulatory peptide is preferentially localized in the plasma membrane where the callose is formed. This question can only be approached experimentally once the catalytic and regulatory subunits of callose synthase are established and antibodies for their immunocytological detection are available. Nevertheless, it appears that a 65-kDa peptide that is able to associate in a cation-dependent manner with callose synthase co-localizes with the plasmodesmata (Delmer *et al.*, 1993b). Even though the function of this peptide is not known, such a result suggests that the presumed callose synthase complex or regulatory components may indeed not be equally distributed on all parts of the plasma membrane.

Another unanswered question is how the deposition of developmentally regulated callose is achieved. It has been reported that growing pollen tubes contain a callose synthase which exhibits properties unlike those of the enzyme in somatic tissue (Schlüpmann *et al.*, 1993). In this case, microsomal membrane preparations contain some GS II activity which does not require Ca^{2+} for activity and can also be rendered 10-fold more active by trypsinization. These results are reminiscent of the proteolytic activation of the Ca^{2+}-dependent GS II (*Table 1*) and of chitin synthase in fungi. The results (Schlüpmann *et al.*, 1993) suggest that the portion of the enzyme that is active *in vitro* is also active *in vivo*, and that the portion activated *in vitro* by trypsinization represents a storage form which might be subject to proteolytic activation in the membrane area where callose is deposited. In pollen tubes this occurs in a subapical zone starting 10 μm from the tip, a region where the Ca^{2+} concentration is not elevated (see Schlüpmann *et al.*, 1993). The way in which these events are regulated is unknown, although the cytoskeleton, via annexins (Andrawis *et al.*, 1993; Clark and Roux, 1995), might provide appropriate signals.

Acknowledgements

The experimental research of the author was financially supported by the Deutsche Forschungsgemeinschaft and the Fonds der Chemischen Industrie.

References

Amor G, Haigler CH, Johnson S, Wainscott M, Delmer DP. (1995) A membrane-associated form of sucrose synthase and its potential role in synthesis of cellulose and callose in plants. *Proc. Natl Acad. Sci. USA* **92:** 9353–9357.

Andrawis A, Solomon M, Delmer DP. (1993) Cotton fiber annexins: a potential role in the regulation of callose synthase. *Plant J.* **3:** 763–772.

Bayles CJ, Ghemawat MS, Aist JR. (1990) Inhibition by 2-deoxy-D-glucose of callose formation, papilla deposition, and resistance to powdery mildew in a *ml-o* barley mutant. *Physiol. Mol. Plant Pathol.* **36:** 63–72.

Benhamou N. (1992) Ultrastructural detection of β-1,3-glucans in tobacco root tissues infected by *Phytophthora parasitica* var. *nicotianae* using a gold-complexed tobacco β-1,3-glucanase. *Physiol. Mol. Plant Pathol.* **41:** 351–370.

Blatt MR, Thiel G. (1993) Hormonal control of ion channel gating. *Annu. Rev. Plant Physiol. Plant Mol. Biol.* **44:** 543–567.

Bolard J. (1986) How do the polyene macrolide antibiotics affect the cellular membrane properties? *Biochem. Biophys. Acta* **864:** 257–304.

Brander U, Wattendorff J. (1987) Staining callose for electron microscopy. Experimental conditions. *La Cellule* **74:** 43–55.

Brown RM, Li L, Okuda K, Kuga S, Kudlicka K, Drake R, Santos R, Clement S. (1994) *In vitro* cellulose synthesis in plants. *Plant Physiol.* **105:** 1–2.

Clark GC, Roux SJ. (1995) Annexins of plant cells. *Plant Physiol.* **109:** 1133–1139.

Cresti M, van Went JL. (1976) Callose deposition and plug formation in *Petunia* pollen tubes *in situ. Planta* **133:** 35–40.

Delmer DP. (1987) Cellulose biosynthesis. *Annu. Rev. Plant Physiol.* **38:** 259–290.

Delmer DP, Solomon M, Read SM. (1991) Direct photolabeling with [^{32}P]UDP-glucose for identification of a subunit of cotton fiber callose synthase. *Plant Physiol.* **95:** 556–563.

Delmer DP, Ohana P, Gonen L, Benziman M. (1993a) *In vitro* synthesis of cellulose in plants: still a long way to go. *Plant Physiol.* **103:** 307–308.

Delmer DP, Volokita M, Solomon M, Fritz U, Delphendahl W, Herth W. (1993b) A monoclonal antibody recognizes a 65 kD higher plant membrane polypeptide which undergoes cation-dependent association with callose synthase *in vitro* and co-localizes with sites of high callose deposition *in vivo*. *Protoplasma* **176:** 33–42.

Dhugga KS, Ray PM. (1991) Isoelectric focusing of plant plasma membrane proteins. Further evidence that a 55 kilodalton polypeptide is associated with β-1,3-glucan synthase activity from pea. *Plant Physiol.* **95:** 1302–1305.

Dhugga KS, Ulvskov P, Gallagher SR, Ray PM. (1991) Plant polypeptides reversibly glycosylated by UDP-glucose. Possible components of Golgi β-glucan synthase in pea cells. *J. Biol. Chem.* **266:** 21977–21984.

Driouich A, Faye L, Staehelin LA. (1993) The plant Golgi apparatus: a factory for complex polysaccharides and glycoproteins. *Trends Biochem. Sci.* **18:** 210–214.

Eschrich W. (1956) Kallose. *Protoplasma* **47:** 487–530.

Eschrich W. (1957) Kallosebildung in plasmolysierten *Allium cepa*-Epidermen. *Planta* **48:** 578–586.

Evans NA, Hoyne PA, Stone BA. (1984) Characteristics and specificity of the interaction of a fluorochrome from aniline blue (Sirofluor) with polysaccharides. *Carbohydr. Polym.* **4:** 215–230.

Fink J, Jeblick W, Blaschek W, Kauss H. (1987) Calcium ions and polyamines activate the plasma membrane-located 1,3-β-glucan synthase. *Planta* **171:** 130–135.

Fink J, Jeblick W, Kauss H. (1990) Partial purification and immunological characterization of 1,3-β-glucan synthase from suspension cells of *Glycine max*. *Planta* **181:** 343–348.

Frederikson K, Kjellbom P, Larsson C. (1991) Isolation and polypeptide composition of 1,3-β-glucan synthase from plasma membrane of *Brassica oleracea*. *Physiol. Plant.* **81:** 289–294.

Gibeaut DM, Carpita NC. (1993) Synthesis of (1→3), (1→4)-β-D-glucan in the Golgi apparatus of maize coleoptiles. *Proc. Natl Acad. Sci. USA* **90:** 3850–3854.

Grotha R. (1986) Tetracyclines, verapamil and nifedipine induce callose deposition at specific cell sites in *Riella helicophylla*. *Planta* **169:** 546–554.

Hächler H, Hohl HR. (1982) Histochemistry of papillae in potato tuber tissue infected with *Phytophthora infestans*. *Bot. Helv.* **92:** 23–31.

Haring V, Gray JE, McClure BA, Anderson MA, Clarke AE. (1990) Self-incompatibility: a self-recognition system in plants. *Science* **250:** 937–941.

Hartland RP, Emerson GW, Sullivan PA. (1991) A secreted β-glucan branching enzyme from *Candida albicans*. *Proc. R. Soc. Lond. B* **246:** 155–160.

Hayashi T, Read SM, Bussel J, Thelen M, Lin FC, Brown RM, Delmer DP. (1987) UDP-glucose: (1→3)-β-glucan synthase from mung bean and cotton. Differential effects of Ca^{2+} and Mg^{2+} on enzyme properties and on macromolecular structure of the glucan product. *Plant Physiol.* **83:** 1054–1062.

Kakimoto T, Shibaoka H. (1992) Synthesis of polysaccharides in phragmoplasts isolated from tobacco BY-2 cells. *Plant Cell Physiol.* **33:** 353–361.

Kamat U, Garg R, Sharma CB. (1992) Purification to homogeneity and characterization of a 1,3-β-glucan (callose) synthase from germinating *Arachis hypogaea* cotyledons. *Arch. Biochem. Biophys.* **298:** 731–739.

Kauss H. (1987) Callose-Synthese. Regulation durch induzierten Ca^{2+}-Einstrom in Pflanzenzellen. *Naturwissenschaften* **74:** 275–281.

Kauss H. (1989) Fluorometric measurement of callose and other 1,3-β-glucans. In: *Modern Methods of Plant Analysis 10* (Linskens HF, Jackson JF, ed.). Berlin: Springer-Verlag, pp. 127–137.

Kauss H. (1990) Role of the plasma membrane in host–pathogen interactions. In: *The Plant Plasma Membrane* (Larsson C, Møller IM, ed.). Berlin: Springer-Verlag, pp. 320–350.

Kauss H, Jeblick W. (1985) Activation by polyamines, polycations, and ruthenium red of the Ca^{2+}-dependent glucan synthase from soybean cells. *FEBS Lett.* **185:** 226–230.

Kauss H, Jeblick W. (1986a) Synergistic activation of 1,3-β-D-glucan synthase by Ca^{2+} and polyamines. *Plant Sci.* **43:** 103–107.

Kauss H, Jeblick W. (1986b) Influence of free fatty acids, lysophosphatidylcholine, platelet-activating factor, acylcarnitine, and Echinocandin B on 1,3-β-D-glucan synthase and callose synthesis. *Plant Physiol.* **80:** 7–13.

Kauss H, Jeblick W. (1991) Induced Ca^{2+} uptake and callose synthesis in suspension-cultured cells of *Catharanthus roseus* are decreased by the protein phosphatase inhibitor okadaic acid. *Physiol. Plant.* **81:** 309–312.

Kauss H, Köhle H, Jeblick W. (1983) Proteolytic activation and stimulation by Ca^{2+} of glucan synthase from soybean cells. *FEBS Lett.* **158:** 84–88.

Kauss H, Jeblick W, Domard A. (1989) The degrees of polymerization and N-acetylation of chitosan determine its ability to elicit callose formation in suspension cells and protoplasts of *Catharanthus roseus*. *Planta* **178:** 385–392.

Kauss H, Waldmann T, Quader H. (1990) Ca^{2+} as a signal in the induction of callose synthesis. In: *Signal Perception and Transduction in Higher Plants* (Ranjeva R, Boudet AM, ed.). Berlin: Springer-Verlag, pp. 117–131.

Kauss H, Waldmann T, Jeblick W, Takemoto TY. (1991) The phytotoxin syringomycin elicits Ca^{2+}-dependent callose synthesis in suspension-cultured cells of *Catharanthus roseus. Physiol. Plant.* **81:** 134–138.

Kessler G. (1958) Zur Charakterisierung der Siebröhrenkallose. *Ber. Schweiz. Bot. Ges.* **68:** 5–43.

Köhle H, Jeblick W, Poten F, Blaschek H, Kauss H. (1985) Chitosan-elicited callose synthesis in soybean cells as a Ca^{2+}-dependent process. *Plant Physiol.* **77:** 544–551.

Krabel D, Eschrich W, Wirth S, Wolf G. (1993) Callase-(1,3-β-D-glucanase) activity during spring reactivation in deciduous trees. *Plant Sci.* **93:** 19–23.

Kuribayashi I, Morita T, Mitsui T, Igaue I. (1993) Purification of β-glucan synthase II from suspension-cultured rice cells. *Biosci. Biotechnol. Biochem.* **57:** 682–684.

Li L, Drake RR, Clement S, Brown RM. (1993) β-glucan synthesis in the cotton fiber. III. Identification of UDP-glucose-binding subunits of β-glucan synthases by photoaffinity labeling with [β-^{32}P]5′-N$_3$-UDP glucose. *Plant Physiol.* **101:** 1149–1156.

Maltby D, Carpita NC, Montezinos D, Kulow C, Delmer DP. (1979) β-1,3-glucan in developing cotton fibers. Structure, localization, and relationship of synthesis to that of secondary wall cellulose. *Plant Physiol.* **63:** 1158–1164.

Meikle PJ, Ng KF, Johnson E, Hoogenraad NJ, Stone BA. (1991a) The β-glucan synthase from *Lolium multiflorum*. Detergent solubilization, purification using monoclonal antibodies, and photoaffinity labeling with a novel photoreactive pyrimidine analogue of uridine 5′-diphosphoglucose. *J. Biol. Chem.* **266:** 22569–22581.

Meikle PJ, Bonig I, Hoogenraad NJ, Clarke AE, Stone BA. (1991b) The location of (1→3)-β-glucans in the walls of pollen tubes of *Nicotiana alata* using a (1→3)-β-glucan-specific monoclonal antibody. *Planta* **185:** 1–8.

Northcote DH, Davey R, Lay J. (1989) Use of antisera to localize callose, xylan and arabinogalactan in the cell-plate, primary and secondary walls of plant cells. *Planta* **178:** 353–366.

Ohana P, Delmer DP, Volman G, Steffens JC, Matthews DE, Benziman M. (1992) β-furfuryl-β-glucoside: an endogenous activator of higher plant UDP-glucose:(1-3)-β-glucan synthase. Biological activity, distribution and *in vitro* synthesis. *Plant Physiol.* **98:** 708–715.

Ohana P, Benziman M, Delmer DP. (1993) Stimulation of callose synthesis *in vivo* correlates with changes in intracellular distribution of the callose synthase activator β-furfuryl-β-glucoside. *Plant Physiol.* **101:** 187–191.

Okuda K, Li L, Kudlicka K, Kuga S, Brown RM. (1993) β-glucan synthesis in the cotton fiber. I. Identification of β-1,4- and β-1,3-glucans synthesized *in vitro. Plant Physiol.* **101:** 1131–1142.

Raskin I. (1992) Role of salicylic acid in plants. *Annu. Rev. Plant Physiol. Plant Mol. Biol.* **43:** 439–463.

Ray PM. (1979) Maize coleoptile cellular membranes bearing different types of glucan synthetase activity. *Meth. Surv. Biochem.* **9:** 135–146.

Ruiz-Herrera J. (1991) Biosynthesis of β-glucans in fungi. *Antonie van Leeuwenhoek* **60:** 73–81.

Schlüpmann H, Bacic A, Read SM. (1993) A novel callose synthase from pollen tubes of *Nicotiana. Planta* **191:** 470–481.

Schmele I, Kauss H. (1990) Enhanced activity of the plasma membrane localized callose synthase in cucumber leaves with induced resistance. *Physiol. Mol. Plant Pathol.* **37:** 221–228.

Sloan ME, Rodis P, Wasserman BP. (1987) CHAPS solubilization and functional reconstitution of β-glucan synthase from red beet (*Beta vulgaris* L.) storage tissue. *Plant Physiol.* **85:** 516–522.

Stone BA, Clarke AE. (1992) *Chemistry and Biology of (1→3)-β-Glucans.* Bundoora, Australia: La Trobe University Press.

Vithanage HIMV, Gleeson PA, Clarke AE. (1980) The nature of callose produced during self-pollination in *Secale cereale. Planta* **148:** 498–509.

Waldmann T. (1992) Untersuchungen zur Regulation der elicitor-induzierten Callose-Synthese bei Suspensionszellkulturen von *Catharanthus roseus* (L.) G. Don. PhD Dissertation, University of Kaiserslautern, Germany.

Waldmann T, Jeblick W, Kauss H. (1988) Induced net Ca^{2+} uptake and callose biosynthesis in suspension-cultured plant cells. *Planta* **173:** 88–95.

Waterkeyn L, Bienfait A. (1987) Localisation et rôle des β-1,3-glucanes (callose et chrysolaminarine) dans le genre pinnularia (Diatomées). *La Cellule* **74:** 199–226.

Wissemeier AH, Klotz F, Horst WJ. (1987) Aluminium induced callose synthesis in roots of soybean (*Glycine max* L.). *J. Plant Physiol.* **129:** 487–492.

Worrall D, Hird DL, Hodge R, Paul W, Draper J, Scott R. (1992) Premature dissolution of the microsporocyte callose wall causes male sterility in transgenic tobacco. *Plant Cell* **4:** 759–771.

Wu A, Wasserman B. (1993) Limited proteolysis of (1,3)-β-glucan (callose) synthase from *Beta vulgaris* L.: topology of protease-sensitive sites and polypeptide identification using pronase E. *Plant J.* **4:** 683–695.

Xu H, Mendgen K. (1994) Endocytosis of 1,3-β-glucans by broad bean cells at the penetration site of the cowpea rust fungus (haploid stage). *Planta* **195:** 282–290.

Zhang G, Taylor GJ. (1993) Callose synthesis induced by aluminium stress. *Plant Physiol.* **102** (Suppl. 1): 174.

Chapter 6

Arabinogalactan-proteins: developmentally regulated proteoglycans of the plant cell surface

J. Paul Knox

1. Introduction

Arabinogalactan-proteins (AGPs) are the only known class of abundant cell surface proteoglycans found in plants. They occur in every taxonomic group of flowering plants tested and also in lower plants, although their distribution has not been explored systematically (Basile and Basile, 1993; Clarke *et al.*, 1979; Fincher *et al.*, 1983). They are increasingly recognized as a remarkably complex and diverse set of molecules that are major components of plant secretions and the cell wall face of all plant cell plasma membranes. To date, we have no clear understanding of AGP function, although various lines of evidence suggest that they may play fundamental roles during plant development in aspects such as cell–cell interactions underlying cell fate, cell expansion and cell proliferation.

2. AGP structure

AGPs have a high proportion of carbohydrate (often over 90%) that is typically acidic and characterized by a backbone of $(1\rightarrow3)$-β-galactan with $(1\rightarrow6)$-β-linked galactosyl side-branches which in turn are substituted with arabinose, uronic acids and other less abundant monosaccharides (Fincher *et al.*, 1983; Gane *et al.*, 1995; Pellerin *et al.*, 1995). The composition of the branched glycan components shows great compositional and structural diversity, and offers potential for an almost limitless heterogeneity. These components have been classified as type-II arabinogalactans by Aspinall (1980) to distinguish them from a second class of arabinogalactans (type I) which occur in plant tissues with a backbone of $(1\rightarrow4)$-β-galactan and arabinose side-chains. These typically form neutral structural elements of pectic polysaccharides (Bacic *et al.*, 1988; Carpita and Gibeaut, 1993; Clarke *et al.*, 1979).

The arabinogalactan polysaccharide is most commonly attached to a protein core by galactosyl *O*-linkages to hydroxyproline (Fincher *et al.*, 1983; Kieliszewski *et al.*, 1992; Strahm *et al.*, 1981). Our understanding of the protein component of AGPs (generally 2–10%) has increased dramatically with the recent cloning of cDNAs for core proteins of soluble AGPs from the style and cultured cells of *Nicotiana alata* and from cultured pear cells (Chasan, 1994; Chen *et al.*, 1994; Du *et al.*, 1994; Mau *et al.*, 1995). These cDNAs indicate that the protein components of AGPs are also highly varied and that the genes can be expressed in a tissue-specific manner. In some cases, the sequences predicted directly from cDNAs indicate Asn-rich domains, in addition to more typical peptide sequences rich in Hyp/Pro, Ala, Thr and Ser (Gleeson *et al.,* 1989; Mau *et al.,* 1995). A comparison of the cDNAs with the corresponding AGPs indicates the loss of Asn-rich domains during AGP biosynthesis and/or secretion (Mau *et al.,* 1995). Another aspect of the predicted core protein is the occurrence of C-terminal hydrophobic regions that may be transmembrane helices (Chen *et al.,* 1994; Du *et al.,* 1994). These may also be lost from the mature protein by processing events. AGP-like characteristics have been reported to occur in proteins that also contain the structural features of a class of cell wall hydroxyproline-rich glycoproteins (HRGPs) known as extensins (Kieliszewski *et al.,* 1992; Lind *et al.,* 1994; Qi *et al.,* 1991). Extensins and AGPs are now thought to belong to a superfamily of HRGPs. The protein structure, known developmental regulation patterns, glycosylational patterns and phylogeny have recently been reviewed (Kieliszewski and Lamport, 1994; Showalter, 1993).

A class of synthetic coloured trivalent phenyl glycosides have become important analytical and diagnostic tools for the study of AGPs (Fincher *et al.,* 1983; Yariv *et al.,* 1962). These compounds, now widely known as Yariv reagents, were originally developed as carbohydrate antigens and were subsequently fortuitously found specifically to precipitate AGPs by a mechanism that is not yet understood (Fincher *et al.,* 1983). As a consequence these reagents have been used extensively as probes for AGP localization and quantification. The most commonly used form is β-glucosyl Yariv reagent (βGlcY). Binding to βGlcY is often used as a diagnostic test for AGPs, but some AGP-like molecules are now known to be unreactive with βGlcY (Komavilas *et al.,* 1991; Smallwood *et al.,* 1996). A recent development of great interest is the discovery and characterization of the biological activity of βGlcY when added to living plant systems, and this is discussed below.

3. AGPs and the plasma membrane

AGPs are most abundant in plant secretions such as those associated with styles and root cap cells. They are also a major component of gums exuded by tree species such as *Acacia,* and are often secreted in large amounts by cells maintained in liquid suspension cultures (Clarke *et al.,* 1979; Fincher *et al.,* 1983). In addition to the soluble AGPs of such secretions, AGPs are also known to be associated with the outer face of the plasma membrane. This was first indicated by the reaction of AGP-binding Yariv reagents with the outer face of protoplasts and with membrane-associated proteins (Larkin, 1978; Nothnagel and Lyon, 1986; Samson *et al.,* 1983). The association of AGPs with the plasma membrane was further indicated by the preparation of monoclonal antibodies (MAC207 and JIM4) that recognized carbohydrate epitopes common to soluble extracellular AGP protoeoglycans secreted by suspension-cultured cells, and to plasma membrane components (Knox *et al.,* 1989; Pennell *et al.,* 1989). The occurrence of glycoproteins with the characteristics of AGPs at the plasma membrane has been confirmed by immunoaffinity purification using the monoclonal antibody PN16.4B4. In this case, tobacco suspension cell membrane glycoproteins with an AGP-like glycan composition (although containing no uronic acids) appeared to be elaborated from a 50-kDa polypeptide rich in Ala, Gly, Ser and Thr (Norman *et al.,* 1990). In an independent study of rose suspension-cultured cells, membrane-associated AGPs were isolated using specific precipitation with Yariv reagents (Komalavilas *et al.,* 1991). A comparison of extracellular and plasma membrane-associated AGPs indicated that the membrane AGPs were characterized by a higher ratio of arabinose to galactose, and they appeared to be more acidic (Komalavilas *et al.,* 1991). Both sets of AGPs were rich in alanine and hydroxyproline, and no clear difference was observed to explain membrane association. It was suggested that the Yariv reagent-reactive glycoproteins accounted for only 9% of the membrane carbohydrate, and that other unreactive arabinose- and galactose-containing glycoproteins were present (Komalavilas *et al.,* 1991).

The nature of the AGP association with the plasma membrane has been investigated in several cases by means of detergent partitioning of membrane fractions. AGPs from tobacco (Norman *et al.,* 1990) and carrot (Pennell *et al.,* 1989; Smallwood *et al.,* 1996) suspension cultures partition into the aqueous phase, suggesting that they are peripherally associated with the plasma membrane. However, AGPs from suspension-cultured rice cells partition with the detergent fraction, indicating that AGPs have a hydrophobic component. In related experiments with rice roots, most of the AGPs partitioned into the detergent fraction, but a high-molecular-weight AGP, as seen on sodium dodecyl sulphate–polyacrylamide gel electrophoresis (SDS–PAGE) analysis, partitioned into the aqueous phase (Smallwood *et al.,* 1996). It was also noted in this study that what appeared to be a soluble form or counterpart of the rice membrane AGP had a slightly increased mobility in the SDS–PAGE gel (Smallwood *et al.,* 1996). This may indicate the loss of a hydrophobic transmembrane domain that correlates with solubilization. It is therefore of interest in this context that the cDNAs encoding soluble AGPs have indicated C-terminal hydrophobic sequences predicted to be transmembrane helices as discussed above (Chen *et al.,* 1994; Du *et al.,* 1994). The possibility arises that, in some cells, plasma membrane-spanning domains of AGPs are cleaved by proteases in cell surface processing events to produce the mature soluble protein counterparts. Such events are well documented at the surface of animal cells (Ehlers and Riordan, 1991). An association of rose cell membrane AGPs with several abundant integral membrane proteins has also been reported (Komavilas *et al.,* 1991).

An important characteristic of the plasma membrane-associated AGPs is their extensive developmental regulation, which is currently most clearly defined by the patterns of carbohydrate AGP epitopes that correlate with a range of developmental parameters (Knox *et al.,* 1991; Knox, 1995; Pennell *et al.,* 1991).

4. AGP localization: the impact of hybridoma technology

As outlined above, the generation of anti-AGP monoclonal antibodies has been instrumental in the confirmation of the occurrence of AGPs at the plasma membrane, and in their characterization. It is through the use of these antibodies in immunolocalization studies that complex and surprising patterns of the developmental regulation of plasma membrane AGPs have become recognized. The restricted distribution patterns of distinct carbohydrate epitopes correlate with fundamental aspects of plant cell development within organs. The use of these probes in taxonomic surveys where epitope occurrence within a family/species could also be mapped with regard to its developmental location has not yet been explored fully, although it is known that the occurrence of a specific epitope within an organ can vary dramatically between species (Knox, 1993, 1995).

Developmental dynamics of AGP epitopes

The localization studies of distinct AGP epitopes have been applied to several developmental systems, including the iterative development at a root apex, floral meristem development and somatic embryogenesis. The common features appear to be the occurrence of epitope patterns in relation to both cell position and cell type during the development of anatomical complexity.

The carrot root apex. The most extensively studied system in terms of epitope number is the developing carrot root apex. Of the five anti-plasma membrane AGP antibodies characterized in this system, two (MAC207 and JIM16) bind to the plasma membrane of all the cells comprising the root, and three further antibodies (JIM4, JIM13 and JIM15) recognize the plasma membrane of sets of cells that reflect the development of the concentric pattern of tissues and the radial diarch symmetry of the stele (Knox *et al.*, 1991; Knox, 1993). The location and appearance/loss of the epitopes correlate with the sequential emergence of identifiable cell shapes during the development of the anatomy prior to the maturation of functional cells. They reflect, in a general way, what can be inferred about the sequence of developmental events from a study of anatomy (Esau, 1940). These observations and the possibility that a limited number of epitopes may be involved in specifying cell identity in a combinatorial manner have been discussed in detail elsewhere (Knox, 1993).

Recently, the patterns of binding of three monoclonal antibodies recognizing epitopes of cell wall extensins have also been described in this system. The occurrence of the extensin epitopes indicates some complementarity with the patterns of AGP epitope expression in terms of groups of cells and the developmental anatomy of the stele (Smallwood *et al.*, 1994).

Floral meristems. The loss of the MAC207 epitope from the plasma membrane of cells that form the progenitors of the germ cells has been observed as an early event in the development of both stamen and carpel primordia of pea (Pennell and Roberts, 1990). The loss affects the gametes, zygote and embryo, although the epitope does appear at the plasma membrane of the vegetative cell of the pollen grain. The epitope reappears in the early heart-stage embryo. The developmental loss and reappearance of this epitope therefore indicates a developmental window in which an AGP structure is absent from the plasma membrane of structures leading to fertilization (Pennell and Roberts, 1990). The developmental regulation of the distinct JIM8 AGP epitope has been characterized during floral development of oilseed rape (Pennell *et al.*, 1991). In some respects mirroring the loss of the MAC207 epitope in pea, the JIM8 epitope in rape appears during the differentiation of both the stamens and the carpels, and shows a complex temporal sequence of appearance and disappearance in the progression from one cell type to an adjacent one. These observations clearly indicate an active metabolism of plasma membrane AGPs, and also suggest the possible involvement of AGPs in interactive events between cell layers.

Somatic embryogenesis. The disruption of the developmental context of plant cells and the preparation of proliferating cell lines with the capacity to express totipotency are important tools in studies of the mechanisms of plant development. AGPs are abundantly secreted by cultured plant cells. The distribution of AGP epitopes in cells in culture and during somatic embryogenesis reflects cell status and again suggests a role in fundamental cell processes. The plasma membrane JIM4 epitope displays a dynamic pattern of expression during maturation of somatic embryos, and correlates with the emergence of distinct tissues (Stacey *et al.*, 1990). Intriguingly, this epitope is also expressed by cells at the surface of clumps of callus — cells that are unorganized in terms of planes of cell division — where it appears as a marker of surface position within the clumps (Stacey *et al.*, 1990). JIM8 recognized the surface of intact cultured carrot cells (i.e. the cell wall) in addition to recognizing the plasma membrane (Pennell *et al.*, 1992). The JIM8 epitope occurred at the surface of only a subset of cells in embryogenic cultures, and its occurrence appeared to be related to the embryogenic potential of the cell cultures (Pennell *et al.*, 1992).

The unadhered plant cell. JIM8 bound to cells with a range of morphologies, but often to larger, more vacuolate unadhered cells in the complex mixture of cells of which an embryogenic carrot cell culture is composed (Pennell *et al.*, 1992). What is significant in this case is that JIM8 and other antibodies bind to unadhered plant cell surfaces. Cells in coherent tissues or larger clumps do not appear to have the AGP epitopes at the equivalent location (i.e. the middle lamella). These observations seem to indicate that AGP distribution — whether in the cell wall or restricted to the plasma membrane — is related to cell context and can be correlated with the presence/absence of adhered neighbours. There is no evidence that AGPs play a role in cell separation, and they rarely appear to be located at developing intercellular spaces (Knox, 1992). It is also of interest that the tip-growing pollen tubes of tobacco (Li *et al.,* 1992) and the protonema of the moss *Physcomitrella patens* (Knox and Knight, unpublished observations) also express AGP epitopes at their surfaces. It is perhaps appropriate to recall here that AGPs are generally most abundant in secretions such as root cap slime, stylar fluids, cell culture conditioned media and wound exudates — developmental locations associated with cell separation and the loss of tissue cohesion. The significance of this is unknown. Possibilities include the involvement of AGPs in the signalling of the nature and proximity of cell neighbours, or in contributing to the required extracellular environments at an interface that is not directly in contact with a neighbouring cell. Many questions concerning the nature and relationship of the cell wall and membrane-associated AGPs and the secretion of AGPs from the cell wall to the extracellular space also remain unanswered.

5. AGPs: metabolism and interactions

The JIM4 epitope can appear and disappear within 12 h when its expression in cultured carrot cells is manipulated by alterations in cell density (Stacey *et al.,* 1990). The transient appearances of the JIM8 epitope in the rape flower also indicate the dynamic metabolism of AGPs (Pennell *et al.,* 1991). These developmental regulations and dynamics of the AGP epitopes raise a number of questions concerning the biochemical basis of epitope appearance and loss. Two broad questions can be asked: (1) What are the structures of the AGP epitopes? (2) Are epitope dynamics related to the modification of the AGPs at the surface, or do they reflect a rapid turnover of the proteoglycans? At this stage we have no precise answers to these questions. The structures of the epitopes are currently being addressed by fragmentation and synthetic approaches for the glycan component of the extracellular AGPs and exudate gums that are recognized by the anti-AGP antibodies (Yates and Knox, 1994), leading towards the identification of antibody-reactive oligosaccharides. Some observations have recently been reported indicating the importance of β-linked glucuronic acid for recognition by several of the antibodies, including MAC207 and JIM4 (Yates *et al.,* 1996).

No substantial body of information is available concerning the biosynthesis and turnover of AGPs (Fincher *et al.,* 1983). We have little understanding of the mechanisms that regulate the retention of AGPs in the wall or their secretion into culture media. There are indications that AGPs of cultured cells are transiently held in the wall prior to their loss to the medium (Takeuchi and Komamine, 1980), but whether this is in association with the plasma membrane is unknown, whereas a further study has indicated a rapid turnover and cycling of the glycan component of AGPs in cultured millet cells, relative to coherent tissues, with only a small proportion of the AGP-associated radiolabel being lost to the medium (Gibeaut and Carpita, 1991). In tobacco leaves, the monoclonal antibody PN16.4B4 locates AGPs to multivesicular bodies associated with the plasma membrane and the cell interior, and also partially degraded structures in the vacuole (Herman and Lamb, 1992). These observations appear to indicate a pathway for the internalization and disposal of cell surface material, including AGPs. Such a mechanism will be required if the basis of AGP modulations involves the transient expression of specific glycan structures at the cell surface and then their removal. Enzymes capable of hydrolysing arabinogalactan structures have been characterized (Hata *et al.,* 1992; Tsumuraya *et al.,* 1990).

Epitope appearance or loss may be due to cleavage events at the cell surface which result in the release of arabinogalactan oligosaccharides. The oligosaccharides released may act as developmental signals in a manner already established for oligosaccharide fragments of pectic polysaccharides, xyloglucan and N-linked glycoproteins (Aldington and Fry, 1993; Fry et al., 1993). Although no arabinogalactan fragments have yet been reported, speculation in this area has increased with the discovery that beet leaf plasma membrane AGPs contain N-acetylglucosamine (Pennell et al., 1995), and the apparent developmental role of chitinases (de Jong et al., 1992).

6. AGPs: possible functions and prospects

The location of AGPs at the outside face of the plasma membrane has led to speculation about their interaction with cell wall components. The plasma membrane is the site where the construction of the cell wall takes place, and as such AGPs are well placed to be components involved in wall assembly or in any mechanisms linking the cell surface and the cell interior across the apoplast. It has been proposed that membrane AGPs interact with cell wall polysaccharides to promote wall–membrane attachment and aid wall assembly and function (Baldwin et al., 1993; Carpita and Gibeaut, 1993; Pennell et al., 1989). The way in which the complex primary wall is constructed is still far from being fully understood, and the role of liquid crystal properties or self-assembly requirements of the wall is uncertain. It has been suggested that plasma membrane AGPs may act as epitaxial matrices, and thus specific AGP glycan structures could be involved in the formation of specific wall assemblies (Kieliszewski and Lamport, 1994).

The biological significance of the reaction of certain AGPs with Yariv reagents is far from clear. It does not seem to indicate a lectin-like capacity to recognize mono/oligosaccharides, as the Yariv reagents are known to self-associate and act in a multimeric form (Komalavilas et al., 1991). The phenyl components are essential for recognition (Fincher et al., 1983), and some naturally occurring monomeric phenyl glycosides can inhibit the interaction (Jermyn, 1978), raising the possibility that AGPs may have a receptor-like role. Alternatively, rather than acting as lectins, AGPs may be bound in carbohydrate recognition events involving lectin-like cell wall proteins. AGPs are recognized by anti-galactose lectins (Knox, 1993; Komavilas et al., 1991), and lectin proteins with the ability to recognize arabinose have been described (Engel et al., 1992). The complex AGPs clearly have immense potential for information presentation and also recognition at plant cell surfaces.

Despite our increasing knowledge of AGP complexity and diversity, the continued use of the term AGP is a sign of the uncertainty surrounding AGP function. An important indicator of AGP function would be the characterization of any biological activity of exogenous AGPs. Such activities have recently been reported. The addition of soluble carrot AGPs isolated from conditioned culture medium and seeds to cultured carrot cells has been reported to have the capacity to promote and inhibit carrot somatic embryogenesis (Kreuger and van Holst, 1993). This work has been extended, and the use of monoclonal antibodies to prepare affinity-purified AGPs has indicated that distinct epitopes are associated with these biological activities (Kreuger and van Holst, 1995). Tomato seed AGPs also influenced the development of carrot cells, and affinity-purified carrot AGPs displayed activity when applied to Cyclamen cells, indicating that the effects are not species-specific (Kreuger and van Holst, 1995; Kreuger et al., 1995). Similar activities have also been reported for spruce AGPs (Egertsdotter and van Arnold, 1995). An AGP-like protein isolated from transmitting tissue of tobacco has been demonstrated to influence the direction and extent of growth of pollen tubes (Cheung et al., 1995; Wu et al., 1995). These are extremely important observations that should allow the development of systems for the full characterization of the cellular and molecular details of the activity of cell surface AGPs.

A second powerful way to probe AGP function has been the recent application of βGlcY to biological systems. Remarkably, βGlcY has emerged as an effective reversible inhibitor of plant growth. The interaction of βGlcY with cell surface AGPs of suspension-cultured rose cells results in the reversible suppression of cell proliferation (Serpe and Nothnagel, 1994). This suggests that the activities of cell surface AGPs are involved in some aspects of the cell division cycle, and it may indicate the prevention of cell expansion (see below) with an indirect effect on cell proliferation. Alternatively, it may reflect the disruption of AGP interaction with cell surface components such as receptor-like protein kinases. Kinases that are likely to signal to the cell interior are known to be associated with plant cell plasma membranes, but with as yet unknown extracellular ligands (Walker, 1994). A case for a role for AGPs in the control of cell proliferation has also been established by studies of leafy liverworts, where the release (or increase in solubility) of distinct AGPs is associated with the de-suppression of cell proliferation in leaf primordia (Basile and Basile, 1993). A similar change in AGP characteristics has also been correlated with the induction of the proliferation of higher plant cells (Mignone and Basile, 1992). At a simple level the anatomical patterns of the plasma membrane epitopes, seen for example at the carrot root apex, may in higher plants relate to the control of cell division in specific cell files or cell groups, although no clear association of a specific AGP or epitope with a stage of the cell cycle has yet been reported.

A role for AGPs as wall-loosening factors involved in cell expansion has been proposed as a result of observations of Yariv reagent binding to the inner face of the wall of the growth-controlling outer epidermal cells of maize coleoptiles (Schopfer, 1990), and the loss of Yariv reagent-reactive material during the adaptation of cultured tobacco cells to high levels of NaCl, an adaptation that correlated with a reduction in cell expansion (Zhu et al., 1993). The use of the anti-AGP monoclonal antibodies JIM13 and JIM14 has revealed no changes in these two epitopes that correlate with cell expansion in maize coleoptiles (Schindler et al., 1995), although, in Arabidopsis seedlings, reduced cell expansion in the diminuto mutant has been correlated with lower levels of βGlcY-reactive AGPs in the hypocotyl (Takahashi et al., 1995). These results may indicate the differing specificities of these probes, and suggest that not all AGPs may be detected by an antibody or βGlcY. It is also possible that different AGPs may have different functions. Further evidence implicating βGlcY-reactive AGPs in cell expansion has also been obtained by growing Arabidopsis seedlings in medium containing βGlcY, which resulted in a severe reduction in cell elongation in the roots (Willats and Knox, 1996). βGlcY had no apparent effect on cell proliferation at the root apex. Defining the AGP(s) that βGlcY binds to in these systems (and also the AGPs that it does not bind to) will be of great interest.

Although biochemically distinct from the proteoglycans which occur in animal systems, the location of AGPs at the cell surface, often in association with the outer face of the plasma membrane, is indicative of possible functional similarities with classes of animal proteoglycans such as the mucins and syndecans. At animal cell surfaces and in the associated extracellular matrices, large proteoglycans are built up from glycosaminoglycan chains covalently linked to protein cores, and they often associate into huge polymeric complexes. They have diverse roles in chemical signalling and cell adhesion, they can bind to and regulate the role of other secreted proteins, and they can act as receptors (Alberts et al., 1994). The plant cell wall is neither biochemically nor functionally the same as the animal extracellular matrix, but AGPs may have roles that are broadly similar to their animal proteoglycan counterparts, in that they function in processes involving maintenance of cellular environments, cell interactions and the development of cell morphology. What we do have at this stage is an indication of the molecular complexity of AGPs. The indications are that knowing more about the structure, biosynthesis, membrane association and regulation of AGPs is likely to be an important component of an understanding of plant morphogenesis and its evolution.

References

Alberts B, Bray D, Lewis J, Raff M, Roberts K, Watson JD. (1994) *Molecular Biology of the Cell,* 3rd edition. New York: Garland Publishing Inc.

Aldington S, Fry SC. (1993) Oligosaccharins. *Adv. Bot. Res.* **19:** 1–101.

Aspinall GO. (1980) Chemistry of cell wall polysaccharides. In: *The Biochemistry of Plants,* Vol. 3 (Preiss J, ed.). New York: Academic Press, pp. 473–500.

Bacic A, Harris PJ, Stone BA. (1988) Structure and function of plant cell walls. In: *The Biochemistry of Plants,* Vol. 14 (Preiss J, ed.). New York: Academic Press, pp. 297–371.

Baldwin TC, McCann MC, Roberts K. (1993) A novel hydroxyproline-deficient arabinogalactan protein secreted by suspension-cultured cells of *Daucus carota. Plant Physiol.* **103:** 115–123.

Basile DV, Basile MR. (1993) The role and control of the place-dependent suppression of cell division in plant morphogenesis and phylogeny. *Mem. Torrey Bot. Club* **25:** 63–84.

Carpita NC, Gibeaut DM. (1993) Structural models of primary cell walls in flowering ·plants: consistency of molecular structure with the physical properties of the walls during growth. *Plant J.* **3:** 1–30.

Chasan R. (1994) Arabinogalactan-proteins: getting to the core. *Plant Cell* **6:** 1519–1521.

Chen CG, Pu ZY, Moritz RL, Simpson RJ, Bacic A, Clarke AE, Mau SL. (1994) Molecular cloning of a gene encoding an arabinogalactan-protein from pear (*Pyrus communis*) cell suspension culture. *Proc. Natl Acad. Sci. USA* **91:** 10305–10309.

Cheung AY, Wang H, Wu H. (1995) A floral transmitting tissue-specific glycoprotein attracts pollen tubes and stimulates their growth. *Cell* **82:** 383–393.

Clarke AE, Anderson RL, Stone BA. (1979) Form and function of arabinogalactans and arabinogalactan-proteins. *Phytochemistry* **18:** 521–540.

de Jong AJ, Cordewener J, LoSchiavo F, Terzi M, Vandekerckhove J, van Kammen A, de Vries SC. (1992) A carrot somatic embryo mutant is rescued by chitinase. *Plant Cell* **4:** 425–433.

Du H, Simpson RJ, Mortiz RL, Clarke AE, Bacic A. (1994) Isolation of the protein backbone of an arabinogalactan-protein from the styles of *Nicotiana alata* and characterization of a corresponding cDNA. *Plant Cell* **6:** 1643–1653.

Egertsdotter U, von Arnold S. (1995) Importance of arabinogalactan proteins for the development of somatic embryos of Norway spruce (*Picea abies*). *Physiol. Plant.* **93:** 334–345.

Ehlers MRW, Riordan JF. (1991) Membrane proteins with soluble counterparts: role of proteolysis in the release of transmembrane proteins. *Biochemistry* **30:** 10065–10074.

Engel M, Bachmann M, Shröder HC, Rinkevich B, Kljajic Z, Uhlenbruck G, Müller WEG. (1992) A novel galactose- and arabinose-specific lectin from the sponge *Pellina semitubulosa*: isolation, characterization and immunobiological properties. *Biochemie* **74:** 527–537.

Esau K. (1940) Developmental anatomy of the fleshy storage organ of *Daucus carota. Hilgardia* **13:** 175–226.

Fincher GB, Stone BA, Clarke AE. (1983) Arabinogalactan-proteins: structure, biosynthesis and function. *Annu. Rev. Plant Physiol.* **34:** 47–70.

Fry SC, Aldington S, Hetherington PR, Aitken J. (1993) Oligosaccharides as signals and substrates in the plant cell wall. *Plant Physiol.* **103:** 1–5.

Gane AM, Craik D, Munro SLA, Howlett GJ, Clarke AE, Bacic A. (1995) Structural analysis of the carbohydrate moiety of arabinogalactan-proteins from stigmas and styles of *Nicotiana alata. Carbohydr. Res.* **277:** 67–85.

Gibeaut DM, Carpita NC. (1991) Tracing cell wall biogenesis in intact cells and plants. Selective turnover and alteration of soluble and cell wall polysaccharides in grasses. *Plant Physiol.* **97:** 551–561.

Gleeson PA, McNamara M, Wettenhall REH, Stone BA, Fincher GB. (1989) Characterization of the hydroxyproline-rich protein core of an arabinogalactan-protein secreted from suspension-cultured *Lolium multiflorum* (Italian ryegrass) endosperm cells. *Biochem J.* **264:** 857–862.

Hata K, Tanaka M, Tsumuraya Y, Hashimoto Y. (1992) α-L-arabinofuranosidase from radish (*Raphanus sativus* L.) seeds. *Plant Physiol.* **100:** 388–396.

Herman EM, Lamb CJ. (1992) Arabinogalactan-rich glycoproteins are localized on the cell surface and in intravacuolar multivesicular bodies. *Plant Physiol.* **98:** 264–272.

Jermyn MA. (1978) Isolation from the flowers of *Dryandra praemorsa* of a flavonol glycoside that reacts with β-lectins. *Aust. J. Plant Physiol.* **5:** 697–705.

Kieliszewski MJ, Lamport DTA. (1994) Extensin: repetitive motifs, functional sites, post-translational codes, and phylogeny. *Plant J.* **5:** 157–172.

Kieliszewski MJ, Kamyab A, Leykam JF, Lamport DTA. (1992) A histidine-rich extensin from *Zea mays* is an arabinogalactan protein. *Plant Physiol.* **99:** 538–547.

Knox JP. (1992) Cell adhesion, cell separation and plant morphogenesis. *Plant J.* **2:** 137–141.

Knox JP. (1993) The role of cell surface glycoproteins in differentiation and morphogenesis. In: *Post-Translational Modification in Plants* (Battey NH, Dickinson HG, Hetherington AM, ed.). SEB Seminar Series 53. Cambridge: Cambridge University Press, pp. 267–283.

Knox JP. (1995) Developmentally regulated proteoglycans and glycoproteins of the plant cell surface. *FASEB J.* **9:** 1004–1012.

Knox JP, Day S, Roberts K. (1989) A set of cell surface glycoproteins forms an early marker of cell position, but not cell type, in the root apical meristem of *Daucus carota* L. *Development* **106:** 47–56.

Knox JP, Linstead PJ, Peart J, Cooper C, Roberts K. (1991) Developmentally regulated epitopes of cell surface arabinogalactan proteins and their relation to root tissue pattern formation. *Plant J.* **1:** 317–326.

Komalavilas P, Zhu JK, Nothnagel EA. (1991) Arabinogalactan proteins from suspension culture medium and plasma membrane of rose cells. *J. Biol. Chem.* **266:** 15956–15965.

Kreuger M, van Holst GJ. (1993) Arabinogalactan-proteins are essential in somatic embryogenesis of *Daucus carota* L. *Planta* **189:** 243–248.

Kreuger M, van Holst GJ. (1995) Arabinogalactan-protein epitopes in somatic embryogenesis of *Daucus carota* L. *Planta* **197:** 135–141.

Kreuger M, Postma E, Brouwer Y, van Holst GJ. (1995) Somatic embryogenesis of *Cyclamen persicum* in liquid-medium. *Physiol. Plant.* **94:** 605–612.

Larkin PJ. (1978) Plant protoplast agglutination by artificial carbohydrate antigens. *J. Cell Sci.* **30:** 283–292.

Li Y, Bruun L, Pierson ES, Cresti M. (1992) Periodic deposition of arabinogalactan epitopes in the cell wall of pollen tubes of *Nicotiana tabacum* L. *Planta* **88:** 532–538.

Lind JL, Bacic A, Clarke AE, Anderson MA. (1994) A style specific hydroxyproline-rich glycoprotein with properties of both extensins and arabinogalactan proteins. *Plant J.* **6:** 491–502.

Mau SL, Chen CG, Pu ZY, Moritz RL, Simpson RJ, Bacic A, Clarke AE. (1995) Molecular cloning of cDNAs encoding the protein backbones of arabinogalactan-proteins from the filtrate of suspension-cultured cells of *Pyrus communis* and *Nicotiana alta*. *Plant J.* **8:** 269–281.

Mignone MP, Basile DV. (1992) The release of high buoyant density AGPs correlated with experimentally induced cell proliferation in lettuce and potato stem pith tissues. *Am. J. Bot.* **79:** S-49.

Norman PM, Kjellbom P, Bradley DJ, Hahn MG, Lamb CJ. (1990) Immunoaffinity purification and biochemical characterization of plasma membrane arabinogalactan-rich glycoproteins of *Nicotiana glutinosa*. *Planta* **181:** 365–373.

Nothnagel EA, Lyon JL. (1986) Structural requirements for the binding of phenylglycosides to the surface of protoplasts. *Plant Physiol.* **80:** 91–98.

Pellerin P, Vidal S, Williams P, Brillouet JM. (1995) Characterization of five type II arabinogalactan-protein fractions from red wine of increasing uronic acid content. *Carbohydr. Res.* **277:** 135–143.

Pennell RI, Roberts K. (1990) Sexual development in the pea is presaged by altered expression of arabinogalactan protein. *Nature* **344:** 547–549.

Pennell RI, Knox JP, Scofield GN, Selvendran RR, Roberts K. (1989) A family of abundant plasma membrane-associated glycoproteins related to the arabinogalactan proteins is unique to flowering plants. *J. Cell Biol.* **108:** 1967–1977.

Pennell RI, Janniche L, Kjellbom P, Scofield GN, Peart JM, Roberts K. (1991) Developmental regulation of a plasma membrane arabinogalactan protein epitope in oilseed rape flowers. *Plant Cell* **3:** 1317–1326.

Pennell RI, Janniche L, Scofield GN, Booji H, de Vries SC, Roberts K. (1992) Identification of a transitional cell state in the developmental pathway to carrot somatic embryogenesis. *J. Cell Biol.* **119:** 1371–1380.

Pennell RI, Cronk QCB, Forsberg LS, Stöhr C, Snogerup L, Kjellbom P, McCabe PF. (1995) Cell-context signalling. *Phil. Trans. R. Soc. Lond. B* **350:** 87–93.

Qi W, Fong C, Lamport DTA. (1991) Gum arabic glycoprotein is a twisted hairy rope. *Plant Physiol.* **96:** 848–855.

Samson MR, Klis FM, Sigon CAM, Stegwee D. (1983) Localization of arabinogalactan proteins in the membrane system of etiolated hypocotyls of *Phaseolus vulgaris* L. *Planta* **159:** 322–328.

Schindler T, Bergfeld R, Schopfer P. (1995) Arabinogalactan-proteins in maize coleoptiles: developmental relationship to cell death during xylem differentiation but not to extension growth. *Plant J.* **7:** 25–36.

Schopfer P. (1990) Cytochemical identification of arabinogalactan protein in the outer epidermal wall of maize coleoptiles. *Planta* **183:** 139–142.

Serpe MD, Nothnagel EA. (1994) Effects of Yariv phenylglycosides on *Rosa* cell suspensions: evidence for the involvement of arabinogalactan-proteins in cell proliferation. *Planta* **193:** 542–550.

Showalter AM. (1993) Structure and function of plant cell wall proteins. *Plant Cell* **5:** 9–23.

Smallwood M, Beven A, Donovan N, Neill SJ, Peart J, Roberts K, Knox JP. (1994) Localization of cell wall proteins in relation to the developmental anatomy of the carrot root apex. *Plant J.* **5:** 237–246.

Smallwood M, Yates EA, Willats WGT, Martin H, Knox JP. (1996) Immunochemical comparison of membrane-associated and secreted arabinogalactan- proteins in rice and carrot. *Planta* **198:** 452–459.

Stacey NJ, Roberts K, Knox JP. (1990) Patterns of expression of the JIM4 arabinogalactan-protein epitope in cell cultures and during somatic embryogenesis in *Daucus carota* L. *Planta* **180:** 285–292.

Strahm A, Amado R, Neukom H. (1981) Hydroxyproline galactoside as a protein polysaccharide linkage in a water soluble arabinogalactan-peptide from wheat endosperm. *Phytochemistry* **20:** 1061–1063.

Takahashi T, Gasch A, Nishizawa N, Chua NH. (1995) The DIMINUTO gene of *Arabidopsis* is involved in regulating cell elongation. *Genes Dev.* **9:** 97–107.

Takeuchi Y, Komamine A. (1980) Turnover of cell wall polysaccharides of *Vinca rosea* suspension culture. III. Turnover of arabinogalactan. *Physiol. Plant.* **50:** 113–118.

Tsumuraya Y, Mochizuki N, Hashimoto Y, Kovác P. (1990) Purification of an exo-β-(1→3)-D-galactanase of *Irpex lacteus* (*Polyporus tulipiferae*) and its action on arabinogalactan-proteins. *J. Biol. Chem.* **265:** 7207–7215.

Walker JC. (1994) Structure and function of the receptor-like protein kinases of higher plants. *Plant Mol. Biol.* **26:** 1599–1609.

Willats WGT, Knox JP. (1996) A role for arabinogalactan-proteins in plant cell expansion: evidence from studies on the interaction of β-glucosyl Yariv reagent with seedlings of *Arabidopsis thaliana. Plant J.* (in press).

Wu H, Wang H, Cheung AY. (1995) A pollen tube growth stimulatory glycoprotein is deglycosylated by pollen tubes and displays a glycosylation gradient in the flower. *Cell* **82:** 395–403.

Yariv J, Rapport MM, Graf L. (1962) The interaction of glycosides and saccharides with antibody to the corresponding phenylazo glycosides. *Biochem. J.* **85:** 383–388.

Yates EA, Knox JP. (1994) Investigations into the occurrence of plant cell surface epitopes in exudate gums. *Carbohydr. Polym.* **24:** 281–286.

Yates EA, Valdor JF, Haslam SM, Morris HR, Dell A, Mackie W, Knox JP. (1996) Characterization of carbohydrate structural features recognized by anti-arabinogalactan-protein monoclonal antibodies. *Glycobiology* **6:** (in press).

Zhu JK, Bressan RA, Hasegawa PM. (1993) Loss of arabinogalactan-proteins from the plasma membrane of NaCl-adapted tobacco cells. *Planta* **190:** 221–226.

Chapter 7

Signal perception at the plasma membrane: binding proteins and receptors

Michael G. Hahn

1. Introduction

Recognition is essential to life. This is true for organisms ranging from the simplest (e.g. viroids) to the most complex (e.g. mammals), and indeed for all levels of cellular organization extending from the whole organism to the biochemical level. Thus, organisms must be able to distinguish self from non-self, friend from foe, food from poison, and be able to respond appropriately. Cells within an organism must interact with other cells both proximally and distally in an integrated manner to give a healthy individual. Different compartments within cells must also communicate with one another in order to maintain cellular viability. The macromolecules that are instrumental for recognition are called receptors. This chapter will focus on one sub-class of receptors in higher plants, namely those that are associated with the plasma membrane, the membrane at which extracellular signals are perceived and across which those signals must be transmitted in order to elicit a physiological response from the cell.

Experimentally, the identification of a receptor proceeds in several stages. To begin with, a signal molecule which triggers the biological response of interest must be purified and its structure elucidated. In addition, structural analogs of the signal molecule, in particular one that can be labeled to high specific radioactivity, must be identified and the biological activities of these analogs determined. The biological response to a particular signal molecule must also be understood, preferably at the cellular and biochemical levels. With these tools in hand, one can then identify and characterize specific binding proteins for the signal molecule.

A set of accepted criteria exists, which a binding protein must fulfill in order to be classified as a biological receptor for a particular ligand. These criteria have been succinctly presented by Venis (1985) and extensively discussed by Hulme and Birdsall (1992), among others. Briefly, the four most important criteria are as follows:

(1) binding of the ligand should be reversible, saturable and of high affinity;
(2) the saturation range of ligand binding should correlate with the concentration range over which the biological response to that ligand saturates;
(3) the pharmacological specificity of the binding must be demonstrated, that is, the affinities of various structural analogs of the ligand should correlate with the abilities of those analogs to induce a biological response;
(4) binding of the ligand by the binding protein should result in a ligand-specific biological response, that is, the functioning of a binding protein in a signal transduction pathway must be demonstrated.

Fulfillment of these criteria is what distinguishes a physiologically relevant receptor from other binding proteins that might be present in cells, but that do not participate in a signal transduction pathway. One example of such a binding protein would be a membrane-localized transporter that is responsible for the uptake of a molecule by a cell. In this example, the formation of the binding protein–ligand complex may not elicit a biological response (e.g. changes in gene expression) in the targeted cell. For the purpose of discussions here, a binding protein is regarded as being functionally a receptor if formation of the binding protein–ligand complex can be causally related to changes in cellular behavior (e.g. growth, differentiation, gene expression) triggered by the signal molecule *in vivo*.

The last few years have seen a dramatic increase in the number of plant ligand-binding protein studies and the diversity of ligands being utilized. For the most part, these studies have focused on the identification, purification and characterization of the binding proteins. Only in a very few cases, such as auxin- and fusicoccin-binding proteins, which will be discussed in detail below, an ethylene-binding protein (Chang *et al.,* 1992; Schaller and Bleecker, 1995; Schaller *at al.,* 1995), a salicylic acid-binding protein (Chen *et al.,* 1993a, b) and binding proteins for the fungal toxins, BmT toxin (Levings and Siedow, 1992) and victorin (Navarre and Wolpert, 1995; Wolpert *et al.,* 1994), has our knowledge progressed to the point where studies are being undertaken with the aim of proving receptor function for the binding proteins and understanding the roles of the binding proteins in signal transduction pathways. This chapter will focus on plant plasma membrane-localized binding proteins for several plant hormones, oligosaccharins, peptide elicitors, and a fungal toxin. Putative receptor protein kinases (Stone and Walker, 1995) that are plasma membrane-localized are discussed in Chapter 8 by Harmon *et al.* and hence will not be covered here. The reader should be aware that several additional plant ligand-binding proteins have been studied which will not be discussed here. These include soluble binding proteins for salicylic acid (Chen *et al.,* 1993a, b) and cytokinins (Brinegar, 1994), endo-membrane-localized binding proteins for ethylene (Chang *et al.,* 1992; Hall *et al.,* 1990; Schaller and Bleecker, 1995; Sisler, 1991) and several fungal toxins (Levings and Siedow, 1992; Wolpert *et al.,* 1994), and binding proteins involved in protein trafficking within plant cells (Kirsch *et al.,* 1994; Wolins and Donaldson, 1994). In addition, receptors for light (Ahmad and Cashmore, 1993; Chory, 1991; Quail, 1991; Short *et al.,* 1993), a very important signal for plants, will not be discussed.

2. Hormone-binding proteins

Plant hormones are among the best studied molecular signals in plants. These molecules can induce a wide variety of responses (Davies, 1995). However, their mechanisms of action have yet to be determined, in particular with regard to signal perception and signal transduction pathways. The plant hormones identified to date are, for the most part, small organic molecules that, in their uncharged form, are able in principle to cross the lipid bilayer and exert their regulatory effects on cellular components directly. Thus, one particular concern to researchers of plant hormone signal transduction is whether plasma membrane-localized receptors for these ligands need even exist. However, fairly conclusive evidence has been obtained for the existence of specific plasma membrane-localized binding proteins for several plant hormones. This evidence will be summarized in the following sections.

Auxin-binding proteins

Auxin [indole-3-acetic acid (IAA); structure shown in *Figure 1*] is involved in various and diverse plant growth and developmental responses, including cell elongation, cell division and

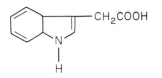

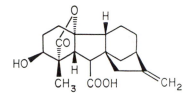

Indole-3-acetic Acid
(an auxin)

Gibberellin A$_4$
(a gibberellin)

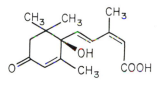

Abscisic Acid

Figure 1. Structures of three plant hormones for which there appear to be plant plasma membrane-localized binding sites.

cell differentiation (Moore, 1989). A number of structural analogs of the natural auxin, indole-acetic acid, have been identified and their biological activities have been characterized.

Specific binding proteins for auxin were among the first plant receptors to be sought, and the first demonstration of specific binding of auxin to a plant membrane fraction was published over 20 years ago (Hertel *et al.*, 1972). Several recent and comprehensive reviews have discussed the studies that led to the identification and characterization of auxin-binding proteins in plants (Barbier-Brygoo, 1995; Goldsmith, 1993; Jones, 1994; Libbenga *et al.*, 1986; Palme *et al.*, 1991; Venis and Napier, 1991, 1995). Thus, only a relatively brief summary of this extensive research record will be provided here, focusing on the membrane-localized auxin-binding proteins. The reader is encouraged to consult those comprehensive reviews for detailed discussions of these and other auxin-binding proteins.

Most of the research on auxin-binding proteins has been generated using maize seedling tissues. Binding of [^{14}C]IAA to membranes prepared from maize coleoptile tissue was found to be reversible, of high affinity (apparent K_d = 3.3 μM), and specific for IAA and biologically active structural analogs (Hertel *et al.*, 1972). In fact, the affinity for the synthetic auxin, naphthyl-1-acetic acid (NAA), was even higher (apparent K_d = 0.1 μM) than that for IAA. Subsequent subfractionation of maize membranes revealed the presence of auxin-binding activity in several subcellular membrane fractions that could be differentiated on the basis of binding parameters (Dohrmann *et al.*, 1978); the bulk of the auxin-binding activity (site 1) is associated with the endoplasmic reticulum. The distribution of auxin-binding proteins among several subcellular membrane fractions has been largely confirmed in subsequent studies (for review see Venis and Napier, 1995).

Photo-affinity labeling has proved to be a powerful, albeit somewhat controversial (see comments by Venis and Napier, 1995) method for identifying auxin-binding polypeptides. This experimental approach was made possible by the synthesis of a tritiated azido-derivative of IAA (Campos *et al.*, 1991; Jones *et al.*, 1984). Half a dozen or more different polypeptides have been tagged using this azido-IAA; many of these are soluble proteins and will not be discussed further here (for review see Jones, 1994; Venis and Napier, 1995). Membrane proteins that can be photo-labeled include a 22-kDa polypeptide that will be discussed in much greater detail below. Other photo-affinity-labeled polypeptides were identified in highly enriched plasma membrane vesicles prepared from maize (Feldwisch *et al.*, 1992) and zucchini (Hicks *et al.*, 1989a, 1993). In maize, a 23-kDa polypeptide is labeled (Feldwisch *et al.*, 1992). However, other results (Zettl *et al.*, 1992) suggest that this polypeptide is likely to be an auxin efflux carrier rather than an auxin receptor. In zucchini, photo-affinity labeling results in the tagging of two polypeptides of 40 and 42 kDa as visualized by sodium dodecyl sulfate–polyacrylamide gel electrophoresis (SDS–PAGE) autoradiography (Hicks *et al.*, 1989a, 1993). Photo-labeling is inhibited in the presence of IAA and biologically active analogs, suggesting that the observed labeling is auxin-specific. Similar photo-affinity labeling experiments carried out in tomato labeled a protein doublet of similar molecular weight (M_r), while greatly reduced labeling of this doublet was observed in membranes prepared from an auxin-insensitive tomato mutant (Hicks *et al.*, 1989b); these results imply physiological relevance for these photo-labeled auxin-binding proteins. Comparison of the properties of these plasma membrane-localized auxin-binding proteins with the 22-kDa ABP1 to be discussed below revealed significant differences (Lomax and Hicks, 1992). Indeed, the properties of the 40- to 42-kDa auxin-binding proteins suggest that they are likely to be involved in the uptake of auxin (Lomax and Hicks, 1992). Further characterization of the 40- to 42-kDa auxin-binding proteins has led to the hypothesis that these polypeptides are subunits of a multimeric channel protein (Hicks *et al.*, 1993), a hypothesis consistent with their presumed function as an auxin uptake carrier. Whether or not the 40- to 42-kDa polypeptides play a role as part of a signal transduction pathway (i.e. function as physiological receptors for auxin) awaits further experimentation.

The bulk of the research on auxin signal perception in plant cells has focused on the site 1 binding proteins that appeared to be localized in the endoplasmic reticulum. These auxin-binding proteins can be solubilized from plant membrane fractions in functional form (as determined by their ligand-binding properties) using nonionic detergents (Cross and Briggs, 1978), although the more prevalent method used for subsequent purification studies involves buffer extraction of an acetone powder prepared from membranes (Venis, 1977). Initial estimates of the M_r of the solubilized auxin-binding proteins ranged from 40 000 to 90 000.

The first purification of a membrane-localized auxin-binding protein utilized a combination of affinity and immuno-affinity columns (Löbler and Klämbt, 1985a). The affinity matrix, consisting of immobilized 2-hydroxy-3,5-diiodobenzoic acid (an auxin analog), was insufficiently specific by itself to achieve purification to homogeneity, hence the need for the second immuno-affinity chromatography step. The purified binding protein has a high affinity (apparent K_d = 0.06 μM) for NAA, is specific for biologically active auxins and their analogs, and shows a predominant polypeptide (*c.* 20 kDa) band on SDS–PAGE. Photo-affinity labeling studies carried out using partially purified preparations of maize auxin-binding protein and a tritiated azido-IAA resulted in the labeling of a 22-kDa polypeptide (Campos *et al.*, 1991; Jones and Venis, 1989); these results are consistent with the data for the purified auxin-binding protein. A native M_r of 40 000 was estimated on the basis of size-exclusion chromatography, suggesting that this auxin-binding protein exists as a dimer. Subsequent purification studies using affinity chromatography on immobilized NAA (Shimomura *et al.*, 1986) or phenylacetic acid (Rademacher and Klämbt, 1993), or combinations of anion-exchange, size-exclusion and affinity chromatographies, and/or native PAGE (Hesse *et al.*, 1989; Napier *et al.*, 1988) have yielded purified auxin-binding proteins with virtually identical properties. This purified auxin-binding protein is a glycoprotein, as evidenced by a decrease in M_r (to 19 000–20 000) following treatment with endoglycosidase H (Inohara *et al.*, 1989; Napier *et al.*, 1988); the oligosaccharide side-chain contains only mannose and *N*-acetylglucosamine (9:2 molar ratio) (Hesse *et al.*, 1989). The properties of the 22-kDa auxin-binding protein (now referred to as ABP1) strongly suggest that it represents the predominantly endoplasmic reticulum-localized binding site (site 1) identified in earlier membrane fractionation studies.

The availability of purified ABP1 made possible the cloning of cDNAs encoding this protein using either oligonucleotide primers, whose design was based on partial amino acid sequences from ABP1 (Hesse *et al.*, 1989; Inohara *et al.*, 1989; Yu and Lazarus, 1991), or antibodies generated against ABP1 (Tillmann *et al.*, 1989). In maize, ABP1 appears to be encoded by a multi-gene family consisting of at least five members (Campos *et al.*, 1994; Hesse *et al.*, 1993; Schwob *et al.*, 1993; Yu and Lazarus, 1991). A single homologous gene has been cloned from *Arabidopsis thaliana* (Palme *et al.*, 1992; Shimomura *et al.*, 1993). Preliminary reports also indicate that highly related sequences are present in tobacco and strawberry (Venis and Napier, 1995). The primary amino acid sequences deduced from these genes share common structural features, namely an N-terminal hydrophobic sequence that may serve as a signal sequence for translocation of ABP1 across the endoplasmic reticulum membrane (Campos *et al.*, 1994), a single potential *N*-glycosylation site, and a C-terminal tetrapeptide (Lys-Asp-Glu-Leu; KDEL) that is a consensus signal for retention of a protein in the endoplasmic reticulum (Pelham, 1990). The latter is consistent with the subcellular localization studies carried out at the binding protein level that were described earlier. Indeed, active ABP1 expressed in insect cells using the baculovirus expression system is targeted to the endoplasmic reticulum (Macdonald *et al.*, 1994). None of the deduced ABP1 sequences contain a hydrophobic domain indicative of an integral membrane protein.

With the sequences of several *ABP1* genes in hand, efforts have been directed at the identification of the amino acid residues that encompass the auxin-binding site. A polyclonal antibody preparation directed against a 10-amino-acid sequence, which is completely conserved among the cloned *ABP1* genes, acts as an auxin agonist, suggesting that the peptide recognized by the antibodies forms at least part of the auxin-binding site (Venis *et al.*, 1992). An arginine residue located within this peptide has been postulated to interact with the carboxylic acid group of IAA (Brown and Jones, 1994). Another peptide has been shown to become specifically photo-labeled using tritiated azido-IAA (Brown and Jones, 1994). It has been suggested that a conserved tryptophan located within this peptide provides the 'hydrophobic platform' postulated on the basis of molecular modeling studies (Edgerton *et al.*, 1994) to be an important feature of the auxin-binding site. Site-directed mutagenesis of specific amino acid residues within these two peptides would substantiate the hypothesis that these sequences encompass the auxin-binding site. The ability to obtain stable expression of active ABP1 in transfected insect cell lines (Henderson *et al.*, 1995; Macdonald *et al.*, 1994) will facilitate such mutagenesis studies.

The lack of a clear transmembrane domain, together with the apparent targeting of ABP1 to the endoplasmic reticulum, has provoked extensive discussion (Cross, 1991; Goldsmith, 1993; Guern, 1987; Hesse and Palme, 1994; Jones, 1990, 1994; Klämbt, 1990; Venis and Napier, 1995) about the dichotomy between the apparent site of action of auxin and the apparent location of the putative receptor for this hormone. Both polyclonal and monoclonal antibodies have been extensively used in experimental efforts to resolve this dichotomy. Purified ABP1 protein has been used to generate antisera in rabbits (Jones *et al.*, 1991; Löbler and Klämbt, 1985a; Napier *et al.*, 1988; Shimomura *et al.*, 1988). The epitopes recognized by these antisera appear to be clustered about, but do not include, the site of glycosylation (Napier and Venis, 1992). These polyclonal antibodies have been used to detect ABP1 in different subcellular membrane fractions (Shimomura *et al.,* 1988) and in different plant species (Napier and Venis, 1992; Venis *et al.,* 1992). More importantly, these antisera have been used to provide evidence that auxin perception occurs at the plasma membrane. Thus, an anti-ABP1 antiserum was found to inhibit auxin-inducible curling of split maize coleoptile sections (Löbler and Klämbt, 1985b), suggesting that the auxin-binding protein is located at the outer face of the plasma membrane. Furthermore, an ABP1-specific antiserum labeled the plasma membrane and cell wall of both maize coleoptiles and suspension-cultured maize cells (Jones and Herman, 1993); the cell wall localization of ABP1 is difficult to reconcile with the results of cell fractionation studies discussed earlier. Auxin-binding sites have also been visualized on the plasma membrane of maize coleoptile protoplasts by means of silver-enhanced immunogold epipolarization microscopy using antisera raised against either whole ABP1 or a synthetic peptide corresponding to the presumed auxin-binding site of ABP1 (Diekmann *et al.,* 1995). Interestingly, clustering of the auxin-binding sites was observed when labeling was carried out in the presence of IAA; this clustering does not appear to be essential for auxin action (Diekmann *et al.,* 1995).

Several electrophysiological studies carried out with either free IAA (Ephritikhine *et al.,* 1987; Marten *et al.,* 1991) or with a cell-impermeant IAA covalently linked to bovine serum albumin (Venis *et al.,* 1990) suggest that auxin acts at the plasma membrane. Electrophysiological experiments carried out using specific antibodies against ABP1 provided additional evidence that the cell surface site of auxin action is ABP1 or a structurally related protein. An antiserum raised against maize ABP1 inhibits the auxin-induced hyperpolarization of tobacco protoplasts, but does not affect the hyperpolarization induced by the fungal toxin fusicoccin (Barbier-Brygoo *et al.,* 1989). Incubation of the protoplasts with an antiserum against the plasma membrane H$^+$-ATPase abolishes the ability of either NAA or fusicoccin to hyperpolarize the protoplasts, suggesting that the ATPase is part of the signal transduction pathway (Barbier-Brygoo *et al.,* 1989). Incubation of tobacco protoplasts with increasing amounts of the anti-ABP1 antiserum increases the concentration of NAA required to induce hyperpolarization (Barbier-Brygoo *et al.,* 1991). In contrast, incubation of tobacco protoplasts with exogenously added maize ABP1 decreases the NAA concentration required to elicit a response. A polyclonal antiserum raised against a conserved peptide thought to encompass the auxin-binding site on ABP1 is able to mimic the ability of NAA to induce the hyperpolarization of tobacco protoplasts (Venis *et al.,* 1992). This antiserum cross-reacts with all known ABP1, including those in tobacco. Finally, patch-clamp experiments carried out using maize protoplasts demonstrated that the auxin-inducible outward-directed current of positive charge can be inhibited by an antiserum generated against whole ABP1 (Rück *et al.,* 1993). The outward-directed current is stimulated in the absence of auxin by the antiserum generated against the conserved peptide encompassing the auxin-binding site of maize ABP1. The results of these experiments also suggest that activation of the plasma membrane H$^+$-ATPase is involved in generation of the observed current across the plasma membrane. In summary, these electrophysiological experiments suggest that ABP1, or antigenically related proteins, are present on the external surface of the plasma membrane and represent a site for auxin perception.

Localization studies using monoclonal antibodies against ABP1 yielded different results to those obtained with the polyclonal sera. Labeling studies carried out with a monoclonal antibody, MAC 256, generated against purified maize ABP1 (Napier *et al.,* 1988), showed only cytoplasmic localization in maize root cells (Napier *et al.,* 1992); this monoclonal antibody recognizes an epitope located at the C-terminal of ABP1 that includes the KDEL sequence (Napier *et al.,* 1992). The same monoclonal antibody did not label the surface of maize coleoptile protoplasts (Diekmann *et al.,* 1995).

The apparent discrepancy between the results of immunolocalization studies using polyclonal versus monoclonal antibodies can be explained if the C-terminal domain of ABP1 is sterically inaccessible when ABP1 is on the plasma membrane, but not when it is in the endoplasmic reticulum. Indeed, the 'docking' of ABP1 with another protein in the plasma membrane has been suggested to explain how a peripheral membrane protein such as ABP1 can transduce the auxin signal to the cell interior (Barbier-Brygoo *et al.,* 1991; Barbier-Brygoo, 1995; Cross, 1991; Klämbt, 1990). Perhaps the docking of ABP1 with the putative 'docking protein', thereby covering the KDEL retention signal, results in the secretion of the docked ABP1 to the plasma membrane; current evidence suggests that ABP1 at the plasma membrane is always 'docked'. Such a hypothesis would predict that the accumulation or rate of synthesis of the 'docking protein' is the rate-limiting step that governs how much ABP1 ends up on the surface of plant cells.

Experimental evidence for the involvement of the C-terminal domain of ABP1 in signal transduction has been reported. Binding of auxin by ABP1 results in a conformational change in the protein, as indicated by an altered circular dichroism spectrum (Shimomura *et al.,* 1986) and the inability of monoclonal antibody MAC 256 to bind liganded ABP1 (Napier and Venis, 1990). The latter result suggests that the auxin-induced conformational change involves the C-terminal domain of ABP1. More significantly, a synthetic dodecapeptide encompassing the C-terminal domain of ABP1 is able to induce auxin-related K$^+$ channel activities in *Vicia faba* stomatal guard cells in the absence of auxin (Thiel *et al.,* 1993). Five other peptides, including one spanning the presumptive auxin-binding site, are not active. Thus, the C-terminal dodecapeptide acts as a molecular mimic of auxin. This dodecapeptide does not appear to interact directly with the K$^+$ channel protein, given the similar charges of the peptide and the mouth of the K$^+$ channel as well as the dependence of the activity of the peptide (and auxin) on changes in cytosolic pH (Thiel *et al.,* 1993). A radiolabeled derivative of this peptide might be a useful probe with which to identify the hitherto elusive 'docking protein' for ABP1 that is postulated to play a central role in auxin signal transduction.

The available experimental evidence strongly suggests that ABP1 is, in fact, an auxin receptor. The most convincing data in support of this conclusion have been obtained from electrophysiological studies on protoplasts. The physiological relevance of this experimental system has been questioned (Hertel, 1995) in the light of the fact that protoplasts have a membrane potential close to zero, in contrast to intact cells whose membrane potential is of the order of –120 mV. Thus, protoplasts are depolarized cells. None the less, it remains clear that the interaction of auxin with ABP1 at the plant cell surface can be tied to rapid electrophysiological responses from those cells. Whether or not the observed electrophysiological responses can, in fact, be causally linked to an auxin-inducible signaling pathway leading to various physiological responses (e.g. cell elongation, mitosis and gene expression) of tissues treated with auxin remains an open question. Thus, while the evidence that ABP1 is a physiological auxin receptor is strong, it is not yet definitive with regard to the definition of a receptor that we have chosen. Experiments seeking to address this question further, such as the study of transgenic plants with altered levels of ABP1 expression, have been reported to be in progress (Hesse and Palme, 1994), but the results of those studies have yet to be reported in detail. The isolation of auxin-insensitive mutants defective in ABP1 would also significantly strengthen the argument that ABP1 is an auxin receptor. Until the experimental evidence becomes more definitive, it is prudent to continue to refer to ABP1 as an 'auxin-binding protein' rather than an 'auxin receptor'.

Abscisic acid-binding proteins

Abscisic acid (structure shown in *Figure 1*) plays a role in several important plant processes (Walton, 1980), including the induction of dormancy in seeds and buds, leaf and fruit abscission, and reduction of drought stress by regulation of the opening and closing of stomata. The specificity of the plant's responses to abscisic acid as determined using structurally related compounds suggests that the carboxylic acid, the presence of a double bond in the α- or β-position in the cyclohexane ring, and a *cis* configuration of the C-2 double bond are the structural features that are essential for biological activity (Van der Meulen *et al.*, 1993). These structural features would also be expected to figure prominently in defining the binding specificity of physiological abscisic acid receptors.

Abscisic acid contains an α,β-unsaturated ketone functionality that can be photoactivated by ultraviolet (UV) light (Bayley and Knowles, 1977), thereby providing a means of photo-affinity labeling abscisic acid-binding protein(s) *in situ*. Using this approach, three polypeptides (14, 19 and 20 kDa) were labeled in guard cell protoplasts of *Vicia faba* (Hornberg and Weiler, 1984). A direct correlation was observed between the abilities of various structurally related compounds to displace the labeled abscisic acid from the protoplast membranes and the abilities of these compounds to inhibit stomatal opening in epidermal strips, thus providing strong evidence for the physiological relevance of the observed binding. Unfortunately, this early study remains unconfirmed, and no further progress toward purification and characterization of abscisic acid-binding proteins has been reported.

Recent papers substantiate the earlier hypothesis of a plasma membrane-localized site for abscisic acid-binding protein(s) by comparing the effects of abscisic acid when applied to the exterior of responsive cells and when injected into those cells. Thus abscisic acid injected into guard cells has no effect on stomatal opening, while externally applied abscisic acid inhibits stomatal opening (Anderson *et al.*, 1994). Similarly, externally applied abscisic acid is able to reverse gibberellic acid-induced synthesis and secretion of α-amylase in barley aleurone protoplasts, while injected abscisic acid has no effect (Gilroy and Jones, 1994). These recent results highlight the importance of identifying and characterizing the plasma membrane-localized binding site(s) for abscisic acid in order to understand how abscisic acid exerts its physiological effects. The generation of monoclonal antibodies against the cell surface of guard cell protoplasts (Knox *et al.*, 1995), some of which inhibit abscisic acid-inducible gene expression (Wang *et al.*, 1995), may provide new tools with which to elucidate abscisic acid perception and signal transduction.

Gibberellin-binding proteins

Gibberellins play significant roles in plant growth and development. Examples of processes in which gibberellins have been shown to be involved include elongation growth of stems (Stoddart, 1987), fruit and seed development (Barendse *et al.*, 1990), and endosperm mobilization in cereals (Fincher, 1989). Aleurone cells have proved to be a particularly useful model system for the study of hormonal regulation of the synthesis and secretion of hydrolytic enzymes involved in the mobilization of endosperm reserves during seed germination (Fincher, 1989; Jones and Jacobsen, 1991). Gibberellins stimulate the synthesis and secretion of the hydrolases (principally α-amylase), an effect that is reversed by abscisic acid. A large number of gibberellins (> 100) have been isolated, structurally characterized and their biological activities determined. These studies have allowed conclusions to be drawn about the structural features of gibberellins (see *Figure 1* for the structure of one gibberellin) that are important for maximum biological activity. These include an intact gibberellane ring system, a carboxyl group at C-6 in the B-ring, and two functionalities in the A-ring, a hydroxyl group at C-3 and a lactone (Hooley *et al.*, 1990). The binding specificity of any physiological gibberellin receptor should be defined by these same structural features.

Experimental investigations using the cereal aleurone have been instrumental in providing evidence that gibberellin perception occurs at the plasma membrane. In one series of experiments, a cell-impermeant derivative of gibberellin A_4 was prepared and shown to induce α-amylase mRNA in *Avena fatua* aleurone protoplasts (Hooley *et al.*, 1991). The hormone had been covalently coupled to Sepharose beads via a spacer arm linked to C-17 of the D-ring of gibberellin A_4. Chemical modification of the gibberellin at this position yields a derivative that still retains its biological activity, albeit reduced by 100-fold compared with the native compound (Hooley *et al.*, 1991). Control experiments demonstrated that intact aleurone cell layers are not responsive to the impermeant gibberellin under conditions where responsiveness to the free hormone was retained; the impermeant gibberellin was unable to pass through the cell wall in intact tissues. Furthermore, the induction of α-amylase gene expression in aleurone protoplasts was not attributable to the release of gibberellin from the beads during the course of the experiments. In another set of experiments, externally applied gibberellin was shown to induce synthesis and secretion of α-amylase in barley aleurone protoplasts, while gibberellin injected into the protoplasts had no effect (Gilroy and Jones, 1994). Taken together, these two studies provide strong evidence that a specific binding site for gibberellins exists in the plasma membrane, although the results do not exclude the possibility of the existence of cell internal binding proteins, particularly in other tissues (Allan and Trewavas, 1994; Hooley, 1994).

Identification of a gibberellin-binding protein in the plasma membrane has proved difficult. Attempts to utilize ligand-binding assays to identify gibberellin-binding proteins in aleurone have been unsuccessful (Hooley *et al.*, 1992). Two other approaches, photo-affinity labeling and anti-idiotypic antibodies, have yielded promising preliminary results. Several photoreactive derivatives of gibberellins have been prepared and shown to retain their biological activity in aleurone (Beale *et al.*, 1992). Radiolabeled photoreactive gibberellin A_4 labels a monoclonal antibody (MAC 182) that specifically recognizes biologically active gibberellins, and whose specificity thus matches that expected of the putative plant gibberellin-binding protein (Walker *et al.*, 1992). Photoreaction of the derivatized gibberellin A_4 with aleurone tissue results in the labeling of numerous polypeptides (Hooley *et al.*, 1993). Labeling of one 60-kDa polypeptide was competitively inhibited by unmodified gibberellin A_4, but not by biologically inactive gibberellins. Subcellular fractionation of the aleurone layers revealed that the photo-tagged 60-kDa polypeptide is recovered in the membrane fraction, although precisely which subcellular membrane is involved remains unknown.

Anti-idiotypic antibodies can act as molecular mimics of a ligand and recognize the ligand-binding site on a receptor. Such antibodies have been used successfully to isolate several mammalian receptors (Linthicum and Farid, 1988). The monoclonal antibody, MAC 182, which specifically recognizes biologically active gibberellins, has been used to generate a polyclonal antiserum that appears to contain anti-idiotypic antibodies against a gibberellin-binding protein (Hooley *et al.*, 1992). Immunoglobin fractions prepared from this antiserum inhibit binding of radiolabeled gibberellin A_4 to MAC 182, and also inhibit gibberellin A_4 induction of α-amylase in aleurone protoplasts. These immunoglobins are also able to agglutinate aleurone protoplasts. Thus, the anti-idiotypic antiserum appears to contain antibody antagonists of gibberellin action and to recognize an epitope on the surface of gibberellin-responsive cells. It remains to be seen whether or not these antibodies can be used successfully to identify or perhaps purify the plasma membrane-localized gibberellin-binding proteins from aleurone cells.

3. Oligosaccharin-binding proteins

Oligosaccharins, defined as 'oligosaccharides with biological regulatory properties' (Albersheim *et al.*, 1983), are a recently identified class of signal molecules in plants. To date, seven oligosaccharins have been purified to homogeneity and their structures characterized (for reviews see Côté and Hahn, 1994; Darvill *et al.*, 1992), although evidence for additional biologically active oligosaccharides has been presented (for review see Aldington and Fry, 1993). The characterized oligosaccharins include oligosaccharide fragments of fungal (hepta-β-glucoside, oligochitin and oligochitosan elicitors) and plant (oligogalacturonides, xyloglucan nonasaccharide) cell wall polysaccharides, glycopeptide elicitors derived from fungal and plant glycoproteins, and lipo-oligosaccharides (nod-factors) synthesized by bacterial symbionts of plants.

Improvements in oligosaccharide purification techniques and contributions from synthetic organic chemists have now made available sufficient quantities of homogeneous oligosaccharins to permit detailed studies on their binding proteins in plants. Recent progress toward the identification and characterization of binding proteins for four oligosaccharins (structures shown in *Figure 2*) is summarized in the following sections.

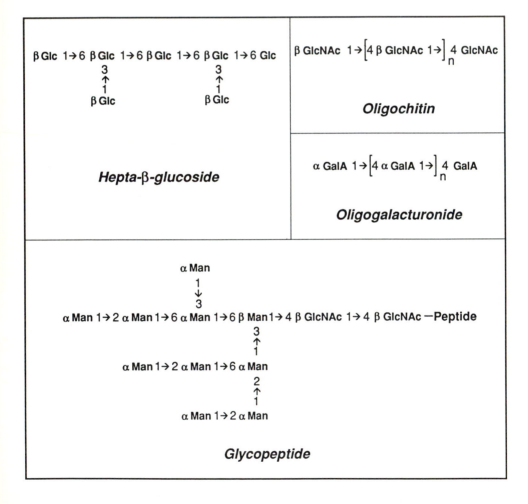

Figure 2. Structures of four oligosaccharins for which there appear to be plant plasma membrane-localized binding sites. The hepta-β-glucoside and oligochitins are fragments of mycelial wall polysaccharides, oligogalacturonides are fragments of a plant cell wall polysaccharide, and the glycopeptide is derived from a fungal glycoprotein. In order to be biologically active, n = 2 or 4 for oligochitins (depending on the plant) and n = 8–10 for oligogalacturonides.

Hepta-β-glucoside elicitor-binding proteins

Glucans with the ability to induce phytoalexin accumulation in soybean tissues were first detected in the culture filtrates of *Phytophthora sojae*, a phytopathogenic oomycete (Ayers *et al.*, 1976a), and were later also purified from commercially available yeast extract (Hahn and Albersheim, 1978). These elicitors were shown to be composed of 3-, 6-, and 3,6-linked β-glucosyl residues (Ayers *et al.*, 1976a; Hahn and Albersheim, 1978), a composition very similar to that of glucans which are major constituents of various mycelial walls (Bartnicki-Garcia, 1968). Similar elicitor-active glucans can be released from the mycelial walls of *P. sojae* by either partial chemical (Ayers *et al.*, 1976b, c; Sharp *et al.*, 1984c) or enzymatic (Ham *et al.*, 1991; Keen and Yoshikawa, 1983; Okinaka *et al.*, 1995; Yoshikawa *et al.*, 1981) hydrolysis. The smallest elicitor-active oligoglucoside purified from the mixture of oligoglucosides generated by partial acid hydrolysis of mycelial walls is a branched hepta-β-glucoside, the structure of which is shown in *Figure 2* (Sharp *et al.*, 1984a, b, c). Structure–activity studies (Cheong *et al.*, 1991, 1993) using synthetic oligosaccharides that are structurally related to the active hepta-β-glucoside have demonstrated that the three non-reducing terminal glucosyl residues are essential for the activity of the elicitor, as is the distribution of the side-chains along the backbone of the molecule. In contrast, the reducing terminal glucosyl residue of the hepta-β-glucoside elicitor does not appear to be essential for activity (Cheong *et al.*, 1991), and it provides a suitable site for radiolabeling of the elicitor for use in experiments designed to identify and characterize glucan elicitor-binding proteins.

Several studies utilizing heterogeneous mixtures of mycelial glucan fragments have indicated that binding sites for glucan fragments exist in plant membranes (Cosio *et al.*, 1988; Peters *et al.*, 1978; Schmidt and Ebel, 1987; Yoshikawa *et al.*, 1983). In particular, binding studies carried out with partially purified acid-released (Cosio *et al.*, 1988; Schmidt and Ebel, 1987) or enzyme-released (Yoshikawa and Sugimoto, 1993) elicitor-active glucans from *P. sojae* mycelial walls demonstrated the presence of high-affinity glucan-binding sites on soybean root plasma membranes (Schmidt and Ebel, 1987) and protoplasts prepared from suspension-cultured soybean cells (Cosio *et al.*, 1988). The glucan preparations utilized in these binding studies were not homogeneous, raising the question of whether the observed binding was specific for the biologically active oligoglucosides.

Subsequent research utilizing homogeneous hepta-β-glucoside elicitor coupled to radio-iodinated aminophenethylamine (Cosio *et al.*, 1990b) or tyramine (Cheong and Hahn, 1991) as the labeled ligand has substantiated the hypothesis that membrane-localized glucan elicitor binding sites exist in soybean cells. The hepta-β-glucoside elicitor binding sites are present in membranes prepared from every major organ of young soybean plants (Cheong and Hahn, 1991). The hepta-β-glucoside elicitor binding sites co-migrate with an enzyme marker (vanadate-sensitive H^+-ATPase) for plasma membranes in isopycnic sucrose density gradients (Cheong *et al.*, 1993), confirming earlier results obtained with partially purified labeled elicitor-active glucan fragments (Schmidt and Ebel, 1987). Binding of the radiolabeled hepta-β-glucoside elicitor to the root membranes is saturable over a concentration range of 0.1 to 5 nM, which is somewhat lower than the range of concentrations (6 to 200 nM) required to saturate the bioassay for phytoalexin accumulation (Cheong *et al.*, 1991; Sharp *et al.*, 1984c). The root membranes possess a single class of high-affinity hepta-β-glucoside binding sites (apparent K_d of *c*.1 nM). These binding sites are inactivated by heat or pronase treatment (Cheong and Hahn, 1991), suggesting that the molecule(s) responsible for the binding are proteinaceous. Binding of the active hepta-β-glucoside to the membrane preparation is reversible, indicating that the elicitor does not become covalently attached to the binding protein(s) (Cheong and Hahn, 1991; Frey *et al.*, 1993). The membrane-localized elicitor-binding proteins exhibit a high degree of specificity with regard to the oligoglucosides that they bind. More importantly, the ability of an oligoglucoside to bind to soybean root membranes correlates with its ability to induce phytoalexin accumulation in soybean tissues (Cheong and Hahn, 1991; Cheong *et al.*, 1993). It is this correlation between biological activity and binding affinity that provides the strongest evidence to date that the binding proteins are physiological receptors for the hepta-β-glucoside elicitor.

Solubilization of glucan elicitor-binding proteins from soybean root membranes has been achieved with the aid of several detergents (Cheong *et al.*, 1993; Cosio *et al.*, 1990a). The nonionic detergent, *n*-dodecanoylsucrose, and the zwitterionic detergent, *N*-dodecyl-*N*,*N*-dimethyl-3-ammonio-1-propane-sulfonate (ZW 3-12) are the principal detergents that have been used, and each solubilizes between 40 and 60% of the elicitor-binding activity from soybean membranes. The detergent-solubilized glucan elicitor-binding proteins appear to be fully functional as defined by retention of their affinity and specificity for the hepta-β-glucoside elicitor. Thus the solubilized binding proteins have an apparent K_d of between 1 and 3 nM (Cheong *et al.*, 1993; Cosio *et al.*, 1992; Frey *et al.*, 1993), and retain the specificity for elicitor-active oligoglucosides characteristic of the membrane-localized proteins (Cheong *et al.*, 1993).

Recently, progress has been made towards the identification and purification of hepta-β-glucoside elicitor-binding proteins. Size-exclusion chromatography of detergent-solubilized hepta-β-glucoside elicitor-binding proteins indicates that elicitor-binding activity is primarily associated with large detergent–protein micelles (M_r > 200 000; Cheong *et al.*, 1993; Cosio *et al.*, 1990a), with detergent–protein micelles having a value of M_r > 400 000 exhibiting the highest specific elicitor-binding activity. Little or no elicitor-binding activity is associated with the smallest detergent–protein micelles (Cheong *et al.*, 1993). Attempts to reduce the size of the large elicitor-binding protein–detergent complexes by altering the solubilization conditions and/or switching to detergents which have higher critical micelle concentrations and smaller micelle sizes have been unsuccessful (F. Côté and M.G. Hahn, unpublished results). These data suggest that the elicitor-binding proteins exist as a multimeric protein complex. Indeed, the fraction of solubilized membrane proteins that is retained on affinity columns carrying either an immobilized mixture of elicitor-active fungal glucans (Cosio *et al.*, 1992; Frey *et al.*, 1993) or immobilized hepta-β-glucoside elicitor (F. Côté and M.G. Hahn, unpublished results) and subsequently eluted with free ligand contains several proteins as visualized on polyacrylamide gels. Photo-affinity labeling experiments carried out on solubilized soybean membrane preparations suggest that three of these polypeptides (*c.* 70, 100 and 170 kDa) carry elicitor-binding domains, of which the 70-kDa species is predominant (Cosio *et al.*, 1992). The affinity-purified hepta-β-glucoside binding proteins retain the same specificity for elicitor-active oligoglucosides that was observed for the membrane-localized and crude detergent-solubilized proteins (F. Côté and M.G. Hahn, unpublished results). Further purification of these proteins and assays of the polypeptides, either alone or in combination, for elicitor-binding activity will be necessary in order to determine whether all of the proteins present in the affinity-purified fraction are essential for elicitor-binding activity. Proof that the hepta-β-glucoside elicitor-binding proteins are functional receptors will require reconstitution of a hepta-β-glucoside elicitor-responsive system using either purified binding proteins or transgenically expressed genes encoding those proteins.

Oligochitin elicitor-binding proteins

Chitin is a major polysaccharide constituent of mycelial walls (Bartnicki-Garcia, 1968), including those of many phytopathogens. Major plant defense responses that can be induced by chitin include lignification in wheat leaves (Pearce and Ride, 1982) and suspension-cultured cells of slash pine (Lesney, 1989), and increased activity of an enzyme in the putative biosynthetic pathway of phytoalexins in rice (Ren and West, 1992). More recently, oligosaccharide fragments of chitin have also been shown to induce defense responses in plant cells (Barber *et al.*, 1989; Koga *et al.*, 1992; Roby *et al.*, 1987; Yamada *et al.*, 1993). Structure–activity studies using purified chitin oligomers indicate that oligomers with a degree of polymerization (DP) of less than four are not active, while oligomers with a DP of six or more are active (Barber *et al.*, 1989; Koga *et al.*, 1992; Ren and West, 1992; Roby *et al.*, 1987; Yamada *et al.*, 1993). Furthermore, the de-*N*-acetylated (chitosan) oligomers are not active in those plants in which the chitin oligomers are active. The availability of purified, biologically active and inactive oligomers of chitin has made possible studies to identify binding sites for these fragments in plant cells.

Several studies have reported results which suggest that perception of chitin fragments occurs at the cell surface of responsive cells. Thus, the elicitation of phytoalexin production in suspension-cultured rice cells by chitin oligomers of DP $\geq$ 6 is accompanied by a rapid (occurring within 1 min) and transient membrane depolarization (Kuchitsu *et al.*, 1993). The induction of reactive oxygen species in suspension-cultured rice cells by chitin oligomers with a DP of 7 and 8 has also been reported (Kuchitsu *et al.*, 1995). Chitin oligomers with a DP $\geq$ 4 also stimulate alkalinization of the medium of suspension-cultured tomato cells (Felix *et al.*, 1993), and this alkalinization is accompanied by changes in the pattern of protein phosphorylation. Biologically inactive, shorter chitin oligomers, as well as chitosan oligomers, do not induce these rapid cellular responses.

The rapid responses of plant cells to oligochitins have yet to be directly linked to a signal transduction pathway leading to activation of a plant defense response. A significant step towards establishing such a link would be the isolation and characterization of specific oligochitin-binding proteins. Studies demonstrating the presence of such binding proteins in the plasma membrane of responsive plant cells were reported recently. Radiolabeled chitin oligomers have been prepared either by reductive amination with tyramine followed by radio-iodination (Shibuya *et al.*, 1993), or by derivatization with *t*-butoxycarbonyl-L-[^{35}S]methionine (Baureithel *et al.*, 1994). Both methods result in incorporation of the radiolabel at the reducing end of the oligosaccharide. These radiolabeled ligands were used to demonstrate the presence of a single class of high-affinity binding sites for biologically active chitin oligomers in membranes prepared from suspension-cultured cells of rice (apparent K_d = 5 nM) (Shibuya *et al.*, 1993) and tomato (apparent K_d = 23 nM) (Baureithel *et al.*, 1994). High-affinity binding (apparent K_d = 1.4 nM) of the radiolabeled pentamer to intact tomato cells was also observed (Baureithel *et al.*, 1994), suggesting that the binding site is localized to the plasma membrane. The observed binding to tomato membranes and cells is reversible (Baureithel *et al.*, 1994). Binding of radiolabeled chitin octamer to rice membranes is abolished by pre-treatment of the membranes with protease (Shibuya *et al.*, 1993), suggesting that the binding site is proteinaceous.

The specificity of these membrane-localized oligochitin-binding sites has been investigated using oligochitins of various degrees of polymerization and oligomers of chitosan, the de-*N*-acetylated form of chitin. In tomato membranes, oligochitins with a DP of 4 and 5 are effective competitors of the radiolabeled pentamer, while smaller chitin oligomers are significantly less effective (Baureithel *et al.*, 1994). Chitosan oligomers are unable to compete for the oligochitin-binding sites in tomato and rice membranes, even at concentrations as high as 100 μM (Baureithel *et al.*, 1994; Shibuya *et al.*, 1993). Thus there is a direct correlation between the ability of chitin oligosaccharides to induce biological responses in treated plant cells and the ability of those oligosaccharides to bind to the membrane-localized binding sites, providing evidence that the oligochitin-binding proteins are physiological receptors. The successful solubilization of chitin-binding proteins from membranes of suspension-cultured tomato cells, with retention of binding affinity and specificity, has recently been reported (Baureithel and

Boller, 1995). It will be of interest to determine the identity and nature of the oligochitin-binding protein(s), and how they connect with the rest of the signal transduction pathway to induce the physiological defense response(s) observed in affected cells.

Glycopeptide elicitor-binding proteins

A variety of glycoproteins have been identified as having biological activity in plants, particularly with regard to the induction of plant defense responses (for review see Anderson, 1989). In some cases, these glycoproteins are enzymes and their ability to induce plant defense responses is dependent on their enzymatic activity (e.g. endopolygalacturonase; Stekoll and West, 1978). In other cases, it has been shown that glycopeptides and/or free N-linked oligosaccharides originating from glycoproteins are also active (Basse *et al.*, 1992; Kogel *et al.*, 1988). These latter glycopeptide elicitors have been the subject of studies designed to identify specific binding proteins for these elicitors in plants.

A glycoprotein elicitor has been purified from cell walls of *Puccinia graminis* f. sp. *tritici* (Beissmann and Reisener, 1990; Kogel *et al.*, 1988), and from apoplastic fluids of infected wheat leaves (Beissmann *et al.*, 1992). The activity of the glycoprotein elicitor is sensitive to periodate oxidation, which suggests that the carbohydrate portion of the molecule is responsible for its activity. It is not yet known which of the oligosaccharide side-chains of the *P. graminis* glycoprotein elicitor, the N-linked high-mannose glycan or the O-linked oligogalactoside, is active. The intact glycoprotein elicitor has been labeled, and high-affinity binding (apparent K_d = 2 μM) of the labeled elicitor to wheat plasma membranes was demonstrated (Kogel *et al.*, 1991). Interestingly, binding of labeled glycoprotein elicitor to two membrane proteins (M_r 30 000 and 34 000) was observed after separation by SDS–PAGE and blotting to nitrocellulose. Binding of the labeled elicitor to a protein fraction that did not enter the resolving gel was also observed, although that binding could not be displaced with unlabeled elicitor. Detailed studies of the specificity of the glycoprotein elicitor-binding site in wheat have not yet been reported.

Another set of glycopeptide elicitors that have been studied in greater detail than the *P. graminis* glycoprotein discussed above are the high mannose-containing glycopeptide elicitors derived from yeast glycoproteins. The yeast glycopeptide elicitors induce ethylene biosynthesis and phenylalanine ammonia-lyase, the first enzyme in the phenylpropanoid pathway, in suspension-cultured tomato cells (IC_{50} of *c.* 5–10 nM) (Basse and Boller, 1992; Basse *et al.*, 1992; Felix *et al.*, 1991; Grosskopf *et al.*, 1991). Structure–activity studies have demonstrated that both the glycan and the peptide portions of the elicitors are important for activity. The most active glycopeptides are those that contain 10–12 mannosyl residues (see *Figure 2*), and a particular α-(1→6)-linked mannosyl residue is essential for maximal activity; high mannose N-linked glycans of this type have not been identified in plant glycoproteins. In contrast, the sequence of the peptide portion does not appear to be important for activity. However, removal of the peptide leads to the loss of elicitor activity, suggesting that the peptide is interacting with the binding proteins in a fairly non-specific manner. In fact, it may be possible to replace the peptide portion altogether with another functional group without altering the activity of the glycopeptide; such a finding has not yet been reported. The free glycans released from the yeast glycopeptides have been shown to suppress the action of the ethylene-inducing glycopeptide elicitors in tomato cells (Basse and Boller, 1992; Basse *et al.*, 1992), that is, the free glycans inhibit the ability of the glycopeptide elicitors to induce plant defense responses.

The most active of the glycopeptides released from yeast invertase was radiolabeled by derivatization with t-butoxycarbonyl-L-[³⁵S]methionine at the amino terminus of the peptide portion of the elicitor (Basse *et al.*, 1993). A single class of reversible, high-affinity binding sites (apparent K_d = 3.3 nM) for the radiolabeled glycopeptide was detected in membranes prepared from suspension-cultured tomato cells (Felix *et al.,* 1993). Saturable binding of the ligand to whole cells (apparent K_d = 0.7 nM) was also observed, suggesting that the glycopeptide elicitor-binding site is located in the plasma membrane. Whether or not the binding site is proteinaceous in nature has not yet been determined. The site concentration is about two orders of magnitude lower (B_{max} = 19 fmol mg^{-1} of membrane protein) than was observed for the hepta-β-glucoside- and oligochitin-binding sites discussed in the previous sections. The glycopeptide elicitor-binding sites have been solubilized from membranes of suspension-cultured tomato cells using nonionic detergents, and they appear to retain their affinity and specificity for the glycopeptide elicitors (Fath and Boller, 1995). Partial purification of the solubilized glycopeptide elicitor-binding sites by ion-exchange chromatography has also been reported (Fath and Boller, 1995).

The specificity of the membrane-localized binding site has been examined using glycopeptides carrying different glycan structures, and with the free glycans (Basse *et al.,* 1993). A direct correlation was observed between the ability of various glycopeptides to compete with the labeled ligand for the binding sites and the ability of the glycopeptides to induce ethylene biosynthesis in tomato cells. A radiolabeled peptide obtained by deglycosylation of the most active glycopeptide elicitor did not bind to the membranes, providing evidence that the peptide portion of the glycopeptide elicitors does not contribute significantly to binding of the elicitor. This conclusion is further substantiated by the observation that the ability of free glycans to compete for the binding site is identical to that of the glycopeptides from which the glycans had been released. Thus, it appears that the glycopeptide elicitors and the glycan suppressors interact with the same binding site in the membranes. This is the first example among the oligosaccharins of an antagonist (free glycan) competing for the same binding site as the agonist (glycopeptide), and presumably thereby preventing the biological response to the elicitor.

Oligogalacturonide-binding proteins

Oligogalacturonides, linear α-(1→4)-linked oligomers of galacturonosyl residues, have been shown to have a variety of activities in plant cells (for review see Darvill *et al.,* 1992). These include important plant defense responses, such as phytoalexin accumulation, induction of glucanases and chitinases, and accumulation of proteinase inhibitors. In addition, roles for oligogalacturonides in normal plant growth and development have been suggested. Structure–activity studies have demonstrated a clear size dependence of the biological activities of oligogalacturonides. In most cases, oligomers shorter than a decamer are not active, while oligomers with a DP of 12 to 14 are the most active.

The results of several studies suggest that oligogalacturonide perception occurs at the plasma membrane of responsive cells. For example, oligogalacturonide mixtures rapidly (within 5 min) depolarize the membranes of tomato leaf mesophyll cells, albeit at relatively high concentrations (1 mg m^{-1}) (Thain *et al.,* 1990). Lower concentrations of size-specific oligogalacturonides (DP of 12 to 15) induce, within 5 min, a transient stimulation of K$^+$ efflux, alkalinization of the extracellular medium, depolarization of the plasma membrane, and a decrease in the external Ca^{2+} concentration in suspension-cultured tobacco cells (Mathieu *et al.,* 1991). These effects are specific for oligogalacturonides; size-heterogeneous mixtures of oligomannuronides and oligoguluronides are approximately 400-fold less effective. Finally, treatment of suspension-cultured soybean cells with heterogeneous oligogalacturonide fractions results in the rapid production of hydrogen peroxide (oxidative burst) (Apostol *et al.,* 1989; Legendre *et al.,* 1993).

The rapid responses at the cell surface described above have not yet been causally linked to any of the known plant responses induced by oligogalacturonides. One critical gap in our knowledge is the absence of information about the molecules responsible for initial recognition of the active oligogalacturonides. Recent improvements in purification techniques (Spiro *et al.,*

1993) allow the preparation of homogeneous samples of biologically active and inactive oligogalacturonides that can be used in experiments to identify and characterize specific binding sites for these oligosaccharins in plants.

Derivatives of biologically active oligogalacturonides have been prepared that would be suitable for the preparation of radiolabeled ligands for use in ligand-binding studies. However, their synthesis is more complicated than that of the other oligosaccharins described in previous sections, due to the occurrence of side-reactions involving the carboxylic acid groups in the oligogalacturonides. Homogeneous tyraminylated and biotinylated derivatives of the tri-decagalacturonide have been prepared (Spiro et al., 1996) and have been shown to retain at least some of their biological activities (M. Spiro and B. Ridley, personal communication). The preparation of a tyrosine hydrazone of the dodecagalacturonide has also been described (Horn et al., 1989), although the purity and structure of the derivative have not been documented. Saturable binding of the radio-iodinated form of the latter derivative to intact soybean cells has been reported (Low et al., 1993), and evidence suggesting uptake of the bound oligogalac-turonide via receptor-mediated endocytosis has been presented (Horn et al., 1989). Most recently, a fluorescein derivative of a tetradecagalacturonide-enriched fraction was prepared, and binding of this derivative to the surface of soybean protoplasts was visualized using silver-enhanced immunogold epipolarization microscopy (Diekmann et al., 1994). However, in no case has the specificity of the observed binding of the labeled oligogalacturonides been demonstrated, raising questions about the significance of these findings with regard to oligogalacturonide signal transduction. These preliminary studies clearly indicate that the molecular tools are now available to pursue the identification and characterization of oligogalacturonide-binding proteins.

4. (Poly)peptide elicitor-binding proteins

Polypeptides and peptides are a prominent group of signal molecules in animal systems. Only recently have polypeptides that are biologically active in plants been purified and characterized. These include the following: the elicitins from *Phytophthora* spp., which can induce systemic leaf necrosis in tobacco (Ricci et al., 1989); harpin, a bacterial polypeptide that induces a hypersensitive necrosis in plants (He et al., 1993); systemin, an 18-amino-acid peptide that induces the systemic accumulation of proteinase inhibitors in tomato and potato (Pearce et al., 1991) and is derived from a larger plant precursor protein (McGurl et al., 1992); a fungal glycoprotein elicitor of phytoalexin accumulation in parsley (Parker et al., 1991), whose activity resides in a non-glycosylated 13-amino-acid peptide (Nürnberger et al., 1994), and AUR9, a 28-amino-acid peptide that induces a hypersensitive neurosis in tomato (Scholtens-Toma and de Wit, 1988) and is derived from a larger fungal precursor protein (Van Kal et al., 1991). The availability of the purified (poly)peptides and information about which amino acids in the molecules are essential for activity (He et al., 1993; Nürnberger et al., 1994; Pearce et al., 1993) have allowed initial studies directed toward the identification of specific binding proteins (candidate receptors) for two of these elicitors in plants.

P. sojae *peptide elicitor-binding proteins*

A proteinaceous elicitor of phytoalexin accumulation in parsley (*Petroselinum crispum*) was first identified as the active component of an elicitor preparation generated from the mycelial walls of *Phytophthora sojae* (Parker et al., 1988). This elicitor was purified to apparent homogeneity from the culture filtrate of the oomycete, and was shown to be a 42-kDa glycoprotein (Parker et al., 1991). The activity of the elicitor resides in the polypeptide portion of the glycoprotein, since it is sensitive to protease digestion and insensitive to deglycosylation (Parker et al., 1988, 1991).

A cDNA encoding the glycoprotein elicitor has been cloned from *P. sojae*, and the amino acid sequences required for elicitor activity were shown to be located in the C-terminal third of the protein (Sacks *et al.*, 1995). Subsequently, a 13-amino-acid oligopeptide with the sequence H_2N-**VWNQPVRGFKVYE**-COOH was shown to be both necessary and adequate for the elicitor activity of the glycoprotein (Nürnberger *et al.*, 1994). Alanine substitution scanning identified two amino acid residues, Trp2 and Pro5, as being essential for the elicitor activity of the oligopeptide.

The availability of the purified glycoprotein elicitor made possible studies directed towards the identification and characterization of specific binding proteins for this elicitor in parsley cells. In initial studies, binding of the *P. sojae* elicitor to the surface of parsley protoplasts was visualized using an antibody against the glycoprotein and silver-enhanced immunogold epipolarization microscopy (Diekmann *et al.*, 1994). Binding of the glycoprotein elicitor to protoplasts prepared from broadbean (*Vicia faba*) leaves and from suspension-cultured cells of soybean was not observed. However, the high level of nonspecific binding and the absence of detailed specificity studies raise questions about whether the observed binding can be attributed to specific elicitor-binding proteins.

Direct binding studies have been carried out using the 13-amino-acid elicitor-active peptide derived from the *P. sojae* glycoprotein elicitor (Nürnberger *et al.*, 1994). The peptide was labeled by radio-iodination of a tyrosine residue in the peptide without any detectable loss in its ability to induce phytoalexin accumulation in parsley protoplasts. Using the labeled peptide, a single class of high-affinity binding sites was found in parsley microsomal membranes and on intact parsley protoplasts (Nürnberger *et al.*, 1994). Apparent K_d values of 2.4 and 11.4 nM were determined for the membranes and protoplasts, respectively. The binding sites are sensitive to heat denaturation and proteolysis, indicating the proteinaceous nature of the sites.

The specificity of the binding proteins for the *P. sojae* peptide elicitor was determined using several deletion and single-amino-acid substitution analogs of the peptide elicitor. The ability of the various peptide analogs to compete with the labeled peptide elicitor for the binding site correlated exactly with the ability of the analogs to induce any of several responses elicited in parsley by the peptide elicitor (Nürnberger *et al.*, 1994, 1995). These responses include the accumulation of coumarin phytoalexins, induction of ion fluxes, and activation of defense-related genes. The correlation between biological activity and binding affinity for the various peptide analogs provides strong evidence for the physiological significance of the peptide elicitor-binding proteins as receptors for this signal molecule. Chemical cross-linking of the radio-iodinated peptide elicitor to its binding site in total membranes and protoplasts prepared from suspension-cultured parsley cells results in the labeling of a 91-kDa polypeptide (Nürnberger *et al.*, 1995). Labeling of this polypeptide can be inhibited by unlabeled elicitor and closely related structural analogs in direct correlation with the elicitor activity of the oligopeptides. These results strongly suggest that the 91-kDa polypeptide is the receptor for the *P. sojae* elicitor, although experimental verification of this possibility awaits purification of the binding protein and/or cloning of its corresponding gene. The peptide elicitor-binding protein in parsley is the first example of a putative receptor for peptide signals in plants; this class of receptors will undoubtedly grow in the future.

Elicitin-binding proteins

Elicitins are a family of highly conserved small proteins that are produced by many *Phytophthora* species (for review see Yu, 1995). These protein elicitors induce a hypersensitive necrosis in tobacco (*Nicotiana tabacum*) leaves and induce systemic acquired resistance (for review see Enyedi *et al.*, 1992; Kessmann *et al.*, 1994) against fungal and bacterial pathogens in tobacco plants (Kamoun *et al.,* 1993; Ricci *et al.,* 1989). Cryptogein, an elicitin from *Phytophthora cryptogea*, is the best characterized of the elicitins, and has been shown to induce several biochemical responses in tobacco cells, including the accumulation of phytoalexins (Blein *et al.,* 1991; Milat *et al.,* 1991b), expression of defense-related genes (Suty *et al.,* 1995), and changes in ion fluxes (Blein *et al.,* 1991; Tavernier *et al.,* 1995b; Viard *et al.,* 1994), lipid composition (Tavernier *et al.,* 1995a) and protein phosphorylation (Viard *et al.,* 1994). In contrast to the glycoprotein elicitor from *Phytophthora sojae* discussed in the previous section, the elicitor activity of elicitins depends on the native conformation of the protein (Ricci *et al.,* 1989), and has not been associated with a specific amino acid sequence within the protein.

Cryptogein has been labeled on tyrosine with ^{125}I and then used to demonstrate the presence of binding sites on the surface of intact suspension-cultured tobacco cells (Blein *et al.,* 1991). Saturable, high-affinity (apparent K_d = 2 nM) binding of radio-iodinated cryptogein to plasma membranes prepared from tobacco leaves has recently been demonstrated (Wendehenne *et al.,* 1995). Binding of the labeled elicitin to the tobacco plasma membrane preparation is reversible. Bound labeled cryptogein can be displaced with unlabeled cryptogein and iodocryptogein, but the specificity of the binding was not examined in further detail (e.g. using other elicitins that have lower activity in tobacco). Binding of cryptogein to the plasma membranes is sensitive to NaCl, suggesting that ionic interactions play a role in the association of the elicitin with its binding site. The apparent affinity of the cryptogein-binding site correlates closely with the effective concentrations of the elicitin required to induce hypersensitive necrosis (Ricci *et al.,* 1989) and phytoalexin accumulation (Milat *et al.,* 1991a) in tobacco leaves, which suggests that the binding site is physiologically relevant. Research is now in progress to identify and purify the molecular species responsible for the elicitin-binding activity.

5. Fusicoccin-binding proteins

A number of phytopathogenic micro-organisms secrete toxins that appear to play a major role in the disease process. These toxins are generally divided into nonspecific (Ballio, 1991; Rudolph, 1976) and host-selective (Scheffer and Livingston, 1984; Walton and Panaccione, 1993) types. The nonspecific toxins, such as fusicoccin, affect a wide variety of plant species, while the host-selective toxins, such as BmT toxin and victorin C, only affect susceptible varieties of a particular plant. These microbial toxins, as their name implies, are damaging to plant cells and tissues, although their site of action varies. Research over the last 20 years has resulted in the identification of specific binding proteins for BmT toxin in plant chloroplasts (Levings and Siedow, 1992) and for victorin C in plant mitochondria (Wolpert *et al.,* 1994). In contrast, the site of action of fusicoccin is clearly the plasma membrane, and will be the focus of our discussion here.

Fusicoccin is the major toxin produced in culture by *Fusicoccum amygdali* Del., a phytopathogen of peach and almond trees (Ballio *et al.,* 1964) that induces wilting in infected trees. This diterpene glucoside, whose structure (Ballio *et al.,* 1968; Barrow *et al.,* 1971) is shown in *Figure 3*, induces most of the symptoms associated with disease when applied to plants; the leaf wilt, in particular, is caused by fusicoccin-induced opening of the stomata (Turner and Graniti, 1969). In addition to the leaf wilt, a variety of other physiological effects are observed in plant tissues treated with fusicoccin (for review see Marrè, 1979), including stimulation of various ion fluxes, induction of cell expansion, and promotion of seed germination. Various structural analogs and derivatives have been isolated and used in structure–activity studies (Ballio *et al.,* 1981a, b, 1991) to determine the essential structural features of the toxin. These include the conformation of the carbotricyclic aglycone, an unsubstituted hydroxyl group at C-8 of the eight-membered ring, and the configuration of the hydroxyl at C-9, which is involved in

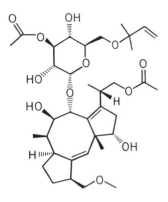

Figure 3. Structure of the fungal toxin, fusicoccin.

the glycosidic linkage. The glucosyl residue is important for maximum activity, but the aglycone is also active, albeit c. 200-fold less so than the native toxin. The 6'-t-pentenyl moiety on the glucosyl residue can be modified without causing major losses in activity, thus providing a potential site for incorporation of a radiolabel into the molecule.

Radiolabeled fusicoccin has been prepared by one of two methods: catalytic hydrogenation of the pentenyl group using tritium gas (Dohrmann *et al.*, 1977), or oxidation of the pentenyl to an aldehyde using osmium tetroxide/periodate followed by reduction with borotritide (Feyerabend and Weiler, 1987). Both derivatives are essentially fully active (Dohrmann *et al.*, 1977; Feyerabend and Weiler, 1988). Tritiated fusicoccin has been used to demonstrate the presence of high-affinity, specific binding sites in membranes prepared from various plants, including maize (Ballio *et al.*, 1981b; Dohrmann *et al.*, 1977; Pesci *et al.*, 1979a), spinach (Ballio *et al.*, 1980), radish (De Michelis *et al.*, 1989), oat (de Boer *et al.*, 1989), *Arabidopsis thaliana* (Holländer-Czytko and Weiler, 1994; Meyer *et al.*, 1989; Stout, 1988), *Vicia faba* (Blum *et al.*, 1988; Feyerabend and Weiler, 1988), *Corydalis sempervirens* (Schulz *et al.*, 1991) and *Commilina communis* (Oecking and Weiler, 1991). The broad distribution of fusicoccin-binding proteins is consistent with the nature of fusicoccin as a nonspecific toxin. In most studies, ligand saturation experiments suggested the presence of two classes of binding sites, namely a high-affinity site with a K_d of c. 1 nM and a low-affinity site with a K_d value 10- to 100-fold higher, although in a few studies only a single class of sites was detected (Abramycheva *et al.*, 1991; De Michelis *et al.*, 1989; Holländer-Czytko and Weiler, 1994; Oecking and Weiler, 1991; Stout, 1988). This apparent discrepancy has not yet been resolved, although differing tissues and/or extraction methods have been suggested as possible reasons. The fusicoccin-binding sites show reversible binding to fusicoccin and are inactivated by proteases, the latter observation suggesting that toxin binding to proteins occurs. The fusicoccin-binding proteins are also specific for fusicoccin and its biologically active structural analogs. Indeed, there is a direct correlation between the ability of a compound to mimic the activity of fusicoccin in several bioassays and its ability to compete with radiolabeled fusicoccin for binding to maize membranes (Ballio *et al.*, 1981a, b). These data support the view that the fusicoccin-binding proteins are physiological receptors for the toxin.

Many of the physiological effects of fusicoccin center around its ability to stimulate ion fluxes, in particular proton efflux, across the plasma membrane (Marrè, 1979). These observations suggested that the toxin exerts its physiological effects via stimulation of the plasma membrane-localized H^+-ATPase. Experimental evidence supporting this hypothesis was provided by the demonstration that fusicoccin is able to stimulate H^+-ATPase in isolated membrane vesicles (Blum *et al.*, 1988; De Michelis *et al.*, 1989; Johansson *et al.*, 1993; Rasi-Caldogno and Pugliarello, 1985; Rasi-Caldogno *et al.*, 1986). Furthermore, fractionation of membranes either on sucrose density gradients (Beffagna *et al.,* 1979; De Michelis *et al.,* 1989; Meyer *et al.,* 1989) or by aqueous two-phase partitioning (Blum *et al.,* 1988; Feyerabend and Weiler, 1988; Johansson *et al.,* 1993; Meyer *et al.,* 1989) demonstrated that H$^+$-ATPase activity and fusicoccin-binding activity always co-fractionate. However, the H$^+$-ATPase is not a fusicoccin-binding protein, since the two activities can be separated after solubilization of the membranes (Stout and Cleland, 1980). The ability of trypsin to inactivate fusicoccin-binding proteins on right-side-out plasma membrane vesicles prepared by aqueous two-phase partitioning (de Boer *et al.,* 1989; Feyerabend and Weiler, 1988) suggests that the binding site faces the apoplast. These results confirm earlier data which demonstrated the agglutination of protoplasts by a cell-impermeant derivative of the toxin consisting of fusicoccin coupled to bovine serum albumin (Aducci *et al.,* 1980).

Solubilization of the fusicoccin-binding proteins from membranes of several plants has been achieved using detergents (Aducci *et al.,* 1984, 1989, 1993b; de Boer *et al.,* 1989; Korthout *et al.,* 1994; Meyer *et al.,* 1989; Oecking and Weiler, 1991; Pesci *et al.,* 1979b; Stout and Cleland, 1980). The nonionic detergents Triton X-100, octylglucoside and nonanoyl-*N*-methylglucamide have been the principal detergents used, with the latter two solubilizing 70 to 90% of the fusicoccin-binding activity from the membranes (de Boer *et al.,* 1989; Feyerabend and Weiler, 1989; Korthout *et al.,* 1994; Meyer *et al.,* 1989; Oecking and Weiler, 1991). An alternative approach, using a cold acetone extraction of the membranes, has also proved successful (Aducci *et al.,* 1984, 1989, 1993b). The solubilized fusicoccin-binding proteins were found to be unstable, principally due to the presence of degradative enzymes (Aducci *et al.,* 1984). The inclusion of glycosidase, phosphatase and proteinase inhibitors in extraction buffers is deemed to be essential for binding-protein stability. Several workers also reported enhanced stability of the fusicoccin-binding proteins if solubilization was carried out in the presence of the toxin ligand (Feyerabend and Weiler, 1989; Korthout *et al.,* 1994; Meyer *et al.,* 1989). Initial characterization of the solubilized fusicoccin-binding proteins suggested that they are phosphorylated glycoproteins (Aducci *et al.,* 1984; Meyer *et al.,* 1989). The behavior of the solubilized fusicoccin-binding proteins after acetone extraction of membranes (Aducci *et al.,* 1989) and during temperature-induced phase partitioning in Triton X-114 (Korthout *et al.,* 1994) suggests that the binding proteins are membrane-associated rather than integral membrane proteins, although a heavily glycosylated integral membrane protein could show a similar partitioning pattern. On the basis of the results of size-exclusion chromatography, a value of M_r in the range 60 000 to 100 000 was estimated for the solubilized fusicoccin-binding proteins (Feyerabend and Weiler, 1989; Pesci *et al.,* 1979; Stout and Cleland, 1980). These results must be interpreted cautiously, however, given that the elution buffers for the chromatography contained detergents at concentrations above the critical micelle concentration; the elution position of membrane proteins is governed by the size of the detergent–lipid–protein micelle under these conditions.

Purification of fusicoccin-binding proteins has been achieved recently using various chromatographic techniques, including traditional (ion-exchange, hydrophobic interaction, size-exclusion) (Aducci *et al.,* 1993b; Feyerabend and Weiler, 1989), ligand-affinity (de Boer *et al.,* 1989) and/or metal-chelate affinity methods (Oecking and Weiler, 1991). A novel variant of ligand-affinity chromatography has recently been applied to the purification of fusicoccin-binding proteins from oat (Korthout *et al.,* 1994); this involved allowing biotinylated fusicoccin to bind to the membrane-localized sites, followed by solubilization and affinity chromatography on an avidin column, after which the fusicoccin-binding proteins were eluted with free biotin. This approach results in improved stability of the binding proteins during membrane solubilization, exploits the tight association between the toxin and its binding protein, and allows the elution of active binding protein from the affinity matrix. The latter was not possible using traditional ligand-affinity chromatography (de Boer *et al.*, 1989). The purified fusicoccin-binding proteins from oat (de Boer *et al.*, 1989; Korthout *et al.*, 1994), *Vicia faba* (Feyerabend and Weiler, 1989) and *Commelina communis* (Oecking and Weiler, 1991) appear as a doublet (29–30 kDa and 31–35 kDa) on SDS–PAGE. Photo-affinity labeling studies carried out on plasma membrane vesicles yielded results that were remarkably consistent with the above-mentioned purification studies: one or two bands of M_r 32 000–35 000 were labeled (Feyerabend and Weiler, 1989; Meyer *et al.*, 1989; Oecking and Weiler, 1991). The single exception to this pattern is the fusicoccin-binding protein purified from acetone-extracted membranes of maize (Aducci *et al.*, 1993b); two protein doublets were observed using SDS–PAGE, one at M_r c. 90 000 and one at M_r c. 32 000. Photo-affinity labeling experiments resulted in the labeling of one polypeptide in each doublet. However, the intensity of labeling of the polypeptides was strongly dependent on the duration of the irradiation: short durations favored labeling of the 90-kDa polypeptide whereas longer durations favored labeling of the 32-kDa band. A monospecific antiserum against the 32-kDa doublet does not recognize the 90-kDa polypeptides, suggesting that the two sets of polypeptides are unrelated (Marra *et al.*, 1994).

Partial amino acid sequences have recently been obtained for the 30-kDa fusicoccin-binding proteins from oat (Korthout and de Boer, 1994), *Commelina communis* (Oecking *et al.*, 1994) and maize (Marra *et al.*, 1994). Comparisons of these partial amino acid sequences with sequences in databases revealed extensive sequence similarity of the fusicoccin-binding proteins with a class of regulatory proteins known as '14-3-3 proteins'. In addition, antibodies generated against known 14-3-3 proteins cross-react with the 30-kDa fusicoccin-binding proteins (Korthout and de Boer, 1994; Oecking *et al.*, 1994). 14-3-3 proteins have been found in a broad range of organisms, including mammals, plants, insects, nematodes and yeast, and they share a high degree of sequence similarity (Aitken *et al.*, 1992). The family of 14-3-3 proteins as a whole are soluble, cytoplasmic proteins that appear to play a variety of roles in cells, principally as regulators of enzymes involved in growth, development and signal transduction (Aitken *et al.*, 1992; Aitken, 1995). The fusicoccin-binding proteins, being membrane-localized and ligand-binding proteins, are unique among the 14-3-3 proteins. It is not currently known whether any of the regulatory mechanisms used by 14-3-3 proteins in other organisms (e.g. regulation of protein kinase C, regulation of Raf protein kinase, regulation of phospholipase A_2) are utilized by the fusicoccin-binding proteins in order to bring about the physiological responses of plant cells after exposure to the toxin.

Recent work suggests that activation of the plasma membrane H+-ATPase by liganded fusicoccin-binding proteins occurs via regulation of the ATPase activity (Johansson *et al.*, 1993; Rasi-Caldogno *et al.*, 1993). Infiltration of spinach leaves with fusicoccin prior to plasma membrane isolation results in alterations in the kinetic properties (increased V_{max} and decreased K_m) of the H+-ATPase (Johansson *et al.*, 1993); similar changes were observed after treatment of *Corydalis sempervirens* membranes with fusicoccin *in vitro* (Schulz *et al.*, 1991). Limited proteolysis of the H+-ATPase, which causes the loss of the C-terminal domain (c. 10 kDa), results in similar changes in the kinetic properties of the ATPase (Johansson *et al.*, 1993; Palmgren *et al.*, 1990, 1991; Rasi-Caldogno *et al.*, 1993). However, fusicoccin does not induce proteolytic cleavage of the ATPase (Johansson *et al.*, 1993; Rasi-Caldogno *et al.*, 1993). Thus, the liganded fusicoccin-binding protein, by interacting with the H+-ATPase, may induce a conformational change in the enzyme that reduces or abolishes the inhibitory effect of its C-terminal domain. A similar proposal had been made earlier to explain the effects of various treatments, including limited proteolysis and altered lipid environment, on the plant plasma membrane H+-ATPases (Palmgren *et al.*, 1991). It has yet to be determined whether the fusicoccin-binding proteins, after toxin binding, directly activate the ATPase or whether other proteins serve as intermediaries.

Examples of such additional intermediary proteins include phospholipase A_2 and heterotrimeric GTP-binding proteins, both of which have been implicated in fusicoccin action (Aducci et al., 1993a; de Boer et al., 1994).

Reconstitution experiments carried out in liposomes have been used in attempts to understand how binding of fusicoccin to its binding protein results in the activation of the plasma membrane H^+-ATPase. Crude solubilized preparations of maize and spinach fusicoccin-binding proteins have been inserted into liposomes, with the retention of ligand affinity and specificity (Aducci et al., 1989). Insertion into liposomes of the crude solubilized fusicoccin-binding proteins, together with a partially purified preparation of the plasma membrane H^+-ATPase, yielded proteoliposomes that transported protons in a fusicoccin-dependent manner (Aducci et al., 1988). Very similar results were obtained when a purified fungal H^+-ATPase (from *Neurospora crassa*) and crude or partially purified preparations of solubilized maize fusicoccin-binding proteins were incorporated into liposomes (Marra et al., 1995). These experiments, carried out with non-purified proteins, clearly support the earlier hypothesis that fusicoccin binding leads to an activation of the plasma membrane H^+-ATPase, and they now need to be followed up with reconstitution studies using purified components. The recent isolation of mutant plants with altered sensitivity to fusicoccin (Gomarasca et al., 1993; Holländer-Czytko and Weiler, 1994) provides additional tools and experimental approaches to demonstrate receptor function for the purified fusicoccin-binding proteins and to elucidate how fusicoccin, after binding to its receptor, activates the plasma membrane H^+-ATPase, and hence brings about its diverse physiological effects in plant cells.

6. Conclusions

The last few years have seen a significant increase in the study of signal-specific binding proteins in plants. This has been brought about in part by the identification and characterization of an increasing diversity of signal molecules that are active in plants. The increased awareness of the central importance of signal transduction pathways in controlling and regulating normal plant growth and development, as well as plant responses to the environment and to other organisms, has also been instrumental in stimulating research on plant receptors. Plant receptor research has made significant progress in identifying and in some cases characterizing ligand-specific binding proteins in plants. Thus, candidate receptors that are plasma membrane-localized have been isolated in the form of specific binding proteins for auxin and fusicoccin. Purification of binding proteins for the hepta-β-glucoside and *P. sojae* oligopeptide elicitors also appears to be within reach. However, we have not yet reached the point where it can be stated that a plant receptor has been identified, although the evidence in support of ABP1 as an auxin receptor is quite strong.

In most cases, the study of plant receptors lags significantly behind equivalent studies of animal receptors. In animals, numerous receptors for various molecular signals have been purified and characterized (for examples, see Drayer and van Haastert, 1994; Hahn, 1989). Studies of these receptors have revealed several important concepts with regard to receptors that are or will be applicable to plant receptors as well. Almost all animal receptors characterized to date can be grouped into families that utilize common mechanisms for signal transmission. Examples include ligand-regulated transcription factors (e.g. steroid receptors; Parker, 1993; Sherman and Stevens, 1984), protein tyrosine kinases (e.g. epidermal growth factor receptor; Adamson, 1993; Iwashita and Kobayashi, 1992), G-protein coupled receptors (e.g. adrenergic receptors; Kobilka, 1992), and the immunoglobin superfamily (e.g. CD2 T-cell adhesion molecules; Springer, 1990; Williams and Barclay, 1988). Closer examination of these superfamilies has revealed that structurally related receptor types differ in the specific ligands that are bound (e.g. insulin, epidermal growth factor and platelet-derived growth factor are ligands for different members of the receptor tyrosine kinase superfamily), while related receptor sub-types differ in the binding properties of agonists and antagonists, and in the nature of the cellular response to binding of the ligand to the receptor. Thus, a given ligand can have different effects on cells depending on the receptor sub-type present in those cells and the signal transduction pathway coupled to those receptors.

Molecular analysis of animal receptors has shown that these proteins consist of linear arrays of functional domains (e.g. ligand-binding domain, protein kinase domain, transmembrane domain, DNA-binding domain). These domains can be viewed as functional cassettes that have been brought together over evolutionary time in order to assemble a functionally useful molecule. Thus, it would be surprising if plant receptors did not show similar structural patterns. Indeed, some of the better characterized ligand-binding proteins (putative receptors) in plants have structural architectures very similar to those of receptors found in other systems, although plants may assemble their receptors using different cassettes than those utilized by other organisms. For example, the putative receptor kinases in plants (Stone and Walker, 1995) utilize a serine/threonine kinase domain in place of the tyrosine kinase domain common in animals (Fantl et al., 1993; Ullrich and Schlessinger, 1990). The putative ethylene receptor in plants, ETR1, has some structural features analogous to the sensor and receiver domains observed in two-component regulatory systems that are widespread in prokaryotes, but ETR1 lacks the output domain (Chang et al., 1993). Thus, molecular characterization of plant receptors may bring to light already familiar architectural and functional themes, although perhaps with different nuances. In addition, the discovery of new structural and functional variants appears likely, given the sessile and phototrophic life-style of plants, the very different organization of plants at the tissue level, and the totipotency of many plant cells (Palme et al., 1991). These unique aspects of plants as experimental organisms, together with the current lack of knowledge of plant signal transduction pathways, should provide ample incentive for further research on signal perception and transmission in plants.

Research on plant receptors, particularly those that are membrane-localized, would benefit from the application and development of additional experimental tools and approaches. One experimental approach that is beginning to bear fruit is the isolation of mutant plants that are unresponsive to a particular signal (Bowler and Chua, 1994). Examples of such mutants are those that are insensitive to particular wavelengths of light (Chory, 1991), or to the hormones ethylene (Kieber and Ecker, 1993) and auxin (Barbier-Brygoo, 1995; Garbers and Simmons, 1994). Characterization of such mutants has the potential to identify not only the ligand-binding proteins, but also other proteins involved in signal transmission and in the regulation of the signal transduction pathway. Thus, the challenge will be to identify, at the biochemical level, the nature of the genetic lesion in the mutant plants. The isolation and biochemical characterization of signal transduction mutants also requires a tractable genetic system for selecting such mutants. This may prove to be problematical for the signaling pathways triggered by some of the signal molecules discussed here (e.g. oligosaccharins).

There is also a need to develop or identify plant systems in which receptor functionality of a binding protein can be tested experimentally. Research on animal receptors has benefited greatly from the ability to achieve functional heterologous expression of ligand-binding protein genes in Xenopus oocytes and mammalian cells, and to confer ligand responsiveness on the transformed cells. Not only can one study the mode of action of binding proteins in response to ligand in such heterologous systems, but it is also possible to use such heterologous systems to clone genes encoding the binding proteins by screening expression libraries (Allen and Seed, 1988; Aruffo and Seed, 1987; Frommer and Ninnemann, 1995; Stengelin et al., 1988). This would allow cloning of binding protein genes without the need to purify the proteins themselves. Thus, it would be useful to identify plant cells that lack signal perception mechanisms for one or more of the molecular signals that have been identified in plants, yet which retain the rest of the signal transduction pathway. Expression of a binding protein for a particular ligand in such a cell would then render the transfected cell responsive to the ligand of interest. Transgenic expression studies will also require more detailed information about the processing and targeting of proteins in both homologous (i.e. plant proteins in plant cells) and heterologous (e.g. plant proteins in animal cells) systems (Bednarek and Raikhel, 1992; Gal and Raikhel, 1993) in order to ensure the functional expression of transgenic binding proteins, particularly those that are plasma membrane-localized. At the same time, additional information is needed about the identity and function of the products of genes that are activated in response to a particular signal (e.g. auxin; Abel *et al.*, 1994; Ballas *et al.*, 1993; Oeller *et al.*, 1993) in order to assess more accurately the activation of a signal transduction pathway in a heterologous expression system.

Acknowledgments

I am grateful to numerous colleagues for making articles available prior to publication. Research in my laboratory on hepta-β-glucoside elicitor-binding proteins is supported by a grant from the United States National Science Foundation (MCB-9206882), and in part by the United States Department of Energy-funded Center for Plant and Microbial Complex Carbohydrates (DE-FG09-93ER20097). The assistance of François Côté and Carol Gubbins Hahn with the preparation of the figures, and of Karen Howard with the preparation of the typescript, is gratefully acknowledged. I also thank many colleagues, both within and outside my research group, for stimulating discussions concerning signal transduction that have helped to sharpen and clarify my thinking about this exciting research area.

References

Abel S, Oeller PW, Theologis A. (1994) Early auxin-induced genes encode short-lived nuclear proteins. *Proc. Natl Acad. Sci. USA* **91:** 326–330.

Abramycheva NY, Babakov AV, Bilushi SV, Danilina EE, Shevchenko VP. (1991) Comparison of the biological activity of fusicoccin in higher plants with its binding to plasma membranes. *Planta* **183:** 315–320.

Adamson ED. (1993) Growth factors and their receptors in development. *Dev. Genet.* **14:** 159–164.

Aducci P, Federico R, Ballio A. (1980) Interaction of a high molecular weight derivative of fusicoccin with plant membranes. *Phytopath. Medit.* **19:** 187–188.

Aducci P, Ballio A, Fiorucci L, Simonetti E. (1984) Inactivation of solubilized fusicoccin-binding sites by endogenous plant hydrolases. *Planta* **160:** 422–427.

Aducci P, Ballio A, Blein J-P, Fullone MR, Rossignol M, Scalla R. (1988) Functional reconstitution of a proton-translocating system responsive to fusicoccin. *Proc. Natl Acad. Sci. USA* **85:** 7849–7851.

Aducci P, Fullone MR, Ballio A. (1989) Properties of proteoliposomes containing fusicoccin receptors from maize. *Plant Physiol.* **91:** 1402–1406.

Aducci P, Ballio A, Donini V, Fogliano V, Fullone MR, Marra M. (1993a) Phospholipase A$_2$ affects the activity of fusicoccin receptors. *FEBS Lett.* **320:** 173–176.

Aducci P, Ballio A, Fogliano V, Fullone MR, Marra M, Proietti N. (1993b) Purification and photoaffinity labeling of fusicoccin receptors from maize. *Eur. J. Biochem.* **214:** 339–345.

Ahmad M, Cashmore AR. (1993) *HY4* gene of *A. thaliana* encodes a protein with characteristics of a blue-light photoreceptor. *Nature* **366:** 162–166.

Aitken A. (1995) 14-3-3 proteins on the MAP. *Trends Biochem. Sci.* **20:** 95–97.

Aitken A, Collinge DB, Van Heusden BPH, Isobe T, Roseboom PH, Rosenfeld G, Soll J. (1992) 14-3-3 proteins: a highly conserved, widespread family of eukaryotic proteins. *Trends Biochem. Sci.* **17:** 498–501.

Albersheim P, Darvill AG, McNeil M *et al.* (1983) Oligosaccharins: naturally occurring carbohydrates with biological regulatory functions. In: *Structure and Function of Plant Genomes* (Ciferri O, Dure L III, ed.). New York: Plenum Publishing Corp., pp. 293–312.

Aldington S, Fry SC. (1993) Oligosaccharins. *Adv. Bot. Res.* **19:** 1–101.

Allan AC, Trewavas AJ. (1994) Abscisic acid and gibberellin perception: inside or out? *Plant Physiol.* **104:** 107–108.

Allen JM, Seed B. (1988) Isolation and expression of functional high-affinity Fc receptor complementary DNAs. *Science* **234:** 378–381.

Anderson AJ. (1989) The biology of glycoproteins as elicitors. In: *Plant–Microbe Interactions. Molecular and Genetic Perspectives* (Kosuge T, Nester E, ed.). New York: McGraw Hill Inc., pp. 87–130.

Anderson BE, Ward JM, Schroeder JI. (1994) Evidence for an extracellular reception site for abscisic acid in *Commelina* guard cells. *Plant Physiol.* **104:** 1177–1183.

Apostol I, Heinstein PF, Low PS. (1989) Rapid stimulation of an oxidative burst during elicitation of cultured plant cells. Role in defense and signal transduction. *Plant Physiol.* **90:** 109–116.

Aruffo A, Seed B. (1987) Molecular cloning of a CD28 cDNA by a high-efficiency COS cell expression system. *Proc. Natl Acad. Sci. USA* **84:** 8573–8577.

Ayers AR, Ebel J, Finelli F, Berger N, Albersheim P. (1976a) Host–pathogen interactions. IX. Quantitative assays of elicitor activity and characterization of the elicitor present in the extracellular medium of cultures of *Phytophthora megasperma* var. *sojae*. *Plant Physiol.* **57:** 751–759.

Ayers AR, Ebel J, Valent B, Albersheim P. (1976b) Host–pathogen interactions. X. Fractionation and biological activity of an elicitor isolated from the mycelial walls of *Phytophthora megasperma* var. *sojae*. *Plant Physiol.* **57:** 760–765.

Ayers AR, Valent B, Ebel J, Albersheim P. (1976c) Host–pathogen interactions. XI. Composition and structure of wall-released elicitor fractions. *Plant Physiol.* **57:** 766–774.

Ballas N, Wong L-M, Theologis A. (1993) Identification of the auxin-responsive element, *AuxRE*, in the primary indoleacetic acid-inducible gene, *PS-IAA4/5*, of pea (*Pisum sativum*). *J. Mol. Biol.* **233:** 580–596.

Ballio A. (1991) Non-host-selective fungal phytotoxins: biochemical aspects of their mode of action. *Experientia* **47:** 783–790.

Ballio A, Chain EB, De Leo P, Erlanger BF, Mauri M, Tonolo A. (1964) Fusicoccin: a new wilting toxin produced by *Fusicoccum amygdali* Del. *Nature* **203:** 297.

Ballio A, Brufani M, Casinovi CG, Cerrini S, Fedeli W, Pellicciari R, Santurbano B, Vaciago A. (1968) The structure of fusicoccin A. *Experientia* **24:** 631–635.

Ballio A, Federico R, Pessi A, Scalorbi D. (1980) Fusicoccin binding sites in subcellular preparations of spinach leaves. *Plant Sci. Lett.* **18:** 39–44.

Ballio A, De Michelis MI, Lado P, Randazzo G. (1981a) Fusicoccin structure–activity relationships: stimulation of growth by cell enlargement and promotion of seed germination. *Physiol. Plant.* **52:** 471–475.

Ballio A, Federico R, Scalorbi D. (1981b) Fusicoccin structure–activity relationships: *in vitro* binding to microsomal preparations of maize coleoptiles. *Physiol. Plant.* **52:** 476–481.

Ballio A, Castellano S, Cerrini S, Evidente A, Randazzo G, Segre AL. (1991) ^{1}H NMR conformational study of fusicoccin and related compounds: molecular conformation and biological activity. *Phytochemistry* **30:** 137–146.

Barber MS, Bertram RE, Ride JP. (1989) Chitin oligosaccharides elicit lignification in wounded wheat leaves. *Physiol. Mol. Plant Pathol.* **34:** 3–12.

Barbier-Brygoo H. (1995) Tracking auxin receptors using functional approaches. *Crit. Rev. Plant Sci.* **14:** 1–25.

Barbier-Brygoo H, Ephritikhine G, Klämbt D, Ghislain M, Guern J. (1989) Functional evidence for an auxin receptor at the plasmalemma of tobacco mesophyll protoplasts. *Proc. Natl Acad. Sci. USA* **86:** 891–895.

Barbier-Brygoo H, Ephritikhine G, Klämbt D, Maurel C, Palme K, Schell J, Guern J. (1991) Perception of the auxin signal at the plasma membrane of tobacco mesophyll protoplast. *Plant J.* **1:** 83–93.

Barendse GWM, Karssen CM, Koornneef M. (1990) Role of endogenous gibberellins during fruit and seed development. In: *Gibberellins* (Takahashi N, Phinney BO, MacMillan J, ed.). New York: Springer-Verlag, pp. 179–187.

Barrow KD, Barton DHR, Chain E, Ohnsorge UFW, Thomas R. (1971) Fusicoccin. Part II. The constitution of fusicoccin. *J. Chem. Soc. (C)* **1971:** 1265–1274.

Bartnicki-Garcia S. (1968) Cell wall chemistry, morphogenesis, and taxonomy of fungi. *Annu. Rev. Microbiol.* **22:** 87–108.

Basse CW, Boller T. (1992) Glycopeptide elicitors of stress responses in tomato cells. *N*-linked glycans are essential for activity but act as suppressors of the same activity when released from the glycopeptides. *Plant Physiol.* **98:** 1239–1247.

Basse CW, Bock K, Boller T. (1992) Elicitors and suppressors of the defense response in tomato cells. Purification and characterization of glycopeptide elicitors and glycan suppressors generated by enzymatic cleavage of yeast invertase. *J. Biol. Chem.* **267:** 10258–10265.

Basse CW, Fath A, Boller T. (1993) High affinity binding of a glycopeptide elicitor to tomato cells and microsomal membranes and displacement by specific glycan suppressors. *J. Biol. Chem.* **268:** 14724–14731.

Baureithel K, Boller T. (1995) Characterization and solubilization of a specific binding site for chitin fragments in suspension-cultured tomato cells and microsomal membranes. (Abstract No. B4-100). *J. Cell. Biochem.* Suppl. 19B: 151.

Baureithel K, Felix G, Boller T. (1994) Specific, high affinity binding of chitin fragments to tomato cells and membranes. Competitive inhibition of binding by derivatives of chitin fragments and a nod factor of *Rhizobium. J. Biol. Chem.* **269:** 17931–17938.

Bayley H, Knowles JR. (1977) Photoaffinity labeling. *Meth. Enzymol.* **46:** 69–114.

Beale MH, Hooley R, Smith SJ, Walker RP. (1992) Photoaffinity probes for gibberellin-binding proteins. *Phytochemistry* **31:** 1459–1464.

Bednarek SY, Raikhel NV. (1992) Intracellular trafficking of secretory proteins. *Plant Mol. Biol.* **20:** 133–150.

Beffagna N, Pesci P, Tognoli L, Marrè E. (1979) Distribution of fusicoccin bound *in vivo* among subcellular fractions from maize coleoptiles. *Plant Sci. Lett.* **15:** 323–330.

Beissmann B, Reisener HJ. (1990) Isolation and purity determination of a glycoprotein elicitor from wheat stem rust by medium-pressure liquid chromatography. *J. Chromatogr.* **521:** 187–197.

Beissmann B, Engels W, Kogel K, Marticke K-H, Reisener HJ. (1992) Elicitor-active glycoproteins in apoplastic fluids of stem-rust-infected wheat leaves. *Physiol. Mol. Plant Pathol.* **40:** 79–89.

Blein J-P, Milat M-L, Ricci P. (1991) Responses of cultured tobacco cells to cryptogein, a proteinaceous elicitor from *Phytophthora cryptogea.* Possible plasmalemma involvement. *Plant Physiol.* **95:** 486–491.

Blum W, Key G, Weiler EW. (1988) ATPase activity in plasmalemma-rich vesicles isolated by aqueous two-phase partitioning from *Vicia faba* mesophyll and epidermis: characterization and influence of abscisic acid and fusicoccin. *Physiol. Plant.* **72:** 279–287.

Bowler C, Chua N-H. (1994) Emerging themes of plant signal transduction. *Plant Cell* **6:** 1529–1541.

Brinegar C. (1994) Cytokinin binding proteins and receptors. In: *Cytokinins — Chemistry, Activity, and Function* (Mok DWS, Mok MC, ed.). Boca Raton: CRC Press, Inc., pp. 217–232.

Brown JC, Jones AM. (1994) Mapping the auxin-binding site of auxin-binding protein 1. *J. Biol. Chem.* **269:** 21136–21140.

Campos N, Feldwisch J, Zettl R, Boland W, Schell J, Palme K. (1991) Identification of auxin-binding proteins using an improved assay for photoaffinity labeling with 5-N$_3$-[7-^{3}H]-indole-3-acetic acid. *Tech. Metab. Res.* **3:** 69–75.

Campos N, Schell J, Palme K. (1994) *In vitro* uptake and processing of maize auxin-binding proteins by ER-derived microsomes. *Plant Cell Physiol.* **35:** 153–161.

Chang C, Bleecker AB, Kwok SF, Meyerowitz EM. (1992) Molecular cloning approach for a putative ethylene receptor gene in *Arabidopsis. Biochem. Soc. Trans.* **20:** 73–75.

Chang C, Kwok SF, Bleecker AB, Meyerowitz EM. (1993) *Arabidopsis* ethylene-response gene *ETR1:* similarity of product to two-component regulators. *Science* **262:** 539–544.

Chen Z, Ricigliano JW, Klessig DF. (1993a) Purification and characterization of a soluble salicylic acid-binding protein from tobacco. *Proc. Natl Acad. Sci. USA* **90:** 9533–9537.

Chen Z, Silva H, Klessig DF. (1993b) Active oxygen species in the induction of plant systemic acquired resistance by salicylic acid. *Science* **262:** 1883–1886.

Cheong J-J, Hahn MG. (1991) A specific, high-affinity binding site for the hepta-β-glucoside elicitor exists in soybean membranes. *Plant Cell* **3:** 137–147.

Cheong J-J, Birberg W, Fügedi P, Pilotti Å, Garegg PJ, Hong N, Ogawa T, Hahn MG. (1991) Structure–activity relationships of oligo-β-glucoside elicitors of phytoalexin accumulation in soybean. *Plant Cell* **3:** 127–136.

Cheong J-J, Alba R, Côté F, Enkerli J, Hahn MG. (1993) Solubilization of functional plasma membrane-localized hepta-β-glucoside elicitor binding proteins from soybean. *Plant Physiol.* **103:** 1173–1182.

Chory J. (1991) Light signals in leaf and chloroplast development: photoreceptors and downstream responses in search of a transduction pathway. *New Biol.* **3:** 538–548.

Cosio EG, Pöpperl H, Schmidt WE, Ebel J. (1988) High-affinity binding of fungal β-glucan fragments to soybean (*Glycine max* L.) microsomal fractions and protoplasts. *Eur. J. Biochem.* **175:** 309–315.

Cosio EG, Frey T, Ebel J. (1990a) Solubilization of soybean membrane binding sites for fungal β-glucans that elicit phytoalexin accumulation. *FEBS Lett.* **264:** 235–238.

Cosio EG, Frey T, Verduyn R, Van Boom J, Ebel J. (1990b) High-affinity binding of a synthetic heptaglucoside and fungal glucan phytoalexin elicitors to soybean membranes. *FEBS Lett.* **271:** 223–226.

Cosio EG, Frey T, Ebel J. (1992) Identification of a high-affinity binding protein for a hepta-β-glucoside phytoalexin elicitor in soybean. *Eur. J. Biochem.* **204:** 1115–1123.

Côté F, Hahn MG. (1994) Oligosaccharins: structures and signal transduction. *Plant Mol. Biol.* **26:** 1375–1411.

Cross JW. (1991) Cycling of auxin-binding protein through the plant cell: pathways in auxin signal transduction. *New Biol.* **3:** 813–819.

Cross JW, Briggs WR. (1978) Properties of a solubilized microsomal auxin-binding protein from coleoptiles and primary leaves of *Zea mays. Plant Physiol.* **62:** 152–157.

Darvill A, Augur C, Bergmann C *et al.* (1992) Oligosaccharins — oligosaccharides that regulate growth, development and defense responses in plants. *Glycobiology* **2:** 181–198.

Davies PJ (ed.). (1995) *Plant Hormones: Physiology, Biochemistry and Molecular Biology,* 2nd edn. Dordrecht, The Netherlands: Kluwer Academic Publishers.

De Boer AH, Watson BA, Cleland RE. (1989) Purification and identification of the fusicoccin binding protein from oat root plasma membrane. *Plant Physiol.* **89:** 250–259.

De Boer AH, van der Molen GW, Prins HBA, Korthout AAJ, van der Hoeven PCJ. (1994) Aluminium fluoride and magnesium, activators of heterotrimeric GTP-binding proteins, affect high-affinity binding of the fungal toxin fusicoccin to the fusicoccin-binding protein in oat root plasma membranes. *Eur. J. Biochem.* **219:** 1023–1029.

De Michelis MI, Pugliarello MC, Rasi-Caldogno F. (1989) Fusicoccin binding to its plasma membrane receptor and the activation of the plasma membrane H+-ATPase. I. Characteristics and intracellular localization of the fusicoccin receptor in microsomes from radish seedlings. *Plant Physiol.* **90:** 133–139.

Diekmann W, Herkt B, Low PS, Nürnberger T, Scheel D, Terschüren C, Robinson DG. (1994) Visualization of elicitor-binding loci at the plant cell surface. *Planta* **195:** 126–137.

Diekmann W, Venis MA, Robinson DG. (1995) Auxins induce clustering of the auxin-binding protein at the surface of maize coleoptile protoplasts. *Proc. Natl Acad. Sci. USA* **92:** 3425–3429.

Dohrmann U, Hertel R, Pesci P, Cocucci SM, Marrè E, Randazzo G, Ballio A. (1977) Localization of 'in vitro' binding of the fungal toxin fusicoccin to plasma-membrane-rich fractions from corn coleoptiles. *Plant Sci. Lett.* **9:** 291–299.

Dohrmann U, Hertel R, Kowalik H. (1978) Properties of auxin binding sites in different subcellular fractions from maize coleoptiles. *Planta* **140:** 97–106.

Drayer AL, Van Haastert PJM. (1994) Transmembrane signalling in eukaryotes: a comparison between higher and lower eukaryotes. *Plant Mol. Biol.* **26:** 1239–1270.

Edgerton MD, Tropsha A, Jones AM. (1994) Modelling the auxin-binding site of auxin-binding protein 1 of maize. *Phytochemistry* **35:** 1111–1123.

Enyedi AJ, Yalpani N, Silverman P, Raskin I. (1992) Signal molecules in systemic plant resistance to pathogens and pests. *Cell* **70:** 879–886.

Ephritikhine G, Barbier-Brygoo H, Muller J-F, Guern J. (1987) Auxin effect on the transmembrane potential difference of wild-type and mutant tobacco protoplasts exhibiting a differential sensitivity to auxin. *Plant Physiol.* **83:** 801–804.

Fantl WJ, Johnson DE, Williams LT. (1993) Signalling by receptor tyrosine kinases. *Annu. Rev. Biochem.* **62:** 453–481.

Fath A, Boller T. (1995) Characterization of a binding site for a fungal glycopeptide elicitor solubilized from tomato microsomal membranes. (Abstract No. B4-106). *J. Cell. Biochem.* Suppl. 19B: 152.

Feldwisch J, Zettl R, Hesse F, Schell J, Palme K. (1992) An auxin-binding protein is localized to the plasma membrane of maize coleoptile cells: identification by photoaffinity labeling and purification of a 23-kDa polypeptide. *Proc. Natl Acad. Sci. USA* **89:** 475–479.

Felix G, Grosskopf DG, Regenass M, Basse CW, Boller T. (1991) Elicitor-induced ethylene biosynthesis in tomato cells. Characterization and use as a bioassay for elicitor action. *Plant Physiol.* **97:** 19–25.

Felix G, Regenass M, Boller T. (1993) Specific perception of subnanomolar concentrations of chitin fragments by tomato cells: induction of extracellular alkalinization, changes in protein phosphorylation, and establishment of a refractory state. *Plant J.* **4:** 307–316.

Feyerabend M, Weiler EW. (1987) Monoclonal antibodies against fusicoccin with binding characteristics similar to the putative fusicoccin receptor of higher plants. *Plant Physiol.* **85:** 835–840.

Feyerabend M, Weiler EW. (1988) Characterization and localization of fusicoccin-binding sites in leaf tissues of *Vicia faba* L. probed with a novel radioligand. *Planta* **174:** 115–122.

Feyerabend M, Weiler EW. (1989) Photoaffinity labeling and partial purification of the putative plant receptor for the fungal wilt-inducing toxin, fusicoccin. *Planta* **178:** 282–290.

Fincher GB. (1989) Molecular and cellular biology associated with endosperm mobilization in germinating cereal grains. *Annu. Rev. Plant Physiol. Plant Mol. Biol.* **40:** 305–346.

Frey T, Cosio EG, Ebel J. (1993) Affinity purification and characterization of a binding protein for a hepta-β-glucoside phytoalexin elicitor in soybean. *Phytochemistry* **32:** 543–550.

Frommer WB, Ninnemann O. (1995) Heterologous expression of genes in bacterial, fungal, animal, and plant cells. *Annu. Rev. Plant Physiol. Plant Mol. Biol.* **46:** 419–444.

Gal S, Raikhel NV. (1993) Protein sorting in the endomembrane system of plant cells. *Curr. Opin. Cell Biol.* **5:** 636–640.

Garbers C, Simmons C. (1994) Approaches to understanding auxin action. *Trends Cell Biol.* **4:** 245–250.

Gilroy S, Jones RL. (1994) Perception of gibberellin and abscisic acid at the external face of the plasma membrane of barley (*Hordeum vulgare* L.) aleurone protoplasts. *Plant Physiol.* **104:** 1185–1192.

Goldsmith MHM. (1993) Cellular signaling: new insights into the action of the plant growth hormone auxin. *Proc. Natl Acad. Sci. USA* **90:** 11442–11445.

Gomarasca S, Vannini C, Venegoni A, Talarico A, Marrè MT, Soave C. (1993) A mutant of *Arabidopsis thaliana* with a reduced response to fusicoccin. I. *Plant Physiol.* **103:** 165–170.

Grosskopf DG, Felix G, Boller T. (1991) A yeast-derived glycopeptide elicitor and chitosan or digitonin differentially induce ethylene biosynthesis, phenylalanine ammonia-lyase and callose formation in suspension-cultured tomato cells. *J. Plant Physiol.* **138:** 741–746.

Guern J. (1987) Regulation from within: the hormone dilemma. *Ann. Bot.* **60:** 75–102.

Hahn MG. (1989) Animal receptors — examples of cellular signal perception molecules. In: *Signal Molecules in Plants and Plant–Microbe Interactions* (Lugtenberg BJJ, ed.). Heidelberg: Springer Verlag, pp. 1–26.

Hahn MG, Albersheim P. (1978) Host–pathogen interactions. XIV. Isolation and partial characterization of an elicitor from yeast extract. *Plant Physiol.* **62:** 107–111.

Hall MA, Connern CPK, Harpham NVJ et al. (1990) Ethylene: receptors and action. In: *Hormone Perception and Signal Transduction in Animals and Plants* (Roberts J, Kirk C, Venis M, ed.). Cambridge: The Company of Biologists Ltd, pp. 87–110.

Ham K-S, Kauffmann S, Albersheim P, Darvill AG. (1991) Host–pathogen interactions. XXXIX. A soybean pathogenesis-related protein with β-1,3-glucanase activity releases phytoalexin elicitor-active heat-stable fragments from fungal walls. *Mol. Plant Microb. Interact.* **4:** 545–552.

He SY, Huang H-C, Collmer A. (1993) *Pseudomonas syringae* pv. *syringae* harpin$_{Pss}$: a protein that is secreted via the Hrp pathway and elicits the hypersensitive response in plants. *Cell* **73:** 1255–1266.

Henderson J, Atkinson AE, Lazarus CM, Hawes CR, Napier RM, Macdonald H, King LA. (1995) Stable expression of maize auxin-binding protein in insect cell lines. *FEBS Lett.* **371:** 293–296.

Hertel R. (1995) Auxin binding protein 1 is a red herring. *J. Exp. Bot.* **46:** 461–462.

Hertel R, Thomson K-S, Russo VE. (1972) *In-vitro* auxin binding to particulate cell fractions from corn coleoptiles. *Planta* **107:** 325–340.

Hesse T, Palme K. (1994) The search for phytohormone receptors: a role for auxin binding proteins. In: *Biochemical Mechanisms Involved in Growth Regulation* (Smith C, Gallon J, Chiatante D, Zocchi G, ed.). Oxford: Clarendon Press, pp. 103–110.

Hesse T, Feldwisch J, Balshüsemann D et al. (1989) Molecular cloning and structural analysis of a gene from *Zea mays* (L.) coding for a putative receptor for the plant hormone auxin. *EMBO J.* **8:** 2453–2461.

Hesse T, Garbers C, Brzobohaty B, Kreimer G, Söll D, Melkonian M, Schell J, Palme K. (1993) Two members of the *ERabp* gene family are expressed differentially in reproductive organs but to similar levels in the coleoptile of maize. *Plant Mol. Biol.* **23:** 57–66.

Hicks GR, Rayle DL, Jones AM, Lomax TL. (1989a) Specific photoaffinity labeling of two plasma membrane polypeptides with an azido auxin. *Proc. Natl Acad. Sci. USA* **86:** 4948–4952.

Hicks GR, Rayle DL, Lomax TL. (1989b) The *Diageotropica* mutant of tomato lacks high specific activity auxin binding sites. *Science* **245:** 52–54.

Hicks GR, Rice MS, Lomax TL. (1993) Characterization of auxin-binding proteins from zucchini plasma membrane. *Planta* **189:** 83–90.

Holländer-Czytko H, Weiler EW. (1994) Isolation and biochemical characterization of fusicoccin-insensitive variant lines of *Arabidopsis thaliana* (L.) Heynh. *Planta* **195:** 188–194.

Hooley R. (1994) Gibberellins: perception, transduction and responses. *Plant Mol. Biol.* **26:** 1529–1555.

Hooley R, Beale MH, Smith SJ. (1990) Gibberellin perception in the *Avena fatua* aleurone. In: *Hormone Perception and Signal Transduction in Animals and Plants* (Roberts J, Kirk C, Venis M, ed). Cambridge: The Company of Biologist Ltd, pp. 79–86.

Hooley R, Beale MH, Smith SJ. (1991) Gibberellin perception at the plasma membrane of *Avena fatua* aleurone protoplasts. *Planta* **183**: 274–280.

Hooley R, Beale MH, Smith SJ, Walker RP, Rushton PJ, Whitford PN, Lazarus CM. (1992) Gibberellin perception and the *Avena fatua* aleurone: do our molecular keys fit the correct locks? *Biochem. Soc. Trans.* **20**: 85–89.

Hooley R, Smith SJ, Beale MH, Walker RP. (1993) *In vivo* photoaffinity labelling of gibberellin-binding proteins in *Avena fatua* aleurone. *Aust. J. Plant Physiol.* **20**: 573–584.

Horn MA, Heinstein PF, Low PS. (1989) Receptor-mediated endocytosis in plant cells. *Plant Cell* **1**: 1003–1009.

Hornberg C, Weiler EW. (1984) High-affinity binding sites for abscisic acid on the plasmalemma of *Vicia faba* guard cells. *Nature* **310**: 321–324.

Hulme EC, Birdsall JM. (1992) Strategy and tactics in receptor binding studies. In: *Receptor–Ligand Interactions: A Practical Approach* (Hulme EC, ed.). Oxford: Oxford University Press, pp. 63–176.

Inohara N, Shimomura S, Fukui T, Futai M. (1989) Auxin-binding protein located in the endoplasmic reticulum of maize shoots: molecular cloning and complete primary structure. *Proc. Natl Acad. Sci. USA* **86**: 3564–3568.

Iwashita S, Kobayashi M. (1992) Signal transduction system for growth factor receptors associated with tyrosine kinase activity: epidermal growth factor receptor signalling and its regulation. *Cell. Signal.* **4**: 123–132.

Johansson F, Sommarin M, Larsson C. (1993) Fusicoccin activates the plasma membrane H$^+$-ATPase by a mechanism involving the C-terminal inhibitory domain. *Plant Cell* **5**: 321–327.

Jones AM. (1990) Do we have the auxin receptor yet? *Physiol. Plant.* **80**: 154–158.

Jones AM. (1994) Auxin-binding proteins. *Annu. Rev. Plant Physiol. Plant Mol. Biol.* **45**: 393–420.

Jones AM, Venis MA. (1989) Photoaffinity labeling of indole-3-acetic acid-binding proteins in maize. *Proc. Natl Acad. Sci. USA* **86**: 6153–6156.

Jones AM, Herman EM. (1993) KDEL-containing auxin-binding protein is secreted to the plasma membrane and cell wall. *Plant Physiol.* **101**: 595–606.

Jones AM, Melhado LL, Ho T-HD, Pearce CJ, Leonard NJ. (1984) Azido auxins: photoaffinity labeling of auxin binding proteins in maize coleoptile with tritiated 5-azidoindole-3-acetic acid. *Plant Physiol.* **75**: 1111–1116.

Jones AM, Cochran DS, Lamerson PM, Evans ML, Cohen JD. (1991) Red light-regulated growth. I. Changes in the abundance of indoleacetic acid and a 22-kilodalton auxin-binding protein in the maize mesocotyl. *Plant Physiol.* **97**: 352–358.

Jones RL, Jacobsen JV. (1991) Regulation of synthesis and transport of secreted proteins in cereal aleurone. *Int. Rev. Cytol.* **126**: 49–88.

Kamoun S, Young M, Glascock CB, Tyler BM. (1993) Extracellular protein elicitors from *Phytophthora*: host-specificity and induction of resistance to bacterial and fungal phytopathogens. *Mol. Plant Microb. Interact.* **6**: 15–25.

Keen NT, Yoshikawa M. (1983) β-1,3-endoglucanase from soybean releases elicitor-active carbohydrates from fungus cell walls. *Plant Physiol.* **71**: 460–465.

Kessmann H, Staub T, Hofmann C, Maetzke T, Herzog J, Ward E, Uknes S, Ryals J. (1994) Induction of systemic acquired disease resistance in plants by chemicals. *Annu. Rev. Phytopathol.* **32**: 439–459.

Kieber JJ, Ecker JR. (1993) Ethylene gas: it's not just for ripening any more! *Trends Genet.* **9**: 356–362.

Kirsch T, Paris N, Butler JM, Beevers L, Rogers JC. (1994) Purification and initial characterization of a potential plant vacuolar targeting receptor. *Proc. Natl Acad. Sci. USA* **91**: 3403–3407.

Klämbt D. (1990) A view about the function of auxin-binding proteins at plasma membranes. *Plant Mol. Biol.* **14**: 1045–1050.

Knox JP, Peart J, Neill SJ. (1995) Identification of novel cell surface epitopes using a leaf epidermal-strip assay system. *Planta* **196**: 266–270.

Kobilka B. (1992) Adrenergic receptors as models for G-protein-coupled receptors. *Annu. Rev. Neurosci.* **15**: 87–114.

Koga D, Hirata T, Sueshige N, Tanaka S, Ide A. (1992) Induction patterns of chitinases in yam callus by inoculation with autoclaved *Fusarium oxysporum*, ethylene, and chitin and chitosan oligosaccharides. *Biosci. Biotech. Biochem.* **56**: 280–285.

Kogel G, Beissmann B, Reisener HJ, Kogel KH. (1988) A single glycoprotein from *Puccinia graminis* f. sp. *tritici* cell walls elicits the hypersensitive lignification response in wheat. *Physiol. Mol. Plant Pathol.* **33**: 173–185.

Kogel G, Beissmann B, Reisener HJ, Kogel K. (1991) Specific binding of a hypersensitive lignification elicitor from *Puccinia graminis* f. sp. *tritici* to the plasma membrane from wheat (*Triticum aestivum* L.). *Planta* **183:** 164–169.

Korthout HAAJ, de Boer AH. (1994) A fusicoccin binding protein belongs to the family of 14-3-3 brain protein homologs. *Plant Cell* **6:** 1681–1692.

Korthout HAAJ, van der Hoeven PCJ, Wagner MJ, Van Hunnik E, de Boer AH. (1994) Purification of the fusicoccin-binding protein from oat root plasma membrane by affinity chromatography with biotinylated fusicoccin. *Plant Physiol.* **105:** 1281–1288.

Kuchitsu K, Kikuyama M, Shibuya N. (1993) *N*-acetylchito-oligosaccharides, biotic elicitor for phytoalexin production, induce transient membrane depolarization in suspension-cultured rice cells. *Protoplasma* **174:** 79–81.

Kuchitsu K, Kosaka H, Shiga T, Shibuya N. (1995) EPR evidence for generation of hydroxyl radical triggered by *N*-acetylchitooligosaccharide elicitor and a protein phosphatase inhibitor in suspension-cultured rice cells. *Protoplasma* **188:** 138–142.

Legendre L, Rueter S, Heinstein PF, Low PS. (1993) Characterization of the oligogalacturonide-induced oxidative burst in cultured soybean (*Glycine max*) cells. *Plant Physiol.* **102:** 233–240.

Lesney MS. (1989) Growth responses and lignin production in cell suspensions of *Pinus elliottii* 'elicited' by chitin, chitosan or mycelium of *Cronartium quercum* f.sp. *fusiforme. Plant Cell Tissue Org. Cult.* **19:** 23–31.

Levings CS III, Siedow JN. (1992) Molecular basis of disease susceptibility in the Texas cytoplasm of maize. *Plant Mol. Biol.* **19:** 135–147.

Libbenga KR, Maan AC, Van der Linde PCG, Mennes AM. (1986) Auxin receptors. In: *Hormones, Receptors and Cellular Interactions in Plants* (Chadwick CM, Garrod DR, ed.). Cambridge: Cambridge University Press, pp. 1–68.

Linthicum DS, Farid NR. (1988) *Anti-Idiotypes, Receptors, and Molecular Mimicry.* New York: Springer-Verlag.

Löbler M, Klämbt D. (1985a) Auxin-binding protein from coleoptile membranes of corn (*Zea mays* L.). I. Purification by immunological methods and characterization. *J. Biol. Chem.* **260:** 9848–9853.

Löbler M, Klämbt D. (1985b) Auxin-binding protein from coleoptile membranes of corn (*Zea mays* L.). II. Localization of a putative auxin receptor. *J. Biol. Chem.* **260:** 9854–9859.

Lomax TL, Hicks GR. (1992) Specific auxin-binding proteins in the plasma membrane: receptors or transporters. *Biochem. Soc. Trans.* **20:** 64–69.

Low PS, Legendre L, Heinstein PF, Horn MA. (1993) Comparison of elicitor and vitamin receptor-mediated endocytosis in cultured soybean cells. *J. Exp. Bot.* **44** (Suppl.): 269–274.

Macdonald H, Henderson J, Napier RM, Venis MA, Hawes C, Lazarus CM. (1994) Authentic processing and targeting of active maize auxin-binding protein in the baculovirus expression system. *Plant Physiol.* **105:** 1049–1057.

McGurl B, Pearce G, Orozco-Cardenas M, Ryan CA. (1992) Structure, expression, and antisense inhibition of the systemin precursor gene. *Science* **255:** 1570–1573.

Marra M, Fullone MR, Fogliano V, Pen J, Mattei M, Masi S, Aducci P. (1994) The 30-kilodalton protein present in purified fusicoccin receptor preparations is a 14-3-3-like protein. *Plant Physiol.* **106:** 1497–1501.

Marra M, Ballio A, Battirossi P, Fogliano V, Fullone MR, Slayman CL, Aducci P. (1995) The fungal H+-ATPase from *Neurospora crassa* reconstituted with fusicoccin receptors senses fusicoccin signal. *Proc. Natl Acad. Sci. USA* **92:** 1599–1603.

Marrè E. (1979) Fusicoccin: a tool in plant physiology. *Annu. Rev. Plant Physiol.* **30:** 273–288.

Marten I, Lohse G, Hedrich R. (1991) Plant growth hormones control voltage-dependent activity of anion channels in plasma membrane of guard cells. *Nature* **353:** 758–762.

Mathieu Y, Kurkdijan A, Xia H, Guern J, Koller A, Spiro M, O'Neill M, Albersheim P, Darvill A. (1991) Membrane responses induced by oligogalacturonides in suspension-cultured tobacco cells. *Plant J.* **1:** 333–343.

Meyer C, Feyerabend M, Weiler EW. (1989) Fusicoccin-binding proteins in *Arabidopsis thaliana* (L.) Heynh. *Plant Physiol.* **89:** 692–699.

Milat M-L, Ducruet J-M, Ricci P, Marty F, Blein J-P. (1991a) Physiological and structural changes in tobacco leaves treated with cryptogein, a proteinaceous elicitor from *Phytophthora cryptogea. Phytopathology* **81:** 1364–1368.

Milat M-L, Ricci P, Bonnet P, Blein J-P. (1991b) Capsidiol and ethylene production by tobacco cells in response to cryptogein, an elicitor from *Phytophthora cryptogea. Phytochemistry* **30:** 2171–2173.

Moore TC. (1989) Auxins. In: *Biochemistry and Physiology of Plant Hormones,* 2nd edn. New York: Springer-Verlag, pp. 28–93.

Napier RM, Venis MA. (1990) Monoclonal antibodies detect an auxin-induced conformational change in the maize auxin-binding protein. *Planta* **182:** 313–318.

Napier RM, Venis MA. (1992) Epitope mapping reveals conserved regions of an auxin-binding protein. *Biochem. J.* **284:** 841–845.

Napier RM, Venis MA, Bolton MA, Richardson LI, Butcher GW. (1988) Preparation and characterisation of monoclonal and polyclonal antibodies to maize membrane auxin-binding protein. *Planta* **176:** 519–526.

Napier RM, Fowke LC, Hawes C, Lewis M, Pelham HRB. (1992) Immunological evidence that plants use both HDEL and KDEL for targeting proteins to the endoplasmic reticulum. *J. Cell Sci.* **102:** 261–271.

Navarre DA, Wolpert TJ. (1995) Inhibition of the glycine decarboxylase multienzyme complex by the host-selective toxin victorin. *Plant Cell* **7:** 463–471.

Nürnberger T, Nennstiel D, Jabs T, Sacks WR, Hahlbrock K, Scheel D. (1994) High-affinity binding of a fungal oligopeptide elicitor to parsley plasma membranes triggers multiple defense responses. *Cell* **78:** 449–460.

Nürnberger T, Nennstiel D, Hahlbrock K, Scheel D. (1995) Covalent cross-linking of the *Phytophthora megasperma* oligopeptide elicitor to its receptor in parsley membranes. *Proc. Natl Acad. Sci. USA* **92:** 2338–2342.

Oecking C, Weiler E. (1991) Characterization and purification of the fusicoccin-binding complex from plasma membranes of *Commelina communis. Eur. J. Biochem.* **199:** 685–689.

Oecking C, Eckerskorn C, Weiler EW. (1994) The fusicoccin receptor of plants is a member of the 14-3-3 superfamily of eukaryotic regulatory proteins. *FEBS Lett.* **352:** 163–166.

Oeller PW, Keller JA, Parks JE, Silbert JE, Theologis A. (1993) Structural characterization of the early indoleacetic acid-inducible genes, *PS-IAA4/5* and *PS-IAA6,* of pea (*Pisum sativum* L.). *J. Mol. Biol.* **233:** 789–798.

Okinaka Y, Mimori K, Takeo K, Kitamura S, Takeuchi Y, Yamaoka N, Yoshikawa M. (1995) A structural model for the mechanisms of elicitor release from fungal cell walls by plant β-1,3-endoglucanase. *Plant Physiol.* **109:** 839–845.

Palme K, Hesse T, Moore I, Campos N, Feldwisch J, Garbers C, Hesse F, Schell J. (1991) Hormonal modulation of plant growth: the role of auxin perception. *Mech. Dev.* **33:** 97–106.

Palme K, Hesse T, Campos N, Garbers C, Yanofsky MF, Schell J. (1992) Molecular analysis of an auxin binding protein gene located on chromosome 4 of *Arabidopsis. Plant Cell* **4:** 193–201.

Palmgren MG, Larsson C, Sommarin M. (1990) Proteolytic activation of the plant plasma membrane H^+-ATPase by removal of a terminal segment. *J. Biol. Chem.* **265:** 13423–13426.

Palmgren MG, Sommarin M, Serrano R, Larsson C. (1991) Identification of an autoinhibitory domain in the C-terminal region of the plant plasma membrane H^+-ATPase. *J. Biol. Chem.* **266:** 20470–20475.

Parker JE, Hahlbrock K, Scheel D. (1988) Different cell-wall components from *Phytophthora megasperma* f. sp. *glycinea* elicit phytoalexin production in soybean and parsley. *Planta* **176:** 75–82.

Parker JE, Schulte W, Hahlbrock K, Scheel D. (1991) An extracellular glycoprotein from *Phytophthora megasperma* f. sp. *glycinea* elicits phytoalexin synthesis in cultured parsley cells and protoplasts. *Mol. Plant Microb. Interact.* **4:** 19–27.

Parker MG. (1993) Steroids and related receptors. *Curr. Opin. Cell Biol.* **5:** 499–504.

Pearce G, Strydom D, Johnson S, Ryan CA. (1991) A polypeptide from tomato leaves induces wound-inducible proteinase inhibitor proteins. *Science* **253:** 895–898.

Pearce G, Johnson S, Ryan CA. (1993) Structure–activity of deleted and substituted systemin, an 18-amino-acid polypeptide inducer of plant defensive genes. *J. Biol. Chem.* **268:** 212–216.

Pearce RB, Ride JP. (1982) Chitin and related compounds as elicitors of the lignification response in wounded wheat leaves. *Physiol. Plant Pathol.* **20:** 119–123.

Pelham HRB. (1990) The retention signal for soluble proteins of the endoplasmic reticulum. *Trends Biochem. Sci.* **15:** 483–486.

Pesci P, Cocucci SM, Randazzo G. (1979a) Characterization of fusicoccin binding to receptor sites on cell membranes of maize coleoptile tissue. *Plant Cell Environ.* **2:** 205–209.

Pesci P, Tognoli L, Beffagna N, Marrè E. (1979b) Solubilization and partial purification of a fusicoccin–receptor complex from maize microsomes. *Plant Sci. Lett.* **15:** 313–322.

Peters BM, Cribbs DH, Stelzig DA. (1978) Agglutination of plant protoplasts by fungal cell wall glucans. *Science* **201:** 364–365.

Quail PH. (1991) Phytochrome: a light-activated molecular switch that regulates plant gene expression. *Annu. Rev. Genet.* **25:** 389–409.

Rademacher E, Klämbt D. (1993) Auxin dependent growth and auxin-binding proteins in primary roots and root hairs of corn (*Zea mays* L.). *J. Plant Physiol.* **141:** 698–703.

Rasi-Caldogno F, Pugliarello MC. (1985) Fusicoccin stimulates the H⁺-ATPase of plasmalemma in isolated membrane vesicles from radish. *Biochem. Biophys. Res. Commun.* **133:** 280–285.

Rasi-Caldogno F, De Michelis MI, Pugliarello MC, Marrè E. (1986) H⁺-pumping driven by the plasma membrane ATPase in membrane vesicles from radish: stimulation by fusicoccin. *Plant Physiol.* **82:** 121–125.

Rasi-Caldogno F, Pugliarello MC, Olivari C, De Michelis MI. (1993) Controlled proteolysis mimics the effect of fusicoccin on the plasma membrane H⁺-ATPase. *Plant Physiol.* **103:** 391–398.

Ren Y-Y, West CA. (1992) Elicitation of diterpene biosynthesis in rice (*Oryza sativa* L.) by chitin. *Plant Physiol.* **99:** 1169–1178.

Ricci P, Bonnet P, Huet J-C, Sallantin M, Beauvais-Cante F, Bruneteau M, Billard V, Michel G, Pernollet J-C. (1989) Structure and activity of proteins from pathogenic fungi (*Phytophthora*) eliciting necrosis and acquired resistance in tobacco. *Eur. J. Biochem.* **183:** 555–563.

Roby D, Gadelle A, Toppan A. (1987) Chitin oligosaccharides as elicitors of chitinase activity in melon plants. *Biochem. Biophys. Res. Commun.* **143:** 885–892.

Rück A, Palme K, Venis MA, Napier RM, Felle HH. (1993) Patch-clamp analysis establishes a role for an auxin binding protein in the auxin stimulation of plasma membrane current in *Zea mays* protoplasts. *Plant J.* **4:** 41–46.

Rudolph K. (1976) Non-specific toxins. In: *Encyclopedia of Plant Physiology, New Series, Vol. 4. Physiological Plant Pathology* (Heitefuss R, Williams PH, ed.). Berlin: Springer Verlag, pp. 270–315.

Sacks W, Nürnberger T, Hahlbrock K, Scheel D. (1995) Molecular characterization of nucleotide sequences encoding the extracellular glycoprotein elicitor from *Phytophthora megasperma*. *Mol. Gen. Genet.* **246:** 45–55.

Schaller GE, Bleecker AB. (1995) Ethylene-binding sites generated in yeast expressing the *Arabidopsis ETR1* gene. *Science* **270:** 1809–1811.

Schaller GE, Ladd AN, Lanahan MB, Spanbauer JM, Bleecker AB. (1995) The ethylene response mediator ETR1 from *Arabidopsis* forms a disulfide-linked dimer. *J. Biol. Chem.* **270:** 12526–12530.

Scheffer RP, Livingston RS. (1984) Host-selective toxins and their role in plant diseases. *Science* **223:** 17–21.

Schmidt WE, Ebel J. (1987) Specific binding of a fungal glucan phytoalexin elicitor to membrane fractions from soybean *Glycine max. Proc. Natl Acad. Sci. USA* **84:** 4117–4121.

Scholtens-Toma IMJ, de Wit PJGM. (1988) Purification and primary structure of a necrosis-inducing peptide from apoplastic fluids of tomato infected with *Cladosporium fulvum* (syn. *Fulvia fulva*). *Physiol. Mol. Plant Pathol.* **33:** 59–67.

Schulz S, Oelgemöller E, Weiler EW. (1991) Fusicoccin action in cell-suspension cultures of *Corydalis sempervirens* Pers. *Planta* **183:** 83–91.

Schwob E, Choi S-Y, Simmons C, Migliaccio F, Ilag L, Hesse T, Palme K, Söll D. (1993) Molecular analysis of three maize 22 kDa auxin-binding protein genes — transient promoter expression and regulatory regions. *Plant J.* **4:** 423–432.

Sharp JK, Albersheim P, Ossowski P, Pilotti Å, Garegg PJ, Lindberg B. (1984a) Comparison of the structures and elicitor activities of a synthetic and a mycelial-wall-derived hexa(β-D-glucopyranosyl)-D-glucitol. *J. Biol. Chem.* **259:** 11341–11345.

Sharp JK, McNeil M, Albersheim P. (1984b) The primary structures of one elicitor-active and seven elicitor-inactive hexa(β-D-glucopyranosyl)-D-glucitols isolated from the mycelial walls of *Phytophthora megasperma* f. sp. *glycinea. J. Biol. Chem.* **259:** 11321–11336.

Sharp JK, Valent B, Albersheim P. (1984c) Purification and partial characterization of a β-glucan fragment that elicits phytoalexin accumulation in soybean. *J. Biol. Chem.* **259:** 11312–11320.

Sherman MR, Stevens J. (1984) Structure of mammalian steroid receptors: evolving concepts and methodological developments. *Annu. Rev. Physiol.* **46:** 83–105.

Shibuya N, Kaku H, Kuchitsu K, Maliarik MJ. (1993) Identification of a novel high-affinity binding site for *N*-acetylchitooligosaccharide elicitor in the membrane fraction from suspension-cultured rice cells. *FEBS Lett.* **329:** 75–78.

Shimomura S, Sotobaya T, Futai M, Fukui T. (1986) Purification and properties of an auxin-binding protein from maize shoot membranes. *J. Biochem.* **99:** 1513–1524.

Shimomura S, Inohara N, Fukui T, Futai M. (1988) Different properties of two types of auxin-binding sites in membranes from maize coleoptiles. *Planta* **175:** 558–566.

Shimomura S, Liu W, Inohara N, Watanabe S, Futai M. (1993) Structure of the gene for an auxin-binding protein and a gene for 7SL RNA from *Arabidopsis thaliana. Plant Cell Physiol.* **34:** 633–637.

Short TW, Reymond P, Briggs WR. (1993) A pea plasma membrane protein exhibiting blue light-induced phosphorylation retains photosensitivity following Triton solubilization. *Plant Physiol.* **101:** 647–655.

Sisler EC. (1991) Ethylene-binding components in plants. In: *The Plant Hormone Ethylene* (Mattoo AK, Suttle JC, ed.). Boca Raton: CRC Press, pp. 81–100.

Spiro MD, Kates KA, Koller AL, O'Neill MA, Albersheim P, Darvill AG. (1993) Purification and characterization of biologically active 1,4-linked α-D-oligogalacturonides after partial digestion of polygalacturonic acid with endopolygalacturonase. *Carbohydr. Res.* **247:** 9–20.

Spiro MD, Ridley BL, Glushka J, Darvil AG, Albersheim P. (1996) Synthesis and characterization of tyramine-derivatized 1→4-linked α-D-oligogalacturonides. *Carbohydr. Res.* (in press).

Springer TA. (1990) Adhesion receptors of the immune system. *Nature* **346:** 425–434.

Stekoll M, West CA. (1978) Purification and properties of an elicitor of castor bean phytoalexin from culture filtrates of the fungus *Rhizopus stolonifer. Plant Physiol.* **61:** 38–45.

Stengelin S, Stamenkovic I, Seed B. (1988) Isolation of cDNAs for two distinct human Fc receptors by ligand affinity cloning. *EMBO J.* **7:** 1053–1059.

Stoddart JL. (1987) Genetic and hormonal regulation of stature. In: *Developmental Mutants in Higher Plants* (Thomas H, Grierson D, ed.). Cambridge: Cambridge University Press, pp. 155–180.

Stone JM, Walker JC. (1995) Plant protein kinase families and signal transduction. *Plant Physiol.* **108:** 451–457.

Stout RG. (1988) Fusicoccin activity and binding in *Arabidopsis thaliana. Plant Physiol.* **88:** 999–1001.

Stout RG, Cleland RE. (1980) Partial characterization of fusicoccin binding to receptor sites on oat root membranes. *Plant Physiol.* **66:** 353–359.

Suty L, Blein JP, Ricci P, Pugin A. (1995) Early changes in gene expression in tobacco cells elicited with cryptogein. *Mol. Plant Microb. Interact.* **8:** 644–651.

Tavernier E, Stallaert V, Blein J-P, Pugin A. (1995a) Changes in lipid composition in tobacco cells treated with cryptogein, an elicitor from *Phytophthora cryptogea. Plant Sci.* **104:** 117–125.

Tavernier E, Wendehenne D, Blein J-P, Pugin A. (1995b) Involvement of free calcium in the action of cryptogein, a proteinaceous elicitor of hypersensitive reactions in tobacco cells. *Plant Physiol.* **109:** 1025–1031.

Thain JF, Doherty HM, Bowles DJ, Wildon DC. (1990) Oligosaccharides that induce proteinase inhibitor activity in tomato plants cause depolarization of tomato leaf cells. *Plant Cell Environ.* **13:** 569–574.

Thiel G, Blatt MR, Fricker MD, White IR, Millner P. (1993) Modulation of K$^+$ channels in *Vicia* stomatal guard cells by peptide homologs to the auxin-binding protein C terminus. *Proc. Natl Acad. Sci. USA* **90:** 11493–11497.

Tillmann U, Viola G, Kayser B, Siemeister G, Hesse T, Palme K, Löbler M, Klämbt D. (1989) cDNA clones of the auxin-binding protein from corn coleoptiles (*Zea mays* L.): isolation and characterization by immunological methods. *EMBO J.* **8:** 2463–2467.

Turner NC, Graniti A. (1969) Fusicoccin: a fungal toxin that opens stomata. *Nature* **223:** 1070–1071.

Ullrich A, Schlessinger J. (1990) Signal transduction by receptors with tyrosine kinase activity. *Cell* **61:** 203–212.

Van der Meulen RM, Heidekamp F, Jastorff B, Horgan R, Wang M. (1993) Effects of abscisic acid analogues on abscisic acid-induced gene expression in barley aleurone protoplasts: relationship between structure and function of the abscisic acid molecule. *J. Plant Growth Reg.* **12:** 13–19.

Van Kan JAL, Van den Ackerveken GFJM, de Wit PJGM. (1991) Cloning and characterization of cDNA of avirulence gene *aur9* of the fungal pathogen *Cladosporium fulvum*, causal agent of tomato leaf mold. *Mol. Plant Microbe-Interact.* **4:** 52–59.

Venis MA. (1977) Solubilization and partial purification of auxin-binding sites of corn membranes. *Nature* **266:** 268–269.

Venis M. (1985) *Hormone Binding Sites in Plants*. New York: Longman.

Venis MA, Napier RM. (1991) Auxin receptors: recent developments. *Plant Growth Reg.* **10:** 329–340.

Venis MA, Napier RM. (1995) Auxin receptors and auxin binding proteins. *Crit. Rev. Plant Sci.* **14:** 27–47.

Venis MA, Thomas EW, Barbier-Brygoo H, Ephritikhine G, Guern J. (1990) Impermeant auxin analogues have auxin activity. *Planta* **182:** 232–235.

Venis MA, Napier RM, Barbier-Brygoo H, Maurel C, Perrot-Rechenmann C, Guern J. (1992) Antibodies to a peptide from the maize auxin-binding protein have auxin agonist activity. *Proc. Natl Acad. Sci. USA* **89:** 7208–7212.

Viard M-P, Martin F, Pugin A, Ricci P, Blein J-P. (1994) Protein phosphorylation is induced in tobacco cells by the elicitor cryptogein. *Plant Physiol.* **104:** 1245–1249.

Walker RP, Beale MH, Hooley R. (1992) Photoaffinity labelling of MAC 182, a gibberellin-specific monoclonal antibody. *Phytochemistry* **31:** 3331–3335.

Walton DC. (1980) Biochemistry and physiology of abscisic acid. *Annu. Rev. Plant Physiol.* **31:** 453–489.

Walton JD, Panaccione DG. (1993) Host-selective toxins and disease specificity: perspectives and progress. *Annu. Rev. Phytopathol.* **31:** 275–303.

Wang M, Heimovaara-Dijkstra S, Van der Meulen RM, Knox JP, Neill SJ. (1995) The monoclonal antibody JIM19 modulates abscisic acid action in barley aleurone protoplasts. *Planta* **196:** 271–276.

Wendehenne D, Binet M-N, Blein J-P, Ricci P, Pugin A. (1995) Evidence for specific, high-affinity binding sites for a proteinaceous elicitor in tobacco plasma membrane. *FEBS Lett.* **374:** 203–207.

Williams AF, Barclay AN. (1988) The immunoglobulin superfamily — domains for cell surface recognition. *Annu. Rev. Immunol.* **6:** 381–405.

Wolins NE, Donaldson RP. (1994) Specific binding of the peroxisomal protein targeting sequence to glyoxysomal membranes. *J. Biol. Chem.* **269:** 1149–1153.

Wolpert TJ, Navarre DA, Moore DL, Macko V. (1994) Identification of the 100 kDa victorin-binding protein from oats. *Plant Cell* **6:** 1145–1155.

Yamada A, Shibuya N, Kodama O, Akatsuka T. (1993) Induction of phytoalexin formation in suspension-cultured rice cells by *N*-acetylchitooligosaccharides. *Biosci. Biotech. Biochem.* **57:** 405–409.

Yoshikawa M, Sugimoto K. (1993) A specific binding site on soybean membranes for a phytoalexin elicitor released from fungal cell walls by β-1,3-endoglucanase. *Plant Cell Physiol.* **34:** 1229–1237.

Yoshikawa M, Matama M, Masago H. (1981) Release of a soluble phytoalexin elicitor from mycelial walls of *Phytophthora megasperma* var. *sojae* by soybean tissues. *Plant Physiol.* **67:** 1032–1035.

Yoshikawa M, Keen NT, Wang M-C. (1983) A receptor on soybean membranes for a fungal elicitor of phytoalexin accumulation. *Plant Physiol.* **73:** 497–506.

Yu LM. (1995) Elicitins from *Phytophthora* and basic resistance in tobacco. *Proc. Natl Acad. Sci. USA* **92:** 4088–4094.

Yu L-X, Lazarus CM. (1991) Structure and sequence of an auxin-binding protein gene from maize (*Zea mays* L.). *Plant Mol. Biol.* **16:** 925–930.

Zettl R, Feldwisch J, Boland W, Schell J, Palme K. (1992) 5′-Azido-[3,6-^{3}H$_2$]-1-naphthylphthalamic acid, a photoactivatable probe for naphthylphthalamic acid receptor proteins from higher plants: identification of a 23-kDa protein from maize coleoptile plasma membranes. *Proc. Natl Acad. Sci. USA* **89:** 480–484.

Chapter 8

Plant membrane-associated protein kinases: proposed great communicators

Alice C. Harmon, Jung-Youn Lee, Byung-Chun Yoo and Jiahong Shao

1. Introduction

One of the characteristics of living organisms is that they respond to stimuli. Mechanisms for coupling stimuli to cellular responses in both prokaryotes and eukaryotes involve protein phosphorylation events in membranes. In animals, many signals are communicated via protein kinases and phosphatases, but in bacteria another type of phosphotransfer system, the two-component system, is most evident (for review see Swanson *et al.,* 1995). In plants, the number of protein kinases that have been described is approaching 100 (for review see Stone and Walker, 1995). They include members of families that are common to all eukaryotes (e.g. cell cycle kinases), and others that are unique to plants (e.g. receptor-like kinases). This review will focus on protein kinases that are associated with membranes, and on recent evidence that plants may also contain a signaling pathway related to the bacterial two-component system.

Protein kinases are enzymes which transfer a phosphoryl group from ATP, or more rarely from GTP, to a serine, threonine or tyrosine residue of a protein substrate and thereby alter the activity of the substrate (Krebs, 1986). The substrate may be an enzyme or a structural protein whose activity or function influences the metabolic or physiological status of the cell. Extensive studies on animal systems have shown that membrane-associated protein kinases are of primary importance in signal transduction pathways, because their location allows them to interact with compartments on both sides of the membrane. Thus these enzymes are poised to communicate information about conditions on one side of the membrane to the metabolic machinery on the other side of the membrane. Plasma membrane-associated protein kinases act as transducers of extracellular signals, while protein kinases associated with the membranes of organelles may be involved in communication between the organelles and the cytoplasm. Phosphotransfer is also employed in transmembrane signal transduction in prokaryotes, but instead of the protein kinases described above, it involves a component that autophosphorylates on histidine and the transfer of the phosphoryl group to an aspartate residue of a second component. This review will examine the evidence for both types of systems in plants.

2. Receptor protein kinases

Animal receptor protein kinases

Receptor protein kinases were first described in animals, and much is known about the structure and function of these proteins. They are integral membrane-spanning proteins that have three distinct domains: an extracellular ligand-binding domain, a membrane-spanning region, and an intracellular protein kinase catalytic domain. These proteins are in direct contact with both cytoplasmic and extracellular spaces, and they perform the dual functions of sensing extracellular signals and activating the intracellular response pathways. Binding of a ligand, which may be a peptide hormone or growth factor, to the receptor domain outside the cell activates both autophosphorylation of the internal protein kinase domain and phosphorylation of substrates inside the cell.

The receptor kinases may be either protein tyrosine kinases or protein serine/threonine kinases, and their catalytic domains contain sequence motifs that typify each class of kinases (Hanks *et al.,* 1988). The receptor tyrosine kinases are the more abundant class, and they can be divided into four subclasses based on the structure of their extracellular domains (Ullrich and Schlessinger, 1990). Subclass I receptors, represented by the epidermal growth factor receptor, are monomeric and have two cysteine-rich repeats in the ligand-binding domain. Subclass II receptors, represented by the insulin receptor, are disulfide-linked heterotetramers with one cysteine-rich region in each monomer. Subclass III, represented by the platelet-derived growth factor receptor, and subclass IV, represented by the fibroblast growth factor receptor, have five or three immunoglobulin-like repeats, respectively, in the ligand-binding domain. Only one class of receptor serine/threonine kinases from animals has been described. These receptor kinases bind various members of the transforming growth factor β family, and have a relatively small ligand-binding domain which contains nine conserved cysteines (Massague, 1992).

Plant receptor-like protein kinases

Plant proteins having the predicted three-domain structure of receptor protein kinases have been identified in both monocots and dicots by molecular genetic approaches. These proteins are called 'receptor-like' because information about the subcellular location, membrane topology, function and ligands is not available for most of them. None of the proteins has been purified from plant sources. Their predicted extracellular domains have little or no similarity to those of the animal receptor kinases; thus no clues about the identity of the ligands that bind to the plant proteins can be gleaned from these data. The catalytic domains contain the 11 conserved subdomains of eukaryotic protein kinases (Hanks *et al.,* 1988). The sequences of subdomains VI and VIII are typical of serine/threonine protein kinases, and all that have been expressed as recombinant proteins phosphorylate serine and/or threonine residues. Thus the plant receptor-like kinases differ from the majority of animal receptor kinases which phosphorylate tyrosine. Furthermore, their kinase domains are not closely related to any of the classes of protein serine/threonine kinases found in animals and yeast, and thus constitute a class of their own (Stone and Walker, 1995).

The plant receptor-like protein kinases can be divided into four groups based on the properties of their extracellular domains. The characteristics of the four family types are discussed below. Other recent reviews on this subject are those by Braun and Walker (1996), Stone and Walker (1995) and Walker (1994).

S-locus receptor-like protein kinases. Proteins encoded by genes of the S-locus of *Brassica* have characteristic features, including a signal peptide sequence at the amino terminus, *N*-glycosylation sites, and clusters of cysteine residues. A class of receptor-like protein kinases with extracellular domains similar to the S-locus proteins is present in *Brassica* and other genera. Members of this group are ZmPK1 (*Zea mays* protein kinase 1) (Walker and Zhang, 1990), the SRKs (S-locus receptor kinases) from *Brassica* (Delorme *et al.,* 1995; Goring and Rothstein, 1992; Kumar and Trick, 1994; Stein *et al.,* 1991; Stein and Nasrallah, 1993), ARK1 (*Arabidopsis* receptor kinase 1) (Tobias *et al.,* 1992), and RLK1 and RLK4 (*Arabidopsis* receptor-like protein kinases) (Walker, 1993). These receptor-like protein kinases range in predicted molecular mass from 91 to 98 kDa. Their catalytic domains are similar to each other, but do not show substantial identity to any other family of protein kinases. For example, SRK910 (Goring and Rothstein, 1992) is 65–84% identical to SRK6 and SRK2 (Stein *et al.*, 1991), 71% identical to ARK1, and 38% identical to ZmPK1. Outside this group, the protein kinase family showing the most similarity to the S-locus receptor-like kinases is the raf serine/threonine kinase family, but the degree of identity is low; ZmPK1 and SRK6 (Stein *et al.*, 1991) have 27 and 29% amino acid sequence identity, respectively, to raf family members. Confirmation that the SRKs are functional serine/threonine protein kinases has come from the expression in *E. coli* of fusion proteins containing the catalytic domains of SRK910 (Goring and Rothstein, 1992) and SRK6 (Stein *et al.*, 1991). These recombinant proteins autophosphorylate on serine and threonine residues, but phosphorylation of other proteins has not been demonstrated. Neither the ligands that bind to the extracellular domains nor the endogenous substrates of these kinases are known, but genetic evidence points to a role in self-incompatibility for the *Brassica* S-locus receptor-like kinase.

(a) *Brassica* SRKs and self-incompatibility. Self-incompatibility is a mechanism by which self-fertilization in androgenous flowers is blocked. Self-incompatibility occurs if identical alleles at the S-locus are expressed in both pollen and papillar cells of the pistils. Proteins encoded by the S-locus include the S-locus glycoprotein (SLG) and the S-locus receptor kinase (SRK). SLG and SRK are specifically expressed in the flower with the highest expression occurring in the pistil and lower expression occurring in the anther (Delorme *et al.*, 1995; Goring and Rothstein, 1992; Stein *et al.*, 1991). The transcription level of SRK is substantially lower than that of the SLG, which is an abundant, soluble, secreted glycoprotein (Stein *et al.*, 1991). A mutant, self-compatible cultivar of Brassica napus contains a normal SLG, but an abnormal SRK (Goring *et al.*, 1993). The gene for the kinase contains one base pair deletion that would result in the expression of a protein truncated in the receptor domain. Because the predicted mutant protein contains no catalytic domain, it would have no protein kinase activity, and this observation supports the hypothesis that SRK activity is required for self-incompatibility.

Stein *et al.* (1991) have suggested that self-incompatibility occurs via a mechanism involving the interaction of SRK on the surface membrane of papillar cells with a putative ligand on self pollen grains. This interaction elicits protein phosphorylation and a cellular response that leads to pollen rejection. Evidence consistent with membrane localization of SRK in stigmas has been presented (Delorme *et al.*, 1995), but little else is known about SRK *in vivo*. It is not known whether the SRKs are monomers, dimers or higher oligomers, or whether SRK acts alone or in a complex with other proteins. It has been suggested, however, that the homologous domains of SLGs and/or SRKs may interact with each other (Goring and Rothstein, 1992; Stein *et al.*, 1991), or that other proteins encoded by alternative transcripts of S-locus genes and expressed in papillar cells may interact directly or indirectly with SRK. These proteins include a truncated form of SRK_3 that contains only the extracellular domain (Giranton *et al.*, 1995), and a membrane-bound form of SLG_2 (Tantikanjana *et al.*, 1993).

(b) S-locus receptor-like kinases of vegetative tissue. In contrast to the localization of SRK to the pistil, ZmPK1, RLK1, RLK4, ARK1–3 and OsPK10 are expressed abundantly in vegetative tissues (Dwyer *et al.*, 1994; Tobias *et al.*, 1992; Walker and Zhang, 1990; Walker, 1993; Zhao *et al.*, 1994). ZmPK1 is highly expressed in shoot and root and less abundantly expressed in the silk of maize. In Arabidopsis, RLK1 and RLK4 are expressed predominantly in rosettes and root, respectively. The transcription level of ARK1 is much higher in stem and leaf than in floral bud, ARK2 transcription is highest in organs of the shoot, and ARK3 transcription is found in both shoot and root. The expression of *OsPK10* in the shoots and roots of rice seedlings is stimulated by light. The presence of these S-locus receptor-like kinase homologs in nonfloral tissues suggests that they are possibly involved in general signal transduction. Information about the activators and substrates of this family of receptor-like kinases is needed in order to confirm this hypothesis.

Leucine-rich repeat (LRR) receptor-like protein kinases. The extracellular domains of TMK1 (transmembrane kinase 1) (Chang *et al.*, 1992), TMKL1 (Valon *et al.*, 1993) and RLK5 (receptor-like kinase 5) (Walker, 1993) from *Arabidopsis,* PRK1 from petunia (Mu *et al.*, 1994) and Xa21 from rice (Song *et al.*, 1995) contain imperfect leucine-rich repeats (for reviews see Braun and Walker, 1996; Walker, 1994). Leucine-rich repeats are found in some tyrosine receptor kinases from animals, and are thought to be involved in dimerization of the receptors. TMK1, RLK5 and PRK1 have been expressed in *E. coli* as fusion proteins and have been shown to have protein serine/threonine kinase activity. TMK1 is immunologically detectable in membrane preparations of *Arabidopsis* and in crude extracts from root, stem, leaf and flower, and PRK1 is expressed specifically in pollen.

One member of this group, Xa21, is known from mutational analysis to have a role in disease resistance. Xa21, which is 36% identical to RLK5, confers resistance to *Xanthomonas oryzae* pv. *oryzae* race 6 (Song *et al.*, 1995). Two proteins from tomato that are also involved in disease resistance are structurally related either to the LRR domain or to the kinase domain of this group of kinases. Cf-9, which is involved in resistance to the fungus *Cladosporium fulvum,* contains a predicted extracellular LRR domain and a transmembrane region, but has no kinase domain (Jones *et al.*, 1994). Resistance to the fungus presumably involves the interaction of Cf-9 with a fungus-encoded 28-residue peptide (Schottens-Toma and de Wit, 1988; Van Kan *et al.*, 1991). Pto, which is involved in resistance to *Pseudomonas syringae* pv. tomato, only has a kinase domain. It has been speculated that Cf-9 may function through interaction with either an LRR-receptor kinase or a cytoplasmic protein kinase.

A protein that has the potential to be a downstream component of an RLK5 signaling pathway has been identified by interaction cloning (Stone *et al.*, 1994). The protein is a type 2C protein phosphatase called KAPP (kinase-associated protein phosphatase). KAPP contains a domain outside its catalytic domain that binds only to autophosphorylated RLK5. This observation opens up the exciting possibility that plants may contain a signaling system based on recognition of phosphorylated serine/threonine residues in receptor kinases rather than the phosphotyrosine-based system of animals. It is not known whether the activity of KAPP is stimulated or inhibited by its interaction with RLK5.

EGF-like repeat receptor-like kinase. The third class of putative receptor protein kinases was serendipitously discovered by a technique designed to identify proteins that interact with the amino terminus of light-harvesting chlorophyll *a/b* binding protein (LHCP) (Kohorn *et al.*, 1992). The cDNA encodes a 595-residue, 70-kDa protein. The predicted extracellular domain contains a repeat of a sequence that is similar to that of epidermal growth factor (EGF). Proteins important in the control of developmental processes in *Drosophila melanogaster*, *Caenorhabditis elegans* and sea urchin contain EGF-like sequences, and similar sequences in several proteins have been shown to be involved in protein–protein interaction (Tepass *et al.*, 1990). The mRNA for this putative receptor protein kinase is expressed in green leaves but not in other plant tissues or in etiolated plants (Kohorn *et al.*, 1992). Recent data from immuno-electron microscopy and biochemical studies suggest that this protein is located in the plasma membrane and that it is also tightly bound to the cell wall, and its name has been changed from PRO25 to WAK1 (for wall-associated kinase) (B.D. Kohorn, personal communication). The function of this receptor-like protein kinase is unknown, but its location shows that its role does not involve interaction with LHCP. This work demonstrates that false, but interesting, positive results may arise from techniques such as the yeast two-hybrid system.

Cysteine-rich repeat receptor-like kinase. The single member of the fourth class of receptor-like protein kinases is Crinkly4 from maize. Genetic analysis has linked this protein to a role in morphogenesis of kernel aleurone and leaf epidermis (Becraft and McCarty, 1995). The extracellular domain of this protein contains cysteine-rich repeats, one of which is homologous to the cysteine-rich motif of the mammalian tumor necrosis factor (TNF) receptor (P.W. Becraft and D.R. McCarty, personal communication). The cysteine-rich repeats of the TNF receptor are involved in oligomerization of the receptor and in interaction with TNF, which is a peptide cytokine (Vandenabeele *et al.*, 1995). The kinase domain of Crinkly4 is very similar to those of the other receptor-like protein kinases.

3. The putative blue-light receptor

Blue light induces phototropic responses in monocots and dicots, and it has been proposed that a receptor for blue light exists in tissues that respond to blue-light irradiation. Recent biochemical evidence suggests that a large plasma-membrane protein from photosensitive tissues of plants could be the blue-light receptor, and that this protein may itself be a protein kinase. The molecular weight of the protein varies from species to species and ranges from 100 000 to 130 000 (Hager and Brich, 1993; Palmer *et al.*, 1993; Reymond *et al.*, 1992a). Blue light dramatically stimulates phosphorylation of these proteins in intact tissue, in isolated membranes and in detergent-solubilized membranes. In corn, induction of phosphorylation of the protein by blue light occurs in the tips of the coleoptile but not in the stem or root. This distribution is consistent with the fact that the tip is the site of maximal sensitivity to phototropic stimuli (Hager and Brich, 1993; Palmer *et al.*, 1993). *Arabidopsis thaliana* mutant JK224 is postulated to be a photoreceptor mutant, since it requires a 20- to 30-fold higher fluence of blue light for the threshold phototropism. The level of blue light-stimulated phosphorylation of a 120-kDa protein in microsomal preparations from this mutant is lower than that in similar preparations of wild-type plants (Reymond *et al.*, 1992b). The lower degree of light-induced phosphorylation could result, in part, from a smaller amount of the protein being present in the mutant, but this question was not addressed. However, this observation does support the hypothesis that the 120-kDa protein is involved in blue-light perception/transduction, but it does not provide any insight into whether the protein is the blue-light receptor, or a component that is downstream of the receptor.

Whether the observed increase in phosphorylation represents autophosphorylation by intrinsic kinase activity or phosphorylation of the 100- to 130-kDa protein by a separate kinase is not clear, since this putative blue-light receptor has not been purified, nor has the gene been cloned. However, since Triton X-100-solubilized membrane fractions from pea (Short *et al.*, 1992) or corn (Hager and Brich, 1993) retain light-induced phosphorylation activity, it has been suggested that the photoreceptor, protein kinase and phosphorylated substrate protein coexist in the same polypeptide or tightly associated protein complex. An experiment that supports this hypothesis was recently reported by Warpeha and Briggs (1993). Triton X-100-solubilized membranes of pea were resolved by nondenaturing electrophoresis in the dark. The gel was cut into lanes and either mock irradiated or given a 30-sec pulse of blue light and incubated with [γ-^{32}P]ATP. Phosphorylation of a protein complex with an estimated molecular weight of 335 kDa was observed in the gel lane irradiated by blue light, but not in the control lane. Upon denaturation of the proteins in the complex in sodium dodecyl sulfate (SDS) and electrophoresis in SDS–polyacrylamide gels, a silver-stained protein of molecular weight 117 kDa, which coincided with the radiolabeled band, was observed.

Liscum and Briggs (1995) have recently reported the identification of *Arabidopsis* mutants that are defective in the perception of phototropic stimuli. Mutations in the *nph*1 (non-phototropic hypocotyl 1) locus lower the abundance of a 120-kDa protein. Although the *nph*1 gene has not yet been isolated, the authors suggest, on the basis of the characteristics of the mutants, that it encodes the apoprotein of the blue-light receptor.

4. Other membrane-associated protein kinases

Signal transduction in animal cells may also occur through the action of soluble protein kinases that transiently associate with the plasma membrane. Two types of growth factor receptor, the interleukin-2 receptor (Taniguchi and Minami, 1993) and T-cell antigen receptor (Weiss, 1993), do not have intrinsic protein kinase activity, but associate with cytoplasmic protein tyrosine kinases. These protein kinases (named lck) bind to the cytoplasmic domain of the activated receptor via SH2 (sarc homology 2) domains, and are activated themselves. Another soluble protein kinase, protein kinase C, translocates from the cytoplasm to the plasma membrane in response to factors which stimulate the phosphatidylinositol signaling pathway. Protein kinases C α and β are dependent on phospholipid and calcium for activity, and are stimulated by diacylglycerol, one of the products of phosphatidylinositol turnover (Nishizuka, 1988). To date, no strong evidence (i.e. purified enzymes or DNA sequences) has been obtained for the presence of either of these types of protein kinase in plants. However, a membrane-associated protein kinase from rice cross-reacts with antibodies directed towards two different synthetic peptides derived from protein kinase C (Abo-El-Saad and Wu, 1995).

Membrane-associated calmodulin-like domain protein kinase

While evidence for protein kinase C homologs in plants is lacking, it is clear that a family of enzymes, the calmodulin-like domain protein kinases (CDPKs), are present in numerous species of plants and algae (Roberts and Harmon, 1992). The activity of these enzymes is greatly stimulated (20- to 100-fold) by micromolar concentrations of free Ca^{2+}. Therefore, these enzymes are capable of sensing intracellular changes in the concentration of the second messenger Ca^{2+}, and of transmitting the message through the action of their protein kinase activities.

Properties of calmodulin-like domain protein kinases. Biochemical and immunological data indicate that members of the family of calmodulin-like domain protein kinases (also called calcium-dependent protein kinases) are tightly associated with plasma membranes isolated from oat roots (Schaller *et al.,* 1992), zucchini hypocotyls (Verhey *et al.,* 1993), silver beet leaves (Klucis and Polya, 1988) and soybean cell cultures (H. Borochov-Neori and A.C. Harmon, unpublished results). Treatment of purified plasma membranes with high concentrations of salt does not remove these enzymes, but detergents are able to solubilize the activity. Only two immunoreactive polypeptides are observed in oat root membranes, but four to six are present in plasma membranes from zucchini and soybean.

The CDPK from oat plasma membranes has been partially purified (Schaller *et al.,* 1992). The activity of the Triton X-100-solubilized enzyme is stimulated by both calcium and a crude lipid preparation. Phosphatidylinositol, platelet-activating factor and lysophosphatidylcholine were each able to stimulate enzyme activity by about five-fold, but none of them achieved the 20-fold stimulation observed with the crude lipid preparation. It is not known whether the effect of the lipids reflects a structural requirement for lipid by this membrane-bound enzyme, or whether it reflects regulation of the enzyme's activity by lipids acting as messengers.

Other membranes in addition to the plasma membrane have been identified as sites of calcium-regulated protein kinase activity. These include peribacteroid membranes of nitrogen-fixing nodules (Weaver et al., 1991), tonoplast (Johnson and Chrispeels, 1992) and chloroplast envelopes (Siegenthaler and Bovet, 1993).

Full-length cDNAs encoding CDPKs from soybean (Harper et al., 1991), Arabidopsis (Harper et al., 1993; Urao et al., 1994a, b; Hraback and Sussman, unpublished results), rice (Breviario et al., 1995; Kawasaki et al., 1993), maize (Estruch et al., 1994) and alfalfa (Bögre et al., 1993), and a partial cDNA from carrot (Suen and Choi, 1991), have been isolated and sequenced. The predicted proteins contain a protein kinase catalytic domain located near the amino terminus, and a calcium-binding regulatory domain that is similar to calmodulin near the carboxyl terminus. The catalytic domains are most similar to those of the family of calmodulin-dependent protein kinases (for reviews see Roberts and Harmon, 1992; Roberts, 1993). Although the amino acid sequence identity of these CDPKs in the catalytic and regulatory domains ranges from 50 to 95%, the size and relative spacing of the domains is constant. The greatest variation between these CDPKs occurs in the length of the sequence at the extreme amino terminus; AK1 has 113 residues preceding the catalytic domain, whereas the rice CDPK has 68 and the soybean CDPK has 40 such residues. None of the proteins have hydrophobic regions that would be predicted to be membrane-spanning regions. One possible mechanism of membrane association is myristoylation. Eight of 12 CDPK sequences from Arabidopsis have potential myristoylation sites (Hraback and Sussman, unpublished results). Carrot CDPK co-expressed in E. coli with a myristoyl transferase incorporates radiolabeled myristate (Farmer and Choi, 1995).

Membrane-bound proteins that are potential endogenous substrates of CDPK. To define the role of membrane-associated CDPK, it is necessary to identify its endogenous substrates. We shall examine the evidence that three membrane-bound proteins, namely the plasma membrane proton pump, nodulin 26, and tonoplast intrinsic protein, are phosphorylated in a calcium-dependent manner, and that they may be endogenous substrates of CDPK. All three of these proteins are involved in transmembrane transport. Other substrates of CDPK include sucrose phosphate synthase and nitrate reductase (Bachmann et al., 1995; McMichael et al., 1995; Bachmann and Huber, unpublished results), but the CDPKs that phosphorylate these proteins are soluble.

(a) Plasma membrane proton pump. The plasma membrane proton pump (H^+-ATPase) is largely responsible for maintaining the difference in electrical potential across the membrane, and transport of many solutes into the cell is dependent on the proton gradient that is generated by the action of this pump. The H^+-ATPase of plasma membranes isolated from oat and potato root cells undergoes Ca^{2+}-dependent phosphorylation. Phosphoamino-acid analysis reveals that serine and threonine but not tyrosine are phosphorylated, and suggests that there are multiple phosphorylation sites (Schaller and Sussman, 1988; Suzuki et al., 1991). CDPK partially purified from oat plasma membranes phosphorylates purified H^+-ATPase (Sussman et al., 1990), but the effect of the phosphorylation on the activity of the pump has yet to be determined. Scherer and co-workers have reported that a protein kinase activity which is stimulated by platelet-activating factor phosphorylates the H^+-ATPase (Nickel et al., 1991), but it is not known whether this protein kinase activity is related to CDPK.

Since removal of the carboxyl-terminal peptide from Arabidopsis H^+-ATPase constitutively activates the enzyme (Palmgren et al., 1991; Palmgren and Christensen, 1993), it has been proposed that the proton pump is regulated by an inhibitory domain located in the carboxyl terminus, and that activation could occur by phosphorylation of a site within the inhibitory domain (Palmgren et al., 1991). Consistent with this hypothesis is the observation that syringomycin, a fungal phytotoxin, stimulates in vitro the H^+-ATPase activity of plasma membranes of red beet storage tissue, and also increases phosphorylation of a 100-kDa protein which cross-reacts with the antibodies raised against synthetic peptides derived from the *Arabidopsis* H⁺-ATPase (Bidwai *et al.,* 1987). However, in conflict with the hypothesis is the observation that phosphorylation of H⁺-ATPase from the isolated microsomal vesicle fraction of corn roots inhibits ATPase activity, and dephosphorylation restores the ATPase activity (Zocchi, 1985). Therefore the relationship between phosphorylation and activity of H⁺-ATPase needs to be clarified.

(b) Nodulin 26 and tonoplast intrinsic protein. Tonoplast intrinsic protein (TIP) and nodulin 26 (NOD 26) are related in sequence to members of a family of membrane transport proteins which include GLpF in bacteria, the *bib* gene product in insects, and MIP and CHIP 28 in mammals. TIP homologues have been reported in bean, *Arabidopsis,* pea, tobacco and soybean, while NOD 26 is expressed specifically in nitrogen-fixing nodules of soybean, and is located in the peribacteroid membrane. Both TIP and NOD 26 are phosphorylated in a calcium-dependent manner (Johnson and Chrispeels, 1992; Weaver *et al.,* 1991).

In bean, a seed-specific form of TIP, called α-TIP, is located in the membrane of protein storage vacuoles and is phosphorylated and dephosphorylated *in vivo* and *in vitro.* Phosphorylation of this 27-kDa integral membrane protein occurs on a single serine residue near the N-terminus. Phosphorylation of TIP in isolated tonoplasts is dependent on calcium and appears to be mediated by a tonoplast-bound calcium-regulated protein kinase (Johnson and Chrispeels, 1992). The protein kinase has not yet been characterized. Experiments in which γ-TIP, the vegetative tissue-specific TIP from *Arabidopsis,* was expressed in *Xenopus* oocytes have suggested that it functions as a water-specific membrane channel (Maurel *et al.,* 1993), but it is not yet known whether the activity of these proteins is regulated by phosphorylation.

Nodulin 26 is phosphorylated *in vitro* and *in vivo* on serine 262, which is in the carboxyl-terminal region of the protein (Weaver and Roberts, 1992) and which faces the cytoplasmic side of the symbiosome membrane (Miao *et al.,* 1992). NOD 26 is phosphorylated in isolated nodules or membranes by a calcium-stimulated protein kinase (Miao *et al.,* 1992; Weaver *et al.,* 1991; Weaver and Roberts, 1992). A calcium-dependent protein kinase that phosphorylates NOD 26 *in vitro* at the same site that is phosphorylated *in vivo* has been purified and characterized (Weaver *et al.,* 1991). The prediction that NOD 26 functions as a channel protein has been tested in experiments in which NOD 26 purified from symbiosome membranes of soybean was reconstituted into liposomes (Weaver *et al.,* 1993). These studies showed that NOD 26 is an ion channel capable of transporting both cations and anions, but with weak selectivity for anions. Recent studies with recombinant NOD 26 suggest that phosphorylation of serine 262 modulates its channel activity by conferring voltage sensitivity (Lee *et al.,* 1995).

In separate studies with isolated symbiosome membranes, the transport of malate across the membranes was enhanced when the membranes were preincubated with ATP, and it was inhibited when they were preincubated with alkaline phosphatase (Ouyang *et al.,* 1991). Since NOD 26 is the single major polypeptide that is labeled when the membranes are incubated with [γ-³²P]ATP (Miao *et al.,* 1992; Ouyang *et al.,* 1991; Weaver *et al.,* 1991), it has been suggested that malate uptake across the symbiosome membrane is controlled by phosphorylation of NOD 26 (Ouyang *et al.,* 1991).

Light-harvesting complex II kinase

Phosphorylation of proteins occurs in both the envelope and thylakoid membranes of the chloroplast, but little is known about the protein kinases present in this organelle. The most well-studied system in chloroplasts is the phosphorylation of two apoproteins of the light-harvesting chlorophyll *a/b*–protein complex of photosystem II (LHCII), with molecular weights in the range

23–27 000, by a thylakoid membrane-associated protein kinase (Allen, 1992; Bennett, 1984, 1991). Phosphorylation of the proteins promotes the physical movement of LHCII complex from a region of the thylakoid membrane close to the photosystem II reaction center to a region near the photosystem I reaction center (Allen *et al.*, 1981). The movement of LHCII serves to balance the rates of the light reactions in photosystems I and II and maximizes the efficiency of noncyclic electron transport. Phosphorylation of LHCII is controlled by the redox state of the plastoquinone pool (Allen *et al.*, 1981). A 64-kDa protein purified from thylakoid membranes and originally thought to be the LHCII kinase (Coughlan and Hind, 1987; Gal *et al.*, 1990) has since been shown to be a polyphenoloxidase (Hind *et al.*, 1995; Race *et al.*, 1995; Sokolenko *et al.*, 1995) that co-solubilizes with the kinase activity.

5. ETR1: a plant homolog of a prokaryotic two-component system?

Characterization of mutant genes in *Arabidopsis* plants defective in their response to ethylene has revealed that plants may contain a bacterial-type phosphotransfer system (Chang *et al.*, 1993; Chang and Meyerowitz, 1994). The prokaryotic regulatory system consists of two components that are present either as separate proteins or combined in a single polypeptide (Stock *et al.*, 1990). The first component, called the sensor, autophosphorylates on a histidine residue in response to a signal. The phosphoryl group is then transferred to an aspartate residue of the second component, called the response regulator. The regulator functions as a positive effector in the regulation of gene expression or cellular physiology.

The *etr*1 (ethylene triple response 1) locus is one of five loci shown to be involved in the ethylene response pathway of *Arabidopsis* (Roman *et al.*, 1995). Plants bearing a mutant allele of this gene (*etr*1-1) exhibit none of the typical responses to ethylene, and have diminished ethylene-binding activity (Chang *et al.*, 1993). Epistatic analysis shows that *etr*1 acts early in the ethylene response pathway (Roman *et al.*, 1995), and indicates that it is either the ethylene receptor or the regulator of the pathway. Schaller and Bleecker (1995) have shown that yeast expressing ETR1 binds significant amounts of [^{14}C]ethylene, while control cells and cells expressing the mutant *etr*1-1 protein do not. This work provides strong evidence that ETR1 is the ethylene receptor.

ETR1 is predicted to be a protein of molecular weight 82 500 (Chang *et al.*, 1993). The amino-terminal 313 residues of ETR1 have no sequence similarity to any protein in the GenBank database, but the carboxyl-terminal portion of the molecule contains a domain with five motifs in the same order and with similar spacing to those found in histidine kinase domains of bacterial sensors. Carboxyl-terminal to the sensor-like domain is a domain having the predicted secondary structure and four conserved residues (three aspartates and one lysine) that are characteristic of response regulator proteins. The sequence of these ETR1 domains is very similar to those of proteins which contain both domains in a single polypeptide (see examples above). ETR1 homologs in *Arabidopsis* (Hua *et al.*, 1995) and tomato (Wilkinson *et al.*, 1995) do not have the aspartate-containing domain. Thus this domain is not necessary for function.

ETR1 has three hydrophobic regions close to the amino terminus, one or more of which may span a membrane (Chang *et al.*, 1993). Recombinant ETR1 expressed in yeast is associated with membrane fractions, but it is not clear in which membrane system it is located (Schaller *et al.*, 1995). ETR1 forms dimers via disulfide bridges between cysteine residues located in the predicted extracytoplasmic/transmembrane region of the protein.

An important question is whether ETR1 is a functional histidine kinase and whether this activity is necessary for the ethylene response pathway. While recombinant ETR1 has been expressed in yeast and shown to bind ethylene (Schaller and Bleecker, 1995), it has not been shown to autophosphorylate on histidine residues. No insight into this question has been gained from study of the available *etr*1 mutants, since all those that have been described contain mutations in the predicted extracytoplasmic/transmembrane region. The question of activity is especially interesting, since other eukaryotic proteins that are related to two-component proteins have been described, and one is a histidine kinase while others are serine/threonine kinases. Yeasts contain a two-component phosphotransfer system consisting of a separate sensor (Sln1) and response regulator (Ssk1) that function in osmoregulation via regulation of a microtubule-associated protein (MAP) kinase cascade (Maeda *et al.,* 1994; Ota and Varshavsky, 1993). A second downstream target of this system is a transcriptional factor (Mcm1) (Yu *et al.,* 1995). In contrast to this yeast system, two mammalian mitochondrial enzymes, branched chain alpha-ketoacid dehydrogenase kinase and pyruvate dehydrogenase kinase, are related to two-component proteins, but are functional serine/threonine kinases (Popov *et al.,* 1992, 1993).

Identification of components in the ethylene response pathway that act downstream of ETR1 is consistent with the hypothesis that ETR1 functions in the regulation of a protein kinase cascade. One of the components is CTR1 (constitutive triple response 1), which is a soluble protein kinase that has limited similarity to the raf kinases (Kieber *et al.,* 1993). In mammals and yeast, the raf kinases act at the beginning of MAP kinase cascades. Questions that remain to be answered concern the identification of the steps between ETR1 and CTR1, and how the binding of ethylene by ETR1 affects the activity of pathway components.

6. Conclusions

The work reviewed here shows that plants have many membrane-associated protein kinases and a potential bacterial-type phosphotransfer system, and that these enzymes are involved in numerous signal transduction pathways. While much information is available for individual components of pathways, all of the details for a complete pathway have not yet been elucidated. It is certain, however, that signal transduction pathways of the same general types as those found in animals exist in plants, but that plants have their own unique variations on the general themes. The discovery of a bacterial type of phosphotransfer system that is involved in the ethylene response pathway implies that there will be others. It will be extremely interesting to determine the physiological function, molecular structures, and mechanism of action of all these systems.

Acknowledgment

This work was supported by grants to A.C.H. from the National Science Foundation (DCB-9117837) and United States Department of Agriculture (91-37304-6654, 94-37304-1177).

References

Abo-El-Saad M, Wu R. (1995) A rice membrane calcium-dependent protein-kinase is induced by gibberellin. *Plant Physiol.* **108:** 787–793.

Allen JF. (1992) Protein phosphorylation in regulation of photosynthesis. *Biochim. Biophys. Acta* **1098:** 275–335.

Allen JF, Bennett J, Steinback KE, Arntzen CJ. (1981) Chloroplast protein phosphorylation couples plastoquinone redox state to distribution of excitation energy between photosystems. *Nature* **291:** 25–29.

Bachmann M, Shiraishi N, Campbell WH, Yoo B-C, Harmon AC, Huber SC. (1996) Identification of the major regulatory phosphorylation site as Ser-543 in spinach leaf nitrate reductase and its phosphorylation by a calcium-dependent protein kinase *in vitro. Plant Cell* (in press).

Bachmann M, McMichael RW Jr, Huber JL, Kaiser WM, Huber SC. (1995) Partial purification and characterization of a calcium-dependent protein kinase and an inhibitor protein required for inactivation of spinach leaf nitrate reductase. *Plant Physiol.* **108:** 1083–1092.

Becraft PW, McCarty DR. (1995) The maize Crinkly Leaf4 gene is involved in the morphogenesis of kernel aleurone and leaf epidermis. *J. Cell. Biochem.* Suppl. 21A: 455.

Bennett J. (1984) Chloroplast protein phosphorylation and the regulation of photosynthesis. *Physiol. Plant.* **60:** 583–590.

Bennett J. (1991) Protein phosphorylation in green plant chloroplasts. *Annu. Rev. Plant Physiol.* **42:** 281–311.

Bidwai AP, Zhang L, Bachmann RC, Takemoto JY. (1987) Mechanism of action of *Pseudomonas syringae* phytotoxin, syringomycin. Stimulation of red beet plasma membrane ATPase activity. *Plant Physiol.* **83:** 39–43.

Bögre L, Harmon AC, Szalay AA, Hirt H, Heberle-Bors E. (1993) Ca²⁺-dependent protein kinase from alfalfa, regulation during the cell cycle and auxin treatment. In: *12th Annual Missouri Plant Biochemistry, Molecular Biology, and Physiology Symposium* (Randall DD, Walker JW, ed.). University of Missouri, Columbia: Interdisciplinary Plant Physiology Group, Abstract 15.

Braun DM, Walker JC. (1996) Plant transmembrane receptors: new pieces in the signaling puzzle. *Trends Biochem. Sci.* **21:** 70–73.

Breviario D, Morello L, Giani S. (1995) Molecular cloning of two novel rice cDNA sequences encoding putative calcium-dependent protein kinases. *Plant Mol. Biol.* **27:** 953–967.

Chang C, Meyerowitz EM. (1994) Eukaryotes have 'two-component' signal transducers. *Res. Microbiol.* **145:** 481–486.

Chang C, Schaller GE, Patterson S, Kwok SF, Meyerowitz EM, Bleecker AB. (1992) The TMK1 gene from *Arabidopsis* codes for a protein with structural and biochemical characteristics of a receptor protein kinase. *Plant Cell* **4:** 1263–1271.

Chang C, Kwok SF, Bleecker AB, Meyerowitz EM. (1993) *Arabidopsis* ethylene-response gene *ETR1*: similarity of product to two-component regulators. *Science* **262:** 539–544.

Coughlan S, Hind G. (1987) Phosphorylation of thylakoid proteins by a purified kinase. *J. Biol. Chem.* **262:** 8402–8408.

Delorme V, Giranton JL, Hatzfeld Y, Friry A, Heizmann P, Ariza MJ, Dumas C, Gaude T, Cock JM. (1995) Characterization of the *S*-locus genes, SLG and SRK, of the *Brassica* S₃ haplotype. Identification of a membrane-localized protein encoded by the S-locus receptor kinase gene. *Plant J.* **7:** 429–440.

Dwyer KG, Kandasamy MK, Mahosky DI, Acciai J, Kudish BI, Miller JE, Nasrallah ME, Nasrallah JB. (1994) A superfamily of S-locus-related sequences in *Arabidopsis*: diverse structures and expression patterns. *Plant Cell* **6:** 1829–1843.

Estruch JJ, Kadwell S, Merlin E, Crossland L. (1994) Cloning and characterization of a maize pollen-specific calcium-dependent calmodulin-independent protein kinase. *Proc. Natl Acad. Sci. USA* **91:** 8837–8841.

Farmer PK, Choi JH. (1995) Expression and potential myristoylation of a calcium-dependent protein-kinase. *J. Cell. Biochem.* Suppl. 21A: 507.

Gal A, Hauska G, Herrmann R, Ohad I. (1990) Interaction between light harvesting chlorophyll-*a/b* protein (LHCII) kinase and cytochrome *b₆/f* complex. *In vitro* control of kinase activity. *J. Biol. Chem.* **265:** 19742–19749.

Giranton J-L, Ariza M, Dumas C, Cock JM, Gaude T. (1995) The S-locus receptor kinase gene encodes a soluble glycoprotein corresponding to the SRK extracellular domain in *Brassica oleracea. Plant J.* **8:** 827–834.

Goring DR, Rothstein SJ. (1992) The S-locus receptor kinase gene in a self-incompatible *Brassica napus* line encodes a functional serine/threonine kinase. *Plant Cell* 4: 1273–1281.

Goring DR, Glavin TL, Schafer U, Rothstein SJ. (1993) An *S* receptor kinase gene in self-compatible *Brassica napus* has a 1-bp deletion. *Plant Cell* **5:** 531–539.

Hager A, Brich M. (1993) Blue-light-induced phosphorylation of a plasma-membrane protein from phototropically sensitive tips of maize coleoptiles. *Planta* **189:** 567–576.

Hanks SK, Quinn AM, Hunter T. (1988) The protein kinase family: conserved features and deduced phylogeny of the catalytic domains. *Science* **241:** 42–52.

Harper JF, Sussman MR, Schaller GE, Putnam-Evans C, Charbonneau H, Harmon AC. (1991) A calcium-dependent protein kinase with a regulatory domain similar to calmodulin. *Science* **252:** 951–954.

Harper JF, Binder BM, Sussman MR. (1993) Calcium and lipid regulation of an *Arabidopsis* protein kinase expressed in *Escherichia coli. Biochemistry* **32:** 3282–3290.

Hind G, Marshak DR, Coughlan SJ. (1995) Spinach thylakoid polyphenol oxidase cloning, characterization, and relation to a putative protein-kinase. *Biochemistry* **34:** 8157–8164.

Hua J, Chang C, Sun Q, Meyerowitz EM. (1995) Ethylene insensitivity conferred by *Arabidopsis* ERS gene. *Science* **269:** 1712–1714.

Johnson KD, Chrispeels MJ. (1992) Tonoplast-bound protein kinase phosphorylates tonoplast intrinsic protein. *Plant Physiol.* **100:** 1787–1795.

Jones DA, Thomas CM, Hammondkosack KE, Balintkurti PJ, Jones JDG. (1994) Isolation of the tomato Cf-9 gene for resistance to *Cladosporium fulvum* by transposon tagging. *Science* **266:** 4789–4793.

Kawasaki T, Hayashida N, Baba T, Shinozaki K, Shimada H. (1993) The gene encoding a calcium-dependent protein kinase located near the *sbe*1 gene encoding starch branching enzyme-I is specifically expressed in developing rice seeds. *Gene* **129:** 183–189.

Kieber JJ, Rothenberg M, Roman G, Feldmann KA, Ecker JR. (1993) *CTR*1, a negative regulator of the ethylene response pathway in *Arabidopsis,* encodes a member of the Raf family of protein kinases. *Cell* **72:** 427–441.

Klucis E, Polya GM. (1988) Localization, solubilization and characterization of plant membrane associated calcium-dependent protein kinases. *Plant Physiol.* **88:** 164–171.

Kohorn BD, Lane S, Smith TA. (1992) An *Arabidopsis* serine/threonine kinase homologue with an epidermal growth factor repeat selected in yeast for its specificity for a thylakoid membrane protein. *Proc. Natl Acad. Sci. USA* **89:** 10989–10992.

Krebs EG. (1986) The enzymology of control by phosphorylation. In: *The Enzymes, XVII* (Boyer P, Krebs EG, ed.). New York: Academic Press, pp. 3–20.

Kumar V, Trick M. (1994) Expression of the S-locus receptor kinase multigene familily in *Brassica oleracea. Plant J.* **6:** 807–813.

Lee JW, Zhang YX, Weaver CD, Shomer NH, Louis CF, Roberts DM. (1995) Phosphorylation of nodulin-26 on serine-262 affects its voltage-sensitive channel activity in planar lipid bilayers. *J. Biol. Chem.* **270:** 27051–27057.

Liscum E, Briggs WR. (1995) Mutations in the NPH1 locus of *Arabidopsis* disrupt the perception of phototropic stimuli. *Plant Cell* **7:** 473–485.

McMichael RW Jr, Bachmann M, Huber SC. (1995) Spinach leaf sucrose-phosphate synthase and nitrate reductase are phosphorylated/inactivated by multiple protein kinases *in vitro. Plant Physiol.* **108:** 1077–1082.

Maeda T, Wurgler-Murphy SM, Saito H. (1994) A two-component system that regulates an osmosensing MAP kinase cascade in yeast. *Nature* **369:** 242–245.

Massague J. (1992) Receptors for the TGF-β family. *Cell* **69:** 1067–1070.

Maurel C, Reizer J, Schroeder JI, Chrispeels MJ. (1993) The vacuolar membrane protein γ-TIP creates water-specific channels in *Xenopus* oocytes. *EMBO J.* **12:** 2241–2247.

Miao G-H, Hong Z, Verma DPS. (1992) Topology and phosphorylation of soybean nodulin 26, an intrinsic protein of the peribacteroid membrane. *J. Cell Biol.* **118:** 481–490.

Mu J-H, Lee H-S, Kao T. (1994) Characterization of a pollen-expressed receptor-like kinase gene of *Petuna inflata* and the activity of its encoded kinase. *Plant Cell* **6:** 709–721.

Nickel R, Schutte M, Hecker D, Scherer GFE. (1991) The phospholipid platelet-activating factor stimulates proton extrusion in cultured soybean cells and protein phosphorylation and ATPase activity in plasma membranes. *J. Plant Physiol.* **139:** 205–211.

Nishizuka Y. (1988) The molecular heterogeneity of protein kinase C and its implications for cellular regulation. *Nature* **334:** 661–665.

Ota IM, Varshavsky A. (1993) A yeast protein similar to bacterial two-component regulators. *Science* **262:** 566–569.

Ouyang L-J, Whelan J, Weaver CD, Roberts DM, Day DA. (1991) Protein phosphorylation stimulates the rate of malate uptake across the peribacteroid membrane of soybean nodules. *FEBS Lett.* **293:** 188–190.

Palmer JM, Short TW, Gallagher S, Briggs WR. (1993) Blue light-induced phosphorylation of a plasma membrane-associated protein in *Zea mays* L. *Plant Physiol.* **102:** 1211–1218.

Palmgren MG, Christensen G. (1993) Complementation *in situ* of the yeast plasma membrane H$^+$-ATPase gene *pma*1 by an H$^+$-ATPase gene from a heterologous species. *FEBS Lett.* **317:** 216–222.

Palmgren MG, Sommarin M, Serrano R, Larsson C. (1991) Identification of an autoinhibitory domain in the C-terminal region of the plant plasma membrane H$^+$-ATPase. *J. Biol. Chem.* **266:** 20470–20475.

Popov KM, Zhao Y, Shimomura Y, Kuntz MJ, Harris RA. (1992) Branched-chain α-ketoacid dehydrogenase kinase. Molecular cloning, expression, and sequence similarity with histidine protein kinases. *J. Biol. Chem.* **267:** 13127–13130.

Popov KM, Kedishvili NY, Zhao Y, Shimomura Y, Crabb DW, Harris RA. (1993) Primary structure of pyruvate dehydrogenase kinase establishes a new family of eukaryotic protein kinases. *J. Biol. Chem.* **268:** 26602–26606.

Race HL, Eatonrye JJ, Hind G. (1995) A 64-kDa protein is a substrate for phosphorylation by a distinct thylakoid protein-kinase. *Photosyn. Res.* **43:** 231–239.

Reymond P, Short TW, Briggs WR. (1992a) Blue light activates a specific protein kinase in higher plants. *Plant Physiol.* **100:** 655–661.

Reymond P, Short TW, Briggs WR, Poff KL. (1992b) Light-induced phosphorylation of a membrane protein plays an early role in signal transduction for phototropism in *Arabidopsis thaliana. Proc. Natl Acad. Sci. USA* **89:** 4718–4721.

Roberts DM. (1993) Protein kinases with calmodulin-like domains: novel targets of calcium signals in plants. *Curr. Opin. Cell Biol.* **5:** 242–246.

Roberts DM, Harmon AC. (1992) Calcium-modulated proteins — targets of intracellular calcium signals in higher plants. *Annu. Rev. Plant Physiol. Mol. Biol.* **43:** 375–414.

Roman G, Lubarsky B, Kieber JJ, Rothenberg M, Ecker JR. (1995) Genetic analysis of ethylene signal transduction in *Arabidopsis thaliana*: five novel mutant loci integrated into a stress response pathway. *Genetics* **139:** 1393–1409.

Schaller GE, Sussman MR. (1988) Phosphorylation of the plasma-membrane H+-ATPase of oat roots by a calcium-stimulated protein kinase. *Planta* **173:** 509–518.

Schaller GE, Bleecker AB. (1995) Ethylene-binding sites generated in yeast expressing the *Arabidopsis ETR1* gene. *Science* **270:** 1809–1811.

Schaller GE, Harmon AC, Sussman MR. (1992) Characterization of a calcium- and lipid-dependent protein kinase associated with the plasma membrane of oat. *Biochemistry* **31:** 1721–1727.

Schaller GE, Ladd AN, Lanahan MB, Spanbauer JM, Bleecker AB. (1995) The ethylene response mediator *ETR1* from *Arabidopsis* forms a disulfide-linked dimer. *J. Biol. Chem.* **270:** 12526–12530.

Schottens-Toma IMJ, de Wit PJGM. (1988) Purification and primary structure of a necrosis-inducing peptide from the apoplastic fluids of tomato infected with *Cladosporium fulvum* (syn. *Fulvia fulva*). *Physiol. Mol. Plant Pathol.* **33:** 59–67.

Short TW, Porst M, Briggs WR. (1992) A photoreceptor system regulating *in vivo* and *in vitro* phosphorylation of a pea plasma membrane protein. *Photochem. Photobiol.* **55:** 773–781.

Siegenthaler P-A, Bovet L. (1993) A unique protein-kinase activity is responsible for the phosphorylation of the 26- and 14-kDa proteins but not of the 67-kDa protein in the chloroplast envelope membranes of spinach. *Planta* **190:** 231–240.

Sokolenko A, Fulgosi H, Gal A, Altschmied L, Ohad I, Herrmann RG. (1995) The 64-kDa polypeptide of spinach may not be the LHCII kinase, but a lumen-located polyphenol oxidase. *FEBS Lett.* **37:** 176–180.

Song W-Y, Wang G-L, Chen L-L et al. (1995) A receptor kinase-like protein encoded by the rice disease resistance gene, *Xa21. Science* **270:** 1804–1806.

Stein JC, Nasrallah JB. (1993) A plant receptor-like gene, the S-locus receptor kinase of *Brassica oleracea* L., encodes a functional serine threonine kinase. *Plant Physiol.* **101:** 1103–1106.

Stein JC, Howlett B, Boyes DC, Nasrallah ME, Nasrallah JB. (1991) Molecular cloning of a putative receptor protein kinase gene encoded at the self-incompatibility locus of *Brassica oleracea. Proc. Natl Acad. Sci. USA* **88:** 8816–8820.

Stock JB, Stock AM, Mottonen JM. (1990) Signal transduction in bacteria. *Nature* **344:** 395–400.

Stone JM, Walker JC. (1995) Plant protein kinase families and signal transduction. *Plant Physiol.* **108:** 451–457.

Stone JM, Collinge MA, Smith RD, Horn MA, Walker JC. (1994) Interaction of a protein phosphatase with an *Arabidopsis* serine–threonine receptor kinase. *Science* **266:** 793–795.

Suen KL, Choi JH. (1991) Isolation and sequence analysis of a cDNA clone for a carrot calcium-dependent protein kinase — homology to calcium calmodulin-dependent protein kinases and to calmodulin. *Plant Mol. Biol.* **17:** 581–590.

Sussman MR, Schaller GE, DeWitt N. (1990) Regulation of the plasma membrane proton pump. In: *Current Topics in Plant Biochemisty and Physiology. Plant Protein Phosphorylation, Protein Kinases, Calcium and Calmodulin, Vol. 9* (Randall DD, Blevins DG, ed.). Columbia: Interdisciplinary Plant Biochemistry and Physiology Program, pp. 183–189.

Suzuki YS, Wang Y, Takemoto JY. (1991) Syringomycin-stimulated phosphorylation of the plasma membrane H+-ATPase from red beet storage tissue. *Plant Physiol.* **99:** 1314–1320.

Swanson RV, Alex LA, Simon MI. (1995) Histidine and aspartate phosphorylation two-component systems and the limits of homology. *Trends Biochem. Sci.* **19**: 485–490.

Taniguchi T, Minami Y. (1993) The IL-2/IL-2 receptor system: a current overview. *Cell* **73**: 5–8.

Tantikanjana T, Nasrallah ME, Stein JC, Chen CH, Nasrallah JB. (1993) An alternative transcript of the S locus glycoprotein gene in a class II pollen-recessive self-incompatibility haplotype of *Brassica oleracea* encodes a membrane-anchored protein. *Plant Cell* **5**: 657–666.

Tepass U, Theres C, Knust E. (1990) *crumbs* encodes an EGF-like protein expressed on apical membranes of *Drosophila* epithelial cells and required for organization of epithelia. *Cell* **61**: 787–799.

Tobias CM, Howlett B, Nasrallah JB. (1992) An *Arabidopsis thaliana* gene with sequence similarity to the S-locus receptor kinase of *Brassica oleracea* — sequence and expression. *Plant Physiol.* **99**: 284–290.

Ullrich A, Schlessinger J. (1990) Signal transduction by receptors with tyrosine kinase activity. *Cell* **69**: 1067–1070.

Urao T, Katagiri T, Mizoguchi T, Yamaguchi-Shinozaki K, Hayashida N, Shinozaki K. (1994a) An *Arabidopsis thaliana* cDNA encoding Ca^{2+}-dependent protein kinase. *Plant Physiol.* **105**: 1461–1462.

Urao T, Katagiri T, Mizoguchi T, Yamaguchi-Shinozaki K, Hayashida N, Shinozaki K. (1994b) Two genes that encode Ca^{2+}-dependent protein kinases are induced by drought and high-salt stresses in *Arabidopsis thaliana. Mol. Gen. Genet.* **244**: 331–340.

Valon C, Smalle J, Goodman HM, Giraudat J. (1993) Characterization of an *Arabidopsis thaliana* gene (TMKL1) encoding a putative transmembrane protein with an unusual kinase-like domain. *Plant Mol. Biol.* **23**: 415–421.

Vandenabeele P, Declercq W, Beyaert R, Fiers W. (1995) Two tumor-necrosis-factor receptors: structure and function. *Trends Cell Biol.* **5**: 392–399.

Van Kan JAL, Van den Ackerveken GFJM, de Wit PJGM. (1991) Cloning and characterization of cDNA of avirulence gene avr9 of the fungal pathogen *Cladosporium fulvum*, causal agent of tomato leaf mold. *Mol. Plant-Microbe Interact.* **4**: 52–59.

Verhey SD, Gaiser JC, Lomax TL. (1993) Protein kinases in zucchini. Characterization of calcium-requiring plasma membrane kinases. *Plant Physiol.* **103**: 413–419.

Walker JC. (1993) Receptor-like protein kinase genes of *Arabidopsis thaliana. Plant J.* **3**: 451–456.

Walker JC. (1994) Structure and function of the receptor-like protein-kinases of higher plants. *Plant Mol. Biol.* **26**: 1599–1609.

Walker JC, Zhang R. (1990) Relationship of a putative receptor protein kinase from maize to the S-locus glycoproteins of *Brassica. Nature* **345**: 743–746.

Warpeha KMF, Briggs WR. (1993) Blue light-induced phosphorylation of a plasma membrane protein in pea: a step in the signal transduction chain for phototropism. *Aust. J. Plant Physiol.* **20**: 393–403.

Weaver CD, Roberts DM. (1992) Determination of the site of phosphorylation of nodulin 26 by the calcium-dependent protein kinase from soybean nodules. *Biochemistry* **31**: 8954–8959.

Weaver CD, Crombie B, Stacey G, Roberts DM. (1991) Calcium-dependent phosphorylation of symbiosome membrane proteins from nitrogen-fixing soybean nodules. *Plant Physiol.* **95**: 222–227.

Weaver CD, Shomer NH, Louis CF, Roberts DM. (1993) Nodulin 26, a nodule-specific symbiosome membrane protein from soybean, is an ion channel. *J. Biol. Chem.* **269**: 17858–17862.

Weiss A. (1993) T-cell antigen receptor signal transduction: a tale of tails and cytoplasmic protein-tyrosine kinases. *Cell* **73**: 209–212.

Wilkinson JQ, Lanahan MB, Yen H-C, Giovannoni JJ, Klee HJ. (1995) An ethylene-inducible component of signal transduction encoded by *Never-ripe. Science* **270**: 1807–1809.

Yu G, Deschenes RJ, Fassler JS. (1995) The essential transcription factor, Mcm1, is a downstream target of Sln1, a yeast 'two-component' regulator. *J. Biol. Chem.* **270**: 8739–8743.

Zhao Y, Feng XH, Watson JC, Bottino PJ, Kung SD. (1994) Molecular cloning and biochemical characterization of a receptor-like serine/threonine kinase from rice. *Plant Mol. Biol.* **26**: 791–803.

Zocchi G. (1985) Phosphorylation–dephosphorylation of membrane proteins controls the microsomal H^+-ATPase activity of corn roots. *Plant Sci.* **40**: 153–159.

Chapter 9

A role for the heterotrimeric G-protein switch in higher plants

Klaus Palme

1. Introduction

Signal perception at the cell surface and information transfer into the cell are important for the continuous adjustment of cellular metabolism and the cell's responsiveness to its environment. Increasing our understanding of the molecular mechanisms underlying these basic processes has been a major challenge. Application of the tools provided by molecular biology and genetics has provided numerous opportunities to visualize and analyse some of the major signalling pathways of eukaryotic cells.

It has been found that protein kinases and GTPases (i.e. guanine-nucleotide-binding regulatory proteins, G-proteins) are among the most important switches utilized intracellularly. They have been found in all eukaryotic systems analysed to date, including higher plants. GTPases are enzymes regarded as molecular switches because they cycle between an active and an inactive state. This mechanism ensures vectorial flow of information at the expense of GTP (Stouten *et al.*, 1993). This molecular switch appears to have been developed very early on, and, through evolution, it has been adapted to a variety of tasks. GTPases perform diverse functions. For example, they mediate hormone and light signalling across plasma membranes, they control accuracy during protein translation and are important during nascent protein translocation. The heterotrimeric G-proteins are mediators in transmembrane signalling through coupling to membrane-spanning receptors, and have been shown to be potent modulators of gene activity. Small GTPases play crucial roles in growth control, cytoskeletal polymerization and vesicle transfer. Detailed knowledge of the many specific systems regulated by GTPases has been accumulating at a rapid pace during the past few years. Information has been collected and summarized in various monographs and reviews and therefore will not be duplicated here (see Bourne, 1989, 1993; Clapham and Neer, 1993; Dickey and Birnbaumer, 1993; Hepler and Gilman, 1992; Stryer and Bourne, 1986; Taylor, 1990). Instead, this review will focus on the role of GTPases, in particular heterotrimeric GTPases, in plants. What is known about plant GTPases, what are the novel insights to be expected from analysis of these plant proteins, and what are the target proteins with which they interact? Before summarizing this information, we shall briefly discuss the GTPase cycle and then consider plant GTPases and various types of interacting proteins.

2. GTPases switch between inactive and active states

In its basal state, the heterotrimeric GTPase exists as a $\alpha\beta\gamma$ heterotrimer composed of three subunits; the α-subunit, which binds and exchanges guanine nucleotides (molecular mass *c*. 39–46 kDa), and the β-subunit (molecular mass *c*. 37 kDa) and the γ-subunit (molecular mass *c*. 8 kDa), which anchor the complex to the membrane. Upon activation of the receptor by an appropriate agonist, the receptor binds to the Gα-subunit, promoting the release of prebound GDP and binding of GTP (i.e. the Gα-subunits cycle between an inactive GDP-bound and an active GTP-bound form) (see *Figure 1* for regulatory cycle). Termination of the active state occurs as a result of the intrinsic GTPase activity of the Gα-subunit, which results in release of the terminal phosphate from bound GTP (Ross, 1992). The action of the ligand–receptor complex on the heterotrimeric G-protein complex has been viewed as a 'guanine nucleotide release protein' (GNRP) activity (Bourne *et al.*, 1990). GNRP activities were noted as being essential for fine-tuning the action of small G-proteins. However, separate GNRP activities have not yet been reported for heterotrimeric GTPases, and may well be presented by ligand-bound receptor proteins. The activated, GTP-bound Gα-subunit dissociates from the βγ-complex and associates with cellular effector proteins. These proteins are the catalytically active compounds responsible for the amplification of the initial signal and the initiation of the final physiological responses as long as the Gα-subunit remains actively bound to them. Termination of effector activity occurs after hydrolysis of Gα-bound GTP, and return of the Gα-subunit to the membrane follows.

The models for the action of heterotrimeric GTPases in different systems are very similar mechanistically. In all cases, the process outlined above was found to take place, resulting in

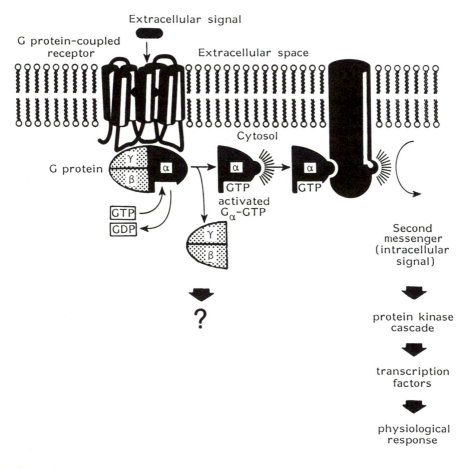

Figure 1. The regulatory cycle of heterotrimeric G-proteins.

interaction of the ligand-bound receptor with the heterotrimeric G-protein complex in guanine exchange and dissociation of the GTP-bound Gα-subunit and the βγ-complex from one another. The various systems differ only in which subunit activates the effector proteins; in some cases the Gα-subunit is the activating component, whereas in others the βγ-subunit is the activating component (for reviews see Clapham and Neer, 1993; Hou *et al.*, 1993; Simon *et al.*, 1991). The ability of both Gα and βγ to activate effectors is of advantage to cells, as it provides flexibility in the capacity of the G-proteins to modulate several effector proteins in response to a single agonist, one by the Gα-subunit and the other by the βγ-complex.

3. Heterotrimeric G-proteins

While purification of the individual components of these heterotrimeric G-protein complexes from mammalian cells has resulted in cloning of their genes and elucidation of their molecular mechanisms of action, it was found that in higher plants the components of these complexes were difficult to isolate to homogeneity, despite intense efforts. Early reports showed that GTPγS, a non-hydrolysable stable GTP analogue, bound to microsomal as well as to highly purified plasma membrane preparations (Blum *et al.*, 1988; Ephritikhine *et al.*, 1993; Jacobs *et al.*, 1987; Perdue and Lomax, 1992; Wise and Hillner, 1992). Interestingly, [α^{32}P]GTP-binding proteins were also found in the outer envelope membrane of pea chloroplasts (Sasaki *et al.*, 1991). All proteins identified were within the size range of small and heterotrimeric GTPases (Bhullar *et al.*, 1990; Bilushi *et al.*, 1991; Blum *et al.*, 1988; Drøbak *et al.*, 1988; Jacobs *et al.*, 1987; Perdue and Lomax, 1992; Sasaki *et al.*, 1991; Wise and Millner, 1992; Zaina *et al.*, 1989, 1990; Zbell *et al.*, 1989). Moreover, photoaffinity labelling with [α^{32}P]GTP-azido-anilide, a photoreactive GTP analogue (Offermans *et al.*, 1990), resulted in identification of proteins of 27 kDa, 43 kDa and 57 kDa that competed with unlabelled guanine nucleotides (Hesse, 1991; Hesse and Palme, unpublished data). Purification of candidate Gα-subunits and βγ-complexes proved to be difficult. This was probably due to detergents present during the purification procedure interfering with the [γ^{35}S]GTP binding assays. Thus, after extraction of proteins from plasma membranes by freeze–thawing and subsequent anion- and size-exclusion chromatography, an apparently homogenous [γ^{35}S]GTP-binding protein complex, consisting of a 34-kDa protein and a 27-kDa protein, was isolated from maize root plasma membranes (Bilushi *et al.*, 1991). Partial purification of Gα-like proteins from *Pisum sativum* and *Arabidopsis thaliana* was achieved using extraction of membranes with neutral detergents such as octyl-glucoside or nonyl-glucamide (White *et al.*, 1992). Similarly, using peptide antibodies raised against the Gα common peptide as an affinity matrix, a 33-kDa polypeptide was isolated from *Sorghum bicolor* (Ricart *et al.*, 1993). An even more rewarding approach involved the use of GTP-biotin derivatives, a molecule that effectively competes with [^{3}H]Gpp(NH)p for binding to GTP-binding proteins from *Avena sativa* (de Boer *et al.*, 1993). Binding of this molecule to solubilized peripheral oat plasma membrane proteins allowed the isolation of 10 different polypeptides through reversible coupling of the biotin moiety to an avidin-bound column. Using Ras and Gα-common peptide antibodies to detect cross-reacting proteins on Western blots, several proteins of molecular mass approximately 26 kDa and 38 kDa were detected. However, as no amino acid sequence data are yet available, the relationship between these proteins and Ras and Gα proteins has yet to be established.

As some of the mammalian Gα-subunits contain short polypeptide motifs that can be ADP-ribosylated by cholera or pertussis toxins (the ADP-ribose acceptor for pertussis toxin is arginine and that for cholera toxin is cysteine), these toxins were also frequently used to identify candidate Gα-subunits (Hasunuma and Funadera, 1987; Hasunuma *et al.*, 1987a,b; Hesse, 1991; Warpeha *et al.*, 1991). While these toxins will ADP-ribosylate several substrate proteins, ADP-ribosylation must be correlated with interference with GTPase function. Indeed, this was demonstrated for a 60-kDa thylakoid protein from *Pisum sativum* (Millner and Robinson, 1989). Another 40-kDa protein was found in highly purified plasma membranes isolated from *P. sativum* leaf tissue, when incubated with cholera toxin and [^{32}P]ADP-ribose (Warpeha *et al.*, 1991). The presence of this signal, as well as the binding of [^{35}S]GTPγS to these membranes, was dependent on exposure of the leaf tissue to blue light before isolation of the membranes. More detailed analysis provided further evidence for a blue light receptor-dependent activation of this putative Gα-subunit protein (Warpeha *et al.*, 1992). All of these data provided circumstantial evidence for the presence of functional small and heterotrimeric G-proteins in plants. As many genes encoding small G-proteins have been cloned over the last few years (for reviews see Palme, 1992a,b; Terryn *et al.*, 1993), we shall concentrate on what is known about the heterotrimeric G-proteins from higher plants.

4. Gα genes

To date, several genes encoding G-proteins homologous to Gα proteins have been reported. The *Arabidopsis thaliana GPA1* gene is localized on chromosome 2, within a distance of approximately 1.2 centimorgan of the *ER* (erecta) and *HY1* locus (Ma *et al.*, 1990). The single-copy gene contains 14 exons that predict a protein of 383 amino acids and molecular mass 44 kDa (GPα1). A related single-copy gene (*TGA1*) with a predicted open reading frame of 384 amino acid residues was isolated from *Lycopersicon esculentum* (Ma *et al.*, 1991). GPα1 and TGα1 are highly homologous proteins (84.6% identity, 93% similarity). They are also highly homologous to ZmGPα1, a protein that is 380 amino acid residues in length and encoded by the maize *ZmGPA1* gene (Hesse, 1991; B. Brzobohaty, F. Hesse and K. Palme, unpublished data). This protein shares 75% identity and 84% similarity with GPα1. Similarity to other Gα proteins is much lower, ranging from 50.6% (rat stimulatory G-protein) to 58% (bovine transducin), with intermediate values for other Gα proteins. Plant Gα-subunit proteins contain all the regions that are well known from other yeast and mammalian Gα proteins. Conserved motifs include the regions involved in binding of the GTPMg^{2+}, lipid modification and ADP-ribosylation by cholera toxin (Landis *et al.*, 1989; Markby *et al.*, 1993; Noel *et al.*, 1993). Moreover, these proteins contain a region of about 100 amino acids that has recently been proposed to function as an intrinsic GTPase-activating-protein-like activity (Landis *et al.*, 1989; Markby *et al.*, 1993). This may help to explain why, compared with small G-proteins like Ras, Gα-subunit proteins have a much higher GTPase activity and therefore apparently do not require a GTPase-activating protein for stimulation of their hydrolysis rates.

In order to investigate GPα1 functions, ectopic expression studies as well as immunolocalization experiments have been performed to study this protein during *Arabidopsis* development (Huang *et al.*, 1994; Weiss *et al.*, 1993). Using antibodies raised against a synthetic peptide of the C-terminal region of GPα1, GPα1 was found to be widely distributed in all organs tested except mature seeds (Weiss *et al.*, 1993). GPα1 levels were higher in growing plant parts such as meristematic zones and elongating tissues. Particularly high levels were detected in shoot and floral meristems, and the leaf and the floral organ primordia (i.e. sepal, petal, stamen and gynoecium). After fertilization, GPα1 expression increased in developing embryos, dividing microspores and growing pollen tubes, as well as in the nectaries. In mature tissues (i.e. leaf and mesophyll cells), the GPα1 protein was found in the vascular tissue. This localization was particularly interesting, as it suggests a possible role for GPα1 in nutrient accumulation and/or transport. Through the control of nutrient transport, it may be possible for GPα1 to influence a signalling chain that could affect the control of cell division.

Whereas in mammals the components of the heterotrimeric GTPases are encoded by large gene families (16 different genes for Gα, four for the Gβ-subunit, and seven for the γ-subunit), it is surprising that to date only one class of genes has been isolated from plants. Does this mean that there is less variability in the requirement of plants for regulatory heterotrimeric GTPases? Or do other plant Gα-subunits differ significantly in their primary amino acid composition and thus their DNA sequence? Answers to this question will probably be obtained either from molecular genetic analysis or, alternatively, from systematic genome DNA-sequencing attempts.

5. Gβ genes

In contrast to the diversity observed among the Gα-subunits, structural and functional differences between the other components of the G-protein heterotrimer are less well understood. The β- and γ-subunits form a tightly associated complex, and studies have indicated that βγ-complexes of various different Gα-subunits are functionally interchangeable. For a long time it was thought that only the Gα-subunit was important for activation of signal transduction pathways, but more recent evidence indicates that the Gβγ-complex also plays a major role in the cell's signal transmission. Effector proteins regulated by the Gβγ-complex include, for example, ion channels and specific subtypes of adenyl cyclases (see Brown, 1991; Birnbaumer et al., 1990; Clapham and Neer, 1993; Iniguez-Lluhi et al., 1993; Simonds et al., 1991). An interesting recent example of target enzyme regulation by Gβγ-complexes is a mammalian phosphoinositide-3-kinase, a key signalling enzyme that is involved in receptor-stimulated mitogenesis, oxidative bursting, membrane ruffling and glucose uptake (Stephens et al., 1994). It will be interesting to examine the data on the regulation of the corresponding plant gene.

The first plant gene encoding a member of the Gβ family has recently been isolated from *Arabidopsis thaliana* (C. Kruse, B. Brzobohaty, J. Schell and K. Palme, submitted for publication). This single-copy gene consists of six exons. The deduced amino acid sequence of AtGβ consists of a polypeptide of 377 amino acid residues with a predicted molecular mass of 37 kDa. Comparison of the AtGβ sequence with protein databases revealed significant homology to the eukaryotic Gβ family. The homology was particularly high towards the Gβ from *Caenorhabditis elegans* (51%), whereas homology to other Gβ proteins from humans, chicken, mouse, chlamydomonas and yeast was lower (in the range of 28–50%). Similar to other Gβ proteins, the AtGβ protein sequence is highly divergent in the N-terminal region, but has increased identity in the C-terminal part. As all Gβ proteins contain several internally repeated homologous segments, it was not surprising to find a similar repeat structure for the AtGβ protein.

Analysis of phylogenetic trees revealed that AtGβ formed a separate branch with the *Saccharomyces cerevisiae STE4* gene, which encodes the yeast Gβ homologue. Although initially the idea that Gβγ-complexes function separately in signalling processes seemed quite controversial, it has now become clear that both G-protein subunits interact with effectors and receptors and play an active role in signal transduction. The finding of a Gβ gene that belongs to the same evolutionary branch as the yeast Gβ gene suggests that this gene may have separate signalling functions in plants as well. Experimental analysis of transgenic *Arabidopsis* plants will resolve this issue.

6. G-protein-coupled receptors

The plasma membrane-located receptors, which bind after activation through external signals to heterotrimeric G-proteins, are integral membrane glycoproteins. They consist of a single polypeptide chain spanning the membrane seven times with α-helical stretches 20–25 amino acid residues long (for reviews see Dohlman *et al.,* 1991; Khorana, 1992; Savarese and Fraser, 1992). For bacteriorhodopsin, a member of the opsin G-protein-coupled receptor family, it has been shown that these transmembrane domains apparently form a hydrophilic pocket surrounded by hydrophobic amino acids, a region that is believed to be involved in ligand binding. Binding of the ligand to these clustered amphipathic α-helical stretches of amino acid residues apparently initiates a conformational shift in the receptor protein, followed by activation of the Gα-subunits. The view that the activation of Gα-subunits occurs by changes in orientation or structure of these amphipathic helices is supported by experiments in which mammalian G-proteins from the G_i and G_o class have been directly activated by small peptides. Similarly, a cationic amphipathic tetradecapeptide amide, originally isolated from wasp venom and known as mastoparan, which apparently mimicks the third cytoplasmic loop, can bind to Gα-subunits and subsequently stimulate GTP binding. It has been shown that over a narrow concentration range this peptide stimulates binding of guanosine-5'-*O*-thiotriphosphate to microsomal and plasma membrane fractions from *Zea mays* and *Pisum sativum,* while an inactive mastoparan analogue, differing by only one amino acid and predicted not to form an amphipathic helix, was unable to increase the binding of this nucleotide (White *et al.,* 1993a,b). As it is known that the C-terminal amino acid residues of Gα-subunits are important for binding to the G-protein-coupled receptor, Millner *et al.* used a peptide corresponding to the 15 C-terminal amino acids of the *Arabidopsis* Gα-subunit for affinity purification of the cognate receptor (Wise *et al.,* 1994). Indeed, a 37-kDa protein was identified which bound to the peptide affinity matrix only in the presence of the G-protein activator AlF_4^-. In addition to the cloning of genes from higher plants encoding members of the Gα and Gβ gene family, these findings provide indirect support for the view that G-protein-coupled receptor proteins are present in higher plants and are functionally relevant. Moreover, the finding that mastoparan stimulated plant Gα-subunits is not surprising, as it is known that this compound activates members of the G_i and the G_o class, and all plant Gα genes cloned to date have been found to encode members of this class (Ma *et al.,* 1990, 1991; White *et al.,* 1992, 1993a,b; B. Brzobohaty and K. Palme, unpublished data).

Recently, Bennett *et al.* succeeded in isolating several cDNAs from *Arabidopsis thaliana* that share similarity to the G-protein-coupled receptor family. Using an insertional mutagenic approach, a T-DNA tagged *aux1* mutant was isolated from K. Feldman's *Arabidopis* T-DNA population that corresponds to the *AUX1* gene, a gene that regulates auxin-dependent root growth in *Arabidopsis* (Bennett *et al.,* unpublished results). Two allelic and recessive mutants, *aux1* and *aux2,* were isolated in an attempt to analyse the various steps involved in gravitropic responses on the basis of their resistance to the auxin herbicide 2,4-dichlorophenoxyacetic acid. Two apparently unlinked loci, named *aux1* and *Dwf,* were identified (Maher and Martindale, 1980). Whereas the *Dwf* allele was lethal when homozygous and displayed dwarf rosettes and unbranched, agravitropic roots, the *aux1* allele gave agravitropic seedling roots with reduced sensitivity to exogenous auxins (Mirza *et al.,* 1984). When a T-DNA tagged line was isolated which displayed the phenotype of the *aux1* mutant, the *Arabidopsis* genomic DNA flanking the T-DNA insert was cloned and used as a probe to screen libraries in order to identify *AUX1* cDNA and genomic clones. Complete sequence determination of the *AUX1* cDNA identified a single open reading frame encoding a highly hydrophobic polypeptide. Database searches identified homology to members of the opsin family of G-protein-coupled receptors. In total, 20% identity and 47% similarity to the opsin family were observed. The AUX1 polypeptide contains many physical features common to other members of the G-protein-coupled receptor family. These include the N-terminal and extracellularly oriented domain, preceded by a putative endoplasmic reticulum (ER) targeting signal. The highly hydrophobic mature polypeptide contains seven predicted transmembrane α-helices. These membrane-spanning α-helical stretches are punctuated by three extra/intracellular loops of varying size, with one of the latter containing a canonical protein kinase C phosphorylation site, while the intracellular, more C-terminally located domains contain several serine and threonine residues. Such sites have also been found in the mammalian G-protein-coupled receptor proteins, and play an important role in the desensitization by regulatory protein kinases. Southern hybridization indicated the presence of a gene family, and in fact several related members of this family have subsequently been isolated. The cloning of several members of a G-protein-coupled receptor family from higher plants has important implications for our understanding of receptor signalling chains in higher plants.

7. Target proteins

G-proteins are destined to activate a variety of cellular effector proteins. These effector proteins are active for as long as they are bound to the G-protein subunit. During this time they amplify the initial signal sensed by the receptor at the plasma membrane. What are the major target proteins and intracellular signalling chains to which the plant G-protein-linked receptors and heterotrimeric G-proteins may be linked?

Among the major target proteins in eukaryotes are enzymes such as adenylcyclase, phosphoinositide lipase C and ion channels. The question now is whether these proteins are also targets of G-proteins in plants. Elegant single-cell assays have been performed (Bowler *et al.,* 1994; Neuhaus *et al.,* 1993), the aim of which was to visualize phytochrome responses after microinjection of 'mammalian' signal chain intermediates into mutant tomato cells that were deficient in phytochrome-dependent transduction of light signals. Two alternative signalling pathways were assayed: calcium/calmodulin-dependent *cab–GUS* reporter gene expression and calcium-independent anthocyanin biosynthesis. As a result of activation of signalling chains, either reporter gene expression, full chloroplast development or pigment biosynthesis was initiated. Indeed, injection of GTPγS (a non-hydrolysable GTP analogue) was able to stimulate all of these responses in the absence of phytochrome, while GDPγS (a compound that binds to and blocks the activation of Gα proteins) was not active. Similarly, the A-subunit of cholera toxin (a known Gα activator) was able to cause activation, whereas pertussis toxin (a known Gα inactivator) was not able to do so. These experiments clearly demonstrate a role for heterotrimeric G-proteins in phytochrome signalling.

Among the major target proteins for Gα-subunits in eukaryotes are the adenylcyclases. Although the presence of cAMP in plant cells has been questioned over the years, convincing evidence for the presence of this compound in plants has been accumulating (Colling *et al.,* 1988; Franco, 1983). Cyclic AMP phosphodiesterases have been studied in various plants and green algae (Feldwisch *et al.,* 1994). However, preliminary evidence in favour of adenylcyclases has only recently been presented by several groups working on *Medicago sativa, Pisum sativa* and *Populus deltoides* (Pacini *et al.,* 1993). Pacini *et al.* (1993) demonstrated a membrane-associated adenylcyclase activity which was inhibited at higher GTP concentrations. However, the microinjection experiments mentioned above were unable to confirm a role for cAMP, at least in phytochrome-dependent phototransduction, but instead suggested a role for cGMP, which acts together with calcium in the regulation of gene expression (Bowler *et al.,* 1994). No evidence of a role for G-proteins in the regulation of plant phospholipase C was found (Yotushima *et al.,* 1993).

Guard cells are another unique model for studying signalling processes (for review see Assmann, 1993). A variety of external stimuli (e.g. hormones, light) induce volume changes in guard cells. For example, light triggers swelling of guard cells and opening of the pore, whereas abscisic acid and darkness reverse this process. As G-proteins have been shown to play a role in light-mediated processes and protoplast swelling (Bossen *et al.,* 1990; Romero *et al.,* 1991a,b;

Romero and Lam, 1993), it was not surprising to find that guard cell opening was affected in epidermal peel of *Commelina communis* treated with cholera or pertussis toxins, or after microinjection of GTPγS and GDPβS into intact guard cells (Lee *et al.*, 1993). While cholera toxin had little effect on stomatal opening, pertussis toxin stimulated stomatal opening in a dose-dependent manner. Similarly, GTPγS released after photolysis of microinjected caged GTPγS induced further stomatal opening.

Good candidates for regulation by G-proteins also include inward- and outward-rectifying potassium channels in *Vicia faba* guard cells and mesophyll cells (Armstrong and Blatt, 1995; Fairely-Grenot and Assmann, 1991; Li and Assmann, 1993). In fact, it has been shown that GTPγS reduces the inward potassium current into guard cells (Fairely-Grenot and Assmann, 1991) as well as the outward potassium current from mesophyll cells (Li and Assmann, 1993). Calcium release from microsomal membranes was stimulated by GTP (Allan *et al.*, 1989), as well as cleavage of inositolphosphate (Dillenschneider *et al.*, 1986). An interesting role for G-proteins may be in defence responses, where a 45-kDa GTP-binding protein has been suggested to play a role (Legendre *et al.*, 1992). The role of auxins in promoting turnover of polyphosphoinositides or the stimulation of auxin binding to rice membrane vesicles has been reported. However, various other research groups have failed to reproduce these results, and so this area needs clarification by additional experiments (Ettinger and Lehle, 1988; Schwendemann and Zbell, 1990; Zaina *et al.*, 1989, 1990; Zbell and Walter, 1987; Zbell *et al.*, 1989). The activation of NADH oxidase of etiolated soybean hypocotyl plasma membranes has been shown to be affected by GTP and by auxins. However, it has been suggested that the effect of auxins is not mediated by common G-protein interaction (Morre *et al.*, 1993).

8. Gβ-like proteins

Gβ-like proteins are a group of proteins that share motifs within their primary amino acid sequence with the Gβ-subunits of the heterotrimeric G-proteins. Members of this rapidly growing protein family are structurally and functionally highly diverse. They contain up to seven Gβ repeat units, typically 40–60 amino acids long and characterized by a core homology block of 20 amino acid residues. This domain contains a cluster of charged amino acids that is followed by a hydrophobic amino acid residue and a characteristic aspartate or arginine residue. Six amino acid residues are typically located upstream of the tryptophan residue (Dalrymple *et al.*, 1989; Duronio *et al.*, 1992). As these blocks contain within their centre the dipeptide Trp-Asp ('WD'), they are often termed WD-repeats (van der Voorn and Ploegh, 1992). Although the precise role of such repeated Trp-containing clusters is not yet understood, it has been proposed that they participate in 'stacking interactions' which may, for example, stabilize the helix–turn–helix interactions of some of these proteins with nucleic acids. A number of different proteins contain such WD repeats, including the original Gβ-subunits of the heterotrimeric G-proteins (Fong *et al.*, 1986, 1987), the yeast *STE4* gene (a homologue of the Gβ-subunit of heterotrimeric G-proteins; Whiteway *et al.*, 1989), the *PRP4* and the *PRP17* genes (a factor involved in yeast RNA splicing; Offermans *et al.*, 1990), a *Drosophila* neurogenic genic locus *E(spl)* (enhancer of split; Hartley *et al.*, 1988), the *RACK* gene (an intracellular receptor of protein kinase C; Ron *et al.*, 1994), the *TUP1* gene (a transcription factor that affects catabolite repression; Williams and Trumbly, 1990) and the yeast *CDC4* gene product (a cell cycle progression factor involved in the initiation of DNA replication; Fong *et al.*, 1986).

Several plant genes belonging to this gene family and representing novel structural features have subsequently been identified. They include the *arcA* gene from *Nicotiana tabacum* (Ishida *et al.*, 1993) and the *COP1* and *PRP1* genes from *Arabidopsis thaliana* (Deng *et al.*, 1992; K. Nemeth and C. Koncz, personal communication). The *arcA* gene product, for example, shares only 25% identity with typical Gβ-subunits; however, it contains seven WD-40 repeat elements similar to the structurally related genes from *Chlamydomonas reinhardii (Cblp;* Schloss, 1990) and chicken (*C12.3*; Guillemot *et al.,* 1989). Despite its striking WD-repeat homologies, not much is known about the function of the *arcA* gene product. Nevertheless, its transcriptional up-regulation after activation of tobacco cell division by hormonal agents such as 2,4-dichlorophenoxyacetic acid may indicate a role for this gene during the control of hormone-regulated cell division (Ishida *et al.,* 1993). More striking are the examples that have arisen from analysis of *Arabidopsis* T-DNA insertional mutants. The COP1 protein appears to represent a novel type of putative transcriptional regulator (Deng, 1994; Deng *et al.,* 1991, 1992; Deng and Quail, 1992; Hou *et al.,* 1993). This protein is encoded by the *Arabidopsis COP1* locus and apparently acts to repress the photomorphogenetic pathway in darkness (Kendrick and Kronenberg, 1986), although the gene does not seem to be directly regulated by light. Recessive mutants in this gene resulted in shoot morphogenesis in the absence of light with many characteristics of light-grown wild-type seedlings (Deng *et al.,* 1991, 1992; Deng and Quail, 1992; McNellis *et al.,* 1994). The COP1 protein, which is 658 amino acid residues long and mostly hydrophilic, contains an array of motifs that have not previously been reported in this arrangement. It contains a zinc-binding domain with several conservatively spaced cysteine and histidine residues which probably constitute the zinc-binding ligands (von Arnim and Deng, 1993), an arrangement of amino acids that is also found in the $TAF_{II}80$ compound of the *Drosophila* TFIID complex. In addition, four WD-repeat units are located closer to the C-terminus of the COP1 protein. Disruption of these WD-repeats or deletion of the 56 C-terminal amino acids led to functional inactivation, suggesting an important role for these domains in COP1 function (McNellis *et al.,* 1994).

9. Perspectives

Over the last few years, numerous biochemical and molecular studies of yeast and mammalian systems have resulted in an explosion of information about eukaryotic signal transduction processes. Although our understanding of related processes in plants has been lagging behind, we are now facing rapid progress which is benefitting greatly from molecular and genetic strategies. While it has become clear from research on diverse organisms that many fundamental cellular processes, including signal transduction, share extensive evolutionary conservation, plant research is now not only unravelling some already known elements and modules, but is also demonstrating novel structural combinations of protein modules in unexpected functional arrangements. Future studies are expected to provide exciting insights into a variety of genetically induced molecular disorders in plants.

Acknowledgements

I gratefully acknowledge advice from and discussions with many colleagues at the Max-Planck Institut für Züchtungsforschung. In particular I am grateful to Dr M. Bennett and Dr C. Koncz for releasing unpublished information, and to Dr C. Redhead for reading the manuscript. Research in our laboratory was supported by the Deutsche Forschungsgemeinschaft (Pa 279/6-1).

References

Allan E, Dawson A, Drøbak B, Roberts K. (1989) GTP causes calcium release from a plant microsomal fraction. *Cell. Signal.* **1:** 23–29.

Armstrong F, Blatt MR. (1995) Evidence for K+ channel control in *Vicia* guard cells coupled by G-proteins to a 7TMS receptor mimetic. *Plant J.* **8:** 187–198.

Assmann SM. (1993) Signal transduction in guard cells. *Annu. Rev. Cell Biol.* **9:** 345–375.

Bhullar RP, Chardin P, Haslam RJ. (1990) Identification of multiple ral gene products in human platelets that account for some but not all of the platelet Gn-proteins. *FEBS Lett.* **260:** 48–52.

Bilushi SV, Shebunin G, Babakov AV. (1991) Purification of a subunit composition of a GTP-binding protein from maize root membranes. *FEBS Lett.* **291:** 219–221.

Birnbaumer L, Abramowitz J, Brown, AM. (1990) Receptor–effector coupling by G proteins. *Biochim. Biophys. Acta* **1031:** 163–224.

Blum W, Hinsch K-D, Schultz G, Weiler EW. (1988) Identification of GTP-binding proteins in the plasma membrane of higher plants. *Biochem. Biophys. Res. Commun.* **156:** 954–959.

Bossen ME, Kendrick RE, Vredenberg WJ. (1990) The involvement of a G-protein in phytochrome-regulated, Ca^{2+}-dependent swelling of etiolated wheat protoplasts. *Physiol. Plant.* **80:** 55–62.

Bourne HR. (1989) G protein subunits: who carries what message? *Nature* **337:** 504–505.

Bourne HR. (1993) A turn-on and a surprise. *Nature* **366:** 628–629.

Bourne HR, Sanders DA, McCormick F. (1990) The GTPase superfamily: conserved structure and molecular mechanism. *Nature* **349:** 117–127.

Bowler C, Neuhaus G, Yamagata H, Chua N-H. (1994) Cyclic GMP and calcium mediate phytochrome phototransdcution. *Cell* **77:** 73–81.

Brown AM. (1991) A cellular logic for G-protein-coupled ion channel pathways. *FASEB J.* **5:** 2175–2179.

Clapham DE, Neer EJ. (1993) New roles for G-protein βγ-dimers in transmembrane signalling. *Nature* **365:** 403–406.

Colling C, Gilles R, Cramer M, Nass N, Moka R, Jaenicke R. (1988) Measurement of 3',5'-cyclic AMP in biological samples using a specific monoclonal antibody. *Second Messenger Phosphoproteins* **12:** 123–133.

Dalrymple MA, Peterson-Bjorn S, Friesen JD, Beggs JD. (1989) The product of the PRP4 gene of *S. cerevisiae* shows homology to β subunits of G proteins. *Cell* **58:** 811–812.

de Boer AH, van Hunnik E, Korthout HAAJ, Sedee NJA, Wang M. (1993) Affinity chromatography of GTPase proteins from oat root plasma membranes using biotinylated GTP. *FEBS Lett.* **337:** 281–284.

Deng LW, Matsui M, Wei N, Wagner C, Chu AM, Feldmann KA, Quail PH. (1992) COP1, an *Arabidopsis* regulatory gene encodes a protein with both a zinc-binding motif and a Gβ-homologous domain. *Cell* **71:** 791–801.

Deng X-W. (1994) Fresh view of light signal transduction in plants. *Cell* **76:** 423–426.

Deng X-W, Quail PH. (1992) Genetic and phenotypic characterization of *cop1* mutants of *Arabidopsis thaliana. Plant J.* **2:** 83–95.

Deng X-W, Caspar T, Quail PH. (1991) COP1: a regulatory locus involved in light-controlled development and gene expression in *Arabidopsis. Genes Dev.* **5:** 1172–1182.

Dickey BF, Birnbaumer L (ed.). (1993) *GTPases in Biology.* Heidelberg: Springer-Verlag.

Dillenschneider M, Hetherington A, Graziana A, Alibert G, Berta P, Haiech H, Ranjeva R. (1986) The formation of inositol phosphate derivatives by isolated membranes from *Acer pseudoplatanus* is stimulated by guanine nucleotides. *FEBS Lett.* **208:** 413–417.

Dohlman HG, Thorner J, Caron MG, Lefkowitz RJ. (1991) Model systems for the study of seven transmembrane-segment receptors. *Annu. Rev. Biochem.* **60:** 655–688.

Drøbak BK, Allan EF, Comerford JG, Roberts K, Dansson AP. (1988) Presence of guanine nucleotide-binding proteins in a plant hypocotyl microsomal fraction. *Biochem. Biophys. Res. Commun.* **150:** 899–903.

Duronio RJ, Gordon JI, Boguski MS. (1992) Comparative analysis of the β-transducin family with identification of several new members including *PWP1,* a nonessential gene of *Saccharomyces cerevisiae* that is divergently transcribed from *NMT1. Proteins: Struct. Function Genet.* **13:** 41–56.

Ephritikhine G, Pradier J-M, Guern J. (1993) Complexity of GTPγS binding to tobacco plasma membranes. *Plant Physiol. Biochem.* **31:** 573–584.

Ettlinger C, Lehle L. (1988) Auxin induces rapid changes in phosphatidylinositol metabolites. *Nature* **331:** 176–178.

Fairely-Grenot K, Assmann SM. (1991) Evidence for G-protein regulation of inward K^+ channel current in guard cells of fava bean. *Plant Cell* **3:** 1037–1044.

Feldwisch O, Lammertz M, Hartmann E, Feldwisch J, Palme K, Jastorff B, Jaenicke L. (1994) Purification and characterization of a cAMP-binding protein of *Volvox carteri* f. *nagariensis* Iyengar. *Eur. J. Biochem.* **228:** 480–489.

Fong HKW, Hurley JB, Hopkins RS, Miake-Lye R, Johnson MS, Doolittle RF, Simon MI. (1986) Repetitive segmental structure of the transducin B-subunit: homology with the CDC4 gene and identification of related mRNAs. *Proc. Natl Acad. Sci. USA* **83:** 2162–2166.

Fong HKW, Amatruda TT III, Birren BW, Simon MI. (1987) Distinct forms of the B-subunit of GTP-binding regulatory proteins identified by molecular cloning. *Proc. Natl Acad. Sci. USA* **84**: 3792–3796.

Franco DA. (1983) Cyclic AMP in photosynthetic organisms: recent developments. *Adv. Cycl. Nucl. Res.* **15**: 97–117.

Guillemot F, Billault A, Auffray C. (1989) Physical linkage of a guanine nucleotide-binding protein-related gene to the chicken major histocompatibility complex. *Proc. Natl Acad. Sci. USA* **86**: 4594–4598.

Hartley DA, Preiss A, Artavanis-Tsakonas S. (1988) A deduced gene product from the *Drosophila* neurogenic locus, Enhancer of split, shows homology to mammalian G-protein β subunit. *Cell* **55**: 785–795.

Hasunuma K, Funadera K. (1987) GTP-binding proteins in a green plant, *Lemna paucicostata*. *Biochem. Biophys. Res. Commun.* **143**: 908–912.

Hasunuma K, Furukawa K, Funadera K, Kubota K, Watanabe W. (1987a) Partial characterization and light-induced regulation of GTP-binding proteins in *Lemna paucicostata*. *Photochem. Photobiol.* **46**: 531–535.

Hasunuma K, Furukawa K, Tomita K, Mukai C, Nakamura T. (1987b) GTP-binding proteins in etiolated epicotyls of *Pisum sativum* (Alaska) seedlings. *Biochem. Biophys. Res. Commun.* **148**: 133–139.

Hepler GR, Gilman AG. (1992) G proteins. *Trends Biochem. Sci.* **17**: 383–387.

Hesse FS. (1991) Untersuchungen zur Signalübertragung von Auxinen in *Zea mays* L. PhD Thesis, Universität Köln.

Hou Y, von Arnim AG, Deng X-W. (1993) A new class of *Arabidopsis* constitutive photomorphogenetic genes involved in regulating cotyledon development. *Plant Cell* **5**: 329–339.

Huang H, Weiss C, Ma H. (1994) Regulated expression of the *Arabidopsis* G protein α-subunit gene GPA1. *Int. J. Plant Sci.* **155**: 3–14.

Iniguez-Lluhi J, Kleuss C, Gilman AG. (1993) The importance of G-protein βγ-subunits. *Trends Cell. Biol.* **3**: 230–236.

Ishida S, Takahashi Y, Nagata T. (1993) Isolation of an auxin-regulated gene encoding a G-protein β-subunit-like protein from tobacco BY-2 cells. *Proc. Natl Acad. Sci. USA* **90**: 11152–11156.

Iyengar R, Birnbaumer L (ed.). (1990) *G-proteins.* San Diego: Academic Press.

Jacobs M, Thelen MP, Farndale RW, Astle MC, Rubery PA. (1987) Specific guanine nucleotide binding by membranes from *Cucurbita pepo* seedlings. *Biochem. Biophys. Res. Commun.* **155**: 1478–1482.

Kendrick RE, Kronenberg GHM (ed.). (1986) *Photomorphogenesis in Plants.* Dordrecht: Martinus Nijhoff Publishers.

Khorana HG. (1992) Rhodopsin, photoreceptor of the rod cell. *J. Biol. Chem.* **267**: 1–4.

Landis CA, Masters SB, Spada A, Pace AM, Bourne HR, Vallar L. (1989) GTPase-inhibiting mutations activate the alpha chain of G_s and stimulate adenylyl cyclase in human pituitary tumours. *Nature* **340**: 692–696.

Lee HJ, Tucker EB, Crain RC, Lee Y. (1993) Stomatal opening is induced in epidermal peels of *Commelina communis* L. by GTP analogs or pertussis toxin. *Plant Physiol.* **102**: 95–100.

Legendre L, Heinstein PF, Low PS. (1992) Evidence for participation of GTP-binding proteins in elicitation of the rapid oxidative burst in cultured soybean cells. *J. Biol. Chem.* **267**: 20140–20147.

Li W, Assmann SM. (1993) Characterization of a G-protein-regulated outward K^+ current in mesophyll cells of *Vicia faba* L. *Proc. Natl Acad. Sci. USA* **90**: 262–266.

Ma H, Yanofsky MF, Meyerowitz EM. (1990) Molecular cloning and characterization of GPA1, a G-protein alpha subunit gene from *Arabidopsis thaliana*. *Proc. Natl Acad. Sci. USA* **87**: 3821–3825.

Ma H, Yanofsky M, Huang H. (1991) Isolation and sequence analysis of TGA1 cDNAs encoding a tomato G-protein subunit. *Gene* **107**: 189–195.

McNellis TW, von Arnim AG, Araki T, Komeda Y, Misera S, Deng X-W. (1994) Genetic and molecular analysis of an allelic series of *cop1* mutants suggests functional roles for the multiple protein domains. *Plant Cell* **6**: 487–500.

Maher EP, Martindale SJB. (1980) Mutants of *A. thaliana* with altered responses to auxin and gravity. *Biochem. Genet.* **18**: 1041–1053.

Markby DW, Onrust R, Bourne HR. (1993) Separate GTP binding on GTPase activating domains of a Gα subunit. *Science* **262**: 1895–1901.

Millner PA, Robinson PS. (1989) ADP-ribosylation of thylakoid membrane polypeptides by cholera toxin is correlated with inhibition of thylakoid GTPase activity and protein phosphorylation. *Cell. Signal.* **1**: 421–433.

Mirza JI, Olsen GM, Iversen T-H, Maher EP. (1984) The growth and gravitropic responses of wild-type and auxin-resistant mutants of *Arabidopsis thaliana*. *Physiol. Plant* **60**: 516–522.

Morre DJ, Brightmann AO, Barr R, Davidson M, Crane FL. (1993) NADH oxidase activity of plasma membranes of soybean hypocotyls is activated by guanine nucleotides. *Plant Physiol.* **102**: 595–602.

Neuhaus G, Bowler C, Kern R, Chua N-H. (1993) Calcium/calmodulin-dependent and -independent phytochrome signal transduction pathways. *Cell* **73**: 937–952.

Noel JP, Hamm HE, Sigler PB. (1993) The 2.2 Å crystal structure of transducin-α complexed with GTP-γS. *Nature* **366**: 654–663.

Offermans S, Schafer R, Hoffmann B, Bombien E, Spicher K, Hinsch K-H, Schultz G, Rosenthal W. (1990) Agonist-sensitive binding of a photoreactive GTP analog to a G-protein alpha-subunit in membranes of HL-60 cells. *FEBS Lett.* **260**: 14–18.

Pacini B, Petrigliano A, Diffley P, Paffetti A, Brown EG, Martelli P, Trabalzini L, Bovalini L, Lusini P, Newton RP. (1993) Adenyl cyclase activity in roots of *Pisum sativum*. *Phytochemistry* **34**: 899–903.

Palme K. (1992a) Molecular analysis of plant signaling elements: relevance of eucaryotic signal transduction models. *Int. Rev. Cytol.* **132**: 223–283.

Palme K. (1992b) The relevance of eucaryotic signalling models to understand the perception of auxin. In: *Control of Plant Gene Expression* (Verma DPS, ed.). Boca Raton: CRC Press, pp. 33–50.

Perdue DO, Lomax TL. (1992) GTP binding and hydrolysis in zuccini PM. *Plant Physiol. Biochem.* **30**: 163–172.

Ricart CA, Morhy L, White IR, Findlay JBC, Millner PA. (1993) Specific immunodetection and purification of G-proteins from *Sorghum bicolor. Biochem. Trans.* **21**: 226S.

Romero LC, Lam E. (1993) Guanine nucleotide binding protein involevement in early steps of phytochrome-regulated gene expression. *Proc. Natl Acad. Sci. USA* **90**: 1465–1469.

Romero LC, Sommer D, Gotor C, Song P-S. (1991a) G-proteins in etiolated *Avena* seedlings. Possible phytochrome regulation. *FEBS Lett.* **282**: 341–346.

Romero LC, Biswal B, Song P-S. (1991b) Protein phosphorylation in isolated nuclei from etiolated *Avena* seedlings. Effects of red/far-red light and cholera toxin. *FEBS Lett.* **282**: 347–350.

Ron D, Chen C-H, Caldwell J, Jamieson L, Orr E, Mocly-Rosen D. (1994) Cloning of an intracellular receptor for protein kinase C: a homolog of the β-subunit of G proteins. *Proc. Natl Acad. Sci. USA* **91**: 839–843.

Ross EM. (1992) Twists and turns on G-protein signalling pathways. *Curr. Opin. Cell Biol.* **2**: 517–519.

Sasaki Y, Sekiguchi K, Nagano Y, Matsuno R. (1991) Detection of small GTP-binding proteins in the outer envelope membrane of pea chloroplasts. *FEBS Lett.* **293**: 124–126.

Savarese TM, Fraser CM. (1992) *In vitro* mutagenesis and the search for structure–function relationships among G-protein-coupled receptors. *Biochem. J.* **283**: 1–19.

Schloss JA. (1990) A *Chlamydomonas* gene encodes a G protein β-subunit-like polypeptide. *Mol. Gen. Genet.* **221**: 443–452.

Schwendemann I, Zbell B. (1990) Detection of a GTP-binding protein possibly involved in auxin transduction on carrot microsomes. *Physiol. Plant* **79**: A188.

Simon MI, Strathmann MP, Gautam N. (1991) Diversity of G-proteins in signal transduction. *Science* **252**: 802–808.

Simonds WF, Butrynski JE, Gautam N, Umson CG, Spiegel AM. (1991) G protein βγ dimers. *J. Biol. Chem.* **266**: 5363–5366.

Stephens L, Smrcka A, Cooke FT, Jackson TR, Sternweis PC, Hawkins PT. (1994) A novel phosphoinositide 3 kinase activity in myeloid-derived cells is activated by G protein βγ subunits. *Cell* **77**: 83–93.

Stouten PFW, Sander C, Wittinghofer A, Valencia A. (1993) How does the switch II region of G-domains work? *FEBS Lett.* **320**: 1–6.

Stryer L, Bourne HR. (1986) G proteins: a family of signal transducers. *Annu. Rev. Cell Biol.* **2**: 391–419.

Taylor CW. (1990) The role of G proteins in transmembrane signalling. *Biochem. J.* **272**: 1–13.

Terryn N, van Montagu M, Inze D. (1993) GTP-binding proteins in plants. *Plant Mol. Biol.* **22**: 143–152.

van der Voorn L, Ploegh HL. (1992) The WD-40 repeat. *FEBS Lett.* **301**: 131–134.

von Arnim AG, Deng X-W. (1993) Ring-finger motif of *Arabidopsis thaliana* COP1 defines a new class of zinc-binding domains. *J. Biol. Chem.* **268**: 19626–19631.

Warpeha KMF, Hamm HE, Rasenick MM, Kaufman LS. (1991) A blue-light-activated GTP-binding protein in the plasma membranes of etiolated peas. *Proc. Natl Acad. Sci. USA* **88**: 8925–8929.

Warpeha KMF, Kaufman LS, Briggs WR. (1992) A flavoprotein may mediate the blue light activated binding of guanosine 5'-triphosphate to isolated plasma membranes of *Pisum sativum* L. *Photochem. Photobiol.* **55**: 595–603.

Weiss CA, Huang H, Ma H. (1993) Immunolocalization of the G protein α-subunit encoded by the *GPA1* gene in *Arabidopsis. Plant Cell* **5**: 1513–1528.

White IR, Wise A, Finan PM, Clarkson J, Millner PA. (1992) GTP-binding proteins in higher plant cells. In: *Transport and Receptor Proteins of Plant Plasma Membranes* (Cooke DT, Clarkson DT, ed.). New York: Plenum Press, pp. 185–192.

White IR, Wise A, Millner PA. (1993a) Evidence for G-protein-linked receptors in higher plants: stimulation of GTP-gamma-S binding to membrane fractions by the mastoparan analogue mas7. *Planta* **191:** 285–288.

White IR, Zamri I, Wise A, Millner PA. (1993b) Use of synthetic peptides to study G-proteins and protein kinases within higher plant cells. In: SEB Seminar Series 53. (Battey NH, Dickinson HG, Hetherington AM, ed.). *Post-Translational Modification in Plants.* Cambridge: Cambridge University Press, pp. 91–108.

Whiteway M, Hougan L, Dignard D, Thomas DY, Bell L, Saari GC, Grant FJ, O'Hara P, MacKay VL. (1989) The STE4 and STE18 genes of yeast encode potential beta and gamma subunits of the mating factor receptor coupled G protein. *Cell* **56:** 467–477.

Williams FE, Trumbly RJ. (1990) Characterization of TUP1, a mediator of glucose repression in *Saccharomyces cerevisiae. Mol. Cell. Biol.* **10:** 6500–6511.

Wise A, Millner PA. (1992) Evidence for the presence of GTP-binding proteins in tobacco leaf and maize hypocotyl plasmalemma. *Biochem. Soc. Trans.* **20:** 7S.

Wise A, Thomas PG, White IR, Millner PA. (1994) Isolation of a putative receptor from *Zea mays* microsomal membranes that interacts with the G-protein, GPalpha1. *FEBS Lett.* **356:** 233–237.

Yotushima K, Mitsui T, Takaoka T, Hayakawa T, Igaue I. (1993) Purification and characterization of membrane-bound inositol phospholipid-specific phospholipase C from suspension-cultured rice (*Oryza sativa* L.) cells. *Plant Physiol.* **102:** 165–172.

Zaina S, Bertani A, Lombardi L, Mapelli S, Torti G. (1989) Membrane-associated binding sites for indoleacetic acid in the rice coleoptile. *Planta* **179:** 222–227.

Zaina S, Reggiani R, Bertani A. (1990) Preliminary evidence for involvement of GTP binding protein(s) in auxin signal transduction in rice (*Oryza sativa* L.) coleoptile. *J. Plant Physiol.* **136:** 653–658.

Zbell B, Walter C. (1987) About the search for the molecular action of high-affinity auxin-binding sites on membrane-localized rapid phosphoinositide metabolism in plant cells. In: *Plant Hormone Receptors* (Klämbt D, ed.). Berlin: Springer-Verlag, pp. 141–153.

Zbell B, Schwendemann I, Bopp M. (1989) High affinity GTP-binding on microsomal membranes prepared from moss protonema of *Funaria hygrometrica. J. Plant Physiol.* **134:** 639–641.

Chapter 10

Lipid transfer proteins: structure, function and gene expression

Jean-Claude Kader, Michèle Grosbois, Françoise Guerbette, Alain Jolliot and Annette Oursel

1. Introduction

Membrane biogenesis and turnover are essential features of plant cell development and growth (Browse and Somerville, 1991; Somerville and Browse, 1991). Among membrane constituents, phospholipids are of considerable interest for various reasons: (1) phosphoinositides appear to be involved in intracellular signalling (Thompson, 1993); (2) the breakdown of phospholipids by phospholipases and lipases provides acyl chains allowing the formation of jasmonate, which is involved in defence reactions against pathogens (Vick, 1993); (3) the nature of chloroplast phospholipids controls chilling and freezing sensitivity (Nishida *et al.*, 1993); and (4) phospholipids are assumed to provide the acyl chains necessary for triacylglycerol formation in oily seeds (Stymne, 1993).

For these reasons, it is important to increase our knowledge of phospholipid biogenesis and renewal in plants. The biosynthesis of the most abundant phospholipids in the extra-chloroplastic compartment, namely phosphatidylcholine (PC) and phosphatidylethanolamine (PE), is principally located in the endoplasmic reticulum (Moore, 1993). Intracellular transport of these phospholipids from the endoplasmic reticulum to membranes that are unable to synthesize them, such as mitochondria, is therefore necessary. A similar requirement has been suggested in animal cells (Bishop and Bell, 1988; Dawidowicz, 1987; Wirtz, 1991). A plant-specific lipid flux has been assumed to occur (Browse and Somerville, 1991; Somerville and Browse, 1991); this flux is linked to the presence of two lipid biosynthetic pathways in higher plants, the so-called prokaryotic (operating entirely within chloroplasts) and eukaryotic (involving co-operation between these organelles and the endoplasmic reticulum) pathways. According to this two-pathway model, fatty acids are exported from the chloroplast as co-enzyme A (CoA) esters and incorporated into the endoplasmic reticulum lipids. They are then imported back into chloroplasts for the biosynthesis of thylakoid lipids by the eukaryotic pathway. It has been suggested that they are imported in this way in the form of PC molecules.

It is thus of interest to know how phospholipids move between membranes. However, the molecular mechanisms involved in phospholipid transport as well as in sorting and targeting are still unclear. Three main hypotheses have been proposed: (1) spontaneous movement of phospholipid monomers through the cytosol; (2) vesicular traffic; and (3) transport by protein vectors. Whereas the first mechanism is unlikely, due to the low solubility of phospholipids with normal long-chain fatty acids in aqueous media, considerable attention has been given to the two other hypotheses. This chapter will be mainly devoted to the description of proteins, initially found in the cytosol of plants, that are able to facilitate phospholipid movements between membranes. They are called phospholipid-transfer proteins, but more recently have been termed lipid-transfer proteins (LTPs) since they have a broad specificity for lipids.

2. Assay and purification of LTPs

Although the purification of LTPs from higher plants is straightforward, the assay of their activity is rather difficult.

Assay

The assay involves the incubation of membranes containing labelled phospholipids ('donor membranes') with non-radioactive membranes ('acceptor membranes') which can be separated from the donor membrane by centrifugation. A routine assay uses liposomes prepared from [^{3}H]-PC (the phospholipid to be transferred) and [^{14}C]cholesteryl-oleate (a non-transferable compound) as donor membranes. The acceptor membranes are mitochondria (easily separated from liposomes by centrifugation at 10 000 g), chloroplasts, plasma membranes or 'microsomes' (endoplasmic-rich fraction).

The increase in ^{3}H/^{14}C ratio of the lipids recovered in the acceptor membranes after incubation with LTP indicates the extent of the transfer of individual PC molecules from donor to acceptor membranes. The low levels of ^{14}C label found in acceptor membranes provide evidence that LTP does not cause a co-sedimentation of donor with acceptor membranes (Kader, 1993) (*Figure 1*).

Alternatively, donor membranes can be natural ones (e.g. microsomes), although in such cases no control can be performed with a non-transferable tracer.

Other assays which do not require a separation of acceptor and donor membranes have recently been used (Moreau *et al.*, 1994). In these assays, the donor membranes are liposomes made from nitrobenzoxadiazol-phosphatidylcholine (NBD-PC). When NBD-PC leaves the donor liposomes to reach acceptor liposomes made from pure PC, the fluorescence increases, due to a transfer of NBD-PC between the two membranes (Moreau *et al.*, 1994). This assay allows

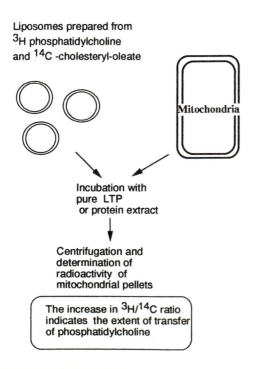

Liposomes prepared from
^{3}H phosphatidylcholine
and ^{14}C -cholesteryl-oleate

Mitochondria

Incubation with
pure LTP
or protein extract

Centrifugation and
determination of
radioactivity of
mitochondrial pellets

The increase in ^{3}H/^{14}C ratio
indicates the extent of transfer
of phosphatidylcholine

Figure 1. Assay of phospholipid transfer activity. Liposomes, prepared from [^{3}H]phosphatidylcholine (270 nmol, 740 Bq) and cholesteryl-[1-^{14}C]-oleate (1 nmol, 700 Bq) were incubated at 30°C for 30 min with mitochondria isolated from maize seedlings (2 mg of protein) and various amounts of protein. The mitochondrial pellets were then collected by centrifugation at 10 000 g for 10 min, and their radioactivity was determined by scintillation counting. The increase in ^{3}H label observed with increasing amounts of protein indicates the extent of phosphatidylcholine, whereas no increase in ^{14}C label was noted.

continuous monitoring of the lipid transfer between membranes. However, it cannot be used with crude protein extracts since several interfering compounds are present. A similar approach has been used for maize LTP and erythrocyte membranes (Geldwerth *et al.,* 1991).

Purification

Purification of LTP from plants is based on their biochemical properties, which have been determined gradually since the initial observations (Kader, 1975). As LTPs are small (9 to 10 kDa) basic proteins, the purification procedure involves a gel-filtration step (Sephadex G75) followed by cation-exchange chromatography (carboxymethyl-Sepharose chromatography or fast protein liquid chromatography (FPLC)-Mono S). The main peak of lipid transfer activity is then subjected to a final gel-filtration step which yields a homogeneous protein. It is important to note that several other peaks of transfer activity are observed, probably corresponding to isoforms of LTP. However, this hypothesis has yet to be validated by determination of the amino-acid sequence of the different proteins found in the active fractions.

Purification of LTPs is aided by the stability of these proteins. For example, LTP from maize seedlings retains its activity after a 5-min incubation at 90°C.

LTPs have been purified to homogeneity from various plants (maize seedlings, spinach leaves, sunflower; see *Table 1*). It is noteworthy that LTPs have been detected in all higher plants studied to date. Furthermore, the fact that a purification factor of about 100 is sufficient to obtain a homogeneous fraction from various seeds or leaves indicates that LTPs are quite abundant.

3. Biochemical properties of LTPs

LTPs purified from higher plants exhibit remarkably similar properties (*Table 1*).

Table 1. Properties of lipid transfer proteins purified from higher plants

Source	Apparent molecular mass (kDa)	Number of residues	Isoelectric point	Substrate specificity
Maize[a]	9	93	8.8	PC, PI, PE
Spinach[b]	9	91	9.3	PC, PI, PE
Castor bean A[c]	7	92	High	PC, PE, MGDG
Castor bean B[c]	8	92	High	PC, PE, MGDG
Castor bean C[c]	9	93	High	PC, PE, MGDG
Castor bean D[c]	7	93	High	PC, PE, MGDG
Sunflower[d]	9	–	9	PC, PI, PE
Rape seed IV[e]	9	–	10.5	PC, acyl binding
Wheat[f]	9	90	10	Only PC studied

MGDG, monogalactosyldiacylglycerol.
[a] Tchang *et al.,* 1988; [b] Kader *et al.,* 1984; [c] Tsuboi *et al.,* 1991; [d] Arondel *et al.,* 1990; [e] Ostergaard *et al.,* 1993; [f] Désormeaux *et al.,* 1992.

Molecular mass

With regard to molecular mass, values varying from 9 to 10 kDa have been found either by gel filtration or by sodium dodecyl sulphate–gel electrophoresis, and these have been confirmed by amino-acid sequence determinations. In contrast, animal cells contain several categories of proteins that specifically transfer PC (PC-TP), preferentially transfer PI (phosphatidylinositol) (PI-TP), or that are non-specific (nsL-TP), and these proteins have a molecular mass ranging from 11 to 33 kDa (Wirtz, 1991). Yeasts contain 35-kDa proteins which preferentially transfer PI (Paltauf and Daum, 1990).

Isoelectric point

The major part of the lipid transfer activity of the crude protein extract is associated with basic proteins (95% in the case of spinach leaves). Only the basic LTPs have been purified. Acidic proteins have been detected in various plants (e.g. maize, castor bean, spinach), but have not been purified to homogeneity (Yamada, 1992). The basic LTPs have an isoelectric point ranging from 8.8 to 10, determined by chromatofocusing or isoelectric focusing.

Specificity

Plant LTPs are able to transfer not only PC from liposomes to mitochondria but also PI, PE or phosphatidylglycerol (PG). In contrast, animal cells contain LTPs with greater specificity.

The possibility that plant LTPs may also transfer sterols needs to be explored. Although cholesteryl-oleate, a highly apolar lipid, is not transferred, it is possible that sterols do interact with plant LTPs. Triacylglycerols are not transferred by maize or castor bean LTPs (Yamada, 1992). In addition, it has been found that LTPs from spinach, sunflower and rape can bind acyl-CoA (Arondel *et al.*, 1990; Ostergaard *et al.*, 1993; Rickers *et al.*, 1984). This is different to the situation in animal cells, which contain fatty acid binding proteins (Veerkamp *et al.,* 1991) or acyl-CoA binding proteins (Knudsen, 1990) that are unable to transfer phospholipids.

4. Structure and mode of action of LTPs

Since several LTPs have been purified to homogeneity, it has been possible to determine their primary structure.

Amino-acid sequence

The complete amino-acid sequence has been determined for various plant LTPs (maize, spinach, wheat, castor bean, rice, barley) (*Table 1*). In the case of barley and rice, LTPs have been primarily referred to as putative amylase–protease inhibitors, due to their homology to a protein of this category isolated from Indian finger millet (Bernhard and Somerville, 1989; Campos and Richardson, 1984; Mundy and Rogers, 1986; Yu *et al.*, 1988). Plant LTPs have a total number of amino acid residues ranging from 91 to 95. They lack tryptophan and they have eight cysteine residues located at conserved positions. According to Yamada's research group, these cysteine residues are engaged in four disulphide bridges (Yamada, 1992). No structural homology has been found between plant and animal lipid transfer proteins.

Structural model

A structural model based on the prediction of the secondary structure of LTP (Tchang *et al.*, 1988) (*Figure 2*) led to the conclusion that the maize LTP was almost entirely composed of β-sheets. In contrast, studies of wheat LTP by Raman and Fourier-transform infrared spectroscopy showed that the protein was composed of 40% α-helices. Interestingly, reduction of the protein by dithiothreitol led to a decrease in the proportion of α-helix from 40% to 25%. The reduction of maize LTP was found to inhibit the lipid transfer activity, confirming the important role of disulphide bridges (Grosbois *et al.*, 1993). These observations suggest that the conformation of LTPs is important for their activity.

Based on these observations, structural models have been proposed (Madrid and Von Wettstein, 1990; Tchang *et al.* 1988) in which the protein appears as a complex with a phospholipid molecule embedded within a cavity. However, no tight binding of phospholipid has yet been demonstrated (Kader, 1993). The only stable complex which has been observed is that

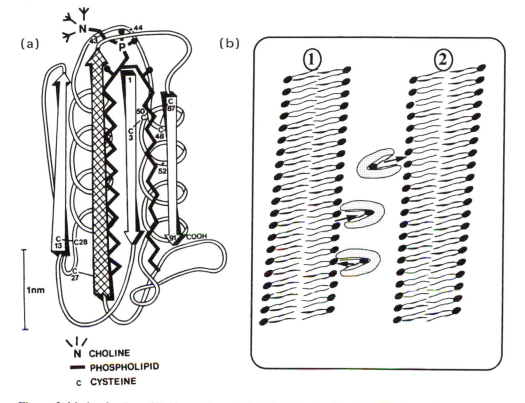

Figure 2. Mode of action of lipid transfer protein. (a) A structural model of lipid transfer protein proposed by Madrid and Von Wettstein (1990), reproduced with permission from Gauthier-Villars Publishers. In this model, a phospholipid molecule is inserted within the protein and interacts with its central domain via its choline part. (b) A suggested mode of action of lipid transfer protein (LTP), which forms a reversible complex with a phospholipid molecule and exchanges its phospholipid with another extracted from a membrane. It has been suggested that LTP facilitates the extraction of phospholipids from membranes. The interaction of LTP with membranes plays an important role in the process.

between wheat LTP and lyso-PC (Désormeaux *et al.*, 1992). This complex was able to form crystals (Pebay-Peyroula *et al.*, 1992) so X-ray diffraction studies will be possible. A nuclear magnetic resonance (NMR) study has recently been performed on wheat LTP in order to construct a three-dimensional structure (Simorre *et al.*, 1991) which will aid understanding of the interaction of LTP with lipids.

Mode of action

The fact that plant LTPs can form complexes with phospholipids led to the suggestion that, as with animal PC-TP, a shuttle mechanism could be responsible for the inter-membrane transfer of phospholipids (Kader, 1993; Wirtz, 1991). According to this hypothesis, the lipid–LTP complex interacts with the membrane and exchanges its bound lipid with a membrane lipid. This involves an interaction of the LTP with the membranes (*Figure 2*).

The fact that the lipid transfer activity can be disturbed when ions (NaCl, MgCl$_2$) are added, or when the acidic phospholipid (PI) content of the acceptor liposomes is varied (Kader, 1993), indicates that electrostatic interactions between LTP and membranes are involved in the functioning of the protein.

However, it has been found that the proportion of lipid–LTP complexes formed with intact phospholipids is very low (less than 1% of the phospholipids are bound) relative to the high proportion of lysophosphatidylcholine–LTP complexes (Désmoreaux *et al.*, 1992).

It has recently been observed, by molecular sieving of the complex formed between PC and maize LTP, that only one in 200 LTP molecules is associated with a PC molecule (Grosbois *et al.*, 1993). This low proportion reflects the low affinity of maize LTP for PC, as well as a lack of stability of the PC–LTP complex.

It has also been shown in animal cells that nsL-TP, in contrast to the other specific proteins, has a low-affinity lipid-binding site (Wirtz, 1991).

The similar properties of plant LTP and animal nsL-TP, namely broad specificity for lipids and low affinity for transferred lipids, suggest that these proteins act by lowering the energy barrier for the lipid monomer–membrane interface equilibration reaction. These proteins may act by facilitating a desorption of lipids from the membrane followed by the formation of a complex with LTP which functions as an intermediate in the transfer reaction.

Are plant LTPs facilitating an exchange or net transfer of phospholipids ? This question is of importance for understanding the role of LTPs. It is clear that a bidirectional transfer of phospholipids is mediated by plant LTP. This leads to a replacement of the phospholipids in the acceptor membranes by those of the donor membranes and vice versa. As a consequence, it is possible to modify the lipid composition of a membrane. This has been done with chloroplast envelope membranes (Miquel *et al.*, 1987), leading to changes in the fatty acid composition. It will be of interest to study the effects of such changes on the functional properties of these membranes. The possibility that plant LTPs promote a net transfer (i.e. an increase in the amount of lipid in a membrane) has not been sufficiently examined, in contrast to animal nsL-TP (Wirtz, 1991).

5.   Cell localization and biogenesis

Although it has not been possible to prepare monoclonal antibodies directed against plant LTPs, polyclonal antibodies prepared in rabbits are available for LTP from maize, spinach, rape seed and castor bean (Grosbois *et al.*, 1989; Kader, 1993). In addition, antibodies prepared against a fusion protein (*Arabidopsis* LTP–maltose binding protein) have been obtained (Thoma *et al.*, 1992). The preparation of these antibodies has allowed studies on the localization and biogenesis of LTP.

Biogenesis

The levels of LTP during the maturation and germination of maize seeds were monitored by immunoblotting protein extracts. A substantial increase in LTP levels was observed during the last stages of maturation and during germination of the seeds, particularly in the aerial part of the plantlets. An increase in lipid transfer activity was found in the protein extracts of the plantlets at various stages of germination. These observations are consistent with the hypothesis that the biosynthesis of LTP could be linked to active biogenesis of membranes occurring during the development of the seedlings (Grosbois *et al.*, 1993). This result was confirmed by *in-vitro* synthesis experiments carried out with RNAs extracted from maize seedlings. By immuno-precipitation with the polyclonal antibody, newly synthesized LTPs were observed, one of which corresponded to a protein of higher molecular mass (Vergnolle *et al.*, 1988).

In castor bean seedlings, active biosynthesis of LTP was noted (Ostergaard *et al.*, 1993). In this case several isoforms (A to D) were detected by immunoblotting and found to be specifically associated with the different organs of the seedlings (Tsuboi *et al.*, 1989). Again, the immuno-precipitated proteins that were newly synthesized from RNAs were of a larger size.

Cell localization

Two surprising conclusions were drawn from studies performed at the light and electron microscope levels in maize, castor bean and *Arabidopsis* (Sossountzov *et al.*, 1991; Thoma *et al.*, 1992). The first of these is a tissue-specific localization in maize: LTP is mainly found in the peripheral cell layers of maize embryos, including the first leaf and the coleoptile. The cells around the vascular bundles also contain large amounts of LTP. These findings led to the concept that LTPs cannot be considered as true 'housekeeping' proteins. The second conclusion, drawn from immuno-electron microscopy, is that LTPs are found in diverse intracellular compartments of plant cells, not always associated with membranes. In the case of castor bean, localization inside glyoxysomes was indicated (Yamada *et al.*, 1990). In this plant and in *Arabidopsis*, localization in the cell wall was also indicated (Thoma *et al.*, 1992; Yamada *et al.*, 1990). This unexpected finding will be discussed later.

6. Gene expression

The availability of antibodies and the determination of amino-acid sequences have provided tools to isolate the genes coding for LTP.

Isolation of cDNAs and genomic clones

Since the first isolation of a cDNA coding for a maize LTP (Tchang *et al.*, 1988), several cDNAs have been isolated and characterized (*Table 2*). These were obtained either by screening cDNA libraries with specific antibodies or with oligonucleotides synthesized according to the amino-acid sequence (Bernhard *et al.*, 1991; Tchang *et al.*, 1988; Thoma *et al.*, 1992; Tsuboi *et al.*, 1991). Other cDNAs have been identified as corresponding to LTP-like proteins by sequence homology (Foster *et al.*, 1992; Hughes *et al.*, 1992; Koltunow *et al.*, 1990; Mundy and Rogers, 1986; Plant *et al.*, 1991; Torres-Schumann *et al.*, 1992).

Table 2. cDNA and genes coding for plant lipid transfer proteins

Source	cDNA or gene	Designation	Predicted mature protein (residues)	Predicted signal peptide (residues)
Maize[a]	cDNA	9 c2	93	27
Spinach[b]	cDNA	pWB3	92	26
Barley[c]	cDNA	PAPI	92	25
Castor bean[d]	cDNA	nsLTP-C	92	24
Rape seed[e]	cDNA	E2 cDNA	95	24
Castor bean[f]	cDNA	nsLTPC1	92	21
Castor bean[f]	cDNA	nsLTPC2	92	23
Carrot[g]	cDNA	EP2	94	26
Tomato[h]	cDNA	PLE16	95	24
Tomato[j]	cDNA	TSW12	91	23
Barley[k]	cDNA	blt4	95	25
Barley[l]	Genomic clone	Ltp1	91	26
Tobacco[m]	Genomic clone	Ltp1	91	23

[a] Tchang *et al.*, 1988; [b] Bernhard *et al.*, 1991; [c] Mundy and Rogers, 1986; [d] Tsuboi *et al.*, 1991; [e] Foster *et al.*, 1992; [f] Weig and Komor, 1992; [g] Sterk *et al.*, 1991; [h] Plant *et al.*, 1991; [j] Torres-Schumann *et al.*, 1992; [k] Hughes *et al.*, 1992; [l] Linnestad *et al.*, 1991 and Skriver *et al.*, 1992; [m] Fleming *et al.*, 1992.

The deduced amino-acid sequences from these cDNAs have several features in common. They all correspond to the sequences determined with the purified LTPs, including the eight cysteine residues. In addition, they contain an N-terminal extrapeptide consisting of 21 to 27 residues. This was suggested to be a signal peptide, involved in the crossing of membranes (Bernhard *et al.*, 1991; Madrid, 1991). Import of the precursor form of spinach or barley LTP, corresponding to larger polypeptides similar to those detected by immunoprecipitation, was demonstrated in experiments performed with endoplasmic reticulum membranes (Bernhard *et al.*, 1991, Madrid, 1991). This import suggested a possible secretion of the LTP, which was shown to occur in the growth medium of somatic embryos from carrot (Sterk *et al.*, 1991) or *Vitis vinifera* (Maes *et al.*, 1994).

This signal sequence was confirmed by the isolation of three genomic clones coding for LTP from barley (Linnestad *et al.*, 1991; Skriver *et al.*, 1992) or tobacco (Fleming *et al.*, 1992). The length of the signal sequence ranged from 23 to 26 amino acids.

Gene expression

The availability of several cDNAs and genes coding for LTPs allowed an extensive study of LTP gene expression in various plants. One major, and unexpected, observation is that LTP genes were not expressed in a constitutive manner as might be predicted from their suggested role in lipid dynamics. On the contrary, tissue- or cell-layer-specific gene expression was found both by Northern blot and by *in-situ* hybridization. An unexpected location in the peripheral cell layers of maize coleoptile (Sossountzov *et al.*, 1991) or in the epidermis of carrot somatic embryos (Sterk *et al.*, 1991) was discovered. In tobacco, the LTP gene is highly expressed in the shoot apex (Fleming *et al.*, 1992), whereas a gene coding for a LTP-like protein was found to be specifically expressed in the tapetum of the inflorescences (Koltunow *et al.*, 1990). A similar high level of expression of a gene was observed in the tapetum of rapeseed anthers (Foster *et al.*, 1992). In barley seeds, a specific expression of LTP genes was found in aleurone cells (Linnestad *et al.*, 1991; Skriver *et al.*, 1992).

LTP gene expression is also time-dependent. The levels of mRNA increase during the germination of maize or castor bean seeds and then decline (Sossountzov *et al.*, 1991; Tsuboi *et al.*, 1989). In addition, the expression of several LTP genes appears to be related to environmental conditions. For example, low temperatures induce an LTP gene in barley (Hughes *et al.*, 1992), whereas drought stress induces LTP genes in tomato (Plant *et al.*, 1991, Torres-Schumann *et al.*, 1992). In this latter case, genes are either expressed in aerial vegetative tissue (Plant *et al.*, 1991) or only in the stem (Torres-Schumann *et al.*, 1992). All these genes also appear to be regulated by abscisic acid.

In conclusion, it seems that a multiple pattern of expression exists for LTP genes, which are switched on in various cellular zones at various stages of plant development. Moreover, in the same plant and at the same stage of development, several genes are expressed in different parts of the plant. This is the case for castor bean (Tsuboi *et al.*, 1989; Weig and Komor, 1992), which has been most intensively studied.

This raises the question of the number of LTP genes in a plant. Based on Southern blotting, values ranging from one in maize (Tchang *et al.*, 1988) or spinach (Bernhard *et al.*, 1991) to four (Tsuboi *et al.*, 1989; Weig and Komor, 1992) in castor bean have been determined. It is of interest to ascertain the roles of the corresponding proteins.

7. Physiological roles

Progress in the knowledge of the biochemistry and molecular biology of plant LTPs has provided tools for understanding their *in-vivo* role in plant cells. Although it was initially proposed that

LTPs act in the cytosol, the recent discovery of their (at least partial) extracellular location has given rise to other hypotheses about their roles in the extracellular matrix.

Intracellular roles

Phospholipid metabolism and fatty acid dynamics. Based on the *in-vitro* properties of LTPs, it has been proposed that these proteins facilitate the movement of phospholipids linked to the biosynthetic pathways, for example, the transfer of PC from the endoplasmic reticulum to the chloroplast (*Figure 3*). However, this remains to be proven. Furthermore, the possibility that the intracellular pools of acyl-CoA can be regulated by LTPs due to their binding ability deserves further study (Ostergaard et al., 1993). It also has yet to be determined whether other apolar ligands can be bound to LTPs.

Lipid signalling. Since PI derivatives are suggested to be involved in intracellular transduction of external signals, it is tempting to suggest a role for LTPs in transporting PI synthesized on the endoplasmic reticulum to the plasma membrane where it is phosphorylated.

The recent demonstration that a wheat basic protein, homologous to LTP, can be phosphorylated by a Ca^{2+}-dependent protein kinase reinforces the idea that LTPs play a role in the dynamics of lipids that are linked to transduction processes (Polya *et al.*, 1992).

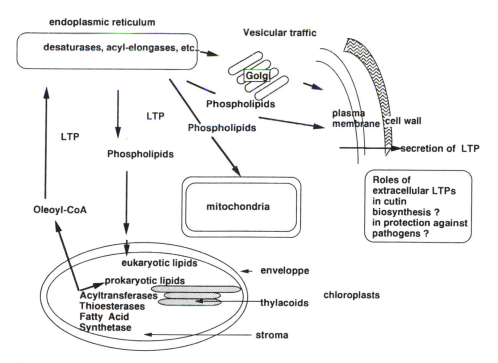

Figure 3. Suggested roles of lipid transfer proteins. According to this simplified model of plant lipid metabolism, lipids are synthesized within chloroplasts by the prokaryotic pathway ('prokaryotic lipids') and are exported as acyl-CoA esters from the chloroplasts and then transferred back for the final steps of the eukaryotic pathway ('eukaryotic lipids'). LTPs are suggested to regulate the pools of oleoyl-CoA and to transfer phospholipids between the endoplasmic reticulum and the organelles. However, movement of phospholipids can occur via vesicular traffic. The fact that LTPs are secreted led to the hypothesis that they are involved in cutin biosynthesis. Other extracellular roles (defence reactions against pathogens) are suggested.

Membrane biogenesis. It is logical to suggest that LTPs, able to facilitate *in-vitro* movements between membranes, behave in the same manner *in vivo* and contribute to the flux of newly synthesized phospholipids necessary for membrane biogenesis or renewal. This hypothesis was proposed several years ago (Kader, 1975; Wirtz, 1991) for animal as well as plant cells. Some indirect evidence favours such a hypothesis: (1) the correlation found between LTP biosynthesis and membrane biogenesis in organs during development and (2) the high level of expression of LTP genes in cellular zones where active membrane biogenesis occurs.

A correlation between the secretory pathway and LTPs has been established in the last few years in yeasts (Bankaitis *et al.*, 1990; Cleves *et al.*, 1991, Skinner *et al.*, 1993). The experiments are based on the use of a *Saccharomyces cerevisiae* mutant, sec 14, which displays impairment of the secretory process and is unable to grow at high temperatures, due to a mutation in the gene coding for a PI-TP. This discovery established for the first time an essential role for an LTP in the normal development of living cells, and, in addition, demonstrated the involvement of an LTP in a secretory process. This could be explained by a need for a suitable lipid composition for a functional Golgi apparatus (Cleves *et al.*, 1991).

Vesicular traffic and secretory pathway. The hypothesis of an endomembrane flow linking the endoplasmic reticulum, the Golgi apparatus and the plasma membrane (Morré *et al.*, 1979) has been suggested to explain the flux of very long chain fatty acids in plant cells (Bertho *et al.*, 1991). According to Cassagne's research group, these fatty acids are transported from the endoplasmic reticulum towards the plasma membranes, where they are accumulated via lipid-rich vesicles. A sorting mechanism of fatty acid flux has also been proposed (Bertho *et al.*, 1991). This hypothesis, based on the use of monensin, which is known to disturb the vesicular traffic at the level of the Golgi apparatus, proposes that, according to the length of their chains, fatty acids can be transported via the Golgi apparatus (very long chain fatty acids) or by another process which may involve LTPs (long chain fatty acids).

However, the hypothesis of essential intracellular roles for LTPs is weakened by the finding that plant LTPs are located partly (Yamada *et al.*, 1990) or mainly (Thoma, *et al.*, 1992) in the extracellular matrix and are secreted (Maes *et al.*, 1994; Sterk *et al.*, 1991). It is thus important to determine precisely the cellular location of all LTP isoforms in plants. The possibility remains that some LTP isoforms are more specifically located in the cytosol, and this must be checked in the future.

A hypothesis that attempts to reconcile an intracellular role of LTPs with their import through the membranes of the endoplasmic reticulum (Bernhard *et al.*, 1991; Madrid, 1991) has been proposed (Madrid and Von Wettstein, 1990). According to this theory, LTPs can act as translocators of phospholipids in the lumen of the endomembrane system (Madrid and Von Wettstein, 1990).

Interestingly, intercellular movements of fluorescent phospholipids have been described recently (Grabski et *al.*, 1993). These movements can occur through plasmodesmata which establish intercellular continuity of the endomembrane system (Lucas and Wolf, 1993).

Extra- or intercellular roles of LTPs

Cutin biosynthesis. It has been suggested that LTPs that are present in the cell wall or secreted are involved in the biosynthesis of cutin by transporting acyl monomers from the cytosol to the sites of cutin formation (Hendricks *et al.*, 1993; Sterk *et al.*, 1991). This interesting hypothesis is supported by the acyl-binding ability of LTPs (Ostergaard *et al.*, 1993), and it would also explain the role of LTPs in the differentiation of the epidermis (Sterk *et al.*, 1991).

Defence reactions. The idea that LTP may be involved in defence reactions against various environmental stress conditions arose from the findings of several experiments which showed that LTP genes are induced by low or high temperatures, high salt concentration, drought or abscisic acid treatments (Hughes *et al.*, 1992; Plant *et al.*, 1991; Torres-Schumann *et al.*, 1992). The observation that an LTP-like protein is secreted by rice suspension cultures treated with salicylic acid suggests the possibility that LTPs are also involved in defence reactions against pathogens (Masuta *et al.*, 1991). The direct demonstration that LTPs act as antifungal proteins (Molina *et al.*, 1993; Terras *et al.*, 1992) opens novel perspectives in this domain. It is conceivable that LTPs act as membrane-permeabilizing agents, probably due to their high isoelectric point. In addition, they belong to a superfamily of cysteine-rich proteins consisting of trypsin inhibitors and thionins (Apel *et al.*, 1990).

8. Conclusions

It is clear that LTPs, initially regarded as cytosolic proteins, are also found in the intercellular matrix. Their role(s) in cell–cell interactions requires investigation in addition to studies of the intracellular compartment.

How can the precise roles of LTPs be determined? One promising approach is the RNA antisense strategy that is being used by Somerville's group in studies of *Arabidopsis* (Somerville, personal communication). However, this approach is problematic since several LTP genes are present. Another approach is to isolate mutants of *Arabidopsis* that are deficient in functional LTP genes. Finally, the complementation of yeast mutants, such as sec 14, can provide indications of the function of LTP genes that have already been isolated, and may lead to the discovery of other as yet unidentified LTPs.

All these projects may lead to exciting developments in our understanding of lipid transfer proteins.

Acknowledgements

The authors thank Ms Marie-France Laforge for the careful preparation of the manuscript.

References

Apel K, Bohlmann H, Reimann-Philipp U. (1990) Leaf thionins, a novel class of putative defence factors. *Physiol. Plant.* **80:** 315–321.

Arondel V, Vergnolle C, Tchang F, Kader JC. (1990) Bifunctional lipid-transfer: fatty acid-binding proteins in plants. *Mol. Cell. Biochem.* **98:** 49–56.

Bankaitis VA, Aitken JR, Cleves AE, Dowhan W. (1990) An essential role for a phospholipid transfer protein in yeast Golgi function. *Nature* **347:** 561–562.

Bernhard WR, Somerville CR. (1989) Coidentity of putative amylase inhibitors from barley and finger millet with phospholipid transfer proteins inferred from amino acid sequence homology. *Arch. Biochem. Biophys.* **269:** 695–697.

Bernhard WR, Thoma S, Botella J, Somerville CR. (1991) Isolation of a cDNA clone for spinach lipid transfer protein and evidence that the protein is synthesized by the secretory pathway. *Plant Physiol.* **95:** 164–170.

Bertho P, Moreau P, Morré DJ, Cassagne C. (1991) Monensin blocks the transfer of very long chain fatty acid-containing lipids to the plasma membrane of leek seedlings. Evidence for lipid sorting based on fatty acyl chain length. *Biochim. Biophys. Acta* **1070:** 127–134.

Bishop WR, Bell RM. (1988) Assembly of phospholipids into cellular membranes: biosynthesis, transmembrane movement and intracellular translocation. *Annu. Rev. Cell. Biol.* **4:** 579–610.

Browse J, Somerville C. (1991) Glycerolipid synthesis: biochemistry and regulation. *Annu. Rev. Plant. Physiol. Plant Mol. Biol.* **42:** 467–506.

Campos FAP, Richardson M. (1984) The complete amino acid sequence of the α-amylase inhibitor 1-2 from seeds of ragi (Indian finger millet, *Eleusine coracana* Gaertn.) *FEBS Lett.* **167:** 221–225.

Cleves A, McGee T, Bankaitis VA. (1991) Phospholipid transfer proteins: a biological debut. *Trends Cell Biol.* **1**: 30–34.

Dawidowicz EA. (1987) Dynamics of membrane lipid metabolism and turnover. *Annu. Rev. Biochem.* **56**: 43–61.

Désormeaux A, Blochet JE, Pézolet M, Marion D. (1992) Amino acid sequence of a non-specific wheat phospholipid transfer protein and its conformation as revealed by infrared and Raman spectroscopy. *Biochim. Biophys. Acta* **1121**: 137–152.

Fleming AJ, Mandel T, Fofmann S, Sterk P, de Vries SC, Kuhlemeier C. (1992) Expression pattern of a tobacco lipid transfer protein gene within the shoot apex. *Plant J.* **2**: 855–862.

Foster GD, Robinson SW, Blundell RP, Roberts MR, Hodge R, Draper J, Scott RJ. (1992) A *Brassica napus* mRNA encoding a protein homologous to phospholipid transfer proteins is expressed specifically in the tapetum and developing microspores. *Plant Sci.* **84**: 184–192.

Geldwerth D, de Kermel A, Zachowski A, Guerbette F, Kader JC, Henry JP, Devaux PF. (1991) Use of spin-labeled and fluorescent lipids to study the activity of the phospholipid transfer protein from maize seedlings. *Biochim. Biophys. Acta* **1082**: 255–264.

Grabski S, De Feijter AW, Schindler M. (1993) Endoplasmic reticulum forms a dynamic continuum for lipid diffusion between contiguous soybean root cells. *Plant Cell* **5**: 25–38.

Grosbois M, Guerbette F, Kader JC. (1989) Changes in level and activity of phospholipid transfer protein during maturation and germination of maize seeds. *Plant Physiol.* **90**: 1560–1564.

Grosbois M, Guerbette F, Jolliot F, Quintin F, Kader JC. (1993) Control of maize lipid transfer protein activity by oxido-reducing conditions. *Biochim. Biophys. Acta* **1170**: 197–203.

Hendricks T, Meije EA, Thoma S, Kader JC, de Vries SC. (1993) The carrot extracellular lipid transfer protein EP2: quantitative respect to its putative role in cutin synthesis. In: *NATO/ASI Seminar on Molecular-Genetic Approaches to Plant Metabolism and Development* (Puigdomenech P, Coruzzi G, ed.). Berlin: Springer-Verlag.

Hughes MA, Dunn MA, Pearce RS, White AJ, Zhang L. (1992) An abscisic-acid-responsive, low temperature barley gene has homology with a maize phospholipid transfer protein. *Plant Cell Environ.* **15**: 961–865.

Kader JC. (1975) Proteins and the intracellular exchange of lipids I. Stimulation of phospholipid exchange between mitochondria and microsomal fractions by proteins isolated from potato tuber. *Biochim. Biophys. Acta* **380**: 31–44.

Kader JC. (1993) Structure, functions and gene expression of lipid transfer proteins. In: *Biochemistry and Molecular Biology of Membrane and Storage Lipids of Plants* (Murata N, Somerville CR, ed.). Rockville: The American Society of Plant Physiologists, pp. 207–214.

Kader JC, Julienne M, Vergnolle C. (1984) Purification and characterization of a spinach leaf protein capable of transferring phospholipids from liposomes to mitochondria or chloroplasts. *Eur. J. Biochem.* **139**: 411–416.

Knudsen J. (1990) Acyl-CoA-binding protein (ACBP) and its relation to fatty-acid binding protein (FABP): an overview. *Mol. Cell. Biochem.* **98**: 217–224.

Koltunow AM, Truettner J, Cox KH, Wallroth M, Goldberg RB. (1990) Different temporal and spatial gene expression patterns occur during anther development. *Plant Cell* **2**: 1201–1224.

Linnestad C, Lönneborg A, Kalla E, Alsen OA. (1991) Promoter of a lipid transfer protein gene expressed in barley aleurone cells contains similar myb and myc recognition sites to the maize Bz-Mcc allele. *Plant Physiol.* **97**: 841–843.

Lucas WJ, Wolf S. (1993) Plasmodesmata: the intercellular organelles of green plants. *Trends Cell Biol.* **3**: 308–315.

Madrid SM. (1991) The barley lipid transfer protein is targeted into the lumen of the endoplasmic reticulum. *Plant Physiol. Biochem.* **29**: 695–703.

Madrid SM, Von Wettstein D. (1990) Reconciling contradictory notions on lipid transfer proteins in higher plants. *Plant Physiol. Biochem.* **29**: 705–711.

Maes O, Coutos-Thevenot P, Jouenne T, Guerbette F, Grosbois M, Le Caer JP, Boulay M, Deloire A, Kader JC, Guern J. (1994) Four 9 kDa proteins excreted by somatic embryos of grapevine are isoforms of lipid transfer proteins. *Eur. J. Biochem.* **217**: 885–889.

Masuta C, Van den Bulcke M, Bauw G, Van Montagu M, Caplan AB. (1991) Differential effects of elicitors on the viability of rice suspension cells. *Plant Physiol.* **97**: 619–629.

Miquel M, Block MA, Joyard J, Dorne AJ, Dubacq JP, Kader JC, Douce R. (1987) Protein mediated transfer of phosphatidylcholine from liposomes to spinach chloroplast envelope membranes. *Biochim. Biophys. Acta* **937**: 219–226.

Molina A, Segura A, Garcia-Olmedo F. (1993) Lipid transfer proteins (nsLTPs) from barley and maize leaves are potent inhibitors of bacterial and fungal plant pathogens. *FEBS Lett.* **316:** 119–122.

Moore TS Jr. (1993) Biosynthesis and utilization of phosphatidylethanolamine and phosphatidylcholine. In: *Biochemistry and Molecular Biology of Membrane and Storage Lipids of Plants* (Murata N, Somerville CR, ed.). Rockville: The American Society of Plant Physiologists, pp. 159–164.

Moreau F, Davy de Virville J, Hoffelt M, Guerbette F, Kader JC. (1994) Use of fluorimetric method to assay the binding and transfer of phospholipids by lipid transfer proteins from maize seedlings and *Arabidopsis* leaves. *Plant Cell Physiol.* **35:** 267–274.

Morré DJ, Kartenbeck J, Franke WW. (1979) Membrane flow and interconversions among membranes. *Biochim. Biophys. Acta* **559:** 71–152.

Mundy J, Rogers JC. (1986) Selective expression of a probable amylase/protease inhibitor in barley aleurone cells: comparison to the barley amylase/subtilisin inhibitor. *Planta* **169:** 51–63.

Nishida I, Imai OI, Nishizawa I, Tasaka H, Shiraishi H, Higashi S, Hayashi H, Beppu T, Matsuo T, Murata N. (1993) Glycerol-3-phosphate acyltranferase, acyl-ACP hydrolase and stearoyl-ACP desaturase: Molecular and physiological studies. In: *Biochemistry and Molecular Biology of Membrane and Storage Lipids of Plants* (Murata N, Somerville CR, ed.). Rockville: The American Society of Plant Physiologists, pp. 79–88.

Ostergaard J, Vergnolle C, Schoentgen F, Kader JC. (1993) Acyl-binding/lipid-transfer proteins from rape seedlings, a novel category of proteins interacting with lipids. *Biochim. Biophys. Acta* **1170:** 109–117.

Paltauf F, Daum G. (1990) Phospholipid transfer in microorganisms. *Subcell. Biochem.* **16:** 279 299.

Pebay-Peyroula E, Cohen-Addad C, Lehmann MS, Marion D. (1992) Crystallographic data for the 9000 dalton wheat non-specific phospholipid transfer protein. *J. Mol. Biol.* **226:** 563–564.

Plant AL, Cohen A, Moses MS, Bray EA. (1991) Nucleotide sequence and spatial expression pattern of a drought- and abscisic acid-induced gene of tomato. *Plant Physiol.* **97:** 900–906.

Polya GM, Chandra S, Chung R, Neumann GM, Hoj PB. (1992) Purification and characterization of wheat and pine small basic protein substrates for plant calcium-dependent protein kinase. *Biochim. Biophys. Acta* **1120:** 273–280.

Rickers J, Tober I, Spener F. (1984) Purification and binding characteristics of a basic fatty acid binding protein from *Avena sativa* seedlings. *Biochim. Biophys. Acta* **784:** 313–319.

Simorre JP, Caille A, Marion D, Marion D, Ptak M. (1991) Two-and three-dimensional ¹H NMR studies of a wheat phospholipid transfer protein: sequential resonance assignments and secondary structure. *Biochemistry* **30:** 11600–11608.

Skinner HB, Alb JG Jr, Whitters EA, Helmkamp GM Jr, Bankaitis VA. (1993) Phospholipid transfer activity is relevant to but not sufficient for the essential function of the yeast SEC 14 gene product. *EMBO J.* **12:** 4775–4784.

Skriver K, Leah R, Müller-Uri F, Olsen FL, Mundy J. (1992) Structure and expression of the barley lipid transfer protein gene Ltp1. *Plant Mol. Biol.* **18:** 585–589.

Somerville C, Browse J. (1991) Plant lipids: metabolism, mutants, and membranes. *Science* **252:** 80–87.

Sossountzov L, Ruiz-Avial L, Vignols F, et al. (1991) Spatial and temporal expression of a maize lipid tranfer protein gene. *Plant Cell* **3:** 923–933.

Sterk P, Booij H, Scheleekens GA, Van Kammen A, de Vries SC. (1991) Cell-specific expression of the carrot EP2 lipid transfer protein gene. *Plant Cell* **3:** 907–921.

Stymne S. (1993) Biosynthesis of 'uncommon' fatty acids and their incorporation into triacylglycerols. In: *Biochemistry and Molecular Biology of Membrane and Storage Lipids of Plants* (Murata N, Somerville CR, ed.). Rockville: The American Society of Plant Physiologists, pp. 150–158.

Tchang F, This P, Stiefel V et al. (1988) Phospholipid transfer protein: full length cDNA and amino acid sequence in maize. Amino acid sequence homologies between plant phospholipid transfer proteins. *J. Biol. Chem.* **263:** 16489–16855.

Terras FRG, Schofs HME, de Bolle MFC, Van Leuven F, Rees SB, Vanderleyden J, Cammue BPA, Broekaert WF. (1992) *In vitro* antifungal activity of a radish (*Raphanus sativus* L.) seed protein homologous to non-specific lipid transfer proteins. *Plant Physiol.* **100:** 1055–1058.

Thoma S, Kaneko Y, Somerville CR. (1992) A non-specific lipid transfer protein from *Arabidopsis* is a cell wall protein. *Plant J.* **3:** 427–436.

Thompson GA. (1993) The dynamic role of inositol phospholipids in plants. In: *Biochemistry and Molecular Biology of Membrane and Storage Lipids of Plants* (Murata N, Somerville CR, ed.). Rockville: The American Society of Plant Physiologists, pp. 17–25.

Torres-Schumann S, Godoy JA, Pintor-Toro JA. (1992) A probable lipid transfer protein gene is induced by NaCl in stems of tomato plants. *Plant Mol. Biol.* **18:** 749–757.

Tsuboi S, Watanabe SI, Ozeki Y, Yamada M. (1989) Biosynthesis of nonspecific lipid transfer proteins in germinating castor bean seeds. *Plant Physiol.* **90:** 841–845.

Tsuboi S, Suga T, Takishima K, Mamiya G, Matsui K, Ozeki Y, Yamada M. (1991) Organ-specific occurrence and expression of the isoforms of nonspecific lipid transfer protein in castor bean seedlings and molecular cloning of a full-length cDNA for a cotyledon-specific isoform. *J. Biochem.* **110:** 823–831.

Veerkamp JH, Peeters RA, Maatman RGHJ. (1991) Structural and functional features of different types of cytoplasmic fatty-acid binding proteins. *Biochim. Biophys. Acta* **1081:** 1–24.

Vergnolle C, Arondel V, Grosbois M, Guerbette F, Jolliot A, Kader JC. (1988) Synthesis of phospholipid transfer proteins from maize seedlings. *Biochem. Biophys. Res. Commun.* **157:** 37–41.

Vick BA. (1993) Oxygenated fatty acids of the lipoxygenase pathway. In: *Lipid Metabolism in Plants* (Moore TS, ed.). Boca Raton: CRC Press, pp. 167–191.

Weig A, Komor E. (1992) The lipid-transfer protein C of *Ricinus communis* L.: isolation of two cDNA sequences which are strongly and exclusively expressed in cotyledons after germination. *Planta* **187:** 367–371.

Wirtz KWA. (1991) Phospholipid transfer proteins. *Annu. Rev. Biochem.* **60:** 73–99.

Yamada M. (1992) Lipid transfer proteins in plants and microorganisms. *Plant Cell Physiol.* **33:** 1–6.

Yamada M, Tsuboi S, Osafune T, Suga T, Takishima K. (1990) Multifunctional properties of non-specific lipid transfer protein from higher plants. In: *Plant Lipid Biochemistry, Structure and Utilization* (Quinn PJ, Harwood JL, ed.). London: Portland Press, pp. 278–280.

Yu YG, Chung CH, Fowler A, Suh SW. (1988) Amino acid sequence of a probable amylase/protease inhibitor from rice seeds. *Arch. Biochem. Biophys.* **265:** 466–475.

Chapter 11

Plant galactolipids and sulfolipid: structure, distribution and biosynthesis

Jacques Joyard, Eric Marechal, Maryse A. Block and Roland Douce

1. Introduction

Glycolipids are a structurally diverse group of membrane constituents (Curatolo, 1987a,b). They are found in numerous bacteria, in algae, and in higher plants as well as in animals. In most cases, complex glycolipids are present as only minor components in cell membranes. In contrast, large amounts of glycolipids, sulfolipid and galactolipids, consisting of structurally simple glycosyl groups, are found in plants. In this case, they are restricted to plastid membranes. The expanding literature on plant glycolipids has been reviewed thoroughly. Molecular aspects of plant lipid biosynthesis, which are in part related to glycolipid biosynthesis, have been described by Somerville and Browse (1991) and Heinz (1993), whereas more specific aspects have been reviewed in several articles by Heinz (1977), Douce and Joyard (1980), Roughan and Slack (1982), Quinn and Williams (1983), Joyard and Douce (1987) and Joyard *et al.* (1993). The purpose of this chapter is to summarize briefly the basic information available about the structure, properties and biosynthesis of plant glycolipids.

2. Structure, distribution and properties of plastid glycolipids

Structure and distribution of galactolipids and sulfolipid

Galactolipids are neutral glycerolipids which contain one or two galactose molecules attached to the *sn*-3 position of the glycerol backbone, corresponding respectively to 1,2-diacyl-3-*O*-(β-D-galactopyranosyl)-*sn*-glycerol (monogalactosyldiacylglycerol or MGDG) and 1,2-diacyl-3-*O*-(α-D-galactopyranosyl-(1→6)-*O*-β-D-galactopyranosyl)-*sn*-glycerol (digalactosyldiacylglycerol or DGDG). MGDG was first isolated as a galactose-containing lipid from wheat flour by Carter *et al.* (1956), and from leaves by Sastry and Kates (1964). A unique feature of galactolipids is their very high content of polyunsaturated fatty acids; in some species, up to 95% of the total fatty acids is linolenic acid (18:3). Therefore, the most abundant molecular species of galactolipids have 18:3 at both the *sn*-1 and *sn*-2 positions of the glycerol backbone (*Figure 1*). Some plants, such as pea, which have almost exclusively 18:3 in MGDG, are called '18:3 plants'. Other plants, such as spinach, which contain large amounts of 16:3 in MGDG, are called '16:3 plants' (Heinz, 1977). The positional distribution of 16:3 in MGDG is highly specific; this fatty acid is only present at the *sn*-2 position of glycerol and is almost excluded from the *sn*-1 position. Therefore, different galactolipid molecular species with either C18 fatty acids at both *sn* positions or with C18 and C16 fatty acids at the *sn*-1 and *sn*-2 positions, respectively, may be present in plastid membranes. The proportions of these two types of MGDG molecular species vary widely among plants (Heinz, 1977). The first structure is typical of 'eukaryotic' lipids (such

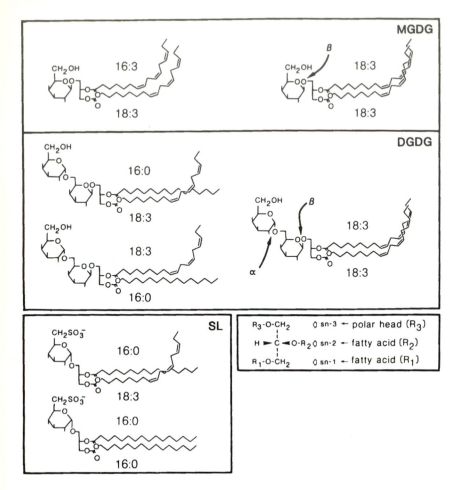

Figure 1. Structure of the major plastid glycolipids. Glycolipids with C16 fatty acids at the *sn*-2 position of the glycerol have a prokaryotic structure.

as phosphatidylcholine) and the second one corresponds to a 'prokaryotic' structure, since it is characteristic of cyanobacterial glycerolipids (Heinz, 1977). In addition, this difference is also valid for other glycerolipids: any membrane lipid containing C16 fatty acids at the *sn*-2 position of glycerol is considered to be prokaryotic. The proportion of eukaryotic to prokaryotic molecular species is not identical for all glycerolipids from a given plant. For instance, although in spinach 50% of MGDG has the prokaryotic structure, this is true for only 10–15% of DGDG (Bishop *et al.*, 1985).

The most important sulfolipid found in higher plants is a 1′,2′-diacyl-3′-*O*-(6-deoxy-6-sulfo-α-D-glucopyranosyl)-*sn*-glycerol (sulfoquinovosyldiacylglycerol or SQDG). The sulfonic residue at C6 of deoxyglucose (quinovose) carries a negative charge at physiological pH. Analyses of the positional distribution of fatty acids (e.g. Siebertz *et al.*, 1979) have demonstrated that a significant proportion of SQDG in higher plants has a dipalmitoyl (16:0/16:0) backbone. However, the major molecular species in SQDG contained both 16:0 and 18:3 fatty acids (Siebertz *et al.*, 1979). In fact, two distinct structures can be observed in higher plants, containing either 18:3/16:0 or 16:0/18:3 (Bishop *et al.*, 1985; Siebertz *et al.*, 1979). The first structures (16:0/16:0 and 18:3/16:0), with 16:0 at the *sn*-2 position, are typical of 'prokaryotic' lipids, whereas the last one (16:0/18:3), with 18:3 at the sn-2 position, is typical of 'eukaryotic' lipids (see above). In a 16:3 plant, such as spinach, a higher proportion of SQDG (compared to the situation for MGDG) has a prokaryotic structure. In contrast, wheat, which is an 18:3 plant, contains almost exclusively SQDG with a eukaryotic structure (Bishop et al., 1985).

Galactolipids and sulfolipid are present in all higher plant tissues, whether or not they are photosynthetic. However, the amount of glycolipids in plant tissues reflects the expansion of plastid membranes within the cell. In photosynthetic tissues, in which chloroplast membranes are the major cell membrane system, with an area that can represent up to 5000 m^2 kg^{-1} of fresh mass (Lawlor, 1987), glycolipids are the major membrane glycerolipids. In nonphotosynthetic tissues, plastid membranes are mostly composed of envelope membranes (representing an area of 180 to 400 m^2 kg^{-1} of fresh mass (Lawlor, 1987) and correspond to only a minor proportion of total cell membranes. Therefore, the amount of phospholipids consistently exceeds the amount of glycolipids (Douce and Joyard, 1980). Galactolipids represent more than 80% of thylakoid and inner envelope membrane lipids (Table 1), of which MGDG constitutes the major part (around 50%). In the outer envelope membrane, DGDG is the major galactolipid (Table 1). SQDG represents 5–10% of plastid membrane glycerolipids (Table 1). Glycolipids are present in the cytosolic leaflet of the outer envelope membrane, since specific antibodies raised against MGDG

Table 1. Glycerolipid composition of plant cell membranes

Organelle	MGDG	DGDG	SL	PC	PG	PI	PE	DPG
Plastids								
Spinach chloroplasts								
Thylakoids[a]	53	27	7	3	7	2	0	0
Recalculated[b]	57	27	7	0	7	1	0	0
Envelope inner membrane[a]	49	30	5	6	8	1	0	0
Recalculated[b]	55	29	5	0	9	1	0	0
Outer membrane	17	29	6	32	10	5	0	0
Total envelope	32	30	6	20	9	4	0	0
Pea etioplasts								
Total envelope	34	31	6	17	5	4	0	0
Cauliflower proplastids								
Total envelope	31.5	27.5	6	20	9	4.5	1	0
Mitochondria								
Sycamore cells[c]								
Total membranes	0	0	0	43	3	6	35	13
Inner membrane	0	0	0	41	2.5	5	37	14.5
Outer membrane	0	0	0	54	4.5	11	30	0
Peroxisomes								
Potato tubers	0	0	0	52	0	0	48	0

[a] The presence of phosphatidylcholine in thylakoids and inner envelope membrane preparations is due to contamination by the outer envelope membrane (Dorne et al., 1985). Therefore, the polar lipid composition of thylakoids and inner envelope membrane was recalculated (data from Block et al., 1983) to account for contamination of these membranes by the outer envelope membrane (Joyard et al., 1991).
[b] Polar lipid composition of plastid membranes was obtained after thermolysin treatment of intact organelles (Dorne et al., 1982).
[c] Data from Bligny and Douce (1980).
Abbreviations used: MGDG, monogalactosyldiacylglycerol; DGDG, digalactosyldiacylglycerol; SL, sulfolipid; PC, phosphatidylcholine; PG, phosphatidylglycerol; PI, phosphatidylinositol; PE, phosphatidylethanolamine; DPG, diphosphatidylglycerol.

and SQDG are able to bind to purified intact chloroplasts (Billecocq *et al.*, 1972). In contrast, membranes of highly purified mitochondria or peroxisomes are devoid of these glycolipids (*Table 1*). The only purified extraplastidial cell membrane preparations that have been shown to contain traces of galactolipids are tonoplast membranes (Haschke *et al.*, 1990; Tavernier *et al.*, 1993). In contrast to plasmalemma or endoplasmic reticulum preparations, relatively pure tonoplast membranes can be obtained, since they can be prepared from purified intact organelles (vacuoles). The presence of galactolipids in such preparations could reflect the autophagic processes characteristic of plants growing under stress conditions (Dorne *et al.*, 1987; Journet *et al.*, 1986; Roby *et al.*, 1987). A marked regression of various cell organelles was observed under these conditions, and it is therefore possible that traces of galactolipids could remain in vacuolar membranes after the stress.

Physical properties of plant glycolipids

Gounaris *et al.* (1983) have demonstrated that pure MGDG with five or more double bonds forms a hexagonal-II phase, whereas MGDG with 4.5 or fewer double bonds per molecule forms a lamellar phase. Therefore MGDG found in plant membranes, which contains six double bonds, is able to form hexagonal-II phase. This phase consists of hexagonally packed cylinders in which water fills the interior, while the acyl chains form the cylinder itself (Luzzati, 1968). In contrast, DGDG and SQDG normally form lamellar phases (Shipley *et al.*, 1973). Electron microscopy of mixtures of MGDG and DGDG indicates the presence of lamellar phases containing lipidic particles arranged in lateral arrays, exhibiting both hexagonal-II and cubic phases (Sen *et al.*, 1981, 1982). Incorporation of SQDG into mixtures of MGDG and DGDG led to the formation of either lamellar phases or paracrystalline arrays of particles (Sakai *et al.*, 1983). However, because MGDG is able to bind to glass walls (Sprague and Staehelin, 1984), the stoichiometry and morphology of MGDG/DGDG or MGDG/DGDG/SQDG phases obtained *in vitro* must be considered with caution. In addition, the physiological significance of such structures is unclear: it is considered that in most cases, nonbilayer structures have no direct biological role in chloroplasts (Gounaris and Barber, 1983; Quinn and Williams, 1983). However, if they do exist *in vivo*, it is possible that hexagonal-II and cubic phases could be restricted in both time and space. For instance, they could be involved in the interactions between the outer and inner envelope membranes, or in the formation of prolamellar bodies in etioplasts. It is also possible that the physical properties of MGDG are essential for the accomodation of specific plastid proteins, such as the light-harvesting chlorophyll *a/b*–protein complex of the thylakoid membranes (Quinn and Williams, 1983; Siefermann-Harms *et al.*, 1982). However, one should bear in mind that the physical properties of glycolipid mixtures, which are the main constituent of plastid membranes, are rather similar to those of the phosphatidylethanolamine/phosphatidylcholine mixture found in most extraplastidial membranes (Curatolo, 1987a). Therefore, the proposition of a specific role for glycolipids in plastid membranes needs further experimental evidence, although the presence of galactolipids in biological membranes is apparently related to the development of oxygenic photosynthesis.

It is also possible to predict the shape of lipid molecules, using minimum conformational energy calculations. Consistent with the possibility that MGDG could form hexagonal-II phases, this molecule has a conical shape with the glycosyl head group at the small end of the cone (Brasseur *et al.*, 1983). Murphy (1982) has proposed that MGDG and DGDG are involved in the formation of the highly curved edges of thylakoids; the cone-shaped MGDG molecules would be present in the inner concave monolayer of the thylakoid membrane, whereas the wedged-shaped DGDG molecules would be found in the convex outer monolayer. However, this hypothesis requires further experimental evidence.

3. Biosynthesis of chloroplast glycerolipids

Plastid envelope membranes play a central role in the biosynthesis of plastid glycerolipids (Douce, 1974; Joyard and Douce, 1977), especially galactolipids and sulfolipid, since they are the site of assembly of fatty acids, glycerol and polar head groups (galactose for galactolipids, and sulfoquinovose for sulfolipid). However, *in-vitro* kinetics of acetate incorporation into chloroplast lipids have demonstrated that, like prokaryotes, isolated intact chloroplasts can synthesize glycerolipids containing almost exclusively a C18/C16 diacylgycerol backbone, but are *apparently* unable to catalyze the formation of phosphatidic acid and diacylglycerol with only C18 fatty acids. Therefore, it is likely that the formation of these two structures proceeds from distinct pathways.

Origin of the diacylglycerol backbone

Formation of C18/C16 diacylglycerol backbone. The first enzyme of the glycerolipid biosynthetic pathway is a soluble glycerol-3-phosphate acyltransferase, closely associated with the inner envelope membrane, which catalyzes the transfer of oleic acid (18:1), from 18:1-ACP (acyl carrier protein) to the *sn*-1 position of glycerol (Frentzen *et al.*, 1983) producing 1-oleoyl-*sn*-glycerol-3-phosphate (lysophosphatidic acid). 1-palmitoyl-*sn*-glycerol-3-phosphate can also be synthesized, but in much lower proportions than 1-oleoyl-*sn*-glycerol-3-phosphate. The proportions are higher in the case of chilling-sensitive plants. The chloroplast glycerol-3-phosphate acyltransferase of chilling-sensitive plants (such as squash) incorporates a higher proportion of palmitic acid at the *sn*-1 position of the glycerol than in chilling-resistant plants (such as pea). Bertrams and Heinz (1981) first provided biochemical data for the purified chloroplast glycerol-3-phosphate acyltransferase from pea and spinach. Depending on the plant species analyzed, the chloroplast stroma contains one to three isomeric forms of glycerol-3-phosphate acyltransferase (Bertrams and Heinz, 1981; Douady and Dubacq, 1987; Nishida *et al.*, 1987). The sequence of the cDNA corresponding to two isoforms of the squash chloroplast protein was determined by Ishizaki *et al.* (1988). An oleate-selective acyl-ACP:*sn*-glycerol-3-phosphate acyltransferase (i.e. one which discriminates efficiently against palmitoyl-ACP) has been purified from chloroplasts of a chilling-resistant plant (pea), and its cDNA was sequenced by Weber *et al.* (1991). The comparison between the selective (pea) and nonselective (squash) acyltransferase did not provide a clue for recognizing the structural differences resulting in different selectivities (Weber *et al.*, 1991). Interestingly, Kunst *et al.* (1988) have characterized an *Arabidopsis* mutant, JB25, in which the chloroplast glycerol-3-phosphate acyltransferase activity was reduced to less than 4% of the activity in the wild type. This mutant could be an interesting tool to investigate possible regulation of glycerolipid biosynthesis in higher plants (see below).

Lysophosphatidic acid is further acylated to form 1,2-diacyl-*sn*-glycerol-3-phosphate (phosphatidic acid) by the action of a 1-acylglycerol-3-phosphate acyltransferase (Joyard and Douce, 1977). In spinach chloroplasts, both the outer and the inner envelope membranes contain this acyltransferase (Block *et al.*, 1983), but in pea chloroplasts it is present only in the inner membrane (Andrews *et al.*, 1985). Since lysophosphatidic acid used for this reaction is esterified at the *sn*-1 position, the enzyme will direct fatty acids, almost exclusively palmitic acid (16:0), to the available *sn*-2 position (Frentzen *et al.*, 1983; Frentzen, 1993). The physiological significance of the outer envelope 1-acylglycerol-3-phosphate acyltransferase is completely unknown, as are its specificity and selectivity for the different substrates.

Therefore the two plastid acyltransferases have distinct specificities and selectivities for acylation of *sn*-glycerol-3-phosphate (*Figure 2*). Together, they lead to the formation of phosphatidic acid with 18:1 (or to a much lesser extent 16:0) and 16:0 fatty acids, respectively,

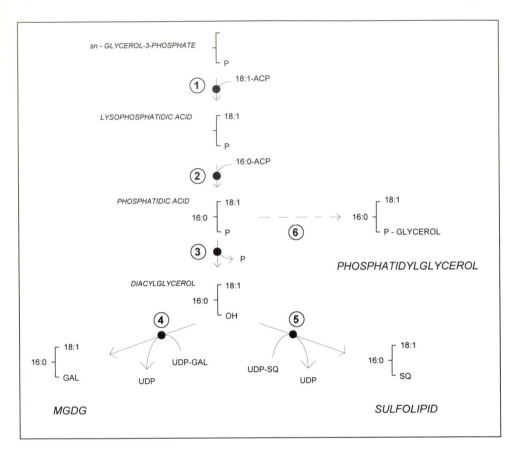

Figure 2. Biosynthesis of plastid glycolipids by plastid envelope membranes. The enzymes involved are (1) glycerol-3-phosphate acyltransferase, (2) 1-acylglycerol-3-phosphate acyltransferase, (3) phosphatidate phosphohydrolase, (4) MGDG synthase and (5) SQDG synthase; (6) indicates phosphatidylglycerol biosynthesis. This biosynthetic pathway leads to the synthesis of MGDG, SQDG and phosphatidylglycerol with C18 and C16 fatty acids at the *sn*-1 and *sn*-2 positions of glycerol, respectively. The fatty acids esterified to these glycerolipid molecules are then desaturated.

at the *sn*-1 and *sn*-2 positions of the glycerol backbone. This structure is typical of the so-called prokaryotic glycerolipids. In contrast, extraplastidial acyltransferases have distinct localization and properties (nature of the acyl donor, specificities, selectivities) as discussed by Frentzen (1993).

Phosphatidic acid synthesized in envelope membranes is further metabolized to either diacylglycerol or phosphatidylglycerol (*Figure 2*). Diacylglycerol (1,2-diacyl-*sn*-glycerol) bio-synthesis occurs in the envelope membrane as a result of the activity of a membrane-bound phosphatidate phosphatase (Joyard and Douce, 1977; 1979) exclusively located on the inner envelope membrane (Andrews *et al.*, 1985; Block *et al.*, 1983). In contrast to chloroplasts from 16:3 plants, those from 18:3 plants have a rather low phosphatidate phosphatase activity (Heinz and Roughan, 1983) and cannot deliver diacylglycerol fast enough to sustain the full rate of glycolipid synthesis. The same is true of nongreen plastids from 18:3 plants (Alban *et al.*, 1989). These results can explain why 18:3 plants contain only small amounts of galactolipids and sulfolipid with C16 fatty acids at the *sn*-2 position, but contain phosphatidylglycerol (which is synthesized from phosphatidic acid) with such a structure. It is not yet known whether the reduced level of phosphatidate phosphatase activity is due to a lower expression (species-specific) of the gene coding for the enzyme, or to the presence of regulatory molecules which control the activity of the enzyme.

The envelope phosphatidate phosphatase exhibits biochemical properties that are clearly different from those of similar enzymes described in the various cell fractions from animals or yeast (for reviews see Bishop and Bell, 1988; Carman and Henry, 1989). Phosphatidate phosphatase activities described in extraplastidial compartments from plant tissues strongly resemble their animal counterparts and are very different from the envelope enzyme (Joyard and Douce, 1987; Stymne and Stobart, 1987). First, the envelope enzyme is tightly membrane-bound, whereas in yeast or animal cells the activity is recovered from both cytosolic and microsomal fractions. Furthermore, the pH optimum for the envelope enzyme is alkaline (pH 9.0), and cations, such as Mg^{2+}, are powerful inhibitors of the enzyme. After solubilization, phosphatidate phosphatase was still sensitive to Mg^{2+} and to a wide range of cations. Mn^{2+}, Cu^{2+} and Zn^{2+} were the most potent inhibitors of the solubilized enzyme (Malherbe *et al.*, 1995). In marked contrast, the yeast or animal phosphatidate phosphatases are active in the pH range 5.5 to 7.5, and their activity is strongly dependent upon the presence of Mg^{2+} (for review see Joyard and Douce, 1987). Malherbe *et al.* (1992) demonstrated that diacylglycerol is a powerful inhibitor (apparent K_i 70 μM) of chloroplast envelope phosphatidate phosphatase, whereas the affinity of this enzyme for its substrate, phosphatidic acid, is rather low (apparent K_m 600 μM). *In vivo*, the steady-state activity of phosphatidate phosphatase is sensitive to the diacylglycerol/phosphatidic acid molar ratio (Malherbe *et al.*, 1992). Feedback inhibition of phosphatidate phosphatase (and consequently of galactolipid and sulfolipid synthesis) by diacylglycerol might lead to the accumulation of phosphatidic acid, and will favor phosphatidylglycerol synthesis. Therefore, the rate of diacylglycerol formation is tightly linked to the rate of its utilization by the envelope enzymes involved in galactolipid and sulfolipid biosynthesis.

Formation of C18/C18 diacylglycerol backbone. As discussed above, the specificities of the envelope acyltransferases do not allow the formation of phosphatidic acid and diacylglycerol with only C18 fatty acids. Prokaryotes are also unable to synthesize such a structure. *In-vivo* kinetics of acetate incorporation into chloroplast lipids have suggested that phosphatidylcholine could provide the diacylglycerol backbone for eukaryotic plastid glycerolipids. The reader is referred to reviews by Heinz (1977), Douce and Joyard (1980), Roughan and Slack (1982) and Joyard and Douce (1987) for detailed presentations of the arguments in favor of this hypothesis. This hypothesis involves (1) the synthesis of phosphatidylcholine in extraplastidial membranes, (2) its transfer to the outer envelope membrane, (3) the formation of diacylglycerol and (4) the integration of diacylglycerol into MGDG or SQDG. Except during the initial steps, the plastid envelope membranes should play a central role in this pathway. However, evidence demonstrating that this pathway indeed operates in plants has not yet been provided. The missing link is the conversion of phosphatidylcholine into diacylglycerol, since envelope membranes apparently lack phospholipase C activity, which would produce diacylglycerol (Joyard and Douce, 1987; Joyard *et al.*, 1991, 1993). If such an enzyme is actually involved in diacylglycerol formation in the outer envelope membrane, its activity is expected to be very tightly regulated.

Glycolipid biosynthesis

MGDG biosynthesis. The inner envelope membrane of spinach chloroplasts is characterized by the presence of a 1,2-diacylglycerol 3-β-galactosyltransferase (or MGDG synthase), which transfers a galactose moiety from a water-soluble donor, UDP-galactose, to a hydrophobic acceptor molecule, diacylglycerol, to synthesize MGDG with the release of UDP (Douce, 1974). MGDG synthase is concentrated in envelope membranes from all plastids and is therefore probably the best marker enzyme for envelope membranes (Douce, 1974). In spinach, this enzyme is exclusively located in the inner envelope membrane (Block *et al.*, 1983).

Covès *et al.* (1986, 1987) first described solubilization of MGDG synthase from spinach chloroplast envelope membranes, and the development of a specific assay for the solubilized activity which allowed a partial purification of this enzyme (Covès *et al.*, 1986). Teucher and Heinz (1991) and Maréchal *et al.* (1991) independently purified several hundredfold MGDG synthase activity from spinach chloroplast envelope, but in both cases, unambiguous characterization of the polypeptide associated with MGDG synthase activity was still very critical. In particular, the final amount of enzyme was so low (of the order of micrograms when starting the purification from 100 mg envelope proteins) that further analyses of the protein were almost impossible. For instance, Teucher and Heinz (1991) proposed that a polypeptide of 22 kDa could be associated with the activity, whereas no 22-kDa polypeptide could be visualized in the purest fraction described by Maréchal *et al.* (1991). In contrast, the purification conducted by Maréchal *et al.* (1991) led to a 90% enrichment of a 19-kDa polypeptide. However, both sets of data led to the conclusion that, despite its importance for chloroplast membrane biogenesis, MGDG synthase is only a minor envelope protein. This observation raised the question of the regulation of MGDG synthase activity within envelope membranes.

The K_m of MGDG synthase for UDP-galactose is 100 μM (Covès *et al.*, 1988; Maréchal *et al.*, 1994). However, no UDP-galactose was detected in spinach chloroplast extracts by [^{31}P] nuclear magnetic resonance (NMR) analyses (Bligny *et al.*, 1990). In contrast, UDP-galactose was characterized in perchloric extracts from whole spinach leaves, corresponding to concentrations in the cytosol of about 0.2 to 0.5 mM (Bligny et al., 1990). These values are high enough to sustain optimal rates of MGDG synthesis, suggesting that the UDP-galactose concentration is probably not a major regulatory factor of the enzyme. Since MGDG synthase is located on the inner membrane of the spinach chloroplast envelope (Block *et al.*, 1983), and since this membrane is impermeable to UDP-galactose (Heber and Heldt, 1981), there is no need for UDP-galactose to accumulate in the plastid stroma if MGDG synthesis occurs on the outer surface of the inner envelope membrane, facing the intermembrane space which is connected to the cytosol by the presence of a porin in the outer envelope membrane (Flügge and Benz, 1984). If this hypothesis is true, then the UDP-galactose binding site of the envelope MGDG synthase could be localized at the outer surface of the inner envelope membrane.

Maréchal *et al.* (1994) carried out kinetic experiments in mixed micelles containing the partially purified enzyme, the substrate diacylglycerol and the detergent 3-((3-cholamidopropyl)-dimethylammonio)-1-propanesulfonate (CHAPS), according to the 'surface dilution' kinetic model proposed by Deems *et al.* (1975). The dependence of kinetic parameters of MGDG synthase on the diacylglycerol mole fraction allows a comparison of the affinity of the enzyme for a wide range of diacylglycerol molecular species. K_m values were obtained ranging from 0.0089-mole fraction for dilinoleoylglycerol and dipalmitoylglycerol to 0.0666-mole fraction for distearoylglycerol, but the differences observed were not really related to the unsaturation of the molecule, since the K_m value for dilinoleoylglycerol (0.0089-mole fraction) was much lower than that for dilinolenoylglycerol (0.040-mole fraction). K_m values for dioleoylglycerol and for 1-oleoyl-2-palmitoylglycerol were in the average range (i.e. lower than 0.030-mole fraction). In addition, this molecule is produced in significant amounts by the envelope phosphatidate phosphatase, explaining why this typical prokaryotic structure is readily incorporated into MGDG. However, the best substrate for MGDG synthase was dilinoleoylglycerol. In mixed micelles, the dilinoleoylglycerol concentration corresponding to the K_m value was as low as one molecule/per micelle. One should question whether this molecule could actually be present (or formed) in envelope membranes. As discussed above, it is assumed that MGDG molecules with the eukaryotic structure could be derived from phosphatidylcholine which is synthesized on the endoplasmic reticulum and transferred to cell organelles such as chloroplasts. Interestingly, phosphatidylcholine, which (in chloroplasts) is concentrated in the outer leaflet of the outer envelope membrane (see above), contains significant amounts of 18:2 at both the *sn*-1 and the *sn*-2 positions of the glycerol backbone (Heinz, 1977), but the enzyme which could generate diacylglycerol from phosphatidylcholine (i.e. phospholipase *c*) has not been found in envelope membranes (see above). Therefore, if phosphatidylcholine is indeed the source of dilinoleoylglycerol in the outer membrane, only limited amounts would be delivered to the inner membrane where MGDG synthase is located. In addition, nothing is known about the possible mechanisms that could be involved in diacylglycerol transfer from the outer to the inner envelope membrane (spontaneous diffusion of free monomers, lateral diffusion of lipids between membranes at regions of direct intermembrane contact). Therefore, only the high affinity of the envelope MGDG synthase for dilinoleoylglycerol would explain the presence of C18 fatty acids at both *sn* positions of glycerol in MGDG.

MGDG does not contain dipalmitoyl species (Siebertz *et al.*, 1979). In addition, dipalmitoylglycerol is only synthesized in small amounts within envelope membranes by the enzymes of the Kornberg–Pricer pathway, whereas 1-oleoyl-2-palmitoylglycerol is the major product (see above). Furthermore, Maréchal *et al.* (1994) have demonstrated that the envelope MGDG synthase has a low affinity for dipalmitoylglycerol. Therefore, the combination of these two major limitations probably explains why the biosynthesis of MGDG molecules containing dipalmitoylglycerol is highly unlikely. As we shall see below, this is in contrast to the situation in SQDG biosynthesis.

The mechanism of MGDG synthase activity was also investigated with two-substrate kinetic studies at different UDP-galactose molar concentrations and different dioleoylglycerol surface concentrations (Maréchal *et al.*, 1994). The families of reciprocal plots obtained were shown to intersect at a single point of the 1/[substrate] axis, thus demonstrating that MGDG synthase was a sequential, either random or ordered, bireactant system, and not a ping-pong mechanism as proposed by Van Besouw and Wintermans (1979). Therefore, MGDG synthase has two distinct and independent substrate-binding sites, a hydrophilic one for UDP-galactose and a hydrophobic one for diacylglycerol (Maréchal *et al.*, 1994, 1995). Further investigations of the pattern of inhibition of MGDG synthase (at varied UDP-galactose molar concentrations and varied dioleoylglycerol surface concentrations) by UDP, and of the effect of various inhibitors (*N*-ethylmaleimide (NEM), citraconic anhydride, etc.) demonstrated that the enzyme mechanism is in fact a random, sequential, bireactant system (Maréchal *et al.*, 1995).

DGDG biosynthesis. The galactolipid:galactolipid galactosyltransferase, which is the only DGDG-forming enzyme clearly described in plastids to date, was first characterized on envelope membranes by Van Besouw and Wintermans (1978). It catalyzes an enzymatic exchange of galactose between galactolipids with the formation of diacylglycerol, but, *in vitro,* unnatural galactolipids with more than two galactose residues can be synthesized. This enzyme is located on the cytosolic side of the outer envelope membrane, as indicated by its destruction during mild proteolytic digestion (with thermolysin) of intact chloroplasts (Dorne *et al.*, 1982). Van Besouw and Wintermans (1978), Heemskerk and Wintermans (1987) and Heemskerk *et al.* (1990) have proposed that the galactolipid:galactolipid galactosyltransferase is indeed responsible for DGDG synthesis. For instance, Heemskerk et al. (1990) analyzed galactolipid synthesis in chloroplasts or chromoplasts from eight species of 16:3 and 18:3 plants and found that digalactosyldiacyl-glycerol formation is never stimulated by UDP-galactose or any other nucleoside 5'-diphosphodigalactoside; in all cases, DGDG formation was reduced by thermolysin digestion of intact organelles. *In vitro*, the galactolipid:galactolipid galactosyltransferase does not show strong specificity for any MGDG molecular species. However, if this enzyme is indeed the DGDG-synthesizing enzyme, it should discriminate *in vivo* between the various MGDG molecular species that are available since (1) the proportion of eukaryotic molecular species is higher in DGDG than in MGDG (Heinz, 1977) and (2) DGDG contains 16:0 fatty acids (up to 10–15%) at both the *sn*-1 and *sn*-2 positions, and very little 16:3 (in 16:3 plants), whereas MGDG contains little 16:0, but (in 16:3 plants) 16:3 at the *sn*-2 position (Heinz, 1977).

Sakaki *et al.* (1990) have proposed another feature of physiological significance for the envelope galactolipid:galactolipid galactosyltransferase. Using ozone-fumigated spinach leaves, they demonstrated *in vivo* that MGDG was converted into diacylglycerol by the galactolipid:galactolipid galactosyltransferase and then to triacylglycerol (by acylation with 18:3-CoA), by a diacylglycerol acyltransferase associated with envelope membranes (Martin and Wilson, 1984). Whether the galactolipid:galactolipid galactosyltransferase is involved in DGDG synthesis is the focus of current investigation in several laboratories, and hopefully a definitive answer to this question will be available in the near future.

Sulfolipid biosynthesis. Intact chloroplasts are able to incorporate SO_4^{2-} into SQDG (Haas *et al.*, 1980; Joyard *et al.*, 1986; Kleppinger-Sparace *et al.*, 1985). Heinz *et al.* (1989) synthesized different nucleoside 5'-diphospho-sulfoquinovoses and demonstrated that both UDP- and GDP-sulfoquinovose significantly increased SQDG synthesis by spinach chloroplasts and by isolated envelope membranes, with UDP-sulfoquinovose being twice as active as the GDP derivative. Therefore, SQDG synthesis in envelope membranes is probably due to a UDP-sulfoquinovose:1,2-diacylglycerol 3-β-sulfoquinovosyltransferase or 1,2-diacylglycerol 3-β-sulfoquinovosyl-transferase (SQDG synthase).

In plants containing SQDG molecular species with C18 fatty acids at the *sn*-2 position, the diacylglycerol used for SQDG synthesis could be derived from eukaryotic lipids, as proposed for MGDG (see above). However, this is only a small proportion of SQDG molecules, except in plants such as wheat or cucumber (Bishop *et al.*, 1985). Because most SQDG molecules have a prokaryotic structure, they should be derived from diacylglycerol molecules formed *de novo* in the inner envelope membrane by the enzymes of the Kornberg–Pricer pathway (Bishop *et al.*, 1985; Kleppinger-Sparace *et al.*, 1985). Of the diacylglycerol molecular species formed within chloroplasts, dipalmitoylglycerol represents only a small proportion, but SQDG contains this structure in significant amounts (Bishop *et al.*, 1985). Furthermore, the two enzymes involved in MGDG and SQDG synthesis compete for the same pool of diacylglycerol molecules (Kleppinger-Sparace *et al.*, 1985). Therefore, one can question how these two enzymes can discriminate between the different molecular species in order to form MGDG and SQDG molecules having distinct backbones. Envelope membranes loaded with 16:0/16:0 and/or 18:1/16:0 diacylglycerol, and incubated in the presence of UDP-galactose or UDP-sulfoquinovose, incorporated 16:0/16:0 diacylglycerol with a much greater efficiency into SQDG than into MGDG, whereas 18:1/16:0 diacylglycerol was incorporated into both MGDG and SQDG with almost the same efficiency (Seifert and Heinz, 1992). Therefore, by comparison with the kinetic studies of MGDG synthase (Maréchal, *et al.*, 1994), one can propose that only a high affinity of SQDG synthase for 16:0/16:0 diacylglycerol would explain its presence in SQDG. Experiments with partially purified SQDG synthase similar to those performed by Maréchal et al. (1994) on MGDG synthase would be of interest to confirm this hypothesis.

Fatty acid desaturation

Glycerolipids that are synthesized on envelope membranes contain 16:0 and 18:1 fatty acids which must be desaturated to polyunsaturated fatty acids. This desaturation of C16 and C18 fatty acids occurs when they are esterified to MGDG (for review see Heinz, 1977). Roughan *et al.* (1979) first demonstrated that isolated intact spinach chloroplasts can synthesize MGDG containing polyunsaturated fatty acids. Heinz and Roughan (1983) further analyzed MGDG synthesized from [^{14}C]acetate by isolated *Solanum nodiflorum* chloroplasts, and found the complete series of radioactive fatty acids, from 16:0 to 16:3 at the *sn*-2 position and from 18:1 to 18:3 at the *sn*-1 position of the glycerol. Desaturation also takes place in nongreen plastids (Alban *et al.*, 1989). Evidence that pure envelope membranes could be used as a source of enzyme for the desaturation of oleic acid to linoleic acid was provided by Schmidt and Heinz (1990a,b). Catalase was required in the incubation medium, apparently to destroy the inhibitory hydrogen peroxide which was formed by auto-oxidation of ferredoxin (Heinz, 1993). A cDNA encoding a 40-kDa polypeptide from chloroplast envelope membranes which could possibly be involved in the n-6 (Δ^{12}) desaturase from spinach was isolated (Heinz, 1993; Schmidt and Heinz, 1992). The sequence data suggest a phylogenetic and functional relationship between this desaturase and the *des*A-coded protein, another Δ^{12} desaturase (Heinz, 1993).

In fact, the nature of the enzymes involved in the desaturation of plastid fatty acids *in vivo* is not yet clearly established. Most of our knowledge of chloroplast desaturases has been limited until recently to the soluble components (ferredoxin, ferredoxin:NADP oxidoreductase, stearoyl-ACP (acyl carrier protein) desaturase). In the case of envelope-bound desaturase, preliminary evidence for the involvement of a ferredoxin:NADPH oxidoreductase (FNR) was obtained by the use of anti-FNR-IgG (Schmidt and Heinz, 1990a,b). Experiments with *Arabidopsis* chloroplasts (Norman *et al.*, 1991) suggest that O_2 is the final electron acceptor, whereas reduced ferredoxin is the source of electrons for the reduction of O_2 to H_2O. Since ferredoxin delivers only one electron at a time, the desaturase must oxidize two reduced ferredoxin molecules, and store the first electron before the double bond is formed (Heinz, 1993). Obviously, the exact mechanism is not yet understood. Another approach was the use of mutants from cyanobacteria or *Arabidopsis*. Browse *et al.* (1985) have initiated a genetic approach to the analysis of lipid composition and synthesis by the isolation of a series of mutants of the small crucifer *Arabidopsis thaliana* with specific alterations in leaf fatty acid or glycerolipid composition. The mutants were isolated without selection by screening a mutagenized population of plants (about 10 000 randomly chosen individuals) by gas chromatography of small leaf samples. Such investigations led to the characterization of several genes associated with membrane-bound fatty acid desaturases (Heinz, 1993; Somerville and Browse, 1991). Most of the mutations with alterations in membrane lipid composition which have been analyzed cause either loss or a reduction in the amount of an unsaturated fatty acid and the corresponding accumulation of a less unsaturated precursor (Heinz, 1993; Somerville and Browse, 1991). Thus, it was inferred that the different *fad* mutants were defective in the desaturation of a lipid-linked fatty acid. Most of these mutants apparently had a normal phenotype and their growth was almost the same as wild-type *Arabidopsis*, thus demonstrating that the mutations which were selected were not lethal. Several mutants affecting fatty acid desaturation in chloroplasts (*fad* A, B, C, D, etc.) have been isolated and characterized. However, it is not yet clear whether the genes characterized from the different mutants correspond to the desaturases themselves or to some regulatory elements. Analysis of the effects of the mutations on membrane lipid composition complements the biochemical studies of leaf lipid metabolism and its regulation. One of the most interesting observations provided by the use of these mutants was the demonstration of the flexibility of the proportions of 16:3 and 18:3 fatty acids observed. Wild-type *Arabidopsis* is a typical 16:3 plant, and it is therefore possible to follow the potential switching between MGDG molecular species that do or do not contain 16:3 fatty acids. Such analyses were performed first using the mutant JB25 (a mutant deficient in chloroplast glycerol-3-phosphate acyltransferase), in which MGDG no longer contains any 16:3, but only 18:3. In addition, the MGDG content of the mutant is nearly the same as that of the wild type (Kunst *et al.*, 1988). Since the acyltransferase involved in the biosynthesis of lysophosphatidic acid is strongly reduced, the normal pathway for the synthesis of glycero-lipids having only the prokaryotic structure is less active and the activity of the enzymes catalyzing the synthesis of the eukaryotic structure is increased to compensate for the loss of pro-karyotic glycerolipids (Kunst *et al.*, 1988). This experiment and others undertaken with *fad* mutants demonstrate the remarkable flexibility of glycerolipid biosynthesis in plants, and provide an insight into the nature of the regulatory mechanisms involved.

4. Conclusions

All the observations summarized in this chapter demonstrate that the inner envelope membrane is the site of synthesis for glycolipids (galactolipids and sulfolipid). The same is true for other plastid constituents that are synthesized in envelope membranes (Kleppinger-Sparace *et al.*, 1985), namely phosphatidylglycerol, prenylquinones (α-tocopherol and plastoquinone-9) and pigments (chlorophyll and carotenoids). Thylakoids represent the main plastid membrane system in the cell, and contain the largest amounts of the plastid lipid constituents. Therefore, extensive transport of molecules from their site of synthesis (envelope membranes) to their site of accumulation (thylakoids) should take place during development. In fact, almost nothing is known about the mechanisms involved in the transfer of lipids. In addition, different strategies for lipid transfer from envelope membranes to thylakoids are probably involved at different stages of chloroplast development. The process which occurs in mature chloroplasts is probably very different from that observed in young developing plastids. In this case, import of thylakoid membrane lipids *en route* to the thylakoids may occur through the stroma via vesicles derived from the inner membrane, but this has yet to be demonstrated. Electron microscopy studies have suggested, however, that numerous vesicles, probably derived from the inner envelope membrane, are formed during development (see for example Carde *et al.*, 1982). Since the inner envelope membrane and the thylakoids do not differ in polar lipid composition (*Table 1*), natural fusion of vesicles produced by the inner envelope membrane with growing thylakoids is possible. In addition, and as discussed above, MGDG can form hexagonal type-II structures (rather than Lα bilayers) which could behave as intermediates in membrane fusion. Therefore, MGDG (which represents almost half of the lipid content of inner envelope vesicles and thylakoids) could favor fusion between plastid membranes. Nothing is known about the way in which the vesicle buds or how its selective fusion with thylakoids is programmed and catalyzed, nor is it known how vesicles derived from the inner membrane select their content while rejecting the bulk constituents of the inner membrane from which they bud. In addition to a putative vesicular transport (see Morré *et al.*, 1991), other possible mechanisms which could be involved include (1) transfer of lipid monomers through the stroma either by protein-facilitated transport (see Nishida and Yamada, 1986) or by spontaneous diffusion of free monomers and (2) lateral diffusion of lipids between membranes at regions of direct intermembrane contact (see Rawyler *et al.*, 1991). Obviously, our understanding of this field is hampered by the lack of conclusive experimental data.

Finally, in recent years substantial progress has been made in understanding the distribution and biosynthesis of plant glycolipids, mostly because of improvements in cell fractionation tech-niques together with the use of modern analytical techniques. However, further improvement of our knowledge of the enzymes involved in MGDG, DGDG and sulfolipid formation requires the purification of the many membrane-bound enzymes, which is very difficult to achieve, especially for such enzymes manipulating hydrophobic substrates. Mutants have been used extensively in this field, but until now their major impact has been restricted to improving our understanding of fatty acid desaturation, with almost no information about gene coding for proteins involved in the formation of typical plastid glycerolipids. Plant glycolipid metabolism is a complex, highly regulated and multicompartmented series of reactions. Obviously, further progress will be made only as a result of joint efforts combining the use of biochemical techniques and molecular biology.

Acknowledgments

We wish to acknowledge financial support of this work as a continuing program by the Commissariat à l'Energie Atomique (CEA) and the Centre National de la Recherche Scientifique (CNRS).

References

Alban C, Dorne AJ, Joyard J, Douce R. (1989) [^{14}C]-acetate incorporation into glycerolipids from cauliflower proplastids and sycamore amyloplasts. *FEBS Lett.* **249**: 95–99.

Alban C, Joyard J, Douce R. (1989) Comparison of glycerolipid biosynthesis in non-green plastids from sycamore (*Acer pseudoplatanus*) cells and cauliflower (*Brassica oleracea*) buds. *Biochem. J.* **259**: 775–783.

Andrews J, Ohlrogge JB, Keegstra K. (1985) Final steps of phosphatidic acid synthesis in pea chloroplasts occurs in the inner envelope membrane. *Plant Physiol.* **78**: 459–465.

Bertrams M, Heinz E. (1981) Positional specificity and fatty acid selectivity of purified *sn*-glycerol-3-phosphate acyltransferases from chloroplasts. *Plant Physiol.* **68**: 653–657.

Billecocq A, Douce R, Faure M. (1972) Structure des membranes biologiques: localisation des galactosyldiglycérides dans les chloroplastes au moyen des anticorps spécifiques. *C. R. Acad. Sci. Paris* **275**: 1135–1137.

Bishop DG, Sparace SA, Mudd JB. (1985) Biosynthesis of sulfoquinovosyldiacylglycerol in higher plants: the origin of the diacylglycerol moiety. *Arch. Biochem. Biophys.* **240**: 851–858.

Bishop WR, Bell RM. (1988) Assembly of phospholipids into cellular membranes: biosynthesis, transmembrane movement and intracellular location. *Annu. Rev. Cell Biol.* **4**: 579–610.

Bligny R, Douce R. (1980) A precise localization of cardiolipin in plant cells. *Biochim. Biophys. Acta* **617**: 254–263.

Bligny R, Gardeström P, Roby C, Douce R. (1990) ^{31}P NMR studies of spinach leaves and their chloroplasts. *J. Biol. Chem.* **265**: 1319–1326

Block MA, Dorne AJ, Joyard J, Douce R. (1983) Preparation and characterization of membrane fractions enriched in outer and inner envelope membranes from spinach chloroplasts. II – Biochemical characterization. *J. Biol. Chem.* **258**: 13281–13286.

Brasseur R, DeMeutter J, Goormaghtigh E, Ruysschaert JM. (1983) Mode of organization of galactolipids: a conformational analysis. *Biochem. Biophys. Res. Commun.* **115**: 666–672.

Browse J, McCourt P, Somerville C. (1985) A mutant of *Arabidopsis* lacking a chloroplast-specific lipid. *Science* **227**: 763–765.

Carde JP, Joyard J, Douce R. (1982) Electron microscopic studies of envelope membranes from spinach plastids. *Biol. Cell* **44**: 315–324.

Carman GM, Henry SA. (1989) Phospholipid biosynthesis in yeast. *Annu. Rev. Biochem.* **58**: 635–669.

Carter HE, McCluer RH, Slifer ED. (1956) Lipids of wheat flour. I. Characterization of galactosylglycerol components. *J. Am. Chem. Soc.* **78**: 3735–3738.

Covès J, Block MA, Joyard J, Douce R. (1986) Solubilization and partial purification of UDP-galactose:diacylglycerol galactosyltransferase activity from spinach chloroplast envelope. *FEBS Lett.* **208**: 401–406.

Covès J, Pineau B, Block MA, Joyard J, Douce R. (1987) Solubilization and partial purification of chloroplast envelope proteins: application to UDP-galactose:diacylglycerol galactosyltransferase. In: *Plant Membranes: Structure, Function, Biogenesis* (Leaver C, Sze H, ed.). New York: Alan R. Liss, pp. 103–112.

Covès J, Joyard J, Douce, R. (1988) Lipid requirement and kinetic studies of solubilized UDP-galactose:diacylglycerol galactosyltransferase activity from spinach chloroplast envelope membranes. *Proc. Natl Acad. Sci. USA* **85:** 4966–4970.

Curatolo W. (1987a) Glycolipid function. *Biochim. Biophys. Acta* **906:** 137–160.

Curatolo W. (1987b) The physical properties of glycolipids. *Biochim. Biophys. Acta* **906:** 111–136.

Deems RA, Eaton BR, Dennis EA. (1975) Kinetic analysis of phospholipase A2 activity towards mixed micelles and its implication for the study of lipolytic enzymes. *J. Biol. Chem.* **250:** 9013–9020.

Dorne A-J, Block MA, Joyard J, Douce R. (1982) The galactolipid:galactolipid galactosyltransferase is located on the outer membrane of the chloroplast envelope. *FEBS Lett.* **145:** 30–34.

Dorne AJ, Joyard J, Block MA, Douce R. (1985) Localization of phosphatidylcholine in outer envelope membrane of spinach chloroplasts. *J. Cell Biol.* **100:** 1690–1697.

Dorne AJ, Bligny R, Rébeillé F, Roby C, Douce R. (1987) Fatty acid disappearance and phosphoryl-choline accumulation in higher plant cells after a long period of sucrose deprivation. *Plant Physiol. Biochem.* **25:** 589–595.

Douady D, Dubacq J-P. (1987) Purification of acyl-CoA:glycerol-3-phosphate acyltransferase from pea leaves. *Biochim. Biophys. Acta* **921:** 615–619.

Douce R. (1974) Site of synthesis of galactolipids in spinach chloroplasts. *Science* **183:** 852–853.

Douce R, Joyard J. (1980) Plant galactolipids. In: *The Biochemistry of Plants, Vol. 4. Lipids: Structure and Function* (Stumpf PK, ed.). New York: Academic Press, pp. 321–362.

Flügge UI, Benz R. (1984) Pore-forming activity in the outer membrane of the chloroplast envelope. *FEBS Lett.* **169:** 85–89.

Frentzen M. (1993) Acyltransferases and triacylglycerol. In: *Lipid Metabolism in Plants* (Moore ST Jr, ed.). Boca Raton: CRC Press, pp. 195-230.

Frentzen M, Heinz E, McKeon TA, Stumpf, PK. (1983) Specificities and selectivities of glycerol-3-phosphate acyltransferase and monoacylglycerol-3-phosphate acyltransferase from pea and spinach chloroplasts. *Eur. J. Biochem.* **129:** 629–636.

Gounaris K, Barber J. (1983) Monogalactosyldiacylglycerol: the most abundant polar lipid in nature. *Trends Biochem. Sci.* **9:** 378–381.

Gounaris K, Mannock DA, Sen A, Brain AP, Williams WP, Quinn PJ. (1983) Polyunsaturated fatty acyl residues of galactolipids are involved in the control of bilayer/non-bilayer lipid transitions in higher plant chloroplasts. *Biochim. Biophys. Acta* **732:** 229–242.

Haas R, Siebertz HP, Wrage K, Heinz E. (1980) Localization of sulfolipid labeling within cells and chloroplasts. *Planta* **148:** 238–244.

Haschke HP, Kaiser G, Martinoia E, Hammer U, Teucher T, Dorne AJ, Heinz E. (1990) Lipid profiles of leaf tonoplasts from plants with different CO_2-fixation mechanisms. *Bot. Acta* **103:** 32–38.

Heber U, Heldt HW. (1981) The chloroplast envelope: structure, function, and role in leaf metabolism. *Annu. Rev. Plant Physiol.* **32:** 139–168.

Heemskerk JHW, Storz T, Schmidt RR, Heinz E. (1990) Biosynthesis of digalactosyldiacylglycerol in plastids from 16:3 and 18:3 plants. *Plant Physiol.* **93:** 1286–1294.

Heemskerk JWM, Wintermans JFGM. (1987) The role of the chloroplast in the leaf acyl-lipid synthesis. *Physiol. Plant.* **70:** 558–568.

Heinz E. (1977) Enzymatic reactions in galactolipid biosynthesis. In: *Lipids and Lipid Polymers* (Tevini M, Lichtenthaler HK, ed.). Berlin: Springer-Verlag, pp. 102–120.

Heinz E. (1993) Biosynthesis of polyunsaturated fatty acids. In: *Lipid Metabolism in Plants* (Moore ST Jr, ed.). Boca Raton: CRC Press, pp. 34–89.

Heinz E, Roughan PG. (1983) Similarities and differences in lipid metabolism of chloroplasts isolated from 18:3 and 16:3 plants. *Plant Physiol.* **72:** 273–279.

Heinz E, Schmidt H, Hoch M, Jung K-H, Binder H, Schmidt RR. (1989) Synthesis of different nucleoside 5′-diphospho-sulfoquinovoses and their use for studies on sulfolipid biosynthesis in chloroplasts. *Eur. J. Biochem.* **184:** 445–453.

Ishizaki O, Nishida I, Agata K, Eguchi G, Murata N. (1988) Cloning and nucleotide sequence of cDNA for the plastid glycerol-3-phosphate acyltransferase from squash. *FEBS Lett.* **238:** 424–430.

Journet EP, Bligny R, Douce R. (1986) Biochemical changes during sucrose deprivation in higher plant cells. *J. Biol. Chem.* **261:** 3193–3199.

Joyard J, Douce R. (1977) Site of synthesis of phosphatidic acid and diacylglycerol in spinach chloroplasts. *Biochim. Biophys. Acta* **486:** 273–285.

Joyard J, Douce R. (1979) Characterization of phosphatidate phosphohydrolase activity associated with chloroplast envelope membranes. *FEBS Lett.* **102:** 147–150.

Joyard J, Douce R. (1987) Galactolipid biosynthesis. In: *The Biochemistry of Plants, Vol. 9. Lipids: Structure and Function* (Stumpf PK, ed.). New York: Academic Press, pp. 215–274.

Joyard J, Blée E, Douce R. (1986) Sulfolipid synthesis from $^{35}SO_4^{2-}$ and $[1-^{14}C]$-acetate in isolated intact spinach chloroplasts. *Biochim. Biophys. Acta* **879:** 78–87.

Joyard J, Block MA, Douce R. (1991) Molecular aspects of plastid envelope biochemistry. *Eur. J. Biochem.* **199:** 489–509.

Joyard J, Block MA, Malherbe A, Maréchal E, Douce R. (1993) Origin and synthesis of galactolipid and sulfolipid head groups. In: *Lipid Metabolism in Plants* (Moore ST Jr, ed.). Boca Raton: CRC Press, pp. 231–258.

Kleppinger-Sparace KF, Mudd JB, Bishop DG. (1985) Biosynthesis of sulfoquinovosyldiacylglycerol in higher plants: the incorporation of $^{35}SO_4$ by intact chloroplasts. *Arch. Biochem. Biophys.* **240:** 859–865.

Kunst L, Browse J, Somerville C. (1988) Altered regulation of lipid biosynthesis in a mutant of *Arabidopsis* deficient in chloroplast glycerol-3-phosphate acyltransferase activity. *Proc. Natl Acad. Sci. USA* **85:** 4143–4147.

Lawlor DW. (1987) *Photosynthesis: Metabolism, Control and Physiology.* Harlow: Longman Scientific and Technical.

Luzzati V. (1968) X-ray diffraction studies of lipid–water systems. In: *Biological Membranes. Physical Fact and Function* (Chapman D, ed.). London: Academic Press, pp. 71–123.

Malherbe A, Block MA, Joyard J, Douce R. (1992) Feedback inhibition of phosphatidate phosphatase from spinach chloroplast envelope membranes by diacylglycerol. *J. Biol. Chem.* **267:** 23546–23553.

Malherbe A, Block MA, Douce R, Joyard J. (1995) Solubilization and biochemical properties of phosphatidate phosphatase from spinach chloroplast envelope membranes. *Plant Physiol. Biochem.* **33:** 97–104.

Maréchal E, Block MA, Joyard J, Douce R. (1991) Purification de l'UDP-galactose:1,2-diacylglycérol galactosyltransferase de l'enveloppe des chloroplastes d'épinard. *C. R. Acad. Sci. Paris* **313:** 521–528.

Maréchal E, Block MA, Joyard J, Douce R. (1994) Kinetic properties of monogalactosyldiacylglycerol synthase from spinach chloroplast envelope membranes. *J. Biol. Chem.* **269:** 5788–5798.

Maréchal E, Block MA, Douce R, Joyard J. (1995) The catalytic site of monogalactosyl diacylglycerol. *J. Biol. Chem.* **270:** 5714–5722.

Martin BA, Wilson RF. (1984) Subcellular localization of triacylglycerol biosynthesis in spinach leaves. *Lipids* **19:** 117–121.

Morré DJ, Morré JT, Morré SR, Sundqvist C, Sandelius AS. (1991) Chloroplast biogenesis. Cell-free transfer of envelope monogalactosylglycerides to thylakoids. *Biochim. Biophys. Acta* **1070:** 437–445.

Murphy DJ. (1982) The importance of non-planar bilayer regions in photosynthetic membranes and their stabilization by galactolipids. *FEBS Lett.* **150:** 19–26.

Nishida I, Yamada M. (1986) Semisynthesis of a spin-labeled monogalactosyl-diacylglycerol and its application in the assay for galactolipid transfer activity in spinach leaves. *Biochim. Biophys. Acta* **813:** 298–306.

Nishida I, Frentzen M, Ishizaki O, Murata N. (1987) Purification of isomeric forms of acyl-[acyl-carrier-protein]:glycerol-3-phosphate acyltransferase from greening squash cotyledons. *Plant Cell Physiol.* **28:** 1071–1079.

Norman HA, Pillai P, StJohn JB. (1991) *In vitro* desaturation of monogalactosyldiacylglycerol and phosphatidylcholine molecular species by chloroplast homogenates. *Phytochemistry* **30:** 2217–2222.

Quinn PJ, Williams WP. (1983) The structural role of lipids in photosynthetic membranes. *Biochim. Biophys. Acta* **737:** 223–266.

Rawyler A, Meylan M, Siegenthaler PA. (1991) Galactolipid export from envelope to thylakoid membranes in intact chloroplasts. I. Characterization and involvement in thylakoid asymmetry. *Biochim. Biophys. Acta* **1104:** 331–341.

Roby C, Martin JB, Bligny R, Douce R. (1987) Biochemical changes during sucrose deprivation in higher plant cells. II. Phosphorus-31 nuclear magnetic resonance studies. *J. Biol. Chem.* **262:** 5000–5007.

Roughan PG, Slack CR. (1982) Cellular organization of glycerolipid metabolism. *Annu. Rev. Plant Physiol.* **33:** 97–132.

Roughan PG, Mudd JB, McManus TT, Slack CR. (1979) Linoleate and α-linolenate synthesis by isolated spinach (*Spinacia oleracea*) chloroplasts. *Biochem. J.* **184:** 571–574.

Sakai WS, Yamamoto HY, Miyazaki T, Ross JW. (1983) A model for chloroplast thylakoid membranes involving orderly arrangements of negatively charged lipidic particles containing sulpho-quinovosyldiacylglycerol. *FEBS Lett.* **158:** 203–207.

Sakaki T, Kondo N, Yamada M. (1990) Pathway for the synthesis of triacylglycerol from mono-galactosyldiacylglycerols in ozone-fumigated spinach leaves. *Plant Physiol.* **94:** 773–780.

Sastry PS, Kates M. (1964) Lipid components of leaves. V. Galactolipids, cerebrosides, and lecithin of runner-bean leaves. *Biochemistry* **3:** 1271–1280.

Schmidt H, Heinz E. (1990a) Involvement of ferredoxin in desaturation of lipid-bound oleate in chloroplasts. *Plant Physiol.* **94:** 214–220.

Schmidt H, Heinz E. (1990b) Desaturation of oleoyl groups in envelope membranes from spinach chloroplasts. *Proc. Natl Acad. Sci. USA* **87:** 9477–9480.

Schmidt H, Heinz E. (1992) n-6 desaturase from chloroplast envelopes: purification and enzymatic characteristics. In: *Metabolism, Structure and Utilization of Plant Lipids* (Cherif A, Miled-Daoud DB, Marzouk B, Smaoui A, Zarrouk M, ed.) Cent. Natl. Ped. Tunis, pp. 140–143.

Seifert U, Heinz E. (1992) Enzymatic characteristics of UDP-sulfoquinovose:diacylglycerol sulfoquinovosyltransferase from chloroplast envelopes. *Bot. Acta* **105:** 197–205.

Sen A, Williams WP, Quinn PJ. (1981) The structure and thermotropic properties of pure 1,2-diacylgalactosylglycerols in aqueous systems. *Biochim. Biophys. Acta* **663:** 380–389.

Sen A, Williams WP, Brain AP, Quinn PJ. (1982) Bilayer and non-bilayer transformation in aqueous dispersion of mixed *sn*-3-galactosyldiacylglycerols isolated from chloroplasts. *Biochim. Biophys. Acta* **685:** 297–306.

Shipley GG, Green JP, Nichols BW. (1973) The phase behavior of monogalactosyl, digalactosyl, and sulphoquinovosyl diglycerides. *Biochim. Biophys. Acta* **311:** 531–544.

Siebertz HP, Heinz E, Linscheid M, Joyard J, Douce R. (1979) Characterization of lipids from chloroplast envelopes. *Eur. J. Biochem.* **101:** 429–438.

Siefermann-Harms D, Ross JW, Kaneshiro KH, Yamamoto HY. (1982) Reconstitution by monogalactosyldiacylglycerol of energy transfer from light-harvesting *a/b*–protein complex to the photosystems in Triton X-100-solubilized thylakoids. *FEBS Lett.* **149:** 191–196.

Somerville C, Browse J. (1991) Plant lipids: metabolism, mutants, and membranes. *Science* **252:** 80–87.

Sprague SG, Staehelin LA. (1984) Effect of reconstitution method on the structural organization of isolated chloroplast membrane lipids. *Biochim. Biophys. Acta* **777:** 306–322.

Stymne S, Stobart AK. (1987) Triacylglycerol biosynthesis. In: *The Biochemistry of Plants, Lipids: Structure and Function Vol. 9* (Stumpf PK, ed.). New York: Academic Press, pp. 175–214.

Tavernier E, Lê Quôc D, Lê Quôc K. (1993) Lipid composition of the vacuolar membrane of *Acer pseudoplatanus* culture cells. *Biochim. Biophys. Acta* **1167:** 242–247.

Teucher T, Heinz E. (1991) Purification of UDP-galactose:diacylglycerol from chloroplast envelopes of spinach (*Spinacia oleracea* L.). *Planta* **184:** 319–326.

Van Besouw A, Wintermans JFGM. (1978) Galactolipid formation in chloroplast envelopes. I. Evidence for two mechanisms in galactosylation. *Biochim. Biophys. Acta* **529:** 44–53.

Van Besouw A, Wintermans JFGM. (1979) The synthesis of galactolipids by chloroplast envelopes. *FEBS Lett.* **102:** 33–37.

Weber S, Wolter FP, Buck F, Frentzen M, Heinz E. (1991) Purification and cDNA sequencing of an oleate-sensitive acyl-ACP:*sn*-glycerol-3-phosphate acyltransferase from pea chloroplasts. *Plant Mol. Biol.* **17:** 1067–1076.

Chapter 12

Metabolism of plant phosphoinositides and other inositol-containing lipids

Bjørn K. Drøbak

1. Inositol-containing lipids in plants

Inositol-containing lipids in plants belong to a heterogenous group of lipids, many of which have yet to be fully characterized. The first inositol-containing lipids were discovered in 1930 by R.J. Anderson, who found that inositol could be isolated from a 'phosphatide' fraction from tubercle bacilli, and the presence of inositol in brain phospholipids was described by Folch and Woolley in 1942. Since this early work, much effort has been devoted to the investigation of these lipids in both mammals and plants. The discovery in the early 1980s that a particular group of inositol-containing lipids (i.e. the phosphoinositides) are involved both in the transmission of a multitude of environmental signals across the plasma membrane and in intracellular signalling has subsequently resulted in this field of research attracting immense interest. The currently known inositol-containing lipids in plants can be divided into three categories: the 4-phosphorylated phosphoinositides, the 3-phosphorylated inositides and the inositol-containing glycolipids. The structure of these general types of lipids is illustrated in *Figure 1*. Further details are given in the figure legend and the sections below.

As several recent reviews have described both the mammalian and plant phosphoinositide signalling systems in detail (mammalian PI system: Berridge, 1993; Irvine, 1992; Michell, 1992; Nishizuka, 1992; Rhee, 1991; plant PI system: Cote and Crain, 1993; Drøbak, 1992, 1993; Rincon and Boss, 1990), only a brief overview will be presented here, and this chapter will concentrate on some recently discovered mechanisms which may play a physiological role in the regulation of the turnover of the phosphoinositide lipids. Many of the hypotheses presented in the second part of this chapter have been strongly influenced by recent progress outside the plant field. Although the extensive 'borrowing' of ideas from, for example, mammalian research is sometimes (and not altogether without justification!) frowned upon by members of the plant (and animal) research community, there are nevertheless powerful reasons for keeping a close eye on progress in a wide range of disciplines when working on cellular signal transduction. The main reason is that many components of signalling cascades are of early evolutionary origin and have remained highly conserved in diverse eukaryotic cells. However, although the main components or basic mechanisms of signal transduction systems may be shared by different cells, the cells have also in several cases been found to have developed their own, often idiosyncratic, ways of utilizing them. One example is cyclic AMP (cAMP), which is used by liver cells as an *intracellular* second messenger for hormones such as glucagon, but is utilized by the slime mould *Dictyostelium* as an *extracellular* signal for aggregation. So whereas cross-species

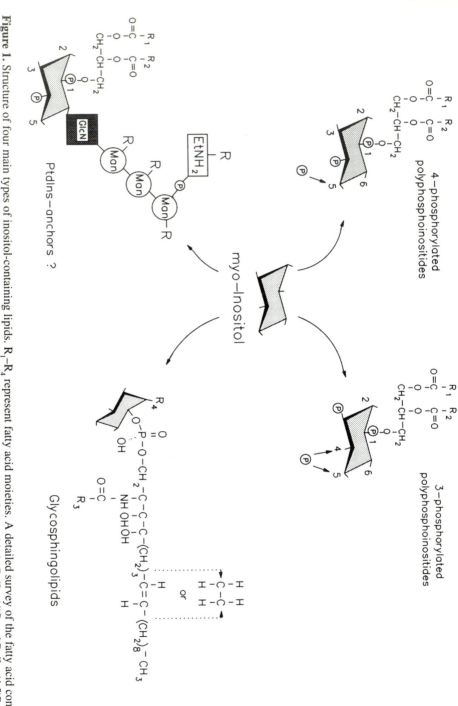

Figure 1. Structure of four main types of inositol-containing lipids. R_1–R_4 represent fatty acid moieties. A detailed survey of the fatty acid composition of polyphosphoinositides has not been carried out; the few data available suggest that the predominant fatty acids in PtdIns(4)P and PtdIns(4,5)P$_2$ are palmitate and linoleate (Drøbak *et al.*, 1988; Rincon and Boss, 1990). This is consistent with a synthetic pathway involving direct phosphorylation of PtdIns. PtdIns anchors have, so far, not been identified in higher plants.

comparisons can be stimulating and important for developing research strategies, critical assessment of experimental data from each species is essential before conclusions can be reached.

2. The plant phosphoinositide system

A schematic and simplified overview of the main components of the phosphoinositide-signalling system in mammalian cells is given in *Figure 2*.

In unstimulated animal and plant cells, the cytosolic Ca^{2+} levels are maintained in the low nanomolar region by active Ca^{2+} transport systems in the plasma membrane and in organelles. Polyphosphoinositides in the inner leaflet of the plasma membrane are continuously being formed and degraded by a set of specific kinases and phosphatases, and are kept in a state of constant turnover. When specific signals (known collectively as agonists) associate with specific receptors, the enzyme phosphoinositidase C (PIC, also known as phosphoinositide-specific phospholipase C) is activated. Many PIC isoforms exist in mammalian cells, and several different modes of activation have been identified. PIC-β-isoform activity is controlled by regulatory, heterotrimeric GTP-binding proteins (G-proteins), whereas γ-isoform activation depends on receptor-associated tyrosine kinase(s). The activation of PIC leads to the hydrolysis of phosphatidylinositol(4,5)bisphosphate (PtdIns(4,5)P_2) and the production of the two second

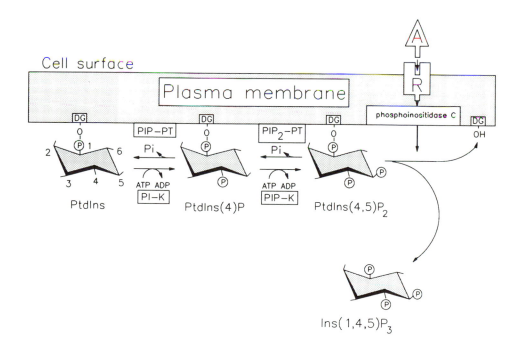

Figure 2. Schematic and simplified model of the main components of the eukaryotic phosphoinositide signalling system. Polyphosphoinositides in the inner leaflet of the plasma membrane are in a state of constant turnover mediated by two sets of lipid kinases (K) and lipid phosphatases (PT). Upon interaction of an appropriate agonist (A) with its receptor (R), the enzyme(s) phosphoinositidase C is activated. This leads to hydrolysis of PtdIns(4,5)P_2 and production of the two second messenger molecules inositol(1,4,5)trisphosphate (Ins(1,4,5)P_3) and diacylglycerol (DG). DG remains associated with the membrane matrix whereas Ins(1,4,5)P_3 is freely diffusable in the cytosol.

messenger molecules inositol(1,4,5)trisphosphate (Ins(1,4,5)P$_3$) and diacylglycerol (DG). Ins(1,4,5)P$_3$ is able specifically to release Ca^{2+} from intracellular stores, whilst DG modulates the activity of the enzyme protein kinase C (PKC). The concomitant increase in cytosolic Ca^{2+} and the switching on of protein kinase activity results in a bifurcated signal which leads to cell activation. When the agonist–receptor complex is dissociated, PIC is inactivated, cellular levels of Ins(1,4,5)P$_3$ and DG decrease, and the cytosolic Ca^{2+} levels are returned to low nanomolar concentrations by Ca^{2+}-transporting systems removing Ca^{2+} from the cytosol. Most of the structural and functional components of the mammalian phosphoinositide system have also been identified in plant cells (see for example Drøbak, 1992; Cote and Crain, 1993), but in the present context the focus will be on the lipid components of the phosphoinositide cycle and the enzymes responsible for their metabolism.

3. Biosynthesis and turnover of plant phosphoinositides

Phosphatidylinositol

Phosphatidylinositol (PtdIns), the precursor for all polyphosphoinositides, is a common phospholipid in plants. In non-photosynthetic plant tissues, PtdIns is the most abundant phospholipid after phosphatidylcholine and phosphatidylethanolamine; it may thus represent as much as 20% of the total phospholipids in various membrane types, and is found in most organellar membranes (Harwood, 1980). Two reactions that are responsible for the incorporation of *myo*-inositol into PtdIns are known to occur in plant cells. The first is mediated by the enzyme CDP-DAG:*myo*-inositol phosphatidyltransferase (EC 2.7.8.11), and the other, the so-called 'head-group exchange reaction', is catalysed by a PtdIns:*myo*-inositol phosphatidyltransferase (see *Figure 3*). Only the first of these two reactions results in net synthesis of PtdIns. Both reactions are thought to occur in the endoplasmic reticulum of plant cells, and also possibly in the Golgi complex (Moore, 1990).

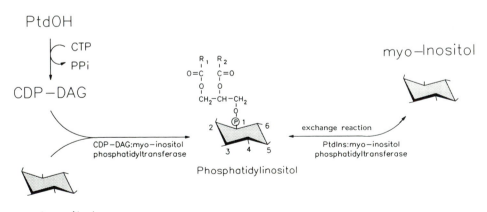

Figure 3. The two routes responsible for phosphatidylinositol synthesis. For further details see main text. CDP-DAG, cytidine diphospho-diacylglycerol.

Phosphatidylinositol 4-phosphate (PtdIns(4)P) and phosphatidylinositol 4,5-bisphosphate

The first reports describing the presence of phosphorylated forms of PtdIns (i.e. poly-phosphoinositides) in plants began to emerge a decade ago (e.g. Boss and Massel, 1985), and work carried out since has led to a general acceptance that polyphosphoinositides are present in all plant cells (e.g. Boss, 1989; Drøbak et al., 1988; Heim and Wagner, 1986; Irvine et al., 1989; Morse et al., 1987; Strasser et al., 1986). There is still some confusion about the chemical levels of polyphosphoinositides in plant cells, because the chemical quantities of polyphosphoinositides in such cells have not so far been determined directly, all current estimates being derived from radiolabelling experiments. Such experiments can be difficult to interpret, and widely different results have been reported, ranging from ratios of 10:1:1 to 300:17:1 for PtdsIns:PtdIns(4)P:PtdIns(4,5)P$_2$ (see Boss, 1989; Drøbak, 1992; Hetherington and Drøbak, 1992; Sandelius and Sommarin, 1990). In most of these experiments, phosphoinositides have been separated by thin-layer chromatography (TLC) and identification has been based on co-chromatography of labelled compounds with authentic mammalian standards. Although this approach, at least until recently, was considered to be adequate for the routine analysis of animal polyphosphoinositides, it has for some time been known that this is not the case when plant polyphosphoinositides are being studied. The reasons for this are several, and include the differences in fatty acid composition between mammalian and plant phosphoinositides (which can lead to 'double running' in commonly used TLC systems), and the presence of considerable amounts of lyso-phosphoinositides and inositol-containing phytoglycolipids in plant tissues (e.g. Wheeler and Boss, 1990). Both of these lipid types show chromatographic behaviour very similar to that of polyphosphoinositides. The recent discovery of the 3-phosphorylated phosphoinositides (see Section 4) further emphasizes the importance of careful identification of polyphosphoinositides – from both plants and other eukaryotes. Experiments in which glycerophosphorylinositol- or inositol-phosphate derivatives of labelled phosphoinositides from plant tissues have been separated by high-pressure liquid chromatography (HPLC) (a procedure which largely overcomes the problems of co-chromatography of other polyphosphoinositide- and inositol-containing lipid derivatives) indicate that PtdIns(4,5)P$_2$ in most plant tissues is unlikely to represent more than approximately 0.05% of total phospholipids or 0.5% of phosphoinositides (Drøbak et al., 1988; Irvine et al., 1989). With the recent advances in methods for determining the mass of very low chemical levels of phosphoinositides (e.g. Divecha et al., 1991), it will hopefully only be a matter of time before a clearer picture of true phosphoinositide levels in plant cells emerges.

Turnover of 4-phosphorylated plant polyphosphoinositides

In mammalian cells, PtdIns(4)P and PtdIns(4,5)P$_2$ are formed by a stepwise phosphorylation of PtdIns. PtdIns is first phosphorylated in the D-4 position of the inositol ring, resulting in the formation of PtdIns(4)P. This lipid is then further phosphorylated in the 5-position, yielding PtdIns(4,5)P$_2$. Two phosphohydrolases work concomitantly with the phosphoinositide kinases to remove the 4 and 5 phosphomonoesters, thus keeping PtdIns(4)P and PtdIns(4,5)P$_2$ in a state of constant turnover. All the current evidence suggests that polyphosphoinositides are synthesized in a similar manner in plants.

Phosphoinositide kinases

The first studies of phosphoinositide kinases in plant cells were carried out by Sommarin and

Sandelius (e.g. Sandelius and Sommarin, 1986; Sommarin and Sandelius, 1988). Using microsomal and purified plasma membrane fractions from wheat as the enzyme source, they demonstrated the presence of PtdIns kinase and PtdInsP kinase activity which could utilize both endogenous and exogenous substrates. Both kinases were highly enriched in plasma membranes and showed complete dependency upon ATP for phosphorylation. The estimated K_m values for ATP for both kinases were around 200 µM, which is considerably higher than the values reported for the mammalian PtdIns 4-kinase (type 2) and PtdIns(4)P 5-kinase (Carpenter and Cantley, 1990), but agrees well with the ATP K_m found for mammalian type 3 PtdIns 4-kinase (Li *et al.*, 1989) and the yeast PtdIns 4-kinase (Belunis *et al.*, 1988). Studies by Kamada and Muto (1991) have shown that the PtdIns kinase of tobacco plasma membranes also resembles the mammalian type 3 kinase with respect to both K_m for ATP and sensitivity to adenosine.

When optimal kinase assay conditions were used, the rate of formation of PtdInsP from endogenous substrates was approximately 175 pmol mg^{-1} protein min^{-1} in plasma membranes from wheat shoots, and 65 pmol mg^{-1} protein min^{-1} in plasma membranes derived from roots. The rates of formation of PtdInsP$_2$ were 18 pmol mg^{-1} protein min^{-1} and 6 pmol mg^{-1} protein min^{-1}, respectively when shoot and root plasma membranes were used. The rates of phosphorylation of both PtdIns and PtdInsP were increased 8- to 20-fold when exogenous lipids were used as substrates.

Little is currently known about the regulation of the plant phosphoinositide kinases *in vivo*, but radiolabelling studies suggest that the rate of turnover of PtdIns(4)P is strikingly rapid. Drøbak *et al.* (1988) thus found that more than 30% of label ($^{32}P_i$) incorporated into phospholipids in suspension-cultured tomato cells after short incubation times (30 min) was found in PtdInsP. Although the position of the label in the PtdIns(4)P molecule was not determined, it is likely that it was predominantly in the 4-phosphomonoester, indicating the presence of a highly active PtdIns 4-kinase. The rapid ^{32}P incorporation, when viewed in conjunction with the low chemical levels of PtdIns(4)P, suggests that PtdIns(4)P is metabolized further with some speed. The apparently very low chemical levels of PtdIns(4,5)P$_2$ and the high rate of turnover of PtdIns(4)P have made the task of obtaining accurate data about PtdIns(4,5)P$_2$ turnover difficult. However, the few available data suggest that either the chemical quantities of PtdIns(4,5)P$_2$ are even lower than is currently assumed, or the turnover rate of the main cellular pools of PtdIns(4,5)P$_2$ is considerably slower than can be predicted from the *in-vitro* values of PtdIns(4)P 5-kinase activity. It is possible that an *in-vivo* inhibitor of PtdIns(4,5)P$_2$ turnover may exist in plant cells, and that its inhibitory effect is overcome only upon cell activation. One potential candidate for such a function is the actin- and polyphosphoinositide-binding protein, profilin, whose function will be discussed in further detail in Section 5.

Polyphosphoinositide phosphatases

Considering the amount of attention that has been given to the role of phosphatases in, for example, cell cycle control and protein phosphorylation/dephosphorylation cascades, the near total neglect of the phosphoinositide phosphatases may seem a little surprising. In the original study by Sandelius and Sommarin (1986), the removal of ATP by addition of glucose/hexokinase did not cause any appreciable decrease in the labelling of PtdInsP, although PtdIns kinase activity was abolished. It is a little difficult to reconcile such data with the apparent rapid turnover of PtdIns(4)P *in vivo*. However, it is possible that optimization of the *in-vitro* assays for phosphoinositide kinase activity may have led to the suppression of phosphatase activity. In recent experiments in the author's laboratory, the presence of active polyphosphoinositide phosphatases in plasma membranes from bean leaves has been demonstrated (P. Xu and B.K. Drøbak, unpublished data).

Phospholipase C and other phospholipases

At least 16 different isoforms of phospholipase C (PLC) have been identified in mammalian cells. These isoforms have been divided into four main families, PLC-β, PLC-δ, PLC-ε and PLC-γ, on the general basis of their amino acid sequences and mode of activation. Relatively little is known about the structure and mode of activation of plant phospholipase C enzymes, but it is likely that several isoforms exist. The first evidence for the presence in plant cells of a phospholipase C that was specific for a phosphoinositide was provided by Irvine *et al.* (1980), who found a soluble activity in celery stems that was capable of hydrolysing PtdIns, with Ins(1)P and cyclic Ins(1:2)P being the resulting products. The presence of a distinct plant PLC which preferred polyphosphoinositides to PtdIns was demonstrated by Melin *et al.* (1987), who found a PLC in the inner leaflet of highly purified plasma membrane vesicles which had a 5- to 20-fold greater affinity for PtdIns(4)P and PtdIns(4,5)P$_2$ than for PtdIns when all lipids were presented as exogenous (liposomal/micellar) substrates. Optimal activity of this PLC required Ca^{2+} concentrations in the low micromolar region, and was abolished in the absence of Ca^{2+}. Several more recent reports have confirmed and extended these findings and it is now evident that at least two distinct PLCs appear to be widespread in plant cells: type I is a predominantly soluble PLC which prefers PtdIns as substrate and requires high Ca^{2+} concentrations (mM) for full activity, whereas the type II PLC is associated with the inner leaflet of the plasma membrane, shows a marked preference for polyphosphoinositides, and is totally dependent on Ca^{2+} in the physiological range (nanomolar to low micromolar). Whether the apparently membrane-associated forms of PtdIns-hydrolysing enzymes reported by Pfaffman *et al.* (1987) and McMurray and Irvine (1988) are distinct, or can be explained by either low-level PtdIns hydrolysis by the type-II enzyme or cross-contamination with the type-I enzyme, has yet to be resolved. However, it is clear that it is the plasma membrane-associated type-II enzyme which most closely resembles the mammalian PIC isoforms involved in signal-induced phosphoinositide hydrolysis. A discussion of the ways in which this group of enzymes can be activated in mammalian and plant cells can be found in Section 5.

Phospholipases A and D

The possibility that phospholipases other than PLC are involved in the metabolism of plant phosphoinositides deserves mention. Phospholipases A$_1$ and A$_2$ remove the fatty acids in the *sn*-1 and *sn*-2 positions of the phospholipid glycerol backbone, respectively. The presence of several lyso-derivatives of the phosphoinositides in certain plant cells suggests the possible presence in these cells of one or more phospholipase(s) A which can specifically utilize phosphoinositides as substrates. Wheeler and Boss (1990) have reported the presence of a Ca^{2+}-dependent, PtdIns-specific phospholipase A$_2$ which is active over the pH range 5.5–7.0, but little else is currently known about this type of enzyme, and most work has been carried out on the non-specific acyl-hydrolases which catalyse the formation of a variety of deacylated lipids (for further details see Hetherington and Drøbak, 1992; Wheeler and Boss, 1990). The fact that lyso-phospholipids can play important roles in cellular physiology has been recognized for some time by researchers in the mammalian field. The demonstration that lyso-phosphatidylcholine can regulate plant H$^+$-ATPase activity (Palmgren *et al.,* 1988), that other plant lyso-lipids affect protein kinase activity (Martiny-Baron and Scherer, 1989) and that membrane-associated phospholipase A$_2$ from soybean cells can be stimulated by auxin (Scherer and Andre, 1993) suggests that further attention to plant phospholipases A could prove very worthwhile.

A recent development in mammalian research is the identification of phospholipase D (PLD) as an important enzyme in transmembrane signal transduction. PLD hydrolyses phospholipids on the 'head-group' side of the phospholipid–phosphodiester bond, and thus gives rise to the products phosphatidic acid and a 'head group' (which depends on the specific type of phospholipid being hydrolysed). Although the preferred substrate for mammalian PLD is phosphatidylcholine (PtdCho) and in some cases phosphatidylethanolamine (PtdEth), there is growing evidence for interactions between the PLD pathway and the phosphoinositide-PLC pathway (e.g. Billah, 1993). In animal cells, PLD exists in both membrane-bound and cytosolic forms and several distinct isozymes have been identified. It appears to be the membrane-associated form which exhibits the narrowest specificity for PtdCho, as the cytosolic forms hydrolyse PtdEth and PtdIns as well as PtdCho. PLD activation is likely to be receptor-linked and occurs rapidly in many cells in response to diverse agonists. The hydrolysis product, phosphatidic acid (PtdOH), and its dephosphorylation product, sn-1,2-diacylglycerol, are both thought to have important second-messenger functions. Whereas the role of DG in PKC activation is well characterized, the precise role of PtdOH is still being debated. However, both PtdOH and its lyso-derivatives have been found to be potent growth factors in cells such as fibroblasts. Fukami and Takenawa (1992) thus showed that the addition of PtdOH and lysoPtdOH to Balb/c fibroblasts leads to a rapid induction of DNA synthesis. Another interesting finding is that PtdOH can also specifically activate PtdIns(4)P 5-kinase (Moritz et al., 1992). PtdOH is furthermore thought to be involved in regulation of the GTP/GDP status of cellular p21[ras] proteins (see Cook and Wakelam, 1992). The presence of active PLD enzymes in plant cells is well documented, and some studies suggest that they may be involved in plant responses to auxin, for example (see Morre, 1989).

4. 3-Phosphorylated phosphoinositides

Several reports have demonstrated that agonist stimulation of mammalian cells not only alters the levels of PtdIns(4)P and PtdIns(4,5)P$_2$, but also leads to the production of a number of polyphosphoinositides containing a monoester in the D-3 position of the inositol ring. The structures of the 3-phosphorylated phosphoinositides are PtdIns(3)P, PtdIns(3,4)P$_2$ and PtdIns(3,4,5)P$_3$, respectively (see *Figure 1*). The chemical levels of this group of lipids are very low in mammalian cells, where they represent only 1–2% of total phosphoinositides. In contrast, much higher levels of 3-phosphorylated phosphoinositides exist in organisms such as yeast. The biosynthetic route leading to the production of 3-phosphorylated phosphoinositides has been the subject of some controversy in the last few years. *In-vitro* studies using a purified PtdIns 3-kinase indicate that both PtdIns(4)P and PtdIns(4,5)P$_2$ can act as kinase substrates, and the following biosynthetic route involving stepwise phosphorylation of phosphoinositides has been proposed: PtdIns $\rightarrow$ PtdIns(3)P $\rightarrow$ PtdIns(3,4)P$_2$ $\rightarrow$ PtdIns(3,4,5)P$_3$ (Majerus, 1992). However, in f-Met-Leu-Phe-activated neutrophils, it has been convincingly demonstrated that the initial step in the formation of 3-phosphorylated phosphoinositides is a direct phosphorylation of PtdIns(4,5)P$_2$ in the 3-position by a PtdIns(4,5)P$_2$ 3-kinase (Stephens et al., 1991). PtdIns(3,4,5)P$_3$ is thus likely, at least in neutrophils, to be the *in-vivo* precursor for the mono- and bisphosphate forms of 3-phosphoinositides, formed by dephosphorylation of phosphatidylinositol trisphosphate.

The physiological roles of the 3-phosphorylated phosphoinositides have not been clarified, but these lipids are currently attracting immense attention since it was found that their synthesis in many cases seems to be causally correlated with progression through the mitotic cycle. The mammalian and yeast PtdIns 3-kinases are heterodimeric proteins with a 85-kDa regulatory subunit and a 110-kDa catalytic subunit. The regulatory subunit contains specific domains which suggest a functional link between regulation of PtdIns 3-kinase activity and tyrosine kinase receptor activation (see Fry and Waterfield, 1993). Surprisingly, the cDNA for the 110-kDa catalytic subunit has been found to have significant sequence homology to a yeast protein, vps34p, which is essential for protein sorting to the lysosome-like vacuole of yeast (Schu *et al.*, 1993). Whether a link exists between PtdIns 3-kinase activation, vacuolar function and mitosis is not yet clear. It is possible that this apparent dual functionality of the PtdIns 3-kinase may turn out to be illusory and that it can be explained by the presence of several isoforms of PtdIns/PtdIns(4,5)P$_2$ 3-kinases in eukaryotic cells, each with its own specialized role.

Recently, the presence of 3-phosphorylated phosphoinositides in plants has been demonstrated, and PtdIns(3)P and PtdIns(3,4)P$_2$ (but so far not PtdIns(3,4,5)P$_3$) have been isolated from duckweed and stomatal guard cells (Brearley and Hanke, 1993; Parmar and Brearley, 1993). The route of synthesis and possible function of PtdIns(3)P in plants is currently unknown, but it will clearly be of considerable interest to investigate whether the 3-phosphorylated plant phosphoinositides are involved in cell cycle control and/or in vacuolar protein sorting, or whether they perhaps have other unique functions.

5. Regulation of phosphoinositide turnover

Formation of signalling complexes: signal-responsive and unresponsive phosphoinositide pools

Although the majority of illustrations depicting the phosphoinositide system resemble that shown in *Figure 2*, a more detailed consideration of the actual situation in the plasma membrane leads to the conclusion that this type of two-dimensional representation, however illustrative, can also be somewhat misleading. The first point to consider is that PtdIns, PtdIns(4)P and PtdIns(4,5)P$_2$ in the membrane are essentially the same molecules, and only the degree of phosphorylation of the inositol head group determines their structure at any given moment. At least four enzymes (two kinases and two phosphatases) are involved in the perpetual interconversion between the different forms of the phosphoinositide molecule in unstimulated cells, so the organization of these enzymes must be highly spatially co-ordinated for the simple reason of substrate accessibility. Upon cell activation, even more enzymes (e.g. PLC and PtdIns 3-kinase) need access to the signal-sensitive phosphoinositide pools. During cell activation there are thus six comparatively large enzymes which must all be within 'striking distance' of the phosphoinositide pools. The assembly of large multi-enzyme complexes (signalling complexes) is therefore likely to be an integral feature of specific signal transduction events, and may also be involved in the control of downstream pleiotropic effects induced by many agonists. The formation of such complexes may help to explain the presence in many cell types of both agonist-responsive and agonist-unresponsive phosphoinositide pools (see Monaco and Gershengorn, 1992). It can be envisaged that only those phosphoinositide molecules which are in close proximity to the site of formation of a signalling complex will be part of the signal-sensitive pool, while phosphoinositide pools which are not in close proximity to such complexes will remain unresponsive to cell activation.

A comprehensive picture of the subcellular localization of plant polyphosphoinositides and phosphoinositide-metabolizing enzymes is still lacking, but data from labelling experiments employing both intact cells and membrane fractions enriched with plasma membrane vesicles suggest that the polyphosphoinositides, as in mammalian cells, are predominantly associated with the plasma membrane (Sandelius and Sommarin, 1990). In isolated membrane fractions from hypocotyls of dark grown soybean, both PtdIns- and PtdInsP-hydroxykinases are thus predominantly associated with the plasma membrane, whereas only negligible activity of these enzymes is found in the tonoplast, mitochondria and plastid fractions (Sandelius and Sommarin, 1990). However, the finding that small amounts of polyphosphoinositides are associated with the nucleus (Hendrix *et al.,* 1989) and the demonstration of the presence of a highly active PtdIns 4-kinase associated with the cytoskeleton (Xu *et al.,* 1992) indicate that much has still to be learned about the subcellular distribution of plant polyphosphoinositides, and their sensitivity to hydrolysis by agonist-activated forms of phospholipase C. (For further details about the cytoskeletal and nuclear phosphoinositide-system components, see Chapter 1 by Lloyd *et al.*).

Activation of mammalian and plant PLC isoforms

As mentioned previously, a large number of mammalian PLC isoforms exist, which differ not only in molecular structure, but also in their mode of activation. The mechanisms of activation of two of the isozymes, PLC-β1 and PLC-γ1, are particularly well understood and involve interactions with heterotrimeric regulatory GTP-binding proteins (G-proteins) and tyrosine kinases respectively. These two modes of activation are not only responsible for the regulation of PLC activity, but also are involved in the control of a range of other signalling enzymes.

G-proteins. The signal transduction pathway which is controlled by G-protein-linked receptors involves three main components: the receptor itself, the regulatory G-protein(s) and the target enzyme (in this case PLC-β). Most of the G-protein-linked receptors identified so far are characterized by having seven membrane-spanning domains connected by extracellular and intracellular loops. It is thought that the transmembrane domains interact with each other to form a 'binding-pocket' where the agonists can bind. Upon agonist binding, the receptor is converted into a configuration which allows activation of the associated G-protein. Mutational studies have revealed that the second and third cytoplasmic loops of G-protein receptors are essential for the activation of the G-proteins, and it is thought that this is the part of the receptor molecule which interacts directly with the αβγ-subunits of the heterotrimeric G-proteins (Simon *et al.,* 1991). When activated, the G-protein α-subunit dissociates from the βγ-subunits, and the α-subunit (which contains the GTP-binding site) exchanges GDP for GTP before interacting with the C-terminal region of PLC-β1. Although the role of the βγ-subunit is still not entirely clear, it is likely to be far more important than just serving as an 'appendix' to the α-subunit. It has been suggested that it may play a role in the activation of certain PLC isoforms (notably PLC-β2; Katz *et al.,* 1992). When agonist activation of G-protein receptors is terminated, the α-subunit-bound GTP is hydrolysed to GDP by an intrinsic GTPase activity, and the α-subunit combines with the βγ-subunits to reform an inactive complex. A very interesting finding is that the intrinsic GTPase activity of the α-subunit is greatly enhanced following its interaction with PLC-β1, which suggests a conserved functional link between PLC-β1 and other GTPase-activating proteins (GAPs), such as the p21[ras]-GAP (Berstein *et al.,* 1992).

Tyrosine kinases. The pathway responsible for modulation of the activity of the PLC-γ isoforms involves activation of tyrosine kinase-linked receptors. One of the most extensively studied examples is the interaction between platelet-derived growth factor (PDGF) and its receptor. The PDGF receptor consists of a single transmembrane protein which contains intrinsic tyrosine kinase activity on its cytoplasmic domain. Perception of the PDGF signal leads to dimerization of the receptor, which allows the kinase domains to phosphorylate each other on specific tyrosine residues. One of the key functions of the phosphorylated receptor domains is to act as attachment sites for signalling molecules such as the γ-isoform of phosphoinositidase C. Other signalling molecules which can bind similarly to activated tyrosine receptors include phosphatidylinositol 3-kinase and GAPs. Both PLC-γ and the other signalling molecules which bind to such receptor-phosphotyrosine domains contain one or more highly conserved domains which are homologous to a non-catalytic region of the c-*src* proto-oncogene, and have been named SH2 (*src* homology region 2). The association of PLC-γ with the PDGF receptor leads to phosphorylation of tyrosine residues of PLC-γ itself by the receptor-associated kinase. The phosphorylation of PLC-γ then results in a conformational change of the enzyme, allowing its catalytic domains access to the membrane-associated lipid substrate, PtdIns(4,5)P$_2$.

Ca^{2+}. It has been known for some time that Ca^{2+} can act as a potent regulator of mammalian PIC, but the precise role of Ca^{2+} in the *in-vivo* modulation of PIC activity remains obscure. The conceptual problem with the Ca^{2+} dependency of PIC isoforms is that activation of PIC in most cells is known to *precede* Ca^{2+} mobilization rather than being a consequence of it. One possibility is that specific isoforms of PIC may exist which are modulated by Ca^{2+} alone, and do not depend on either G-proteins or tyrosine phosphorylation. The activation of such PIC isoforms could be envisaged either as a downstream event to PtdIns(4,5)P$_2$ hydrolysis by other PICs, or by changes in their Ca^{2+} sensitivity. The discovery that PLC-δ isoforms contain the canonical Ca^{2+}-binding EF-hand motif (Bairoch and Cox, 1990) begins to lend credence to the hypothesis of Ca^{2+}-activated forms of PIC.

Plant phospholipase C (type II)

The possibility that the polyphosphoinositide-specific plant phospholipase C may be regulated by one of the pathways outlined above has not escaped attention, but so far no final conclusions have been reached. The presence in higher plants of tyrosine kinases in general, and tyrosine kinase receptors in particular, remains contentious. There are, however, some reports which provide biochemical evidence in favour of the presence of tyrosine kinases in plants (e.g. Torruella *et al.,* 1986), and genetic information suggests the presence of plant genes with homology to v-*src* (Zabulionis *et al.,* 1988). The lack of a general acceptance of the presence of tyrosine kinases in plant cells may have several causes, including the generally very low levels of phosphotyrosines in any eukaryotic cells, the presence of highly active phosphotyrosine phosphatases in plants (Cheng and Tao, 1989), and the transient nature of signal-induced tyrosine phosphorylation.

As most of the evidence for and against the involvement of G-protein in the control of plant PLC activity has been discussed recently (Drøbak, 1992), only the main conclusions will be presented here. In favour of G-protein involvement are the findings that the plasma membrane polyphosphoinositide-specific phospholipase C from the alga *Dunaliella salina* can be markedly stimulated by 100 μM GTPγS over a range of free Ca^{2+} concentrations (Einspahr *et al.,* 1989), and that guanine nucleotides are able to stimulate the release of inositol phosphates from membrane fractions isolated from cultured sycamore cells (Dillenschneider *et al.,* 1986). However, in another series of experiments using both crude membrane fractions and highly purified plasma membrane vesicles, no evidence has been found for any involvement of GTP-binding proteins (e.g. Biffen and Hanke, 1990; McMurray and Irvine, 1988; Melin *et al.,* 1987, 1992; Pica *et al.,* 1992; Tate *et al.,* 1989; Yotsushima *et al.,* 1993). One problem which has been pointed out by several authors is that G-proteins may be lost during the preparation and purification of plant membrane fractions. Genetic and immunological evidence for the presence in plants of proteins which have many similarities to mammalian heterotrimeric G-proteins (e.g. Blum *et al.,* 1988; Ishida *et al.,* 1993; Jacobs *et al.,* 1988; Ma *et al.,* 1990) nevertheless suggests that further experimentation, perhaps on intact/permeabilized cells, may help to resolve the question of whether or not G-proteins are involved in the modulation of plant PLC activity. A recent elegant paper by Legendre *et al.* (1993) provides evidence for the activation of plant PLC by the G-protein activator, mastoparan, and thus lends further support to the hypothesis of G-protein coupling of plant signalling enzymes.

The absolute dependency of plant PLC-II on Ca^{2+} in the high nanomolar to low micromolar range suggests that cellular Ca^{2+} changes could play an important role in the *in-vivo* regulation of this enzyme. If this should turn out to be the case, it clearly raises important questions about the proposed physiological roles of PLC-II in agonist-induced $Ins(1,4,5)P_3$ formation. Elucidation of the mechanism(s) involved in PLC-II activation would be considerably aided if information were available about its amino acid structure. Hopefully such data will emerge in the near future.

It is worth mentioning that PLC-II is not the only enzyme of the plant phosphoinositide cycle which is Ca^{2+}-dependent. In isolated plasma membranes from suspension-cultured tobacco cells, both of the enzymes responsible for $PtdIns(4,5)P_2$ generation (i.e. the PtdIns 4-kinase and PtdIns(4)P 5-kinase) are strongly inhibited by nanomolar Ca^{2+} concentrations (Kamada and Muto, 1991). The PtdIns 4-kinase is by far the most sensitive, with a reduction in activity of more than 90% in the presence of 100 nM Ca^{2+}. The reduction in PtdIns(4)P 5-kinase activity is less dramatic, and this enzyme still shows 30–50% of full activity in the concentration range 1–100 µM Ca^{2+}.

Interactions between the phosphoinositide pathway and components of the cytoskeleton

The actin cytoskeleton. One of the key responses which characterizes growth-factor stimulation (used here in its widest sense) of many cell types, including plants, is rapid reorganization of the actin cytoskeleton. It is thus not surprising that links between growth factor modulation of the phosphoinositide pathway and cytoskeletal reorganization have been sought. The cytoskeleton in eukaryotic cells has three main components: actin, tubulin and intermediate filaments. In the present context the focus will be on the actin component (for further details about the other cytoskeletal components see Chapter 1 by Lloyd *et al.*). The chemico-physical state of cellular actin is regulated by three reversible cycles. The first involves the interconversion between monomeric actin (globular actin or G-actin) and the polymerized form of actin (filamentous or F-actin). This interconversion between G- and F-actin is known to be regulated by at least six classes of actin-associated proteins, including profilin, villin, fragmin, β-actinin, gelsolin and depactin. Another factor which is involved in determining the state of cellular actin is the ability of F-actin to cross-link and form actin bundles. This bundling of actin is also controlled by several proteins, such as villin, α-actinin and spectrin. The third reaction involved in actin metabolism is the reversible anchoring of actin structures to membranes. This latter process has been shown to be regulated by a member of the phosphoinositide signalling system, namely protein kinase C.

Protein kinase C and the cytoskeleton. It is believed that protein kinase C may play a key role in the control of this part of the actin cycle via its effect on a class of actin-binding proteins which include the MARCKS (myristoylated, alanine-rich, C-kinase substrates). MARCKs are specific protein kinase C substrates which are targeted to membranes by the amino-terminal myristoylated membrane-attachment domain. Phosphorylation of MARCKS results in their translocation from membranes to the cytosol, where they associate with actin. The following model for the action of MARCKS during signal transduction has been proposed. In non-stimulated cells the unphosphorylated form of MARCKS associates with the cytoplasmic face of the plasma membrane (probably by association with receptor molecules) where it cross-links with actin. This cross-linking is thought to render the actin meshwork, associated with the plasma membrane via MARCKS, relatively rigid. Upon cell activation, MARCKS are phosphorylated by PKC and released from the membrane.

Although cytosolic phosphorylated MARCKS can still associate with actin, they are no longer able to cross-link with actin. The spatial separation of MARCKS-bound actin from the membrane, as well as its increased plasticity, is thought to mediate reorganization of the actin cytoskeleton in events such as secretion, phagocytosis and mitogenesis (for review see Aderem, 1992). Although there is currently no direct evidence for the presence of MARCKS in plant cells, the presence of several PKC-like (i.e. DG-activated, phospholipid- and Ca^{2+}-dependent) enzymes (for references see Drøbak, 1992) and the reported dramatic effects of DG on mitotic progression (Larsen and Wolniak, 1990) suggest that a link between DG-activated plant kinases and mitotic events (such as reorganization of the actin cytoskeleton) deserves further attention.

Actin-binding proteins and phospholipase C. Another actin-binding protein which has been shown to interact closely with components of the phosphoinositide system is profilin. Profilin is a small (12–15 kDa) protein which, in addition to its actin-binding properties, also has specific binding sites for polyphosphoinositides. Profilin has a 10-fold higher affinity for $PtdIns(4,5)P_2$ than for actin, so $PtdIns(4,5)P_2$ is highly effective in disrupting profilin–actin (profilactin) complexes. Polyphosphoinositides are thus able to exert control over the ability of profilin to interact with actin, but at the same time profilin controls the availability of polyphosphoinositides for second-messenger production. In fact, the association of profilin with clusters of $PtdIns(4,5)P_2$ molecules is so strong that it completely abolishes the activity of purified mammalian PLC-γ (Goldschmidt-Clermont *et al.,* 1990). A very interesting finding is that the activated γ-forms of PIC (i.e. the tyrosine-phosphorylated forms, see above) are capable of hydrolysing $PtdIns(4,5)P_2$ even in its profilin-bound form (Goldschmidt-Clermont *et al.,* 1991). This has led to the hypothesis that activation of certain phosphotyrosine kinase receptors (such as the PDGF receptor) *in vivo* not only results in the production of $Ins(1,4,5)P_3$ and DG, but also leads to profilin release from clusters of $PtdIns(4,5)P_2$, causing an increase in the cellular pool of free profilin that is capable of interacting with actin.

Until recently, the favoured view of profilin's function was that it acts primarily as a sequestering protein, and through binding to actin monomers promotes actin depolymerization. However, recent evidence has led to reconsideration of this model. In addition to its sequestering function, profilin is also capable of accelerating the exchange of adenine nucleotides on G-actin (Goldschmidt-Clermont *et al.,* 1992). This means that, at cellular levels of ADP and ATP, profilin can catalyse the conversion of ADP-actin to ATP-actin, a form which polymerizes much faster, and thus profilin promotes net polymer assembly and actin polymerization.

Recently, a plant profilin homologue was identified in a search for pollen allergens from the dicotyledonous birch tree (Valenta *et al.,* 1991), and additional plant profilin genes have been characterized from maize and timothy grass (Staiger *et al.,* 1993; Valenta *et al.,* unpublished data). The overall amino acid sequence homology between the plant profilins and profilins from other eukaryotes is relatively low (30–40% identity), and only nine amino acids are completely conserved in all profilins (Staiger *et al.,* 1993). However, the ability of plant profilin to form complexes with phosphoinositides (in particular $PtdIns(4,5)P_2$) and thereby to affect their turnover dramatically has recently been demonstrated in the author's laboratory (Drøbak *et al.,* 1994). The precise molecular mechanism responsible for the interaction between plant profilin and $PtdIns(4,5)P_2$ is currently being studied, but it is a clear possibility that the highly polar head group of $PtdIns(4,5)P_2$ interacts with a specific, localized region of positive charge on plant profilin. One such region, rich in basic residues and highly conserved in all plant profilins, is IRGKKGSGGITIKKT (Staiger *et al.,* 1993). Other actin-binding proteins, such as villin and gelsolin, which also interact with polyphosphoinositides have similar (poly)basic motifs and their ability to function in this context has been demonstrated directly using synthetic peptides (Janmey *et al.,* 1992).

Another interesting recent development is the partial purification and characterization of a novel activator of the plant plasma membrane PtdIns 4-kinase (Yang *et al.*, 1993). The activator, named PIK-A49, was found to have a relative molecular mass of 49 kDa, and sequence analysis of seven peptide fragments (totalling 142 amino acid residues) indicated that PIK-A49 was 69% identical to the actin-binding protein ABP-50 from *Dictyostelium*. Surprisingly, it was found that PIK-A49 had more than 90% homology with the eukaryotic elongation factor 1α (EF-1α). It is clearly an intriguing possibility that a link may exist between EF-1α activity, protein synthesis, phosphoinositide turnover and cytoskeletal reorganization — an idea which deserves further attention.

6. Glycophosphosphingolipids and other complex inositol-containing lipids

The general structure of glycophosphosphingolipids or phytosphingolipids is shown in *Figure 1*. These lipids are complex inositol- and phosphate-containing glycolipids which all contain a phytosphingosine base. It should be noted that phytosphingosine differs from sphingosine with regard to the position of a carbon–carbon double bond. Pioneering work on these lipids was carried out by Professor Herbert Carter and his colleagues at the University of Illinois, and they were the first to partially characterize several types of glycophosphosphingolipid from plants (Carter *et al.*, 1958). Since Carter's early work, progress in our understanding of this enigmatic group of lipids has been modest, with the result that relatively little is known about the structure of many of the glycophosphosphingolipids, their route of biosynthesis and their possible functions (for reviews of the current views about this class of plant lipids see e.g. Hetherington and Drøbak, 1992; Laine and Hsieh, 1987). It is clear that quite a large number of these lipids exist in a wide variety of tissues; in tobacco at least 10 different types have been identified (Kaul and Lester, 1975). The presence in higher plants of glycosylated forms of phosphatidylinositols, including glycosyl-phosphatidylinositol anchors (for review see McConville and Ferguson, 1993) is still not established, but since such structures are very widespread amongst eukaryotes, a search for their occurrence in plants could prove successful.

More than 150 different forms of sphingoglycolipids have been identified in mammalian cells to date, and this has recently begun to attract considerable attention among researchers in a wide variety of fields. This sudden upsurge in interest is due to the discovery that many members of the sphingoglycolipid family play key roles in processes such as the regulation of cell growth, transmembrane signal transduction and cell–cell recognition (for a recent review see Hakomori, 1993). Although the plant glycophosphosphingolipids differ somewhat in structure from their mammalian sphingoglycolipid counterparts, it is certainly possible that functional similarities may exist. That this could indeed be the case is illustrated by the recent finding that expression of a novel plant inositol-containing glycolipid appears to be developmentally regulated in plant membranes involved in the process of infection of legume root cells by the nodule-forming bacterium *Rhizobium* (Brewin *et al.*, 1993; see also Chapter 29 by Brewin). Further research on the plant phytosphingolipids is clearly needed, and one can only hope that the breakthroughs made by, for example, researchers in the mammalian field will provide the necessary impetus for plant scientists to pursue this line of enquiry.

7. Conclusions and future research

A great deal has been learned about the occurrence, function and metabolism of inositol-containing lipids in the last decade, but hand in hand with the increase in our knowledge of these lipids goes the growing realization of the immense complexity which underlies their biological functions.

The modest number of plant scientists who until recently have been involved in inositol lipid research and, perhaps of even greater importance, the lack of appropriate techniques adapted to the specific problems associated with research on plant cells have prevented a rate of progress comparable to that seen, for example, in the mammalian field. However, this situation seems to be changing rapidly. The growing numbers of plant scientists who are turning their attention toward this enigmatic field of research, and the recent development and adaption for plant cell studies of new and powerful techniques, are already beginning to bear fruit. Two particularly welcome additions to the technical arsenal are the recently developed highly sensitive receptor assays for inositol-phosphate production, and the increased use of laser-scanning confocal microscopy (and aequorin-transformed plants) in the elucidation of both temporal and spatial aspects of agonist-induced activation of the plant phosphoinositide system. Many of the necessary tools are now available, so research over the next few years should lead to significant progress in our understanding not only of the plant phosphoinositide system, but also of plant cell signalling in general.

8. Notes added in proof — update

Since preparation of the original manuscript for this chapter (in early January 1994), a number of new developments in the field of plant phosphoinositide research have occurred. Limitations of space prevent me from reviewing all the new data which have emerged during the last 2 years, so I will merely point out a few selected pieces of work which are particular relevant to the present chapter. I apologize to authors whose work has not been included in this brief section.

Phospholipase C

Studies on phosphoinositide-specific phospholipases C in oat have demonstrated that multiple isoforms are likely to exist (Huang et al., 1995), and several cDNAs encoding plant phosphoinositide-specific phospholipase C have been identified and characterized (Hirayama et al., 1995; Shi et al., 1995; Yamamoto et al., 1995). The cDNA (cAtPLC1) cloned from Arabidopsis by Hirayama et al. (1995) encoded a polypeptide of 561 amino acids with a calculated molecular mass of 64 kDa. The deduced polypeptide contained the X and Y domains found in all polyphosphoinositide-PLCs so far identified in eukaryotic cells. The overall structure of the putative AtPLC1 protein was found to be most similar to that of mammalian PLC-δ, although the plant protein was found to be much smaller than polyphosphoinositide-PLCs from other organisms. Somewhat surprisingly, Hirayama et al. (1995) observed that the AtPLC1 gene was expressed at very low levels under normal conditions, whereas it was significantly induced by various environmental stresses such as dehydration, salinity and low temperature. It is clearly an attractive hypothesis that polyphosphoinositide-PLC activity is linked to the perception of and adaption to environmental stress signals, but further work is needed to investigate the links between expression and cellular activity. Unfortunately, the PLC-δ isoforms of PLC are the least well understood isoforms of eukaryotic PLCs with regard to mode of activation, but the absolute requirement for Ca^{2+}, combined with the identification of a putative EF-hand motif in the plant AtPLC1 protein, may well point to a potential mode of regulation. The presence of multiple isoforms of plant polyphosphoinositide-PLCs obviously raises the question of whether other isoforms are regulated in a similar fashion, and indeed whether other PLC isoforms are also involved in phosphoinositide-mediated signalling.

Ins(1,4,5)P_3 production

The effects of the wasp venom toxin, mastoparan, on $Ins(1,4,5)P_3$ production described by Legendre et al. (1993) have recently been confirmed and extended by several groups. Drøbak and Watkins (1994) showed that mastoparan induced a rapid and transient increase in $Ins(1,4,5)P_3$

levels in cultured carrot cells, and a similar effect was also observed when the bee venom toxin, mellittin, was substituted for mastoparan. Interestingly, Drøbak and Watkins (1994) also found that Ins(1,4,5)P$_3$ levels in protoplasts were comparable to the levels found in intact cells after stimulation by mastoparan and/or mellittin. These results suggest that the phosphoinositide system is activated by the protoplast preparation procedure and that great caution should be exercised when protoplasts are used for studies of this signalling pathway. The studies of Legendre *et al.* (1993) and Drøbak and Watkins (1994) both employed an Ins(1,4,5)P$_3$-binding assay based on bovine adrenal gland preparations for the quantification of Ins(1,4,5)P$_3$, and the specificity of this preparation for studies on plant extracts has occasionally been questioned. However, a subsequent report from Cho *et al.* (1995), who used a combined enzymatic–HPLC approach for the analysis of Ins(1,4,5)P$_3$, confirms the conclusion that mastoparan does lead to rapid production of the 1,4,5-trisphosphate isomer of *myo*-inositol when added externally to plant cells.

Phosphatidylinositol 3-kinases

In addition to the radiolabelling data mentioned in Section 4, more recent work has led to the identification of a plant enzyme capable of *in-vitro* phosphorylation of PtdIns in the D-3 position (Dove *et al.*, 1994). Interestingly, a considerable proportion of this enzyme is associated with the cytoskeleton. Studies in *Chlamydomonas eugametos* further demonstrate that PtdIns(3)P, at least in this organism, is metabolized with sufficient speed for there to be a signalling function for this lipid (Munnik *et al.*, 1994). Perhaps the most significant recent development in this new area of phosphoinositide research has been the cloning and characterization of two cDNAs encoding plant PtdIns 3-kinases (Hong and Verma, 1994; Welters *et al.*, 1994). The study by Hong and Verma (1994) further points to a specific role for PtdIns 3-kinase(s) in membrane proliferation and formation of the peribacteroid membrane in soybean nodules.

Novel phosphoinositides

Further evidence of possible role(s) for inositol-containing glycolipid(s) in nodule development has recently been published (Perotto *et al.*, 1995), and the purification and determination of the precise chemical structure of one of these lipids is in progress (Hernandez *et al.*, 1995).

Tyrosine kinases

Further evidence for the occurrence of tyrosine phosphorylation was provided by Suzuki and Shinshi (1995), who showed that a 47-kDa myelin basic protein kinase in tobacco cells was phosphorylated on a tyrosine residue in response to treatment of the cells with a fungal elicitor.

Cytosolic Ca^{2+}-fluxes

The prediction made in Section 7 that single-cell Ca^{2+}-imaging and the use of aequorin-transformed plants would become invaluable tools for the further investigation of phosphoinositide-mediated and other events involving intracellular Ca^{2+} transients has indeed proved to be valid. The reader is referred to several elegant examples of the genre (Franklin-Tong *et al.*, 1995; Haley *et al.*, 1995; McAinsh *et al.*, 1995; Malho *et al.*, 1995; Subbaiah *et al.*, 1994).

References

Aderem A. (1992) Signal transduction and the actin cytoskeleton: the roles of MARCKS and profilin. *Trends Biochem. Sci.* **17:** 438–443.
Bairoch A, Cox JA. (1990) EF-hand motifs in inositol phospholipid-specific phospholipase C. *FEBS Lett.* **269:** 454–456.

Belunis CJ, Bae-Lee M, Kelley MJ, Carman GM. (1988) Purification and characterization of phosphatidylinositol kinase from *Saccharomyces cerevisiae. J. Biol.Chem.* **263:** 18897–18903.

Berridge MJ. (1993) Inositol trisphosphate and calcium signalling. *Nature* **361:** 315–325.

Berstein G, Blank JL, Jhon D-Y, Exton JH, Rhee SG, Ross, EM. (1992) Phospholipase C-β1 is a GTPase-activating protein for G$_{q/11}$, its physiological regulator. *Cell* **70:** 411–418.

Biffen M, Hanke DE. (1990) Polyphosphoinositidase activity in soybean membranes is Ca^{2+} dependent and shows no requirement for guanine-nucleotides. *Plant Sci.* **69:** 147–155.

Billah MM. (1993) Phospholipase D and cell signalling. *Curr. Opin. Immunol.* **5:** 114–123.

Blum W, Hinsch K-D, Schultz G, Weiler EW. (1988) Identification of GTP-binding proteins in the plasma membrane of higher plants. *Biochem. Biophys. Res. Commun.* **156:** 954–959.

Boss WF. (1989) Phosphoinositide metabolism: its relation to signal transduction in plants. In: *Second Messengers in Plant Growth and Development* (Boss WF, Morre DJ, ed.). New York: Alan R Liss, pp. 29–56.

Boss WF, Massel, MO. (1985) Polyphosphoinositides are present in plant tissue culture cells. *Biochem. Biophys. Res. Commun.* **132:** 1018–1023.

Brearley CA, Hanke DE. (1993) Pathway of synthesis of 3,4- and 4,5-phosphorylated phosphatidyl-inositols in the duckweed *Spirodela polyrhiza* L. *Biochem. J.* **290:** 145–150.

Brewin NJ, Perotto S, Kannanberg EL, *et al.* (1993) Mechanisms of cell and tissue invasion by *Rhizobium leguminosarum*: the role of cell surface interaction. In: *Advances in Molecular Genetics of Plant–Microbe Interaction, Vol. 2* (Nester EW, Verma DPS, ed.). Dordrecht: Kluwer Academic Publishers, pp. 369–380.

Carpenter CL, Cantley LC. (1990) Phosphoinositide kinases. *Biochemistry* **29:** 11147–11156.

Carter HE, Gigg RH, Law JH, Nakayama T, Weber E. (1958) Biochemistry of the sphingolipids. XI. Structure of phytoglycolipid. *J. Biol. Chem.* **233:** 1309–1314.

Cheng HF, Tao M. (1989) Purification and characterization of a phosphotyrosyl-protein phosphatase from wheat seedlings. *Biochem. Biophys. Acta* **998:** 271–276.

Cho MH, Tan Z, Erneux C, Shears SB, Boss WF. (1995) The effects of mastoparan on the carrot cell plasma membrane polyphosphoinositide phospholipase C. *Plant Physiol.* **107:** 845–856.

Cook SJ, Wakelam MJ. (1992) Phospholipases C and D in mitogenic signal transduction. *Rev. Physiol. Biochem. Pharmacol.* **119:** 13–45.

Cote GG, Crain RC. (1993) Biochemistry of phosphoinositides. *Annu. Rev. Plant Physiol. Plant Mol. Biol.* **44:** 333–356.

Dillenschneider M, Hetherington A, Graziana A, Alibert G, Berta P, Haiech J, Ranjeva R. (1986) The formation of inositol phosphate derivatives by isolated membranes from *Acer pseudoplatanus* is stimulated by guanine nucleotides. *FEBS Lett.* **208:** 413–417.

Divecha N, Banfic H, Irvine RF. (1991) The polyphosphoinositide cycle exists in the nuclei of Swiss 3T3 cells under the control of a receptor (for IGF-1) in the plasma membrane, and the stimulation of the cycle increases nuclear diacylglycerol and apparently induces translocation of protein kinase C to the nucleus. *EMBO J.* **10:** 3207–3214.

Dove SK, Lloyd CW, Drøbak BK. (1994) Identification of a phosphatidylinositol 3-hydroxy kinase in plant cells: association with the cytoskeleton. *Biochem. J.* **303:** 347–350.

Drøbak BK. (1992) The plant phosphoinositide system. *Biochem. J.* **288:** 697–712.

Drøbak BK. (1993) Phosphoinositides and intracellular signalling. *Plant Physiol.* **102:** 705–709.

Drøbak BK, Watkins PAC. (1994) Inositol(1,4,5)trisphosphate production in plant cells: stimulation by the venom peptides, melittin and mastoparan. *Biochem. Biophys. Res. Commun.* **205:** 739–745.

Drøbak BK, Ferguson IB, Dawson AP, Irvine RF. (1988) Inositol-containing lipids in suspension cultured plant cells. *Plant Physiol.* **87:** 217–222.

Drøbak BK, Watkins PAC, Valenta R, Dove SK, Lloyd CW, Staiger CJ. (1994) Inhibition of plant plasma membrane phosphoinositide phospholipase C by the actin-binding protein, profilin. *Plant J.* **6:** 389–400.

Einspahr KJ, Peeler TC, Thompson GA. (1989) Phosphatidylinositol 4,5-bisphosphate phospholipase C and phosphomonoesterase in *Dunaliella salina* membranes. *Plant Physiol.* **90:** 1115–1120.

Franklin-Tong VE, Ride JP, Franklin FCH. (1995) Recombinant stigmatic self-incompatibility (S-) protein elicits a Ca^{2+} transient in pollen of *Papaver rhoeas. Plant J.* **8:** 299–307.

Fry MJ, Waterfield MD. (1993) Structure and function of phosphatidylinositol 3-kinase: a potential second messenger system involved in growth control. *Phil. Trans. R. Soc. Lond. B.* **340:** 337–344.

Fukami K, Takenawa T. (1992) Phosphatidic acid that accumulates in platelet-derived growth factor-stimulated Balb/c 3T3 cells is a potential mitogenic signal. *J. Biol. Chem.* **267:** 10988–10993.

Goldschmidt-Clermont PJ, Machesky LM, Baldassare JJ, Pollard TD. (1990) The actin-binding protein profilin binds to PIP$_2$ and inhibits its hydrolysis by phospholipase C. *Science* **247:** 1575–1578.

Goldschmidt-Clermont PJ, Kim JW, Machesky LM, Rhee SG, Pollard TD. (1991) Regulation of phospholipase C-γ1 by profilin and tyrosine phosphorylation. *Science* **251:** 1231–1233.

Goldschmidt-Clermont PJ, Furman MI, Wachsstock D, Safer D, Nachmias VT, Pollard TD. (1992) The control of actin nucleotide exchange by thymosin β4 and profilin. A potential regulatory mechanism for actin polymerization in cells. *Mol. Biol. Cell* **3:** 1015–1024.

Hakomori S-I. (1993) Structure and function of sphingoglycolipids in transmembrane signalling and cell–cell interactions. *Biochem. Soc. Trans.* **21:** 583–595.

Haley A, Russell AJ, Wood N, Allan AC, Knight M, Campbell AK, Trewavas AJ. (1995) Effects of mechanical signalling on plant cell cytosolic calcium. *Proc. Natl Acad. Sci. USA* **92:** 4124–4128.

Harwood JL. (1980) Plant acyl lipids: structure, distribution and analysis. In: *The Biochemistry of Plants, Vol. 4.* (Stumpf PK, ed.). New York: Academic Press, pp. 1–55.

Heim S, Wagner KG. (1986) Evidence of phosphorylated phosphatidylinositols in the growth cycle of suspension cultured plant cells. *Biochem. Biophys. Res. Commun.* **134:** 1175–1181.

Hendrix KW, Qasefa HA, Boss WF. (1989) The polyphosphoinositides, phosphatidylinositol monophosphate and phosphatidylinositol bisphosphate are present in nuclei isolated from carrot protoplasts. *Protoplasma* **151:** 62–72.

Hernandez LE, Perotto S, Brewin NJ, Drøbak BK. (1995) A novel inositol-lipid in plant–bacteria symbiosis. *Biochem. Soc. Trans.* **23:** 582S.

Hetherington AM, Drøbak BK. (1992) Inositol-containing lipids in higher plants. *Prog. Lipid Res.* **31:** 53–63.

Hirayama T, Ohto C, Mizoguchi T, Shinozaki K. (1995) A gene encoding a phosphatidylinositol-specific phospholipase C is induced by dehydration and salt stress in *Arabidopsis thaliana. Proc. Natl Acad. Sci. USA* **92:** 3903–3907.

Hong Z, Verma DPS. (1994) A phosphatidylinositol 3-kinase is induced during soybean nodule organogenesis and is associated with membrane proliferation. *Proc. Natl Acad. Sci. USA* **91:** 9617–9621.

Huang C-H, Tate BF, Crain RC, Cote GG. (1995) Multiple phosphoinositide-specific phospholipases C in oat roots: characterization and partial purification. *Plant J.* **8:** 257–267.

Irvine RF. (1992) Inositol lipids in cell signalling. *Curr. Opin. Cell Biol.* **4:** 212–219.

Irvine RF, Letcher AJ, Dawson RMC. (1980) Phosphatidylinositol phosphodiesterase in higher plants. *Biochem. J.* **192:** 279–283.

Irvine RF, Letcher AJ, Lander DJ, Drøbak BK, Dawson AP, Musgrave A. (1989) Phosphatidyl-inositol(4,5)bisphosphate and phosphatidylinositol(4)phosphate in plant tissues. *Plant Physiol.* **89:** 888–892.

Ishida S, Takahashi Y, Nagata T. (1993) Isolation of cDNA of an auxin-regulated gene encoding a G-protein β-subunit-like protein from tobacco BY-2 cells. *Proc. Natl Acad. Sci. USA* **90:** 11152–11156.

Jacobs M, Thelen MP, Farndale RW, Astle MC, Rubery PH. (1988) Specific nucleotide binding by membranes from *Cucurbita pepo* seedlings. *Biochem. Biophys. Res. Commun.* **155:** 1478–1484.

Janmey PA, Lamb J, Allen PG, Matsudaira PT. (1992) Phosphoinositide-binding peptides derived from the sequences of gelsolin and villin. *J. Biol. Chem.* **267:** 11818–11823.

Kamada Y, Muto S. (1991) Ca^{2+} regulation of phosphatidylinositol turnover in the plasma membrane of tobacco suspension culture cells. *Biochim. Biophys. Acta* **1093:** 72–79.

Katz A, Wu DQ, Simon MI. (1992) Subunits of βγ of heterotrimeric G-proteins activate β2-isoform of phospholipase C. *Nature* **360:** 686–689.

Kaul K, Lester RL. (1975) Characterization of inositol-containing phosphosphingolipids from tobacco leaves. *Plant Physiol.* **55:** 120–129.

Laine RA, Hsieh TC-Y. (1987) Inositol-containing sphingolipids. *Methods Enzymol.* **138:** 186–195.

Larsen PM, Wolniak SM. (1990) 1,2-dioctanoylglycerol accelerates or retards mitotic progression in *Tradescantia* stamen hair cells as a function of the time of its addition. *Cell Motil. Cytoskel.* **16:** 190–203.

Legendre L, Yueh YG, Crain RC, Haddock N, Heinstein PF, Low PS. (1993) Phospholipase C activation during elicitation of the oxidative burst in cultured plant cells. *J. Biol. Chem.* **268:** 24559–24563.

Li Y-S, Porter FD, Hoffman RM, Deuel TF. (1989) Separation and identification of two phosphatidyl-inositol 4-kinase activities in bovine uterus. *Biochem. Biophys. Res. Commun.* **160:** 202–209.

Ma H, Yanofsky MF, Meyerowitz EM. (1990) Molecular cloning and characterization of GPA1, a G protein α subunit gene from *Arabidopsis thaliana. Proc. Natl Acad. Sci. USA* **87:** 3821–3825.

McAinsh MR, Webb AAR, Taylor JE, Hetherington AM. (1995) Stimulus-induced oscillations in guard cell cytosolic free calcium. *Plant Cell* **7:** 1207–1219.

McConville MJ, Ferguson MAJ. (1993) The structure, biosynthesis and function of glycosylated phosphatidylinositols in the parasitic protozoa and higher eukaryotes. *Biochem. J.* **294:** 305–324.

McMurray WC, Irvine RF. (1988) Phosphatidylinositol 4,5-bisphosphate phosphodiesterase in higher plants. *Biochem. J.* **249:** 877–881.

Majerus PW. (1992) Inositol phosphate biochemistry. *Annu. Rev. Biochem.* **61:** 225–250.

Malho R, Read ND, Trewavas AJ, Pais MS. (1995) Calcium channel activity during pollen tube growth and reorientation. *Plant Cell* **7:** 1173–1184.

Martiny-Baron G, Scherer GFE. (1989) Phospholipid-stimulated protein kinase in plants. *J. Biol. Chem.* **264:** 18052–18059.

Melin P-M, Sommarin M, Sandelius AS, Jergil B. (1987) Identification of Ca^{2+}-stimulated polyphosphoinositide phospholipase C in isolated plant plasma membranes. *FEBS Lett.* **223:** 87–91.

Melin P-M, Pical C, Jergil B, Sommarin M. (1992) Polyphosphoinositide phospholipase C in wheat root plasma membranes. Partial purification and characterization. *Biochim. Biophys. Acta* **1123:** 163–169.

Michell RH. (1992) Inositol lipids in cellular signalling mechanisms. *Trends Biochem. Sci.* **17:** 274–276.

Monaco ME, Gershengorn MC. (1992) Subcellular organization of receptor-mediated phosphoinositide turnover. *Endocr. Rev.* **13:** 707–718.

Moore TS. (1990) Biosynthesis of phosphatidylinositol. In: *Inositol Metabolism in Plants* (Morre DJ, Boss WF, Loewus FA, ed.). New York: John Wiley, pp. 107–112.

Moritz A, De Graan PNE, Gispen WH, Wirtz KWA. (1992) Phosphatidic acid is a specific activator of phosphatidylinositol 4-phosphate kinase. *J. Biol. Chem.* **267:** 7207–7210.

Morre DJ. (1989) Stimulus–response coupling in auxin regulation of plant cell elongation. In: *Second Messengers in Plant Growth and Development* (Boss WF, Morre DJ, ed.). New York: Alan R. Liss, pp. 29–56.

Morse MJ, Crain RC, Satter RL. (1987) Phosphatidylinositol cycle metabolites in *Samanea saman* pulvini. *Plant Physiol.* **83:** 640–644.

Munnik T, Irvine RF, Musgrave A. (1994) Rapid turnover of phosphatidylinositol 3-phosphate in the green alga *Chlamydomonas eugametos* — signs of a phosphatidylinositol 3-kinase signalling pathway in lower plants. *Biochem. J.* **298:** 269–273.

Nishizuka Y. (1992) Intracellular signalling by hydrolysis of phospholipids and the activation of protein kinase C. *Science* **258:** 607–614.

Palmgren MG, Sommarin M, Ulvskov P, Jorgensen PL. (1988) Modulation of plasma membrane H^+-ATPase from oat roots by lysophosphatidylcholine, free fatty acids and phospholipase A_2. *Physiol. Plant.* **74:** 11–19.

Parmar PN, Brearley CA. (1993) Identification of 3- and 4-phosphorylated phosphoinositides and inositol phosphates in stomatal guard cells. *Plant J.* **4:** 255–263.

Perotto S, Donovan N, Drøbak BK, Brewin NJ. (1995) Differential expression of a glycosyl inositol phospholipid antigen on the peribacteroid membrane during pea nodule development. *Mol. Plant Microb. Interact.* **8:** 560–568.

Pfaffmann H, Hartmann E, Brightman AO, Morre DJ. (1987) Phosphatidylinositol specific phospholipase C of plant stems. *Plant Physiol.* **85:** 1151–1155.

Pical C, Sandelius AS, Melin P-M, Sommarin M. (1992) Polyphosphoinositide phospholipase C in plasma membranes of wheat (*Triticum aestivum* L.). *Plant Physiol.* **100:** 1296–1303.

Rhee SG. (1991) Inositol phospholipid-specific phospholipase C: interaction of the γ1 isoform with tyrosine kinase. *Trends Biochem. Sci.* **16:** 297–301

Rincon M, Boss WF. (1990) Second-messenger roles of phosphoinositides. In: *Inositol Metabolism in Plants* (Morre DJ, Boss WF, Loewus FA, ed.). New York: Wiley-Liss, pp. 173–200.

Sandelius AS, Sommarin M. (1986) Phosphorylation of phosphatidylinositols in isolated plant membranes. *FEBS Lett.* **201:** 282–286.

Sandelius AS, Sommarin M. (1990) Membrane-localized reactions involved in polyphosphoinositide turnover in plants. In: *Inositol Metabolism in Plants* (Morre DJ, Boss WF, Loewus FA, ed.). New York: Wiley-Liss, pp. 139–161.

Scherer GFE, Andre B. (1993) Stimulation of phospholipase A_2 by auxin in microsomes from suspension-cultured soybean cells is receptor-mediated and influenced by nucleotides. *Planta* **191:** 515–523.

Schu PV, Takegawa K, Fry MJ, Stack JH, Waterfield, MD, Emr SD. (1993) A yeast gene essential for protein sorting in the secretory pathway encodes a phosphatidylinositol 3-kinase. *Science* **260:** 88–91.

Shi J, Dixon RA, Gonzales RA, Kjellbom P, Bhattacharyya MK. (1995) Identification of cDNA clones encoding valosin-containing protein and other plant plasma membrane-associated proteins by a general immunoscreening strategy. *Proc. Natl Acad. Sci. USA* **92:** 4457–4461.

Simon MI, Strathmann MP, Gautam N. (1991) Diversity of G proteins in signal transduction. *Science* **252:** 802–808.

Sommarin M, Sandelius AS. (1988) Phosphatidylinositol and phosphatidylinositolphosphate kinases in plant plasma membranes. *Biochim. Biophys. Acta* **958:** 268–278.

Staiger CJ, Goodbody KC, Hussey PJ, Valenta R, Drøbak BK, Lloyd CW. (1993) The profilin multigene family of maize: differential expression of 3 isoforms. *Plant J.* **4:** 631–641.

Stephens LR, Hughes KT, Irvine RF. (1991) Pathway of phosphatidylinositol (3,4,5)trisphosphate synthesis in activated neutrophils. *Nature* **351:** 33–39.

Strasser H, Hoffman C, Grisebach H, Matern U. (1986) Are polyphosphoinositides involved in signal transduction of elicitor-induced phytoalexin synthesis in cultured plant cells? *Z. Naturforsch.* **41:** 717–724.

Subbaiah CC, Bush DS, Sachs MM. (1994) Elevation of cytosolic calcium preceded anoxic gene expression in maize suspension-cultured cells. *Plant Cell* **6:** 1747–1762.

Suzuki K, Shinshi H. (1995) Transient activation and tyrosine phosphorylation of a protein kinase in tobacco cells treated with a fungal elicitor. *Plant Cell* **7:** 639–647.

Tate BF, Schaller GE, Sussman MR, Crain RC. (1989) Characterization of a polyphosphoinositide phospholipase C from the plasma membrane of *Avena sativa. Plant Physiol.* **91:** 1275–1279.

Torruella M, Casano LM, Vallejos RH. (1986) Evidence of activity of tyrosine kinase(s) and of the presence of phosphotyrosine in pea plantlets. *J. Biol. Chem.* **261:** 6651–6653.

Valenta R, Duchene M, Pettenburger K, *et al.* (1991) Identification of profilin as a novel pollen allergen; IgE autoreactivity in sensitized individuals. *Science* **109:** 619–626.

Welters P, Takegawa K, Emr SD, Chrispeels MJ. (1994) ATVPS34, a phosphatidylinositol 3-kinase of *Arabidopsis thaliana*, is an essential protein with homology to a calcium-dependent lipid-binding domain. *Proc. Natl Acad. Sci. USA* **91:** 11398–11402.

Wheeler JJ, Boss WF. (1990) Inositol lysophospholipids. In: *Inositol Metabolism in Plants* (Morre DJ, Boss WF, Loewus FA, ed.). New York: Wiley-Liss, pp. 163–172.

Yamamoto YT, Conkling MA, Sussex IM, Irish VF. (1995) An *Arabidopsis* cDNA related to animal phosphoinositide-specific phospholipase C genes. *Plant Physiol.* **107:** 1029–1030.

Yang W, Burkhart W, Cavallius J, Merrick WC, Boss WF. (1993) Purification and characterization of a phosphatidylinositol 4-kinase activator in carrot cells. *J. Biol. Chem.* **268:** 392–398.

Yotsushima K, Mitsui T, Takaoka T, Hayakawa T, Igaue I. (1993) Purification and characterization of membrane-bound inositol phospholipid-specific phospholipase C from suspension-cultured rice (*Oryza sativa* L.) cells. *Plant Physiol.* **102:** 165–172.

Xu P, Lloyd CW, Staiger CJ, Drøbak BK. (1992) Association of phosphatidylinositol 4-kinase with the plant cytoskeleton. *Plant Cell* **4:** 941–951.

Zabulionis RB, Atkinson BG, Procunier JD, Walden DB. (1988) Maize (*Zea mays* L.) DNA sequences homologous to the oncogenes *myb, ras* and *src. Genome* **30:** 820–824.

Jasmonates: global regulators of plant gene expression

Gary J. Loake

1. Role of membranes in cell signalling

In addition to delineating the boundary of the cytoplasm, the plasma membrane plays a central role in the interaction of the cell with its external environment. It provides a matrix for a plethora of specific protein receptors that constitute the first components of cellular signalling pathways, which allow the cell to respond to a profusion of diverse external cues. In response to some stimuli, the plasma membrane itself may play a direct role in the signalling process, via the receptor-dependent release of intracellular signals derived from plasma membrane lipids. These signals may initiate a complex series of cellular events, often resulting in the modulation of gene expression.

Lipid-based signals have been studied extensively in animal systems, where they orchestrate a wide spectrum of cellular responses. Perhaps the best characterized is the release of inositol 1,4,5-triphosphate ($InsP_3$) from phosphatidylinositol-4,5-biphosphate present within the plasma membrane, which results in Ca^{2+} release following $InsP_3$ binding to an endomembrane Ca^{2+} channel (Berridge, 1993). Other lipid-derived signals include the prostaglandins and leuko-trienes, a large group of signal molecules which stimulate a spectrum of cellular responses, including smooth muscle contraction and platelet aggregation (Bailey, 1985; Sigal, 1991). In contrast, lipid-based signalling in plant cells is a relatively nascent field. However, rapidly accumulating evidence suggests that lipid-derived signals may play a central role in plant growth and development and responses to environmental stimuli. Advances in our understanding of lipid-based signalling pathways in plant cells may therefore have a major impact within plant biology.

2. Jasmonates as global regulators of gene expression

Jasmonates are naturally occurring compounds that have been identified in a wide variety of plant species (*Figure 1*). These C12 fatty acids, derived from membrane lipids, are now widely regarded as endogenous plant growth substances, which play key roles in plant growth, development and responses to environmental stresses.

The signal for induction of potato tuber development has been proposed to be a jasmonate family member. Tuberonic acid β-glucoside is synthesized in the leaves and transported to the stolons, where it is thought to activate tuberization (Yoshikara *et al.*, 1989). The touch-induced coiling of the climbing plant *Bryonia dioica*, in addition to some of the developmental changes induced by coiling, are thought to be transduced through jasmonates (Falkenstein *et al.*, 1991). Moreover, the application of relatively high concentrations of exogenous jasmonic acid (JA) has been shown to induce senescence and growth inhibition in a variety of plant species (Ueda and Kato, 1980). Recently, a mechanism to explain this phenomenon has been proposed which

215

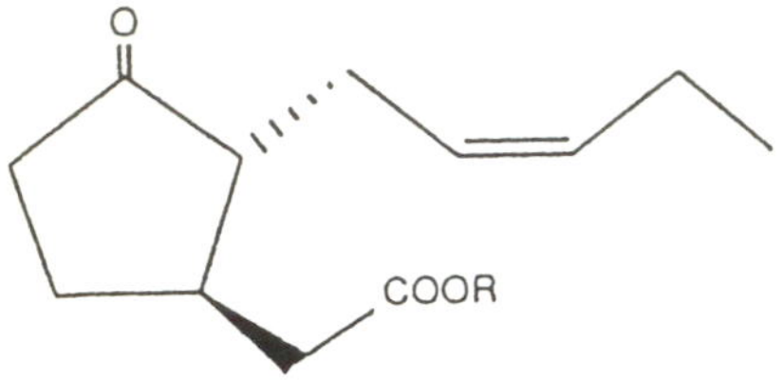

Figure 1. Structure of jasmonic acid (JA) and methyl jasmonate (Me-JA). For JA, R is H, for Me-JA, R is CH_3.

suggests that extended accumulation of JA results in 'marking' of ribosomes for cleavage and inactivation (Reinbothe *et al.*, 1994).

Jasmonates are also thought to underpin a variety of plant stress responses. In response to water deficit, JA rapidly accumulates in dehydrating tissue, where it is proposed to transduce transient responses to reduced turgor (Creelman and Mullet, 1995). Moreover, JA accumulation occurs in response to mechanical wounding of leaf tissue. In this context there is strong evidence that JA and methyl jasmonate (Me-JA) are important local wound signals (Farmer and Ryan, 1990).

The accumulating experimental data are therefore consistent with a central role for jasmonates in the biology of plants. The biosynthesis and mode of action of these compounds therefore provide an important paradigm for the study of lipid-based signalling in plant cells.

3. Jasmonate biosynthesis

The precursor for the biosynthesis of jasmonates is presumed to be 18:3 linolenic acid (*Figure 2*). Lipid analysis of plant cell membranes has revealed this fatty acid to be present in large quantities (Roy *et al.*, 1995). The presence of free fatty acid within the cytoplasm is negligible. Therefore, the first step in JA biosynthesis must be the release of the unsaturated fatty acid precursor, linolenic acid, from the cell membrane. The identity of the cellular membrane providing the source of linolenic acid is at present unclear, but it is thought to originate from either the plasma membrane, the chloroplast membranes, or both. Linolenic acid is characteristically found attached at the *sn*-2 position of plant plasma membrane phospholipids, requiring the activity of either a membrane-associated lipolytic acyl hydrolase or phospholipase A_2 activity for release (Croft *et al.*, 1990). The mechanism underlying the activation of phospholipase activity in response to external stimuli is at present unknown.

The released fatty acids are oxidized by the action of lipoxygenase (LOX). The expression of genes encoding this enzyme is activated by a number of environmental stimuli, preceding increases in enzyme activity (Creelman *et al.*, 1992). LOX catalyses the incorporation of molecular oxygen into polyunsaturated fatty acids (Vick and Zimmerman, 1984). Fatty acid substrates for LOX must contain a *cis,cis*-1,4-pentadiene structure. In the case of JA biosynthesis, a specific 13-LOX incorporates oxygen into the fatty acid at the n-13 position of linolenic acid. The product of the reaction is 13(S)-hydroperoxylinolenic acid, a conjugated hydroperoxydiene, in which the *cis* double bond attacked by oxygen moves into conjugation with a neighbouring *cis* double bond (Vick and Zimmerman, 1980). During this process, the migrating double bond assumes the *trans*configuration. The carbon bearing the hydroperoxide group has the S configuration.

Lipoxygenases are widely distributed among plant organs. Seeds are a particularly rich source of LOX activity, which presumably reflects the massive mobilization of lipid reserves following germination. Vegetative lipoxygenases are found in the cotyledons of germinating

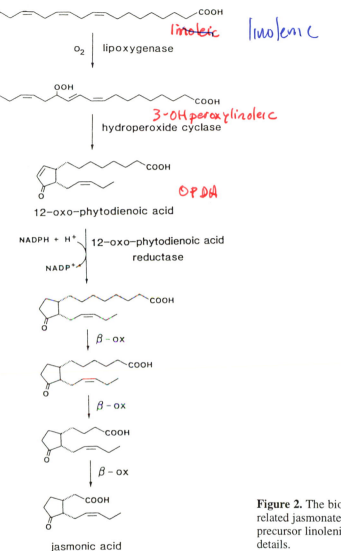

Figure 2. The biosynthesis of JA and related jasmonates from the fatty acid precursor linolenic acid. See text for details.

seedlings and in young leaves and stems (Galliard and Chan, 1980). Immunofluorescent studies have demonstrated the presence of LOX activity in the parenchyma cells below the epidermis, in the cortex parenchyma and, intriguingly, in the vascular cylinder (Vernooygerritsen *et al.*, 1982). The location of LOX activity in a variety of plant organ and cell types suggests a high division of labour among plant lipoxygenases. Therefore, there are likely to be multiple isoforms of this enzyme encoded by a relatively large gene family.

Despite a number of detailed studies, the precise subcellular location of LOX is still ambiguous, reflecting conflicting experimental results. However, it is likely that different LOX isoforms occur in different cellular locations, possibly reflecting their potentially different biochemical roles.

Hydroperoxide cyclase activity was first detected in flax seed by Zimmerman and Feng in 1978 (Vick and Zimmerman, 1979). These investigators reported the formation of a cyclic fatty acid from the reaction of linolenic acid with flax seed extract. An analogous cyclic product formed from linoleic acid was not found. The enzyme responsible for catalysing the reaction was called hydroperoxide cyclase, and the product formed was given the common name 12-oxo-phytodienoic acid (12-oxo-PDA). This enzyme is found throughout the plant kingdom. Hydroperoxide cyclase activity is present in a variety of tissues including the leaves of developing seedlings and ripening fruits.

Oxygen isotope labelling studies have shown that the 12-oxo oxygen atom arises from the 13-hydroperoxy oxygen atom (Vick and Zimmerman, 1980). The proposed mechanism involves the formation of an epoxy-cation intermediate, followed by abstraction of a proton at carbon 12 to generate an enolate anion. The rearrangement of this intermediate would produce cyclization between carbon 9 and carbon 13.

In plants, 12-oxo-PDA reductase catalyses the reduction of the double bond in the cyclo-pentenone ring to form 3-oxo-2-(2'-pentenyl) cyclopentaneoctanoic acid. It is thought that NADPH provides the necessary reductant (Vick and Zimmerman, 1983). 12-oxo-PDA reductase from corn kernels has a molecular weight of approximately 54 000, and has a broad pH range, from 6.8 to 9.0.

The identification of intermediates in the pathway which have hexanoic and butanoic acid side-chains led to the proposal that chain shortening to produce the 12-carbon product, JA, occurs by three cycles of β-oxidation. The final product, (1R,2S)-3-oxo-2-(2'-pentenyl)cyclopentane-acetic acid (jasmonic acid), retains the same *cis* stereoconfiguration of the side-chains as the parent 12-oxo-PDA. The *cis* form of JA is easily isomerized to the *trans* form, indeed originally JA was the term used to denote the *trans* form of the compound, which, unknown to the observers, had isomerized from the *cis* form during its isolation (Demole *et al.*, 1962).

4. Jasmonate-induced developmental responses

JA and Me-JA are powerful inhibitors of germination and plant growth. The application of Me-JA to *Avena* and *Lactuca* seeds inhibits germination (Satler and Thimann, 1981). Moreover, the exogenous application of relatively high concentrations of JA (10^{-3} M) to rice, lettuce and wheat seedlings induces a growth inhibition of over 90% (Yamane *et al.*, 1981a). In this regard, the physiological events induced by JA resemble those of the plant hormone abscisic acid (ABA), although ABA is biologically active at significantly lower concentrations.

Me-JA at 10^{-6} M is a powerful inhibitor of primary root growth in *Arabidopsis* (Staswick *et al.*, 1992). This phenomenon facilitated the development of a rapid screen to identify Me-JA-resistant mutants in this species.

Pollen germination is also influenced by JA. In *Impatiens,* 10^{-5} M JA actively inhibited pollen tube growth (Yamane *et al.*, 1981b). In contrast, however, *Lilium* pollen tube growth was strongly stimulated by 10^{-7} M JA (Sembdner and Gross, 1986).

The induction of leaf senescence by JA produces a rapid degradation of chlorophyll and ribulose-1,5-bisphosphate carboxylase (Rubisco), and a concomitant decrease in photosynthetic activity (Koda, 1992). These events follow an increase in protease and peroxidase activities. The expression of genes encoding proteins in photosynthetic carbon assimilation is suppressed by Me-JA. In both the nucleocytoplasmic and the plastidic compartments of barley leaf cells, Me-JA affects various steps in the formation of chloroplast proteins, such as the small and large subunits of Rubisco, the light-harvesting chlorophyll *a/b* binding proteins, and the 65- and 68-kDa proteins of photosystem II (Reinbothe *et al.*, 1994). Presumably, it is these effects which promote symptoms of senescence within the plastid compartment characterized by chlorophyll loss and Rubisco degradation.

Transcription rate determinations using nuclear run-off assays from isolated nuclei have demonstrated that Me-JA suppresses the expression of *rbcs* and *cab* genes at the transcriptional level (Reinbothe *et al.*, 1994). Less is known about the effect of jasmonates on post-transcriptional events. However, the *rbcl* transcript of barley encoding the Rubisco large subunit has been shown to be modified at its 5′ end. In JA-treated tissue, two shorter transcript sizes can be identified. Interestingly, the induction of leaf senescence by JA can be counteracted by the application of cytokinins, which are known to delay the onset of senescence in leaf tissue (Weidhase *et al.*, 1987).

JA has been shown to promote the biosynthesis of ethylene during fruit ripening in tomato (Saniewski *et al.*, 1987). In this context it has been demonstrated to increase the activity of 1-aminocyclopropane-1-carboxylic acid (ACC) synthase and ACC oxidase, key enzymes in the biosynthesis of ethylene. Whether jasmonates play a more general role in fruit ripening has yet to be determined.

There is some evidence that JA and Me-JA are involved in thigmotropism (Falkenstein *et al.*, 1991). *Bryonia dioica* is a dioecious climbing plant belonging to the Cucurbitaceace. Both sexes produce tendrils which coil around objects so providing support for plant growth. A touch stimulus applied to the ventral side of the tendril promotes coiling; in contrast, a stimulus to the dorsal side inhibits this response. The application of JA and jasmonate biosynthetic pathway intermediates induces the coiling response. Lignification of a zone of sclerenchyma cells termed the Bianconi plate, in addition to proliferation of a layer of collenchyma cells around this plate, is promoted during the coiling response. These developmental changes are also activated by JA.

Data have accumulated which suggest a physiological role for JA in tuber development in a number of plant species, including potato, yam and *Helianthus tuberosus*. It is postulated that JA may elaborate tuber development by disrupting intracellular microtubules, facilitating the radial cell expansion which precedes the development of the tuber (Abe *et al.*, 1990). The tuber-inducing stimulus in potato is thought to be a jasmonate family member (Yoshikara *et al.*, 1989). Even low levels of tuberonic acid β-glucoside provide a powerful signal for tuber formation. This compound is synthesized in the leaves and transported downwards to the stolons. It is interesting that the β-glucoside conjugate rather than the free acid is the most active compound. Conjugated fatty acids may therefore also be important signalling compounds in plant cells.

5. Jasmonate-induced genes and proteins

The exogenous application of jasmonic acid to plant tissues induces the synthesis of a plethora of proteins in all plant species tested thus far. These proteins are known as jasmonate-induced proteins (JIPs) (Parthier, 1991). The identity of the vast majority of JIPs is still unknown, and even some of the most abundant members of this class of proteins are characterized only by their molecular weights. Most of the JIPs of known function can be placed into one of two classes: plant storage proteins, or proteins involved in the defence of the plant cell against microbial pathogens or insect herbivores.

Among the storage protein class, the soybean vegetative storage proteins (VSPs) have been most extensively studied (Anderson *et al.*, 1989). These proteins were originally discovered by Wittenbach as 28-kDa and 32-kDa proteins that accumulate as temporary sources of nitrogen in response to various stresses, including wounding, seed pod removal and water deficit (Wittenbach, 1983). VSPs accumulate in vegetative organs, for example in the epidermis and paraveinal mesophyll cells of leaves (Staswick, 1990). Me-JA can induce VSPs in all cell types, therefore overriding cell-type-specific gene expression (Huang *et al.*, 1991). During vegetative organ maturation, VSPs are gradually degraded and the nitrogen is translocated into developing organs, such as flower buds and ripening fruits, which act as sinks. Recently, specific VSPs and their corresponding genes have been sequenced, revealing that at least some VSPs are lipoxy-genases and acid phosphatases (Aarts *et al.*, 1991; Tranbarger *et al.*, 1991). Elevated expression of the soybean *vsp* in seedling hypocotyl hooks and lower expression in roots was paralleled by higher concentrations of Me-JA in hooks and lower levels in roots. Moreover, inhibitors of lipoxygenase that have the potential to inhibit jasmonate synthesis also block *vsp* mRNA accumulation. The experimental evidence therefore strongly suggests the modulation of these genes by endogenous Me-JA (Staswick *et al.*, 1991).

Zenk and co-workers have shown that the addition of JA to plant cell cultures derived from a wide variety of different species results in the rapid accumulation of species-specific phyto-alexins (Gundlach *et al.*, 1992). These secondary plant products are antimicrobial compounds of central importance to the plant defence response (Loake *et al.*, 1992a). Species-specific phyto-alexins rapidly accumulate in response to pathogen infection. A rapid and transient increase in phenylalanine ammonia-lyase (PAL) activity in response to exogenous JA application has also been demonstrated. This key enzyme catalyses the first committed step in the phenylpropanoid biosynthetic pathway, from which all phytoalexins are derived. The induction of PAL and the accumulation of the phytoalexin end-products in response to JA clearly suggests that other pathogen-inducible biosynthetic enzymes integral to the pathway of phytoalexin production must also be induced in response to jasmonates.

A further important observation in this study was that fungal cell wall fragments, the classical elicitors of phytoalexin biosynthesis, induced a massive and transient increase in intracellular JA levels, which preceded PAL induction and the subsequent accumulation of phytoalexins. These observations suggest that JA may function as an intracellular signal inducing the biosynthesis of phytoalexins.

In soybean, jasmonates are potent inducers of a gene encoding chalcone synthase (CHS), which catalyses the first committed step in flavonoid biosynthesis. Flavonoid secondary plant products are found in all higher plants and function as UV-protectants, signal molecules in plant–microbe interactions, antimicrobial compounds, and pigments in flowers, fruits and seed coats (Schmid *et al.*, 1990). Moreover, hydroxymethylglutaryl CoA reductase, a key enzyme in the biosynthesis of sesquiterpenoid secondary products, is also activated in response to JA. Signals that induce the biosynthesis of numerous secondary plant products are therefore thought to be transduced through jasmonates.

Other stress-related proteins, such as antifungal thionins (Andresen *et al.*, 1992), insecticidal proteinase inhibitors (Farmer *et al.*, 1992) and proline-rich cell wall proteins, in addition to antimicrobial secondary plant products, have also been shown to accumulate in response to JA and Me-JA. Jasmonates may therefore play a central role in the intracellular signalling cascades that activate inducible plant defences.

6. Modulation of jasmonate signals

The application of gas chromatography–mass spectroscopy (GC–MS) techniques has facilitated the quantitation of JA accumulation in response to environmental stimuli (Creelman and Mullet, 1995). Recently, a monoclonal antibody has been raised against JA and an enzyme-linked immunosorbent assay (ELISA) detection method developed (Albrecht *et al.*, 1993). The accumulation of JA has been measured for a number of environmental cues. *Rauvolfia* suspension cultures treated with a yeast cell wall elicitor produced a relatively rapid and transient rise in JA (Gundlach *et al.*, 1992). In contrast, JA accumulation continued for 24 h in wounded soybean hypocotyl tissue (Creelman *et al.*, 1992). The profile of JA accumulation is therefore stimulus- and possibly tissue-dependent.

The metabolism of jasmonates has been followed using [^{14}C]dihydro-JA in barley shoots (Meyer *et al.*, 1989). Specific conjugates have been detected with glucose introduced at carbons 11 or 12, suggesting the presence of a JA-selective glucosyltransferase. Moreover, conjugates of dihydro-JA with a number of other molecules, including the hydrophobic amino acids leucine, isoleucine and valine, have also been detected (Meyer *et al.*, 1989).

Conjugation is a characteristic of many signal molecules, including auxin, cytokinin, gibberellin and salicylic acid, which is an important signal in the activation of systemic acquired resistance (Ryals *et al.*, 1994). The rapid metabolism of plant signal molecules is obviously crucial for maintaining their transient nature. The rapid release of free JA from glucose-conjugated forms, by the action of a JA-selective glucosidase, may play a key functional role in the augmentation of *de novo* JA biosynthesis in response to environmental cues. Therefore, specific glucosyltransferase and glucosidase enzymes may play central roles in the modulation of jasmonate signals.

7. Genetic dissection of the jasmonate-signalling pathway

The isolation of mutants within a genetically tractable system such as that of *Arabidopsis thaliana* provides a powerful technique for the genetic dissection of a specific signalling pathway. Moreover, the recent advances in genome mapping and chromosome walking techniques, coupled with the availability of T-DNA tagged lines, are facilitating the cloning of the corresponding genes. A number of mutants compromised in JA signalling have recently been described (Feys *et al.*, 1994; Staswick *et al.*, 1992). Staswick *et al.,* (1992) employed a mutant selection scheme based on the ability of JA to inhibit primary root growth. When *Arabidopsis* seedlings were grown on an agar medium containing 0.1 µM Me-JA, root growth was inhibited by 50%. Screening of 35 000 seedlings identified four mutants which did not show root growth inhibition; these were designated jasmonate resistant (*jar*) mutants. Progeny analysis revealed that *jar-1* segregated as a single recessive gene. ABA and Me-JA can mediate similar effects on some developmental responses, including germination inhibition. However, *jar-1* seeds were more sensitive rather than resistant to germination inhibition by ABA, demonstrating that they were distinct from ABA-insensitive mutants. Moreover, the *Arabidopsis* ABA-insensitive mutants *abi-1*, *abi-2* and *abi-3* had little or no resistance to ABA, again suggesting that ABA and Me-JA modulate plant development via independent signalling pathways (Staswick, *et al.*, 1992). Analysis of adult plants demonstrated that the *jar-1* mutant was compromised in the activation of Me-JA inducible polypeptides. This mutant therefore exerted pleiotropic effects.

A phenotypically distinct JA-insensitive mutant was described by Feys *et al.* (1994). This mutant, designated *coi-1*, was identified by screening *Arabidopsis* seedlings for their resistance to the *Pseudomonas syringae* phytotoxin, coronatine. This chlorosis-inducing toxin, produced by several pathovars of *P. syringae,* is important for infection because mutants that do not produce it show reduced virulence. Coronatine has diverse effects on plants: it promotes senescence in tobacco (Kenyon and Turner, 1992), inhibits root growth in wheat (Sakai, 1980) and stimulates stomatal opening in ryegrass (Mino *et al.*, 1987). All of these responses are stimulated by JA. Moreover, there is also some similarity between the structure of jasmonates and that of coronatine. Both compounds possess a cyclopentanone ring, and the keto group at C3 of JA is also found in coronatine. Mutants of *Arabidopsis* that are resistant to growth inhibition by coronatine (*coi*) were therefore characterized with respect to their responses to JA. The pleiotropic, recessive *coi-1* mutant was resistant to root inhibition by JA and was compromised in the activation of JA-inducible polypeptides. These experiments suggested that the phytotoxin coronatine mediated at least some of its responses by transducing signals through the JA-signalling pathway. Intriguingly, *coi-1* mutants were male sterile. Stamens of *coi-1* flowers had shorter filaments than

those of wild-type flowers. In contrast, the stigma and style appeared normal and siliques developed normally when wild-type *Arabidopsis* flowers donated pollen. The anthers from *coi-1*
flowers were perturbed in anther dehiscence, and protein profiles from *coi-1* and wild-type
flowers revealed the absence of a number of polypeptides from *coi-1* flowers. These observations
suggest that jasmonates may play an important role in orchestrating the development of structures that are integral to successful plant reproduction.

The further characterization of *jar*, *coi* and other JA-insensitive *Arabidopsis* mutants and the
subsequent isolation of the corresponding genes should significantly contribute to our
understanding of the role of JA in plant development and environmental interactions.

8. Role of jasmonates in wound signalling

Insect grazing on a tomato leaf results in local leaf injury, which serves as a signal to activate a
plethora of inducible defence responses. One such response is the transcriptional activation of
genes encoding serine proteinase inhibitors (*pin*), which perturb the digestive system of the
feeding insects (Farmer *et al.*, 1992). This response occurs not only in leaves under direct local
attack, but also in unwounded systemic leaves. Both local and long-range systemic signals are
therefore elaborated, which results in the activation of *pin* gene expression throughout the plant.

The action of cutting a tomato stem did not induce the appearance of proteinase inhibitors in
the leaves. This facilitated the development of a simple bioassay for compounds that, when
applied to the cut stem or petiole, would enter the transpiration stream and trigger the expression
of *pin* gene expression in the leaves. Using this bioassay a number of chemicals have been
proposed to function as wound signals. These include plant cell wall pectic fragments (Bishop *et
al.*, 1981), especially oligogalacturonides, the plant growth regulator ABA (Pena-Cortes *et al.*,
1989), a small 18-amino-acid peptide termed systemin, derived from a larger prosystemin
precursor (Pearce *et al.*, 1991) and JA and Me-JA (Farmer and Ryan, 1990).

The application of 10^{-6} M JA to the cut stem of a tomato plant induces a massive
accumulation of *pin* mRNA in the leaves, to a level significantly higher than that achieved by
mechanical wounding itself. This observation suggests that JA may be a key signalling molecule
in the activation of *pin* gene expression. The significance of JA in wound signalling was
highlighted when it was demonstrated that spraying of 10^{-3} M Me-JA directly on to the surface of
tomato leaves activated high levels of *pin* gene expression (Farmer and Ryan, 1992). Moreover,
fatty acid intermediates of JA biosynthesis, such as linolenic acid and phytodienoic acid, were
also found to activate *pin* gene expression when applied in a similar fashion. The specificity of
this response was confirmed by the failure of stearic and arachidonic acids to elaborate *pin*
induction, thus demonstrating that the response was not merely a general effect of fatty acids
per se. Furthermore, a structural specificity of hydroperoxides and linolenic acid was shown.
Only 13(S)-hydroperoxylinolenic acid, the intermediate in JA biosynthesis, and not 13(S)-
hydroperoxylinoleic acid, a structurally similar fatty acid, could induce *pin* mRNA accumulation.
In a similar manner, linolenic acid was shown to induce *pin* gene expression, but the structural
analogue γ-linolenic did not (Farmer and Ryan, 1992).

These results suggested that the applied JA biosynthetic intermediates were being converted
directly to JA within the unwounded tomato leaf. The enzymes responsible for JA biosynthesis
may therefore be continually present, at least at a basal level, within leaf tissue. Hence the rate-
determining step in jasmonate biosynthesis is likely to be the release of the fatty acid precursor
linolenic acid following the wound-induced activation of a membrane-associated lipase.

A number of studies have revealed that mechanical leaf injury results in a massive and
transient accumulation of JA and Me-JA in leaf tissue of tomato and soybean (Creelman *et al.*,
1992; Pena-Cortes *et al.*, 1993). These observations are particularly significant, as they suggest

a genuine physiological role for JA as a signalling molecule following mechanical wounding (*Figure 3*). Direct evidence that JA is an essential component of this signal transduction pathway was provided by an elegant series of experiments employing mammalian anti-inflammatory drugs. The cyclo-oxygenase inhibitors aspirin and propyl gallate, and the lipoxygenase inhibitors salicylhydroxamic acid (SHAM) and ZK139817, can effectively block the accumulation of JA following mechanical wounding (Pena-Cortes *et al.*, 1993). Importantly, when JA biosynthesis is blocked by application of these enzyme inhibitors and the plant is subsequently injured, no *pin* gene activation occurs in the leaves. These experiments demonstrate a direct correlation between JA accumulation and subsequent *pin* gene activation, providing compelling evidence that JA is an essential component of the wound-signalling pathway.

Studies have suggested that the systemic wound signal may not be chemical in nature. An electrical activation potential has been shown to exit the leaf following wounding, and the appearance of this electrical signal was found to correlate with the systemic induction of *pin* gene expression in distal leaves (Wildon *et al.*, 1992).

Although translocation of jasmonates has been shown to take place in the phloem (Ryan, 1992), the mode of biosynthesis of JA suggests that it is unlikely to be a systemic wound signal. An integrated model combining the putative wound signal candidates is now beginning to

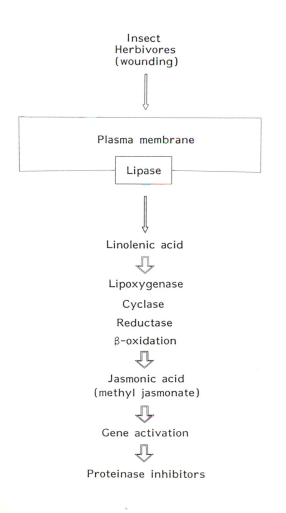

Figure 3. Model showing a possible mechanism for the production of intracellular jasmonate signals in response to mechanical injury.

emerge. Mechanical wounding may generate changes in plasma membrane potential, creating electrical signals which could exit the injured leaf and rapidly travel to distal unwounded leaves. Within systemic leaves, the propagated electrical signal may activate a mechanism for the release of systemin from prosystemin, possibly via the action of a specific protease. Released systemin may subsequently stimulate local ABA accumulation, which could activate the release of linolenic acid from the plasma membrane, initiating jasmonate biosynthesis, with the resulting rise in the intracellular concentration of jasmonates ultimately activating the expression of wound-inducible genes (*Figure 4*).

Jasmonates are therefore thought to function as intracellular signals near the terminus of the wound-signalling pathway in systemic leaves. Interestingly, the induction of wound-inducible genes by JA and Me-JA is inhibited by cyclohexamide treatment. This result suggests that *de*

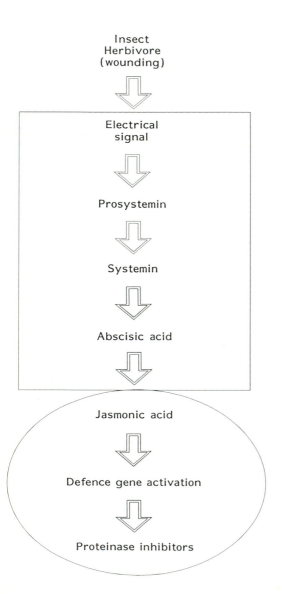

Figure 4. Emerging model integrating a number of different signals which have been implicated in the systemic signalling pathway coupling mechanical injury to the activation of proteinase inhibitor gene expression. The extracellular systemic signalling pathway is shown within the rectangle. Intracellular signalling events are shown within the circle.

novo protein synthesis is required subsequent to JA accumulation for transduction of the wound signal. This may be explained by the rapid activation of specific transcription factors following JA accumulation, which could bind conserved *cis* elements within the promoters of wound-inducible genes, activating their expression in response to mechanical injury.

9. Role of jasmonates in disease resistance

The role of jasmonates in plant disease resistance is at present not well understood. Fungal elicitor preparations induce both JA and Me-JA accumulation in a large number of cell cultures (Gundlach *et al.*, 1992). Many elicitor-inducible genes are also activated by JA and Me-JA (Creelman *et al.*, 1992; Gundlach *et al.*, 1992). The data are therefore consistent with the proposal that the jasmonate-signalling pathway transduces the signal following elicitor perception. Moreover, a number of key plant defence genes, such as *pal*, *chs*, *lox* and those which encode proline-rich cell wall proteins (PRPs) are induced following both microbial pathogen attack and exogenous application of JA and Me-JA, further supporting a role for jasmonate signalling in the response to microbial plant pathogens.

The application of jasmonate sprays for the protection of plants from microbial attack has been investigated in some systems. Spraying jasmonic acid on potato plants induced local and systemic protection against subsequent infection by the late blight fungus *Phytophthora infestans*. Further studies showed that JA had a direct effect on the germination of *P. infestans* sporangia. However, systemic leaves which were not directly sprayed with JA were also protected, suggesting that the mechanism of resistance was at least in part conveyed by defence gene induction (Cohen *et al.*, 1993).

Intriguingly, a JA-inducible protein in barley, JIP60, has been shown to cleave 'marked' ribosomes isolated from tissues exposed to JA for long periods (Reinbothe *et al.*, 1994). It has been proposed that the accumulation of JA resulting from an incompatible interation could trigger ribosome cleavage by JIP60. This mechanism could augment the proposed programmed cell death response which results in the phenotypic expression of the hypersensitivity which functions to isolate the pathogen in a zone of dead host cells, so preventing further pathogen spread. It will be interesting to test this hypothesis further experimentally.

An incompatible interaction is often followed by the subsequent induction of a durable resistance against a broad spectrum of pathogens throughout the plant. This phenomenon has been termed systemic acquired resistance (SAR) (Ryals *et al.*, 1994). The phenylpropanoid derivative salicylic acid (SA) has been shown to play a key role in orchestrating the local development of SAR. Moreover, SA accumulation has recently been demonstrated to be essential for successful gene-for-gene mediated resistance (Delaney *et al.*, 1994). Recent evidence suggests that SA and other related hydroxybenzoic acids are powerful inhibitors of both the biosynthesis of JA and its subsequent perception. SA is thought to inhibit selectively the enzyme 13(S)-hydroperoxide dehydrase, which converts 13(S)-hydroperoxylinolenic acid to 12-oxo-phytodienoic acid (Pena-Cortes *et al.*, 1993). Moreover, SA has been shown to inhibit the induction of the JA-inducible gene *pin-II* following exogenous JA application; therefore SA also inhibits JA perception (Doares *et al.*, 1995). These results suggest that SA may block JA signalling during the SA-dependent development of SAR and the later stages of gene-for-gene mediated resistance. SA accumulation may therefore provide a mechanism for 'cross-talk' between the SA and JA signalling pathways. The expression of plant defence genes, which are induced in response to both wounding and pathogen challenge (e.g. those encoding the enzymes of central phenylpropanoid metabolism), may therefore be under the control of at least two distinct defence signalling pathways.

Limited pathogen infection studies employing the *Arabidopsis coi-1* mutant have been undertaken. This mutant is insensitive to both JA and the *P. syringae* phytotoxin coronatine, which mimics the action of JA (Feys *et al.*, 1994). Infection of *coi-1* plants with a virulent *P. syringae* pathovar resulted in suppression of the colonizing bacterial population by two orders of magnitude compared to wild-type plants. Significant suppression of bacterial growth can therefore occur in coronatine-resistant plants which are unable to activate defence genes via the JA-signalling pathway, raising questions about the importance of jasmonates in disease resistance. Careful studies are now required to assess the response of *coi-1* and *jar-1 Arabidopsis* plants to a range of microbial pathogens which do not produce coronatine.

10. Methyl jasmonate: an airborne interplant signal

Until recently, the plant hormone ethylene was the only chemical signal known to travel through the atmosphere and activate gene expression. However, experiments have now indicated that Me-JA may also function as an airborne signal during interplant communication (Farmer and Ryan, 1990).

The presence of Me-JA on cotton swabs, placed in airtight chambers containing plants from three species of two families, the Solanaceae and Fabaceae, resulted in the accumulation of protease inhibitors in the leaves of all three species. Moreover, when sagebrush (*Artemisia tridentata*), a plant known to possess Me-JA in leaf surface structures, is incubated in airtight chambers with tomato plants, protease inhibitors accumulate in the tomato leaves. This observation has profound implications because it demonstrates that interplant communication can occur from the leaves of one plant species to those of another.

Although the exact mechanism is still unclear, it has been proposed that Me-JA may enter the vascular system via leaf stomata and activate *pin* gene expression through a receptor-mediated signal transduction pathway. Alternatively, Me-JA may diffuse into the leaf cell cytoplasm, functioning as an intracellular signal.

It is possible that Me-JA or other volatile signals could be released by plants such as sagebrush, which would have profound biological effects upon their immediate neighbours. The responses evoked could range from the induction of defence gene expression to any of the diverse physiological responses mediated by JA, such as leaf senescence or growth control. If this mechanism of communication is widespread in nature, it may be of central significance when evaluating the ecological interactions within plant communities.

11. Transcriptional activation by jasmonates

The intracellular accumulation of JA can elicit the transcriptional activation of specific sets of genes. Although the mechanism remains unclear, intracellular JA may either directly or indirectly activate gene-specific transcription factors. Binding of these activated transcription factors to their cognate *cis* elements within the promoters of JA-inducible genes may lead to transcriptional activation.

The unravelling of the molecular mechanisms underlying the activation of gene expression by JA and Me-JA has only recently been initiated. It will be essential to define the promoter *cis* elements which are responsible for co-ordinating JA transcriptional activation. The identification of such motifs will be a prerequisite for the subsequent isolation and characterization of the cognate *trans* factors which specifically bind these motifs. These transcription factors would constitute the terminal components of the JA-signalling pathway and would provide a point of entry for molecular studies to define more upstream signalling components.

Wound-activated binding of a putative transcription factor to a JA-inducible gene promoter has been described by Palm *et al.* (1990). Using deletion analysis of the potato *pin-II* gene promoter, these authors identified a nucleotide sequence between positions –165 and –151 that bound *in vitro* a nuclear protein from wounded potato leaves approximately fivefold more than a similar protein from unwounded leaves. An imperfect repeat of this motif was also identified between positions –620 and –520 within the same promoter (Ryan, 1992). This element lies within a region of the promoter between positions –642 and –506, which has been implicated in mediating gene activation following wounding of mutant *pin-II* promoter reporter gene fusions in transgenic plants (Lorbeth *et al.*, 1992). Deletion of the –700 to –500 region of the *pin-II* promoter completely abolished wound inducibility in these experiments.

The use of mutant *pin-II* promoter–reporter gene fusions in transgenic plants has been applied directly to define JA-responsive *cis* elements. Deletion analysis showed that the promoter region between positions –625 and –520 was sufficient for JA responsiveness (Kim *et al.*, 1992). A G-box-like element (TCACGT) was identified within this region. This motif is conserved in a number of distinct plant promoters which respond to a variety of cues, including white light, ABA and *para*-coumaric acid (Loake *et al.*, 1992b). A family of basic leucine zipper transcription factors termed GBFs, which bind G-box-like elements, have been cloned and well characterized (Oeda *et al.*, 1991; Schindler *et al.*, 1992). However, further studies are required to confirm a functional role in wound signalling for the G-box present within the *pin-II* gene promoter. Site-specific mutational analysis needs to be employed to confirm that the G-box hexamer is necessary, let alone sufficient, for JA-mediated transcriptional activation. There is still much detailed work to be undertaken in order to define and characterize the *cis* elements and *trans* factors which constitute the terminal components in the JA-signalling pathway.

12. Summary

A class of membrane lipid-derived molecules, termed jasmonates, has been identified in plants, and these may play profound roles in plant development and interactions with the environment. Intriguingly, the structure and biosynthesis of these compounds are similar to those of mammalian eicosonoids synthesized from C_{20} trienoic acids. Jasmonates are synthesized from the 18:3 unsaturated fatty acid linolenic acid, which is thought to be released from cellular membranes following activation of a specific lipase. Linolenic acid is subsequently converted to JA and related jasmonate compounds via a lipoxygenase-dependent pathway (Vick and Zimmerman, 1984). Jasmonates have been implicated in a diverse range of developmental processes, including senescence (Ueda and Kato, 1980), growth control (Staswick *et al.*, 1992), fruit development (Saniewski *et al.*, 1987) and thigmotropism (Falkenstein *et al.*, 1991). The molecular mechanism underlying the role of JA in leaf senescence is now beginning to emerge. Accumulating JA and Me-JA in leaf tissue have been shown to repress the transcription and translation of key genes involved in photosynthesis and carbon assimilation (Reinbothe *et al.*, 1994). Moreover, jasmonates are thought to initiate ribosome inactivation in tissues exposed to such signals for extended periods. These compounds may therefore play a key role in the promotion of leaf senescence.

Jasmonates have been shown to transduce stimuli resulting from exposure to diverse environmental stresses including water deficit (Creelman and Mullet, 1995), mechanical injury (Farmer *et al.*, 1992) and microbial pathogen attack (Gundlach *et al.*, 1992). In all cases, these compounds have been shown to activate rapidly appropriate sets of defensive genes. The role of jasmonates in the transduction of the wound signal has been characterized in most detail. These compounds have been shown to function as intracellular signals residing at the terminus of the wound-signalling pathway (Pena-Cortes *et al.*, 1995). Perhaps in response to stress stimuli, jasmonates channel the available cellular resources into mobilization of the appropriate defensive strategy. This may be achieved by suppressing the expression of genes involved in photosynthesis and carbon assimilation, while simultaneously inducing the transcription of stress-related genes. In this context, jasmonates may function as global regulators of plant gene expression.

References

Aarts JMMJG, Hontelez JGJ, Fisher P, Verkerk R, Van Kammen A, Zabel P. (1991) Phosphatase-11, a tightly linked molecular marker for root-knot nematode resistance in tomato — from protein to gene, using PCR and degenerate primers containing deoxyinosine. *Plant Mol. Biol.* **16:** 647–661.

Abe M, Shibaoka H, Yamane H, Takahashi N. (1990) Cell cycle dependent disruption of microtubules by methyl jasmonate in tobacco BY-2-cells. *Protoplasm* **1561:** 1–8.

Albrecht T, Kehlen A, Stahl K, Knofel H-D. (1993) Quantitation of rapid and transient increases in jasmonic acid in wounded plants using a monoclonal antibody. *Planta* **191:** 86–94.

Anderson JM, Spilatro SR, Klauer SF, Franceshi VR. (1989) Jasmonic acid dependent increase in vegetative storage proteins in soybean. *Plant Sci.* **62:** 45–52.

Andresen I, Beccker W, Schulter K, Burges J, Parthier B, Apel K. (1992) The identification of leaf thionin as one of the main jasmonate-induced proteins of barley (*Hordeum vulgare*). *Plant Mol. Biol.* **19:** 193–204.

Bailey M, ed. (1985) *Prostaglandins, Leukotrienes and Lipoxins: Biochemistry, Mechanism of Action and Clinical Applications.* London: Plenum Press.

Berridge MJ. (1993) Inositol triphosphate and calcium signalling. *Nature* **361:** 315–325.

Bishop PD, Makus DJ, Pearce G, Ryan CA. (1981) Proteinase inhibitor-inducing factor activity in tomato leaves resides in oligosaccharides enzymically released from cell walls. *Proc. Natl Acad. Sci. USA* **78:** 3526–3540.

Cohen Y, Grisi U, Mosinger E. (1993) Local and systemic protection against *Phytophthora infestans* induced in potato and tomato by jasmonic acid and methyl jasmonate. *Phytopathology.* **83:** 1054–1062.

Creelman R, Mullet JE. (1995) Jasmonic acid action and distribution in plants: regulation during development and responses to biotic and abiotic stresses. *Proc. Natl Acad. Sci. USA* **92:** 4114–4119.

Creelman R, Tierney ML, Mullet JE. (1992) Jasmonic acid/methyl jasmonate accumulate in wounded soybean hypocotyls and modulate wound gene expression. *Proc. Natl Acad. Sci. USA* **89:** 4938–4941.

Croft KPC, Voisey CR, Slusarenko AJ. (1990) Mechanism of hypersensitive cell collapse: correlation of increased lipoxygenase activity with membrane damage in leaves of *Phaseolus vulgaris* (L.) inoculated with an avirulent race of *Pseudomonas syringae* pv. *phaseolicola. Physiol. Mol. Plant Path.* **36:** 49–62.

Delaney T, Uknes S, Vernooij B et al. (1994) A central role of salicylic acid in plant disease resistance. *Science* **266:** 1247–1250.

Demole E, Lederer E, Mercier D. (1962) Isolement et determination de la structure du jasmonic acid. *Helv. Chim. Acta.* **45:** 675–685.

Doares S, Narvaez-Vasquez J, Conconi A, Ryan CA. (1995) Salicylic acid inhibits synthesis of proteinase inhibitors in tomato leaves induced by systemin and jasmonic acid. *Plant Physiol.* **108:** 1741–1746.

Falkenstein E, Groth B, Mithofer A, Weiler EW. (1991) Methyl-jasmonate and linolenic acid are potent inducers of tendril coiling. *Planta* **185:** 316–322.

Farmer EE, Ryan CA. (1990) Interplant communication: air-borne methyl jasmonate induces synthesis of proteinase inhibitors in plant leaves. *Proc. Natl Acad. Sci. USA* **87:** 7713–7716.

Farmer EE, Ryan CA. (1992) Octadeconoid precursors of jasmonic acid activate the synthesis of wound-inducible proteinase inhibitors. *Plant Cell* **4:** 129–134.

Farmer EE, Johnson RR, Ryan CA. (1992) Regulation of expression of proteinase inhibitor genes by methyl jasmonate and jasmonic acid. *Plant Physiol.* **98:** 995–1002.

Feys BJF, Benedetti CE, Penfold CN, Turner JG. (1994) *Arabidopsis* mutants selected for resistance to the phytoxin coronatine are male sterile, insensitive to methyl jasmonate and resistant to a bacterial pathogen. *Plant Cell* **6:** 751–759.

Galliard T, Chan HWS. (1980) Lipids: structure and function. In: *The Biochemistry of Plants: a Comprehensive Treatise, Vol. 4* (Stumpf PK, Conn EE, ed.). London: Academic Press, pp. 131–161.

Gundlach H, Muller MJ, Kutchan TM, Zenk MH. (1992) Jasmonic acid is a signal transducer in elicitor-induced plant cell cultures. *Proc. Natl Acad. Sci. USA* **89:** 2389–2393.

Huang J-F, Bantroch DJ, Greenwood JS, Staswick PE. (1991) Methyl jasmonate treatment eliminates cell-specific expression of vegetative storage protein genes in soybean leaves. *Plant Physiol.* **97:** 1512–1520.

Kenyon JS, Turner JG. (1992) The stimulation of ethylene synthesis in *Nicotiana tabacum* leaves by the phytotaxin coronatine. *Plant Physiol.* **100:** 219–224.

Kim SR, Choi J-L, Costa MA, An G. (1992) Identification of G-box sequence as an essential element for methyl jasmonate response of potato proteinase inhibitor II promoter. *Plant Physiol.* **99:** 627–631.

Koda Y. (1992) The role of jasmonic acid and related compounds in plant development. *Int. Rev. Cytol.* **135:** 155–199.

Loake GJ, Choudhary AD, Harrison MJ, Mavandad M, Lamb CJ, Dixon RA. (1992a) Phenylpropanoid pathway intermediates regulate transient expression of a chalcone synthase gene promoter. *Plant Cell* **3:** 829–840.

Loake GJ, Faktor O, Lamb CJ, Dixon RA. (1992b) Combination of H-box [CCTACC(N)$_7$CT] and G-box (CACGTG) *cis*-elements is necessary for feed-forward stimulation of a chalcone synthase promoter by the phenylpropanoid-pathway intermediate *p*-coumaric acid. *Proc. Natl Acad. Sci. USA* **89:** 9230–9234.

Lorbeth R, Dammann C, Ebneth M, Amati S, Sanchez-Serrano JJ. (1992) Promoter elements involved in environmental and developmental control of potato proteinase inhibitor II expression. *Plant J.* **2:** 477–486.

Meyer A, Gross D, Vorkefeld S, Kummer M, Schmidt J, Sembder G, Schreiber K. (1989) Metabolism of the plant growth regulator dihydrojasmonic acid in barley shoots. *Phytochemistry* **28:** 1007–1011.

Mino Y, Matasushita Y, Sakai R. (1987) Effect of coronatine on stomatal opening in leaves of broad bean and Italian ryegrass. *Ann. Phytopath. Japan* **53:** 53–55.

Oeda K, Salinas J, Chua N-H. (1991) A tobacco bZip transcriptional activator (TAF-1) binds to a G-box-like motif conserved in plant genes. *EMBO J.* **10:** 1793–1802.

Palm CJ, Costa MJ, An G, Ryan CA. (1990) Wound-inducible nuclear protein binds DNA fragments that regulate a proteinase inhibitor II gene from potato. *Proc. Natl Acad. Sci. USA* **87:** 603–607.

Parthier B. (1991) Jasmonates, new regulators of plant growth and development: many facts and few hypotheses on their actions. *Bot. Acta* **104:** 446–454.

Pearce G, Strydom D, Johnson S, Ryan CA. (1991) A polypeptide from tomato leaves induces wound-inducible proteinase inhibitor proteins. *Science* **253:** 895–898.

Pena-Cortes H, Sanchez-Serrano JJ, Mertens R, Willmitzer L, Prat S. (1989) Abscisic acid is involved in the wound-induced expression of the proteinase inhibitor II gene in potato and tomato. *Proc. Natl Acad. Sci. USA* **86:** 9851–9855.

Pena-Cortes H, Albrecht T, Prat S, Weiler EW, Willmitzer L. (1993) Aspirin prevents wound-induced gene expression in tomato leaves by blocking jasmonic acid biosynthesis. *Planta* **191:** 123–128.

Pena-Cortes H, Fisahn J, Willmitzer L. (1995) Signals involved in wound-induced proteinase inhibitor II gene expression in tomato and potato plants. *Proc. Natl Acad. Sci. USA* **92:** 4106–4113.

Reinbothe S, Mollenhauer B, Reinbothe C. (1994) JIPS and RIPS: the regulation of plant gene expression in response to environmental cues and pathogens. *Plant Cell* **6:** 1197–1209.

Roy S, Pouenat M-L, Caumont C, Cariven C, Prevost M-C, Esquerre-Tugaye M-T. (1995) Phospholipase activity and phospholipid patterns in tobacco cells treated with fungal elicitor. *Plant Sci.* **107:** 17–25.

Ryals J, Uknes S, Ward E. (1994) Systemic acquired resistance. *Plant Physiol.* **104:** 1109–1112.

Ryan CA. (1992) The search for the proteinase inhibitor inducing factor, PIIF. *Plant Mol. Biol.* **19:** 123–133.

Sakai R. (1980) Comparison of the physiological activities between coronatine and indole-3-acetic acid to some plant tissues. *Ann. Phytopath. Soc. Japan* **46:** 499–503.

Saniewski M, Czapski J, Nowacki J, Lange E. (1987) The effect of methyl jasmonate on ethylene and 1-amino-cyclopropane-1-carboxylic acid production in apple fruits. *Biol. Plant.* **29:** 199–203.

Satler SO, Thimann KV. (1981) Le jasmonate de methyle: nouveauet puissant promoteur de la senescence des fleuilles. *R. Acad. Sci. Paris Series III* **293:** 735–740.

Schindler U, Menkens AE, Beckman H, Ecker JR, Cashmore AR. (1992) Heterodimerization between light-regulated and ubiquitously expressed *Arabidopsis* GBF bZip proteins. *EMBO J.* **11:** 1261–1273.

Schmid J, Doerner PW, Clouse SD, Dixon RA, Lamb CJ. (1990) Developmental and environmental regulation of a chalcone synthase gene promoter in transgenic tobacco. *Plant Cell* **2:** 619–631.

Sembdner G, Gross D. (1986) Plant growth substances of plant and microbial origin. In: *Plant Growth Substances* (Bopp M, ed.). Berlin: Springer-Verlag, pp. 139–147.

Sigal E. (1991) The molecular biology of mammalian arachidonic acid metabolism. *J. Physiol.* **260:** L13–L28.

Staswick PE. (1990) Novel regulation of vegetative storage protein genes. *Plant Cell* **2:** 1–6.

Staswick PE, Huang J-F, Rhee Y. (1991) Nitrogen and methyl jasmonate induction of soybean vegetative storage protein genes. *Plant Physiol.* **96:** 130–136.

Staswick PE, Wenpei S, Howell SH. (1992) Methyl jasmonate inhibition of root growth and induction of a leaf protein are decreased in an *Arabidopsis thaliana* mutant. *Proc. Natl Acad. Sci. USA* **89:** 6837–6840.

Tranbarger TJ, Franceschi VR, Hildebrand DF, Grimes HD. (1991) The soybean 94-kilodalton vegetative storage protein is a lipoxygenase that is localised in paraveinal mesophyll cell vacuoles. *Plant Cell* **3**: 973–987.

Ueda J, Kato J. (1980) Identification of a senescence promoting substance from wormwood (*Artemisia absinthium*). *Plant Physiol.* **66**: 246–249.

Vernooygerritsen M, Veldink GA, Vliegenthart JFG. (1982) Specificities of antisera directed against soybean lipoxygenase-1 and lipoxygenase-2 and purification of lipoxygenase-2 by affinity chromatography. *Biochem. Biophys. Acta* **708**: 330–334.

Vick BA, Zimmerman DC. (1979) Distribution of fatty acid cyclase activity in plants. *Plant Physiol.* **64**: 203–205.

Vick BA, Zimmerman DC. (1980) Oxidative systems for modification of fatty acids: the lipoxygenase pathway. In: *The Biochemistry of Plants: a Comprehensive Treatise, Vol. 9* (Stumpf PK, Conn EE, ed.). London: Academic Press, pp. 53–89.

Vick BA, Zimmerman DC. (1983) Characterisation of 12-oxo-PDA reductase activity in plants. *Plant Physiol.* **57**: 780–788.

Vick BA, Zimmerman DC. (1984) Biosynthesis of jasmonic acid by several plant species. *Plant Physiol.* **75**: 458–461.

Weidhase RA, Krammel H-M, Lehmann J, Liebisch W, Lerbs W, Parthier B. (1987) Methyl jasmonate induced changes in the polypeptide pattern of senescing barley leaf segments. *Plant Sci.* **51**: 177–186.

Wildon DC, Thain JF, Minchin PEH, Gubb IR, Reilly AJ, Skipper YD, Doherty HM, O'Donnell PJ, Bowles DJ. (1992) Electrical signalling and systemic proteinase inhibitor induction in the wounded plant. *Nature* **360**: 62–65.

Wittenbach VA. (1983) Purification and characterisation of a soybean leaf storage glycoprotein. *Plant Physiol.* **73**: 125–129.

Yamane H, Takagi H, Abe H, Yokota T, Takahashi N. (1981a) Identification of jasmonic acid in three species of higher plants and its biological activities. *Plant Cell Physiol.* **22**: 689–697.

Yamane H, Abe H, Takahashi N. (1981b) Jasmonic acid and methyl jasmonate in pollen and anthers of three *Camellia* species. *Plant Cell Physiol.* **23**: 1125–1127.

Yoshikara T, Omer E-SA, Koshino H, Sakamura S, Kikuta Y, Koda Y. (1989) Structure of a tuber inducing stimulus from potato leaves. *Agric. Biol. Chem.* **53**: 2835–2837.

Chapter 14

Functions of ion channels in plant cells

[handwritten: + exocytotic vesiculation]

Mark Tester

1. Introduction

The rate of passive movement of solutes across membranes is almost entirely determined by the chemical properties of both the lipids and the proteins embedded in the lipid bilayer, in particular by the so-called transport proteins. Passive transport can take place across the lipid component of the bilayer, or through the proteins embedded within the lipid.

Movement across the lipid bilayer can be a significant component of transmembrane transport for molecules which have a high solubility in the lipid component of the membrane. Furthermore, small (molecular weight less than *c.* 60) neutral molecules which are relatively insoluble in lipid, such as water and urea, are more permeant than would be expected from their lipid solubility; this may be due to movement of these small molecules through transient pores in the lipid structure. Thus, membranes are highly permeable to a wide range of non-polar molecules (such as the dissolved gases, O_2 and CO_2), and also to small polar molecules such as water and urea. Larger molecules which have polar side groups (such as glucose and sucrose) are not particularly permeant, and ions are almost completely impermeant. As a result, most transmembrane movement of ions must be catalysed by proteins. The passive movements of ions are largely catalysed by carriers and by ion channels.

Carriers are differentiated from channels by their much slower rate of turnover, by thermodynamic differences (carriers have one infinite energy barrier in contrast to finite barriers at all times in channels) and by their primary structure (carriers have structural homologies to solute-coupled transporters, whereas channels have distinct structures). However, there are no known confirmed reports regarding the presence of carriers in plant membranes, although there is some evidence for their existence (e.g. for NH_4^+ uptake; Walker *et al.,* 1979). This chapter will deal solely with passive transport catalysed by ion channels.

2. What are ion channels?

Ion channels are large, membrane-spanning proteins found in most, if not all, biological membranes. By homology with channels from animal cells, channels in plants are likely to be oligomeric glycoproteins, a factor to be remembered when considering the expression of cloned genes for channels in heterologous systems. They effectively form a hydrophilic pore through the membrane, and allow very high rates of solute movement. Channels have the greatest turnover rate of all enzymes, catalysing the net movement of 10^6 to 10^8 molecules per second; by comparison, the fastest water-soluble enzyme known, catalase, can perform 10^5 reactions per second, and most enzymes have much slower turnover rates. Transport catalysed by ion channels can be many thousands of times faster than that catalysed by pumps and carriers.

3. Measurement of ion channels

Movements due to the activity of ion channels can be measured using a range of techniques which reflect varying compromises between (a) physiological reality and (b) biochemical definition and biophysical resolution. These techniques include:

(1) unidirectional radioactive tracer fluxes (e.g. Reid and Tester, 1992; Smith *et al.,* 1987);
(2) intracellular impalements with fine-tipped microelectrodes (e.g. Blatt *et al.,* 1990);
(3) patch-clamp microelectrodes in the 'whole cell' mode (e.g. Schroeder and Fang, 1991);
(4) patch-clamp microelectrodes containing excised patches of membrane (e.g. Johannes *et al.,* 1992); and
(5) incorporation of channels into artificial lipid bilayers (e.g. White and Tester, 1992).

In addition, net changes in intracellular ionic activities resulting from fluxes through ion channels can also be measured, for example with ratiometric fluorescence techniques (Fricker *et al.,* 1993; Read *et al.,* 1992). The use of transgenic plants expressing apoaequorin enables, on addition of coelenterazine, the qualitative measurement and even imaging of changes in intracellular free Ca^{2+} in intact whole plants (Knight *et al.,* 1991, 1993).

The choice of a particular technique depends, at least in part, on the question being addressed (Tester, 1990), but techniques can also be chosen for more pragmatic reasons. For example, there have until recently been very few records of Ca^{2+}-selective channels in the plant plasma membrane, and certainly there has been no thorough characterization of such channels. This may be due to a very low density of these channels, making it difficult to find one of these proteins when probing the plasma membrane with a patch-clamp microelectrode. However, when incorporating plasma membrane vesicles into artificial planar lipid bilayers, it has proved relatively easy to measure these channels (Piñeros and Tester, 1995); this is presumably because the investigator 'simply' needs to wait until a vesicle containing one of these proteins fuses with the bilayer, rather than probing (with much effort) with a patch electrode until the (apparently rare) channel happens to occur in the area of membrane enclosed by the patch pipette.

The first three techniques listed above measure (either chemically or electrically) the fluxes of ions through many channel proteins. However, one exciting aspect of research on ion channels is the ability conferred by the last two techniques to measure the movements of ions through a single channel protein. Such measurements have shown that channels can change conformation rapidly between catalytic ('open') and non-transporting ('closed') states. This movement between open and closed states is called 'gating'. When the channels are open, for a given substrate concentration and driving force (i.e. ion activity and electrochemical potential difference), the protein will move ions at a characteristic rate, known as the unitary current. The ease of ion movement is the unitary conductance. An example of the electrical traces resulting from such protein activity is shown in *Figure 1*. Many channels will exhibit more than one open state, with smaller conductance conformations being referred to as sub-conductance states (Tyerman *et al.,* 1992). Although of interest at the molecular biophysical level, the significance of such states in the intact plant remains unknown.

Such traces represent the most sensitive enzyme assay known. For example, the clear channel event indicated in *Figure 1* by an arrow represents the movement of 4.2 pA of current for 100 msec, which is equivalent to the movement of 390 000 ions, of the order of 10^{-19} moles. The technique can resolve events less than one-tenth of this size. Such remarkable resolution is aided by the very high turnover rates of ion channels. The activity of a single pump, for example, cannot be resolved, because their turnover rate is too low, and the resulting currents cannot be resolved above the noise of the electrical circuitry.

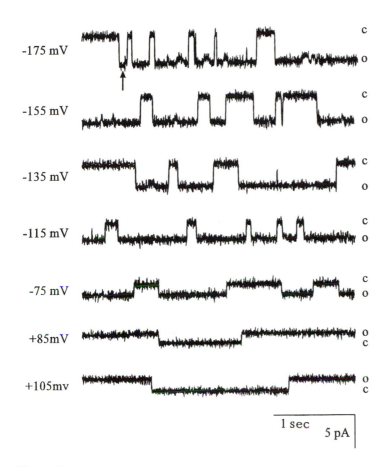

-175 mV

-155 mV

-135 mV

-115 mV

-75 mV

+85mV

+105mv

1 sec

5 pA

Figure 1. Single channel currents of a plasma membrane Ca^{2+} channel incorporated into a planar lipid bilayer. Membrane with 1 mM $CaCl_2$, pH 5.5 on each side. The voltages are equivalent to those used in intact cells, namely cytosol with respect to the outside. C and O represent closed and open states, respectively. Reproduced with permission from Piñeros and Tester (1995). © 1995 Springer-Verlag.

Using a combination of the above techniques, we are now able to study the activity of single proteins responsible for significant movements of solutes across the membrane, the control of their opening and closing, their selectivity of transport and their physiological role in the intact plant.

4. Primary structure of ion channels

Molecular biological techniques are providing insights into the primary structure of ion channels in plants, and will surely prove to be very powerful tools in the investigation of the function of these proteins, as well as their evolutionary relationships with analogous proteins in other organisms (Strong *et al.,* 1993). Two similar genes from *Arabidopsis,* which have been cloned by complementation of a K^+ transport mutant in yeast, have quite high sequence homology and a similar overall primary structure to K^+-selective channels from animal cells, particularly in the area containing the membrane-spanning amino acids and the so-called H5 sequence that is apparently responsible for the binding and transport of the ion (Anderson *et al.,* 1992; Sentenac *et al.*, 1992). The open reading frames can encode proteins of 78 kDa (Anderson *et al.*, 1992) and 95.4 kDa (Sentenac *et al.*, 1992). Expression of the smaller clone in *Xenopus* gave K^+ channel activity (Schachtman *et al.*, 1992). In addition, both clones contain sequences homologous to cyclic nucleotide binding sites, and the larger clone contains ankyrin repeats, suggesting that the protein may interact with the cytoskeleton or with regulatory proteins (Sentenac *et al.*, 1992). Although the function of these proteins in the intact plant is as yet unknown, site-directed mutagenesis of the putative ion-binding site has been shown to alter the ability of the channel to discriminate between various cations (Anderson *et al.*, 1994). This provides exciting possibilities for the insertion of such altered proteins back into growing plants, in order to provide the first opportunities for the analytical investigation of channel function *in vivo*.

5. Properties of ion channels

Mechanism of ion movement: conduction

Despite their very high rates of transport, ion channels can show typical enzymic saturation kinetics; in this respect, the transport is distinct from simple diffusion across the lipid bilayer (where the flux increases linearly as the concentration of substrate increases). Permeation of ions through channels should not simply be considered as a 'ball rolling through a drain-pipe' (Tester, 1990), but involves binding and de-binding of ions to a specific site (which can now even be identified to the level of particular amino acids: Kumpf and Dougherty, 1993; Yang *et al.*, 1993), together with conformational changes of the protein (Eisenberg, 1990). So long as these conformational changes are not too large, such properties can explain the ability of channels to maintain both high rates of transport and a high degree of selectivity.

Selectivity of ion movement

Ion channels can be very specific, showing a high degree of selectivity with regard to the solutes transported. For example, certain channels will mainly catalyse the movement of the monovalent cations K^+ and NH_4^+, while other ions, such as Na^+, are transported at approximately one-tenth of the rate of K^+ (Tester, 1988). The high selectivity of channels is all the more remarkable given their very high rates of transport.

Not all channels exhibit such high selectivity. For example, a channel from the plasma membrane of rye roots discriminates poorly between a wide range of monovalent and divalent cations (White, 1993), and the so-called 'SV' channel found in the tonoplast of a diverse range of cells (Hedrich *et al.*, 1988) apparently allows the movement of a wide range of both cations and anions. In fact, given the large differences in ionic composition between the cytoplasm and vacuole, the low selectivity of this channel suggests that it must be closed for much of the time in the intact cell. The imperfection of ion channel transport may be very important, and may well be a key to the entry of many 'undesirable' ions into plants. Due to the high rate of flux through ion channels, even if selectivity for the desired ion is 10 times or even 100 times that for a different ion, it may be enough to allow significant 'leaks' of undesirable ions. For example, it seems likely that a major pathway for the entry of Na^+ into the cortical cells of plant roots is through K^+-selective channels (Schachtman *et al.*, 1991); also, the entry of some toxic metal ions into plants may occur through Ca^{2+}-selective channels. Mutagenesis of the putative ion-binding site (Anderson *et al.*, 1994) should give insights into the molecular basis for ionic selectivity.

It is notable that anion channels appear to discriminate poorly between NO_3^- and Cl^- (e.g. Terry *et al.*, 1991); given the approximately similar cytosolic concentrations of NO_3^- and Cl^-, this would appear to allow the loss of NO_3^- from the cytoplasm to the soil when plasma membrane anion channels are open. How does this anticipated efflux affect the net uptake of this important plant nutrient? Similarly, anion channels in guard cells appear to allow the rapid movement of malate (in addition to that of Cl⁻; Hedrich *et al.,* 1990), which must have important consequences for the fluxes of organic solutes during stomatal movements.

Control of ion movement: gating

A cell cannot allow rapid, dissipative fluxes over long periods of time, and most channels are not transporting for most of the time. The control of their opening is complex, and since gating is a random process, the probability of the protein being open must be considered. Some of the factors which may control the probability of channel opening in plants include voltage (Thiel *et al.,* 1992), cytosolic Ca^{2+} (Schroeder and Hagiwara, 1989) and pH (Blatt, 1992), external Ca^{2+} (Fairley-Grenot and Assmann, 1992) and vacuolar Ca^{2+} (Johannes *et al.,* 1992), redox potential (Bertl and Slayman, 1990), stretch (Cosgrove and Hedrich, 1991; see also Chapter 15 by Garrill *et al.* in this volume), phosphorylation and dephosphorylation (Blatt and Armstrong, 1993), plant growth regulators (Blatt and Thiel, 1993; Marten *et al.,* 1991) and a range of possible secondary messenger metabolites such as inositol-1,4,5-trisphosphate (InsP$_3$: Alexandre *et al.,* 1990) and GTP-binding proteins (Fairley-Grenot and Assmann, 1991). There is some evidence from work on algal cells for the control of Ca^{2+}-channel gating by cAMP (Zherelova and Grishchenko, 1987); it will be interesting to see whether such control can also be found in the *Arabidopsis* K^+ channel, given the deduction of a cyclic nucleotide-binding domain from its primary sequence (Sentenac *et al.,* 1992).

There have been fewer studies of anion-selective channels compared to research on cation channels (Tyermann, 1992), perhaps because they do not appear as frequently during patch-clamp experiments. This may be partly due to the so-called 'run down' of channel activity upon perturbation of the cell during the (rather disruptive) process of making a patch-clamp recording (Tyermann, 1992), and with planar lipid bilayers (an even more disruptive technique) very few anion channels have been observed (unpublished results). Control of anion channel activity appears to be complex (e.g. Hedrich *et al.,* 1990; Tyermann, 1992), so it may be susceptible to disruption.

The factors controlling channel activity are many, and this does not seem unreasonable given the capacity of these proteins to move ions at potentially useful but, equally, dangerous speeds. However, such complexity also makes it difficult to unravel the functions of channels in the intact plant. Therefore, physiological interpretations of function derived from experiments based on one technique should be qualified until confirmatory results are obtained using less physiologically disruptive techniques.

Pharmacology of ion channels

In addition to the 'natural' factors mentioned above, a range of pharmacological and other agents can affect ion channel activity. A wide range of compounds which are known to affect ion channels in animal cells have also been found to affect such channels in plants, although not necessarily with the same specificity or affinity (Terry *et al.,* 1992; Tester and MacRobbie, 1990). This suggests that, although there is some similarity between plant and animal ion channels, significant differences also exist between them, differences which will, of course, become more defined as further primary sequences of plant channels are obtained (Strong *et al.,* 1993). These differences also mean that the effects of putative ion channel effectors on physiological processes in plants should not be assumed to be mediated via an effect on a particular channel, without checking directly whether the compound does in fact affect channel activity in plants.

6. Classification of ion channels

Given the above properties of ion channels, we are able to classify channels on the basis of their selectivity, gating, unitary conductance and pharmacology. For example, the channel illustrated in *Figure 1* has a characteristic unitary conductance of 27 pS (in 1 mM Ca^{2+}), is selective for Ca^{2+}, has an open probability strongly affected by voltage, and is sensitive to the drug verapamil (Piñeros and Tester, 1995); it could therefore be conveniently referred to as a verapamil-sensitive, voltage-activated 27-pS Ca^{2+} channel. Such classifications are inevitably anthropocentric, based on what is of particular interest at the time, rather than on what may be of more importance for the plant. Thus a so-called 'outward rectifying K^+ channel' may, in saline soils, be of more importance to the plant as a protein which is allowing the 'leaking' in of Na^+, rather than the efflux of K^+ (Schachtman *et al.*, 1991).

The main types of channel which are found in plant cells fall into three main classes (Tester, 1990):

(1) Ca^{2+} channels, which catalyse Ca^{2+} entry into the cytoplasm from outside and from intracellular stores such as the vacuole and (possibly) the endoplasmic reticulum;
(2) Cl^- channels, which catalyse the loss of a range of anions from the cell; not only is Cl^- lost, but also NO_3^-, organic anions and possibly $H_2PO_4^-$. Under saline conditions, where external Cl^- concentrations are high and membrane potentials are small, these channels may be responsible for Cl^- uptake into the root (Skerrett and Tyerman, 1993);
(3) K^+ channels which, depending on the voltages at which they open, can catalyse the influx or efflux of K^+ into or out of the cytoplasm and vacuole. Other monovalent cations can also move, most notably Na^+, as mentioned above.

7. Roles of ion channels in plants

Ion channels have a wide range of functions in plants. By virtue of the remarkable speed with which channels transport solutes, these proteins not only have a role in transport, but can also be used by cells in processes where the rapid transmission of a message is crucial. In this review, a range of processes in which the role of ion channels is becoming clear is discussed, from the relatively leisurely uptake of nutrients by roots, to the role of ion channels in charge balance, a process which can occur over periods of a few milliseconds.

Role of ion channels in nutrient uptake

Although channels catalyse passive movement, they can have a very important role in accumulation, because they are acting in parallel with the pumps. Due to the potential difference across the plasma membrane, with the cytoplasm negative with respect to the soil solution, there can be large accumulations of cations inside the cell. Another basis for accumulation is the ability of the cytoplasm to chelate and/or sequester ions, most notably Ca^{2+}, but also other divalent cations such as Zn^{2+} (O'Halloran, 1993). Thus Ca^{2+} uptake at the root tip (Huang *et al.*, 1992) is likely to occur through Ca^{2+}-selective channels in the plasma membrane of root cells (Piñeros and Tester, 1995). Such channels may also be the pathway for the uptake of micronutrients such as Mg^{2+}, Zn^{2+} and Cu^{2+} (M. Piñeros, unpublished data), and could allow the entry of heavy metals such as Cd^{2+}, Ni^{2+} and Co^{2+}.

Accumulation of monovalent cations can also occur through ion channels, and for K^+ this appears to be likely to occur only at soil K^+ concentrations greater than about 1 mM (Maathius and Sanders, 1993). Although concentrations of the order of 1 mM are commonly found in the bulk soil solution (Edmeades *et al.*, 1985; Marschner, 1986, p. 414), they may be depleted to below this level near the root surface. Therefore the inwardly rectifying K^+ channel may often not be involved in the uptake of K^+ from the soil solution. Similarly, NH_4^+ uptake, at least from higher concentrations of NH_4^+ in the soil solution, may occur through plasma membrane cation channels, but active transport must be invoked for uptake from lower soil concentrations (Wang et al., 1993).

An example where anion channels may be involved in the uptake of nutrients is during the movement of phosphate from the soil solution to the plant, when that movement occurs via mycorrhizal fungi. For phosphate to be transferred from the fungus to the plant in the mycorrhizal symbiosis, phosphate must be lost from the fungus across the fungal plasma membrane at the mycorrhizal interface. This intuitively unusual loss of phosphate from the fungus is likely to be passive and catalysed by ion channels (Tester et al., 1992). The steady-state rate of phosphate efflux from cultured ericaceous mycorrhizal fungus is of the order of 10 pmol m^{-2} sec^{-1} (M. Tester, C. Cherry, R.J. Reid and F.A. Smith, unpublished data), which is 100 to 1000 times less than the fluxes necessary for phosphate transfer in an intact symbiosis, as calculated by Smith et al. (1994). There must, therefore, be a large stimulation of phosphate efflux, either by greatly increasing efflux through existing transporters or by stimulating the insertion of more (or novel) transporters into the fungal plasma membrane (Tester et al., 1994). Such a stimulation must be effected by the plant, but the mechanism underlying this stimulation is unknown.

Ion channels may also be involved in the adaptation of plants to nutritional extremes, such as low soil pH or high soil salinity. The toxic effect of low soil pH is primarily due to high levels of free Al^{3+}; the resulting inhibition of root elongation is apparently due to the binding of Al^{3+} to cell walls (Kinraide et al., 1994), with these ions perhaps replacing Ca^{2+} in the developing wall structure of the growing root tip (Reid et al., 1995). The ability of acid-tolerant lines of wheat to maintain root elongation in acidic soils appears to be due at least in part to the ability to exclude Al from the root tip (Delhaize et al., 1993a) through an increase in secretion of malic acid (which chelates Al) from the root tips in response to increased soil Al (Delhaize et al., 1993b). This feature appears to be controlled by a single genetic locus, Alt1 (Delhaize et al., 1993b). The pathway and control of the malic acid efflux is of much interest. Is malate lost through anion channels (with H^+ moving via another pathway), or could efflux occur by exocytosis? How are these pathways controlled? How does Al stimulate the efflux, and what is the basis for the difference in Al stimulation controlled by Alt1?

The ability of some graminaceous crops to tolerate high soil salinity appears to be at least partly due to the ability to exclude Na^+ from the shoot (e.g. Munns, 1993; although compare Cramer et al., 1994). However, there is no evidence that increased tolerance of high Na^+ is due to the increased selectivity of cation channels from root plasma membrane for K^+ over Na^+ (Schachtman et al., 1991), nor for an increased ability of vacuoles to compartmentalize Na^+ within the root (Maathius and Prins, 1990). It should be noted that such studies have been carried out with root cortical cells, and the role of ion channels in different cell types within the root (e.g. stelar parenchyma cells) has not yet been investigated.

Turgor adjustment, especially for cells placed in solutions of less negative osmotic potential, where rapid loss of solutes may be necessary, is a process which is likely to involve ion channels. In the giant-celled brackish water alga, Lamprothamnium, the process appears to be dependent on an increase in cytosolic Ca^{2+} (Okazaki and Iwasaki, 1991) due to channel-mediated influx across the plasma membrane (Okazaki and Tazawa, 1990). A scenario has been proposed whereby an initial rise in cytosolic Ca^{2+} activates Cl^- channels, and the depolarization due to the Ca^{2+} influx and Cl^- efflux activates outwardly rectifying K^+ channels, so causing a net efflux of KCl (Okazaki and Tazawa. 1990). The fluxes are analogous to those which occur during the plant action potential (e.g. Tester, 1990) but, as with the action potential, the mechanism for initiation of these fluxes is unknown. As suggested below for movements of plant parts, stretch-activated channels (see Chapter 15 by Garrill *et al.*) may have an important role in the transduction of increased turgor to ion efflux.

Role of ion channels in intracellular malate movements

Uptake of malate^{2-} into the vacuole during nocturnal CO_2 fixation by plants exhibiting Crassulacean acid metabolism photosynthesis is likely to occur via an anion channel acting in parallel with the tonoplast H^+-translocating pumps (Iwasaki *et al.*, 1992; Smith and Bryce, 1992). The problem of malate release from the vacuole during the daytime has yet to be resolved, although Iwasaki *et al.* (1992) claim that this could occur through the so-called 'SV' channel (see above). However, other transport processes (H^+ efflux from the vacuole) would need to be activated in order to enable charge balance across the tonoplast and pH homeostasis in the cytosol. Furthermore, we do not know the nature of the (kinetic) control(s) necessary to inhibit loss of malate from the vacuole at night.

Another interesting system where large transmembrane movements of metabolites are required is in the 'peribacteroid' membrane surrounding bacteroids in leguminous nodules. The large fluxes of organic compounds from the plant to the bacteroids are quite specific, with dicarboxylates such as malate and succinate moving preferentially to other organic compounds (Day *et al.*, 1989). This movement may be catalysed by ion channels.

Role of ion channels in movements of guard cells

Guard cells are highly specialized cells in which rapid changes in turgor occur in order to regulate gas exchange between the leaf and the atmosphere. There has been a large amount of research on these cells, both because of their central role in the growth of plants, and because they are cells in which rapid, controlled movements of large amounts of solutes occur (MacRobbie, 1988). Many of the controls of channel gating listed above have been elucidated with various channels from the plasma membrane of guard cells. There have been many excellent reviews over the past few years which have discussed in much detail the activity of ion channels in controlling guard cell movements (e.g. Blatt and Thiel, 1993; Kearns and Assmann, 1993; Thiel *et al.*, 1992); the reader is referred to these works and to the references therein. In this review, other systems in which channel activity has been less thoroughly discussed will be covered.

Role of ion channels in movements of other plant parts

Rapid turgor changes in a limited number of cells can cause the movement of specialized plant parts, such as the leaves of *Mimosa,* the trap lobes of the Venus fly-trap (*Dionaea*) and the reproductive parts of a range of flowers (Hill and Findlay, 1981). These turgor changes are due to loss of solutes (KCl), apparently through ion channels that open during action potentials (Stoeckel and Takeda, 1993). It seems likely that Cl$^-$ channels are opened by elevated cytosolic Ca^{2+} due to Ca^{2+} influx through plasma membrane Ca^{2+} channels, and that K^+ channels are opened by depolarization of the membrane potential due to a combination of the Ca^{2+} influx and the Cl$^-$ efflux. The degree of difference between the mechanism and control of the loss of solutes in these processes and those which occur in guard cells is unknown. It would seem likely that some components of the processes will be similar, but clearly there will also be differences, if only because the movements occur in response to different environmental cues.

The action potentials responsible for the large movements of plant parts also appear to initiate the movement, propagating through the plant tissue from the site of the stimulus (Sibaoka, 1991).

However, the mechanism of initiation of the action potentials is still unknown; perhaps this is one function of stretch-activated ion channels (see also Chapter 15 by Garrill *et al.*).

Turgor changes underlying slower plant movements, such as heliotropism and other nyctinastic movements, may well be driven by solute movements through ion channels.

Role of ion channels in long-distance signalling within plants: systemic wound responses

Fragments of plant cell walls released upon damage to plant tissue (oligogalacturonides) alter membrane potential and ion fluxes (Mathieu *et al.*, 1991; Thain *et al.*, 1990), consistent with the effects of these compounds on ion channels. Perhaps related to this effect is the stimulation of ion fluxes in cultured cells of parsley by fungal glycopeptides (Scheel *et al.*, 1991). Thus the proposal has been made that these effects play a central role in the response of plants to wounding. It has been suggested that the depolarization induced by wounding can be propagated through the plant, perhaps by action potentials along the vascular tissue, to distant, undamaged tissues where induction of wound-response proteins (such as proteinases) can occur (Wildon *et al.*, 1992; see also Chapter 18 by Wildon and Thain). Such long-distance signalling may be similar to that which stimulates plant movements, as discussed above.

Needless to say, any comparison of such long-distance communication in plants with nervous impulses in animals (Simons, 1992) needs to be tempered with the realization that plants lack a discernible specialized central nervous system.

Role of ion channels in signal transduction and control of development

The role of ion channels in signal transduction and the control of cellular development is an area in which there is still much room for speculation. Although there has been much research which has provided many very interesting observations, there is still a great deal of work to be done before a meaningful picture can emerge of exactly what processes are occurring between an environmental change and a physiological or developmental response by the plant. As a result, this is an exciting and complex area, in which important advances continue to be made.

Increases in cytosolic Ca^{2+} activity, presumably due to fluxes through Ca^{2+}-permeable channels, have been implicated in the response of cells to increases in a range of plant growth regulators, such as abscisic acid (Gilroy *et al.*, 1991; McAinsh *et al.*, 1992), cytokinins (Hahm and Saunders, 1991), auxins (Gehring *et al.*, 1990; Irving *et al.*, 1992) and gibberellic acid (Bush and Jones, 1988; Montague, 1993), and to environmental changes, such as red light (Shacklock *et al.*, 1992), increased salinity (Lynch *et al.*, 1989) and temperature, touch, wind, wounding and fungal elicitors (Knight *et al.*, 1991, 1992, 1993). Some effectors appear to stimulate Ca^{2+} movement from external sources (e.g. salinity: Reid *et al.*, 1993; red light: Mehta *et al.*, 1993), whereas other stimuli may activate Ca^{2+} movement across endomembranes (e.g. wind: Knight *et al.*, 1992). The effect of growth regulators on a range of channels in addition to those which are permeable to Ca^{2+} is reviewed by Blatt and Thiel (1993), and Schroeder (1992) reviews effects of abscisic acid in particular.

As part of a possible signal transduction chain (Blatt *et al.*, 1990; Gilroy *et al.*, 1990), $InsP_3$ has been shown to activate Ca^{2+} channels in the tonoplast of red beet (Alexandre and Lassalles, 1992; Alexandre *et al.*, 1990). Ca^{2+} fluxes into the cytoplasm through tonoplast-bound, Ca^{2+}-selective channels which are insensitive to $InsP_3$ may also occur (Gelli and Blumwald, 1993; Johannes *et al.*, 1992), although the physiological role of these fluxes is unclear. The ability of cells to maintain localized areas of elevated Ca^{2+} within the cell (e.g. near the spindle poles during cell division; Keith *et al.*, 1985) suggests a high level of spatial control of Ca^{2+} movements. This implies a role for the endoplasmic reticulum (ER) in intracellular Ca^{2+}

regulation, but Ca^{2+} channels in ER-enriched membranes have yet to be found (Niemietz and Tester, unpublished results). Localized elevation of Ca^{2+}, such as that which occurs at growing tips (Berger and Brownlee, 1993; Miller *et al.*, 1992), may also be due to localized activation of plasma membrane Ca^{2+} channels (Taylor and Brownlee, 1992), although how such fine control of channel activity is maintained is not known.

Although many environmental stimuli appear to increase cytosolic Ca^{2+}, they can cause a wide range of physiological and developmental responses in the plant. This suggests that, for many stimuli, there must be more to the signal transduction process than mere elevation of cytosolic Ca^{2+}. These other responses could include changes in cytosolic pH; there is evidence for this modulation upon addition of a range of growth regulators to coleoptiles (Gehring *et al.*, 1990) and guard cells (Irving *et al.*, 1992). In addition, there is good evidence that the effects of both abscisic acid and auxin on the outward and inward rectifying K^+ channels, respectively, in the plasma membrane of guard cells are at least partly due to changes in cytosolic pH (Blatt and Armstrong, 1993; Blatt and Thiel, 1994; Thiel *et al.*, 1993).

Role of ion channels in charge balance

The electrical potential difference across membranes is an important component of the electrochemical gradient of ions, controlling the direction of ion movements. It is also an important regulator of the activity of proteins that catalyse ion movement (whether through ion channels or pumps). As a result, membrane potential must be carefully regulated; this is often achieved by controlling the activity of ion channels. Upon depolarization of the plasma membrane, K^+ channels can open to allow K^+ efflux (the 'outward rectifier') or, under conditions of high salinity, Cl^- channels may open to allow Cl^- influx (Skerrett and Tyerman, 1993). Upon hyperpolarization, such as occurs in solutions of relatively high pH, Cl^- channels may open, allowing Cl^- efflux (Tyerman *et al.*, 1986). These hyperpolarization-activated anion channels may also have a role in the uptake of some anions. Although the permeability of these channels to OH^- is not known, it is conceivable that they are responsible for the high passive flux of OH^- across the plasma membrane of charophytes in high pH solutions (Bisson and Walker, 1980). Consequently, such channels could have a central role in maintaining the alkaline 'bands' which are characteristic of some charophytes and appear to be central to the uptake of HCO_3^- for photosynthesis (Lucas, 1976). It should be noted that the high passive conductance could equally well be due to movement of H^+, and a model for passive H^+ movement through the H^+-translocating ATPase has been proposed by Fisahn *et al.* (1992).

The potential difference across endomembranes must also be controlled. Various channels may exist in thylakoids, permeable to Mg^{2+}, K^+ and Cl^-, and these could balance the charge build-up due to H^+ movement in response to illumination (Enz *et al.*, 1993). K^+ channels from the chloroplast inner envelope (Wang *et al.*, 1993) could balance the charge resulting from H^+ fluxes occurring as part of stromal pH regulation during photosynthesis (Wu and Berkowitz, 1992). Channels have also been reported from the inner membrane of mitochondria of various animal tissues (e.g. Zorov *et al.*, 1992); these may function in short-term charge balancing, but roles in volume regulation and protein translocation have also been proposed (for review see Manella, 1992). Although none have yet been reported from plant cells, these channels are likely to be present and serving similar functions to those in animal cells.

Channels also exist in a range of endomembranes with high conductance and low selectivity, such as the chloroplast outer envelope (Flügge and Benz, 1984; Pottosin, 1993), the mitochondrial outer membrane (Smack and Colombini, 1985) and the nuclear envelope (Matzke *et al.*, 1992). It would seem unlikely that these are involved in charge balancing, but rather that they play a role in the controlled transport of a range of solutes.

8. Conclusions

Ion channels are responsible for the rapid movement of a wide range of solutes, and their activity is subject to complex controls. As a result, they are central to a large diversity of functions in plants, including nutrient uptake, signal transduction and long-distance communication. However, given the many complex manifestations arising from malfunctions of single channels in animal cells, such as cystic fibrosis (Riordan, 1993) and myotonias (Koch *et al.,* 1992; Steinmeyer *et al.*, 1991), as well as the apparent role of ion channels in multi-drug resistance (Valverde *et al.*, 1992), it seems reasonable to expect that many more functions of ion channels will be discovered in plants over the next few years. For example, the possible roles of ion channels in multiple herbicide resistance (Devine *et al.,* 1993; Häusler *et al.,* 1991), exocytosis (Zorec and Tester, 1992, 1993) and adaptation to soils contaminated with pollutants such as heavy metals and various organic compounds remain largely unexplored, and could provide fruitful areas for future research. Such work will, of course, benefit from an integration of molecular biological approaches with a range of electrophysiological, radiotracer and fluorescence techniques, and will build on knowledge of the mechanism of action of the channel proteins (e.g. Kumpf and Dougherty, 1993; Yang *et al.,* 1993). In addition, the rigorous testing of the hypotheses for channel function presented in this paper is needed. Now that the genes coding for some ion channels in plants have been cloned, experiments using transgenic plants with altered channel activity will soon be possible, and will surely advance our understanding of the roles of ion channels in plants.

Acknowledgements

The author would like to thank Professor E.A.C. MacRobbie and Dr A. Kaile, Dr C. Morris and Dr S.K. Roberts for comments on the manuscript, various colleagues for access to results prior to publication, and the Australian Research Council for their financial support of research cited in this paper.

References

Alexandre J, Lassalles JP. (1992) Intracellular Ca^{2+} release by $InsP_3$ in plants and effect of buffers on Ca^{2+} diffusion. *Phil. Trans. R. Soc. Lond.* **338:** 53–61.

Alexandre J, Lassalles JP, Kado RT. (1990) Opening of Ca^{2+} channels in isolated red beet root vacuole membrane by inositol 1,4,5-trisphosphate. *Nature* **343:** 567–570.

Anderson JA, Huprikar SS, Kochian LV, Lucas WJ, Gaber RF. (1992) Functional expression of a probable *Arabidopsis thaliana* potassium channel in *Saccharomyces cerevisiae. Proc. Natl Acad. Sci. USA* **89:** 3736–3740.

Anderson JA, Nakamura RL, Gaber RF. (1994) Heterologous expression of K+ channels in *Saccharomyces cerevisiae*: strategies for molecular analysis of structure and function. In: *Membrane Transport in Plants and Fungi: Molecular Mechanisms and Control,* SEB Symposium Series (Blatt MR, Leigh RA, Sanders D, ed.). Cambridge: Company of Biologists, pp. 85–97.

Berger F, Brownlee C. (1993) Ratio confocal imaging of free cytoplasmic calcium gradients in polarising and polarised *Fucus* zygotes. *Zygote* **1:** 9–15.

Bertl A, Slayman CL. (1990) Cation-selective channels in the vacuolar membrane of *Saccharomyces:* dependence on calcium, redox state and voltage. *Proc. Natl Acad. Sci. USA* **87:** 7824–7828.

Bisson MA, Walker NA. (1980) The *Chara* plasmalemma at high pH. Electrical measurements show a rapid passive uniport of H^+ or OH^-. *J. Memb. Biol.* **56:** 1–7.

Blatt MR. (1992) K^+ channels of stomatal guard cells: characteristics of the inward rectifier and its control by pH. *J. Gen. Physiol.* **99:** 615–644.

Blatt MR, Armstrong F. (1993) K^+ channels of stomatal guard cells: abscisic-acid-evoked control of the outward rectifier mediated by cytoplasmic pH. *Planta* **191:** 330–341.

Blatt MR, Thiel G. (1993) Hormonal control of ion channel gating. *Annu. Rev. Plant Physiol. Plant Mol. Biol.* **44:** 543–567.

Blatt MR, Thiel G. (1994) K⁺ channels of stomatal guard cells: bimodal control of the K⁺ inward-rectifier evoked by auxin. *Plant J.* **5**: 55–68.

Blatt MR, Thiel G, Trentham DR. (1990) Reversible inactivation of K⁺ channels of *Vicia* stomatal guard cells following the photolysis of caged inositol 1,4,5-trisphosphate. *Nature* **346**: 766–769.

Bush DS, Jones RL. (1988) Cytoplasmic calcium and α-amylase secretion from barley aleurone protoplasts. *Eur. J. Cell Biol.* **46**: 466–469.

Cosgrove DJ, Hedrich R. (1991) Stretch-activated chloride, potassium, and calcium channels coexisting in plasma membranes of guard cells of *Vicia faba* L. *Planta* **186**: 143–153.

Cramer GR, Alberico GJ, Schmidt C. (1994) Salt tolerance is not associated with the sodium accumulation of two maize hybrids. *Aust. J. Plant Physiol.* **21**: 675–692.

Day DA, Price GD, Udvardi MK. (1989) Membrane interface of the *Bradyrhizobium japonicum–Glycine max* symbiosis: peribacteroid units from soybean nodules. *Aust. J. Plant Physiol.* **16**: 69–84.

Delhaize E, Craig S, Beaton CD, Bennet RJ, Jagadish VC, Randall PJ. (1993a) Aluminum tolerance in wheat (*Triticum aestivum* L.). I. Uptake and distribution of aluminum in root apices. *Plant Physiol.* **103**: 685–693.

Delhaize E, Ryan PR, Randall PJ. (1993b) Aluminum tolerance in wheat (*Triticum aestivum* L.). II. Aluminum-stimulated excretion of malic acid from root apices. *Plant Physiol.* **103**: 695–702.

Devine MD, Hall JC, Romano ML, Marles MAS, Thomson LW, Shimabukuro RH. (1993) Diclofop and fenoxaprop resistance in wild oats is associated with an altered effect on the plasma membrane electric potential. *Pest. Biochem. Physiol.* **45**: 167–177.

Edmeades DC, Wheeler DM, Clinton OE. (1985) The chemical composition and ionic strength of soil solutions from New Zealand topsoils. *Aust. J. Soil Res.* **23**: 151–165.

Eisenberg RS. (1990) Channels as enzymes. *J. Memb. Biol.* **115**: 1–12.

Enz C, Steinkamp T, Wagner R. (1993) Ion channels in the thylakoid membrane (a patch-clamp study). *Biochim. Biophys. Acta* **1143**: 67–76.

Fairley-Grenot K, Assmann SM. (1991) Evidence for G-protein regulation of inward K⁺ current in guard cells of fava bean. *Plant Cell* **3**: 1037–1044.

Fairley-Grenot K, Assmann SM. (1992) Permeation of Ca²⁺ through K⁺-selective channels in the plasma membrane of *Vicia faba* guard cells. *J. Memb. Biol.* **128**: 103–113.

Fisahn J, Hansen U-P, Lucas WJ. (1992) Reaction kinetic model of a proposed plasma membrane two-cycle H⁺-transport system of *Chara corallina*. *Proc. Natl Acad. Sci. USA* **89**: 3261–3265.

Flügge UI, Benz R. (1984) Pore-forming activity in the outer membrane of the chloroplast envelope. *FEBS Lett.* **169**: 85–89.

Fricker MD, Tester M, Gilroy S. (1993) Fluorescence and luminescence techniques to probe ion activities in living plant cells. In: *Fluorescent and Luminescent Probes for Biological Activity* (Mason WT, ed.), London: Academic Press, pp. 360–377.

Gehring CA, Irving HR, Parish RW. (1990) Effects of auxin and abscisic acid on cytosolic calcium and pH in plant cells. *Proc. Natl Acad. Sci. USA* **87**: 9645–9649.

Gelli A, Blumwald E. (1993) Calcium retrieval from vacuolar pools. Characterization of a vacuolar calcium channel. *Plant Physiol.* **102**: 1139–1146.

Gilroy S, Read ND, Trewavas AJ. (1990) Elevation of cytoplasmic calcium by caged calcium or caged inositol trisphosphate initiates stomatal closure. *Nature* **346**: 769–771.

Gilroy S, Fricker MD, Read ND, Trewavas AJ. (1991) Role of calcium in signal transduction of *Commelina* guard cells. *Plant Cell* **3**: 333–344.

Hahm SH, Saunders MJ. (1991) Cytokinin increases intracellular Ca²⁺ in *Funaria*: detection with indo-1. *Cell Calcium* **12**: 675–681.

Häusler RE, Holtum JAM, Powles SB. (1991) Cross-resistance to herbicides in annual ryegrass (*Lolium rigidum*). IV. Correlation between membrane effects and resistance to graminicides. *Plant Physiol.* **97**: 1035–1043.

Hedrich R, Barbier-Brygoo H, Felle H *et al.* (1988) General mechanism for solute transport across the tonoplast of plant vacuoles: a patch-clamp survey of ion channels and proton pumps. *Bot. Acta* **101**: 7–13.

Hedrich R, Busch H, Raschke K. (1990) Ca²⁺ and nucleotide dependent regulation of voltage dependent anion channels in the plasma membrane of guard cells. *EMBO J.* **9**: 3889–3892.

Hill BS, Findlay GP. (1981) The power of movement in plants: the role of osmotic machines. *Q. Rev. Biophys.* **14**: 173–222.

Huang JW, Grunes DL, Kochian LV. (1992) Aluminum effects on the kinetics of calcium uptake into cells of the wheat root apex. Quantification of calcium fluxes using a calcium-selective vibrating microelectrode. *Planta* **188**: 414–421.

Irving HR, Gehring CA, Parish RW. (1992) Changes in cytosolic pH and calcium of guard cells precede stomatal movements. *Proc. Natl Acad. Sci. USA* **89:** 1790–1794.

Iwasaki I, Arata H, Kijima H, Nishimura M. (1992) Two types of channels involved in the malate ion transport across the tonoplast of a Crassulacean acid metabolism plant. *Plant Physiol.* **98:** 1494–1497.

Johannes E, Brosnan JM, Sanders D. (1992) Parallel pathways for intracellular Ca^{2+} release from the vacuole of higher plants. *Plant J.* **2:** 97–102.

Kearns EV, Assmann SM. (1993) The guard cell–environment connection. *Plant Physiol.* **102:** 711–715.

Keith CH, Ratan R, Maxfield FR, Bajer A, Shelanski ML. (1985) Local cytoplasmic calcium gradients in living mitotic cells. *Nature* **316:** 848–850.

Kinraide TB, Ryan PR, Kochian LV. (1994) Al^{3+}–Ca^{2+} interactions in aluminum rhizotoxicity. II. Evaluating the Ca^{2+}-displacement hypothesis. *Planta* **192:** 104–109.

Knight MR, Campbell AK, Smith SM, Trewavas AJ. (1991) Transgenic plant aequorin reports the effect of touch and cold-shock and elicitors on cytoplasmic calcium. *Nature* **352:** 524–526.

Knight MR, Smith SM, Trewavas AJ. (1992) Wind-induced plant motion immediately increases cytosolic calcium. *Proc. Natl Acad. Sci. USA* **89:** 4967–4971.

Knight MR, Read ND, Campbell AK, Trewavas AJ. (1993) Imaging calcium dynamics in living plants using semi-synthetic recombinant aequorins. *J. Cell Sci.* **121:** 83–90.

Koch MC, Steinmeyer K, Lorenz C et al. (1992) The skeletal muscle chloride channel in dominant and recessive human myotonia. *Science* **257:** 797–800.

Kumpf RA, Dougherty DA. (1993) A mechanism for ion selectivity in potassium channels: computational studies of cation-π interactions. *Science* **261:** 1708–1710.

Lucas WJ. (1976) Plasmalemma transport of HCO_3^- and OH^- in *Chara corallina*: non-antiporter systems. *J. Exp. Bot.* **27:** 19–31.

Lynch J, Polito VS, Läuchli A. (1989) Salinity stress increases cytoplasmic calcium activity in maize root protoplasts. *Plant Physiol.* **90:** 1271–1274.

Maathius FJM, Prins HBA. (1990) Patch clamp studies on root cell vacuoles of a salt-tolerant and a salt-sensitive *Plantago* species. Regulation of channel activity by salt stress. *Plant Physiol.* **92:** 23–28.

Maathius FJM, Sanders D. (1993) Energization of potassium uptake in *Arabidopsis thaliana*. *Planta* **191:** 302–307.

McAinsh MR, Brownlee C, Hetherington AM. (1992) Visualizing changes in cytosolic free Ca^{2+} during the response of stomatal guard cells to abscisic acid. *Plant Cell* **4:** 1113–1122.

MacRobbie EAC. (1988) Control of ion fluxes in stomatal guard cells. *Bot. Acta* **101:** 140–148.

Mannella C. (1992) The 'ins' and 'outs' of mitochondrial membrane channels. *Trends Biochem. Sci.* **17:** 315–320.

Marschner H. (1986) *Mineral Nutrition of Higher Plants.* London: Academic Press.

Marten I, Lohse G, Hedrich R. (1991) Plant growth hormones control voltage-dependent activity of anion channels in plasma membrane of guard cells. *Nature* **353:** 758–762.

Mathieu Y, Kurkdjian A, Xia H, Guern J, Koller A, Spiro MD, O'Neill M, Albersheim P, Darvill A. (1991) Membrane responses induced by oligogalacturonides in suspension-cultured tobacco cells. *Plant J.* **1:** 333–343.

Matzke AJM, Behensky C, Weiger T, Matzke MA. (1992) A large conductance ion channel in the nuclear envelope of a higher plant cell. *FEBS Lett.* **302:** 81–85.

Mehta M, Malik MK, Khurana JP, Maheshwari SC. (1993) Phytochrome modulation of calcium fluxes in wheat (*Triticum aestivum* L.) protoplasts. *Plant Growth Reg.* **12:** 293–302.

Miller DD, Callaham DA, Gross DJ, Hepler PK. (1992) Free Ca^{2+} gradient in growing pollen tubes of *Lilium*. *J. Cell Sci.* **101:** 7–12.

Montague MJ. (1993) Calcium antagonists inhibit sustained gibberellic acid-induced growth of *Avena* (oat) stem segments. *Plant Physiol.* **101:** 399–405.

Munns R. (1993) Physiological processes limiting plant growth in saline soils: some dogmas and hypotheses. *Plant Cell Environ.* **16:** 15–24.

O'Halloran TV. (1993) Transition metals in control of gene expression. *Science* **261:** 715–725.

Okazaki Y, Tazawa M. (1990) Calcium ion and turgor regulation in plant cells. *J. Memb. Biol.* **114:** 189–194.

Okazaki Y, Iwasaki N. (1991) Injection of a Ca^{2+}-chelating agent into the cytoplasm retards the progress of turgor regulation upon hypotonic treatment in the alga, *Lamprothamnium*. *Plant Cell Physiol.* **32:** 185–194.

Piñeros M, Tester M. (1995) Characterization of a voltage dependent Ca^{2+}-selective channel in the plasma membrane of wheat roots. *Planta* **195:** 478–488.

Pottosin I. (1993) One of the chloroplast envelope ion channels is probably related to the mitochondrial VDAC. *FEBS Lett.* **330:** 211–214.

Read ND, Allan WTG, Knight H, Knight MR, Malhó R, Russell A, Shacklock PS, Trewavas AJ. (1992) Imaging and measurement of cytosolic free calcium in plant and fungal cells. *J. Microscopy* **166:** 57–86.

Reid RJ, Tester M. (1992) Measurements of Ca^{2+} fluxes in intact plant cells. *Phil Trans. R. Soc. Lond. B* **338:** 73–82.

Reid RJ, Tester M, Smith FA. (1993) Effects of salinity and turgor on calcium influx in *Chara. Plant Cell Environ.* **16:** 547–554.

Reid RJ, Tester M, Smith FA. (1995) Calcium/aluminium interactions in the cell wall and plasma membrane of *Chara. Planta* **195:** 362–368.

Riordan JR. (1993) The cystic fibrosis transmembrane conductance regulator. *Annu. Rev. Physiol.* **55:** 609–630.

Schachtman DP, Tyerman SD, Terry BR. (1991) The Na^+/K^+ selectivity of a cation channel in the plasma membrane of root cells does not differ in salt-tolerant and salt-sensitive wheat species. *Plant Physiol.* **97:** 598–605.

Schachtman DP, Schroeder JI, Lucas WJ, Anderson JA, Gaber RF. (1992) Expression of an inward-rectifying potassium channel by the *Arabidopsis KAT1* cDNA. *Science* **258:** 1654–1658.

Scheel D, Colling C, Hedrich R, Kawalleck P, Parker JE, Sacks WR, Somssich IE, Hahlbrock K. (1991) Signals in plant defense gene activation. *Adv. Mol. Genet. Plant-Microbe Interact.* **1:** 373–380.

Schroeder JI. (1992) Plasma membrane ion channel regulation during abscisic acid-induced closing of stomata. *Phil. Trans. R. Soc. Lond. B* **338:** 83–89.

Schroeder JI, Hagiwara S. (1989) Cytosolic calcium regulates ion channels in the plasma membrane of *Vicia* guard cells. *Nature* **338:** 427–430.

Schroeder JI, Fang HH. (1991) Inward-rectifying K^+ channels in guard cells provide a mechanism for low affinity K^+ uptake. *Proc. Natl Acad. Sci. USA* **88:** 11583–11587.

Sentenac H, Boonneaud N, Minet M, Lacroute F, Salmon J-M, Gaymard F, Grignon C. (1992) Cloning and expression in yeast of a plant potassium ion transport system. *Science* **256:** 663–665.

Shacklock PS, Read ND, Trewavas AJ. (1992) Cytosolic free calcium mediates red light-induced photomorphogenesis. *Nature* **358:** 753–755.

Sibaoka T. (1991) Rapid plant movements triggered by action potentials. *Bot. Mag.* **104:** 73–95.

Simons P. (1992) *The Action Plant.* Oxford: Blackwell Science Ltd.

Skerrett M, Tyerman SD. (1993) A channel that allows inwardly directed fluxes of anions in protoplasts derived from wheat roots. *Planta* **102:** 295–305.

Smack DP, Colombini M. (1985) Voltage-dependent channels found in the membrane fraction of corn mitochondria. *Plant Physiol.* **79:** 1094–1097.

Smith JAC, Bryce JH. (1992) Metabolite compartmentation and transport in CAM plants. In: *Plant Organelles,* SEB Seminar Series 50 (Tobin A, ed.). Cambridge: Cambridge University Press, pp. 141–167.

Smith JR, Smith FA, Walker NA. (1987) Potassium transport across the membranes of *Chara.* I. The relationship between radioactive tracer influx and electrical conductance. *J. Exp. Bot.* **38:** 731–751.

Smith SE, Dickson S, Morris C, Smith FA. (1994) Transfer of phosphate from fungus to plant in VA mycorrhizas: calculation of the area of symbiotic interface and of fluxes of P from two different fungi to *Allium porrum* L. *New Phytol.* **127:** 93–99.

Steinmeyer K, Klocke R, Ortland C, Gronemeier M, Jockusch H, Gründer S, Jenntsch TJ. (1991) Inactivation of muscle chloride channel by transposon insertion in myotonic mice. *Nature* **354:** 304–308.

Stoeckel H, Takeda K. (1993) Plasmalemmal, voltage-dependent ionic currents from excitable pulvinar motor cells of *Mimosa pudica. J. Memb. Biol.* **131:** 179–192.

Strong M, Chandy KG, Gutman GA. (1993) Molecular evolution of voltage-sensitive ion channel genes: on the origins of electrical excitability. *Mol. Biol. Evol.* **10:** 221–242.

Taylor AR, Brownlee C. (1992) Localized patch-clamping of plasma membrane of a polarized plant cell. Laser microsurgery of the *Fucus spiralis* rhizoid cell wall. *Plant Physiol.* **99:** 1686–1688.

Terry BR, Tyerman SD, Findlay GP. (1991) Ion channels in the plasma membrane of *Amaranthus* protoplasts: one cation and one anion channel dominate the conductance. *J. Memb. Biol.* **121:** 223–236.

Terry BR, Findlay GP, Tyerman SD. (1992) Direct effects of Ca^{2+}-channel blockers on plasma membrane cation channels of *Amaranthus tricolor* protoplasts. *J. Exp. Bot.* **43:** 1457–1473.

Tester M. (1988) Blockade of potassium channels in the plasmalemma of *Chara corallina* by tetraethylammonium, Ba^{2+}, Na^+ and Cs^+. *J. Memb. Biol.* **105:** 77–85.

Tester M. (1990) Plant ion channels: whole-cell and single channel studies. *New Phytol.* **114:** 305–340.

Tester M, MacRobbie EAC. (1990) Cytoplasmic calcium affects the gating of potassium channels in the plasma membrane of *Chara corallina*: a whole-cell study using calcium channel effectors. *Planta* **180:** 569–581.

Tester M, Smith FA, Smith SE. (1992) The role of ion channels in controlling solute exchange in mycorrhizal associations. In: *Mycorrhizas in Ecosystems* (Read DJ, Lewis DH, Fitter AH, Alexander IJ, ed.). Wallingford: CAB International, pp. 348–351.

Thain JF, Doherty HM, Bowles DJ, Wildon DC. (1990) Oligosaccharides that induce proteinase inhibitor activity in tomato plants cause depolarization of tomato leaf cells. *Plant Cell Environ.* **13:** 569–574.

Thiel G, MacRobbie EAC, Blatt MR. (1992) Membrane transport in stomatal guard cells: the importance of voltage control. *J. Memb. Biol.* **126:** 1–18.

Thiel G, Blatt MR, Fricker MD, White IR, Millner P. (1993) Modulation of K^+ channels in *Vicia* stomatal guard cells by peptide homologs to the auxin-binding protein C terminus. *Proc. Natl Acad. Sci. USA* **90:** 11493–11497.

Tyerman SD. (1992) Anion channels in plants. *Annu. Rev. Plant Physiol. Plant Mol. Biol.* **43:** 351–373.

Tyerman SD, Findlay GP, Patterson GJ. (1986) Inward membrane current in *Chara inflata*. II. Effects of pH, Cl^- channel blockers and NH_4^+, and significance for the hyperpolarised state. *J. Memb. Biol.* **89:** 153–161.

Tyerman SD, Terry BR, Findlay GP. (1992) Multiple conductances in the large K^+ channel from *Chara corallina* shown by a transient analysis. *Biophys. J.* **61:** 736–749.

Valverde MA, Díaz M, Sepúlveda FV, Gill DR, Hyde SC, Higgins CF. (1992) Volume-regulated chloride channels associated with the human multidrug-resistance P-glycoprotein. *Nature* **355:** 830–833.

Walker NA, Beilby MJ, Smith FA. (1979) Amine uniport at the plasmalemma of charophyte cells. I. Current–voltage curves, saturation kinetics, and effects of unstirred layers. *J. Memb. Biol.* **49:** 21–55.

Wang MY, Glass ADM, Shaff JE, Kochian LV. (1993) Ammonium uptake by rice roots. III. Electrophysiology. *Plant Physiol.* **104:** 899–906.

Wang X-C, Berkowitz GA, Peters JS. (1993) K^+ conducting ion channel of the chloroplast inner envelope: functional reconstitution into liposomes. *Proc. Natl Acad. Sci. USA* **90:** 4981–4985.

White PJ. (1993) Characterization of a high-conductance, voltage-dependent cation channel from the plasma membrane of rye roots in planar lipid bilayers. *Planta* **191:** 541–551.

White PJ, Tester M. (1992) Potassium channels from plasma membrane of rye roots characterized following incorporation into planar bilayers. *Planta* **186:** 188–202.

Wildon DC, Thain JF, Minchin PEH, Gubb IR, Reilly AJ, Skipper YD, Doherty HM, O'Donnell PJ, Bowles DJ. (1992) Electrical signalling and systemic proteinase inhibitor induction in the wounded plant. *Nature* **360:** 62–65.

Wu W, Berkowitz GA. (1992) Stromal pH and photosynthesis are affected by electroneutral K^+ and H^+ exchange through chloroplast envelope ion channels. *Plant Physiol.* **98:** 666–672.

Yang J, Ellinor PT, Sather WA, Zhang J-F, Tsien RW. (1993) Molecular determinants of Ca^{2+} selectivity and ion permeation in L-type Ca^{2+} channels. *Nature* **366:** 158–161.

Zherelova OM, Grishchenko VM. (1987) Effect of a calmodulin-like protein on activation of Ca^{2+} channels in *Nitellopsis obtusa*. *Doklady Bot. Sci.* **293:** 47–49.

Zorec R, Tester M. (1992) Cytoplasmic calcium stimulates exocytosis in a plant secretory cell. *Biophys. J.* **63:** 864–867.

Zorec R, Tester M. (1993) Rapid pressure driven exocytosis–endocytosis cycle in a single plant cell. Capacitance measurements in aleurone protoplasts. *FEBS Lett.* **333:** 283–286.

Zorov DB, Kinnally KW, Petrini S, Tedeschi H. (1992) Multiple conductance levels in rat heart inner mitochondrial membranes studied by patch-clamping. *Biochim. Biophys. Acta* **1105:** 263–270.

Chapter 15

1996

taxane

Osmo-

Mechanosensitive ion channels

A. Garrill, G.P. Findlay and S.D. Tyerman

1. Introduction

Ion channels in cell membranes can generally be classified into three groups, voltage-gated, ligand- (or receptor-) gated and mechanosensitive, reflecting the cues that cause them to open and close. This chapter deals with the third group, the mechanosensitive (MS) channels. These channels, first described by Guharay and Sachs (1984) and Brehm *et al.* (1984), appear to be ubiquitous, having been described in numerous systems, both pro- and eukaryotic, as well as in sensory and non-sensory cells (Morris, 1990; Sachs, 1988, 1989, 1992). At present they are perhaps the least well understood of the three major groups of ion channels.

Mechanosensitive channels include stretch-activated channels, stretch-inactivated channels, displacement-sensitive channels and the putative shear-stress-sensitive channels (Morris, 1990). We will consider the MS channels of plant membranes together with those of bacterial and fungal membranes. For the latter two groups, it is reasonable to surmise that as walled cells which generate turgor pressure, the MS channels of these different types of organisms are likely to have similar properties and functions to those in higher plant cells. Where relevant, we shall also draw comparisons with animal systems.

It is really not surprising that MS channels have been identified in plant cells, considering that many of the cues in morphogenesis are mechanical. It is thought that the channels may act as mechanoreceptors or transducers converting mechanical stimuli into ionic fluxes that are then likely to set in motion a variety of biochemical reactions, leading to a cellular response to the initial stimuli. To date, several MS channels have been described in plant cells, that differ in ion selectivity, conductance and voltage sensitivity. General features are their activation by tension, blockage by gadolinium and greater permeability to cations than to anions, apart from channels in *Nicotiana* (Falke *et al.*, 1988) and *Samanea* (Moran, 1990).

2. Studying MS channels

MS channels have mainly been studied with the patch-clamp technique (Hamill *et al.*, 1981), using whole-cell preparations, in which the activities of an ensemble of channels from the whole plasma membrane are measured, and detached patch preparations in which it is possible to resolve the currents flowing through single ion channels. Activation or deactivation are usually measured as changes in the channel open probability (P_o) or mean current, while the mechanical force is applied as negative or positive pressure via the patch pipette (*Figure 1a*). The pressure can best be applied using a pressure clamp, an apparatus in which a balance between negative and positive pressures enables stepwise changes in pressure that are both rapid (*c.* 10 msec) and precise (McBride and Hamill, 1992). More recently, the coupling of electrophysiology with molecular biology has enabled advances such as those described in Section 3.

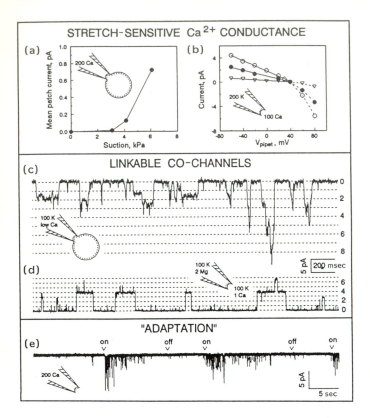

Figure 1. The properties of MS Ca²⁺-selective channels from onion bulb scale parenchyma. (a) Increasing the suction via the pipette gives an increase in the mean patch current across a cell-attached patch, suggesting the presence of MS channels. The pipette contained 200 mM CaCl₂ and the pipette potential was clamped to 90 mV. (b) The selectivity of the channels for Ca²⁺ over K⁺ can be shown by measuring the current flowing across an outside-out patch as a function of clamped voltage when electrochemically comparable concentrations of the two ions are present on opposite sides of the membrane (triangles, filled circles and open circles represent current through one, two and three co-channels, respectively). The movement of cations into the pipette when the membrane potential is clamped at 0 mV is suggestive of Ca²⁺ selectivity. (c) Current flowing through suspected linkable co-channels in a cell-attached patch with 100 mM K⁺ + low Ca²⁺ outside and a pipette potential of 60 mV. (d) As for (c), but in an outside-out patch with 2 mM Mg²⁺ + 100 mM K⁺ in the pipette vs. 1 mM Ca²⁺ + 10 mM K⁺ in the bath and a membrane potential of –90 mV. (e) During periods of more prolonged suction, the MS channels show 'adaptive' behaviour, becoming less sensitive to the stimulus. The sensitivity may be regained after a period with no suction. The membrane potential is –100 mV and suction is 2.5 kPa. Reproduced from Pickard and Ding (1993) with permission from CSIRO Editorial Services.

3. MS channel genes

An important breakthrough has been the recent cloning of the gene encoding the large conductance channel of *Escherichia coli* (Sukharev *et al.*, 1994). The authors initially purified an envelope fraction that displayed mechanosensitive activity *in vitro*. The protein responsible for this activity was isolated and partially sequenced, and the information thus obtained was used to identify and clone the *mscL* gene. Knockout, rescue and overexpression experiments, as well as the ability to use *mscL* as a template in a reticulocyte expression system, provide further evidence that *mscL* encodes the large conductance channel and that 'mscL alone is not only necessary but sufficient for the entire structure responsible for ion conduction and mechanosensitivity' (Sukharev *et al.*, 1994). It is interesting to note that *mscL* is located only 133 nucleotides downstream from a gene encoding a protein responsible for K^+ uptake, a fact that may have implications when considering the functions of the channel (see below).

The open reading frame of *mscL* indicates a protein of 136 amino acids with a molecular mass of *c.* 15 kDa. The functional channel is thought to be a homomultimer (possibly a tetramer (Huse *et al.*, 1995) of several 15-kDa subunits (Sukharev *et al.*, 1995)). Hydropathy plots of the protein sequence suggest a hydrophobic core containing two membrane-spanning domains and a hydrophilic C-terminus. Only approximately 75% of the amino acids are required for activity, as mutants with 25 C-terminal residues deleted are able to form functional channels (Sukharev *et al.*, 1995).

In eukaryotic systems the picture is not quite as clear, indeed to the best of our knowledge there have been no reports of MS channel genes in plant or fungal cells. There has been speculation that the *mec-4*, *mec-6* and *mec-10* genes encode a heteromultimeric ion channel in the nematode *Caenorhabditis elegans* (Hong and Driscoll, 1994; Huang and Chalfie, 1994). If this is the case, then the channel could be mechanosensitive, as mutations in the above genes give worms that are insensitive to touch. While there is as yet no direct evidence to suggest that the *mec* genes actually encode an ion channel, they do show considerable sequence homology with the rat epithelial Na^+ channel (Canessa *et al.*, 1994) (none of these show any homology with *mscL*). This channel is composed of three subunits, and the same channel from bovine kidney has recently been reported to show mechanosensitive activity (Awayda *et al.*, 1995).

4. Channel gating

It is thought that tension, and not other consequences of applying pressure, gates MS channels. Gustin *et al.* (1988) utilized Laplace's law which states that, for a thin-walled sphere at equilibrium, the tension (T) is proportional to the pressure (P) multiplied by the diameter (d); $T = Pd/4$. For the MS channels in yeast (described in more detail below), it was found that P_o was a function of both the applied pressure and cell size, as expected if the channels were gated by tension.

How then is the tension transferred to the channels? Sokabe *et al.* (1991) have demonstrated that, in response to a step in pressure, the area of a patch increases by more than 10%, far greater than the maximum possible increase of 2% due to elastic or intrinsic stretching of the lipids (Wolfe and Steponkus, 1981; Wolfe *et al.*, 1985). The increase in area is accounted for by the incorporation of lipid into the membrane from some internal reservoir (Sokabe *et al.*, 1991; Wolfe and Steponkus, 1983; Wolfe *et al.*, 1985). This would suggest that, in the steady state, patch lipid is stress-free and the channels must be gated via tension in some parallel structure, possibly the cytoskeleton. The presence of a membrane-attached cytoskeleton is implicated by the observation that the tensions of 6 to 12 kN m^{-1} required to rupture whole cells (Gustin *et al.*, 1988) are two to three times greater than that required to rupture phospholipid vesicles (3 to 4 kN m^{-1}).

The nature of these cytoskeletal components is unknown at present, although there have been suggestions that in animal cells they may be members of the spectrin/fodrin family (Sachs, 1988). In plant cells, emphasis has been placed on the involvement of glycoproteins, possibly integrins and the extracellular matrix (ECM), focusing mechanical stress from the wall on MS channels in the plasma membrane (Edwards and Pickard, 1987; Pont-Lezica *et al.*, 1993; Roberts, 1990; Wayne *et al.*, 1990). Significantly, there have been demonstrations of proteins similar to fibronectin, vitronectin, vinculin and subunits of integrin in plant and fungal cells (Faraday and Spanswick, 1993; Kaminskyj and Heath, 1995; Odani *et al.*, 1987; Quatrano *et al.*, 1991; Sanders *et al.*, 1991). Wayne *et al.* (1992) have shown that a mechanoreceptor which is likely to be gravisensing in *Chara* internodal cells is inhibited in a concentration-dependent manner by the tetrapeptide Arg-Gly-Asp-Ser (RGDS), a compound that inhibits interactions between integrins and proteins of the ECM (Schindler *et al.*, 1989). A second mechanoreceptor in *Chara*, responsible for a touch-induced action potential, does not contain RGDS-binding integrin-like proteins, and cytoskeletal inhibitor studies suggest that this system may be analogous to animal systems involving spectrin-like molecules (Staves and Wayne, 1993). *Figure 2*, which shows a modification of a model first proposed by Pickard and Ding (1993), suggests how the various cytoskeletal components may be arranged relative to MS channels in a plant cell.

It is possible that local distortions in the lipid bilayer itself may in some instances also transmit sufficient tension to activate MS channels (Hamill and McBride, 1994). Opsahl and Webb (1994) have reconstituted alamethicin (a channel-forming peptide of 20 amino acids pro-duced as an antibiotic by the fungus *Trichoderma veride*) into pure lipid membranes devoid of cytoskeletal components, and have demonstrated mechanosensitivity. The MscS and MscL channels of *E. coli* (described below) can be reconstituted in liposomes (by analogy, devoid of the peptidoglycan layer) and retain their mechanosensitivity (Sukharev *et al.*, 1993). Further-

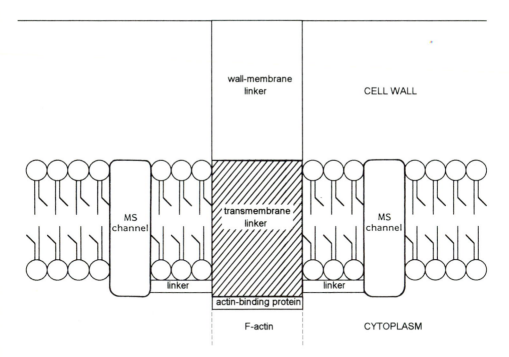

Figure 2. Hypothetical arrangement of MS channels and cytoskeletal elements in the plasma membrane of a plant cell, modified from an original model proposed by Pickard and Ding (1993). Stress may be transferred from the cell wall to the plasma membrane via a wall–membrane linker which is attached to a transmembrane linker analogous to the integrins of animal cells (in the current model this is shown as a direct attachment, but alternatively it may be via some attachment protein). The transmembrane linker may be attached to MS channels directly, or via linker proteins which lie just at the periphery of the cytoplasm (as shown in the current example), or alternatively within the membrane. In addition, the transmembrane linker may be attached to cytoskeletal elements such as F-actin via actin-binding proteins. The actin may determine the cellular location of the channels and also transfer force to the transmembrane linker and thus the MS channels.

more, activation of MS channels through the addition of various amphipathic compounds (that insert preferentially into either the inner or outer leaflet of the lipid bilayer, depending on their charge, and thus result in contour changes in the bilayer) would suggest that a mechanical gating force can come from the surrounding lipids (Martinac *et al.*, 1990). More recently, Sukharev *et al.* (1995) have reported that the opening of the MscL channel can be influenced by agents that affect hydrophobic interactions, and they suggest that channel opening may be a result of stretch-induced breakage of the hydrophobic interactions between certain protein domains. The osmotic sensitivity of certain MS channels may be indicative of a gating mechanism that involves an osmotic gradient between the internal volume of the channel and the medium, generated by large molecules which are excluded from channel entry (Parsegian *et al.*, 1992).

5. Clustering of channels

There have been several reports which suggest that MS channels may occur in clusters of as many as nine conductance units (Alexandre and Lassalles, 1991; Ding and Pickard, 1993a; Garrill *et al.*, 1993; Szabo *et al.*, 1990; Zoratti *et al.*, 1990) (Figure 1c, d). In fungal cells it is possible that these clusters may arise because of an interaction between the channels and peripheral actin arrays (Garrill *et al.*, 1993; Levina *et al.*, 1994). Ding and Pickard (1993a) suggest that such arrangements may offer sensory benefits to the cells, such as permitting quantitative responses to qualitatively different signals.

6. Pharmacology

One of the more frustrating aspects of the work on MS channels has been the lack of a specific inhibitor for these channels. Hamill *et al.* (1992) have investigated the effects of amiloride on the MS channel of *Xenopus* oocytes, and report an IC_{50} of 500 μM (at a potential difference of 100 mV). Certain amiloride analogues, in particular bromohexamethylene amiloride (BrHMA), show greater potency. Many of the studies cited in this chapter utilize the trivalent lanthenide gadolinium. It is important to note that this ion is far from specific, and may also block calcium channels (Lansman, 1990), endplate channels (Yang and Sachs, 1989) and, in *Amaranthus tricolor*, the time-dependent K^+ outward rectifier (Terry *et al.*, 1992). Caution must therefore be exercised in the attribution of a physiological role to stretch-activated channels on the basis of inhibitor effects on cells, although it must also be stated that it is currently the best pharmacological tool available.

7. MS channels in walled cells

We shall first describe the various MS channels that have been described in protoplasts or membrane patches from walled cells (summarized in *Table 1*), and then review their respective physiological functions.

Higher plant cells

MS channels have been observed in the plasma membrane of cells derived from various tissues of several species of plant. In most cell types studied so far the MS channels are cation-selective. However, guard cells contain three types of MS channel, specific for K^+, Cl^- and Ca^{2+}, respectively, probably reflecting the importance of the turgor/volume relationship of these cells in the control of stomatal aperture (Cosgrove and Hedrich, 1991). These channels were distinguished from previously described non-MS channels by their conductances, kinetics, voltage dependence and pressure sensitivity.

Table 1. MS channels of walled cells

Species and membrane origin	Channel conductance (pS)	Permeability
Plant		
Vicia faba guard cell plasma membrane (Cosgrove and Hedrich, 1991)	6	Ca^{2+}
Vicia faba guard cell plasma membrane (Cosgrove and Hedrich, 1991)	27 (outward current) 13 (inward current)	Cl^-
Vicia faba guard cell plasma membrane (Cosgrove and Hedrich, 1991)	50 (outward current) 25 (inward current)	K^+
Nicotiana leaf mesophyll plasma membrane (Falke *et al.,* 1988)	86 (outward current) 146 (inward current)	Anions
Samanea saman pulvinar motor cell plasma membrane (Moran, 1990)	15–20	Anions
Arabidopsis leaf plasma membrane (Spalding and Goldsmith, 1993)	Not determined	Unselective
Allium bulb scale inner epidermis plasma membrane (Ding and Pickard, 1993a, b)	6.5	$Ca^{2+} > K^+$
Zostera meulleri leaf epidermis plasma membrane (Garrill *et al.,* 1994)	100	K^+ (not yet determined for other ions)
Beta tonoplast (Alexandre and Lassalles, 1991)	20	$K^+ > Cl^-$
Allium bulb scale parenchyma tonoplast (Badot *et al.,* 1992)	Not determined	Not determined
Fungal		
Saccharomyces cerevisiae plasma membrane (Gustin *et al.,* 1988)	36	Cations, anions
Schizosaccharomyces pombe plasma membrane (Zhou and Kung, 1992)	180	Cations
Uromyces appendiculatus plasma membrane (Zhou *et al.,* 1991)	600	Cations
Saprolegnia ferax plasma membrane (Garrill *et al.,* 1992)	Not determined	K^+, Ca^{2+}
Saprolegnia ferax plasma membrane (Garrill *et al.,* 1992)	Not determined	Mg^{2+}
Neurospora crassa plasma membrane (Levina *et al.,* 1995)	Not determined	Ca^{2+}
Bacterial		
Escherichia coli inner membrane (Berrier *et al.,* 1989, 1992; Buechner *et al.,* 1990; Cui *et al.,* 1995; Delcour *et al.,* 1989; Martinac *et al.,* 1987; Sukharev *et al.,* 1993, 1994, 1995; Szabo *et al.,* 1990)	970 and several lower conductances	Anions > cations
Escherichia coli inner membrane (Berrier *et al.,* 1989, 1992; Buechner *et al.,* 1990; Delcour *et al.,* 1989; Martinac *et al.,* 1987; Sukharev *et al.,* 1993, 1994, 1995)	3 nS (MscL)	Non-selective
Steptococcus faecalis cytoplasmic membrane (Zoratti and Petronilli, 1988)	Numerous from 10 pS to several nS	Non-selective
Bacillis subtilis cytoplasmic membrane (Zoratti *et al.,* 1990)	Numerous in the nS range	Not determined

Voltage-, temperature- and auxin-sensitive MS channels of onion bulb epidermal cells have been extensively studied by Ding and Pickard (1993a,b) (*Figure 1a–e*), and Pickard and Ding (1993) suggest that they serve as a major component of a proposed plasmalemmal control centre integrating signals from mechanical, electrical, thermal and chemical sources. It is the high Ca^{2+} permeability of these channels that has aroused most interest with regard to the possible functions of such channels, given the involvement of Ca^{2+} in intracellular signalling and the fact that increases in intracellular Ca^{2+} often accompany mechanical stimulation (*Figure 1b*). Knight *et al.* (1991) have shown that *Nicotiana phonbaginifolia* plants expressing aequorin, a Ca^{2+} sensitive luminescent protein, luminesce when gently touched, thus indicating a rise in cytosolic Ca^{2+}, although the only MS channel described in this genus so far is anion-selective (Falke *et al.*, 1988). When prodded, animal cells loaded with Ca^{2+}-sensitive dyes show waves of Ca^{2+} which emanate from the site of stimulus (Sigurdson and Sachs, 1991). The MS channels of onion epidermal cells lose their sensitivity with continual stimulation, yet are able to regain sensitivity after a period of rest (Figure 1e). Pickard and Ding (1993) argue that these responses indicate an accommodation to ambient stress which allows the channels to respond to stresses over a wide range of background levels.

To date there have been two reports of MS channels in the plant tonoplast (Alexandre and Lasalles, 1991; Badot *et al.*, 1992). Alexandre and Lasalles (1991) found a 20-pS channel in tonoplast from red beet that was stretch-activated. They speculate that this channel may be the same as the 15-pS channel previously identified but not characterized by Coyaud *et al.* (1987) in sugar beet tonoplast.

Fungi

Several mechanosensitive ion channels have been described in fungal membranes. The channels in the plasma membrane of the yeasts *Saccharomyces cerevisiae* (Gustin *et al.*, 1988), *Schizosaccharomyces pombe* (Zhou and Kung, 1992) and the filamentous species *Uromyces appendiculatus* (Zhou *et al.*, 1991) are all blocked by submillimolar concentrations of Gd^{3+}, yet differ in several important characteristics. Channel conductances of 36 pS for *S. cerevisiae*, 180 pS for *S. pombe* and 600 pS for *U. appendiculatus* have been reported, yet surprisingly the greater the conductance the higher the ionic selectivity, with P_K/P_{Cl} values of 1.6, 3.6 and 19.3, respectively. There appear to be differences in the kinetics of the channels, for example the open-channel current of *S. pombe*, in contrast to that of *U. appendiculatus*, flickers rapidly. In addition, the channels differ in their voltage sensitivity, which may have important implications in the regulation of the channels in their open states. The MS channels of *S. pombe*, in contrast to those of *S. cerevisiae* and *U. appendiculatus,* become less activated at depolarized membrane potentials. As suggested by Zhou and Kung (1992), this voltage sensitivity could form part of a negative feedback loop, where mechanical stimulation of MS channels leads to depolarization that in turn inactivates the channel, thereby reducing the time that the channel is open *in vivo*.

MS channels have been observed in *Neurospora crassa* and in the oomycete *Saprolegnia ferax* (whether this latter organism is in fact a fungus is a point of contention for many mycologists), although in each of these studies seal resistances of the order of 0.5 gigaohms precluded detailed characterization of the channels (Garrill *et al.*, 1992, 1993; Levina *et al.,* 1994, 1995). The protoplasting regime used did enable a description of the channel distribution along the hyphae and thus an evaluation of the role of the channel in polarized growth (discussed in more detail below). These channels are permeable to Ca^{2+}.

Bacteria

Bacterial cells are amenable to the patch-clamp technique through the production of giant spheroplasts (with diameters up to 6 μM), obtained by treating cells with septum formation inhibitors such as cephalexin followed by lysozyme and ethylenediamine tetra-acetic acid (EDTA) (Martinac *et al.*, 1987). Studies with both Gram-positive species (Zoratti and Petronilli, 1988; Zoratti *et al.*, 1990) and Gram-negative species (Berrier *et al.*, 1989; Buechner *et al.*, 1990; Delcour *et al.*, 1989; Martinac *et al.*, 1987) have demonstrated the existence of MS channels with several conductances in the nS range, although it is possible that these conductances arise from the co-operative gating in multiplexes of one or more channel types (Szabo *et al.*, 1990; Zoratti and Petronilli, 1988). Following repeated stretch cycles these channels may begin to act in a spontaneous manner. There has been debate as to the location of such channels (Berrier *et al.*, 1989; Delcour *et al.*, 1989). In Gram-positive species, they can, of course, only be located in the cytoplasmic membrane. As for *E. coli*, Western blots performed with MscL-specific antibodies have now specifically located this channel to the inner membrane (P. Blount *et al.*, personal communication).

8. Functions

The question of the physiological significance of MS channels was raised by Morris and Horn (1991), who were unable to find the expected MS whole-cell currents despite various stimuli, yet were readily able to detect mechanosensitive currents in single-channel studies on cultured molluscan neurones and growth cones. Their findings raised the possibility that the significant currents observed in single-channel studies were a manifestation of, for example, cytoskeletal disruption during patch formation, which simply did not occur in undisturbed cells. The argument has been countered by Gustin (1991), who showed that a current could indeed be activated in whole-cell preparations (upon cell inflation) which was of a magnitude expected from single-channel studies. Similar currents are also apparent in studies on other fungal species (Zhou *et al.*, 1991; Zhou and Kung, 1992). Furthermore, it was suggested (Sachs et al., 1991) that the whole-cell currents reported by Morris and Horn (1991) were not as expected because of difficulties in uniformly stimulating a whole cell and therefore estimating the expected current. It is clear that checks should be made in other systems along the lines of those made for yeast, especially in view of how little we know of the mechanism of seal formation in patch clamping, and the fact that apparent channel events can result from simply pushing patch electrodes against hydrophobic substrates (Sachs and Qin, 1993).

Having established that MS channels are unlikely simply to be a patch-clamping artefact, it is important to show that the pressures required for activation are in the physiological range. In the studies by Cosgrove and Hedrich (1991) on guard cells, the pressures needed to induce channel opening in a patch were of the order of 2–10 kPa. Although these pressures are much smaller than the turgor pressure of intact guard cells (which may be in excess of 1000 kPa when the stoma is open), it should be remembered that almost all of the turgor-generated force is borne by the cell wall, which has a greater elastic modulus than the parallel membrane–cytoskeletal system. Cosgrove and Hedrich (1991) have calculated that the membrane cytoskeletal system need bear only 0.05% of the total mechanical stress generated by turgor to bring a pressure of 2.5 kPa into the physiological range.

Osmoregulation

MS channels have been implicated as osmo-receptors/regulators in bacterial (Berrier *et al.*, 1989, 1992; Delcour *et al.*, 1989; Sukharev *et al.*, 1994), fungal (Gustin *et al.*, 1988; Zhou and Kung,

1992), plant (Cosgrove and Hedrich, 1991; Garrill *et al.*, 1994; Moran, 1990) and animal cell membranes (Christensen, 1987; Cornet *et al.*, 1993; Falke and Misler, 1989; Filipovic and Sackin, 1992; Morris *et al.,* 1989; Ollet and Bourque, 1993; Ubl *et al.*, 1988), as well as the tonoplast of plants (Alexandre and Lassalles, 1991; Badot *et al.*, 1992). In many of the above studies MS channels were shown to activate in response to reductions in the osmotic potential of the cell bathing media.

Berrier *et al.* (1992) have shown MS channels to be blocked with submillimolar concentrations of Gd^{3+}, a treatment that also has the effect of inhibiting the release of metabolites such as lactose and ATP in *E. coli* and ATP in *S. faecalis* when the cells are subjected to hypotonic osmotic stress. It is therefore possible that MS channels provide a means for the cells to unload solutes. As stated above, *mscL,* the gene encoding the large conductance MS channel of *E. coli,* is located only 133 nucleotides downstream from the gene that encodes a K^+ transporter (both genes are suspected of being transcribed clockwise) (Sukharev *et al.*, 1994). Potassium is the most important osmoticum in *E. coli,* and a number of K^+ transport systems are regulated by the osmolarity of the medium. A null mutation in one of these transport systems (KefA) has recently been shown to affect MS channel activity (Cui *et al.*, 1995).

Cosgrove and Hedrich (1991) have proposed that the MS channels in guard cells may function in the relief of excess turgor pressure. In the case of MS K^+ and Cl^- channels this may occur directly as well as indirectly. The efflux of K^+ and Cl^- that occurs when the channels activate (in response to the excess pressure) would lead directly to water efflux and a decrease in volume. The efflux of Cl^- may depolarize the membrane potential, thus activating voltage-dependent non-MS channels such as the outward rectifying K^+ and anion channels (Keller *et al.*, 1989; Schroeder *et al.*, 1987), with a consequent further efflux of ions. In addition, as suggested by Cosgrove and Hedrich (1991), an influx of Ca^{2+} through the MS Ca^{2+} channel could increase cytoplasmic Ca^{2+} and activate Ca^{2+}-dependent anion channels (Schroeder and Hagiwara, 1989). This could give rise to anion efflux, depolarization of membrane potential, inhibition of inward-rectifying K^+ channels and activation of outward-rectifying K^+ channels, with a resultant efflux of ions and water. The guard cell MS channels appear to have an important role in leaf gas exchange as a result of their role in guard cell turgor regulation.

Tropic responses

The observation that Gd^{3+} can block gravitropism (Millet and Pickard, 1988) has led to the suggestion that MS channels may be involved in the response of plant cells to gravity. These channels would act as the gravireceptor, the cellular component that transduces the 'signal' from the susceptor, that is acted upon directly by gravity, into physiological information. Suggested candidates for the susceptor include amyloplasts, and various non-plastid components such as the entire protoplasm and the ECM (Sack, 1991). Wayne *et al.* (1990) suggest that the entire mass of the protoplasm can provide sufficient force to activate membrane channels; alternatively, subplasmalemmal networks of cytoskeletal elements could amplify and focus force from the susceptor to activate MS channels (Edwards and Pickard, 1987). It should be noted, however, that the channels need not necessarily be mechanosensitive (see above discussion of the specificity of Gd^{3+}), and Hepler and Wayne (1985) have discussed alternative models based on the involvement of voltage-gated Ca^{2+} channels.

More recently, Wayne *et al.* (1992) have presented evidence suggesting that the junction of the cell–extracellular matrix is essential for gravity perception in *Chara* internodal cells, with the possible involvement of integrin-like proteins that link the plasmalemma to the cell wall (see above and *Figure 2*). The authors propose a model in which 'tension-opened channels' and/or 'tension-closed channels' at the top of the cells and 'compression-opened channels' and/or 'compression-closed channels' at the base of the cell trigger a Ca^{2+}-dependent signal transduction chain that mediates a graviresponse.

The MS channel in *U. appendiculatus* has been implicated in transducing mechanical stress into cation influx that triggers differentiation of the tips of germlings into appressoria, structures that are required for the successful invasion of bean leaf stomata. Appressoria formation and channel activity are inhibited in the presence of Gd^{3+} (Zhou *et al.*, 1991). Hoch *et al.* (1987) had previously shown that the tips of germlings will form these infection structures when they come into contact with chemically inert plastic ridges of similar dimensions to the stomatal ridge. Importantly, Zhou *et al.* (1991) have calculated that the tension experienced by the germ tube upon such a contact is comparable to that which activated MS channels in excised patches. It is also worth noting that thigmotropism in higher plants may be blocked with Gd^{3+} (Millet and Pickard, 1988).

Generation of Ca^{2+} gradient in polarized growth

MS channels permeable to Ca^{2+} appear to be present at higher densities in the apical plasma membrane than in the more distal regions of the hypha of the oomycete *Saprolegnia ferax* (Garrill *et al.*, 1992, 1993; Levina *et al.*, 1994). Channel activity could be blocked with Gd^{3+}, which was also shown to inhibit hyphal growth reversibly and dissipate a tip high gradient of cytoplasmically free Ca^{2+}. The Ca^{2+} gradient was only present in growing hyphae (Garrill *et al.*, 1993). The data suggest that the MS channels play a fundamental role in the process of tip growth, generating a tip high gradient of free Ca^{2+} that may be required to regulate tip extensibility, cytoplasmic migration and vesicle fusion. Similar processes may occur in other tip-growing systems. For example, pollen tubes of *Lilium* (Miller *et al.*, 1992) and growing rhizoids of *Fucus* (Brownlee *et al.*, 1993) display elevated Ca^{2+} levels at the apex. In this latter study, hypo-osmotic stress produced transients of elevated Ca^{2+} at the apex, and preliminary patch-clamp studies revealed higher average channel activities in this region. Furthermore, in yeast it is possible that MS channels permit a localized influx of Ca^{2+} that determines a new site of budding (Gustin *et al.*, 1988; Zhou and Kung, 1992).

A tip high gradient of cytoplasmic Ca^{2+} is necessary for growth in the ascomycete fungus *Neurospora crassa* (Levina *et al.*, 1995). In contrast to those systems described above, it appears that MS channels may not be involved in the generation and maintenance of such a gradient. Levina *et al.* (1995) have found that MS channels are not preferentially located at the tip and that Gd^{3+} has no effect on growth, and they hypothesize that the Ca^{2+} gradient may be generated from internal stores.

Cell cycle regulation

MS channels permeable to Ca^{2+} may function in regulating the cell cycle. Zhou and Kung (1992) have suggested that MS channels may be subjected to varying membrane tensions that occur as microfilaments and microtubules are dis- and reassembled during different stages of the cell cycle.

9. Channel evolution

The ubiquity of MS channels points to a common and very early origin. Sachs (1992) has suggested that an MS channel in the early protocells may have acted as an osmoregulator to prevent osmotic lysis. Bearing in mind that it is likely that the internal constituents of the earliest cells were similar to those of the external medium, the channel would presumably need to leak osmotically active organic ions generated by metabolic processes. This could account for the low selectivity of MS channels, as originally these channels were required to pass a variety of large ions. With time, in cells such as guard cells which require a very fine control of cell turgor, these channels may have evolved with increasing selectivity.

10. Non-channel MS proteins

There have been numerous reports (e.g. Morre *et al.*, 1973; Reinhold *et al.*, 1984; Tomos, 1989) that reductions in cell turgor cause acidification of the external medium. This acidification appears to be due to an increase in the activity of the plasma membrane H^+-ATPase (Curti *et al.*, 1993). It is at present unclear whether this is a direct effect on the H^+-ATPase or whether it occurs indirectly via changes in ion concentrations or membrane potential through the activation of channels. Van Wees *et al.* (personal communication) have observed effects of stretching on the H^+-ATPase in isolated vesicles, although it is possible that these vesicles may also contain channels. Changes in the osmotic potential of the medium may also affect the cytoskeleton (Chowdury *et al.*, 1992). Adenylate cyclase has been shown to respond directly to mechanical forces (Watson, 1991), and it will not be surprising if, in future studies, other proteins prove to be stress-sensitive.

11. Future work

There is clearly a pressing need for the development of a pharmacological agent which will specifically block MS channels. Sachs (1988) states that there have been 'promising leads' in screening antibodies, venoms and organic reagents for such a drug, and it can only be hoped that one or more of these leads may eventually enable us to describe the cellular functions of MS channels with more certainty. Of equal importance is the use of molecular biology techniques, particularly the development of MS channel mutants. Clearly, the recent report of the *mscL* gene in *E. coli* is a very important step forward. With the demonstration of MS channels in yeast and *Arabidopsis* (systems in which the application of molecular biology techniques is well established), it can only be hoped that similar advances are soon realized in fungal and higher plants. The involvement of the cytoskeleton and ECM in channel activation and distribution is an exciting area which is deservedly creating much interest at the present time. If MS channels act as mechanical transducers, there needs to be a better understanding of the biochemical train of events that they set in motion. It can only be hoped that advances in the above areas will lead to future reviews on MS channels of plant cells that are more definitive and less speculative.

References

Alexandre J, Lassalles J-P. (1991) Hydrostatic and osmotic pressure activated channel in plant vacuole. *Biophys. J.* **60**: 1326–1336.

Awayda MS, Ismailov II, Berdiev BK, Benos DJ. (1995) A cloned renal epithelial Na^+ channel protein displays stretch activation in planar lipid bilayers. *Am. J. Physiol.* **268**: C1450–C1459.

Badot P-M, Ding JP, Pickard BG. (1992) Mechanically activated ion channels occur in vacuoles of onion bulb scale parenchyma. *C. R. Acad. Sci. [III]* **315**: 437–443.

Berrier C, Coulombe A, Houssin C, Ghazi A. (1989) A patch clamp study of ion channels of inner and outer membranes and of contact zones of *E. coli*, fused into giant liposomes. Pressure-activated channels are localised in the inner membrane. *FEBS Lett.* **259**: 27–32.

Berrier C, Coulombe A, Szabo I, Zoratti M, Ghazi A. (1992) Gadolinium ion inhibits loss of metabolites induced by osmotic shock and large stretch-activated channels in bacteria. *Eur. J. Biochem.* **206**: 559–565.

Brehm P, Kidokoro Y, Moody-Corbett F. (1984) Properties of non-junctional acetylcholine receptor channels on innervated muscle of *Xenopus laevis*. *J. Physiol.* **350**: 631–648.

Brownlee C, Berger F, Taylor AR. (1993) Laser microsurgery, ion channels and intracellular signalling in *Fucus* zygotes. *J. Exp. Bot.* **44** (Suppl.): 55.

Buechner M, Delcour AH, Martinac B, Adler J, Kung C. (1990) Ion channel activities in the *Escherichia coli* outer membrane. *Biochim. Biophys. Acta* **1024:** 111–121.

Canessa, CM, Schlid L, Buell G, Thorens B, Gautschl J, Horisberger J-D, Rossier BC. (1994) Amiloride-sensitive epithelial Na$^+$ channel is made of three homologous subunits. *Nature* **367:** 463–467.

Chowdury S, Smith KW, Gustin MC. (1992) Osmotic stress and the yeast cytoskeleton: phenotype-specific suppression of an actin mutation. *J. Cell Biol.* **118:** 561–571.

Christensen O. (1987) Mediation of cell volume regulation by Ca^{2+} influx through stretch-activated channels. *Nature* **330:** 66–68.

Cornet M, Ubl J, Kolb H-A. (1993) Cytoskeleton and ion movements during volume regulation in cultured PC12 cells. *J. Memb. Biol.* **133:** 161–170.

Cosgrove DJ, Hedrich R. (1991) Stretch-activated chloride, potassium and calcium channels coexisting in plasma membranes of guard cells of *Vicia faba* L. *Planta* **186:** 143–153.

Coyaud L, Kurkdjian A, Kado RT, Hedrich R. (1987) Ion channels and ATP-driven pumps involved in ion transport across the tonoplast of sugar beet vacuoles. *Biochim. Biophys. Acta* **902:** 263–268.

Cui C, Smith DO, Adler J. (1995) Characterization of mechanosensitive channels in *Escherichia coli* cytoplasmic membrane by whole-cell patch clamp recording. *J. Memb. Biol.* **144:** 31–42.

Curti G, Massardi F, Lado P. (1993) Synergistic activation of plasma membrane H$^+$-ATPase in *Arabidopsis thaliana* cells by turgor decrease and by fusicoccin. *Physiol. Plant.* **87:** 592–600.

Delcour AH, Martinac B, Adler J, Kung C. (1989) Modified reconstruction method used in patch-clamp studies of *Escherichia coli* ion channels. *Biophys. J.* **56:** 631–636.

Ding JP, Pickard BG. (1993a) Mechanosensory calcium selective channels in onion epidermis. *Plant J.* **3:** 83–110.

Ding JP, Pickard BG. (1993b) Modulation of mechanosensory calcium channels by temperature. *Plant J.* **3:** 713–720.

Edwards KL, Pickard BG. (1987). Detection and transduction of physical stimuli in plants. In: *The Cell Surface in Signal Transduction* (Wagner E, Greppin H, Millet B, ed.). Berlin: Springer-Verlag, pp. 41–66.

Falke L, Misler S. (1989) Activity of ion channels during volume regulation by clonal N1E115 neuroblastoma cells. *Proc. Natl Acad. Sci. USA* **86:** 3919–3923.

Falke L, Edwards KL, Pickard BG, Misler S. (1988) A stretch-activated anion channel in tobacco protoplasts. *FEBS Lett.* **237:** 141–144.

Faraday CD, Spanswick RM. (1993) Evidence for a membrane cytoskeleton in higher plants: a spectrin-like polypeptide co-isolates with rice root plasma membranes. *FEBS Lett.* **318:** 313–316.

Filipovic D, Sackin H. (1992) Stretch- and volume-activated channels in isolated proximal tubule cells. *Am. J. Physiol.* **262:** 857–870.

Garrill A, Lew RR, Heath IB. (1992) Stretch-activated Ca^{2+} and Ca^{2+}-activated K$^+$ channels in the hyphal tip plasma membrane of the oomycete *Saprolegnia ferax. J. Cell Sci.* **101:** 721–730.

Garrill A, Jackson SL, Lew RR, Heath IB. (1993) Ion channel activity and tip growth: tip-localised stretch-activated channels generate an essential Ca^{2+} gradient in the oomycete *Saprolegnia ferax. Eur. J. Cell Biol.* **60:** 358–365.

Garrill A, Tyerman SD, Findlay GP. (1994) Ion channels in the plasma membrane of protoplasts from the halophytic angiosperm *Zostera muelleri. J. Memb. Biol.* **142:** 381–393.

Guharay F, Sachs F. (1984) Stretch-activated single ion channel currents in tissue-cultured embryonic chick skeletal muscle. *J. Physiol.* **352:** 685–701.

Gustin MC. (1991) Single channel mechanosensitive currents. *Science* **253:** 800.

Gustin MC, Zhou X-L, Martinac B, Kung C. (1988) A mechano-sensitive ion channel in the yeast plasma membrane. *Science* **242:** 762–765.

Hamill OP, McBride DW. (1994) The cloning of a mechano-gated membrane ion channel. *Trends Neurosci.* **17:** 439–443.

Hamill OP, Marty A, Neher E, Sakmann B, Sigworth FJ. (1981) Improved patch clamp techniques for high-resolution current recording from cells and cell-free membrane patches. *Pflügers Archiv.* **391:** 85–100.

Hamill OP, Lane JW, McBride DW. (1992) Amiloride: a molecular probe for mechanosensitive channels. *Trends Pharmacol. Sci.* **13:** 373–375.

Hase CC, Le Dain AC, Martinac B. (1995) Purification and functional reconstitution of the recombinant large mechanosensitive ion channel (MscL) of *Escherichia coli. J. Biol. Chem.* **270:** 18329–18334.

Hepler PK, Wayne RO. (1985) Calcium and plant development. *Annu. Rev. Plant Physiol.* **36:** 397–439.

Hoch HC, Staples RC, Whitehead B, Comeau J, Wolf ED. (1987) Signalling for growth orientation and cell differentiation by surface topography in *Uromyces. Science* **235:** 1659–1662.

Hong K, Driscoll M. (1994) A transmembrane domain of the putative channel subunit MEC-4 influences mechanotransduction and neurodegradation in *Caenorhabditis elegans*. *Nature* **367**: 470–473.

Huang M, Chalfie M. (1994) Gene interactions affecting mechanosensory transduction in *Caenorhabditis elegans*. *Nature* **367**: 467–470.

Kaminskyj SGW, Heath IB. (1995) Integrin and spectrin homologues and cytoplasm–wall adhesion in tip growth. *J. Cell Sci.* **108**: 849–856.

Keller BU, Hedrich R, Raschke K. (1989) Voltage-dependent anion channels in the plasma membrane of guard cells. *Nature* **341**: 450–453.

Knight MR, Campbell AK, Smith SM, Trewavas AJ. (1991) Transgenic plant aequorin reports the effects of touch and cold-shock and elicitors in cytoplasmic calcium. *Nature* **352**: 524–526.

Lansman JB. (1990) Blockage of current through single calcium channels by trivalent lanthenide cations. Effect of ionic radius on the rates of ionic entry and exit. *J. Gen. Physiol.* **95**: 679–696.

Levina NN, Lew RR, Heath IB. (1994) Cytoskeletal regulation of ion channel distribution in the tip-growing organism *Saprolegnia ferax*. *J. Cell Sci.* **107**: 127–134.

Levina NN, Lew RR, Hyde GJ, Heath IB. (1995) The roles of Ca^{2+} and plasma membrane ion channels in hyphal tip growth of *Neurospora crassa*. *J. Cell Sci.* **108**: 3405–3417.

McBride DW, Hamill OP. (1992) Pressure-clamp: a method for rapid step perturbation of mechanosensitive ion channels. *Pflügers Archiv.* **421**: 606–612.

Martinac B, Saimi Y, Gustin MC, Kung C. (1987) Pressure sensitive ion channel in *Escherichia coli*. *Proc. Natl Acad. Sci. USA* **84**: 1–5.

Martinac B, Adler J, Kung C. (1990) Mechanosensitive ion channels of *E. coli* activated by amphipaths. *Nature* **348**: 261–263.

Miller DD, Callaham DA, Gross DJ, Hepler PK. (1992) Free Ca^{2+} gradient in growing pollen tubes of *Lilium*. *J. Cell Sci.* **101**: 7–12.

Millet B, Pickard BG. (1988) Gadolinium ion as an inhibitor for testing the putative role of stretch-activated ion channels in geotropism and thigmotropism. *Biophys. J.* **53**: 155a.

Moran N. (1990) Stretch-activated channels in plasmalemma of pulvinar motor cells. *Plant Physiol.* **93**: 17 (abstract).

Morre E, Lado P, Rasi-Caldogno F, Colombo R. (1973) Correlation between cell enlargement in pea internode segments and decrease in the pH of the medium of incubation. I. Effects of fusicoccin, natural and synthetic auxins and mannitol. *Plant Sci.* **1**: 179–184.

Morris CE. (1990) Mechanosensitive ion channels. *J. Memb. Biol.* **113**: 93–107.

Morris CE, Horn R. (1991) Failure to elicit neuronal macroscopic mechanosensitive currents anticipated by single channel studies. *Science* **251**: 1246–1249.

Morris CE, Williams B, Sigurdson WJ. (1989) Osmotically-induced volume changes in isolated cells of a pond snail. *Comp. Biochem. Physiol.* **92A**: 479–483.

Odani S, Takehiko K, Ono T. (1987) Amino acid sequence of a soybean (*Glycine max*) seed polypeptide having a poly (L-aspartic acid) structure. *J. Biol. Chem.* **262**: 10502–10505.

Ollet SHR, Bourque CW. (1993) Mechanosensitive channels transduce osmosensitivity supraoptic neurons. *Nature* **364**: 341–343.

Opsahl LR, Webb WW. (1994) Transduction of membrane tension by the ion channel alamethicin. *Biophys. J.* **66**: 71–74.

Parsegian VA, Rand RP, Rau DC. (1992) Swelling from the perspective of molecular assemblies and single functioning biomolecules. In: *Mechanics of Swelling, From Clays to Living Cells and Tissues* (Karalis TK, ed.). Berlin: Springer-Verlag, pp. 623–647.

Pickard BG, Ding J-P (1993) The mechanosensory calcium-selective ion channel: key component of a plasmalemmal control centre. *Aust. J. Plant Physiol.* **20**: 439–459.

Pont-Lezica RF, McNally JG, Pickard BG. (1993) Wall-to-membrane linkers in onion epidermis: some hypotheses. *Plant Cell Environ.* **16**: 111–123.

Quatrano RS, Brian L, Aldridge J, Schutz T. (1991) Polar axis fixation in *Fucus* zygotes: components of the cytoskeleton and extracellular matrix. *Development* **1** (Suppl.): 11–16.

Reinhold L, Seiden A, Volokita M. (1984) Is modulation of the rate of proton pumping a key event in osmoregulation? *Plant Physiol.* **75**: 846–849.

Roberts K. (1990) Structures at the plant cell surface. *Curr. Opin. Cell Biol.* **2**: 920–928.

Sachs F. (1988) Mechanical transduction in biological systems. *CRC Crit. Rev. Biomed. Eng.* **16**: 141–169.

Sachs F. (1989) Ion channels as mechanical transducers. In: *Cell Shape: Determinants, Regulation and Regulatory Role* (Bronner F, Stein W, ed.). New York: Academic Press, pp. 63–92.

Sachs F. (1992) Stretch activated ion channels: an update. In: *Sensory Transduction* (Corey DP, Roper RS, ed.). New York: The Rockefeller University Press, pp. 464–509.

Sachs F, Qin F. (1993) Gated ion channels observed with patch pipettes in the absence of membranes. *Biophys. J.* **65:** 1101–1107.

Sachs F, Sigurdson WJ, Rudnikin A, Bowman C. (1991) Single channel mechanosensitive currents. *Science* **253:** 800–801.

Sack FD. (1991) Plant gravity sensing. *Int. Rev. Cytol.* **127:** 193–252.

Sanders LC, Wang C-S, Walling LL, Lord EM. (1991) A homolog of the substrate adhesion molecule vitronectin occurs in four species of flowering plants. *Plant Cell* **3:** 629–635.

Schindler M, Meiners S, Cheresh DA. (1989) RGD-dependent linkage between plant cell wall and plasma membrane: consequences for growth. *J. Cell Biol.* **108:** 1955–1965.

Schroeder JI, Hagiwara S. (1989) Cytosolic calcium regulates ion channels in the plasma membrane of *Vicia faba* guard cells. *Nature* **338:** 427–430.

Schroeder JI, Raschke K, Neher E. (1987) Voltage dependence of K⁺ channels in guard cell protoplasts. *Proc. Natl Acad. Sci. USA* **84:** 4108–4112.

Sigurdson WJ, Sachs F. (1991) Mechanical stimulation of cardiac myocytes increases intracellular calcium. *Biophys. J.* **59**: 469a (Abstract).

Sokabe M, Sachs F, Jing Z. (1991) Quantitative video microscopy of patch clamped membranes: stress, strain, capacitance and stretch channel activation. *Biophys. J.* **59:** 722–728.

Spalding EP, Goldsmith MHM. (1993) Activation of K⁺ channels in the plasma membrane of *Arabidopsis* by ATP produced photosynthetically. *Plant Cell* **5:** 477–484.

Staves MP, Wayne R. (1993) The touch-induced action potential in *Chara*: inquiry into the ionic basis and the mechanoreceptor. *Aust. J. Plant Physiol.* **20:** 471–488.

Sukharev SI, Martinac B, Arshavsky VY, Kung C. (1993) Two types of mechanosensitive channels in *Escherichia coli* cell envelope: solubilization and functional reconstitution. *Biophys. J.* **65:** 177–183.

Sukharev SI, Blount P, Martinac B, Blattner FR, Kung C. (1994) A large conductance mechanosensitive channel in *E. coli* encoded by *mscL* alone. Nature **368:** 265–268.

Sukharev SI, Blount P, Nagle S, Kung C. (1995) Experimental evidence that mechanosensitivity of the large conductance MS channel (MscL) is determined by hydrophobic interactions within the channel complex. *Biophys. J.* **68:** A132.

Szabo I, Petronilli V, Guerra L, Zoratti M. (1990) Cooperative mechanosensitive ion channels in *Escherichia coli.* Biochem. Biophys. Res. Commun. **171:** 280–286.

Terry BR, Findlay GP, Tyerman SD. (1992) Direct effects of Ca²⁺-channel blockers on plasma membrane cation channels of *Amaranthus tricolor* protoplasts. *J. Exp. Bot.* **43:** 1457–1473.

Tomos AD. (1989) Turgor pressure and membrane transport. In: *Plant Membrane Transport: The Current Position* (Dainty J, DeMichelis MI, Marre E, Rasi-Caldogno F, ed.). Amsterdam: Elsevier, pp. 559–562.

Ubl J, Murer H, Kolb H-A. (1988) Ion channels activated by osmotic and mechanical stress in membranes of opossum kidney cells. *J. Memb. Biol.* **104:** 223–232.

Watson PA. (1991) Function follows form: generation of intracellular signals by cell deformation. *FASEB J.* **5:** 2014–2019.

Wayne R, Staves MP, Leopold AC. (1990) Gravity-dependent polarity of cytoplasmic streaming in *Nitellopsis*. Protoplasma **155:** 43–57.

Wayne R, Staves MP, Leopold AC. (1992) The contribution of the extracellular matrix to gravisensing in characean cells. *J. Cell Sci.* **101:** 611–623.

Wolfe J, Steponkus PL. (1981) The stress–strain relation of the plasma membrane of isolated plant protoplasts. *Biochim. Biophys. Acta* **643:** 662–668.

Wolfe J, Steponkus PL. (1983) Mechanical properties of the plasma membrane of isolated plant protoplasts. Mechanism of hyperosmotic and extracellular freezing injury. *Plant Physiol.* **71:** 276–285.

Wolfe J, Dowgert MF, Steponkus PJ. (1985) Dynamics of membrane exchange of the plasma membrane and the lysis of isolated protoplasts during rapid expansions in area. *J. Memb. Biol.* **86:** 127–138.

Yang X-C, Sachs F. (1989) Block of stretch-activated ion channels in *Xenopus* oocytes by gadolinium and calcium ions. *Science* **243:** 1068–1071.

Zhou X-L, Kung C. (1992) A mechanosensitive ion channel in *Schizosaccharomyces pombe*. EMBO J. **11:** 2869–2875.

Zhou X-L, Stumpf MA, Hoch HC, Kung C. (1991) A mechanosensitive cation channel in membrane patches and in whole cells of the fungus *Uromyces*. *Science* **253:** 1415–1417.

Zoratti M, Petronilli V. (1988) Ion-conducting channels in a Gram-positive bacterium. *FEBS Lett.* **240:** 105–109.

Zoratti M, Petronilli V, Szabo I. (1990) Stretch-activated composite ion channels in *Bacillus subtilis.* Biochem. Biophys. Res. Commun. **168:** 443–450.

Chapter 16

Proton-translocating ATPases of the plasma membrane: biological functions, biochemistry and molecular genetics

Baudouin Michelet and Marc Boutry

1. Introduction

General aspects of transport and ATPases

Living cells constantly exchange energy, matter and information with their environment. The plasma membrane surrounding them cannot, therefore, be completely impermeable, and is indeed better viewed as a selective barrier through which many solutes can be transported, sometimes against a concentration gradient. This active transport requires the plasma membrane to be energized by cation-translocating ATPases and possibly, to a lesser extent, by redox reactions.

Cation-translocating ATPases reside in the plasma membrane. They actively pump cations against their concentration gradient. This process is powered by hydrolysis of ATP to ADP and inorganic phosphate (P_i), whatever the type of transported cation (H^+, Ca^{2+}, Na^+/K^+, H^+/K^+). The catalytic cycle of plasma-membrane ATPases involves the formation of an aspartyl-phosphate residue. This is why these enzymes are referred to as P-type ATPases. They are also called E_1E_2-ATPases, referring to the fact that they pass through two different conformational states during each catalytic cycle. Other types of ATPase are the V-type or vacuolar ATPases, and the F-type enzymes, which include mitochondrial and chloroplastic ATP synthases. All these enzymes have quite different structures. This review discusses only the main P-type ATPase found in higher plant plasma membranes, namely the proton-translocating ATPase.

Figure 1 illustrates the primary role of plant H^+-ATPase in the transport of solutes. This theoretical diagram helps us to understand how solutes (charged or uncharged) can be accumulated on one side of the membrane by carrier proteins (uniport, symport and antiport systems), and how massive fluxes of ions, passing through channels, are allowed. In each case, the process is powered either by the pH gradient (ΔpH) or by the electrical potential (E_m) of the proton-motive force created by the H^+-ATPase.

It must be understood that, in addition to its single primary physiological function of pumping protons, H^+-ATPase promotes very diverse biological processes, each requiring active transport of a specific solute. It has been postulated, for instance, and in some cases actually shown, that plasma-membrane ATPase plays an essential role in regulating cytoplasmic pH, in acidifying the cell wall during cell elongation, in mineral nutrition, stomatal opening, and sap loading and unloading in vascular tissues.

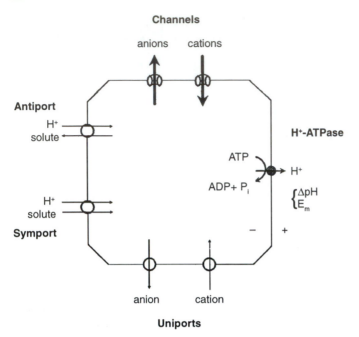

Figure 1. Energization of secondary transport across the plasma-membrane by H[+]-ATPase. The electrochemical gradient (ΔpH and E_m) created by ATPase across the plasma membrane is used by transport proteins (channels, symports, antiports and uniports) to accumulate ions and solutes against their concentration gradients.

Demonstration of the existence of the plasma-membrane H[+]-ATPase

Early evidence for the involvement of proton fluxes in solute movements across the plant plasma membrane came from studies of potassium absorption in roots (Jacobson *et al.*, 1950). Later, proton fluxes were also found to accompany the transport of organic molecules (reviewed in Bush, 1993) as well as guard cell ion fluxes (Assmann *et al.*, 1985; Shimazaki *et al.*, 1986).

Root plasma membrane preparations have long been known to exhibit ATPase activity (see, for example, Hodges *et al.*, 1972). This activity was assumed to be coupled with H[+] transport (Fisher *et al.*, 1970; reviewed in Poole, 1978; Spanswick, 1981), but biochemical proof of this coupling was not obtained until more recently, by direct measurement of ATP-dependent H[+] transport by purified plasma-membrane ATPase incorporated into liposomes (Vara and Serrano, 1982). It should be emphasized here that studies of plant H[+]-ATPase have been, and still are, facilitated by knowledge gained from other P-type ATPases, especially the yeast enzyme, which is also an H[+]-ATPase (see, for example, Serrano, 1989).

Our knowledge of plant H[+]-ATPase has dramatically increased with the sequencing and characterization of the corresponding genes in several species (Boutry *et al.*, 1989; Ewing *et al.*, 1990; Ewing and Bennett, 1994; Harms *et al.*, 1994; Harper *et al.*, 1989, 1990, 1994; Houlné and Boutry, 1994; Jin and Bennetzen, 1994; Michelet and Boutry, 1995; Moriau *et al.*, 1993; Nakijama *et al.*, 1995; Ookura *et al.*, 1994; Pardo and Serrano, 1989; Perez *et al.*, 1992; Sussman, 1994; Wada *et al.*, 1992). The discovery of multiple H[+]-ATPase-encoding genes in all plants studied so far has certainly changed our perspective in this area of research; H[+]-ATPase can no longer be viewed as a single enzyme. While this complicates investigations, it also makes it easier to explain the many conflicting reports in the literature. Most probably, 'the' ATPase studied in different systems actually represents different isozymes or mixtures of isozymes which may have different biochemical properties. As this review will stress, the molecular characterization of H⁺-ATPase has created a rich terrain for physiological, biochemical and molecular studies.

We shall first review the biochemical and molecular aspects of H⁺-ATPase, and then discuss its physiological roles.

2. Biochemistry and molecular biology

Properties of the enzyme

The plasma-membrane H⁺-pumping ATPase of plants energizes the transport of H⁺ by hydrolysing ATP. It has a strict requirement for Mg^{2+} and is ATP-specific. Different reports place the K_m for MgATP in the range 0.3 to 1.4 mM and the pH optimum at about 6.6 (Becker *et al.,* 1993; Brauer *et al.,* 1989; Vara and Serrano, 1982). The enzyme's specific activity in purified plasma membranes is usually of the order of 1 to 2 µmol P_i min⁻¹ mg⁻¹ protein, but it can reach 13 µmol P_i min⁻¹ mg⁻¹ protein (Palmgren and Christensen, 1993; reviewed in Marrè and Ballarin-Denti, 1985). This activity is inhibited by vanadate (K_i = 1 µM), dicyclohexylcarbodiimide (K_i = 20 µM) and diethylstilbestrol (K_i = 40 µM). It is resistant to 5 mM NaN_3 and 5 µg ml⁻¹ oligomycin (two mitochondrial ATPase inhibitors), to 0.1 M KNO_3 (a vacuolar ATPase inhibitor), and to 0.1 mM molybdate (an inhibitor of non-specific phosphatases present in plasma membrane preparations) (Briskin and Leonard, 1982; Palmgren and Christensen, 1993; Vara and Serrano, 1982, reviewed in Sussman, 1992). H⁺-ATPase forms an acyl-phosphate intermediate with a K_m for MgATP of 0.5 mM (Briskin and Leonard, 1982; Scalla *et al.,* 1983; Vara and Serrano, 1983). It is stimulated *in vitro* by K⁺ but, as reviewed in Briskin and Hanson (1992), K⁺ acts as a simple effector and is not transported by the H⁺-ATPase. By analogy with what is observed in yeast, however, it has been postulated that K⁺ transport could be indirectly coupled to H⁺ transport by close association between the ATPase and a K⁺ channel (Briskin and Hanson, 1992; Serrano, 1989).

Molecular cloning of H⁺-ATPase genes has made it possible to deduce its putative primary structure. In *Arabidopsis thaliana,* there are at least 10 H⁺-ATPase genes (Harper *et al.,* 1994), at least four of which are expressed. The corresponding proteins range in size from 948 to 954 amino acids and in predicted molecular mass from 104 to 105 kDa (Harper *et al.,* 1989, 1990, 1994; Houlné and Boutry, 1994; Pardo and Serrano, 1989). In *Nicotiana plumbaginifolia,* nine genes have been isolated (Boutry *et al.,* 1989; Michelet and Boutry, 1995; Moriau *et al.,* 1993; Perez *et al.,* 1992; M. Oufattole, M. Arango and M. Boutry, unpublished results), at least six of which are expressed, giving rise to proteins of 952 to 957 amino acids, all with a predicted molecular mass of about 105 kDa (Boutry *et al.,* 1989; Moriau *et al.,* 1993; Perez *et al.,* 1992). For tomato, seven genes have been characterized, at least three of which are expressed relatively abundantly at the mRNA level (Ewing *et al.,* 1990; Ewing and Bennett, 1994). Finally, sequences of H⁺-ATPase genes from rice (Ookura *et al.,* 1994), maize (Jin and Bennetzen, 1994), potato (Harms *et al.,* 1994) and broad bean (Nakijama *et al.,* 1995) have been published. They all encode a protein of *c.* 105 kDa.

Sequence comparisons of plant H⁺-ATPases suggest the existence of two sub-families which emerged before the appearance of dicotyledonous species (Moriau *et al.,* 1993).

Structure and function

The topology and conformation of an enzyme are revealed by X-ray crystallography. Crystallography of membrane proteins is extremely difficult. It is therefore of interest to gather as much information as possible about the structure of ATPases by other techniques, such as secondary structure prediction or sequence analogy studies.

Hydropathy studies of plant H$^+$-ATPases have led to different models predicting eight (Harper *et al.*, 1989, 1990), nine (Pardo and Serrano, 1989; Serrano, 1989) or 10 (Boutry *et al.*, 1989; Michelet *et al.*, 1989) membrane spans. By analogy to yeast, where the N- and C-termini are cytoplasmic, one would expect an even number of spans (reviewed in Serrano, 1993; Wach *et al.*, 1992). For the reasons explained in Wach *et al.* (1992), we still favour a 10-span model.

The membrane spans of the enzyme delimit hydrophilic stretches which constitute about 80% of the protein. Only 7% of the protein is predicted to be exposed on the outer face of the plasma membrane. Secondary structure predictions, for what they are worth, are described in Michelet *et al.* (1989).

Putative catalytic functions can be assigned to different conserved regions of the ATPase by analogy to other P-type ATPases (Ewing *et al.*, 1990; Michelet *et al.*, 1989; Serrano, 1989). Recently, the 17 predicted H$^+$-ATPase sequences known to date were aligned, highlighting residues conserved either in all P-type ATPases or specifically in all H$^+$-ATPases (Wach *et al.*, 1992). The latter residues have been proposed for a site-directed mutagenesis study aimed at elucidating the specificity of the enzyme for H$^+$.

Plant H$^+$-ATPase has been shown to exist naturally in an autoinhibited state. *In vitro*, limited proteolysis of H$^+$-ATPase cleaves a 7- to 10-kDa portion from its C-terminus, thus activating proton pumping by 70 to 590%, depending on the plant species used (Palmgren *et al.*, 1990, 1991). This result has been confirmed in radish (Rasi-Caldogno *et al.*, 1993) and *Arabidopsis* (Olivari *et al.*, 1993), and by expression of a modified *Arabidopsis* H$^+$-ATPase gene in yeast (Palmgren and Christensen, 1993). We do not expect proteolytic activation of H$^+$-ATPase to take place *in vivo*, but, as discussed further, it provides an interesting insight into the mechanisms of enzyme modulation by other factors.

At present, the expression and analysis of a mutated H$^+$-ATPase in plants is hindered by the fact that the same organ may co-express many isozymes (Harper *et al.*, 1990; Moriau *et al.*, 1993; Perez *et al.*, 1992). Heterologous expression seems much more promising, and interesting results have been obtained in yeast. In an initial study involving the expression of a plant ATPase in yeast, little plant enzyme reached the plasma membrane, but structures resembling endoplasmic reticulum (ER) filled with active plant ATPase accumulated massively in the yeast cells (Villalba *et al.*, 1992). The procedure used made it possible to purify 70 mg of ATPase per kg of yeast (i.e. nothing less than Ali Baba's cave for plant H$^+$-ATPase enzymologists). In this case, the plant ATPase could not restore growth of yeast cells whose own H$^+$-ATPase was deficient, but removal of various lengths of the carboxy-terminal region of the plant H$^+$-ATPase resulted in efficient targeting of the enzyme to the yeast plasma membrane, increased specific activity, and complementation of the defective endogenous yeast ATPase (Palmgren and Christensen, 1993; Regenberg *et al.*, 1995). We used a similar approach with *pma2* and *pma4*, two *N. plumbaginifolia* H$^+$-ATPase genes, and found that they can both sustain yeast growth when the yeast gene for H$^+$-ATPase is either repressed or deleted. However, they differ in conferring distinct pH sensitivity to growth. PMA2 can replace the yeast H$^+$-ATPase provided that the pH of the medium is kept above 5.0, whereas the expression of PMA4 still allows yeast growth at pH 4.0 (de Kerchove d'Exaerde *et al.*, 1995; Luo *et al.*, unpublished results). Interestingly, point mutations of *pma2* were found to increase H$^+$ pumping and allow yeast growth at lower pH, suggesting that single amino acid substitutions are sufficient to activate the enzyme (Morsomme *et al.*, unpublished results).

Activity modulation: auxin, fusicoccin, light, phospholipids and kinases

The multiplicity of biological functions related to H⁺-ATPase and its concomitant role in integrating environmental variations suggest a complex pattern of activity modulation and regulation. In the following discussion, the term 'modulation' will be used to refer to any effect on the ATPase itself and the term 'regulation' will be used to refer to earlier effects (i.e. on gene expression, mRNA stability or translation, or protein targeting). We expect modulation to be more rapid than regulation.

Several plant pathogens produce H⁺-ATPase-modulating toxins, such as syringomycin, which is produced by *Pseudomonas syringae* (Bidway *et al.,* 1987), fusicoccin (FC), which is produced by *Fusicoccum amygdali* (Rasi-Caldogno and Pugliarello, 1985), and the NIP1 and NIP3 peptides produced by *Rhynchosporium secalis* (Wevelsiep *et al.,* 1993). FC is the most well studied of these toxins. It activates H⁺-ATPase in many plants and even in the alga *Dunaliella acidophila* (Sekler and Pick, 1993). FC increases the V_{max} of the enzyme and there are conflicting reports as to whether it decreases the K_m for MgATP (Johansson *et al.,* 1993; Marra *et al.,* 1992; Rasi-Caldogno *et al.,* 1986, 1993). The effect of FC is manifest only at physiological and thus suboptimal pH values (7–7.5). Sustained activation of H⁺-ATPase by FC is explained by a shift of the optimal pH towards more alkaline values, enabling the ATPase to keep working as it renders the cytoplasmic pH more alkaline (Rasi-Caldogno *et al.,* 1993, and references therein).

The mechanism of modulation is indirect. The receptor which binds FC is not the ATPase itself (Aducci *et al.,* 1988; Marra *et al.,* 1992; Meyer *et al.,* 1989). FC modulation would appear to involve the autoinhibitory C-terminus of H⁺-ATPase, since the stimulatory effects of FC and proteolysis are not additive (Johansson *et al.,* 1993; Rasi-Caldogno *et al.,* 1993). FC-insensitive variant lines of *Arabidopsis thaliana* have been isolated (Holländer-Czytko and Weiler, 1994), and the FC-receptor has been purified and identified as belonging to the 14-3-3 family of regulating proteins (Korthout and de Boer, 1994; Korthout *et al.,* 1994; Marra *et al.,* 1994; Decking *et al.,* 1994).

Like the effect of proteolysis, the ability of lysophospholipids to modulate other P-type ATPases has been known for some time. When the effects of these compounds were tested on plant ATPase, they were shown to activate the enzyme by increasing its V_{max} without altering the K_m for MgATP or the optimal pH (Palmgren and Sommarin, 1989). Here again, modulation seems to involve the C-terminal domain of H⁺-ATPase. This is evidenced by the similarity between proteolytic and phospholipid activation (Palmgren *et al.,* 1990, 1991), and by the fact that a truncated plant H⁺-ATPase expressed in yeast is no longer stimulated by phospholipids (Palmgren and Christensen, 1993).

The effect of light on stomata is known to involve activation of H⁺-ATPase in guard cells (Assmann *et al.,* 1985; Serrano *et al.,* 1988; Shimazaki *et al.,* 1986, reviewed in Serrano and Zeiger, 1989). The mechanism of this modulation is unknown, but interestingly, H⁺-ATPase activation is paralleled by light-induced modification of phosphoinositide in the plasma membrane (Memon and Boss, 1990), and by light-induced activation of a Ca^{2+}/calmodulin-dependent kinase (Shimazaki *et al.,* 1992).

This brings us to the much discussed subject of kinase-catalysed phosphorylation of H⁺-ATPase (reviewed in Serrano, 1989; Sussman, 1992, 1994). Oat root H⁺-ATPase is phosphorylated by a Ca^{2+}-stimulated kinase associated with the plasma membrane (Schaller and Sussman, 1988). This kinase exhibits maximal activity at 7 µM Ca^{2+}, but is active at Ca^{2+} concentrations as low as 1 µM. As Ca^{2+} concentrations are known to vary within this range and by about this order of magnitude during some signal transduction events, it has been argued that this kinase might modulate the ATPase. A kinase resembling the enzyme characterized by Schaller and Sussman (1988) has been cloned. When expressed in *E. coli*, it displays lipid- and

Ca^{2+}-dependent activity (Harper *et al.*, 1993). Furthermore, oat root ATPase is a good substrate for this kinase.

A phosphorylation site has been identified in a trypsin-released carboxy-terminal 12-kDa fragment of the H$^+$-ATPase from the alga *Dunaliella acidophila* (Sekler *et al.*, 1994). It has recently been shown that dephosphorylation of a tomato H$^+$-ATPase is induced by a fungal elicitor and results in activation of the enzyme activity (Vera-Estrella *et al.*, 1994).

Modulation of ATPase activity by auxin has been reviewed (Barbier-Brygoo *et al.*, 1992), but the action of auxin in plant cells is complex, suggesting that it might regulate ATPase at various levels. This point will be discussed in a later section.

This chapter would not be complete if it failed to mention two plant hormones that are believed to modulate H$^+$-ATPase, namely abscisic acid and brassinolide. As we are not aware of any detailed or recent work on the subject, we refer the reader to the review by Marrè and Ballarin-Denti (1985), where a few relatively old references can be found.

Gene regulation and tissue-specific expression

Plants possess a fairly large number of H$^+$-ATPase-encoding genes. This raises the question of the specialization of these genes and the corresponding isozymes. Given the diversity of biological functions of ATPase and the multiplicity of factors affecting its activity, it is tempting to speculate that certain genes and isozymes have become specialized to function in specific cell types or under specific environmental conditions. According to this scenario, factors known to affect 'the' ATPase may actually modulate only a subpopulation of ATPase isoforms or regulate the expression of some ATPase genes only.

H$^+$-ATPase is one of the most abundant plasma-membrane proteins, but even so it accounts for, at most, only 1% of the total plasma-membrane protein (reviewed in Sussman and Harper, 1989). This, however, is an average figure and some cell types probably have higher levels of ATPase. Studies of the tissue distribution of H$^+$-ATPase can shed some light on the multiple H$^+$-ATPase isoforms, their encoding genes, and their respective roles in plant physiology. Two main techniques are currently being used for such studies: immunocytolocalization and reporter gene analyses.

Immunological studies suggest that H$^+$-ATPase accumulates in particular cell types. The root cap, epidermis and central cylinder display a high level of reactivity towards an anti-ATPase antibody (Parets-Soler *et al.*, 1990). In another study achieving higher resolution (Samuels *et al.*, 1992), reactivity was found in the root cap, the root hairs, the epidermis, and all the cells of the stele. Stelar cells can be classified according to their level of reactivity. Companion cells exhibit the highest reactivity, followed by the pericycle, xylem, parenchyma and endodermis cells (Samuels *et al.*, 1992). Both of the studies just mentioned tend to confirm some of the biological functions attributed to the ATPase. In both studies, however, mature xylem vessels clearly react with the antiserum. As these vessels are known to be composed exclusively of dead cells devoid of plasma membrane (Esau, 1977), it is to be feared that artefactual results were obtained. It is also important to bear in mind that there are distinct ATPase isoforms, and it is impossible to know whether the antibodies used in the above studies revealed all or only some of the H$^+$-ATPases.

Reporter gene studies make it possible to analyse the expression of single genes. The principle of such studies is to fuse the transcriptional promoter of a gene of interest with an exogenous reporter gene encoding an easily assayed enzyme (such as β-glucuronidase) which wild-type plants lack. The fusion gene is introduced into the plant by genetic transformation. The reporter enzyme is monitored and considered to demonstrate expression of the gene of interest. This technique showed the *aha3* gene of *Arabidopsis* to be specifically expressed in the phloem of all organs analysed (DeWitt *et al.,* 1991). This confirms both the hypothesis of gene specialization and the primary role played by H⁺-ATPase in phloem loading and unloading. Expression was also found in pollen and ovules. Another *Arabidopsis* isoform, *aha10*, was found to be expressed only in developing seeds (Harper *et al.,* 1994).

A similar study of six *N. plumbaginifolia* H⁺-ATPase genes points, once again, to their differential specialization and to regulation of one of them by environmental factors (Michelet *et al.,* 1994; Moriau *et al.,* unpublished results). Some of the tissues identified in this study (root epidermis and stomata) are consistent with previous physiological and biochemical results, but others (tapetum, transmitting tissue and ovule) point to previously unsuspected biological roles for H⁺-ATPase. Our study also shows that the expression of different H⁺-ATPase genes may partly overlap, indicating that at least two different H⁺-ATPases are expressed at the same developmental stage in some cell types. As mentioned above, the expression of H⁺-ATPase isoforms in yeast has shown that they might have distinct kinetic properties.

ATPase gene expression would appear to be regulated at yet another level. This is suggested by the peculiar structure of the corresponding mRNAs (Harper *et al.,* 1989, 1990). Several ATPase transcripts display an unusually long 5′ untranslated region (more than 250 nucleotides) containing a small open reading frame (three to nine residues). These features are typical of translational regulation. A detailed expression analysis of two *N. plumbaginifolia* genes (*pma1* and *pma3*) led to the hypothesis that the small open reading frame is translated, but that ribosome reinitiation occurs at the level of the ATPase reading frame (Michelet *et al.,* 1994; Lukaszewicz *et al.,* unpublished results). The significance of these observations and the modes of possible regulation have yet to be elucidated.

Modification of H⁺-ATPase expression

It is possible to characterize the enzymatic properties of single isoforms by heterologous expression of plant H⁺-ATPases in yeast. However, this method is ill suited for elucidating how H⁺-ATPase activity relates to plant physiology. Although no plant ATPase mutants have been identified so far, molecular biology tools are available for modifying the expression of an ATPase isoform directly in the plant. When an ATPase gene of *N. plumbaginifolia* was overexpressed in tobacco, some transgenic plants were found to overproduce H⁺-ATPase, but in others expression of the introduced and resident ATPase genes was turned off in leaf tissues (the so-called co-suppression phenomenon) (Bogaerts *et al.,* unpublished results). Any consequences of a modified ATPase level have yet to be analysed at the physiological level.

Another recently initiated approach involves modifying ATPase transcript levels by expressing anti-sense constructs in a plant. Combined with the use of transcription promoters that are tissue-specific or inducible by external factors, this approach should make it possible to modify at will the physiological traits that depend on a given H⁺-ATPase.

3. Physiological functions of H⁺-ATPase

A housekeeping enzyme?

Plasma-membrane H⁺-ATPase has mainly been studied in tissues that are available in large amounts, and in which it is expected to play some specialized function. It is probable, however, that every single cell, alive or dead, needs or has needed H⁺-ATPase activity. During the growth phase, a cell must indeed import all its nutrients, and when it enlarges it has to maintain its turgor (for review see Cosgrove, 1986). Active nutrient transport and turgor maintenance by osmotic pressure adjustment are both thought to be energized by H⁺-ATPase, which can therefore be viewed as a housekeeping enzyme.

This reasoning is theoretical and hard to prove. Moreover, it again presents H$^+$-ATPase as a single enzyme, and we know that an individual plant expresses several isozymes. So even if we believe that a H$^+$-ATPase is active in all plant cells at some stage, it is still possible that there is no single housekeeping ATPase gene.

Figure 1 shows how transport processes can be energized by the proton gradient and electrical potential created by H$^+$-ATPase. This kind of diagram is common in textbooks, but most books barely mention what exactly is known about these mechanisms and what is purely hypothetical. We therefore feel that this issue is worth discussing in detail here, and we shall therefore review the molecular identification of carriers and channels of plant plasma membranes.

Mineral nutrition

Mineral nutrition of the plant takes place in the roots. Minerals are absorbed by root hair, epidermal and cortical cells (Taiz and Zeiger, 1991). Once in the cell, the nutrients are transported to the stele. The endodermis which delimits the stele forms a barrier to all apoplastic ion movements (Peterson *et al.*, 1993, and references therein). Apoplastic ion movements occur through the intercellular space, as opposed to symplastic movements which take place through the intracellular spaces of a series of cells interconnected by plasmodesmata. Once in the stele, ions are transported to the xylem, where they are eventually carried to the whole plant by bulk water movement. It thus seems that H$^+$-ATPase activity is needed not only in the cortex and epidermis of roots, for ion absorption, but also in the stele, for loading ions into the conductive tissues. This barely discussed point has been confirmed by a recent study of H$^+$-ATPase activity in the stele and cortex of maize roots (Cowan *et al.*, 1993), and by immunolocalization (Parets-Soler *et al.*, 1990; Samuels *et al.*, 1992).

Potassium uptake and transport. The uptake of potassium by plant roots can produce major concentration differences. Under natural conditions the concentration of potassium in the soil is often below 100 µM and the intracellular concentration is of the order of 100 mM (Maathuis and Sanders, 1993). Uptake occurs through two transport systems with different affinities for K$^+$ (first demonstrated by Epstein *et al.*, 1963). A low-affinity transport system (LATS) works when the soil potassium concentration is high (typically > 1 mM) and a high-affinity transport system (HATS) operates at low K$^+$ concentrations (< 1 mM), which may occur more commonly in nature.

When K$^+$ is abundant, the mechanism of transport is simple facilitated diffusion allowed by the membrane potential created by H$^+$-ATPase. This transport is energetically downhill at normal potentials (Cheeseman and Hanson, 1979; Romani *et al.*, 1985). It has therefore been suggested that the LATS is a K$^+$ channel. Two *Arabidopsis thaliana* genes coding for such channels have indeed been cloned by complementation of yeast K$^+$ transport mutants (AKT1, Sentenac *et al.*, 1992; KAT1, Anderson *et al.*, 1992). Homologies with known channels and hydropathy analysis led to the prediction that these K$^+$ channels may contain six membrane spans and form a pore in the membrane when assembled as a tetramer. In conclusion, a K$^+$ channel can be included in *Figure 1* with confidence.

With regard to the HATS, the situation is not so clear. There is indeed controversy about the nature of this system and the ions it transports. Cheeseman and Hanson (1979) have proposed that the HATS is a H$^+$/K$^+$ exchange ATPase. Kochian *et al.* (1989) propose either a K$^+$-ATPase or a K$^+$–H$^+$ co-transport system. There is no clear evidence for any one type of ATPase or co-transporter (Hedrich and Schroeder, 1989; Maathuis and Sanders, 1993). Interestingly, the *Arabidopsis* KAT1 K$^+$ channel expressed in yeast allows growth at micromolar K$^+$ concentrations, thus working under conditions typical of the HATS. However, as noted both by Maathuis and Sanders (1993) and, in their original paper, by Sentenac *et al.* (1992), a simple diffusion facilitator (channel) can promote growth under such conditions because yeasts can attain higher membrane potentials than plants. At micromolar K^+ concentrations, K^+ transport may thus be energetically downhill in yeast, but it is uphill in plants. Therefore, the HATS cannot be a K^+ channel. Furthermore, the properties of the KAT1 channel expressed in oocytes and yeast differ from those described for the HATS in maize roots (Kochian and Lucas, 1993). The cloning and characterization of a high-affinity K^+ uptake transporter from wheat have shown that this enzyme uses K^+–H^+ co-uptake (Schachtman and Schroeder, 1994).

This section on K^+ transport would not be complete without a mention of the channels found elsewhere in plants (for review see Hedrich and Schroeder, 1989). These are outwardly conducting channels found in guard cells, pulvini motor cells, and cells involved in repolarizing plant action potentials. Inwardly conducting channels are found not only in root cells, as discussed above, but also in aleurone-layer cells, suspension-culture cells, coleoptile cells, tumour cells, epidermal cells and guard cells.

Nitrogen uptake and transport. Nitrogen is usually taken up from the soil as nitrate, but is sometimes absorbed in the form of ammonium (Gabathuler and Cleland, 1985). Nitrate transport must be energized, because plants accumulate nitrate against a concentration gradient and also because the entrance of anions into a cell is impeded by the negative inside membrane potential. Within the plant, nitrate is reduced and assimilated. Depending on the species, reductive assimilation takes place either in the roots themselves or, more commonly, in the aerial parts of the plant. Even if nitrate reduction occurs in the roots, as in barley for instance, its transport into those cells remains an energetically uphill process (Glass *et al.,* 1992).

Like K^+, nitrate is transported by two different systems, which have been reviewed by Crawford (1995). While the HATS is constitutively expressed, the LATS is induced by growth at high nitrate concentrations. As both transport systems depolarize the membrane, they are both proposed to be proton–nitrate symport systems which transport more than one proton per nitrate molecule (Glass *et al.,* 1992; McClure *et al.,* 1990, and references therein).

In many studies of nitrate transport *in vitro*, chlorate is used as a tracer element because it is readily transported by the nitrate transport systems. When applied to living plants, chlorate acts as a herbicide and defoliant. This toxic effect is due to the chlorite released by chlorate reduction in the cells. A chlorate-resistant mutant of *Arabidopsis* has been used to isolate the gene of a nitrate transporter (Tsay *et al.,* 1993). Functional expression of this gene *(CHL1)* in oocytes causes nitrate-induced membrane depolarization under acidic conditions. This confirms that two or more protons are transported per nitrate molecule. Like many transporters, the CHL1 protein is predicted to contain 12 membrane spans. Gene expression is induced in roots by nitrate and low pH, so it is probably a gene for the LATS. Members of the high-affinity transport system have also been cloned in barley (Crawford, 1995). A yeast mutant deficient in two NH_4^+ uptake systems was used for isolating, by complementation, a high-affinity NH_4^+ transporter from *Arabidopsis* (Ninnemann *et al.,* 1994).

In conclusion, nitrate and ammonium transport systems can confidently be included in *Figure 1*.

Other minerals. The transport of several other ions is thought to be energized by the electrochemical gradient created by the H^+-ATPase. A proton–sulphate symport system has been characterized in plasma membrane vesicles from *Brassica* roots (Hawkesford *et al.,* 1993). This transport system has a high affinity for sulphate, depends on the pH gradient, and is induced in plants grown on a sulphur-depleted medium. Recently, three cDNAs encoding different sulphate transporters have been cloned in *Stylosanthes hamata*, a legume (Smith *et al.,* 1995). This symport system can thus reliably be included in *Figure 1*.

Phosphate carriers are better known in mitochondrial and chloroplast membranes, but at least one is probably present in the plasma membrane as well. There is some evidence for the existence of chloride channels in plasma membranes of higher plants (Hedrich and Schroeder, 1989).

Finally, calcium ions constitute a special case because they act as second messengers in signal transduction pathways. Therefore, cells are equipped with rapidly inducible transport systems (channels) which enable the cytosolic calcium concentration to change rapidly. The flux of calcium through these well-characterized channels is energized by the membrane potential. Moreover, there are Ca^{2+}-ATPases in the plasma membrane and endoplasmic reticulum (ER) whose function is to deplete the cytosol of its calcium between signal transduction events (Askerlund and Evans, 1992; Wimmers *et al.,* 1992, and references therein).

Transport of organic molecules

Photosynthetic plants are autotrophs, but when single organs or tissues of a plant are considered, some are heterotrophic. A plant thus consists of source and sink organs, the former being net exporters of reduced carbon or nitrogen, and the latter net consumers. Carbon source organs are typically mature leaves and green stems; all other organs can be classified as carbon sinks. Nitrogen source organs can be roots or, more commonly, mature leaves, the other organs being nitrogen sinks.

Reduced carbon and nitrogen are transported throughout the plant as organic solutes. The solutes leave the photosynthetic and nitrogen-reducing cells, transit through the phloem (or through the xylem and phloem for reduced nitrogen when roots are the nitrogen sources; Frommer *et al.,* 1993), and end up in sink tissues and cells. In most plant species, reduced carbon is transported as sucrose, and reduced nitrogen is transported as amino acids. Specific organic nutrient transport occurs during seed filling, germination, seedling growth, filling and mobilization in storage organs, and secretion in nectaries. As the subject of this section has been discussed previously by Bush (1993), we shall focus here only on the molecular aspects of organic nutrient transport and its relationship to H^+-ATPase.

Sugar transport. In mature leaves, sucrose follows a symplastic route from the mesophyll cells that produce it towards the sieve element/companion cell complex. It then probably enters this complex from the apoplast and 8- to 10-fold accumulation occurs (Riesmeier *et al.,* 1992). This accumulation is energized by H^+-ATPase, being achieved by a ΔpH- and membrane-potential-dependent proton–sucrose symport system (for review see Bush, 1993). The stoichiometry of transport is 1:1.

A molecular approach to the study of sucrose transport was adopted by purification of the activity and obtention of specific antibodies inhibiting the transport in plasma membrane vesicles (Gallet *et al.,* 1992). A gene encoding a sucrose carrier has been isolated by complementation of a specially tailored yeast mutant (Riesmeier *et al.,* 1992, 1993). More recently, similar genes were isolated from *Arabidopsis* (Sauer and Stolz, 1994). The protein is predicted to contain 12 membrane spans, a typical structure for carrier proteins. There is a predicted hydrophilic loop in the middle of the protein which may be involved in substrate recognition (six spans–loop–six spans structure). Antisense strategy has shown that the sucrose transporter located at the phloem plasma membrane is the primary route for sugar uptake from the apoplast (Riesmeier *et al.,* 1994).

In sink organs, sucrose can reach individual cells by either a symplastic or an apoplastic route. In the latter case it is hydrolysed by an extracellular invertase, and fructose and glucose are then transported into the sink cells. H+–hexose symport protein genes have been isolated from the alga *Chlorella* and from *Arabidopsis* (Sauer and Tanner, 1989, 1993; Sauer *et al.*, 1990). The encoded proteins are analogous to a series of known H+–glucose symport proteins and are predicted to display a six spans-loop-six spans topology. A general discussion of sugar transport in plants can be found in the review by Sauer *et al.* (1994).

Active transport of sugars is thus energized by H+-ATPase. To date, only H+-sucrose symports have been characterized in plant plasma membranes, but theoretical arguments suggest that antiport systems could exist as well. In any case, both a proton–sucrose and a proton–hexose symport can be confidently included in *Figure 1*.

Amino acid transport. A comprehensive study of amino acid transport in plasma membrane vesicles from *Beta vulgaris* has led to the conclusion that four distinct transport systems exist in this species (Li *et al.*, 1990, 1991). One system is specific for acidic amino acids, and another for basic ones. Neutral system I is specific for isoleucine, threonine, valine, proline and hydroxyproline. Neutral system II is more abundant, and is specific for the other 11 neutral amino acids. All of these transport systems are H+–amino acid symport systems, as revealed by their ΔpH dependence (for review see Bush, 1993).

Neutral amino acid carrier genes from *Arabidopsis* have been isolated independently by two groups (Frommer *et al.*, 1993; Hsu *et al.*, 1993; Kwart *et al.*, 1993). The strategy adopted by both groups was identical, and once again based on complementation of defects in yeast mutants, in this case strains deficient in amino acid transport. The carrier is predicted to exhibit a 12-membrane-span topology, with hydrophilic loops between spans 2 and 3 and spans 9 and 10. Amino acid uptake by mutant yeasts expressing this carrier is ΔpH-dependent.

An *Arabidopsis* histidine-transporting protein was found to be a high-affinity, low-specificity oligopeptide carrier. This enzyme may represent a nitrogen translocation system when amino acid export is limiting (Frommer *et al.*, 1994a,b; Rentsch *et al.*, 1995).

Thus amino acid transport in plants is energized by H+-ATPase and involves H+–amino acid symport systems. These can be included with confidence in *Figure 1*. Clearly, major differences must exist in transport systems and fluxes between plants that reduce nitrogen in their roots and plants that reduce it in their leaves. For instance, the vessels used for transport are not the same, and when the xylem is used, its loading must be indirect, since functional xylem is composed solely of dead cells.

Salinity tolerance and turgor regulation

Exposure to a saline environment leads to the accumulation of toxic NaCl in the cytoplasm and to loss of cell turgor (i.e. decrease in osmotic pressure). H+-ATPase is reported to be involved in the response of plants to both of these effects. The rationale behind this is that enhanced H+-ATPase activity is needed firstly to energize Na+ efflux from the cytoplasm, for instance by Na+/H+ antiport systems in the tonoplast and possibly the plasma membrane (Spickett *et al.*, 1993), and secondly to stimulate K+ influx to compensate for loss of turgor. The loss of turgor is also compensated by the accumulation of organic solutes in the cytoplasm (Niu *et al.*, 1993).

Response to osmotic stress in relation to H+-ATPase has been studied in carrot cells (Reuveni *et al.*, 1987). A slight hyperpolarization of the plasmalemma was observed, as well as a decrease in external pH. Fusicoccin, a H+-ATPase activator, simulates the osmoticum effect. The same experimental system was used later to show that it was tonoplast pyrophosphatase (PPase), and not the plasma-membrane H+-ATPase, that was responding to the stress (Colombo and Cerana, 1993).

A study quite similar in its design to that of Reuveni *et al.* (1987) has been conducted in *Arabidopsis* cells (Curti *et al.*, 1993). Enhanced proton extrusion was measured upon osmotic stress. The effect was only slightly inhibited when protein synthesis was blocked, but it was completely prevented by erythrosin B, a plasma-membrane H^+-ATPase inhibitor. H^+ extrusion required the presence of K^+ in the medium. This evidence, although indirect, is convincing and demonstrates the importance of H^+-ATPase in osmotic pressure regulation.

H^+-ATPase activity and gene expression have been studied in relation to NaCl stress and adaptation (Niu *et al.*, 1993a, b; Reuveni *et al.*, 1993). In *Nicotiana tabacum* and the halophyte *Atriplex nummularia*, total H^+-ATPase mRNA accumulates during salt stress, a finding observed in both roots and expanded leaves. Furthermore, the response of *A. nummularia* was 20-fold greater than that of tobacco, prompting the authors to suggest a primary role for H^+-ATPase in salinity tolerance (Niu *et al.*, 1993a). Once adapted to salt, the halophyte regained its pre-stress H^+-ATPase mRNA levels (Niu *et al.*, 1993b). Clearly, the next step in this line of research should be a more detailed analysis, taking into account the existence of several H^+-ATPase genes, not just one. Accordingly, it is interesting to note that, after salt adaptation, the kinetic properties of the H^+-ATPase of tobacco change, while the total amount of ATPase remains unaltered (Reuveni *et al.*, 1993). A more detailed study extending to the single-gene level could yield the first evidence in plants of different kinetic properties and maybe different regulatory properties for distinct H^+-ATPase isoforms.

Intracellular pH regulation

This subject has been extensively reviewed by Kurkdjian and Guern (1989). Here we shall simply raise a few points of general interest and mention some specific aspects of H^+-ATPase.

The cytoplasmic pH is kept at a remarkably constant level in natural and even challenging experimental conditions. The following environmental factors are known to induce cytoplasmic pH responses: anaerobiosis, light-to-dark transitions, and temperature changes (for review see Kurkdjian and Guern, 1989). Cytoplasmic pH modification and regulation result firstly from the production and consumption of H^+ and secondly from the traffic of H^+ through the boundary membranes of the cytoplasm. Therefore H^+-ATPase must certainly participate in cytoplasmic pH regulation. Acidification of the cytoplasm does indeed activate H^+-ATPase, which then hyperpolarizes the membrane. The hyperpolarization is counterbalanced by K^+ uptake. The mechanism for H^+-ATPase activation seems to be straightforward, since the enzyme has a slightly acidic pH optimum (6.6) in comparison with the normal cytoplasmic pH value of 7.5 (reviewed in Kurkdjian and Guern, 1989).

The involvement of H^+-ATPase in cytoplasmic pH regulation is well illustrated by the case of *Dunaliella acidophila*. The H^+-ATPase of this acidophilic alga has a more acidic pH optimum (6.0) than do those of higher plants, and its affinity for MgATP is 10 times greater ($K_m = 40\ \mu M$). The enzyme is estimated to represent 5% of the *total* membrane protein (Sekler and Pick, 1993), as compared to no more than 1% of the *plasma* membrane protein in higher plants (Sussman and Harper, 1989). It is therefore not surprising that this strange little ATPase-doped alga is in peak form when growing at pH 1.0, and that it happily maintains a cytoplasmic pH of 7.0 (Sekler and Pick, 1993, and references therein).

Acid growth

The so-called acid growth theory of auxin action assumes that auxin promotes growth by causing apoplastic acidification which in turn increases the extendibility of the cell wall (for review see Rayle and Cleland, 1992). This is followed by turgor-driven cell growth. Underlying this theory is the hypothesis that auxin activates H^+-ATPase. The enzyme would thus serve as an early intermediate in a growth-signalling event initiated by an increased auxin concentration.

There is controversy about this theory as to whether H^+ extrusion is really an obligatory step in growth promotion by auxin, or just some kind of side-effect (see, for example, Schopfer, 1989). In fact, what seems to differ between the pH and auxin effects is the long-term action of auxin, but as has been shown in a careful and convincing study (Lüthen *et al.,* 1990), the effects of auxin, FC and pH are almost exactly parallel during the early phase of growth induction.

An interesting study has reported a rapid auxin effect on the amount, rather than the specific activity, of H^+-ATPase in maize coleoptile plasma membrane (Hager *et al.,* 1991). Auxin was shown to enhance exocytosis of internal vesicles and *de novo* synthesis of ATPase; the authors further suggest that auxin does not affect H^+-ATPase specific activity. This is demonstrated by the inhibition of auxin-induced growth after treatment with inhibitors of protein and RNA synthesis (Hager *et al.,* 1991). However, these findings do not rule out the possibility of a fast auxin effect at enzyme level. Indeed, several studies show an *in-vitro* effect of auxin on proton pumping by tobacco H^+-ATPase, under conditions which preclude exocytosis and protein synthesis (François *et al.,* 1992; Gabathuler and Cleland, 1985; Stantoni *et al.,* 1991, 1993). Clearly, auxin effects on the cell are multiple and can occur through different pathways, since it is known that there are auxin receptors on the plasma membrane, in the cytoplasm and in the nucleus of plant cells (see for instance Barbier-Brygoo *et al.,* 1989, 1991; Prasad and Jones, 1991; for review see Barbier-Brygoo *et al.,* 1992).

It is not impossible that some of the remaining conflicting reports on acid growth result from differences in H^+-ATPase isoform content among the different tissues and species studied (see for example Kokubo *et al.,* 1993). Taking into account the existence of several isoforms, combined with the use of sensitive methods such as that reported by Lüthen *et al.* (1990), should help to resolve some of the remaining controversies. For us, the conclusion of this section is that H^+-ATPase, in a number of cases, seems to be the target of growth-regulating signals, and that the enzyme plays an early role by acidifying and promoting loosening of the cell wall. Considering the primary role of H^+-ATPase in maintaining turgor pressure, it is clear that this enzyme is a key participant in cell growth.

Other less well-documented functions

Apical cell growth is often accompanied by a polarized growth current. Such currents have been detected in growing root hairs and pollen tubes. In the latter case, a growth current supported by entry of K^+ along the tube and exit of H^+ from the grain was observed (Weisenseel and Jaffe, 1976), and a high membrane potential was measured (Weisenseel and Wenish, 1980). Moreover, ATPase has been found to accumulate in grains of germinated pollen, while immunodetection experiments failed to reveal any ATPase in the tube (Obermeyer *et al.,* 1992). Using reporter gene analysis, however, some expression of a tobacco H^+-ATPase gene was detected in pollen tubes, although the level of expression was much lower than in the grain. The asymmetrical distribution of H^+-ATPase in germinated pollen suggests the possible involvement of H^+-ATPase in determining the direction of apical growth by governing the growth current.

Pulvini movements are often mentioned as examples of the biological functions of H^+-ATPase. The rationale behind this is that the movement is caused by a change in turgor of the pulvini motor cells. The changes in turgor result from massive ion movements across the plasma membrane, and are thus energized by the H^+-ATPase. However, we are not aware of any recent work on this subject.

In contrast, stomatal opening, which relies on the same mechanism of turgor change, has been intensively studied. Stomata are known to open or close in response to various stimuli such as light, relative humidity, CO_2 and hormones. These stimuli trigger a cascade of signals which eventually increase or decrease the turgor of guard cells, with subsequent opening or closure of the stomata (for review see Serrano and Zeiger, 1989). H$^+$-ATPase is involved in some of these cascades (Assmann et al., 1985; Assmann and Schwartz, 1992; Marten et al., 1991; Serrano et al., 1988; Shimazaki et al., 1986, 1992). The enzyme seems to be more abundant in guard cells than in mesophyll cells, but H$^+$-ATPase preparations of both origins appear to have identical kinetic properties (Becker et al., 1993; Villalba et al., 1991).

4. Conclusions

Several participants in the transport processes which take place across the plant plasma membrane have been identified at the protein and/or gene levels. Among them, H$^+$-ATPase is currently the best characterized enzyme. The existence of a multigene family encoding H$^+$-ATPase has made the picture more complex than was originally thought. The next step will thus consist of relating single isozymes and the expression of their corresponding genes to the various roles of the ATPase in different cell types and at different developmental stages. A future challenge will require the identification of H$^+$-ATPase regulatory mechanisms at various levels and their integration with plant transport physiology. Reverse physiology should become feasible with the development of tools aimed at altering the expression level of the ATPase, either at the single-gene or sub-family level.

Acknowledgements

Research carried out in this laboratory was supported by grants from the Commission of the European Communities, the Belgian Services de la Politique Scientifique and the National Fund for Scientific Research (Belgium).

References

Aducci P, Ballio A, Blein J-P, Fullone MR, Rossignol M, Scalla R. (1988) Functional reconstitution of a proton-translocating system responsive to fusicoccin. Proc. Natl Acad. Sci. USA 85: 7849–7851.

Anderson JA, Huprikar SS, Kochian LV, Lucas VJ, Gaber RF. (1992) Functional expression of a probable Arabidopsis thaliana potassium channel in Saccharomyces cerevisiae. Proc. Natl Acad. Sci. USA 89: 3736–3740.

Askerlund P, Evans DE. (1992) Reconstitution and characterization of a calmodulin-stimulated Ca^{2+}-pumping ATPase purified from Brassica oleracea L. Plant Physiol. 100: 1670–1681.

Assmann SM, Schwartz A. (1992) Synergistic effect of light and fusicoccin on stomatal opening. Plant Physiol. 98: 1349–1355.

Assmann SM, Simoncini L, Schroeder JI. (1985) Blue light activates electrogenic ion pumping in guard cell protoplasts of Vicia faba. Nature 318: 285–287.

Barbier-Brygoo H, Ephritkine G, Klämbt D, Ghislain M, Guern J. (1989) Functional evidence for an auxin receptor at the plasmalemma of tobacco mesophyll protoplasts. Proc. Natl Acad. Sci. USA 86: 891–895.

Barbier-Brygoo H, Ephritkine G, Klämbt D, Maurel C, Palme K, Schell J, Guern J. (1991) Perception of the auxin signal at the plasmalemma of tobacco mesophyll protoplasts. Plant J. 1: 83–93.

Barbier-Brygoo H, Ephritkine G, Maurel C, Guern J. (1992) Perception of the auxin signal at the plasma membrane of tobacco mesophyll protoplasts. Biochem. Soc. Trans. 20: 59–63.

Becker D, Zeilinger C, Lohse G, Depta H, Hedrich R. (1993) Identification and biochemical characterization of the plasma-membrane H$^+$-ATPase in guard cells of Vicia faba L. Planta 190: 44–50.

Bidway AP, Zhang L, Bachmann RC, Takemoto JY. (1987) Mechanism of action of Pseudomonas syringae phytotoxin syringomycin. Stimulation of red beet plasma-membrane ATPase activity. Plant Physiol. 83: 39–43.

Boutry M, Michelet B, Goffeau A. (1989) Molecular cloning of a family of plant genes encoding a protein homologous to plasma-membrane H$^+$-translocating ATPases. Biochem. Biophys. Res. Commun. 162: 567–574.

Brauer D, Tu S-I, Hsu A-F, Thomas CE. (1989) Kinetic analysis of proton transport by the vanadate-sensitive ATPase from maize root microsomes. Plant Physiol. 89: 464–471.

Briskin DP, Leonard RT. (1982) Partial characterization of a phosphorylated intermediate associated with the plasma membrane ATPase of corn roots. *Proc. Natl Acad. Sci. USA* **79**: 6922–6926.

Briskin DP, Hanson JB. (1992) How does the plant plasma-membrane H⁺-ATPase pump protons? *J. Exp. Bot.* **43**: 269–289.

Bush DR. (1993) Proton coupled sugar and amino acid transporters in plants. *Annu. Rev. Plant Physiol. Plant Mol. Biol.* **44**: 513–542.

Cheeseman JM, Hanson JB. (1979) Energy-linked potassium influx as related to cell potential in corn roots. *Plant Physiol.* **64**: 842–845.

Colombo R, Cerana R. (1993) Enhanced activity of tonoplast pyrophosphatase in NaCl-grown cells of *Daucus carota. J. Plant Physiol.* **142**: 226–229.

Cosgrove D. (1986) Biophysical control of plant cell growth. *Annu. Rev. Plant Physiol.* **37**: 377–405.

Cowan DSC, Clarkson DT, Hall JL. (1993) A comparison between the ATPase and proton pumping activities of plasma membranes isolated from the stele and cortex of *Zea mays* roots. *J. Exp. Bot.* **44**: 983–989.

Crawford NM. (1995) Nitrate: nutrient and signal for plant growth. *Plant Cell* **7**: 859–868.

Curti G, Massardi F, Lado P. (1993) Synergistic activation of plasma-membrane H⁺-ATPase in *Arabidopsis thaliana* cells by turgor decrease and by fusicoccin. *Physiol. Plant.* **87**: 592–600.

de Kerchove d'Exaerde A, Supply P, Dufour JP, Bogaerts P, Thinès D, Goffeau A, Boutry M. (1995) Functional complementation of a null mutation of the yeast *Saccharomyces cerevisiae* plasma membrane H⁺-ATPase by a plant H⁺-ATPase gene. *J. Biol. Chem.* **270**: 23828–23837.

DeWitt ND, Harper JF, Sussman MR. (1991) Evidence for a plasma membrane proton pump in phloem cells of higher plants. *Plant J.* **1**: 121–128.

Epstein E, Rains DW, Elzam OE. (1963) Resolution of dual mechanisms of potassium absorption by barley roots. *Proc. Natl Acad. Sci. USA* **49**: 684–692.

Esau K. (1977) *Anatomy of Seed Plants*, 2nd edition. New York: John Wiley and sons.

Ewing NN, Bennett AB. (1994) Assessment of the number and expression of P-Type H⁺-ATPase genes in tomato. *Plant Physiol.* **106**: 547–557.

Ewing NN, Wimmers LE, Meyer DJ, Chetelat RT, Bennett AB. (1990) Molecular cloning of tomato plasma membrane H⁺-ATPase. *Plant Physiol.* **94**: 1874–1881.

Fisher JD, Hansen D, Hodges TK. (1970) Correlation between ion fluxes and ion-stimulated adenosine triphosphatase activity of plant roots. *Plant Physiol.* **46**: 812–814.

François J-M, Berville A, Rossignol M. (1992) Development and line dependent variations of *Petunia* plasma membrane H⁺-ATPase sensitivity to auxin. *Plant Sci.* **87**: 19–27.

Frommer WB, Hummel S, Riesmeier JW. (1993) Expression cloning in yeast of a cDNA encoding a broad specificity amino acid permease from *Arabidopsis thaliana. Proc. Natl Acad. Sci. USA* **90**: 5944–5948.

Frommer WB, Hummel S, Rentsch D. (1994a) Cloning of an *Arabidopsis* histidine transporting protein related to nitrate and peptide transporters. *FEBS Lett.* **347**: 185–189.

Frommer WB, Kwart M, Hirner B, Fischer WN, Hummel S, Ninnemann O. (1994b) Transporters for nitrogenous compounds in plants. *Plant Mol. Biol.* **26**: 1651–1670.

Gabathuler R, Cleland RE. (1985) Auxin regulation of a proton translocating ATPase in pea root plasma membrane vesicles. *Plant Physiol.* **79**: 1080–1085.

Gallet O, Lemoine R, Gaillard C, Larsson C, Delrot S. (1992) Selective inhibition of active uptake of sucrose into plasma membrane vesicles by polyclonal sera directed against a 42 kilodalton plasma membrane polypeptide. *Plant Physiol.* **98**: 17–23.

Glass ADM, Shaff JE, Kochian LV. (1992) Studies of the uptake of nitrate in barley. IV. Electrophysiology. *Plant Physiol.* **93**: 281–289.

Hager A, Debus G, Edel H-G, Stransky H, Serrano R. (1991) Auxin induces exocytosis and the rapid synthesis of a high-turnover pool of plasma membrane H⁺-ATPase. *Planta* **185**: 527–537.

Harms K, Wöhner RV, Schulz B, Frommer WB. (1994) Isolation and characterization of P-type H⁺-ATPase genes from potato. *Plant Mol. Biol.* **26**: 979–988.

Harper JF, Surowy TK, Sussman MR. (1989) Molecular cloning and sequence of cDNA encoding the plasma membrane proton pump (H⁺-ATPase) of *Arabidopsis thaliana. Proc. Natl Acad. Sci. USA* **86**: 1234–1238.

Harper JF, Manney L, DeWitt ND, Yoo MH, Sussman MR. (1990) The *Arabidopsis thaliana* plasma membrane H⁺-ATPase multigene family. *J. Biol. Chem.* **265**: 13601–13608.

Harper JF, Binder BM, Sussman MR. (1993) Calcium and lipid regulation of an *Arabidopsis* protein kinase expressed in *Escherichia coli. Biochemisty* **32**: 3282–3290.

Harper JF, Manney L, Sussman MR. (1994) The plasma membrane H⁺-ATPase gene family in *Arabidopsis*: genomic sequence of *AHA10* which is expressed primarily in developing seeds. *Mol. Gen. Genet.* **244:** 572–587.

Hawkesford MJ, Davidian J-C, Grignon C. (1993) Sulphate/proton cotransport in plasma-membrane vesicles isolated from roots of *Brassica napus* L.: increased transport in membranes isolated from sulphur-starved plants. *Planta* **190:** 297–304.

Hedrich R, Schroeder JI. (1989) The physiology of ion channels and electrogenic pumps in higher plants. *Annu. Rev. Plant Physiol.* **40:** 539–569.

Hodges TK, Leonard RT, Bracker CE, Keenan TW. (1972) Purification of an ion-stimulated ATPase from plant roots: association with plasma membranes. *Proc. Natl Acad. Sci. USA* **69:** 3307–3311.

Holländer-Czytko H, Weiler EW. (1994) Isolation and biochemical characterization of fusicoccin-insensitive variant lines of *Arabidopsis thaliana* (L.) Heynh. *Planta* **195:** 188–194.

Houlné G, Boutry M. (1994) Identification of an *Arabidopsis thaliana* gene encoding a plasma membrane H⁺ATPase whose expression is restricted to anther tissues. *Plant J.* **5:** 311–317.

Hsu L, Chiou T, Chen L, Bush DR. (1993) Cloning a plant amino acid transporter by functional complementation of a yeast amino acid transport mutant. *Proc. Natl Acad. Sci. USA* **90:** 7441–7445.

Jacobson L, Overstreet R, King H, Handley R. (1950) A study of potassium absorption by barley roots. *Plant Physiol.* **25:** 639–647.

Jin YK, Bennetzen JL. (1994) Integration and non-random mutation of a plasma membrane proton ATPase gene fragment within the *Bs1* retroelement of maize. *Plant Cell* **6:** 1177–1186.

Johansson F, Sommarin M, Larsson C. (1993) Fusicoccin activates the plasma membrane H⁺-ATPase by a mechanism involving the C-terminal inhibitory domain. *Plant Cell* **5:** 321–327.

Kochian LV, Lucas WJ. (1993) Can K⁺ channels do it all? *Plant Cell* **5:** 720–721.

Kochian LV, Shaff JE, Lucas WJ. (1989) High affinity K⁺ uptake in maize roots. A lack of coupling with H⁺ efflux. *Plant Physiol.* **91:** 1202–1211.

Kokubo A, Yamamoto R, Masuda Y. (1993) Galactose prevents proton excretion but not IAA-induced growth in segments of azuki bean epicotyls. *Planta* **190:** 284–287.

Korthout HAAJ, de Boer AH. (1994) A fusicoccin binding protein belongs to the family of 14-3-3 brain protein homologs. *Plant Cell* **6:** 1681–1692.

Korthout HAA, van der Hoeven PCJ, Wagner MJ, Van Hunnik E, De Boer AH. (1994) Purification of the fusicoccin-binding protein from oat root plasma membrane by affinity chromatography with biotinylated fusicoccin. *Plant Physiol.* **105:** 1281–1288.

Kurkdjian A, Guern J. (1989) Intracellular pH: measurement and importance in cell activity. *Annu. Rev. Plant Physiol. Plant Mol. Biol.* **40:** 271–303.

Kwart M, Hirner B, Hummel S, Frommer WB. (1993) Differential expression of two related amino acid transporters with differing substrate specificity in *Arabidopsis thaliana. Plant J.* **4:** 993–1002.

Li Z-C, Bush DR. (1990) ΔpH-dependent amino acid transport into plasma membrane vesicles isolated from sugar beet (*Beta vulgaris* L.) leaves. I. Evidence for carrier-mediated, electrogenic flux through multiple transport systems. *Plant Physiol.* **94:** 268–277.

Li Z-C, Bush DR. (1991) ΔpH-dependent amino acid transport into plasma membrane vesicles isolated from sugar beet (*Beta vulgaris* L.) leaves. II. Evidence for multiple aliphatic, neutral amino acid symports. *Plant Physiol.* **96:** 1338–1344.

Lüthen H, Bigdon M, Böttger M. (1990) Re-examination of the acid growth theory of auxin action. *Plant Physiol.* **93:** 931–939.

Maathuis FJM, Sanders D. (1993) Energization of potassium uptake in *Arabidopsis thaliana. Planta* **191:** 302–307.

McClure PR, Kochian LV, Spanswick RM, Shaff JE. (1990) Evidence for cotransport of nitrate and proton in maize roots. I. Effects of nitrate on the membrane potential. *Plant Physiol.* **93:** 281–289.

Marra M, Ballio A, Fullone MR, Aducci P. (1992) Some properties of a reconstituted plasmalemma H⁺-ATPase activated by fusicoccin. *Plant Physiol.* **98:** 1029–1034.

Marra M, Fullone MR, Fogliano V, Pen J, Mattei M, Masi S, Aducci P. (1994) The 30-kilodalton protein present in purified fusicoccin receptor preparations is a 14-3-3-like protein. *Plant Physiol.* **106:** 1497–1501.

Marrè E, Ballarin-Denti A. (1985) The proton pumps of the plasmalemma and the tonoplast of higher plants. *J. Bioenerg. Biomembr.* **17:** 1–21.

Marten I, Lohse G, Hedrich R. (1991) Plant growth hormones control voltage-dependent activity of anion channels in plasma membrane of guard cells. *Nature* **353:** 758–762.

Memon AR, Boss WF. (1990) Rapid light-induced changes in phosphoinositide kinase and H⁺-ATPase in plasma membrane of sunflower hypocotyls. *J. Biol. Chem.* **265:** 14817–14821.

Meyer C, Feyerabend M, Weiler EW. (1989) Fusicoccin-binding proteins in *Arabidopsis thaliana* (L.) Heynh. Characterization, solubilization and photoaffinity labeling. *Plant Physiol.* **89:** 692–699.

Michelet B, Boutry M. (1995) The plasma membrane H⁺-ATPase. A highly regulated enzyme with multiple physiological functions. *Plant Physiol.* **108:** 1–6.

Michelet B, Perez C, Goffeau A, Boutry M. (1989) The plasma membrane H⁺-ATPase of *Nicotiana plumbaginifolia*. In: *Plant Membrane Transport: the Current Position* (Dainty J, De Michelis MI, Marrè E, Rasi-Caldogno F, ed.). Amsterdam: Elsevier Science Publishers, pp. 455–460.

Michelet B, Lukaszewicz M, Dupriez V, Boutry M. (1994) A plant plasma membrane proton-ATPase gene is regulated by development and environment and shows signs of a translational regulation. *Plant Cell* **6:** 1375–1389.

Moriau L, Bogaerts P, Jonniaux J-L, Boutry M. (1993) Identification and characterization of a second plasma membrane H⁺-ATPase gene subfamily in *Nicotiana plumbaginifolia*. *Plant Mol. Biol.* **21:** 955–963.

Nakijama N, Saji H, Aono M, Kondo N. (1995) Isolation of cDNA for a plasma membrane H⁺-ATPase from guard cells of *Vicia faba* L. *Plant Cell Physiol.* **36:** 919–924.

Ninnemann O, Jauniaux JC, Frommer WB. (1994) Identification of a high affinity NH₄⁺ transporter from plants. *EMBO J.* **15:** 3464–3471.

Niu X, Narasimhan ML, Salzman RA, Bressan RA, Hasegawa PM. (1993a) NaCl regulation of plasma membrane H⁺-ATPase gene expression in a glycophyte and a halophyte. *Plant Physiol.* **103:** 713–718.

Niu X, Zhu J-K, Narasimhan ML, Bressan RA, Hasegawa PM. (1993b) Plasma-membrane H⁺-ATPase gene expression is regulated by NaCl in cells of the halophyte *Atriplex nummularia* L. *Planta* **190:** 433–438.

Obermeyer G, Lützelschwab M, Heumann H-G, Weisenseel MH. (1992) Immunolocalization of H⁺-ATPases in the plasma membrane of pollen grains and pollen tubes of *Lilium longiflorum*. *Protoplasma* **171:** 55–63.

Oecking C, Eckerskorn C, Weiler EW. (1994) This fusicoccin receptor of plants is a member of the 14-3-3 superfamily of eukaryotic regulatory proteins. *FEBS Lett.* **352:** 163–166.

Olivari C, Pugliarello MC, Rasi-Caldogno F, De Michelis MI. (1993) Characteristics and regulatory properties of the H⁺-ATPase in a plasma membrane fraction purified from *Arabidopsis thaliana*. *Bot. Acta* **106:** 13–19.

Ookura T, Wada M, Sakakibara Y, Jeong KH, Maruta I, Kawamura Y, Kasamo K. (1994) Identification and characterization of a family of genes for the plasma membrane H⁺-ATPase of *Oryza sativa* L. *Plant Cell Physiol.* **35:** 1251–1256.

Palmgren MG, Sommarin M. (1989) Lysophosphatidylcholine stimulates ATP dependent accumulation in isolated oat root plasma membrane vesicles. *Plant Physiol.* **90:** 1009–1014.

Palmgren MG, Christensen G. (1993) Complementation *in situ* of the yeast plasma membrane H⁺-ATPase gene *pma*1 by an H⁺-ATPase gene from a heterologous species. *FEBS Lett.* **317:** 216–222.

Palmgren MG, Larsson C, Sommarin M. (1990) Proteolytic activation of the plant plasma membrane H⁺-ATPase by removal of a terminal segment. *J. Biol. Chem.* **265:** 13423–13426.

Palmgren MG, Sommarin M, Serrano R, Larsson C. (1991) Identification of an autoinhibitory domain in the C-terminal region of the plant plasma membrane H⁺-ATPase. *J. Biol.Chem.* **266:** 20470–20475.

Pardo JM, Serrano R. (1989) Structure of a plasma membrane H⁺-ATPase gene from the plant *Arabidopsis thaliana*. *J. Biol. Chem.* **264:** 8557–8562.

Parets-Soler A, Pardo JM, Serrano R. (1990) Immunocytolocalization of plasma membrane H⁺-ATPase. *Plant Physiol.* **93:** 1654–1658.

Perez C, Michelet B, Ferrant V, Bogaerts P, Boutry M. (1992) Differential expression within a three-gene subfamily encoding a plasma membrane H⁺-ATPase in *Nicotiana plumbaginifolia*. *J. Biol. Chem.* **267:** 1204–1211.

Peterson CA, Murrmann M, Steudle E. (1993) Location of the major barriers to water and ion movement in young roots of *Zea mays* L. *Planta* **190:** 127–136.

Poole RJ. (1978) Energy coupling for membrane transport. *Annu. Rev. Plant Physiol.* **29:** 437–460.

Prasad PV, Jones AM. (1991) Putative receptor for the plant growth hormone auxin identified and characterized by anti-idiotypic antibodies. *Proc. Natl Acad. Sci. USA* **88:** 5479–5483.

Rasi-Caldogno F, Pugliarello MC. (1985) Fusicoccin stimulates the H⁺-ATPase of plasmalemma in isolated membrane vesicles from radish. *Biochem. Biophys. Res. Commun.* **133:** 280–285.

Rasi-Caldogno F, De Michelis MI, Pugliarello MC, Marrè E. (1986) H+-pumping driven by the plasma membrane ATPase in membrane vesicles from radish: stimulation by fusicoccin. *Plant Physiol.* **82:** 121–125.

Rasi-Caldogno F, Pugliarello MC, Olivari C, De Michelis MI. (1993) Controlled proteolysis mimics the effect of fusicoccin on the plasma membrane H+-ATPase. *Plant Physiol.* **103:** 391–398.

Rayle DL, Cleland R. (1992) The acid growth theory of auxin-induced cell elongation is alive and well. *Plant Physiol.* **99:** 1271–1274.

Regenberg B, Villalba JM, Lanfermeijer FC, Palmgren MG. (1995) C-terminal deletion analysis of plant plasma membrane H+-ATPase: yeast as a model system for solute transport across the plant plasma membrane. *Plant Cell* **7:** 1655–1666.

Rentsch D, Laloi M, Rouhara I, Schmelzer E, Delrot S, Frommer WB. (1995) *NTR1* encodes a high affinity oligopeptide transporter in *Arabidopsis. FEBS Lett.* **370:** 264–268.

Reuveni M, Colombo R, Lerner HR, Pradet A, Poljakoff-Mayber A. (1987) Osmotically induced proton extrusion from carrot cells in suspension culture. *Plant Physiol.* **85:** 383–388.

Reuveni M, Bressan RA, Hasegawa PM. (1993) Modification of proton transport kinetics of the plasma membrane H+-ATPase after adaptation of tobacco cells to NaCl. *J. Plant Physiol.* **142:** 312–318.

Riesmeier JW, Willmitzer L, Frommer WB. (1992) Isolation and characterization of a sucrose carrier cDNA from spinach by functional expression in yeast. *EMBO J.* **11:** 4705–4713.

Riesmeier JW, Hirner B, Frommer WB. (1993) Potato sucrose transporter expression in minor veins indicates a role in phloem loading. *Plant Cell* **5:** 1591–1598.

Riesmeier JW, Willmitzer L, Frommer WB. (1994) Evidence of an essential role of the sucrose transporter in phloem loading and assimilate partitioning. *EMBO J.* **1:** 1–7.

Romani G, Marrè MT, Bellando M, Alloatti G, Marrè E. (1985) H+ extrusion and potassium uptake associated with potential hyperpolarization in maize and wheat root segments treated with permeant weak acids. *Plant Physiol.* **76:** 734–739.

Samuels AL, Fernando M, Glass ADM. (1992) Immunofluorescent localization of plasma membrane H+-ATPase in barley roots and effects of K nutrition. *Plant Physiol.* **99:** 1509–1514.

Santoni V, Vansuyt G, Rossignol M. (1991) The changing sensitivity to auxin of the plasma membrane H+-ATPase: relationship between plant development and ATPase content of membranes. *Planta* **185:** 227–232.

Santoni V, Vansuyt G, Rossignol M. (1993) Indoleacetic acid pretreatment of tobacco plants *in vivo* increases the *in vitro* sensitivity to auxin of the plasma membrane H+-ATPase from leaves and modifies the polypeptide composition of the membrane. *FEBS Lett.* **326:** 17–20.

Sauer N, Tanner W. (1989) The hexose carrier from *Chlorella:* cDNA cloning of a eucaryotic H+-cotransporter. *FEBS Lett.* **259:** 43–46.

Sauer N, Tanner W. (1993) Molecular biology of sugar transporters in plants. *Bot. Acta* **106:** 277–286.

Sauer N, Stolz J. (1994) SUC1 and SUC2: two sucrose transporters from *Arabidopsis thaliana;* expression and characterization in baker's yeast and identification of the histidine-tagged protein. *Plant J.* **6:** 67–77.

Sauer N, Friedländer K, Gräml-Wicke U. (1990) Primary structure, genomic organization and heterologous expression of a glucose transporter from *Arabidopsis thaliana. EMBO J.* **9:** 3045–3050.

Sauer N, Baier K, Gahrtz M, Stadler R, Stolz J, Truernit E. (1994) Sugar transport across the plasma membranes of higher plants. *Plant Mol. Biol.* **26:** 1671–1679.

Scalla R, Amory A, Rigaud J, Goffeau A. (1983) Phosphorylated intermediate of a transport ATPase and activity of protein kinase in membranes from corn roots. *Eur. J. Biochem.* **132:** 525–530.

Schachtman DP, Schroeder JI. (1994) Structure and transport mechanism of a high-affinity potassium uptake transporter from higher plants. *Nature* **370:** 655–658.

Schaller GE, Sussman MR. (1988) Phosphorylation of the plasma-membrane H+-ATPase of oat roots by a calcium-stimulated protein kinase. *Planta* **173:** 509–518.

Schopfer P. (1989) pH dependence of extension growth in *Avena* coleoptiles and its implications for the mechanism of auxin action. *Plant Physiol.* **90:** 202–207.

Sekler I, Pick U. (1993) Purification and properties of a plasma membrane H+-ATPase from the extremely acidophilic alga *Dunaliella acidophila. Plant Physiol.* **101:** 1055–1061.

Sekler I, Weiss M, Pick U. (1994) Activation of the *Dunaliella acidophila* plasma membrane H+-ATPase by trypsin cleavage of a fragment that contains a phosphorylation site. *Plant Physiol.* **105:** 1125–1132.

Sentenac H, Bonneaud N, Minet M, Lacroute F, Salmon JF, Gaymard F, Grignon C. (1992) Cloning and expression in yeast of a plant potassium ion transport system. *Science* **256:** 663–665.

Serrano R. (1989) Structure and function of plasma membrane ATPase. *Annu. Rev. Plant Physiol. Plant Mol. Biol.* **40:** 61–94.

Serrano R. (1993) Structure, function and regulation of plasma membrane H⁺-ATPase. *FEBS Lett.* **325:** 108–111.

Serrano EE, Zeiger E. (1989) Sensory transduction and electrical signaling in guard cells. *Plant Physiol.* **91:** 795–799.

Serrano EE, Zeiger E, Hagiwara S. (1988) Red light stimulates an electrogenic proton pump in *Vicia* guard cell protoplasts. *Proc. Natl Acad. Sci. USA* **85:** 436–440.

Shimazaki K, Lino M, Zeiger E. (1986) Blue light-dependent proton extrusion by guard cell protoplasts of *Vicia faba. Nature* **319:** 324–326.

Shimazaki K, Kinoshita T, Nishimura M. (1992) Involvement of calmodulin and calmodulin-dependent myosin light chain kinase in blue light-dependent H⁺ pumping by guard cell protoplasts from *Vicia faba* L. *Plant Physiol.* **99:** 1416–1421.

Smith FW, Ealing PM, Hawkesford MJ, Aarkson DT. (1995) Plant members of a family of sulfate transporters reveal functional subtypes. *Proc. Natl Acad. Sci. USA* **92:** 9373–9377.

Spanswick RM. (1981) Electrogenic ion pumps. *Annu. Rev. Plant Physiol.* **32:** 267–289.

Spickett CM, Smirnoff N, Ratcliffe RG. (1993) An *in vivo* magnetic resonance investigation of ion transport in maize (*Zea mays*) and *Spartina anglica* roots during exposure to high salt concentrations. *Plant Physiol.* **102:** 629–638.

Sussman MR. (1992) A plethora of plant plasmalemma proton pumps. In: *Transport and Receptor Proteins of Plant Membranes. Molecular Structure and Function* (Cooke DT, Clarkson DT ed.). New York: Plenum Press, pp. 5–11.

Sussman MR. (1994) Molecular analysis of proteins in the plant plasma membrane. *Annu. Rev. Plant Physiol. Plant Mol. Biol.* **45:** 211–234.

Sussman MR, Harper JF. (1989) Molecular biology of the plasma membrane of higher plants. *Plant Cell* **1:** 953–960.

Taiz L, Zeiger E. (1991) *Plant Physiology.* Amsterdam: The Benjamin Kumming Publishing Company Inc.

Tsay Y-F, Schroeder JI, Feldmann KA, Crawford NM. (1993) The herbicide sensitivity gene *CHL1* of *Arabidopsis* encodes a nitrate-inducible nitrate transporter. *Cell* **72:** 705–713.

Vara F, Serrano R. (1982) Partial purification and properties of the proton-translocating ATPase of plant plasma membrane. *J. Biol. Chem.* **257:** 12826–12830.

Vara F, Serrano R. (1983) Phosphorylated intermediate of the ATPase of plant plasma membranes. *J. Biol. Chem.* **258:** 5334–5336.

Vera-Estrella R, Barkla BJ, Higgins VJ, Blumwald E. (1994) Plant defense response to fungal pathogens. Activation of host-plasma membrane H⁺-ATPase by elicitor-induced enzyme dephosphorylation. *Plant Physiol.* **104:** 209–215.

Villalba JM, Lützelschwab M, Serrano R. (1991) Immunocytolocalization of plasma-membrane H⁺-ATPase in maize coleoptiles and enclosed leaves. *Planta* **185:** 458–461.

Villalba JM, Palmgren MG, Berberian GE, Ferguson C, Serrano R. (1992) Functional expression of plant plasma membrane H⁺-ATPase in yeast endoplasmic reticulum. *J. Biol. Chem.* **267:** 12341–12349.

Wach A, Schlesser A, Goffeau A. (1992) An alignment of 17 deduced protein sequences from plant, fungi, and protozoa H⁺-ATPase genes. *J. Bioenerg. Biomembr.* **24:** 309–317.

Wada M, Takano M, Kasamo K. (1992) Nucleotide sequence of a complementary DNA encoding plasma membrane H⁺-ATPase from rice (*Oryza sativa* L.) *Plant Physiol.* **99:** 794–795.

Weisenseel MH, Jaffe LF. (1976) The major growth current through lily pollen tubes enters as K⁺ and leaves as H⁺. *Planta* **133:** 1–7.

Weisenseel MH, Wenish HH. (1980) The membrane potential of growing lily pollen. *Z. Pflanzenphysiol.* **99:** 313–323.

Wevelsiep L, Rüpping E, Knogge W. (1993) Stimulation of barley plasmalemma H⁺-ATPase by phytotoxic peptides from the fungal pathogen *Rhynchosporium secalis. Plant Physiol.* **101:** 297–301.

Wimmers LE, Ewing NN, Bennett AB. (1992) Higher plant Ca²⁺-ATPase: primary structure and regulation of mRNA abundance by salt. *Proc. Natl Acad. Sci. USA* **89:** 9205–9209.

Calcium efflux transporters in higher plants

Per Askerlund and Marianne Sommarin

1. Introduction

Many stimuli (e.g. light, gravity, salinity, temperature, wind, touch, elicitors and plant hormones) which control different physiological processes in plants have been shown to induce rapid changes in cytosolic free Ca^{2+} concentration, demonstrating that a change in cytosolic Ca^{2+} is a primary event in these signal transduction pathways, and establishing that Ca^{2+} acts as an intracellular messenger in plants. The change in cytosolic Ca^{2+} concentration, which may be a transient and large increase, a modest steady-state increase/decrease or an oscillatory change, may in turn modulate Ca^{2+}-dependent intracellular processes such as protein phosphorylation, eventually giving rise to a physiological response (for reviews see Bush, 1993, 1995; Gilroy *et al.*, 1993; Poovaiah and Reddy, 1993).

Like all other living organisms, plants maintain their cytosolic free Ca^{2+} concentration at very low levels under resting conditions. Measurements using a variety of different techniques have determined that the concentration of cytosolic free Ca^{2+} in a resting plant cell is between 30 and 900 nM (Fricker *et al.*, 1993; Read *et al.*, 1992). A stimulus-induced increase in cytosolic Ca^{2+} is accomplished either by an influx of extracellular Ca^{2+} via Ca^{2+} channels in the plasma membrane (PM), or by the release of Ca^{2+} from intracellular Ca^{2+} stores. Alternatively, a change in cytosolic Ca^{2+} may occur due to altered efflux activities against a relatively constant background influx (Bush, 1993). Influx of Ca^{2+} into the cytosol from outside the cell and release of Ca^{2+} from intracellular stores is an energetically downhill process and it occurs through Ca^{2+} channels. In mature cells, the vacuole is the most significant intracellular pool of Ca^{2+}, although the endoplasmic reticulum (ER) can represent a major Ca^{2+} store in young, rapidly developing cells (Sanders *et al.*, 1990). Voltage-gated Ca^{2+} channels are present both in the PM and in the vacuole membrane (tonoplast), stretch-operated Ca^{2+} channels have been identified in the PM, and inositol trisphosphate- and cyclic ADP-ribose-operated Ca^{2+} channels are present in the vacuole membrane (Allen *et al.*, 1995; Gilroy *et al.*, 1993; Johannes *et al.*, 1991, 1992). The presence of voltage-dependent Ca^{2+} release channels in the ER from touch-sensitive tendrils of *Bryonica dioica* was recently demonstrated (Klüsener *et al.*, 1995). Animal cells contain inositol trisphosphate- and cyclic ADP-ribose-gated Ca^{2+} channels in their ER/sarcoplasmic reticulum (Berridge, 1993; Mészáros *et al.*, 1993), but as yet there is no evidence for such channels in plant ER (Canut *et al.*, 1993). The cytosolic Ca^{2+} concentration is restored and maintained at the low level by extrusion of Ca^{2+} from the cell or by sequestration into intracellular organelles. In plants, primary Ca^{2+}-pumping ATPases and secondary nH^+/Ca^{2+} antiporters are responsible for these functions (*Figure 1*; Chanson, 1993; Evans *et al.*, 1991). In addition to regulating the cytosolic Ca^{2+} concentration, these Ca^{2+} transporters may have an important role in regulating Ca^{2+} levels inside organelles (Bush, 1993). Several reviews about the role of Ca^{2+} as an intracellular messenger in plants have appeared (Bush, 1993; Gilroy *et al.,* 1993; Poovaiah and Reddy, 1993), but only a few concentrate on the characteristics of Ca^{2+} transporters (Chanson, 1993; Evans *et al.,* 1991). The present review is intended to give an updated overview of Ca^{2+} efflux transporters in higher plants. Comparisons with animal Ca^{2+} efflux transporters will be made, but for detailed information on mammalian systems the reader is referred to several excellent reviews (Carafoli, 1987, 1991, 1994; Crompton, 1990; Garrahan and Rega, 1990; McCormack *et al.,* 1990; Pietrobon *et al.,* 1990; Schatzmann, 1989).

2. Plasma membrane

A number of studies on several different species suggest that a primary Ca^{2+}-ATPase is present in plant PM (Briskin, 1990; De Michelis *et al.,* 1992; Evans *et al.,* 1991). This pump probably operates continuously to export Ca^{2+} from the cell. The amount of Ca^{2+}-ATPase in the PM is very small compared to the much more abundant H^+-ATPase (Serrano, 1990). This makes it difficult

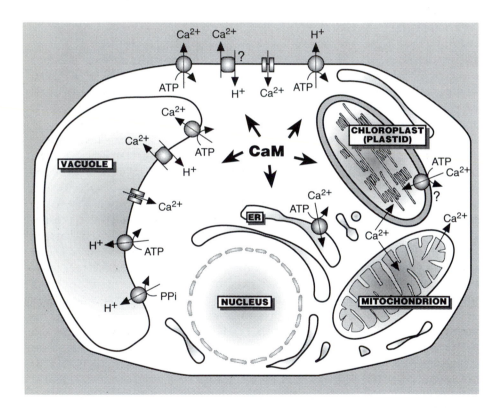

Figure 1. Schematic overview of Ca^{2+} transporters in a higher plant cell. Calcium efflux transporters such as primary Ca^{2+}-pumping ATPases and secondary nH^+/Ca^{2+} antiporters are shown to pump Ca^{2+} out from the cell and into intracellular compartments. The existence of a nH^+/Ca^{2+} antiporter in the PM as well as a Ca^{2+}-pumping ATPase in the plastid inner envelope is highly controversial. nH^+/Ca^{2+} antiporters are energized by proton electrochemical gradients established by primary H^+ pumps in the PM and vacuolar membrane. In addition, Ca^{2+} is shown to be accumulated in the chloroplast stroma and mitochondrial matrix via uniporter mechanisms energized by photosynthetic and oxidative electron flow, respectively. Rapid influx of Ca^{2+} into the cytoplasm can occur via the opening of Ca^{2+} channels in the PM and vacuolar membrane. A mitochondrial Ca^{2+} efflux mechanism is also indicated. Calmodulin (CaM) regulates Ca^{2+}-ATPases in one or several types of membranes, and may also regulate nH^+/Ca^{2+} antiporters and channels.

to distinguish the hydrolytic activity representing the Ca^{2+}-ATPase from the total ATP hydrolytic activity in the PM. This problem may be partly overcome by substituting either ITP or GTP for ATP (Carnelli *et al.,* 1992; Williams *et al.,* 1990), since the Ca^{2+}-ATPase is less substrate-specific than the H^+-ATPase. More commonly, however, the Ca^{2+}-ATPase activity is measured as ATP-dependent accumulation of $^{45}Ca^{2+}$ by inside-out PM vesicles (Olbe and Sommarin, 1991).

The accumulation of Ca^{2+} by PM vesicles is insensitive to protonophores, suggesting that it is catalysed by a primary Ca^{2+}-ATPase and not by a nH^+/Ca^{2+} antiport mechanism driven by a proton gradient (Gräf and Weiler, 1989, 1990; Malatialy *et al.,* 1988; Olbe and Sommarin, 1991). One exception to this general observation was demonstrated for maize leaf PM vesicles, where both systems were present (Kasai and Muto, 1990, 1995). However, the PM Ca^{2+}-ATPase may itself translocate protons in the opposite direction to Ca^{2+} by means of charge neutralization, and thus act as a direct ATP-fuelled nH^+/Ca^{2+} antiporter in a similar manner to animal Ca^{2+}-ATPases (Da Costa and Madeira, 1994; Felle *et al.,* 1992; Rasi-Caldogno *et al.,* 1987). Further investigations are necessary to resolve the exact mechanism of Ca^{2+} transport by the plant PM Ca^{2+}-ATPase, and to determine whether in some cases a true antiport mechanism is present in the plant PM. The ATP-driven Ca^{2+} uptake shows simple saturation kinetics when measured as a function of either ATP or Ca^{2+} concentration (Briskin, 1990). A biphasic response showing high and low affinity components for ATP is a common property of cation-transport ATPases in animals (Garrahan and Rega, 1990), but has not been reported for plant Ca^{2+}-ATPases. One reason for this may be that measurements have rarely been conducted in the presence of an ATP-regenerating system, even though accurate determinations at low concentrations of ATP are difficult to achieve in the absence of such a system. The apparent K_m values estimated for ATP for the plant PM Ca^{2+}-ATPase range between 4 and 300 µM; the lower values were obtained when assays were performed in the presence of excess Mg^{2+} and only ATP was varied, and the higher values were obtained when both Mg^{2+} and ATP were varied as the 1:1 concentration ratio (Gräf and Weiler, 1989; Kasai and Muto, 1991; Olbe and Sommarin, 1991; Rasi-Caldogno *et al.,* 1989; Williams *et al.,* 1990). The apparent K_m for free Ca^{2+} is in the micromolar range (0.4 to 8 µM; Carnelli *et al.,* 1992; Giannini *et al.,* 1987b; Kasai and Muto, 1990; Olbe and Sommarin, 1991; Williams *et al.,* 1990; see also Section 8). ATP is the preferred nucleotide substrate. GTP and ITP give activities 30–70% of the uptake rate observed with ATP, while other nucleotides are less effective (De Michelis *et al.,* 1993; Gräf and Weiler, 1990; Olbe and Sommarin, 1991; Rasi-Caldogno *et al.,* 1989; Thomson *et al.,* 1993). This relatively low nucleotide specificity differs from that of the animal PM Ca^{2+}-ATPase, which is highly specific for ATP (Garrahan and Rega, 1990). The PM Ca^{2+}-ATPase has a broad pH profile with an optimum between pH 7.2 and 7.8 (Carnelli *et al.,* 1992; De Michelis *et al.,* 1993; Gräf and Weiler, 1989, 1990; Kasai and Muto, 1991; Olbe and Sommarin, 1991).

The PM Ca^{2+}-ATPase is characterized by a high sensitivity to the iodinated fluorescein derivative, erythrosin B. This inhibitor is assumed to modify a lysine residue close to the ATP-binding site, causing inhibition of ATP binding (Garrahan and Rega, 1990). The PM enzyme is completely inhibited by 1 µM erythrosin B (De Michelis *et al.,* 1993; Gräf and Weiler, 1989, 1990; Olbe and Sommarin, 1991; Rasi-Caldogno *et al.,* 1989), while Ca^{2+}-ATPases in intracellular membranes appear to be less sensitive (see Section 3 and *Figure 3*). Erythrosin B inhibits other membrane-bound ATPases as well, but the concentrations required are much higher. The mycotoxin cyclopiazonic acid, and hydrophobic reagents such as thapsigargin and benzohydroquinones, did not have specific effects on the red beet PM Ca^{2+}-ATPase (Thomson *et al.,* 1993, 1994). These reagents are specific inhibitors of the animal ER and sarcoplasmic reticulum Ca^{2+}-ATPases, and interact with reaction cycle intermediates (Seidler *et al.,* 1989; Thastrup *et al.,* 1990). The PM Ca^{2+} pump is sensitive to vanadate, indicating that it forms a phosphorylated intermediate during the reaction cycle. Vanadate acts as a non-competitive inhibitor against ATP, and is a transition-state analogue of phosphate; it binds to the aspartyl residue that forms the phosphoenzyme intermediate at the active site (Carafoli, 1991; Serrano, 1990). The potency of vanadate depends on the ionic composition of the medium and on the concentration of ATP (Carafoli, 1991), which may explain why the vanadate concentration giving 50% inhibition ranges from 2 to 500 µM in different investigations (Carnelli *et al.*, 1992; Gräf and Weiler, 1989; Kasai and Muto, 1990; Olbe and Sommarin, 1991; Williams *et al.*, 1990).

Two distinct phosphorylated intermediates representing the Ca^{2+}-ATPase and H^+-ATPase, respectively, were detected in PM from cauliflower inflorescences (*Figure 2*; Askerlund and Evans, 1993). A 116-kDa phosphoprotein was identified as the Ca^{2+}-ATPase by its Ca^{2+}-dependent phosphorylation, which was enhanced by La^{3+}. The effect of La^{3+} results from an increase in the steady-state phosphorylation level and is characteristic of PM Ca^{2+}-ATPases in animals. In contrast, the phosphorylated intermediates of all intracellular Ca^{2+} pumps in animal cells are either inhibited or unaffected by La^{3+} (Carafoli, 1991; Wuytack *et al.*, 1982). In plants, however, the effect of La^{3+} cannot be used to distinguish between PM and intracellular Ca^{2+}-ATPases, since La^{3+} also enhances the phosphorylated intermediates of intracellular Ca^{2+}-ATPases (Askerlund, 1996; Chen *et al.*, 1993). Possibly this is because La^{3+} is effective with

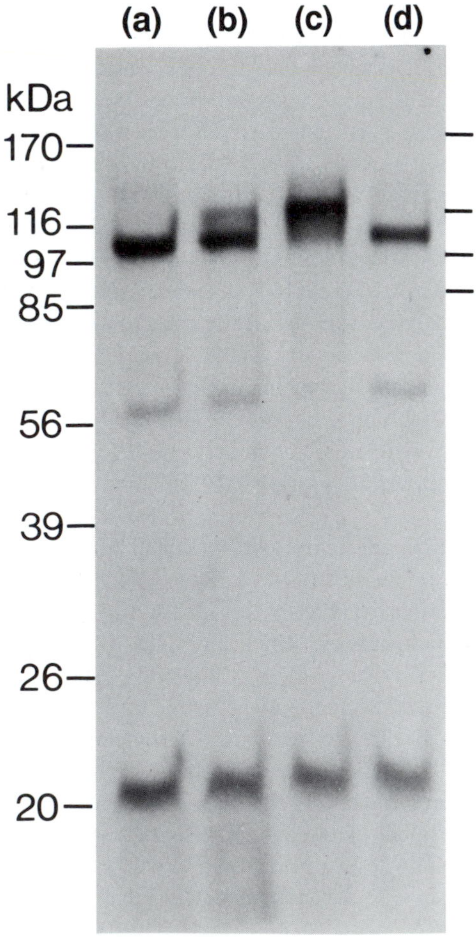

Figure 2. Autoradiogram showing phosphorylated intermediates of Ca^{2+}-ATPase and H^+-ATPase in PM from cauliflower inflorescences after phosphorylation with $[\gamma$-$^{32}P]ATP$ (5 µM) at 0°C for 15 sec and separation of polypeptides by means of SDS–PAGE under acidic conditions. The phosphorylation medium contained 20 mM Mes-Tris (pH 6.5), 12.5 µM $MgCl_2$ (a), or the same medium with the addition of 50 µM $CaCl_2$ (b); 50 µM $CaCl_2$ plus 50 µM $LaCl_3$ (c); or 50 µM $CaCl_2$ plus 0.5 mM EGTA (d). Phosphorylation of the Ca^{2+}-ATPase (*c.* 116 kDa; visible in b and c) required Ca^{2+} and was enhanced by La^{3+}. Phosphorylation of the H^+-ATPase (*c.* 105 kDa; visible in a, b and d) did not require Ca^{2+} and was inhibited by La^{3+}. Adapted from Askerlund and Evans (1993) with permission from Gauthier-Villars Publishers.

calmodulin (CaM)-stimulated rather than necessarily PM-bound Ca^{2+}-ATPases (see Section 8). The phosphointermediate of the cauliflower PM Ca^{2+}-ATPase was detected after phosphorylation at both pH 6.5 and pH 7.4, in contrast to the H^+-ATPase phosphointermediate, which was only detected after phosphorylation at pH 6.5 (Askerlund and Evans, 1993). Two phosphorylated intermediate bands with molecular masses of about 119 and 124 kDa, respectively, were identified in PM from red beet after phosphorylation at pH 7.4 (Thomson *et al.*, 1993). Both of these bands were suggested to represent Ca^{2+}-ATPases since they were Ca^{2+}-dependent, sensitive to hydroxylamine and showed rapid turnover. Molecular masses based on phosphorylated intermediates should be considered as relatively approximate estimates, since the acidic gel systems used often result in diffuse bands and sometimes cause atypical migration. However, the Ca^{2+}-ATPase in cauliflower PM was also estimated to have a molecular mass of 116 kDa when identified by other methods, such as immunoblotting with a cross-reacting antiserum against an intracellular Ca^{2+}-ATPase from cauliflower, and in [^{125}I]CaM overlays (Askerlund and Evans, 1993; P. Askerlund, unpublished data). In addition, antibodies against a conserved region of the putative Ca^{2+}-ATPase from *Arabidopsis thaliana* (Huang *et al.*, 1993) recognized a polypeptide of 116 kDa in cauliflower PM (P. Askerlund, L. Huang and N. Hoffman, unpublished data). With PM from spinach leaves, the same antiserum recognized a polypeptide of 118 kDa (M. Olbe and M. Sommarin, unpublished data). A molecular mass of 133 kDa was estimated for the Ca^{2+}-ATPase in radish PM using FITC (fluorescein isothiocyanate) labelling and [^{125}I]CaM overlay (Rasi-Caldogno *et al.*, 1995). Taken together, these data suggest that PM Ca^{2+}-ATPases have apparent molecular masses ranging from 116 to 133 kDa. A molecular mass of 270 kDa was estimated by radiation-inactivation analysis, indicating that the Ca^{2+}-ATPase may function as a dimer (Rasi-Caldogno *et al.*, 1990).

There appears to be variation between plant species in the sensitivity of the PM Ca^{2+}-ATPase to CaM. Calmodulin-stimulated Ca^{2+} uptake has been identified in PM from radish (Carnelli *et al.*, 1992; Rasi-Caldogno *et al.*, 1992, 1993), *Arabidopsis thaliana* (De Michelis *et al.*, 1993), spinach leaves (Malatialy *et al.*, 1988), cucumber roots (Erdei and Matsumoto, 1991), red beet (Williams *et al.*, 1990) and carrot cells (Kurosaki and Kaburaki, 1994). In PM from leaves of *Commelina communis*, no CaM stimulation was observed (Gräf and Weiler, 1989), and reports for PM from maize are conflicting (Kasai and Muto, 1991; Robinson *et al.*, 1988). Plasma membrane preparations contain large amounts of CaM, suggesting that the absence of CaM-stimulated activity in PM from some species may be due to saturating levels of PM-bound CaM rather than to the presence of a CaM-insensitive Ca^{2+} pump (Briskin, 1990; Evans *et al.*, 1992). In some species, removal of most of the endogenous CaM from the PM by EDTA/EGTA and/or salt washing, or by Triton X-114 fractionation, resulted in improved CaM sensitivity (Evans *et al.*, 1992; Rasi-Caldogno *et al.*, 1992; Williams *et al.*, 1990). However, with PM isolated from leaves of maize and *Commelina communis*, the Ca^{2+} pump activity was not activated by CaM, even after solubilization and reconstitution of the Ca^{2+} pump (Gräf and Weiler, 1990; Kasai and Muto, 1991). In addition to differences between species, factors such as proteolysis, lipid environment, or varying degrees of contamination by intracellular membranes may explain differences in sensitivity to CaM. Recently, Rasi-Caldogno *et al.* (1993) demonstrated that controlled proteolysis by trypsin resulted in activation of the PM Ca^{2+} pump. Removal of a 15-kDa fragment rendered the enzyme insensitive to further activation by CaM, indicating that activation by CaM and proteolysis proceeded by similar mechanisms (Rasi-Caldogno *et al.*, 1995). Similar results were obtained with the spinach PM Ca^{2+} pump (M. Olbe and M. Sommarin, unpublished data). Thus, in a similar manner to the animal PM Ca^{2+}-ATPase, limited proteolysis of the plant PM Ca^{2+}-ATPase results in loss of a regulatory autoinhibitory domain (see Section 7).

3. Endoplasmic reticulum

Calcium transport into the ER is driven by a primary Ca^{2+}-ATPase which shows similar properties to those of the Ca^{2+}-ATPase of PM (Briskin, 1990; Chanson, 1993; Evans *et al.*, 1991). The Ca^{2+}- ATPase in ER from garden-cress roots showed nucleotide specificities similar to those of PM Ca^{2+}-ATPases, with GTP and ITP giving rates that were 25–50% of those observed with ATP (Buckhout, 1983). In red beet and wheat root ER, the Ca^{2+}-ATPases were more specific for ATP (Giannini *et al.*, 1987a; Thomson *et al.*, 1993; M. Olbe, S. Widell and M. Sommarin, unpublished data). The K_m for ATP was between 70 and 400 µM (Buckhout, 1984; Bush and Sze, 1986; Bush and Wang, 1995; Giannini *et al.*, 1987a). Reported K_m values for Ca^{2+} range between 0.1 and 7.5 µM, similar to the affinity of the PM Ca^{2+}-ATPase (Buckhout, 1983; Bush and Sze, 1986; Bush and Wang, 1995; Giannini *et al.*, 1987a; Thomson *et al.*, 1993; M. Olbe, S. Widell and M. Sommarin, unpublished data). The pH profile of the ER Ca^{2+}-ATPase shows an optimum between pH 6.8 and 7.5 (e.g. Buckhout, 1983; Bush and Sze, 1986; Giannini *et al.*, 1987a). In a similar manner to the PM Ca^{2+}-ATPase, the ER enzyme is inhibited by vanadate and erythrosin B. However, the sensitivity to erythrosin B of ER preparations from different sources varies, probably because preparations are often contaminated to varying degrees, in particular by vacuole membranes. For this reason, it is necessary to design assays that make it possible to distinguish between different types of transporters present on different membranes. As demonstrated in *Figure 3*, the activity of the ATP-dependent H^+ pump in the vacuole membrane interferes with the sensitivity of Ca^{2+}-ATPases to erythrosin B. When bafilomycin, a specific inhibitor of the vacuolar H^+ pump (Bowman *et al.*, 1988), was included in the assay, the sensitivity of the Ca^{2+}-ATPase in the ER fraction increased by about 10-fold (*Figure 3*). In red beet, the PM and ER Ca^{2+}-ATPases exhibited similar sensitivities to erythrosin B (Thomson *et al.*, 1994). Thapsigargin and cyclopiazonic acid, inhibitors of the animal ER- and sarcoplasmic reticulum-type Ca^{2+}-ATPases (Seidler *et al.*, 1989; Thastrup *et al.*, 1990), had no specific effects on the ER Ca^{2+}-ATPase in red beet (Thomson *et al.*, 1993, 1994) and wheat roots (M. Olbe, S. Widell and M. Sommarin, unpublished data), while a significant inhibitory effect was

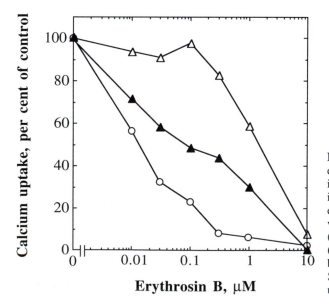

Figure 3. The effect of erythrosin B on the Ca^{2+} uptake in PM (O) and in an intracellular membrane fraction enriched with ER (△). Assays were performed in the absence (open symbols) or presence (filled symbols) of 1 µM bafilomycin. (M. Olbe, S. Widell and M. Sommarin, unpublished data).

observed with carrot membranes (Hsieh *et al.*, 1991). The gravitropic response in cress roots was inhibited by treatment with cyclopiazonic acid. It was postulated that the ER Ca^{2+}-ATPase of statocytes was involved in transduction of the gravity stimulus, and that cyclopiazonic acid disrupted the cytosolic Ca^{2+} signal necessary for graviperception (Sievers and Busch, 1992). A rapidly turning-over 96-kDa phosphoprotein, possibly representing a Ca^{2+}-ATPase, was demonstrated in ER from red beet (Giannini *et al.*, 1987a). In another study, the ER Ca^{2+}-ATPase in red beet was detected as a 119-kDa phosphoprotein (Thomson *et al.*, 1993). The 119-kDa band showed all the characteristics of the phosphorylated intermediate of a Ca^{2+}-ATPase, including Ca^{2+} dependency, hydroxylamine sensitivity and rapid turnover.

Calcium uptake in ER from garden-cress roots was not stimulated by CaM (Buckhout, 1984). However, in other species the ER Ca^{2+}-ATPase was found to be CaM-stimulated. Fractionation of carrot membranes on a continuous sucrose gradient resulted in two peaks of CaM-stimulated Ca^{2+} uptake, one co-migrating with ER and the other co-migrating with PM markers (Hsieh *et al.*, 1991). Calmodulin-stimulated Ca^{2+} uptake in barley aleurone membranes has also been localized to the ER (Gilroy and Jones, 1993; see Section 7). In contrast, Ca^{2+} uptake in ER isolated from wheat aleurone layers was not stimulated by CaM (Bush and Wang, 1995). A three- to fourfold stimulation of the Ca^{2+}-ATPase by CaM was observed in wheat root ER (M. Olbe, S. Widell and M. Sommarin, unpublished data). The CaM-stimulated ATPase in maize roots was localized to the ER on the basis of density gradient centrifugation in the absence and presence of Mg^{2+} (Brauer *et al.*, 1990).

4. Vacuole membrane

Calcium is accumulated in the vacuole via a nH^+/Ca^{2+} antiporter which is energized by a proton electrochemical gradient generated by the vacuolar H^+-ATPase and the vacuolar pyrophosphatase (PP_iase) (Blumwald and Poole, 1986; Chanson, 1993; Schumaker and Sze, 1990) as illustrated in *Figure 1*. Calcium uptake via this antiporter can therefore be inhibited by compounds that dissipate the proton gradient across the vacuole membrane and compounds that inhibit either the vacuolar H^+-ATPase (nitrate, bafilomycin; Bowman *et al.*, 1988) or the PP_iase (aminomethylene-diphosphonate; Zhen *et al.*, 1994). Inhibitors that act directly on the antiporter are N,N'-dicyclohexyl-carbodiimide (DCCD), ruthenium red and La^{3+} (Schumaker and Sze, 1990). The nH^+/Ca^{2+} antiporter has a high capacity but a low affinity for Ca^{2+} ($K_m > 10$ µM; Andreev *et al.*, 1990; Blumwald and Poole, 1986; Bush and Sze, 1986; Muto, 1992). However, CaM has been suggested to increase the affinity of the antiporter for Ca^{2+} in sugar beet tap roots by almost 1000-fold (Andreev *et al.*, 1990). In addition to the nH^+/Ca^{2+} antiporter, there appears to be a primary Ca^{2+}-pumping ATPase in the vacuole membrane displaying a high affinity for Ca^{2+}. Fukumoto and Venis (1986) demonstrated that the accumulation of Ca^{2+} by tonoplast vesicles from apple fruit was CaM-stimulated and directly coupled to ATP hydrolysis. Additional reports supporting the presence of a Ca^{2+}-ATPase in the vacuole membrane have appeared during the last few years (DuPont *et al.*, 1990; DuPont and Morrissey, 1992; Gavin *et al.*, 1993; Pfeiffer and Hager, 1993). The CaM-stimulated, protonophore-insensitive, ATP-dependent Ca^{2+} uptake in maize root membranes was found to be associated with markers for vacuole membranes (Gavin *et al.*, 1993). In other studies with maize roots and shoots, the same activity was proposed to be located in the ER (Brauer *et al.*, 1990; Logan and Venis, 1995), but data that excluded a vacuole membrane location were not presented. The distribution of CaM-stimulated Ca^{2+}-ATPase in cauliflower (as detected by activity and immunoblotting) after sucrose gradient centrifugation correlated very closely with vacuole membrane markers (PP_i-dependent H^+ pumping and antibodies against the vacuolar H^+-ATPase). In contrast, the distribution of ER markers (antimycin A-insensitive NADH-cytochrome c reductase activity and antibodies against the ER luminal binding protein,

BiP) was significantly different from that of the Ca^{2+}-ATPase (P. Askerlund, unpublished data). This is in contrast with the earlier report in which the major part of the CaM-stimulated Ca^{2+} uptake was suggested to be located in the ER (Askerlund and Evans, 1992). In the earlier experiments, PP_i hydrolysis rather than PP_i-dependent H^+ pumping was measured. The current view is that PP_i hydrolysis is a much less specific marker for the vacuole membrane than PP_i-dependent H^+ pumping, possibly due to binding of soluble PP_iases to membranes. In addition, fractionation of a low-density membrane fraction from cauliflower by two-phase partitioning, a method which separates membranes on the basis of surface properties rather than density, strongly supports the view that a CaM-stimulated Ca^{2+}-ATPase in cauliflower is present in vacuole membranes, but not in the ER (P. Askerlund, unpublished data).

5. Mitochondria and plastids

Very little is known about Ca^{2+} transport systems in plant mitochondria and their role in cellular Ca^{2+} metabolism. In mammalian systems, mitochondria passively accumulate Ca^{2+} by a uniporter mechanism in the inner membrane. The uptake is driven by the membrane potential (about 180 mV, negative inside) generated by proton extrusion by the respiratory chain (Crompton, 1990; McCormack *et al.*, 1990; Pietrobon *et al.*, 1990). However, the nature of the uptake mechanism (carrier or channel) has not been established with certainty, and the molecular components involved have not been identified. Mammalian mitochondria also possess one or two Ca^{2+} efflux pathways in the inner membrane: a $2Na^+/Ca^{2+}$ and/or a $2H^+/Ca^{2+}$ antiporter. The uptake pathway has a potential capacity about 10-fold higher than the efflux mechanisms (Crompton, 1990; McCormack *et al.*, 1990; Pietrobon *et al.*, 1990).

The mechanism of Ca^{2+} uptake in plant mitochondria seems to differ according to species, age and type of tissue (Zottini and Zannoni, 1993). In some species, uptake is by a simple uniport mechanism, while in others uptake has an absolute requirement for inorganic phosphate and probably occurs via a ruthenium red- and mersalyl-sensitive Ca^{2+}/P_i-cotransporter. As in the case of mammalian mitochondria, the plant mitochondrial Ca^{2+} uptake systems are sensitive to the respiratory state of the organelle. The matrix free Ca^{2+} concentration differs considerably between species. Pea stem mitochondria (phosphate-independent uptake) were found to have a free matrix Ca^{2+} concentration of 60 to 100 nM, while in Jerusalem artichoke mitochondria (phosphate-dependent uptake) this concentration was 400 to 600 nM (Zottini and Zannoni, 1993). By using the fluorescent Ca^{2+} indicator fura-2, it was shown that some plant mitochondria can accumulate external free Ca^{2+} by an electrophoretic uniporter mechanism with a high apparent affinity for Ca^{2+} (K_m approximately 150 nM; Zottini and Zannoni, 1993). This is in contrast to earlier investigations, using isotopes and Ca^{2+}-sensitive electrodes to estimate free Ca^{2+} concentrations, which showed that uptake required high exogenous Ca^{2+} concentrations (K_m approximately 30 µM; Martins *et al.*, 1986; Muto, 1992). Our knowledge about Ca^{2+} efflux mechanism(s) in plant mitochondria is limited: a sodium- and phosphate-independent and mersalyl-insensitive Ca^{2+} efflux pathway operating at high membrane potential was recently demonstrated in maize mitochondria. In a similar manner to the situation in vertebrates, this efflux pathway may be either directly or indirectly coupled to the influx of protons (Rugolo *et al.*, 1990; Silva *et al.*, 1992).

Under normal cellular conditions the vertebrate mitochondrial Ca^{2+} transport system is thought to have a role in controlling matrix Ca^{2+} levels and consequently in the regulation of Ca^{2+}-dependent intramitochondrial processes. It is only under some pathophysiological conditions that mitochondria are believed to act as buffers of extramitochondrial Ca^{2+}. Calcium transport across the inner membrane is assumed to play a critical role in coordinating the regulation of mitochondrial processes with those of the cytoplasm (Crompton, 1990;

McCormack *et al.,* 1990). A change in cytoplasmic Ca^{2+} can be relayed to the matrix, and the oxidative metabolism may thus respond to an increased demand for energy. In plants, the regulation of matrix Ca^{2+} concentration is likely to be essential for regulation of intra-mitochondrial processes. Only two mitochondrial enzymes that are regulated by the matrix Ca^{2+} concentration are known at present: the NAD(H)-glutamate dehydrogenase (Itagaki *et al.,* 1990) and the rotenone-insensitive NADPH dehydrogenase (Rasmusson and Møller, 1991); both these enzymes are regulated by micromolar Ca^{2+} concentrations.

Calcium uptake into isolated chloroplasts increases in response to illumination. This light-dependent uptake is electrogenic, and is thought to be mediated by a uniport-type carrier that is driven by the change in membrane potential caused by photosynthetic electron transport (Kreimer *et al.,* 1985). The stromal free Ca^{2+} concentration has been estimated to be around 2 to 6 µM (Kreimer *et al.,* 1988). The importance of the Ca^{2+} transport system(s) in chloroplasts is not clear, but the low affinity for Ca^{2+} uptake (K_m = 100 to 200 µM; Kreimer *et al.,* 1985) suggests that they are not involved in the regulation of cytoplasmic Ca^{2+} concentration under normal conditions, but possibly during periods or locations of increased cytosolic Ca^{2+} (Bush, 1993). Stromal enzymes regulated by micromolar Ca^{2+} include NAD kinase (Muto and Miyachi, 1986) and fructose-1,6-bisphosphatase (Kreimer *et al.,* 1988). Other Ca^{2+}-regulated processes are the gating between localized and delocalized proton gradient coupling in the thylakoid, and oxygen evolution by photosystem II (Huang *et al.,* 1993, and references therein).

6. Purification and cloning of Ca^{2+} transporters

A CaM-stimulated Ca^{2+}-ATPase was partially purified from maize microsomes as early as 1981 using CaM-affinity chromatography (Dieter and Marmé, 1981), a method that was first developed for purification of the CaM-stimulated Ca^{2+}-ATPase from erythrocytes (Carafoli, 1987, 1991). The maize enzyme was not characterized in detail, probably because of its low stability after solubilization. It was later shown by Briars *et al.* (1988) to be a P-type ATPase with an apparent molecular mass of 140 kDa, a value which seems high in comparison with molecular masses reported by other investigators (Askerlund and Evans, 1992; Logan and Venis, 1995). The CaM-stimulated Ca^{2+}-ATPase from maize microsomes was functionally reconstituted in liposomes (Theodoulou *et al.,* 1994). A CaM-stimulated Ca^{2+}-ATPase was purified *c.* 120-fold from cauliflower microsomes, again using CaM-affinity chromatography (Askerlund and Evans, 1992). The fraction eluted from the CaM-affinity column showed a dominant Coomassie-stained band at 115 kDa, and a few weaker bands at lower molecular weights. The 115-kDa band was identified as the Ca^{2+}-ATPase by its ability to form a Ca^{2+}-dependent phosphorylated intermediate (Askerlund and Evans, 1992). The partly purified cauliflower Ca^{2+}-ATPase was reconstituted into liposomes using CHAPS (3-[(3-cholamidopropyl)dimethylammonio]-1-propanesulphonate) dialysis. The resulting proteoliposomes catalysed an ATP-dependent accumulation of $^{45}Ca^{2+}$ which was strongly stimulated by CaM (Askerlund and Evans, 1992). The activity had a pH optimum of about 7.0 and could also be supported with GTP (at a rate 50% of that obtained with ATP). The purified and reconstituted Ca^{2+}-ATPase was inhibited by vanadate (K_i = 20 µM) and erythrosin B (K_i =12 µM), but not by thapsigargin or cyclopiazonic acid (Askerlund and Evans, 1992). The K_m for Ca^{2+} was about 7 µM both in the absence and in the presence of CaM. In this respect the activity of the purified ATPase differed from that measured with membranes from cauliflower, which usually show a significantly higher affinity for Ca^{2+} in the presence of CaM (*Figure 4;* K_m = 0.6 µM) than in its absence. The CaM-stimulated Ca^{2+}-ATPase in intracellular membranes from cauliflower probably originates from the vacuole membrane, as discussed above (Section 5). The apparent molecular mass appears to be slightly lower than the value reported previously, *c.* 111 kDa (*Figure 5;* Askerlund, 1996). A partial amino acid sequence has been obtained from the 111-kDa Ca^{2+}-ATPase (purified from low-density cauliflower membranes), and an antiserum has been raised against it. Polymerase chain reaction of a combination of primers from this amino acid sequence and a sequence (DKTGTL) conserved among P-type ATPases resulted in amplification of a 1200-bp fragment (S. Malmström, P. Askerlund and M.G. Palmgren, unpublished data). The deduced amino acid sequence shows high homology with the putative *Arabidopsis thaliana* Ca^{2+}-ATPase (Huang *et al.*, 1993) and CaM-stimulated PM Ca^{2+}-ATPases from various animal species.

A CaM-stimulated Ca^{2+}-ATPase was partially purified from spinach chloroplast envelope membranes using CaM-affinity chromatography (Nguyen and Siegenthaler, 1985). The eluted fraction showed a single polypeptide of 65 kDa when resolved by sodium dodecyl sulphate–polyacrylamide gel electrophoresis (SDS–PAGE). This molecular mass is low compared to that of other CaM-stimulated ATPases (Askerlund and Evans, 1992; Carafoli, 1991, 1992; Garrahan and Rega, 1990; Logan and Venis, 1995; Rasi-Caldogno *et al.*, 1995), and may therefore represent a proteolytic product of the ATPase. A cDNA clone encoding a putative Ca^{2+}-ATPase was isolated by screening an *Arabidopsis thaliana* expression library with an antiserum against spinach chloroplast envelope proteins (Huang *et al.*, 1993). The corresponding genomic clone was also isolated. The predicted polypeptide (designated PEA1p) had a molecular mass of 111 kDa and showed the highest degree of homology with the CaM-stimulated Ca^{2+}-ATPase in animal PM. In contrast to the animal PM Ca^{2+}-ATPase, however, PEA1p lacked the C-terminal CaM-binding domain, suggesting that it is not identical to the CaM-stimulated ATPase in spinach, identified by Nguyen and Siegenthaler (1985). In this context it is of interest that one of the two Ca^{2+}-ATPase genes that have been cloned from *Saccharomyces cerevisiae* (PMC1, encoding a vacuole membrane protein) is also highly homologous to Ca^{2+}-ATPases in animal PM, but lacks the C-terminal CaM-binding domain (Cunningham and Fink, 1994). The Ca^{2+}-ATPase identified in cauliflower (*Figure 5*; Askerlund, 1996; Askerlund and Evans, 1992) has an apparent molecular mass close to that of PEA1p, but should likewise not be identical to PEA1p, since it is CaM-stimulated. Antibodies raised against a portion of PEA1p recognized a single 90–95 kDa polypeptide in the inner chloroplast envelope preparation from peas, indicating that a Ca^{2+}-ATPase was present in these membranes (Huang *et al.*, 1993). For unknown reasons, attempts to measure Ca^{2+}-ATPase activity (as ATP hydrolysis or ATP-dependent Ca^{2+} uptake) in these preparations were unsuccessful. Because Ca^{2+} uptake into the stroma is not thought to be

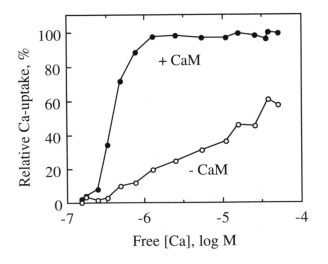

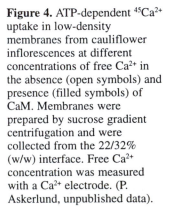

Figure 4. ATP-dependent $^{45}Ca^{2+}$ uptake in low-density membranes from cauliflower inflorescences at different concentrations of free Ca^{2+} in the absence (open symbols) and presence (filled symbols) of CaM. Membranes were prepared by sucrose gradient centrifugation and were collected from the 22/32% (w/w) interface. Free Ca^{2+} concentration was measured with a Ca^{2+} electrode. (P. Askerlund, unpublished data).

mediated by an ATPase (Evans *et al.*, 1991; Kreimer *et al.*, 1985; Muto, 1992), the function of the putative chloroplast envelope Ca^{2+}-ATPase may instead be to pump Ca^{2+} in the opposite direction, but as all known higher-plant Ca^{2+}-ATPases pump Ca^{2+} out of the cytoplasm, this would be an unusual orientation of the enzyme (Huang *et al.*, 1993). It was also speculated that the Ca^{2+}-ATPase may be involved in the energy-dependent import of proteins into the stroma (Nguyen and Siegenthaler, 1985). Peptide sequences deduced from a genomic DNA clone isolated from tomato (Wimmers *et al.*, 1992) and from a cDNA clone from tobacco (Perez-Prat *et al.*, 1992) show greatest homology with Ca^{2+}-ATPases of the animal ER/sarcoplasmic reticulum type. These putative Ca^{2+}-ATPases were therefore suggested to be located in the ER. The predicted tomato ATPase has a calculated molecular mass of *c.* 116 kDa, close to the value of 119 kDa estimated for the Ca^{2+}-ATPase phosphorylated intermediate in red beet ER (Thomson *et al.*, 1993).

The plant PM Ca^{2+} pump has not yet been extensively purified. Maize leaf PM was solubilized with $C_{12}E_8$ (octaethyleneglycol monododecylether), and the extract was fractionated on a diethylaminoethyl (DEAE) anion-exchange column by high-performance liquid chromatography. A Ca^{2+}-ATPase peak was resolved and was successfully reconstituted into liposomes by detergent dilution. Studies on the reconstituted Ca^{2+}-ATPase demonstrated that it functioned as a primary Ca^{2+} pump (Kasai and Muto, 1991). A Ca^{2+}-ATPase was also solubilized from PM of *Commelina communis* leaves using CHAPS, and the Ca^{2+} pump was functionally reconstituted into proteoliposomes with characteristics similar to those of the PM-bound enzyme (Gräf and Weiler, 1990).

The nH^+/Ca^{2+} antiporter has been solubilized from oat root tonoplasts and reconstituted into proteoliposomes in an active form (Schumaker and Sze, 1990). A nH^+/Ca^{2+} antiporter from *Arabidopsis thaliana* was recently cloned by complementation in yeast, and was found to be 40% identical to the yeast vacuolar nH^+/Ca^{2+} antiporter at the deduced amino acid level (K. Hirschi and G.R. Fink, unpublished data).

7. Regulation of Ca^{2+}-ATPases

In animal cells the PM Ca^{2+}-ATPase is activated by CaM. Calmodulin interacts with the C-terminal region of the pump, resulting in both a decreased K_m for Ca^{2+} and an increased V_{max}. Mild proteolysis with trypsin mimics the effects of CaM on the affinity for Ca^{2+} and on the value of V_{max} (Carafoli, 1991, 1992; Garrahan and Rega, 1990). Proteolysis interferes with the stimulatory effect of CaM by removing a regulatory autoinhibitory domain that, in the absence of CaM, inhibits the system. Proteolysis may also be important *in vivo* since calpain, an intracellular protease, has similar effects to trypsin and is preferentially targeted against CaM-binding domains. In contrast, Ca^{2+}-ATPases in the ER and sarcoplasmic reticulum are not directly stimulated by CaM, but may be indirectly stimulated via phosphorylation of the regulatory proteolipid phospholamban by CaM-dependent protein kinases (Pietrobon *et al.*, 1990). The mammalian PM Ca^{2+} pump activity is controlled by a large number of regulators in addition to CaM. It is stimulated by acidic phospholipids, polyunsaturated fatty acids, phosphorylation by cyclic AMP-dependent protein kinases and by protein kinase C, and by induction of oligomerization (Carafoli, 1992; Wuytack and Raeymaekers, 1992). All of these effectors act at different sites on the enzyme and form an intricate regulatory network which allows the PM Ca^{2+} pump to adapt to the prevailing situation. In plants, the presence of directly CaM-stimulated Ca^{2+}-ATPases is not restricted to the PM, as discussed above and below in more detail. With similarity to the situation in animal cells, CaM affects the plant pump both by decreasing its K_m for Ca^{2+} and by increasing its V_{max} (*Figure 4*; Dieter and Marmé, 1983; Rasi-Caldogno *et al.*, 1992), but in many cases the effect on the K_m has been difficult to demonstrate (Gilroy and Jones,

1993). In a similar manner to the mammalian PM Ca^{2+}-ATPase, the CaM-stimulated Ca^{2+}-ATPase in radish PM was activated by limited proteolysis, and the proteolyzed enzyme was no longer sensitive to CaM activation (Rasi-Caldogno *et al.*, 1993, 1995). Treatment of PM from spinach leaves with CaM or trypsin resulted in a two- and threefold increase in Ca^{2+} affinity and V_{max}, respectively, of the Ca^{2+} pump (M. Olbe and M. Sommarin, unpublished data). Trypsin digestion was also found to activate a Ca^{2+}-ATPase in low-density intracellular membrane vesicles, probably of vacuolar origin, from cauliflower inflorescences (*Figure 5*; Askerlund, 1996). Trypsin treatment of vesicles resulted in an approximately threefold activation of Ca^{2+} uptake and loss of CaM sensitivity (*Figure 5a*). Immunoblotting experiments with an antiserum raised against the Ca^{2+}-ATPase showed that the trypsin activation was accompanied by a decrease in the amount of intact Ca^{2+}-ATPase (111 kDa), and by successive appearances of polypeptides of 102 and 99 to 84 kDa (*Figure 5b*). [125I]calmodulin overlays showed that only the intact Ca^{2+}-ATPase bound CaM. The data suggested that trypsin digestion and CaM activated the cauliflower Ca^{2+}-ATPase by at least partially different mechanisms, since removal of the CaM binding domain (*c.* 9 kDa) by trypsin was not sufficient to obtain full activation (*Figure 5*), and since trypsin proteolysis resulted in a significantly higher Ca^{2+} affinity than addition of CaM (Askerlund, 1996). Whether limited proteolysis of plant Ca^{2+} pumps is of physiological significance is not yet known.

Treatment of a variety of plant tissues with fusicoccin, the major toxin produced by the fungus *Fusicoccum amygdali*, gives rise to increased acidification of the apoplast, and this effect has been ascribed to an activation of the PM H^+-ATPase (Marrè, 1979). It was recently shown that fusicoccin activates the PM H^+-ATPase in spinach leaves by a mechanism involving displacement of the C-terminal inhibitory domain (Johansson *et al.*, 1993). In *Corydalis sempervirens* both the H^+-ATPase (Schulz *et al.*, 1990) and a Ca^{2+}-ATPase (Liß *et al.*, 1991) were shown to be activated by fusicoccin. As in the case of the spinach PM H^+-ATPase, fusicoccin activated the Ca^{2+}-ATPase by lowering the K_m for ATP, but was only effective when added prior to homogenization, indicating that the effect was not the result of a direct interaction between the toxin and the enzyme, but involved components that were inactivated or lost during cell fractionation (Liß *et al.*, 1991). In contrast, no effect on ATP-dependent Ca^{2+} uptake was observed in PM-enriched membranes isolated from fusicoccin-treated maize coleoptiles (Zocchi and Rabotti, 1993).

Treatment of barley aleurone layers with gibberellic acid led to a drastic increase in both ATP-dependent Ca^{2+} uptake and CaM or CaM-like protein levels in ER isolated from this tissue (Bush *et al.*, 1993; Gilroy and Jones, 1993), while abscisic acid treatment prevented or reversed the gibberellic acid-induced increase in Ca^{2+} transport activity, and lowered the level of Ca^{2+} binding proteins in the ER (Bush *et al.*, 1993). Gibberellic acid and abscisic acid thus regulated Ca^{2+} transport into barley aleurone ER in an antagonistic way which was shown to correlate very closely with the effects on α-amylase production, and it was suggested that the hormonally regulated levels of Ca^{2+} in the ER may in turn regulate Ca^{2+}-dependent α-amylase synthesis in the lumen of the ER (Bush *et al.*, 1993; Gilroy and Jones, 1993). With maize coleoptile segments, treatment with indoleacetic acid prior to homogenization increased ATP-dependent Ca^{2+} uptake in PM-enriched membranes by 50% (Zocchi and Rabotti, 1993). The stimulation by CaM was lower in the membrane vesicles prepared from indoleacetic acid-treated segments than in the vesicles isolated from control segments, but in the presence of CaM the activity was the same in both preparations. One explanation for this observation could be that indoleacetic acid treatment increased the amount of membrane-bound CaM, so that exogenously added CaM could not stimulate the activity further (Zocchi and Rabotti, 1993). The CaM-stimulated Ca^{2+}-ATPase activity in PM isolated from Ca^{2+}-deficient cucumber roots was lower than that in PM from roots

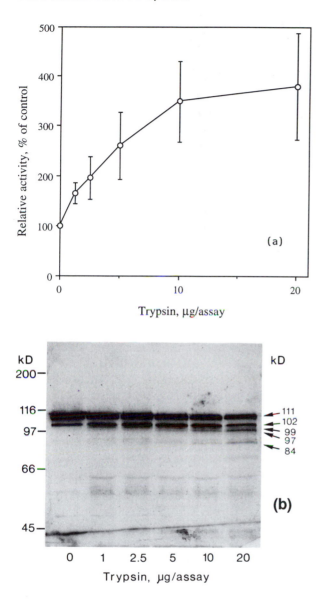

Figure 5. Effect of different concentrations of trypsin on (a) ATP-dependent Ca^{2+} uptake and (b) antibody binding pattern of Ca^{2+}-ATPase in low-density intracellular membrane vesicles from cauliflower inflorescences. Membranes were prepared by sucrose gradient centrifugation and were collected from the 10/32% (w/w) interface. Samples were collected from the assay mixture exactly 2 min after the addition of trypsin, and were analysed by immunoblotting with antiserum against cauliflower Ca^{2+}-ATPase. The 102-kDa degradation product was also present in small amounts in untreated membranes. Adapted from Askerlund (1996) with permission from the American Society of Plant Physiologists.

of Ca^{2+}-sufficient plants. *In-vivo* treatment of the Ca^{2+}-deficient plants with the synthetic cytokinin benzyladenine increased the activity to the level observed in the Ca^{2+}-sufficient plants (Erdei and Matsumoto, 1991). Benzyladenine has also been shown to increase the activity and the affinity towards Ca^{2+} and CaM of the microsomal Ca^{2+}-ATPase in wheat seedlings (Oláh *et al.*, 1983). Triacontanol is a potent plant growth promoter of many agronomic and horticultural crops (Lesniak *et al.*, 1986). Treatment of barley roots with physiologically active concentrations of triacontanol prior to membrane isolation resulted in a 64 to 85% stimulation of the membrane-associated Ca^{2+}/Mg^{2+}-dependent ATPase activity (Lesniak *et al.*, 1986). The effect of triacontanol did not seem to depend on *de-novo* synthesis of proteins, since stimulation was also observed in the presence of cycloheximide. A smaller effect on the activity was seen when PM-enriched vesicles from barley roots were treated directly with triacontanol (Lesniak *et al.*, 1989).

Exposure to high concentrations (50–400 mM) of NaCl increased the levels of a putative Ca^{2+}-ATPase mRNA by up to fourfold in tobacco cells and in tomato (Perez-Prat *et al.,* 1992; Wimmers *et al.,* 1992). In contrast, the abundance of the tomato H^+-ATPase mRNA was not affected by NaCl, suggesting that the induction of Ca^{2+}-ATPase mRNA in response to salt stress was relatively specific. Salt stress results in increased intracellular levels of Ca^{2+} (Läuchli, 1990), and an increased amount of Ca^{2+}-ATPase sequestering Ca^{2+} into intracellular compartments may represent a way for the plant cell to adapt to a saline environment (Perez-Prat *et al.,* 1992; Wimmers *et al.,* 1992). Mild oxidant stress in the form of ozone was suggested to lead to an increased influx of Ca^{2+} into cells of pinto bean leaves, but also to an activation of active Ca^{2+} extrusion mechanisms in the PM (Castillo and Heath, 1990). The regulation of different phytochrome-affected processes involves Ca^{2+} fluxes, but the molecular mechanisms involved are not understood (Tretyn *et al.,* 1991). Irradiation of maize seedlings with far-red light reduced CaM-stimulated- and total Ca^{2+} uptake activities in microsomes as compared to dark-grown tissue, suggesting that a CaM-stimulated Ca^{2+}-ATPase may be involved in phytochrome-regulated processes (Dieter and Marmé, 1983).

In summary, the Ca^{2+}-ATPase activity in plants can be directly influenced by CaM and proteolysis. In addition, Ca^{2+}-ATPases can be affected by growth regulators as well as by environmental factors such as light, toxins, salt and oxidant stress, but all these effects seem to be of an indirect nature, and may be linked to intracellular messenger production leading to, for example, altered levels of CaM in the cell.

8. Concluding remarks

The subcellular location of the CaM-stimulated Ca^{2+}-ATPases in plants has long been a matter of controversy (Chanson, 1993; Evans, 1994; Evans *et al.,* 1991). The analogy with animals would suggest that CaM-stimulated Ca^{2+}-ATPases should be present only in the PM. It is now clear, however, that plants differ from animals in this respect and that CaM-stimulated Ca^{2+}-ATPases are located in intracellular membranes as well as in the PM (Askerlund and Evans 1992; Gavin *et al.,* 1993; Gilroy and Jones, 1993). At present it is not possible to ascertain the exact subcellular location of CaM-stimulated Ca^{2+}-ATPases. Convincing reports suggesting the ER (Gilroy and Jones, 1993), PM (Malatialy *et al.,* 1988; Rasi-Caldogno *et al.,* 1993, 1995), vacuole membrane (Fukumoto and Venis, 1986; Gavin *et al.,* 1993) and chloroplast inner envelope (Nguyen and Siegenthaler, 1985) as subcellular locations have appeared. It seems most likely that Ca^{2+}-ATPases (sensitive or insensitive to CaM) can be present in several types of membrane, the relative amount and location depending on the species, organ and developmental state of the plant. In yeast (*Saccharomyces cerevisiae*), Ca^{2+}-ATPases are present in the vacuole membrane, the PM, the Golgi apparatus and the secretory system (Antebi and Fink, 1992; Cunningham and Fink, 1994; Rudolph *et al.,* 1989). The possibility that the Golgi apparatus and secretory systems harbour Ca^{2+} transporters in plants as well has yet to be investigated. The PM Ca^{2+}-ATPase appears to have a slightly higher molecular mass than the Ca^{2+}-ATPase(s) in intracellular membranes. In cauliflower inflorescences, for example, Ca^{2+}-ATPases located in the PM and in presumptive vacuole membranes have apparent molecular masses of 116 and 111 kDa, respectively (Askerlund, 1996; Askerlund and Evans, 1993; P. Askerlund, unpublished data). Both of these Ca^{2+}-ATPases are sensitive to CaM. In addition to differences in molecular mass, the PM Ca^{2+}-ATPase in many species is usually more sensitive to the inhibitor erythrosin B than Ca^{2+}-ATPases in intracellular membranes (*Figure 3*). In contrast to the situation in animal cells, where several specific inhibitors of the ER/sarcoplasmic reticulum Ca^{2+}-ATPase are available, at present no specific inhibitors of intracellular Ca^{2+}-ATPases have been identified in plants.

The relative roles of different Ca^{2+} transporters in the regulation of Ca^{2+} homeostasis in plants are not understood (*Figure 1*). However, a working hypothesis is that the PM Ca^{2+}-ATPase operates continuously to expel Ca^{2+} from the cell. All the available data point to the vacuole rather than the ER as the major intracellular pool for transient release of Ca^{2+} into the cytoplasm during signal transduction (Canut *et al.*, 1993). The vacuolar nH^+/Ca^{2+} antiporter is therefore likely to be involved in short-term regulation of cytoplasmic Ca^{2+} levels (Sanders *et al.*, 1990). Intracellular Ca^{2+}-ATPases may also function in the short-term regulation of cytosolic Ca^{2+} concentration; their higher affinity for Ca^{2+} may be necessary to deplete the cytosol of Ca^{2+}. The K_m values (Ca^{2+}) reported for the Ca^{2+}-ATPases are about one order of magnitude higher than the concentration of free Ca^{2+} in the cytosol (Fricker *et al.*, 1993; Read *et al.*, 1992). One could therefore argue that the Ca^{2+}-ATPases are virtually inactive except under stress conditions, when the concentration of free Ca^{2+} in the cytosol increases. However, the K_m values (Ca^{2+}) reported in the literature may be overestimated, since regulatory factors (*Figure 4*; CaM and others) that are present in the cell, but absent in the assay, may increase the Ca^{2+} binding affinities of the enzymes. The relative roles and abundance of different Ca^{2+} efflux transporters may also vary between different cell types and/or developmental stages. In addition to regulating the cytosolic Ca^{2+} concentration, Ca^{2+} transporters may have an important role in regulating Ca^{2+} levels inside organelles and during secretory processes (Antebi and Fink, 1992; Bush, 1993; Bush *et al.*, 1993; McCormack *et al.*, 1990; Rudolph *et al.*, 1989).

During the last few years, genes coding for plant Ca^{2+}-ATPases have been cloned (Huang *et al.*, 1993; Perez-Prat *et al.*, 1992; Wimmers *et al.*, 1992) and a few Ca^{2+} transporters have been purified and characterized. These achievements and future research in this area will lead to a deeper understanding of how different Ca^{2+} transporters function and interact, and of plant Ca^{2+} homeostasis and signal transduction in general.

Acknowledgements

We would like to acknowledge the financial support of the Swedish Natural Science Research Council and the Swedish Council for Forestry and Agricultural Research.

References

Allen GJ, Muir SR, Sanders D. (1995) Release of Ca^{2+} from individual plant vacuoles by both $InsP_3$ and cyclic ADP-ribose. *Science* **268:** 735–737.

Andreev IM, Koren'kov V, Molotkovsky YG. (1990) Calmodulin stimulation of Ca^{2+}/nH^+ antiport across the vacuolar membrane of sugar beet tap root. *J. Plant Physiol.* **136:** 3–7.

Antebi A, Fink GR. (1992) The yeast Ca^{2+}-ATPase homologue, PMR1, is required for normal Golgi function and localizes in a novel Golgi-like distribution. *Mol. Biol. Cell* **3:** 633–654.

Askerlund P. (1996) Modulation of an intracellular calmodulin-stimulated Ca^{2+}-pumping ATPase in cauliflower by trypsin. The use of Calcium Green-5N to measure Ca^{2+} transport in membrane vesicles. *Plant Physiol.* (in press).

Askerlund P, Evans DE. (1992) Reconstitution and characterization of a calmodulin-stimulated Ca^{2+}-pumping ATPase purified from *Brassica oleracea* L. *Plant Physiol.* **100:** 1670–1681.

Askerlund P, Evans DE. (1993) Detection of distinct phosphorylated intermediates of Ca^{2+}-ATPase and H^+-ATPase in plasma membranes from *Brassica oleracea*. *Plant Physiol. Biochem.* **31:** 787–791.

Berridge MJ. (1993) Inositol trisphosphate and calcium signalling. *Nature* **361:** 315–325.

Blumwald E, Poole RJ. (1986) Kinetics of Ca^{2+}/H^+ antiport in isolated tonoplast vesicles from storage tissue of *Beta vulgaris* L. *Plant Physiol.* **80:** 727–731.

Bowman EJ, Siebers A, Altendorf K. (1988) Bafilomycins: a class of inhibitors of membrane ATPases from microorganisms, animal cells, and plant cells. *Proc. Natl Acad. Sci. USA* **85:** 7972–7976.

Brauer D, Schubert C, Tsu S-I. (1990) Characterization of a Ca^{2+}- translocating ATPase from corn root microsomes. *Physiol. Plant.* **78:** 335–344.

Briars SA, Kessler F, Evans DE. (1988) The calmodulin-stimulated ATPase of maize coleoptiles is a 140 000-Mr polypeptide. *Planta* **176:** 283–285.

Briskin DP. (1990) Ca²⁺-translocating ATPase of the plant plasma membrane. *Plant Physiol.* **94:** 397–400.

Buckhout TJ. (1983) ATP-dependent Ca²⁺ transport in endoplasmic reticulum isolated from roots of *Lepidium sativum* L. *Planta* **159:** 84–90.

Buckhout TJ. (1984) Characterization of Ca²⁺ transport in purified endoplasmic reticulum membrane vesicles from *Lepidium sativum* L. roots. *Plant Physiol.* **76:** 962–967.

Bush DR, Sze H. (1986) Calcium transport in tonoplast and endoplasmic reticulum vesicles isolated from cultured carrot cells. *Plant Physiol.* **80:** 549–555.

Bush DS. (1993) Regulation of cytosolic calcium in plants. *Plant Physiol.* **103:** 7–13.

Bush DS. (1995) Calcium regulation in plant cells and its role in signaling. *Annu. Rev. Plant Physiol. Plant Mol. Biol.* **46:** 95–122.

Bush DS, Wang T. (1995) Diversity of calcium-efflux transporters in wheat aleurone cells. *Planta* **197:** 19–30.

Bush DS, Biswas AK, Jones RL. (1993) Hormonal regulation of Ca²⁺ transport in the endomembrane system of the barley aleurone. *Planta* **189:** 507–515.

Canut H, Carrasco A, Rossignol M, Ranjeva R. (1993) Is the vacuole the richest store of IP₃-mobilizable calcium in plant cells? *Plant Sci.* **90:** 135–143.

Carafoli E. (1987) Intracellular calcium homeostasis. *Annu. Rev Biochem.* **56:** 395–433.

Carafoli E. (1991) Calcium pump of the plasma membrane. *Physiol. Rev.* **71:** 129–153.

Carafoli E. (1992) The Ca²⁺ pump of the plasma membrane. *J. Biol. Chem.* **267:** 2115–2118.

Carafoli E. (1994) Biogenesis: plasma membrane calcium ATPase: 15 years of work on the purified enzyme. *FASEB J.* **8:** 993–1002.

Carnelli A, De Michelis MI, Rasi-Caldogno F. (1992) Plasma membrane Ca-ATPase of radish seedlings. I. Biochemical characterization using ITP as a substrate. *Plant Physiol.* **98:** 1196–1201.

Castillo FJ, Heath RL. (1990) Ca²⁺ transport in membrane vesicles from pinto bean leaves and its alteration after ozone exposure. *Plant Physiol.* **94:** 788–795.

Chanson A. (1993) Active transport of proton and calcium in higher plant cells. *Plant. Physiol. Biochem.* **31:** 943–955.

Chen FH, Ratterman DM, Sze H. (1993) A plasma membrane-type Ca²⁺-ATPase of 120 kilodaltons on the endoplasmic reticulum from carrot (*Daucus carota*) cells. Properties of the phosphorylated intermediate. *Plant Physiol.* **102:** 651–661.

Crompton M. (1990) Role of mitochondria in intracellular calcium regulation. In: *Intracellular Calcium Regulation* (Bronner F, ed.).New York: Wiley-Liss, pp. 181–209.

Cunningham KW, Fink GR. (1994) Calcineurin-dependent growth control in *Saccharomyces cerevisiae* mutants lacking *PMC1*, a homolog of plasma membrane Ca²⁺ ATPases. *J. Cell Biol.* **124:** 351–363.

Da Costa AG, Madeira VMC. (1994) Proton ejection as a major feature of the Ca²⁺-pump. *Biochim. Biophys. Acta* **1189:** 181–188.

De Michelis MI, Rasi-Caldogno F, Pugliarello MC. (1992) The plasma membrane Ca²⁺ pump: potential role in Ca²⁺ homeostasis. In: *Progress in Plant Growth Regulation* (Karssen CM, van Loon LC, Vreugdenhil D, ed.). Dordrecht: Kluwer Academic Publishers, pp. 675–685.

De Michelis MI, Carnelli A, Rasi-Caldogno F. (1993) The Ca²⁺ pump of the plasma membrane of *Arabidopsis thaliana*: characteristics and sensitivity to fluorescein derivatives. *Bot. Acta* **106:** 20–25.

Dieter P, Marmé D. (1981) A calmodulin-dependent, microsomal ATPase from corn (*Zea mays* L.). *FEBS Lett.* **125:** 245–248.

Dieter P, Marmé D. (1983) The effect of calmodulin and far-red light on the kinetic properties of the mitochondrial and microsomal calcium-ion transport system from corn. *Planta* **159:** 277–281.

DuPont FM, Morrissey PJ. (1992) Subunit composition and Ca²⁺-ATPase activity of the vacuolar ATPase from barley roots. *Arch. Biochem. Biophys.* **294:** 341–346.

DuPont FM, Bush DS, Windle JJ, Jones RL. (1990) Calcium and proton transport in membrane vesicles from barley roots. *Plant Physiol.* **94:** 179–188.

Erdei L, Matsumoto H. (1991) Mitigation of symptoms of Ca²⁺ deficiency by benzyladenine in cucumber: ion levels, polyamines and Ca²⁺-Mg²⁺-ATPase. *Biochem. Physiol. Pflanzen* **187:** 177–188.

Evans DE. (1994) Calmodulin-stimulated calcium pumping ATPases located at higher plant intracellular membranes: a significant divergence from other eukaryotes? *Physiol. Plant.* **90:** 420–426.

Evans DE, Briars SA, Williams LE. (1991) Active calcium transport by plant cell membranes. *J. Exp. Bot.* **42:** 285–303.

Evans DE, Askerlund P, Boyce JM, Briars S-A, Coates J, Cooke DT, Theodoulou FL. (1992) Studies on higher plant calmodulin-stimulated ATPase. In: *Transport and Receptor Proteins of Plant Membranes* (Cooke DT, Clarkson DT, ed.). New York: Plenum Press, pp. 39–53.

Felle HH, Tretyn A, Wagner G. (1992) The role of the plasma-membrane Ca^{2+}-ATPase in Ca^{2+} homeostasis in *Sinapis alba* root hairs. *Planta* **188:** 306–313.

Fricker M, Tester M, Gilroy S. (1993) Fluorescence and luminescence techniques to probe ion activities in living plant cells. In: *Fluorescent and Luminescent Probes for Biological Activity. A Practical Guide to Technology for Quantitative Real-Time Analysis* (Mason WT, ed.). London: Academic Press Ltd, pp. 360–377.

Fukumoto M, Venis MA. (1986) ATP-dependent Ca^{2+} transport in tonoplast vesicles from apple fruit. *Plant Cell Physiol.* **27:** 491–497.

Garrahan PJ, Rega AF. (1990) Plasma membrane calcium pump. In: *Intracellular Calcium Regulation* (Bronner F, ed.). New York: Wiley-Liss, pp. 271–303.

Gavin O, Pilet P-E, Chanson A. (1993) Tonoplast localization of a calmodulin-stimulated Ca^{2+}-pump from maize roots. *Plant Sci.* **92:** 143–150.

Giannini JL, Gildensoph LH, Reynolds-Niesman, Briskin DP. (1987a) Calcium transport in sealed vesicles from red beet (*Beta vulgaris* L.) storage tissue. I. Characterization of a Ca^{2+}-pumping ATPase associated with the endoplasmic reticulum. *Plant Physiol.* **85:** 1129–1136.

Giannini JL, Ruiz-Cristin J, Briskin DP. (1987b) Calcium transport in sealed vesicles from red beet (*Beta vulgaris* L.) storage tissue. II. Characterization of $^{45}Ca^{2+}$ uptake into plasma membrane vesicles. *Plant Physiol.* **85:** 1137–1142.

Gilroy S, Jones RL. (1993) Calmodulin stimulation of unidirectional calcium uptake by the endoplasmic reticulum of barley aleurone. *Planta* **190:** 289–296.

Gilroy S, Bethke PC, Jones RL. (1993) Calcium homeostasis in plants. *J. Cell Sci.* **106:** 453–462.

Gräf P, Weiler EW. (1989) ATP-driven Ca^{2+} transport in sealed plasma membrane vesicles prepared by aqueous two-phase partitioning from leaves of *Commelina communis. Physiol. Plant.* **75:** 469–478.

Gräf P, Weiler EW. (1990) Functional reconstitution of an ATP-driven Ca^{2+}-transport system from the plasma membrane of *Commelina communis* L. *Plant Physiol.* **94:** 634–640.

Hsieh WL, Pierce WS, Sze H. (1991) Calcium-pumping ATPases in vesicles from carrot cells. Stimulation by calmodulin or phosphatidylserine, and formation of a 120 kilodalton phosphoenzyme. *Plant Physiol.* **97:** 1535–1544.

Huang L, Berkelman T, Franklin AE, Hoffman NE. (1993) Characterization of a gene encoding a Ca^{2+}-ATPase-like protein in the plastid envelope. *Proc. Natl Acad. Sci. USA* **90:** 10066–10070 (and Correction (1994) *Proc. Natl. Acad. Sci. USA* **91:** 9664).

Itagaki T, Dry IB, Wiskich JT. (1990) Effects of calcium on NAD(H)-glutamate dehydrogenase from turnip (*Brassica rapa*) mitochondria. *Plant Cell Physiol.* **31:** 993–997.

Johannes E, Brosnan JM, Sanders D. (1991) Calcium channels and signal transduction in plant cells. *BioEssays* **13:** 331–336.

Johannes E, Brosnan JM, Sanders D. (1992) Parallel pathways for intracellular Ca^{2+} release from the vacuole of higher plants. *Plant J.* **2:** 97–102.

Johansson F, Sommarin M, Larsson C. (1993) Fusicoccin activates the plasma membrane H^+-ATPase by a mechanism involving the C-terminal inhibitory domain. *Plant Cell* **5:** 321–327.

Kasai M, Muto S. (1990) Ca^{2+} pump and Ca^{2+}/H^+ antiporter in plasma membrane vesicles isolated by aqueous two-phase partitioning from corn leaves. *J. Memb. Biol.* **114:** 133–142.

Kasai M, Muto S. (1991) Solubilization and reconstitution of Ca^{2+} pump from corn leaf plasma membrane. *Plant Physiol.* **96:** 565–570.

Kasai M, Muto S. (1995) Effects of Mg^{2+} on the activities of Ca^{2+}/H^+ antiporter and Ca^{2+} pump in maize-leaf plasma membranes. *J. Plant Physiol.* **145:** 450–452.

Klüsener B, Boheim G, Liß H, Engelberth J, Weiler EW. (1995) Gadolinium-sensitive, voltage-dependent calcium release channels in the endoplasmic reticulum of a higher plant mechanoreceptor organ. *EMBO J.* **14:** 2708–2714.

Kreimer G, Melkonian M, Latzko E. (1985) An electrogenic uniport mediates light-dependent Ca^{2+} influx into intact spinach chloroplasts. *FEBS Lett.* **180:** 253–258.

Kreimer G, Melkonian M, Holtum JAM, Latzko E. (1988) Stromal free calcium concentration and light-mediated activation of chloroplast fructose-1,6-bisphosphatase. *Plant Physiol.* **86:** 423–428.

Kurosaki F, Kaburaki, H. (1994) Calmodulin-dependency of a Ca^{2+}-pump at the plasma membrane of cultured carrot cells. *Plant Sci.* **104:** 23–30.

Läuchli A. (1990) Calcium, salinity and the plasma membrane. In: *Calcium in Plant Growth and Development* (Leonard RT, Hepler PK, ed.). Rockville, MD: American Society of Plant Physiologists, pp. 26–35.

Lesniak AP, Haug A, Ries SK. (1986) Stimulation of ATPase activity in barley (*Hordeum vulgare*) root plasma membrane after treatment of intact tissues and cell-free extracts with triacontanol. *Physiol. Plant.* **68:** 20–26.

Lesniak AP, Haug A, Ries SK. (1989) Stimulation of ATPase activity in barley (*Hordeum vulgare*) root plasma membranes after treatment with triacontanol and calmodulin. *Physiol. Plant.* **75:** 75–80.

Liß H, Siebers B, Weiler EW. (1991) Characterization, functional reconstitution and activation by fusicoccin of a Ca^{2+}-ATPase from *Corydalis sempervirens* Pers. cell suspension cultures. *Plant Cell Physiol.* **32:** 1049–1056.

Logan DC, Venis MA. (1995) Characterisation and immunological identification of a calmodulin-stimulated Ca^{2+}-ATPase from maize shoots. *J. Plant Physiol.* **145:** 702–710.

McCormack JG, Halestrap AP, Dento RM. (1990) Role of calcium ions in regulation of mammalian intramitochondrial metabolism. *Physiol. Rev.* **70:** 391–425.

Malatialy L, Greppin H, Penel C. (1988) Calcium uptake by tonoplast and plasma membrane vesicles from spinach leaves. *FEBS Lett.* **233:** 196–200.

Marrè E. (1979) Fusicoccin: a tool in plant physiology. *Annu. Rev. Plant Physiol.* **30:** 273–288.

Martins IS, Carnieri EGS, Vercesi E. (1986) Characteristics of Ca^{2+} transport by corn mitochondria. *Biochim. Biophys. Acta* **850:** 49–56.

Mészáros LG, Bak J, Chu A. (1993) Cyclic ADP-ribose as an endogenous regulator of the non-skeletal type ryanodine receptor Ca^{2+} channel. *Nature* **364:** 776–779.

Muto S. (1992) Intracellular Ca^{2+} messenger system in plants. *Int. Rev. Cytol.* **142:** 305–345.

Muto S, Miyachi S. (1986) Roles of calmodulin dependent and independent NAD kinases in regulation of nicotinamide coenzyme levels of green plant cells. In: *Molecular and Cellular Aspects of Calcium in Plant Development* (Trevawas AJ, ed.). New York: Plenum Press, pp. 107–114.

Nguyen TD, Siegenthaler PA. (1985) Purification and some properties of an Mg^{2+}, Ca^{2+}- and calmodulin-stimulated ATPase from spinach chloroplast envelope membranes. *Biochim. Biophys. Acta* **840:** 99–106.

Oláh Z, Bérczi A, Erdei L. (1983) Benzylaminopurine induced coupling between calmodulin and Ca-ATPase in wheat root microsomal membranes. *FEBS Lett.* **154:** 395–399.

Olbe M, Sommarin M. (1991) ATP-dependent Ca^{2+} transport in wheat root plasma membrane vesicles. *Physiol. Plant.* **83:** 535–543.

Perez-Prat E, Narasimhan ML, Binzel ML, Botella MA, Chen Z, Valpuesta V, Bressan RA, Hasegawa PM. (1992) Induction of a putative Ca^{2+}-ATPase mRNA in NaCl-adapted cells. *Plant Physiol.* **100:** 1471–1478.

Pfeiffer W, Hager A. (1993) A Ca^{2+}-ATPase and a Mg^{2+}/H^+-antiporter are present on tonoplast membranes from roots of *Zea mays* L. *Planta* **191:** 377–385.

Pietrobon D, Di Virgilio F, Pozzan T. (1990) Structural and functional aspects of calcium homeostasis in eukaryotic cells. *Eur. J. Biochem.* **193:** 599–622.

Poovaiah BW, Reddy ASN. (1993) Calcium and signal transduction in plants. *Crit. Rev. Plant Sci.* **12:** 185–211.

Rasi-Caldogno F, Pugliarello MC, De Michelis MI. (1987) The Ca^{2+} transport ATPase of plant plasma membrane catalyzes a nH^+/Ca^{2+} exchange. *Plant Physiol.* **83:** 994–1000.

Rasi-Caldogno F, Pugliarello MC, Olivari C, De Michelis MI. (1989) Identification and characterization of the Ca^{2+}-ATPase which drives active transport of Ca^{2+} at the plasma membrane of radish seedlings. *Plant Physiol.* **90:** 1429–1434.

Rasi-Caldogno F, Pugliarello MC, Olivari C, De Michelis MI, Gambarini G, Colombo P, Tosi G. (1990) The plasma membrane Ca^{2+} pump of plant cells: a radiation inactivation study. *Bot. Acta* **103:** 39–41.

Rasi-Caldogno F, Carnelli A, De Michelis MI. (1992) Plasma membrane Ca-ATPase of radish seedlings. II. Regulation by calmodulin. *Plant Physiol.* **98:** 1202–1206.

Rasi-Caldogno F, Carnelli A, De Michelis MI. (1993) Controlled proteolysis activates the plasma membrane Ca^{2+} pump of higher plants. Comparison with the effect of calmodulin in plasma membrane from radish seedlings. *Plant Physiol.* **103:** 385–390.

Rasi-Caldogno F, Carnelli A, De Michelis MI. (1995) Identification of the plasma membrane Ca^{2+}-ATPase and of its autoinhibitory domain. *Plant Physiol.* **108:** 105–113.

Rasmusson AG, Møller IM. (1991) Effect of calcium ions and inhibitors on internal NAD(P)H dehydrogenases in plant mitochondria. *Eur. J. Biochem.* **202:** 617–623.

Read N, Allan WTG, Knight H, Knight MR, Malhó R, Russel A, Shacklock PS, Trevawas AJ. (1992) Imaging and measurement of cytosolic free calcium in plant and fungal cells. *J. Microscopy* **166:** 57–86.

Robinson C, Larsson C, Buckhout TJ. (1988) Identification of a calmodulin-stimulated (Ca^{2+} + Mg^{2+})-ATPase in a plasma membrane fraction isolated from maize (*Zea mays*) leaves. *Physiol. Plant.* **72:** 177–184.

Rudolph HK, Antebi A, Fink GR, Buckley CM, Dorman TE, Levitre J, Davidow LS, Mao JI, Moir DT. (1989) The yeast secretory pathway is perturbed by mutations in PMR1, a member of a Ca^{2+} ATPase family. *Cell* **58:** 133–145.

Rugolo M, Pistocchi R, Zannoni D. (1990) Calcium ion transport in higher plant mitochondria (*Helianthus tuberosus*). *Physiol. Plant.* **79:** 297–302.

Sanders D, Miller AJ, Blackford S, Brosnan JM, Johannes E. (1990) Cytosolic free calcium homeostasis in plants. In: *Current Topics in Plant Biochemistry and Physiology, Vol. 9* (Randall DD, Blevins DG, ed.). Columbia: Interdisciplinary Plant Group, University of Missouri, pp. 20–37.

Schatzmann HJ. (1989) The calcium pump of the surface membrane and of the sarcoplasmic reticulum. *Annu. Rev. Physiol.* **51:** 473–485.

Schulz S, Oelgemöller E, Weiler EW. (1990) Fusicoccin action in cell suspension cultures of *Corydalis sempervirens*. *Planta* **183:** 83-91.

Schumaker KS, Sze H. (1990) Solubilization and reconstitution of the oat root vacuolar H^+/Ca^{2+} exchanger. *Plant Physiol.* **92:** 340–345.

Seidler NW, Jona I, Vegh M, Martonosi A. (1989) Cyclopiazonic acid is a specific inhibitor of the Ca^{2+}-ATPase of sarcoplasmic reticulum. *J. Biol. Chem.* **264:** 17816–17823.

Serrano R. (1990) Plasma membrane ATPase. In: *The Plant Plasma Membrane. Structure, Function and Molecular Biology* (Larsson C, Møller IM, ed.). Berlin: Springer-Verlag, pp. 127–153.

Sievers A, Busch MM. (1992) An inhibitor of the Ca^{2+}-ATPases in the sarcoplasmic and endoplasmic reticula inhibits transduction of the gravity stimulus in cress roots. *Planta* **188:** 619–622.

Silva MAP, Carnieri EGS, Vercesi AE. (1992) Calcium transport by corn mitochondria. *Plant Physiol.* **98:** 452–457.

Thastrup O, Cullen PJ, Drøbak BK, Hanley MR, Dawson AP. (1990) Thapsigargin, a tumor promoter discharges intracellular Ca^{2+} stores by specific inhibition of the endoplasmic reticulum Ca^{2+}-ATPase. *Proc. Natl Acad. Sci. USA* **87:** 2466–2470.

Theodoulou FL, Dewey FM, Evans DE. (1994) Calmodulin-stimulated ATPase of maize cells: functional reconstitution, monoclonal antibodies and subcellular localization. *J. Exp. Bot.* **45:** 1553–1564.

Thomson LJ, Xing T, Hall JL, Williams LE. (1993) Investigation of the calcium-transporting ATPases at the endoplasmic reticulum and plasma membranes of red beet (*Beta vulgaris*). *Plant Physiol.* **102:** 553–564.

Thomson LJ, Hall JL, Williams LE. (1994) A study of the effect of inhibitors of the animal sarcoplasmic/endoplasmic reticulum-type calcium pumps on the primary Ca^{2+}-ATPases of red beet. *Plant Physiol.* **104:** 1295–1300.

Tretyn A, Kendrick RE, Wagner G. (1991) The role(s) of calcium ions in phytochrome action. *Photochem. Photobiol.* **54:** 1135–1155.

Williams LE, Schueler SB, Briskin DP. (1990) Further characterization of the red beet plasma membrane Ca^{2+}-ATPase using GTP as an alternative substrate. *Plant Physiol.* **92:** 747–754.

Wimmers LE, Ewing NN, Bennett AB. (1992) Higher plant Ca^{2+}-ATPase: primary structure and regulation of mRNA abundance by salt. *Proc. Natl Acad. Sci. USA* **89:** 9205–9209.

Wuytack F, Raeymaekers L. (1992) The Ca^{2+}-transport ATPases from the plasma membrane. *J. Bioenerg. Biomemb.* **24:** 285–300.

Wuytack F, Raeymaekers L, De Schutter G, Casteels R. (1982) Demonstration of the phosphorylated intermediates of the Ca^{2+}-transport ATPase in a microsomal fraction and in a (Ca^{2+} + Mg^{2+})-ATPase purified from smooth muscle by means of calmodulin affinity chromatography. *Biochim. Biophys. Acta* **693:** 45–52.

Zhen R-G, Baykov AA, Bakuleva NP, Rea PA. (1994) Aminomethylene-diphosphonate: a potent type-specific inhibitor of both plant and phototrophic bacterial H^+-pyrophosphatases. *Plant Physiol.* **104:** 153–159.

Zocchi G, Rabotti G. (1993) Calcium transport in membrane vesicles isolated from maize coleoptiles. Effect of indoleacetic acid and fusicoccin. *Plant Physiol.* **101:** 135–139.

Zottini M, Zannoni D. (1993) The use of fura-2 fluorescence to monitor the movement of free calcium ions into the matrix of plant mitochondria (*Pisum sativum* and *Helianthus tuberosus*). *Plant Physiol.* **102:** 573–578.

Chapter 18

Electrical signalling in plants

J.F. Thain and D.C. Wildon

1. Introduction

Although the first well-attested example of electrical signalling in a plant species was reported over 100 years ago (Burdon-Sanderson, 1873), electrical signalling has generally been regarded as a phenomenon of very minor importance in plants, and one restricted to a role in leaf movements in a very small number of species.

This view has persisted in spite of the gradual accumulation of evidence (Pickard, 1973, 1974) to suggest that it may be wrong, or at least over-stated. A major problem has continued to be the paucity of reproducible evidence linking electrical events in plants to functional roles, and especially to functional roles of a different type to the leaf movements mentioned above.

Recently, however, new evidence has been obtained which suggests that the traditional view described above may need revision. The purpose of this chapter is to bring up to date the reviews by Pickard (1973) and Thain and Wildon (1992), especially with regard to our own results on electrical signalling in relation to the wound-induced systemic induction of proteinase inhibitor proteins in tomato seedlings. First, however, we must define what we mean by electrical signalling.

2. The characteristics of electrical signals

The characteristics of an electrical signalling system are best described by reference to a well-understood system such as the giant nerve axon of the squid (Aidley, 1989), which is a very long, cylindrical cytoplasm-filled extension of the nerve cell body. An electrical potential difference (p.d.) exists across the axon's plasma membrane such that the interior is about 60 mV negative compared with the external medium. Stimulation (e.g. by an electric current) of a small region of the plasma membrane causes the excitation of an action potential at that region, i.e. the production of a large transient depolarization of the membrane p.d., so that the cytoplasm becomes electrically positive with respect to the external medium for a short period (1 msec) before the p.d. returns to its usual value.

The action potential occurs because the stimulus causes the transient opening of selective ion channels in the membrane so that enhanced fluxes of ions can occur down their electrochemical gradients. In order for excitation to occur, it is necessary for the stimulus to exceed a certain threshold value so that excitation is an all-or-nothing response. After excitation, that region of the membrane is inexcitable for a short period, known as the refractory period.

The ionic currents that flow during an action potential at one part of the axon membrane cause local currents to flow over neighbouring regions of the membrane, and these local currents can in turn excite action potentials in those regions by the opening of voltage-sensitive ion channels. In this way, excitation of one region of the axon can lead to propagation of the action potential along the axon to a site (e.g. a muscle) where an appropriate response is produced.

While the description given above is greatly simplified and not typical in detail of all nerve systems, it serves to show that any electrical signalling system requires (a) cells that can be excited to produce action potentials by appropriate chemical or physical stimuli, (b) a pathway consisting of excitable plasma membrane along which the action potential can propagate by the transient opening of voltage-sensitive ion channels, and (c) cells that can be stimulated by the arrival of an action potential to produce some physiological, biochemical or developmental response.

It is worth noting here that the pathway need not be a structure such as a nerve axon. In the phenomenon of epithelial conduction in animals (reviewed by Anderson, 1980), excitation propagates in two dimensions through epithelia in which the cells are essentially isodiametric in the plane of the epithelium. Propagation of excitation from any cell to its neighbours is made possible by the presence of gap junctions (Robards and Pitts, 1991) which provide cytoplasmic pathways through which excitation currents can pass from cell to cell. Nerves are not necessary for electrical signalling, although they do obviously confer advantages in terms of definition of the pathway and speed of signal transmission.

3. Experimental methods

Ideally, studies of electrical signalling should be made with the aid of intracellular microelectrodes (Purves, 1981) which permit direct measurement of changes in electrical p.d. across the plasma membranes of cells of the signalling pathway. However, the practical difficulties involved in applying these methods to studies on whole plants have meant that, in the few cases where they have been used, they have been employed in the later stages of the investigations, when evidence for the existence of electrical signalling has already been obtained by the use of surface-contact electrodes (Thain, 1995), and in order to carry out more detailed investigations of the signalling pathway.

In all established and potential cases of electrical signalling in plants, the initial and often the only studies have involved the use of surface-contact electrodes which make contact with the surface of the plant via a thread or wick of absorbent material soaked in a dilute salt solution such as 10 mM KCl. These electrodes make electrical contact with the extracellular aqueous phase (apoplast) of the plant tissue. Since only electrical potential differences between two electrodes can be measured, any observed electrical events represent transient changes in electrical p.d. between two points in the apoplast due to ionic currents flowing in that region of the apoplast. The most probable origin of ionic currents in the apoplast is changes in ion fluxes across the plasma membranes of cells in the tissue, because it is at the plasma membranes that the major electrical and concentration-driving forces on ions are to be found. While it is possible that some electrical events recorded in this way could be streaming potentials caused by bulk movement of the aqueous medium in the apoplast through regions of cell wall containing fixed charges (Dainty et al., 1963), such effects are likely to be small because of short-circuiting by uncharged regions. In fact, recent theoretical analysis (Due, 1993) and experiments with plant tissues (Frachisse-Stoilskovic and Julien, 1993; Zawadzki and Trebacz, 1985) both indicate that measurements with surface-contact electrodes can give good representations of changes in plasma membrane p.d.

One possible problem with the use of surface-contact electrodes is that the electrical events thus observed represent the sum of all the ionic currents flowing in the apoplast between the two electrodes (Geddes, 1972). There may, for example, be more than one signalling pathway in the tissue, or there may be simultaneous changes in plasma membrane p.d. at different points in the tissue between the two electrodes. Thus although the observed electrical events have their origin in changes in plasma membrane p.d., their interpretation in those terms is often not easy. In practice, the shapes of the electrical events observed in this way can vary from single spikes, whose duration typically ranges from a few seconds to tens of seconds, to more complex forms which can be of somewhat indefinite shape and longer duration (often termed variation potentials), or which may consist of a number of overlapping spikes.

4. Electrical signalling in plants: cases and candidates

Here we shall briefly review the situations where electrical signalling plays, or may play, a role in plants. In each case the application of a stimulus is followed by the observation of transient electrical events that propagate through the plant to sites where physiological, biochemical or developmental changes then occur. The strength of the evidence that these cases are true examples of electrical signalling, as defined above, will be discussed in the next section.

Leaf movements

The earliest and best established examples of electrical signalling in plants involve various kinds of leaf movement: the insect-trapping leaf closure of the Venus fly-trap, *Dionaea muscipula*, and the closely related but aquatic species, *Aldrovanda vesiculosa*; the insect-trapping bending of mucilage-secreting tentacles on the leaves of sundew (*Drosera*) species; and the seismonastic movements of leaves of *Mimosa pudica*, at least in response to electrical stimulation or stimulation by cold. The general features of these systems have been reviewed elsewhere (Iijima and Sibaoka, 1981; Pickard, 1973, 1974; Sibaoka, 1966, 1969; Williams and Pickard, 1980). In *Dionaea*, *Aldrovanda* and *Drosera*, distances of about 1 cm or less are involved in the signalling pathway, but in *Mimosa* distances of over 10 cm can be involved.

Respiration and photosynthesis

Several reports indicate a role for electrical signalling in causing changes in the rate of respiration of plant tissues. Stimulation of the basal regions of pumpkin stems with heat or with a high concentration of KCl caused electrical transients to propagate acropetally along the stem. This was followed, a few minutes later, by transient changes in rates of respiration and photosynthesis in the leaves (Gunar and Sinyukhin, 1963). Pollination of the stigma of *Incarvillea grandiflora* caused the propagation of an electrical transient down the style to the ovary, where it was followed about a minute later by an increased rate of respiration (Sinyukhin and Britikov, 1967). Electrical blockage of the propagation of the electrical transient down the style inhibited the increase in respiration. Similarly, Fromm *et al.* (1995) have shown that pollination of *Hibiscus rosa-sinensis* caused transient (2-sec duration) depolarizations of cell membrane p.d. to propagate rapidly (3.5 cm sec^{-1}) down the style to the ovary. The rate of respiration of the ovary increased transiently for about 15 min after pollination. In the liverwort *Conocephalum conicum,* both cutting and electrical stimulation of the thallus caused electrical transients to propagate across the thallus, followed by a transient increase in the rate of respiration of the thallus (Dziubinska *et al.*, 1989). Fromm and Eschrich (1993) have reported that treatment of the roots of willow plants with various nutrients, hormones or pH changes caused electrical transients to propagate from the roots to the leaves; the arrival of these electrical transients at the leaves was followed within 3 min by changes in the rates of transpiration and photosynthesis.

Transport of assimilates and phloem unloading

The arrival of an action potential at the pulvinus of *Mimosa pudica* is followed by unloading of sucrose into the apoplast (Fromm, 1991). Stimulation of the hypocotyl of beet seedlings with cold water caused electrical transients to propagate acropetally, and this was followed by an increase in basipetal transport of ^{14}C-labelled photosynthate (Opritov, 1978).

With maize, Fromm and Bauer (1994) reported that electrical stimulation and stimulation by cold shock both initiated electrical transients that propagated along the leaf; at the same time phloem translocation was inhibited.

Plant development

A very interesting correlation between electrical events and plant development has been described by Desbiez and co-workers (Frachisse *et al.*, 1985). Excision of the shoot apex of a seedling of *Bidens pilosus* leads to growth of one or other of the two buds in the axils of the cotyledons. Pricking of a cotyledon before the shoot apex is excised causes propagation of electrical transients down the petiole of that cotyledon to the axil, and bias in favour of growth of the bud in the axil of the other cotyledon when the apex is later excised.

Protein synthesis

Wounding one cotyledon or leaflet of a tomato seedling, by crushing or by heat, leads to the following sequence of events (Wildon *et al.*, 1992): the propagation of electrical transients down the petiole of the wounded organ and into the other leaves; the production, within 4 h, of detectable quantities of mRNA for the Pin 2 proteinase inhibitor protein in the unwounded leaves; and the production, within 24 to 48 h, of significant amounts of proteinase inhibitor activity in the unwounded leaves.

5. Mechanism: the nature of the signal

Possible mechanisms

In each of the examples described above the evidence indicates the existence of a long-distance signal, and is consistent with the signal being an electrical one. In fact, most of the workers studying these systems have used the term 'action potential' to describe the travelling electrical events that they observe. However, this interpretation is not universally accepted, at least for some of the cases, and other mechanisms have been proposed.

For example, the actual long-distance signal could be a chemical one that travels from the site of stimulation to the site of response, and that produces secondary local electrical effects *en route*. Movement could take place in the phloem, which raises the question of entry into the translocation pathway, in the xylem, which should impose some directionality on the signal, or through the symplast.

A wounding stimulus could cause the release of tension in the xylem and a consequent hydraulic shock that would propagate very rapidly through the plant, and could produce secondary local electrical effects as it travelled (Malone, 1993). Alternatively, the release of tension in the xylem at the site of wounding could set up a reversed flow of xylem contents; the reversed flow could carry with it a wound-released chemical signal that could have secondary local electrical effects (Malone, 1993; Malone and Alarcon, 1995).

In the rest of this section we shall discuss the evidence concerning the nature of the long-distance signal in the cases described in the preceding section. First, we shall consider whether plants possess the necessary machinery for electrical signalling, that is electrically excitable membranes that form a pathway along which action potentials can propagate.

Action potentials in plants: evidence obtained with intracellular microelectrodes

There is definite evidence that at least some kinds of plant cell are excitable and will generate action potentials in response to a range of stimuli. The best evidence comes from studies of the giant cells of various species of characean algae (Hope and Walker, 1975; Wayne, 1994), for which a wider range of electrophysiological techniques is available (Beilby, 1989; Shimmen *et al.*, 1994) than can be used on most other types of plant cell. The characean algae belong to the order Charales of the class Charophyceae, which is regarded as the algal group having the most recent common ancestor with the land plants (Graham, 1993; McCourt, 1995). Thus it seems reasonable to use characean algae, such as *Chara* and *Nitella*, as model systems for higher plants.

Of particular importance is the fact that these characean cells can be stimulated by the passage of depolarizing electric currents across their plasma membranes, as well as by other physical and chemical stimuli. The resulting action potentials (*Figure 1*) have profiles very similar to those of animal nerve action potentials, which they also resemble in being 'all-or-nothing' responses that require a minimum-threshold stimulating depolarization, and in having a post-excitation refractory period. The main differences are that the plant action potentials are much slower than the animal ones, with durations of the order of seconds rather than milliseconds, and that the nature of the ionic current is not the same. In the characean cells, stimulation results in an increase in the cytosolic Ca^{2+} concentration, which causes the opening of Ca^{2+}-sensitive Cl^- channels. The depolarizing phase of the action potential is due to the consequent efflux of Cl^- ions down their electrochemical gradient (Lunevsky *et al.*, 1983), whereas in most animal nerve cells the depolarizing phase results from the opening of Na^+ channels, which allows Na^+ ions to flow into the cell (Aidley, 1989). Near the peak of the action potential these channels close again, and in both cases K^+ channels open to allow K^+ ions to flow out of the cell down their electrochemical gradient to bring the membrane p.d. back to its original value. With the characean cells, which can be several centimetres long, stimulation of one end of the cell causes an action potential to propagate along the length of the cell at velocities of a few centimetres per second.

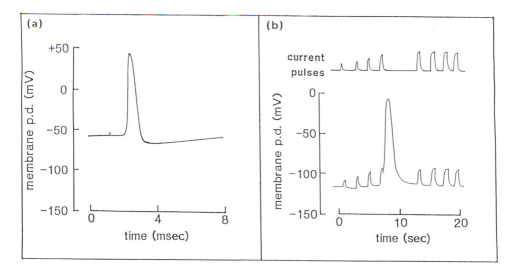

Figure 1. Action potentials recorded with intracellular microelectrodes from (a) a squid axon and (b) a cell of the alga *Chara corallina*. In the latter case the action potential was stimulated electrically, the current pulses being shown in the upper trace; these demonstrate that a minimum theshold stimulation is required to cause excitation and that, during the refractory period after the action potential, pulses of the same magnitude as that which caused excitation are not effective. Note the difference between the time scales of the two action potentials. The *Chara* trace was recorded in the authors' laboratory; the squid axon trace is reproduced from A.L. Hodgkin (1964) *The Conduction of the Nervous Impulse* with permission from Liverpool University Press.

Experiments with intracellular microelectrodes on a range of other plant species show that they too possess cells that are electrically excitable. The most detailed studies have been made on the species described above in the paragraph on leaf movements: *Dionaea muscipula* (Hodick and Sievers, 1988), *Drosera* spp. (Williams and Spanswick, 1976), *Mimosa pudica* (Opritov, 1978; Sibaoka, 1962) and *Aldrovanda vesiculosa* (Iijima and Sibaoka, 1981). Other species that have been studied are the willow, *Salix viminalis* (Fromm and Spanswick, 1993; Fromm and Bauer, 1994), and the liverwort, *Conocephalum conicum* (Zawadzki and Trebacz, 1985). Because the depolarizing transients observed in these cases are electrically stimulated, there can be no doubt that they are action potentials with durations of a few seconds or, in the case of *Conocephalum*, a few tens of seconds.

In many of these experiments the stimulating electrical pulse was applied to cells distant from those in which the action potentials were recorded. These experiments thus show not only that the tissues contained excitable cells, but also that the action potentials produced could propagate through the tissue. The microelectrode experiments also demonstrated that depolarizing electrical transients identical to those produced by electrical stimulation were produced by other kinds of stimuli: mechanical stimuli in *Dionaea* (Hodick and Sievers, 1988), *Aldrovanda* (IIjima and Sibaoka, 1981) and the *Mimosa* pulvinus (Oda and Abe, 1972); cold water/ice in the *Mimosa* pulvinus (Abe and Oda, 1976) and the *Mimosa* petiole (Samejima and Sibaoka, 1983); cutting in *Conocephalum* (Zawadzki and Trebacz, 1985); change of bathing solution, application of plant hormones in *Salix* (Fromm and Eschrich, 1993); and pollination in *Hibiscus* (Fromm *et al.*, 1995). This comparison indicates that these more natural stimuli also initiate action potentials.

Propagating action potentials in plants: evidence obtained with surface-contact electrodes

When the species mentioned in the preceding paragraph were stimulated electrically and by the other methods described there, measurements with surface-contact electrodes detected propagating transients that were similar in shape and duration to the action potentials observed with intracellular microelectrodes (Dziubuiska *et al.*, 1983; Fromm and Spanswick, 1993; Houwink, 1935; Sibaoka, 1966, 1969, 1980; Williams and Pickard, 1972a, b, 1980). Furthermore, the transients detected by surface-contact electrodes in response to electrical stimulation were found to be 'all-or-nothing' responses requiring stimuli greater than minimum threshold values and with post-excitation refractory periods (Dziubuiska *et al.*, 1983; Fromm and Spanswick, 1993; Roblin, 1979; Sibaoka, 1966; Williams and Pickard, 1972a, b, 1980). These results show that the propagating electrical transients observed with surface-contact electrodes in these species are the surface manifestations of propagating action potentials within the tissues.

The responses of *Lupinus angustifolius* (Paszewski and Zawadzki, 1973, 1974, 1976), *Helianthus annuus* (Zawadzki *et al.*, 1991) and *Luffa cylindrica* (Shiina and Tazawa, 1986) to electrical stimulation have been measured only with surface-contact electrodes. In each case (e.g. *Figure 2*) the response was a propagating electrical transient similar to those observed in the species described above. Furthermore, in all three species, the propagating responses showed the 'all-or-nothing', threshold and refractory properties characteristic of action potentials.

Surface-contact electrodes have also been used to study the electrical responses of the vine *Clematis zeylanica* to stimulation by local cooling (Houwink, 1938), and the responses of various cucurbits to local raised concentrations of KCl and local cooling (Karmanov *et al.*, 1972; Mamulashvili *et al.*, 1972, 1973; Opritov *et al.*, 1982; Sinyukhin and Gorchakov, 1966, 1968). Silver/silver chloride wires inserted into the stems of potato plants have been used to investigate propagating electrical events produced by a variety of stimuli (Volkov and Haack, 1995). In all cases the response consisted of propagating electrical transients that were very similar in shape and duration to those described in the previous paragraphs.

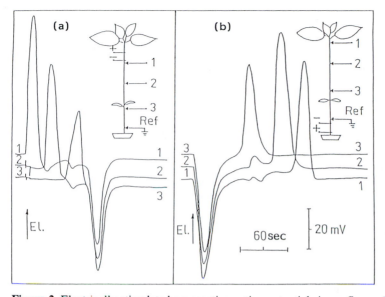

Figure 2. Electrically stimulated propagating action potentials in sunflower (*Helianthus annuus*). The stimulating electrodes are indicated by + and –, and the time of stimulation by the arrow labelled E1. Traces 1, 2 and 3 are records of the potential differences between electrode 1 and the reference electrode (Ref), between electrode 2 and the reference electrode, and between electrode 3 and the reference electrode. The time and voltage scales are the same for both panels. (a) A basipetally propagating action potential passing electrodes 1, 2, 3 and the common reference electrode in turn. (b) An acropetally propagating action potential passing the electrodes in the opposite sequence to (a). Reproduced from Zawadzki *et al.* (1991) with permission from *Physiologica Plantarum*.

On the basis of the evidence described in this and the previous section, we believe that any observation, with surface-contact electrodes, of propagating electrical transients of a simple shape ('spike') and of durations ranging from several seconds to several tens of seconds is likely to be indicative of action potentials propagating through the tissues of the plant. This conclusion is supported by studies of the ion fluxes that accompany the electrical transients observed in several of the species mentioned above (Fromm and Spanswick, 1993; Hodick and Sievers, 1988; Kumon and Suda, 1984; Trebacz *et al.*, 1994). The results provide evidence for effluxes of K^+ and Cl^- and increases in cytosolic Ca^{2+} concentrations, a pattern that is the same as that observed during action potentials in characean cells.

As we mentioned earlier, damaging stimuli such as crushing or heat can produce electrical responses of less definite shape and much longer duration than those described above. At least in some cases, these 'variation potentials' are the secondary local consequences of substances released at the site of wounding into the transpiration stream. Often, however, associated with these variation potentials and superimposed on them there are spike transients that are very similar in shape and duration to those described earlier in this section. It seems likely that these, too, represent propagating action potentials.

Individual cases

In Section 4 we described a number of phenomena involving long-distance signalling in plants, and in the last section we reviewed the evidence that many species have the properties required for long-distance propagation of action potentials (i.e. for electrical signalling). Here we shall consider the evidence that the long-distance signals involved in the various phenomena described earlier are electrical ones.

In the cases of the species which show leaf movements (*Dionaea, Drosera, Aldrovanda,* and *Mimosa* when it responds to electrical stimulation or cold shock), there is no doubt that the long-distance signal is electrical. The similarity of the responses observed with intracellular micro-electrodes and with surface-contact electrodes following either electrical stimulation or mechanical stimulation shows that these plants contain excitable membranes which form a pathway along which action potentials can propagate, and that the natural mechanical stimuli can initiate propagating action potentials. The fact that these are indeed the long-distance signal is confirmed by the observations that electrical stimulation leads to the natural leaf movement response (Sibaoka, 1962; Williams and Pickard, 1972a). With more injurious forms of stimulation, however, there is evidence that xylem-borne chemical signals may also play a role in long-distance signalling in *Mimosa* (Malone, 1994).

In the willow, *Salix viminalis,* Fromm and Eschrich (1993) showed that changes in the medium bathing the roots caused electrical transients to propagate through the plant, and that their passage was followed by changes in the rates of photosynthesis and transpiration. Subsequently, Fromm and Spanswick (1993) showed that the phloem tissue of the bark contains electrically excitable cells, and that action potentials can propagate along that tissue. Electrical stimulation and cold shock both initiated action potentials that propagated through the plant and were accompanied by a reduction in phloem translocation (Fromm and Bauer, 1994). Taken together, these separate observations provide evidence for the presence of an electrical signalling system in this species.

The very detailed studies on the liverwort *Conocephalum conicum* (Zawadzki and Trebacz, 1985) show that the thallus contains cells which are excitable, and that action potentials can propagate through the thallus. The fact that the propagating action potentials initiated by a cutting stimulus are the signal causing the subsequent transient increase in respiration (Dziubinska *et al.,* 1989) is confirmed by the fact that an electrical stimulus produced the same sequence of events.

Intracellular microelectrodes and electrical stimulation were not used in the study of the effects of pollination in *Incarvillea grandiflora* (Sinyukhin and Britikov, 1967), but it was found that electrical blockage prevented both the propagation of the electrical transients down the style and the subsequent increase in the rate of respiration of the ovary. This is strong evidence for a long-distance electrical signal, and supports the authors' description of the electrical transients observed with surface-contact electrodes as propagating action potentials. Intracellular micro-electrodes were used by Fromm *et al.* (1995) in their study of the electrical and biochemical events in the pistil of *Hibiscus* plants following pollination. The short duration (about 2 sec) and large magnitude of the pollination-induced depolarizations of cell membrane p.d., their speed (3.5 cm sec^{-1}) of propagation down the style, and the fact that they occur within about 2 min after pollination, before the transient increase in respiration of the ovary, are all indicative of an electrical signalling system.

Electrical stimulation of a cotyledon of *Bidens pilosus* causes the production of an electrical transient, detected with surface-contact electrodes, that travels down the petiole of the cotyledon. This transient has all the properties of a propagating action potential as detected by surface-contact electrodes: it is a simple spike with a duration of about 15 sec, its initiation is an 'all-or-nothing' event requiring at least a minimum threshold stimulus, and its occurrence is followed by a refractory period (Frachisse *et al.,* 1990). Pricking of a cotyledon is also followed by the propagation down the petiole of a rapid simple electrical transient of about 10–15 sec duration, which the authors identify as an action potential, followed by a slower transient with a duration of about 1 min, which the authors identify as a 'slow wave' or variation potential. It is the arrival of the slow wave at the axil of the cotyledon that is correlated with the inhibition of outgrowth of that axillary bud when the shoot apex is subsequently excised. Both the movement of the slow wave and the degree of bias between the two axillary buds for growth are dependent on the water status of the plant. It seems, therefore, that this may not be a case of long-distance electrical signalling; the long-distance signal could be a chemical one with secondary local electrical effects.

In our studies of wound-induced systemic synthesis of proteinase inhibitor in tomato seedlings, we have tried to distinguish between the various possible long-distance signalling mechanisms (Wildon et al., 1992). Experiments in which the wounded cotyledon was excised at various times after wounding showed that the period during which the systemic signal passed from the cotyledon into the stem coincided with the passage of the electrical transients caused by wounding. The signal reached the stem between 0.5 and 5 min after wounding, which is too long a period for the signal to be a hydraulic shock due to tension release in the xylem, and too fast for the signal to be a chemical moving in the symplast. Chilling of a length of the petiole of the cotyledon stopped phloem translocation out of the cotyledon, but did not stop the passage of the wound-induced electrical transients or of the systemic signal for proteinase inhibitor synthesis. The fact that similar results were obtained both from excised shoots placed with their bases in water and from intact seedlings suggests that the signal is not a chemical carried in a reverse flow of xylem contents. These observations, together with the very high correlation between the occurrence of the wound-induced electrical transients and systemic proteinase inhibitor synthesis, provide strong evidence that an electrical signal can act as the long-distance signal for proteinase inhibitor synthesis. This conclusion is reinforced by recent evidence (Herde et al., 1995; Stankovic and Davies, 1995) that local electrical stimulation can lead to systemic induction of genes for proteinase inhibitor synthesis in tomato plants.

Evidence has been presented for the involvement of other long-distance signalling mechanisms in the tomato proteinase inhibitor response. Malone and Alarcon (1995) have produced evidence which suggests that a chemical signal, as yet unidentified, is carried in a reverse flow of xylem sap from the wounded leaflet to the rest of the plant. Narvaez-Vasquez et al. (1995) have argued for a phloem-borne chemical signal, the signal being the naturally occurring oligopeptide systemin (Pearce et al., 1991) that is an elicitor of proteinase inhibitor synthesis. Thus there is evidence to support at least three different long-distance signalling mechanisms in the tomato proteinase inhibitor response. It is possible that all three may operate, perhaps in different circumstances. In particular, it is difficult to see how the systemic induction of proteinase inhibitor synthesis in response to local electrical stimulation by low applied voltages (10 V over several centimetres of leaf; Herde et al., 1995) could be mediated by a reverse flow of xylem sap.

6. Pathways for electrical signalling

From the evidence presented above it is clear that a number of plant species can be excited to produce action potentials which then propagate through the tissues, either as part of a natural electrical signalling mechanism or in response to experimental stimuli such as electric currents, ice/water, etc. The tissues of these species must therefore contain continuous pathways of excitable membrane along which action potentials can propagate. In this section we shall review the evidence concerning which cell types constitute these pathways.

Pathways for short distances

In *Drosera,* the action potential that is initiated near the top of the tentacle propagates down to its base via the epidermal cells of the tentacle stalk (Williams and Spanswick, 1976); it does not spread into the lamina of the leaf. In *Dionaea,* the cells of all the major tissues of the leaf are excitable (Hodick and Sievers, 1988), so that action potentials excited in the sensory cells at the bases of the trigger hairs can spread throughout the leaf lamina (Sibaoka, 1966). A similar situation occurs in the thallus of the liverwort *Conocephalum* (Zawadzki and Trebacz, 1985), in which all cells appear to be excitable. In all three species, the propagation of action potentials occurs over distances of about 1 cm or less, and involves relatively unspecialized cells whose plasma membranes are continuous through the plasmodesmata that connect adjacent cells.

Pathways for longer distances

In cases where action potentials propagate over longer distances (e.g. several centimetres), there is much evidence that the pathways consist of vascular tissues. This has been shown for the propagation of action potentials along petioles of *Mimosa* by various surgical treatments (Sibaoka, 1966). The cortical tissue between the vascular bundles is not excitable, but electrical currents from action potentials in the large central vascular bundle can spread across the cortical tissue sufficiently well to excite action potentials in the two smaller vascular bundles. In *Hibiscus* plants, the action potentials induced by pollination propagate down the vascular tissue of the style (Fromm *et al.*, 1995).

Stimulation of one end of an excised segment of gourd stem by exposure to 1 M KCl initiated action potentials that then travelled along the whole length of the stem segment, a distance of about 15 cm, but if the vascular bundles were removed along the length of the segment, no propagation occurred (Mamulashvili *et al.*, 1973). However, if the vascular bundles were removed over a 1-cm length halfway along the segment, action potentials could be observed as normal at the far end of the segment from the site of stimulation. Here, too, it seems that electrical currents from action potentials arriving at one side of the 1-cm stripped zone could spread sufficiently well through the inexcitable tissues of that zone to excite action potentials in the vascular tissue on its far side. The same authors (Mamulashvili *et al.*, 1973) also found that excitation in one vascular bundle of sunflower hypocotyl caused excitation of neighbouring vascular bundles, so that action potentials could propagate along the hypocotyl even through staggered incisions meant that no one vascular bundle was continuous along the whole length.

Similar evidence for the role of vascular bundles in long-distance propagation of action potentials was obtained from experiments with pumpkin stem segments. Removal of the vascular bundles over a distance of 1 to 2 cm blocked propagation of action potentials, but removal of all other tissues except vascular bundles did not have this effect (Sinyukhin and Gorchakov, 1968).

Excitable cells of the vascular pathway

The vascular bundles that serve as the pathway for propagation of action potentials over distances of several centimetres or more consist of a number of different cell types. Elucidation of which cell types actually constitute the signalling pathway requires detailed investigation with intracellular microelectrodes. Such a study was carried out on *Mimosa pudica* by Samejima and Sibaoka (1983), who identified the phloem parenchyma as the only excitable tissue. Phloem sieve tubes were not identified as being excitable, but this may reflect the difficulty encountered in obtaining accurate measurements from these very narrow and highly turgid cells. In a later study on *Mimosa*, Fromm and Eschrich (1988) used the aphid stylet technique in an attempt to make recordings from phloem sieve tubes, and they concluded that the sieve tubes are excitable.

In our most recent studies of long-distance signalling in the systemic wound response in tomato (J.D. Rhodes *et al.*, unpublished results), we have used intracellular microelectrodes to investigate which cells in the petiole of an unwounded leaf are involved in the electrical events

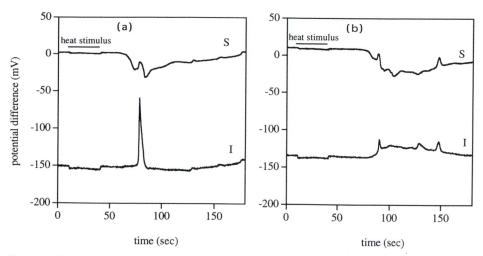

Figure 3. Electrical recordings from the petiole of leaf 1 of a tomato seedling subsequent to stimulation of one cotyledon by heat. In both parts of the figure, symbol I indicates the record from an intracellular microelectrode and S indicates the record from a surface electrode positioned about 2 cm from the microelectrode and basal to it. (a) Microelectrode inserted into a phloem sieve-tube element/companion cell complex. (b) Microelectrode inserted into a phloem parenchyma cell. The line labelled 'heat stimulus' indicates the period during which the cotyledon was being stimulated.

that are detected by surface electrodes on that petiole after another leaf has been wounded. The cell types investigated were epidermal cells, cortical cells outside the vascular tissue, phloem parenchyma cells, other parenchyma cells in the vascular tissue, and cells of the sieve-tube element/companion cell complex. Each cell studied was identified by injection of the fluorescent dye Lucifer Yellow CH into the cell from the microelectrode at the end of the recording period, but it was not possible to discriminate between sieve tube elements and their companion cells. The results showed a very clear and consistent pattern. When the electrical signal passed along the petiole, all the recordings from sieve tube elements/companion cells showed a large transient depolarization of the membrane potential (*Figure 3a*). Recordings from all the other cell types examined showed much smaller depolarizations of the membrane potential, with more variable shapes (*Figure 3b*). These results indicate that it is depolarization of the plasma membranes of the sieve tube elements/companion cells that is responsible for the propagating electrical events detected with surface electrodes. Furthermore, the magnitude, shape and duration of the transient depolarizations recorded from sieve tube elements/companion cells are very similar to those of known plant action potentials (e.g. *Figure 1*).

These results are clearly consistent with an electrical signalling system whereby action potentials propagate along the plasma membrane of the sieve-tube element/companion cell complex. The results could be consistent with a mechanism involving a chemical signal which is translocated in the phloem and causes local depolarizations of the plasma membranes of the sieve tube elements/companion cells. However, our earlier results have shown that, in our experimental system, phloem translocation is not necessary for the passage of the long-distance signals. Furthermore, in the experiments described here, the fluorescent dye injected into the sieve-tube element/companion cell complex was seen to be translocated basipetally (i.e. in the opposite direction to that of the propagation of the electrical signal). With regard to the possibility that the long-distance signal is a chemical one moving via reverse flow in the xylem (Malone, 1993) and causing depolarization of cells in the petiole, our results would require that the cells of the sieve-tube element/companion cell complex be much more sensitive to the chemical signal than the other cells. In addition, there should be a correlation between the timing of the depolarization of a cell and its distance from the xylem vessels, but we did not observe such an association.

Propagation of excitation in plant tissues

The phloem sieve tubes satisfy all the geometrical requirements described earlier for a long-distance electrical signalling pathway: they consist of a series of contiguous sieve tube elements whose plasma membranes are continuous through the modified plasmodesmata that form the sieve pores of the intervening sieve plates; there should be good electrical conductivity between successive sieve tube elements via the sieve pores; and the sieve-tube element/companion cell complex has very few plasmodesmatal connections to the surrounding cells (Van Bel and Kempers, 1990). This gives a pathway with a continuous membrane over long distances, low longitudinal resistance and minimal loss of excitation current. Thus it is not surprising that, as described above, propagation of electrical events over long distances within plants is associated with a continuous vascular pathway.

The propagation of electrical signals over shorter distances (c. 1 cm or less), as in *Dionaea*, *Drosera* or *Conocephalum*, does not involve a vascular pathway (Hodick and Sievers, 1988; Williams and Spanswick, 1976; Zawadzki and Trebacz, 1985). In those systems, where the pathway consists of cells that do not possess the features described above for sieve tubes, propagation of electrical signals probably occurs by a mechanism similar to that of epithelial conduction (Anderson, 1980) in animals. In plants, plasmodesmata would have the role played by gap junctions in animals. Directionality of propagation would be determined partly by the shapes of the cells involved (e.g. whether they are nearly isodiametric or whether they are long cylinders) and partly by the frequencies of plasmodesmata connecting each cell to its various neighbours. Over longer distances the gross architecture of the vascular system would be important in determining directionality.

Propagation of excitation in the whole plant

The architecture of the vascular system may also provide a system of controls on the pathway for the propagation of electrical signals in plants. For example, action potentials stimulated electrically in the stem of *Lupinus angustifolius* propagated along the stem but not into the stem apex, the root system or the petioles (Zawadzki, 1980). Similarly, electrical stimulation of a stem internode of *Luffa cylindrica* caused an action potential to propagate along that internode but not into adjacent internodes or petioles, and action potentials stimulated in petioles did not spread into adjacent stem internodes (Shiina and Tazawa, 1986). Stimulation of the roots of gourd plants with increased concentrations of KCl caused action potentials to propagate along the stem, but not in the roots themselves.

Electrical stimulation of action potentials was found to be much more difficult in young plants or in younger organs than in older stems of both *Lupinus* and *Luffa* (Shiina and Tazawa, 1986; Zawadzki, 1980). We have found a similar difference between tomato seedlings and older plants (unpublished data). In other unpublished experiments we have not been able to interfere with the wound-induced electrical signals by applying substances such as ion-channel blockers to the petiole, even with abrasion of the surface or infiltration of the tissue. Such protection of a signalling pathway is, perhaps, not surprising.

Explanations of these gross features of the pathway will probably involve symplastic domains (Erwee and Goodwin, 1985; Van der Schoot and Van Bel, 1989, 1990) and apoplastic barriers (Canny, 1990).

7. Initiation and transduction of electrical signals

Besides requiring a cellular pathway suitable for the transmission of propagating action potentials, an electrical signalling system also needs a mechanism for initiating action potentials at the site of stimulation, and mechanisms for signal transduction (i.e. conversion of the electrical signal into a biochemical or physiological response at the distal end of the pathway, or along it). While these aspects of electrical signalling have been widely studied and are well understood in many animal systems, our understanding of plant systems is meagre. However, our knowledge of the ion fluxes involved in plant action potentials and of signalling mechanisms in plant cells has reached a stage at which some reasonable ideas can be formulated.

Signal initiation

As we have described above (Section 5), the evidence obtained from characean cells indicates that the depolarizing phase of a plant action potential is caused by the opening of Cl^- channels in the plasma membrane so that Cl^- ions flow out of the cell. The Cl^- channels open in response to an increased Ca^{2+} concentration in the cytoplasm as a result of the opening, at least in the experimental systems where electrical stimulation is used, of voltage-gated Ca^{2+} channels in the plasma membrane. Furthermore, as we have already seen (Section 5), evidence consistent with this mechanism is available from studies on other plants.

According to the mechanism described above, the first step in initiation of a plant action potential would be the creation of a generator potential (Aidley, 1989) (i.e. a depolarization of the plasma membrane p.d. to a value sufficient to cause opening of the voltage-sensitive calcium channels). One mechanism by which the generator potential could be produced would involve inhibition of the outward H^+-pump of the plasma membrane, a pump that is responsible for a large part of the plasma membrane p.d. in actively metabolizing plant cells. The activity of the plasma membrane H^+-pump is sensitive to a variety of modulating factors (Michelet and Boutry, 1995; Serrano, 1989), and the results of Pyatigin *et al.* (1992) indicate that inhibition of the H^+-pump causes the generator potential that gives rise to the action potential produced by cooling of pumpkin stems. In this context, it is interesting that certain oligosaccharide elicitors depolarize the membrane potential of tomato leaf mesophyll cells, apparently by inhibiting the proton pump of the plasma membrane (Thain *et al.*, 1995).

However, the mechanism involving voltage-dependent Ca^{2+} channels and Ca^{2+}-dependent Cl^- channels is probably not the only one whereby plant action potentials can be initiated. A variety of stimuli, including elicitors of defence responses, plant hormones and mechanical stimulation, can cause increased cytosolic concentrations of Ca^{2+}, in some cases by direct effects on Ca^{2+} channels in the plasma membrane (Assmann, 1993), and in others by the release of Ca^{2+} from intracellular stores such as the endoplasmic reticulum and vacuole (Bush, 1993). Furthermore, several different types of anion channel have been found in the plant plasma membrane (Assmann, 1993), and the opening of these channels appears to be controlled not by the cytosolic Ca^{2+} concentration but by other factors such as the plasma membrane p.d., auxins and membrane stretching. Thus it is possible that a variety of different mechanisms may, in different systems or in response to different stimuli, bring about the opening of the Cl^- channels that is characteristic of the depolarizing phase of plant action potentials.

Signal transduction

Recent patch-clamp studies (Piñeros and Tester, 1995; Thuleau *et al.*, 1994) have confirmed the presence of voltage-dependent Ca^{2+} channels in the plasma membrane of higher plant cells. Furthermore, the Ca^{2+} ion is an important intracellular signal in plant cells, with regulatory roles in a variety of physiological and biochemical processes, including gene expression (De Silva and Mansfield, 1993/94; Poovaiah and Reddy, 1993). It is important to bear in mind that it is possible that very local changes in Ca^{2+} concentration in the cytoplasm are important in regulating other processes in the cell. In this context, Pickard and Ding (1993) have proposed an interesting model for a 'plasmalemmal control centre' consisting of mechanosensory Ca^{2+} channels located in the plasma membrane and intimately associated with other plasma membrane proteins (e.g. the proton pump, indole-3-acetic acid (IAA) porter, NADH oxidase, etc.) and with components of the cytoskeleton. Similar models could be constructed around voltage-dependent Ca^{2+} channels. Voltage-dependent Ca^{2+} channels are thus obvious candidates for involvement in the transduction of electrical signals (action potentials) into physiological and biochemical responses.

Transduction of electrical signals may also operate via other pathways. Davies (1987) argues that transduction may involve other events that follow directly from an action potential, such as the effluxes of Cl^- and K^+ ions and of water, and there is evidence that anion channels play important roles in signal transduction in plant cells (Ward et al., 1995).

Clearly, the processes we have described here as electrical signal initiation and electrical signal transduction are both examples of signal transduction in the wider sense. Both processes may involve similar mechanisms, such as control by voltage-dependent Ca^{2+} channels, but, equally, both processes may occur by a variety of mechanisms in different situations. There is no difficulty in constructing plausible hypotheses, but detailed experimental evidence is still in short supply.

8. Conclusions

It is now clear that a wide range of plant species possess cells that can generate action potentials in response to electrical and other stimuli, and that these action potentials can propagate through the plant's tissues. In some cases the evidence comes only from studies using surface-contact electrodes, but in others intracellular recording has also been used.

The evidence for long-distance electrical signalling in plants is still most convincing for those phenomena that involve the leaf movements of species such as *Dionaea muscipula, Drosera rotundifolia, Mimosa pudica* and *Aldrovanda vesiculosa*. With these systems, the use of a variety of techniques, especially intracellular recording of plasma membrane p.d. and electrical stimulation, have established that action potentials propagating through the plant's tissues from the site of stimulation do act as long-distance signals, although in some of these systems other signalling mechanisms may also operate. Furthermore, studies of these systems give us a clear picture of propagating action potentials as they appear to surface-contact electrodes, so that we can then more easily identify electrical events recorded with surface-contact electrodes in other plants.

Several other cases have been described where the evidence is at least consistent with long-distance electrical signalling leading to a variety of responses, including changes in rates of respiration and photosynthesis, changes in assimilate transport and phloem unloading, regulation of plant development and the initiation of proteinase inhibitor synthesis. In all of these cases the stimulus and response are linked by travelling electrical transients, and in some the evidence in favour of these transients being electrical signals is much stronger. In particular, electrical stimulation of distant responses (e.g. systemic induction of proteinase inhibitor synthesis in tomato) and electrical blocking (e.g. of the pollination signal in *Incarvillea*) are both highly indicative of electrical signalling.

References

Abe T, Oda K. (1976) Resting and action potentials in excitable cells in the main pulvinus of *Mimosa pudica*. *Plant Cell Physiol.* **17:** 1343–1346.

Aidley DJ. (1989) *The Physiology of Excitable Cells,* 3rd edn. Cambridge: Cambridge University Press.

Anderson PAV. (1980) Epithelial conduction: its properties and functions. *Prog. Neurobiol.* **15:** 161–203.

Assmann SM. (1993) Signal transduction in guard cells. *Annu. Rev. Cell Biol.* **9:** 345–375.

Beilby MJ. (1989) Electrophysiology of giant algal cells. *Meth. Enzymol.* **174:** 403–443.

Burdon-Sanderson J. (1873) Note on the electrical phenomena which accompany stimulation of the leaf of *Dionaea muscipula*. *Proc. R. Soc. Lond.* **21:** 495–496.

Bush DS. (1993) Regulation of cytosolic calcium in plants. *Plant Physiol.* **103:** 7–13.

Canny MJ. (1990) What becomes of the transpiration stream? *New Phytol.* **114:** 341–368.

Dainty J, Croghan PC, Fensom DS. (1963) Electro-osmosis, with some applications to plant physiology. *Can. J. Bot.* **41:** 953–966.

Davies E. (1987) Action potentials as multifunctional signals in plants: a unifying hypothesis to explain apparently disparate wound responses. *Plant Cell Environ.* **10:** 623–631.

De Silva DLR, Mansfield TA. (1993/94) Calcium in plants: control problems associated with a diversity of roles. *Sci. Prog.* **77:** 233–251.

Due G. (1993) Interpretation of the electrical potential on the surface of plant roots. *Plant Cell Environ.* **16:** 501–510.

Dziubinska H, Paszewski A, Trebacz K, Zawadzki T. (1983) Electrical activity of the liverwort *Conocephalum conicum*: the all-or-nothing law, strength–duration relation, refractory periods and intracellular potentials. *Physiol. Plant.* **57:** 279–284.

Dziubinska H, Trebacz K, Zawadzki T. (1989) The effect of excitation on the rate of respiration in the liverwort *Conocephalum conicum*. *Physiol. Plant.* **75:** 417–423.

Erwee MG, Goodwin PB. (1985) Symplast domains in extrastelar tissue of *Egeria densa* Planch. *Planta* **163:** 9–19.

Frachisse JM, Desbiez MO, Champagnat P, Thellier M. (1985) Transmission of a traumatic signal via a wave of electric depolarization, and induction of correlations between the cotyledonary buds in *Bidens pilosus*. *Physiol. Plant.* **64:** 48–52.

Frachisse JM, De Jaegher G, Desbiez MO. (1990) Possible role of the wave of electrical depolarization induced by wounding: basis for an intracellular study of the wave. In: *Intra- and Intercellular Communications in Plants* (Millet B, Greppin H, ed.). Paris: INRA, pp. 91–101.

Frachisse-Stoilskovic JM, Julien JL. (1993) The coupling between extra- and intracellular electric potentials in *Bidens pilosa* L. *Plant Cell Environ.* **16:** 633–641.

Fromm J. (1991) Control of phloem unloading by action potentials in *Mimosa. Physiol. Plant.* **83:** 529–533.

Fromm J, Eschrich W. (1988) Transport processes in stimulated and non-stimulated leaves of *Mimosa pudica*. II. Energesis and transmission of seismic stimulations. *Trees* **2:** 18–24.

Fromm J, Eschrich W. (1993) Electric signals released from roots of willow (*Salix viminalis* L.) change transpiration and photosynthesis. *J. Plant Physiol.* **141:** 673–680.

Fromm J, Spanswick R. (1993) Characteristics of action potentials in willow (*Salix viminalis* L). *J. Exp. Bot.* **44:** 1119–1125.

Fromm J, Bauer T. (1994) Action potentials in maize sieve tubes change phloem translocation. *J. Exp. Bot.* **45:** 463–469.

Fromm J, Hajirezaei M, Wilke I. (1995) The biochemical response of electrical signaling in the reproductive system of *Hibiscus* plants. *Plant Physiol.* **109:** 375–384.

Geddes LA. (1972) *Electrodes and the Measurement of Bioelectric Events*. New York: Wiley Interscience.

Graham LE. (1993) *Origin of Land Plants*. New York: John Wiley & Sons, Inc.

Gunar I, Sinyukhin AM. (1963) Functional significance of action currents affecting the gas exchange of higher plants. *Sov. Plant Physiol.* **10:** 219–226.

Herde O, Fuss H, Pena-Cortés H, Fisahn J. (1995) Proteinase inhibitor II gene expression induced by electrical stimulation and control of photosynthetic activity in tomato plants. *Plant Cell Physiol.* **36:** 737–742.

Hodick D, Sievers A. (1988) The action potential of *Dionaea muscipula* Ellis. *Planta* **174:** 8–18.

Hope AB, Walker NA. (1975) *The Physiology of Giant Algal Cells*. Cambridge: Cambridge University Press.

Houwink AL. (1935) The conduction of excitation in *Mimosa pudica*. *Rec. Trav. Bot. Néerl.* **32:** 51–91.

Houwink AL. (1938) The conduction of excitation in *Clematis zeylanica* and in *Mimosa pudica*. *Ann. Jard. Bot. Buitenzorg* **48:** 10–16.

Iijima T, Sibaoka T. (1981) Action potential in the trap-lobes of *Aldrovanda vesiculosa. Plant Cell Physiol.* **22:** 1595–1601.

Karmanov VG, Lyalin OO, Mamulashvili GG. (1972) Form of action potentials and co-operativeness of excited elements in winter squash stems. *Sov. Plant Physiol.* **19:** 354–359.

Kumon K, Suda S. (1984) Ionic fluxes from pulvinar cells during the rapid movement of *Mimosa pudica* L. *Plant Cell Physiol.* **25:** 975–979.

Lunevsky VZ, Zherelova OM, Vostrikov IY, Berestovsky GN. (1983) Excitation of Characeae cell membranes as a result of activation of calcium and chloride channels. *J. Memb. Biol.* **72:** 43–58.

McCourt RM. (1995) Green algal phylogeny. *Tree* **10:** 159–163.

Malone M. (1993) Hydraulic signals. *Phil. Trans. R. Soc. Lond. B* **341:** 33–39.

Malone M. (1994) Wound-induced hydraulic signals and stimulus transmission in *Mimosa pudica* L. *New Phytol.* **128:** 49–56.

Malone M, Alarcon J-J. (1995) Only xylem-borne factors can account for systemic wound signalling in the tomato plant. *Planta* **196:** 740–746.

Mamulashvili GG, Krasavina, MS, Lyalin OO. (1972) Comparative study of electrical activity of the root and stem of plants. *Sov. Plant Physiol.* **19:** 462–467.

Mamulashvili GG, Krasavina, MS, Lyalin OO. (1973) The role of different stem tissues in transmission of excitation. *Sov. Plant Physiol.* **20:** 365–371.

Michelet B, Boutry M. (1995) The plasma membrane H$^+$-ATPase. *Plant Physiol.* **108:** 1–6.

Narvaez-Vasquez J, Pearce G, Orozco-Cardenas ML, Franceschi VR, Ryan CA. (1995) Autoradiographic and biochemical evidence for the systemic translocation of systemin in tomato plants. *Planta* **195:** 593–600.

Oda K, Abe T. (1972) Action potentials and rapid movement in the main pulvinus of *Mimosa pudica. Bot. Mag. Tokyo* **85:** 135–145.

Opritov VA. (1978) Propagating excitation and assimilate transport in the phloem. *Sov. Plant Physiol.* **25:** 828–837.

Opritov VA, Pyatygin SS, Retivin VG. (1982) Emergence of action potentials in higher plants in response to insignificant local cooling. *Sov. Plant Physiol.* **29:** 260–266.

Paszewski A, Zawadzki T. (1973) Action potentials in *Lupinus angustifolius* L. shoots. *J. Exp. Bot.* **24:** 804–809.

Paszewski A, Zawadzki T. (1974) Action potentials in *Lupinus angustifolius* L. shoots. II. Determination of the strength–duration relation and the all-or-nothing law. *J. Exp. Bot.* **25:** 1097–1103.

Paszewski A, Zawadzki T. (1976) Action potentials in *Lupinus angustifolius* L. shoots. III. Determination of the refractory periods. *J. Exp. Bot.* **27:** 369–374.

Pearce G, Strydom D, Johnson S, Ryan CA. (1991) A polypeptide from tomato leaves induces wound-inducible proteinase inhibitor proteins. *Science* **253:** 895–898.

Pickard BG. (1973) Action potentials in higher plants. *Bot. Rev.* **39:** 172–201.

Pickard BG. (1974) Electrical signals in higher plants. *Naturwisseuschaften* **61:** 60–64.

Pickard BG, Ding JP. (1993) The mechanosensory calcium-selective ion channel: key component of a plasmalemmal control centre? *Aust. J. Plant Physiol.* **20:** 439–459.

Piñeros M, Tester M. (1995) Characterization of a voltage-dependent Ca^{2+}-selective channel from wheat roots. *Planta* **195:** 478–488.

Poovaiah BW, Reddy ASN. (1993) Calcium and signal transduction in plants. *Crit. Rev. Plant Sci.* **12:** 185–211.

Purves RD. (1981) *Microelectrode Methods for Intracellular Recording and Ionophoresis.* London: Academic Press.

Pyatigin SS, Opritov VA, Khudyakov VA. (1992) Subthreshold changes in excitable membranes of *Cucurbita pepo* L. stem cells during cooling-induced action-potential generation. *Planta* **186:** 161–165.

Robards AW, Pitts JD. (1991) Parallels in cell to cell communication in plants and animals. In: *Cell to Cell Signals in Plants and Animals* (Neuhoff V, Friend J, ed.). Berlin: Springer, pp. 63–81.

Roblin G. (1979) *Mimosa pudica*: a model for the study of excitability in plants. *Biol. Rev.* **54:** 135–153.

Samejima M, Sibaoka T. (1983) Identification of excitable cells in the petiole of *Mimosa pudica* by intracellular injection of procion yellow. *Plant Cell Physiol.* **24:** 33–39.

Serrano R. (1989) Structure and function of plasma membrane ATPase. *Annu. Rev. Plant Physiol. Plant Mol. Biol.* **40:** 61–94.

Shiina T, Tazawa M. (1986) Action potential in *Luffa cylindlica* and its effects on elongation growth. *Plant Cell Physiol.* **27:** 1081–1089.

Shimmen T, Mimura T, Kikuyama M, Tazawa M. (1994) Characean cells as a tool for studying electrophysiological characteristics of plant cells. *Cell Struct. Funct.* **19:** 263–278.

Sibaoka T. (1962) Excitable cells in *Mimosa. Science* **137**: 226.

Sibaoka T. (1966) Action potentials in plant organs. *Symp. Soc. Exp. Biol.* **20**: 49–73.

Sibaoka T. (1969) Physiology of rapid movements in higher plants. *Annu. Rev. Plant Physiol.* **20**: 165–184.

Sibaoka T. (1980) Action potentials and rapid plant movements. In: *Plant Growth Substances 1979* (Skoog F, ed.). Berlin: Springer, pp. 462–469.

Sinyukhin AM, Gorchakov VV. (1966) Characteristics of the action potentials of the conducting system of pumpkin stems evoked by various stimuli. *Sov. Plant Physiol.* **13**: 727–733.

Sinyukhin AM, Britikov EA. (1967) Action potentials in the reproductive system of plants. *Nature* **215**: 1278–1280.

Sinyukhin AM, Gorchakov VV. (1968) Role of the vascular bundles of the stem in long-distance transmission of stimulation by means of bioelectric impulses. *Sov. Plant Physiol.* **15**: 400–407.

Stahlberg R, Cosgrove DJ. (1992) Rapid alterations in growth rate and electrical potentials upon stem excision in pea seedlings. *Planta* **187**: 523–531.

Stankovic B, Davies E. (1995) Direct electrical induction of gene expression in tomato plants. *J. Cell Biochem.* (Suppl. 21A): 503.

Thain JF. (1995) Electrophysiology. In: *Methods in Cell Biology, Vol. 49. Methods in Plant Cell Biology, Part A* (Gallbraith DW, Bohnert HJ, Bourque DP, ed.). San Diego: Academic Press Inc., pp. 259–274.

Thain JF, Wildon DC. (1992) Electrical signalling in plants. *Sci. Prog.* **76**: 553–564.

Thain JF, Gubb IR, Wildon DC. (1995) Depolarization of tomato leaf mesophyll cells by oligogalacturonide elicitors. *Plant Cell Environ.* **18**: 211–214.

Thuleau P, Ward JM, Ranjeva R, Schroeder JI. (1994) Voltage-dependent calcium-permeable channels in the plasma-membrane of a higher-plant cell. *EMBO J.* **13**: 2970–2975.

Trebacz K, Simonis W, Schönknecht G. (1994) Cytoplasmic Ca^{2+}, K^+, Cl^- and NO_3^- activities in the liverwort *Conocephalum conicum* L. at rest and during action potentials. *Plant Physiol.* **106**: 1073–1084.

Van Bel AJE, Kempers R. (1990) Symplastic isolation of the sieve element–companion cell complex in the phloem of *Ricinus communis* and *Salix alba* stems. *Planta* **183**: 69–76.

Van der Schoot C, Van Bel AJE. (1989) Glass microelectrode measurements of sieve tube membrane potentials in internode discs and petiole strips of tomato (*Solanum lycopersicum* L.). *Protoplasma* **149**: 144–154.

Van der Schoot C, Van Bel AJE. (1990) Mapping membrane potential differences and dye-coupling in internodal tissues of tomato (*Solanum lycopersicum* L.). *Planta* **182**: 9–21.

Volkov AG, Haack RA. (1995) Bioelectrochemical signals in potato plants. *Russian J. Plant Physiol.* **42**: 17–23.

Ward JM, Pei Z-M, Schroeder JI. (1995) Roles of ion channels in initiation of signal transduction in higher plants. *Plant Cell* **7**: 833–844.

Wayne R. (1994) The excitability of plant cells: with a special emphasis on characean internodal cells. *Bot. Rev.* **60**: 265–367.

Wildon DC, Thain JF, Minchin PEH, Gubb IR, Reilly AJ, Skipper YD, Doherty HM, O'Donnell PJ, Bowles DJ. (1992) Electrical signalling and systemic proteinase inhibitor induction in the wounded plant. *Nature* **360**: 62–65.

Williams SE, Pickard BG. (1972a) Receptor potentials and action potentials in *Drosera* tentacles. *Planta* **103**: 193–221.

Williams SE, Pickard BG. (1972b) Properties of action potentials in *Drosera* tentacles. *Planta* **103**: 222–240.

Williams SE, Spanswick RM. (1976) Propagation of the neuroid action potential of the carnivorous plant *Drosera. J. Comp. Physiol.* **108**: 211–223.

Williams SE, Pickard BG. (1980) The role of action potentials in the control of capture movements of *Drosera* and *Dionaea*. In: *Plant Growth Substances 1979* (Skoog F, ed.). Berlin: Springer, pp. 470–480.

Zawadzki T. (1980) Action potentials in *Lupinus angustifolius* L. shoots. V. Spread of excitation in the stem, leaves and root. *J. Exp. Bot.* **31**: 1371–1377.

Zawadzki T, Trebacz K. (1985) Extra- and intracellular measurements of action potentials in the liverwort *Conocephalum conicum. Physiol. Plant.* **64**: 477–481.

Zawadzki T, Davies E, Dziubinska H, Trebacz K. (1991) Characteristics of action potentials in *Helianthus annuus. Physiol. Plant.* **83**: 601–604.

Chapter 19

Sugar transport in higher plants

Wolf B. Frommer, Brigitte Hirner, Christina Kühn, Karsten Harms, Thomas Martin, Jörg W. Riesmeier and Burkhard Schulz

1. Introduction

The myriad reactions that proceed in parallel in an organism are only possible due to the compartmentation created by intracellular and surrounding membranes. Membranes consist of two basic types of molecule: lipids which prevent free exchange and proteins which control the exchange of molecules between compartments and ensure the exchange of information. Carbohydrates synthesized during photosynthesis must cross several membranes on their way from the site of production to the site of consumption. Production and consumption must be coordinated. The primary fixation of CO_2 takes place in the chloroplasts, from which export to the cytoplasm can occur via two possible routes, during the day preferably via the triose-phosphate translocator (TPT) and during the night via glucose transporters (GT). Sucrose, the major transport molecule, is formed in the cytoplasm of the green parts of the plant and provides both carbon skeletons and energy for the non-photosynthetic organs and cells. During the past few years molecular tools have enabled significant progress in the study of plant biochemical and physiological aspects of assimilate partitioning (Frommer and Sonnewald, 1995). In this review, we shall focus on this progress and the still unresolved questions in the area of sugar transport, especially with regard to current understanding of phloem loading.

2. Plastidic sugar transport

The primary products of photosynthesis are triose phosphates (triose-P), representing the net products of the Calvin cycle. Triose-P can either be shuttled into various biosynthetic pathways within the chloroplast, such as starch formation or lipid and amino acid biosynthesis, or it can be exported. In the cytosol it is diverted into sucrose and amino acids and a variety of other pathways. To enter the cytosol, metabolites must cross the two membranes that enclose the plastidic lumen: (1) the inner membrane that controls which substances can enter or leave the plastid, and (2) the outer membrane which is permeable, with an exclusion limit of 10 kDa (Flügge and Benz, 1984).

The main route for the export of assimilates is assumed to be via the TPT, an integral membrane protein located at the inner membrane of the chloroplast envelope that catalyses the export of triose-P from the chloroplast. The TPT represents the major protein of the inner envelope, accounting for up to 15% of the total protein content of this membrane. It counter-exchanges triose-P, 3-phosphoglycerate (3-PGA) and inorganic phosphate (P_i) (for a recent review see Heldt and Flügge, 1992). During the metabolism of triose-P in the cytoplasm, phosphate is released and made available as the counter-ion for triose-P export. More importantly, the TPT serves as a means of communication between chloroplast and cytosol, enabling adjustment of the rate of photosynthesis to the demands of various parts of the plant for photoassimilates. When the export of triose-P from the chloroplasts is limited by shortage of P_i, (e.g. due to decreased sucrose synthesis), fixed carbohydrates can be deposited in the chloroplasts in the form of transitory starch (Stitt, 1989). The TPT gene is nuclear-encoded and respective cDNAs have been cloned from a number of species, including spinach, pea, potato and tobacco (Flügge et al., 1989; Schulz et al., 1993; Willey et al., 1991; and Schulz and Frommer, unpublished results). Structural analyses have predicted that the protein consists of six membrane-spanning domains, with the N- and C-termini pointing to the intermembrane space of the chloroplast (Wallmeier et al., 1992). Expression of the TPT gene was observed predominantly in green tissues, indicating that a second unrelated protein is involved in the import of carbohydrates into the amyloplasts in sink organs (Schulz et al., 1993).

In order to assess the in-vivo function of the translocator, potato plants were transformed with an antisense construct consisting of the TPT cDNA under control of the constitutive CaMV 35S promoter (Riesmeier et al., 1993a). Several transformants showed a reduction in the amounts of both TPT mRNA and protein. The decrease in TPT protein is consistent with the 20–30% reduction in phosphate transport activity. Even a 20% reduction in transport activity was sufficient to have marked effects, such as a dwarf phenotype and altered metabolite and starch levels in leaves. Maximal rates of photosynthesis were reduced by up to 60%. The retarded development of the antisense plants is clearly dependent on the growth conditions (Kühn and Frommer, unpublished results). In the stroma, the level of 3-PGA was increased fourfold, and there was a corresponding decrease in P_i, whereas cytosolic triose-P and glucose-6-phosphate were reduced (Heineke et al., 1994). Owing to the allosteric activation of ADP-glucose pyrophosphorylase by phosphoglycerate, and its inhibition by P_i, the elevated stromal ratio of 3-PGA/P_i can explain the increase in starch formation. As a consequence of the reduced TPT activity, less triose-P is available in the cytosol for the biosynthetic pathways (e.g. for the synthesis of sucrose and amino acids). In wild-type plants, 43% of the assimilate, compared to a much larger proportion (61–89%) in antisense plants, is partitioned into starch. At night, normal potato plants translocate approximately 75% of the assimilate exported from leaves during the daytime. In antisense plants, the ratio translocated at night compared to the day is much higher (Heineke et al., 1994). These mechanisms appear to compensate for the reduced availability of carbohydrates, and probably prevent stronger phenotypic effects. The changes are very similar to the responses of detached leaves to phosphate limitation (i.e. a reduction in photosynthesis, an increase in the partitioning of photoassimilates towards starch, and a decrease in phosphorylated intermediates).

Taken together, these data indicate that the activity of the TPT may become rate-limiting for maximal photosynthesis. This might be due to the fact that TPT activity is largely occupied by synonymous exchanges. It would thus be interesting to test the effect of TPT overexpression on the flow of carbon. However, as the contribution of the TPT to the total protein content of the inner envelope is already very large, a further increase might be difficult to achieve. It is well known that the overexpression of integral membrane proteins, at least in bacteria, is problematic for technical reasons.

Several other sugar transport activities have been identified in plastids. There is evidence for a glucose transporter which represents the main route of carbohydrate export during the night (Schäfer et al., 1977). This carrier may also be responsible for the compensatory effects observed in the TPT antisense plants. The role of such a transporter is further supported by the finding that, despite the absence of chloroplastidic fructose-1,6-bisphosphatase (FBPase) activity during the night, chloroplasts can synthesize starch when leaves are supplied with sucrose through the petiole (Koßmann et al., 1992). A mutant with altered regulation of starch degradation might be a potential candidate for a mutation in this protein (Caspar et al., 1991). A model system for studying the role of the TPT and the glucose carrier could be provided by cyanelles (Schlichting and Bothe, 1993). Such a transporter might be a more effective tool for increasing the flow of carbon from chloroplasts by overexpression. Modification of the expression in either of these transporters might represent a means whereby plants can adapt to specific environments, such as differences in day length.

Another important area with regard to carbohydrate partitioning is the import into plastids in sink tissues. Sugars or sugar phosphates must be imported into the amyloplasts. Expression studies have indicated that the TPT gene is not or is only marginally expressed in non-green sink organs. Moreover, under conditions of reduced stringency, no related gene could be detected (Schulz et al., 1993). Uptake studies with amyloplasts have shown that a translocator with a different specificity (i.e. glucose 6-phosphate) is responsible for the uptake of precursors for starch biosynthesis (Borchert et al., 1989). As starch biosynthesis also requires energy, adenylates must be imported as well, possibly via a protein similar to the mitochondrial ATP/ADP exchanger. A mutant that is potentially affected at this step might be a useful tool for manipulating this protein (Sullivan et al., 1992).

These findings demonstrate that new insights into the role of transporters and the overall function and regulation of carbon metabolism in plastids will be gained from experiments that involve developing mutants with regard to specific transport functions.

3. Plasma membrane sugar transport

Many organisms are unicellular and exchange solutes across the plasma membrane via carrier-mediated processes. In higher organisms, the functional differentiation of the cells has made the development of specific transport vessels necessary. Especially in higher land plants, remarkable distances must be travelled in order to allow the exchange of metabolites between different cells at the distal ends of the plant (e.g. leaves and roots or flowers). For the long-distance transport (of sugars, amino acids and ions) from source (net exporting tissues) to sink (net importing tissues), plants have developed a veinal network, namely the phloem. The most abundant compound transported in the phloem of most species is sucrose, whereas hexose concentrations are very low (Zimmermann and Ziegler, 1975). A central question therefore is how such metabolites enter the long-distance distribution pathways at the sites of synthesis, and how they are unloaded at the sites where they are needed. Two main routes can be envisaged: (1) carrier-mediated transport across the plasma membrane and diffusion through the cell wall (apoplastic route); and (2) direct diffusional cell-to-cell transport via the plasmodesmata (symplastic route).

Symplastic transport

Apart from the possibility of transporting nutrients across the plasma membrane by carrier-mediated mechanisms, higher organisms have developed a second system that allows direct transfer from cell to cell via cytoplasmic connections such as gap junctions or plasmodesmata (Robards and Lucas, 1990). Fluorescent dyes with a molecular mass of up to 800 Da can move freely between mesophyll cells. Transport via the plasmodesmata is especially important between the companion cells and the sieve tube elements because the latter have lost their nuclei and must therefore be supplied from the neighbouring companion cells. Depending on the species, however, the continuity between the mesophyll and the sieve element/companion cell complex varies. In some species, classified as potential symplastic loaders, the density of plasmodesmata at the interface between the mesophyll/vascular parenchyma and the companion cells is high, whereas, in species with only a few plasmodesmata in this region, apoplastic transport is required. This hypothesis has stimulated an analysis of numerous species of plant with regard to the density of plasmodesmata, as well as the classification into different types of plant according to the density of plasmodesmata at this interface (Gamalei, 1986; Van Bel, 1993).

Several lines of evidence have been used as criteria for demonstrating symplastic routes of phloem loading in certain species (Van Bel *et al.*, 1994).

(1) Anatomical analyses using the electron microscope have revealed the structure of plasmodesmata. The permeability of plasmodesmata to fluorescent dyes has revealed that molecules below a certain molecular weight threshold can move from cell to cell. However, these studies have been criticized (e.g. Delrot, 1989).

(2) Carrier-mediated sucrose transport appears to be sensitive to thiol-group-modifying agents. Two methods were used to determine the effects of *p*-chloromercuri benzyl sulphonic acid (PCMBS) on phloem loading. The efflux of sugars from the petiole of detached leaves can be measured, and differences in the sensitivity of exudation to PCMBS correlate with the classification based on anatomical studies, indicating that symplastic loaders are not affected by PCMBS.

(3) Microautoradiographs of leaves were analysed in order to determine the accumulation of sugars in the veins after labelling with $^{14}CO_2$ (e.g. Fritz *et al.*, 1983). Again, the lack of sensitivity to PCMBS was regarded as an indication of symplastic loading.

All these methods are indirect, and their interpretation may be influenced by the argument that pits are simple holes through which metabolites can diffuse. However, recent studies have elegantly demonstrated that plasmodesmata are complex structures, the 'open' status of which can be regulated not only by external factors but possibly also by endogenous factors (Lucas *et al.*, 1994). Furthermore, it is difficult to explain symplastic transport from a thermodymanic viewpoint as transport must occur against a concentration gradient. Nevertheless, active transport through plasmodesmata is conceivable. An interesting, mechanistic model for active loading through plasmodesmata was developed by Turgeon and Gowan (1990). The synthesis of raffinose takes place mainly in intermediary or companion cells (Beebe and Turgeon, 1992). The concept of active transport through plasmodesmata is based on the size exclusion limit, which is sufficient for the passage of sucrose but insufficient to allow a backflow of raffinose. It is quite possible that both pathways are present within a single species (for review see Van Bel, 1993). An uncontrolled flow may not be desirable under all conditions. A useful assumption could therefore be that under a variety of conditions the plant transports small molecules in a non-selective manner through the plasmodesmata. However, under certain conditions the plasmodesmata might be closed, and then only those metabolites for which the appropriate carriers are present could be transported. The regulation of the permeability of plasmodesmata by changes in turgor pressure supports this hypothesis (Oparka and Prior, 1992). The use of viral movement proteins has proved to be an invaluable tool for studying plasmodesmal transport (Citovsky, 1993; Lucas *et al.*, 1993). It has been found that overexpression of viral movement proteins has significant effects on carbohydrate partitioning in tobacco (Lucas *et al.*, 1993).

Apoplastic transport

At least in species with low levels of symplastic continuity, apoplastic phloem loading seems highly probable. As mentioned previously, a pair of specific carrier proteins, one of which is responsible for export into the cell wall and the other responsible for import into the veins, must be involved (see also *Figure 1*). The hypothesis of apoplastic transport is supported by an analysis of plants that overexpress invertase in the cell wall (Dickinson *et al.*, 1991; Heineke *et al.*, 1992; von Schaewen *et al.*, 1990). In the case of symplastic transport, sucrose should be present at high concentrations in the cell wall compartment. Metabolite analysis showed that the cell wall contained only marginal concentrations of sucrose (less than 5 mM sucrose in contrast to around 10 to 20-fold higher concentrations in the cytosol of the mesophyll and a 100 to 200-fold higher concentration in the phloem sap) (Delrot, 1989; Lohaus *et al.*, 1994; Riens *et al.*, 1991). If symplastic transport occurs, the invertase in the apoplast should not have substantial effects. However, if the major route of sucrose transport is carrier-mediated and thus sucrose must cross the apoplastic compartment, the invertase should dramatically affect assimilate partitioning, because hexoses do not appear to be translocated efficiently into the phloem. As strong phenotypic effects such as leaf curling and local bleaching, reduced root growth and tuber yield, and strong physiological effects such as the accumulation of soluble sugars, amino acids and starch occur, we can conclude that at least in the cases examined so far, apoplastic and therefore carrier-mediated loading of the phloem appears to be predominant.

Sucrose transporters. Transport activities have been studied in recent decades in leaf tissues, protoplasts and, most recently of all, since the development of efficient methods for purification, in plasma membrane vesicles (Larsson, 1985). Transport activities were mainly detected in leaves, although some activity was also found in sink tissues (Delrot, 1989). Normally the uptake consists of several kinetic components (e.g. in *Vicia,* two saturable (low and high affinity) and one linear component were identified; Delrot and Bonnemain, 1981). The best studied systems are proton symporters (Buckhout, 1989; Bush, 1989, 1993; Delrot, 1981; Giaquinta, 1977, 1983; Komor *et al.*, 1977; Lemoine and Delrot, 1989; Williams *et al.*, 1992). The transport is active and has been described as a sucrose/proton co-transport with a 1:1 stoichiometry (Bush, 1990). The substrate specificity is restricted to sucrose and phenylglucosides (Delrot, 1989; Fondy and Geiger, 1977; Hitz *et al.*, 1986; Lucas and Madore, 1988), and the activity is sensitive to thiol-group-modifying agents and to diethylpyrocarbonate. Comparison of the transport activity in developing and mature leaves has shown that the sucrose/proton cotransport is differentially active and develops during maturation of the leaf (Lemoine *et al.*, 1992). In sugar beet, a protein with an apparent molecular mass of 42 kDa was identified as a potential candidate for the sucrose transporter, and it was partially purified (Li *et al.*, 1992). An antiserum directed against the 42-kDa fraction from the plasma membrane was able specifically to block sucrose transport in vesicles (Gallet *et al.*, 1992). Attempts to isolate the carrier gene by screening cDNA expression libraries with this antiserum have failed (Frommer and Delrot, unpublished results). In plants, molecular studies of metabolite transport across the plasma membrane have been neglected for many years because of the problems associated with the identification and purification of the respective proteins.

Sucrose transporter genes. Complementation of yeast mutants has proved to be an effective method for isolating K⁺-channel and amino acid permease genes from *Arabidopsis* (Anderson *et al.*, 1992; Frommer *et al.*, 1993; Sentenac *et al.*, 1992). At first sight, yeast complementation does not appear to be suitable for isolating sucrose transporters owing to the ability of budding yeast to metabolize sucrose extracellularly. However, a modified strain deficient in secreted invertase, but able to metabolize ingested sucrose due to the expression of a sucrose-cleaving activity, was used as an artificial complementation system to isolate a sucrose transporter cDNA from spinach and potato (Riesmeier *et al.*, 1992, 1993b). Owing to the high levels of expression in leaves and the high degree of conservation at the DNA level, heterologous screening is a rather simple tool for the isolation of respective genes from other species (e.g. tobacco, tomato, *Arabidopsis* and *Plantago)* (Bürkle *et al.*, 1996; Gahrtz *et al.*, 1994; Sauer and Stolz, 1994; Kühn, Bürkle and Frommer, unpublished results). The yeast expression system has enabled analysis of the biochemical properties of the transporters by direct measurement of the uptake of radiolabelled sucrose under different conditions in the presence of competitors and inhibitors. The two proteins from spinach and potato are very similar with regard to the inhibition of transport by protono-phores, thiol-group-modifying agents and diethylpyrocarbonate (DEPC). The K_m value of the sucrose carriers was estimated to be between 0.3 and 1 mM, and the specificity towards other sugars is in close agreement with the data for the sucrose carrier as determined in plants (for reviews see Bush, 1993; Delrot, 1989).

The transporter genes from spinach (*SoSUT1*) and potato (*StSUT1*) encode highly hydrophobic proteins that belong to the class of metabolite transporters which consist of two sets of six membrane-spanning regions separated by a large cytoplasmic loop. At the protein level the two proteins are 68% identical. The areas with the highest variability are the N- and C-terminal extensions and the large central loop, whereas the putative membrane-spanning regions are highly conserved. Although no sequence homologies were found to the prototype of sugar-transporting proteins, namely the lactose permease from *E. coli*, the transporters appear to be related not only in being disaccharide transporters but also in their structure, with 12 membrane-spanning regions separated by a large central loop. Interestingly, a conserved motif, [RK]X_{2-3}[RK], is located in the second and eighth loop of the sucrose transporters, and is one of the characteristic features of the major facilitator superfamily (MFS) of transporters (Marger and Saier, 1993). Mutational studies of the sucrose transporter in conjunction with the yeast expression system represent efficient tools for analysing which regions of the proteins are important for function and which amino acids are involved in substrate recognition. Active sucrose transport activity in the leaves of sugar beet develops upon maturation of the leaves (Lemoine *et al.*, 1992). Two-dimensional gel electrophoresis of plasma membranes from sink and source leaves of sugar beet has allowed us to identify a number of polypeptides that are specifically expressed in source leaves of sugar beet, and which are within the size and pI range predicted for the sucrose transporters from spinach and potato (Frommer *et al.*, 1994). As might have been expected, the expression profile of the sucrose transporter mRNA from potato follows the sink-to-source transition (Riesmeier *et al.*, 1993b). RNA *in-situ* hybridization experiments localize the expression of the carrier in the phloem, suggesting a role for StSUT1 in phloem loading. *StSUT1* RNA is also found in stems, although at lower levels than in the leaves. These data are supported by an analysis of SUT promoters from tomato and *Arabidopsis* (Truernit and Sauer, 1995; Hirner and Frommer, unpublished results). This is interpreted as an argument against a single role for StSUT1 in retrieval along the translocation pathway. Low levels of expression were also found in sink tissues, where the carrier might play a role in unloading processes. Further experiments will be necessary in order to define in more detail the expression profile of the carrier at the cellular level (e.g. by immunolocalization studies).

Using photoaffinity labelling techniques, a sucrose binding protein (SBP) has been isolated from soybean cotyledons (Grimes *et al.* 1992; Ripp *et al.*, 1988; Warmbrodt *et al.*, 1989). As the protein is localized in the phloem, the authors have speculated that it might be involved in sucrose transport. At present it is unclear whether this protein constitutes a regulatory subunit of the sucrose transporter, or whether it serves other functions. Heteromeric structures have been reported for mammalian amino acid transporters (Bertran *et al.*, 1992). Further analysis (e.g. by coexpression in yeast or in oocytes) might allow this hypothesis to be examined in more detail. Interestingly, the SBP is able to mediate sucrose transport on its own in the yeast system described above (Overvoorde *et al.*, 1996). The biochemical properties indicate that *SBP* may encode the linear component of sucrose uptake observed in many transport studies.

Transgenic plants with reduced sucrose transport activity. The best method for studying the actual function of a protein appears to be the analysis of mutants. In order to generate mutants with reduced sucrose transport activity, potato plants were transformed with the *StSUT1* gene in antisense orientation under the control of the constitutive CaMV 35S promoter (Riesmeier *et al.*,

1994). If sucrose transport mediated by this transporter is essential for phloem loading, a reduction in transport activity should affect carbon partitioning and photosynthesis. As expected, the antisense plants display dramatic phenotypic effects. The plants show retarded growth, crinkled leaves, local bleaching and the accumulation of anthocyanins. Development of the phenotype is dependent on light period and intensity (Kühn, Riesmeier and Frommer, unpublished results). An analysis of metabolites shows a five- to tenfold increase in leaf sucrose and starch content, and an even greater increase in hexoses. A similar accumulation of soluble carbohydrates was observed when petioles of potato leaves were cold-girdled, a treatment which is thought to block phloem translocation (Krapp *et al.*, 1993). Enhanced partitioning into insoluble carbohydrates was also found in a number of studies where attempts were made to block the export by heat girdling (Grusak *et al.*, 1990, and references therein). Efflux measurements on excised leaves from the antisense plants indicate a substantial reduction in phloem transport. The reduced efflux strongly influences the supply of sucrose to sink organs, as the plants have a reduced root system and decreased tuber yield. The similarities to the phenotype of transgenic potato plants overexpressing a yeast invertase in the cell wall of leaves are striking (Heineke *et al.*, 1992; Riesmeier *et al.*, 1994). Comparable effects were observed in potato plants in which *SUT1* was expressed in antisense orientation under the control of the companion cell-specific rolC promoter (Kühn and Frommer, unpublished results). However, it was not possible to estimate the control coefficient of SUT1, since low levels of SUT expression in mesophyll cells masked the actual amount of SUT1 protein and sucrose transport activity in the phloem (Lemoine *et al.*, unpublished results). Due to the large biomass transferred to potato tubers, loading in *Solanum tuberosum* might represent a special case. The effect of an antisense inhibition was therefore analysed in tobacco, in which the ratio of dry matter in seeds to that in leaves is at least 10-fold lower than the ratio of dry matter in tubers and leaves in potato. Furthermore, in tobacco (minor vein configuration type IIa) the inhibition of SUT1 led to dramatic growth retardation and the accumulation of carbohydrates in the leaves (Bürkle *et al.*, unpublished results). Thus carrier-mediated phloem loading appears to be a more general property, at least in solanaceous species.

The observed effects are thus in agreement with the expected results if the transporter is essential for phloem loading. It seems that no alternative pathway that can compensate is present in potato. The increase in carbohydrates in the leaves is surprising, as a reduced efflux from the leaves should lead to feedback inhibition of photosynthesis via fructose-2,6-bisphosphate (Stitt, 1990). We interpret these results as a sign either that the regulation capacity of the pathway has been overridden or that a signal from the sink maintains carbohydrate production in the leaves. Possibly the high levels of soluble sugars in the mesophyll allow at least a small amount of carbohydrate to enter the phloem by diffusion. Analysis of the level of reduction and determination of whether only one of the kinetic components of sucrose transport (see above) is affected will be important future objectives. Due to possible changes in the osmotic potential, one might also expect that amino acid transport and overall translocation rates would be reduced. However, it should be borne in mind that modern potato varieties are plants selected for high yield and therefore probably for high translocation rates. Tuber dry weight as a proportion of plant dry weight has increased from 7 to 81% (Inoue and Tanaka, 1978). A significant role in this process was ascribed to the shift from short- to long-day plants. It would therefore be important, for the purposes of generalization, to study sucrose transport by similar methods in wild-type species (e.g. potato varieties that do not produce tubers).

Other proteins involved in phloem loading

The phloem consists of a number of different cell types. The actual transport vessels are sieve tubes, which do not contain nuclei and which are supplied with nutrients by the neighbouring companion cells. In some species the companion cells are characterized by an increased membrane surface area (transfer cells; Wimmers and Turgeon, 1991). The sieve element/companion (transfer) cell complex is in close contact with parenchymatous cells, and the whole complex is surrounded by a bundle sheath. Potato veins down to the sixth order are bicollateral and consist of the major abaxial phloem and the minor adaxial phloem strands (McCauley and Evert, 1989). Active loading is assumed to take place at the border between the mesophyll and the phloem. Companion cells contain a comparatively large number of mitochondria, indicating that they might have a role in supplying energy for loading and transport. This view is supported by the findings that plasma membrane H^+-ATPases are specifically expressed in the phloem, and that antibodies detect the highest levels of ATPase protein in the phloem (Parets-Soler et al., 1990). The high concentration of sucrose in the phloem renders very attractive the hypothesis that a small proportion of sucrose serves as substrate to supply the ATP for the H^+-ATPase responsible for generating the proton gradient across the plasma membrane. The only sucrose-cleaving activity detected in the phloem is sucrose synthase (Geigenberger et al., 1993). If this enzyme is actually involved in phloem loading, it might be induced by blocking of ATP synthesis (e.g. by anaerobiosis). Using a sucrose synthase promoter–GUS fusion, it was demonstrated that, under ATP-limiting conditions (i.e. during anaerobiosis), the gene upregulated specifically in the phloem. Furthermore, under cold conditions, in which translocation rates and several glycolytic enzymes show a reduction in activity, the sucrose synthase gene is upregulated in the phloem (Martin et al., 1993). This may mean that other proteins also involved in energy transduction could have regulatory functions.

A number of glucose transporter genes with differing expression patterns have been identified. These proteins do not appear to play a role in phloem loading, but might be involved in a number of other processes, including phloem unloading (Sauer and Tanner, 1993).

Amino acid transporters. A major part of nitrogen fixation into amino acids takes place in the leaves, using the energy provided by photosynthesis. Therefore leaves are also a source of amino acids that can be exported to the sink tissues. By analogy with the description of sucrose transport in the previous section, it is assumed that amino acids enter the phloem by carrier-mediated processes. At least one basic, one acidic and two neutral proton/amino-acid transport systems have been described in the leaves of higher plants (Li and Bush, 1990). In order to study the role of membrane transporters in this distribution process, the respective genes were isolated using complementation of yeast mutants (Frommer et al., 1993; Hsu et al., 1993; Kwart et al., 1993; Rentsch et al., 1995). All 10 transporters isolated to date from *Arabidopsis* are differentially expressed and appear to serve different functions (Fischer et al., 1995; Frommer et al., 1995; Rentsch and Frommer, unpublished results). On the basis of this complexity one might speculate that plants also contain further sucrose carriers serving different functions in loading (e.g. in pollen) or unloading processes.

Plasma membrane H^+-ATPase. Proton ATPase genes have been isolated from a variety of species (Boutry et al., 1989; Ewing et al., 1990; Pardo and Serrano, 1989; Perez et al., 1992; Sussman and Harper, 1989). Both immunolocalization and the analysis of GUS fusions have shown that they are mainly localized in the phloem (DeWitt et al., 1991; Parets-Soler et al., 1990). Thus a major contribution to the energy requirements for phloem loading must probably be made by the phloem. In order to study the role of ATPase genes in this process in parallel to the studies on sucrose transport in potato, cDNAs encoding H^+-ATPase genes were isolated from a cDNA library of mature potato leaves by heterologous screening (Harms et al., 1994). The cDNAs are full length and encode proteins with an estimated molecular mass of 105 kDa. *PHA1* was mapped to two loci on chromosomes III and VI, and therefore is in a similar location to the tomato genes *LHA1* and *LHA2*, and is related to the tobacco gene *PMA1-3* (Boutry *et al.*, 1989; Ewing *et al.*, 1990; Gebhardt *et al.*, 1993). The second potato gene, *PHA2,* is more closely related to the tobacco gene *PMA4* and to the *Arabidopsis* genes that map to chromosome VII (Moriau *et al.*, 1993). Both genes are highly expressed in leaves, and this is also the case in the sucrose transporter antisense plants (Harms *et al.*, 1994). Attempts to reduce the ATPase by antisense repression were unsuccessful (Harms, Wöhner and Frommer, unpublished results).

K^+-*transport.* In order to compensate the H^+-flux and regulate the membrane potential, it has been proposed that K^+-transporters also play an important role in phloem loading (Cho and Komor, 1980). This view is supported by the finding that exogenous application of potassium strongly affects phloem loading (Doman and Geiger, 1979). Several genes encoding potassium channels have been identified, again by yeast complementation (Anderson *et al.*, 1992; Sentenac *et al.*, 1992). Further analysis of their expression pattern and the analysis of transgenic plants will be necessary in order to obtain a better understanding of K^+ transport.

Regulation of phloem loading

In principle, phloem loading can be regulated at three different levels: (1) the transporters; (2) the H^+-ATPase; or (3) the metabolic pathways associated with phloem loading. Various physical and chemical treatments have been used to inhibit phloem loading. Typical examples include anaerobiosis, which probably acts at the level of the ATP supply to the ATPase, and uncouplers which directly inhibit the energization (Giaquinta, 1977; Maynard and Lucas, 1982; Servaites *et al.*, 1979; Sowonick *et al.*, 1974; Thorpe *et al.*, 1979).

Phytohormones have multiple effects on plant growth and development. One possible mechanism is an action exerted through stimulation of transport. For a long time phytohormones such as auxins and cytokinins have been known to increase phloem loading (Lepp and Peel, 1970; Patrick, 1976). Fusicoccin or auxins can rapidly promote sucrose uptake, while abscisic acid acts as an inhibitor (Malek and Baker, 1978; Sturgis and Rubery, 1982; Vreugdenhil, 1983). In broad bean, direct promotion of assimilate export by the application of gibberellic acid was reported (Aloni *et al.*, 1986). Furthermore, in isolated bundles of celery, phloem loading appears to be directly influenced by gibberellic acid and auxins (Daie *et al.*, 1986). However, the main problem with such studies is that it is difficult to differentiate between the effects operating at the different levels of regulation. Analysis at the transcriptional level has shown that the sucrose transporter mRNA can be induced by the addition of auxins and cytokinins to detached leaves, whereas no such effects could be observed on the two major H^+-ATPase genes expressed in leaves (Harms *et al.*, 1994). Thus at least one possible regulation mechanism appears to operate directly at the mRNA level. Other levels of regulation are intracellular targeting and post-translational modification which, now that the proteins are accessible, can be studied in more detail.

A number of environmental factors have been identified that also affect phloem loading. Metals at high concentrations reduce sugar transport (Delrot, 1989; Rauser and Samarakoon, 1980). SO_2 can also strongly inhibit phloem loading, possibly by interfering directly with sucrose transport due to its sensitivity to sulphydryl-modifying agents (Maurousset *et al.*, 1992; Minchin and Gould, 1986; Teh and Swanson, 1982).

A hypothetical model for phloem loading

Geiger (1975) suggested that the release of assimilates from photosynthetically active cells occurs in the vicinity of the phloem. This hypothesis is attractive for two reasons: (1) export from all cells *en route* from synthesis to the site of loading would take place in the opposite direction to the transpiration stream, and (2) the ubiquitous high levels of sugar in the apoplast could serve as a substrate for the growth of micro-organisms. Despite the high concentrations of sucrose in the mesophyll (20–200 mM) and the phloem (500–1500 mM), only very low levels of sugars were found in the apoplast (0.07–5 mM; Delrot, 1989; Ntsika and Delrot, 1986). The reason for this could be that apoplastic transport is restricted to the interface between the phloem parenchyma and the companion cells, and that otherwise sugar moves along the symplast. This view is supported by the finding of a steep osmotic gradient at the boundary of the conducting complex (for review see Delrot, 1989). In addition, certain species classified as apoplastic loaders (type 2b) have developed specific companion cells with an enlarged surface area which were defined as transfer cells. The existence of a correlation between increase in membrane surface area and sucrose transport activity indicates a functional role for these cells in phloem loading (Wimmers and Turgeon, 1991). This may indicate that the carriers are localized only at the interface membranes. Such asymmetrical targeting of sugar transport proteins is well established in mammalian systems, where different types of glucose transporter are targeted differentially to either the basolateral or the apical membranes (e.g. Kong *et al.*, 1993; Silverman, 1991).

We therefore postulate that the proton sucrose symporter StSUT1 is localized either on the companion cell membrane or on the sieve element membrane adjacent to the neighbouring cells, such as the phloem parenchyma (*Figure 1*). According to this model, an increase in the membrane surface area of the companion cells might permit an increase in transport rates due to increased amounts of transporter on the surface. Alternatively, sucrose uptake directly into the sieve element could take place. One might speculate that the companion cell is electrically coupled to the sieve elements via the plasmodesmata, and that the membrane potential and the proton gradient generated by the H^+-ATPase located either at the companion cell or at the sieve element membrane should be able to drive the transport across the common outer membrane against a concentration gradient.

Juxtaposed to this transporter is a second 'release carrier' that requires more detailed characterization. This carrier is thought to oppose the proton symporter so that sucrose transfer through the cell wall is restricted to the interface. It is also possible that the nature of the cell wall might be different at this interface, in order to facilitate diffusion from one carrier to the other. A sensitive screening of the artificial complementation system has not resulted in the identification of other sucrose transporters (Kühn and Frommer, 1995). However, an efflux system was characterized by Delrot (1989) and by Laloi *et al.* (1993). Mechanistically the release carrier could be a facilitator or an antiporter. Alternatively, a model has been presented which suggests the presence of proton co-transporters at both membranes (Lohaus *et al.*, 1995).

Phloem unloading

Despite the important role of unloading in sink organs, the study of this process has been neglected. The main reason is that unloading can differ not only between species, but also between different tissues within the same plant. The possibility cannot be excluded that several pathways coexist in the same cell. Several models exist for unloading, as described below.

(1) Symplastic unloading. Unloading can occur along a concentration gradient, as the concentration of assimilates in the phloem is higher than that in the surrounding sink cells. Subcellular compartmentation or further use of the imported metabolites (e.g. the storage of sugars in the form of insoluble starch or the storage of amino acids in the form of storage proteins) ensures that the gradient can be maintained.

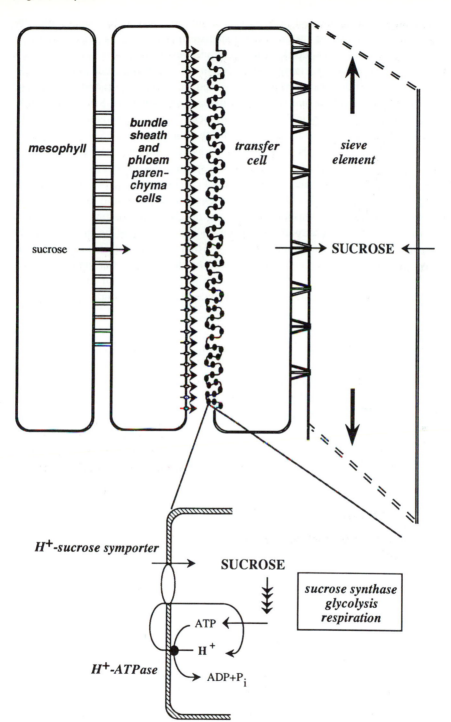

Figure 1. Speculative model for phloem loading in type IIa plants. A combination of symplastic routes (mesophyll to bundle sheath and phloem parenchyma, and transfer cell to sieve element) and polar apoplastic transport via an as yet unknown sucrose carrier at the bundle sheath/phloem parenchyma border and an active proton–sucrose co-transporter (SUT) on the convoluted plasma membrane of the transfer cells or the sieve element.

(2) Sucrose can be unloaded into the apoplast via a sucrose carrier. The uptake into the parenchyma can then either be mediated by a sucrose transporter or, in cases where an apoplastic invertase is present, uptake can occur via hexose transport systems. In this context the expression of sucrose transporter mRNA in the roots is of particular interest. The respective hexose transporter genes have been isolated from *Arabidopsis* and tobacco, and some of them are specifically expressed in sink tissues (for review see Sauer and Tanner, 1993). The unloading of amino acids from the phloem in seeds appears to be a simpler process. Amino acids are transported through both the phloem and the xylem, but due to xylem-to-phloem transfer along the translocation pathway, the assimilates reach the seeds mainly via the phloem (Pate *et al.*, 1977). In many species, the embryo is symplastically isolated from the maternal tissue and assimilates must cross the apoplast before entering the developing seeds (Thorne, 1985). Thus at least two membranes must be crossed and possibly two different sets of transport systems are necessary. We assume that a similar situation occurs for sucrose transport, and again an invertase might be present and the second uptake system could then be one of the hexose transporters. Mutants might represent the best means of analysing the contribution made by the different transporters and the invertase. Parallel studies of the expression of sucrose and amino acid transporters will certainly result in a clearer understanding of assimilate partitioning in higher plants.

4. Conclusions

On the one hand molecular tools have allowed identification of the proteins responsible for the transfer of assimilates across the chloroplast envelope and the plasma membrane, while on the other hand it has been possible to study the *in-vivo* function of these proteins in engineered mutants with altered transport capacity. Due to the variability of the mechanisms involved in phloem loading and unloading and the complexity of these processes, the data reviewed here provide only initial evidence. Progress in this area will certainly stimulate the further research that is needed for a better understanding of assimilate partitioning and allocation, which are important parameters of the complex character yield (Gifford *et al.*, 1984; Gifford and Evans, 1988).

Acknowledgements

We are much indebted to Professor Ulf-Ingo Flügge (Würzburg) and Dr D. Heineke and Professor Hans-Walter Heldt (Göttingen) for their collaboration with the research on triose-phosphate translocation. The work described was supported by grants to W.B.F. from BHFT, DFG and EC.

References

Aloni B, Daie J, Wyse RE. (1986) Enhancement of [^{14}C]sucrose export from source leaves of *Vicia faba* by gibberellic acid. *Plant Physiol.* **82**: 962–967.

Anderson JA, Huprikar SS, Kochian LV, Lucas WJ, Gaber RF. (1992) Functional expression of a probable *Arabidopsis thaliana* potassium channel in *Saccharomyces cerevisiae*. *Proc. Natl Acad. Sci. USA* **89**: 3736–3740.

Beebe DU, Turgeon R. (1992) Localization of galactinol, raffinose and stachyose synthesis in *Cucurbita pepo* leaves. *Planta* **188**: 354–361.

Bertran J, Werner A, Moore ML, *et al.* (1992) Expression cloning of a cDNA from rabbit kidney cortex that induces a single transport system for cystine and dibasic and neutral amino acids. *Proc. Natl Acad. Sci. USA* **89**: 5601–5605.

Borchert S, Große H, Heldt HW. (1989) Specific transport of phosphate, glucose-6-phosphate, dihydroxyacetonephosphate and 3-phosphoglycerate into amyloplasts from pea roots. *FEBS Lett.* **253**: 183–186.

Boutry M, Michelet B, Goffeau A. (1989) Molecular cloning of a family of plant genes encoding a protein homologous to plasma membrane H⁺-translocating ATPases. *Biochem. Biophys. Res. Commun.* **162:** 567–574.

Buckhout TJ. (1989) Sucrose transport into isolated membrane vesicles from sugar beet. *Planta* **178:** 393–399.

Bush DR. (1989) Proton-coupled sucrose transport in plasma membrane vesicles isolated from sugar beet leaves. *Plant Physiol.* **89:** 1318–1323.

Bush DR. (1990) Electrogenicity, pH-dependence, and stoichiometry of the proton–sucrose symport. *Plant Physiol.* **93:** 1590–1596.

Bush DR. (1993) Proton-coupled sugar and amino acid transporters in plants. *Annu. Rev. Plant Physiol. Plant Mol. Biol.* **44:** 513–542.

Caspar T, Lin T, Kakefuda G, Benbow L, Preiss J, Somerville C. (1991) Mutants in *Arabidopsis* with altered starch degradation. *Plant Physiol.* **95:** 1181–1188.

Cho BH, Komor E. (1980) The role of potassium in charge compensation for sucrose-proton symport by cotyledons of *Ricinus communis. Plant Sci. Lett.* **17:** 425–435.

Citovsky V. (1993) Probing plasmodesmal transport with plant viruses. *Plant Physiol.* **102:** 1071–1076.

Daie J, Watts M, Aloni B, Wyse RE. (1986) *In vitro* and *in vivo* modification of sugar transport and translocation in celery by phytohormones. *Plant Sci.* **46:** 35–41.

Delrot S. (1981) Proton fluxes associated with sugar uptake in *Vicia faba* leaf tissue. *Plant Physiol.* **68:** 706–711.

Delrot S. (1989) Phloem loading. In: *Transport of Photoassimilates* (Baker DA, Milburn JA, ed.). London: Longman Scientific, pp. 167–205..

Delrot S, Bonnemain JL. (1981) Involvement of protons as a substrate for the sucrose carrier during phloem loading in *Vicia faba* leaves. *Plant Physiol.* **67:** 560–564.

DeWitt ND, Harper JF, Sussman MR. (1991) Evidence for a plasma membrane proton pump in phloem cells of higher plants. *Plant J.* **1:** 121–128.

Dickinson CD, Altabella T, Chrispeels M. (1991) Slow-growth phenotype of transgenic tomato expressing apoplastic invertase. *Plant Physiol.* **95:** 420–425.

Doman DC, Geiger DR. (1979) Effect of exogenously supplied foliar potassium on phloem loading in *Beta vulgaris* L. *Plant Physiol.* **64:** 528–533.

Ewing NN, Wimmers LE, Meyer DJ, Chetelat RT, Bennett AB. (1990) Molecular cloning of tomato plasma membrane H⁺-ATPase. *Plant Physiol.* **94:** 1874–1881.

Fischer WN, Kwart M, Hummel S, Frommer WB. (1995) Substrate specificity and expression profile of amino acid transporters (AAPs) in *Arabidopsis. J. Biol. Chem.* **270:** 16315–16320.

Flügge UI, Benz R. (1984) Pore forming activity in the outer membrane of the chloroplast envelope. *FEBS Lett.* **169:** 85–89.

Flügge UI, Fischer K, Gross A, Sebald W, Lottspeich F, Eckerskorn C. (1989) The triose phosphate–3-phosphoglycerate-phosphate translocator from spinach chloroplasts: nucleotide sequence of a full-length cDNA clone and import of the *in vivo* synthesized precursor protein into chloroplasts. *EMBO J.* **8:** 39–46.

Fondy BR, Geiger DR. (1977) Sugar selectivity and other characteristics of phloem loading in *Beta vulgaris* L. *Plant Physiol.* **59:** 953–960.

Fritz E, Evert RF, Heyser W. (1983) Microautoradiographic studies of phloem loading and transport in the leaf of *Zea mays* L. *Planta* **159:** 193–206.

Frommer WB, Sonnewald U. (1995) Control of assimilate partitioning in solanaceous plants. *J. Exp. Bot.* **46:** 587–607.

Frommer WB, Hummel S, Riesmeier JW. (1993) Yeast expression cloning of a cDNA encoding a broad specificity amino-acid permease from *Arabidopsis thaliana. Proc. Natl Acad. Sci. USA* **90:** 5944–5948.

Frommer WB, Hummel S, Lemoine R, Delrot S. (1994) Developmental changes in the two-dimensional protein pattern of plasma membrane vesicles between sink and source leaves from sugar beet. *Plant Physiol. Biochem.* **32:** 205–209.

Frommer WB, Hummel S, Unseld M, Ninnemann O. (1995) Seed and vascular expression of a high affinity transporter for cationic amino acids in *Arabidopsis. Proc. Natl Acad. Sci. USA* **92:** 12036–12040.

Gahrtz M, Stolz J, Sauer N. (1994) A phloem-specific sucrose-H⁺ symporter from *Plantago major* L. supports the model of apoplastic phloem loading. *Plant J.* **6:** 697–706.

Gallet O, Lemoine R, Gaillard C, Larsson C, Delrot S. (1992) Selective inhibition of active uptake of sucrose into plasma membrane vesicles by polyclonal antisera directed against a 42 kD plasma membrane polypeptide. *Plant Physiol.* **98:** 17–23.

Gamalei YV. (1986) Characteristics of phloem loading in woody and herbaceous plants. *Sov. Plant Physiol.* **32:** 656–665.

Gebhardt C, Ritter E, Salamini F. (1993) RFLP map of the potato. In: *Advances in Cellular and Molecular Biology of Plants, Vol. 1, DNA-Based Markers in Plants* (Vasil IK, Phillips R, ed.). Dordrecht: Kluwer Academic Publishers.

Geigenberger P, Langenberger S, Wilke I, Heineke D, Heldt HW, Stitt M. (1993) Sucrose is metabolized by sucrose synthase and glycolysis within the phloem of *Ricinus communis* L. seedlings. *Planta.* **190:** 446–453.

Geiger DR. (1975) Phloem loading. In: *Transport in Plants. I. Phloem Transport, Encyclopedia of Plant Physiology* (Zimmermann MH, Milburn JA, ed.). Berlin: Springer-Verlag, pp. 395–431.

Giaquinta RT. (1977) Possible role of pH gradient and membrane ATPase in the loading of sucrose into the sieve tubes. *Nature* **267:** 369–370.

Giaquinta RT. (1983) Phloem loading of sucrose. *Annu. Rev. Plant Physiol.* **34:** 347–387.

Gifford RM, Evans LT. (1981) Photosynthesis, carbon partitioning and yield. *Annu. Rev. Plant Physiol.* **32:** 485–509.

Gifford RM, Thorne JH, Hitz WD, Giaquinta RT. (1984) Crop productivity and photoassimilate partitioning. *Science* **225:** 801–808.

Grimes HD, Overvoorde PJ, Ripp K, Franceschi VR, Hitz WD. (1992) A 62-kD sucrose binding protein is expressed and localized in tissues actively engaged in sucrose transport. *Plant Cell* **4:** 1561–1574.

Grusak MA, Delrot S, Ntsika G. (1990) Short-term effects of heat-girdles on source leaves of *Vicia faba*: analysis of phloem loading and carbon partitioning parameters. *J. Exp. Bot.* **41:** 1371–1377.

Harms K, Wöhner RV, Schulz B, Frommer WB. (1994) Regulation of two p-type H$^+$-ATPase genes from potato. *Plant Mol. Biol.* **26:** 979–988.

Heineke D, Sonnewald U, Büssis G, Günter G, Leitreiter K, Wilke I, Raschke K, Willmitzer L, Heldt HW. (1992) Apoplastic expression of yeast-derived invertase in potato. *Plant Physiol.* **100:** 301–308.

Heineke D, Kruse A, Flügge UI, Frommer WB, Riesmeier J, Willmitzer L, Heldt HW. (1994) Effect of antisense repression of the triose phosphate translocator on photosynthesis metabolism in transgenic potato plants. *Planta* **193:** 174–180.

Heldt HW, Flügge UI. (1992) Metabolite transport in plant cells. In: *Plant Organelles,* SEB Seminar Series 50 (Tobin AK, ed.). Cambridge: Cambridge University Press, pp. 21–47.

Hitz WD, Card PJ, Ripp KG. (1986) Substrate recognition by a sucrose transporting protein. *J. Biol. Chem.* **261:** 11986–11991.

Hsu L, Chiou T, Chen L, Bush DR. (1993) Cloning a plant amino acid transporter by functional complementation of a yeast amino acid transport mutant. *Proc. Natl Acad. Sci. USA* **90:** 7441–7445.

Inoue H, Tanaka A. (1978) Comparison of source and sink potentials between wild and cultivated potatoes. *J. Sci. Soil Mgmt Jpn.* **49:** 321–327.

Komor E, Rotter M, Tanner W. (1977) A proton-cotransport system in a higher plant: sucrose transport in *Ricinus communis. Plant Sci Lett.* **9:** 153–162.

Kong C, Varde A, Lever J. (1993) Targeting of recombinant Na$^+$/glucose transporter (SGLT1) to the apical membrane. *FEBS Lett.* **333:** 1–4.

Koßmann J, Müller-Röber B, Dyer TA, Raines CA, Sonnewald U, Willmitzer L. (1992) Cloning and expression analysis of plastidic fructose-1,6-bisphosphatase coding sequence from potato: circumstantial evidence for the import of hexoses into chloroplasts, *Planta* **188:** 7–12.

Krapp A, Hofmann B, Schäfer C, Stitt M. (1993) Regulation of the expression of *rbcS* and other photosynthetic genes by carbohydrates: a mechanism for the 'sink regulation' of photosynthesis. *Plant J.* **3:** 817–828.

Kühn C, Frommer WB. (1995) The couch potato gene PCP1 from potato enables a sucrose transport deficient yeast strain to grow on sucrose. *Mol. Gen. Genet.* **247:** 759–763.

Kwart M, Hirner B, Hummel S, Frommer WB. (1993) Differential expression of two related amino-acid transporters with differing substrate specificity in *Arabidopsis thaliana. Plant J.* **4:** 993–1002 .

Larsson C. (1985) Plasma membranes. In: *Modern Methods of Plant Analysis. Vol. 1. Cell Components* (Linskens HF, Jackson JF, ed.). Berlin: Springer Verlag, pp. 85–104.

Lemoine R, Delrot S. (1989) PMV-driven sucrose uptake in sugar beet plasma membrane vesicles. *FEBS Lett.* **249:** 129–133.

Lemoine R, Gallet O, Gaillard C, Frommer WB, Delrot S. (1992) Plasma membrane vesicles from source and sink leaves. *Plant Physiol.* **100:** 1150–1156.

Lepp NW, Peel AJ. (1970) Some effects of IAA and kinetin on the movement of sugars in the phloem of willow. *Planta* **90:** 230–235.

Li Z, Bush DR. (1990) ΔpH-dependent amino acid transport into plasma membrane vesicles isolated from sugar beet leaves. *Plant Physiol.* **94:** 268–277.

Li Z, Gallet O, Gaillard C, Lemoine R, Delrot S. (1992) The sucrose carrier of the plant plasmalemma. III. Partial purification and reconstitution of active sucrose transport in liposomes. *Biochim. Biophys. Acta* **1103:** 259–267.

Lohaus G, Winter H, Riens B, Heldt HW. (1995) Further studies of the phloem loading process in leaves: comparison of metabolite concentrations in the apoplastic compartment with those in the cytosolic compartment and in sieve tubes. *Bot. Acta* **108:** 270–275.

Lucas WJ, Madore MA. (1988) Recent advances in sugar transport. In: *The Biochemistry of Plants, Vol. 14.* (Preiss J, ed.). Orlando: Academic Press, pp. 35–84.

Lucas W, Olesinski A, Hull RJ, Haudenshield JS, Deom CM, Beachy RN, Wolf S. (1993) Influence of the tobacco mosaic virus 30-kDa movement protein on carbon metabolism and photosynthate partitioning in transgenic tobacco plants. *Planta* **190:** 88–96.

Lucas WJ, Ding B, van der Schoot C. (1993) Plasmodesmata and the supracellular nature of plants. *New Phytol.* **125:** 435–476.

McCauley MM, Evert RF. (1989). Minor veins of the potato *(Solanum tuberosum* L.) leaf: ultrastructure and plasmodesmatal frequency. *Bot. Gazette* **150:** 351–368.

Malek F, Baker DA (1978) Effect of fusicoccin on proton co-transport of sugars in the phloem loading of *Ricinus communis* L. *Plant Sci. Lett.* **11:** 233–239.

Marger MD, Saier MH Jr. (1993) A major superfamily of transmembrane facilitators that catalyse uniport, symport and antiport. *Trends Biochem. Sci.* **18:** 13–20.

Martin T, Frommer WB, Salanoubat M, Willmitzer L. (1993) The expression of an *Arabidopsis* sucrose synthase gene indicates a role in the metabolism of sucrose both during phloem loading and in sink organs. *Plant J.* **4:** 367–377.

Maurousset L, Lemoine R, Gallet O, Delrot S, Bonnemain JL. (1992) Sulfur dioxide inhibits the sucrose carrier of the plant plasma membrane. *Biochem. Biophys. Acta* **1105:** 230–236.

Maynard JW, Lucas WJ. (1982) Sucrose and glucose uptake in *Beta vulgaris* leaf tissue. A case for a general (apoplastic) retrieval system. *Plant Physiol.* **70:** 1436–1443.

Minchin PEH, Gould R. (1986) Effect of SO$_2$ on phloem loading. *Plant Sci. Lett.* **43:** 179–183.

Moriau L, Bogaerts P, Jonniaux J, Boutry M. (1993) Identification and characterization of a second plasma membrane H$^+$-ATPase gene subfamily in *Nicotiana plumbaginifolia. Plant Mol. Biol.* **21:** 955–963.

Ntsika G, Delrot S. (1986) Changes in apoplastic and intracellular leaf sugars induced by the blocking of export in *Vicia faba. Physiol. Plant.* **68:** 145–153.

Oparka KJ, Prior DAM. (1992) Direct evidence for pressure-generated closure of plasmodesmata. *Plant J.* **2:** 741–750.

Overvoorde PJ, Frommer WB, Grimes HD. (1996) Heterologous expression demonstrates that the sucrose binding protein mediates linear, non-saturable uptake of sucrose in yeast. *Plant Cell,* (in press).

Pardo JM, Serrano R. (1989) Structure of a plasma membrane H$^+$-ATPase gene from *Arabidopsis thaliana. J. Biol. Chem.* **264:** 8557–8562.

Parets-Soler A, Pardo JM, Serrano R. (1990) Immunolocalization of plasma membrane H$^+$-ATPase. *Plant Physiol.* **93:** 1654–1658.

Pate JS, Sharkey PJ, Atkins CA. (1977) Nutrition of a developing legume fruit. *Plant Physiol.* **59:** 506–510.

Patrick JW. (1976) Hormone directed transport of metabolites. In: *Transport and Transfer Processes in Plants* (Wardlaw IF, Passioura JB, ed.). New York: Academic Press, pp. 433–446.

Perez C, Michelet B, Ferrant V, Bogaerts P, Boutry M. (1992) Differential expression within a three-gene subfamily encoding a plasma membrane H$^+$-ATPase in *Nicotiana plumbaginifolia. J. Biol Chem.* **267:** 1204–1211.

Rauser WE, Samarakoon AB. (1980) Vein loading in seedlings of *Phaseolus vulgaris* exposed to excess cobalt, nickel, and zinc. *Plant Physiol.* **65:** 578–583.

Rentsch D, Laloi M, Rouhara I, Schmelzer E, Delrot S, Frommer WB. (1995) *NTR1* encodes a high affinity oligopeptide transporter in *Arabidopsis. FEBS Lett.* **370:** 264–268.

Riens B, Lohaus G, Heineke D, Heldt HW. (1991) Amino acid and sucrose content determined in the cytosolic, chloroplastic, and vacuolar compartments and in the phloem sap of spinach leaves. *Plant Physiol.* **97:** 227–233.

Riesmeier JW, Willmitzer L, Frommer WB. (1992) Isolation and characterization of a sucrose carrier cDNA from spinach by functional expression in yeast. *EMBO J.* **11:** 4705–4713.

Riesmeier JW, Flügge UI, Schulz B, Heineke D, Heldt HW, Willmitzer L, Frommer WB. (1993a) Antisense repression of the chloroplast triose phosphate translocator affects carbon partitioning in transgenic potato plants. *Proc. Natl Acad. Sci USA* **90:** 6160–6164.

Riesmeier JW, Hirner B, Frommer WB. (1993b) Expression of the sucrose transporter from potato correlates with the sink-to-source transition in leaves. *Plant Cell* **5:** 1591–1598.

Riesmeier JW, Willmitzer L, Frommer WB. (1994) Evidence for an essential role of the sucrose transporter in phloem loading and assimilate partitioning. *EMBO J.* **13:** 1–7.

Ripp KG, Viitanen PV, Hitz WD, Franceschi VR. (1988) Identification of a membrane protein associated with sucrose transport into cells of developing soybean cotyledons. *Plant Physiol.* **88:** 1435–1445.

Robards AW, Lucas WJ. (1990) Plasmodesmata. *Annu. Rev. Plant Physiol. Plant Mol. Biol.* **41:** 369–419.

Sauer N, Tanner W. (1993) Molecular biology of sugar transporters in plants. *Bot. Acta* **106:** 277–286.

Sauer N, Stolz J. (1994) SUC1 and SUC2: two sucrose transporters from *Arabidopsis thaliana*; expression and characterization in baker´s yeast and identification of the histidine-tagged protein. *Plant J.* **6:** 67–77.

Schäfer G, Heber U, Heldt HW. (1977) Glucose transport into spinach chloroplasts. *Plant Physiol.* **60:** 286–289.

Schlichting R, Bothe H. (1993) The cyanelles (organelles of a low evolutionary scale) possess a phosphate translocator and a glucose-carrier in *Cyanophora paradoxa. Bot. Acta* **106:** 428–434.

Schulz B, Frommer WB, Flügge UI, Hummel S, Fischer K, Willmitzer L. (1993) Expression of the triose phosphate translocator gene from potato is light dependent and restricted to green tissues. *Mol. Gen. Genet.* **238:** 357–361.

Sentenac H, Bonneaud N, Minet M, Lacroute F, Salmon JM, Gaymard F, Grignon C. (1992) Cloning and expression in yeast of a plant potassium ion transport system. *Science* **256:** 663–665.

Servaites JC, Schrader LE, Jung DM. (1979) Energy-dependent loading of amino acids and sucrose into the phloem of soybean. *Plant Physiol.* **64:** 546–550.

Silverman M. (1991) Structure and function of hexose transporters. *Annu. Rev. Biochem.* **60:** 757–794.

Sowonick SA, Geiger DR, Fellows RJ. (1974) Evidence for active phloem loading in the minor veins of sugar beet. *Plant Physiol.* **54:** 886–891.

Sturgis JN, Rubery PH. (1982) The effects of indole 3-indolyl-acetic acid and fusicoccin on the kinetic parameters of sucrose uptake by dics from expanded primary leaves of *Phaseolus vulgaris. Plant Sci. Lett.* **24:** 319–326.

Sullivan TD, Strelow LI, Illingworth CA, Phillips RL, Nelson OE Jr. (1992) Analysis of maize *Brittle-1* alleles and a defective *suppressor-mutator*-induced mutable allele. *Plant Cell* **3:** 1337–1348.

Sussman MR, Harper JF. (1989) Molecular biology of the plasma membrane of higher plants. *Plant Cell* **1:** 953–960.

Stitt M. (1990) Fructose-2,6-bisphosphate as a regulatory molecule in plants. *Annu. Rev. Plant Physiol. Plant Mol. Biol.* **41:** 153–158.

Teh KH, Swanson CA. (1982) Sulfur dioxide inhibition of translocation in bean plants. *Plant Physiol.* **69:** 88–92.

Thorne JH. (1985) Phloem unloading of C and N assimilates in developing seeds. *Annu. Rev. Plant Physiol. Plant Mol. Biol.* **36:** 317–343.

Thorpe MR, Minchin PEH, Dye EA. (1979) Oxygen effects on phloem loading. *Plant Sci. Lett.* **15:** 345–350.

Truernit E, Sauer N. (1995) The promoter of the *Arabidopsis thaliana* SUC2 sucrose-H^+ symporter gene directs expression of β-glucuronidase to the phloem: evidence for phloem loading and unloading by SUC2. *Planta* **196:** 564–570.

Turgeon R. (1989) The sink–source transition in leaves. *Annu. Rev. Plant Physiol. Plant Mol. Biol.* **40:** 119–138.

Turgeon R, Gowan E. (1990) Phloem loading in *Coleus blumei* in the absence of export sugar from the apoplast. *Plant Physiol.* **94:** 1244–1249.

Van Bel AJE. (1993) Strategies of phloem loading. *Annu. Rev. Plant Physiol. Plant Mol. Biol.* **44:** 253–281.

Van Bel AJE, Ammerlaan A, van Dijk AA. (1994) A three-step screening procedure to identify the mode of phloem loading in intact leaves. *Planta* **192:** 31–39.

von Schaewen A, Stitt M, Schmidt R, Sonnewald U, Willmitzer L. (1990) Expression of a yeast-derived invertase in the cell wall of tobacco and *Arabidopsis* plants leads to accumulation of carbohydrates and inhibition of photosynthesis, and strongly influences growth and phenotype of transgenic tobacco plants. *EMBO J.* **9:** 3033–3044.

Vreugdenhil D. (1983) Abscisic acid inhibits phloem loading of sucrose. *Physiol. Plant.* **57:** 463–467.

Wallmeier H, Weber A, Gross A, Flügge UI. (1992) Insights into the structure of the chloroplast phosphate translocator protein. In: *Transport and Receptor Proteins of Plant Membranes* (Cooke DT, Clarkson DT, ed.). New York: Plenum Press, pp. 77–89.

Warmbrodt RD, Buckhout TJ, Hitz WD. (1989) Localization of a protein, immunologically similar to a sucrose binding protein from developing soybean cotyledons, on the plasma membrane of sieve tube members of spinach leaves. *Planta* **180**: 105–115.

Willey DL, Fischer K, Wachter E, Link TA, Flügge U-I. (1991) Molecular cloning and structural analysis of the phosphate translocator from pea chloroplasts and its comparison to the spinach phosphate translocator. *Planta* **183**: 451–461.

Williams LE, Nelson SJ, Hall JL. (1992) Characterization of a solute/proton cotransport in plasma membrane vesicles from *Ricinus* cotyledon, and comparison with other tissues. *Planta* **186**: 541–550.

Wimmers LE, Turgeon R. (1991) Transfer cells and solute uptake in minor veins of *Pisum sativum* leaves. *Planta* **186**: 2–12.

Zimmermann MH, Ziegler H. (1975) List of sugars and sugar alcohols in sieve-tube exudates. In: *Transport in Plants. I. Phloem Transport, Encyclopedia of Plant Physiology* (Zimmermann MH, Milburn JA, ed.). Berlin: Springer-Verlag, pp. 245–271.

Chapter 20

The Golgi apparatus and pathways of vesicle trafficking

Chris Hawes, Loïc Faye and Béatrice Satiat-Jenemaitre

1. Introduction — the pathways of vesicle traffic in plant cells

Compartmentalization of the cytoplasm is one of the key features that enabled the successful evolution of eukaryotic cells from their protoeukaryote ancestors. The movement of macro-molecules in membrane-bounded vesicles is a rapid and efficient mechanism for transfer between the various compartments in the secretory pathway. Such molecules are transported from sites of synthesis to sites of action or deposition, from sites of action to sites of degradation, or are endocytosed and delivered to sites of action/degradation. Cargoes in carrier vesicles can range from secretory and storage products to signal molecules and toxic waste. The trafficking of such vesicles requires efficient recognition and sorting mechanisms to load them with the correct products, as well as separate guidance and recognition systems to ensure that they and their enclosed cargoes arrive at the right destinations at the right time and fuse successfully with their acceptor membrane. Concomitant with the vesicular transport of macromolecules is the inevitable flow of membrane within the cell, and to and from the cell surface. Thus the production and cycling of macromolecules within the cytoplasm must be balanced by a parallel turnover and/or recycling of many of the cytoplasmic membranes.

The organelles involved in the various pathways of vesicle traffic in plant cells are clearly defined and together comprise the 'endomembrane system' as previously described by Morré and co-workers (Morré and Mollenhauer, 1974). However, the role of vesicles themselves in some of the routes within this pathway has yet to be elucidated. A map of the membrane flow pathway in plant cells is given in *Figure 1*, and the various stages are detailed in the different sections of this chapter.

In mammalian and yeast cells, structural and molecular dissection of the various vesicle trafficking pathways has generated important new information about the mechanisms of vesicle transport, targeting and fusion, and about the plethora of genes and proteins involved in these processes (Goud and McCaffrey, 1991; Gruenberg and Clague, 1992; Kreis, 1992: Rothman and Orci, 1992; Rothman, 1994; Schekman, 1992). Such a systematic exploration of vesicle trafficking in plants has only recently been started (Satiat-Jenemaitre and Hawes, 1993b), but preliminary data indicate the possibility of a unified concept of vesicle-mediated transport in eukaryotic cells. However, it is important to remember that there are key differences between the structural and functional make-up of the endomembrane systems of plant and animal cells, due to both the physical nature of plant cells and the different products that they synthesize.

In this chapter we shall concentrate on the recent advances in our understanding of vesicle transport pathways in plant cells, and on the central role of the Golgi apparatus in the membrane flow phenomenon. More specific accounts of the structure and function of the endoplasmic reticulum, of the targeting of proteins transported in vesicle vectors, and the conservation of the

337

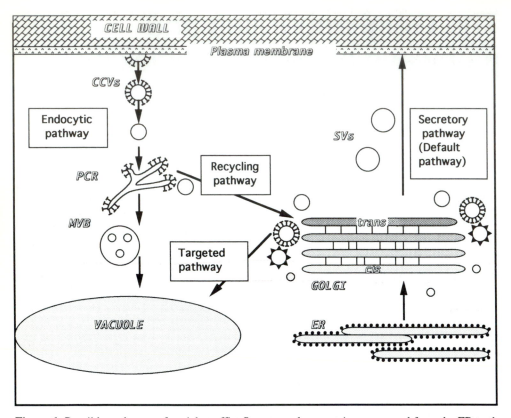

Figure 1. Possible pathways of vesicle traffic. Secretory glycoproteins are passed from the ER to the Golgi (possibly via transition vesicles). These macromolecules, together with newly synthesized polysaccharides, are passed through the Golgi stack (possibly by a shuttle of non-clathrin-coated vesicles (▲) and then either targeted to the vacuole (possibly via clathrin-coated vesicles) (T) or transferred to the plasma membrane in smooth secretory vesicles (SVs). Endocytosis is by means of clathrin-coated vesicles (CCVs) which deliver to the partially coated reticulum (PCR), which may in turn route vesicles to the vacuole via multivesicular bodies (MVBs) or directly to the Golgi apparatus.

plasma membrane may be found elsewhere in this volume (see, respectively, Chapters 21 by Denecke, 22 by Lichtscheidl and Hepler, 23 by Reynolds and Raikhel, and 3 by Oparka *et al.*).

2. Budding, targeting and fusion of vesicles

With regard to membrane trafficking, vesicles can be divided into two categories: those surrounded by a protein coat during their formation and scission from a donor membrane, and larger uncoated (smooth) vesicles. The latter are generally formed towards the *trans*-face of the Golgi apparatus, are involved in the bulk flow of membrane and product through the secretory pathway to the cell surface, and are discussed further in Section 5 of this chapter.

It is now becoming apparent that the production and delivery of a coated vesicle is a well-defined operation with various control points to ensure efficient coating, scission, uncoating and fusion with the correct acceptor membrane. This apparently simple operation may involve a plethora of different proteins and protein/protein interactions, the complexity of which is only now becoming apparent. A combination of biochemical and molecular research on animal and yeast cells has resulted in the emergence of a concept of vesicle operation that may prove to be ubiquitous across the kingdoms (Rothman and Orci, 1992; Rothman, 1994). It has long been known that plants produce clathrin-coated vesicles which are involved in endocytosis and intracellular transport (Fowke *et al.*, 1991). However, it is only recently that the other components involved in the mediation of vesicle trafficking, such as small GTP-binding proteins of the rab family (Palme *et al.*, 1992), annexins involved in vesicle fusion (Blackbourn and Battey, 1993) and non-clathrin-coated vesicles, have been reported (Staehelin and Moore, 1995).

Vesicle budding and coats

In the animal kingdom, two distinct forms of coated vesicle have been characterized: (1) those surrounded by clathrin and involved in receptor-mediated transport of macromolecules between post-Golgi compartments; and (2) those surrounded by the coatomer complex of protein and involved in the non-selective transport of material from the endoplasmic reticulum (ER) to the *cis* Golgi and between the cisternae of the Golgi stack. In plants, the former are now well characterized (Coleman *et al.*, 1991), but although non-clathrin-coated vesicles have been reported from microscopy studies, there is to date no biochemical characterization of a coatomer homologue in plants.

It is now becoming clear that a common set of rules have evolved across the kingdoms which enable the production of a coated vesicle from a donor membrane and the fusion of that vesicle with the correct acceptor membrane. This apparently simple feat follows a clearly defined set of events (Stammes and Rothman, 1993), which can be summarized as follows:

(1) GTP-mediated binding of a small G-protein to the donor membrane;
(2) recognition and binding of a cytoplasmic coat protein to the G-protein;
(3) final assembly of coat complex on the budding vesicle;
(4) scission of the vesicle;
(5) release of the vesicle coat by a cytosolic ATP/GTPase;
(6) receptor-mediated docking of the vesicle on an acceptor membrane, which is mediated by cytosolic docking proteins;
(7) formation of a fusion pore and fusion of lipid bilayers of vesicle and acceptor membrane.

Clathrin-coated vesicles. In plant cells, clathrin-coated vesicles (CCVs) are associated with the plasma membrane, partially coated reticulum, multivesicular bodies and *trans* Golgi membranes. The first three organelles are involved in the endocytic pathway (see Section 6), whilst CCVs associated with the Golgi apparatus have been implicated in transport to the vacuole (Harley and Beevers, 1989; Hoh *et al.*, 1991; Robinson *et al.*, 1989; see Section 5).

The plant clathrin molecule is homologous with its mammalian counterpart, being triskelial in nature, although the predominant heavy chain has a molecular mass of 190 kDa in comparison with that of 180 kDa in animals (Coleman *et al.*, 1988). The clathrin triskelions possess the property of being able to self-assemble to form the characteristic clathrin cage of CCVs composed of a polygonal basket of pentagons and hexagons. Associated with the clathrin heavy chain are a range of proteins. Several of these, in the molecular mass range 30 kDa to 50 kDa, have been reported to possess properties similar to those attributed to the light chains of animal clathrin (i.e. heat stability, Ca^{2+} binding, elastase sensitivity) and as such may be involved in maintaining the integrity of the clathrin cage (Demmer *et al.*, 1993; Lin *et al.*, 1992).

Perhaps, in terms of function, the most important group of clathrin-associated proteins are the adaptor complexes which, in animal cells, are involved in binding the clathrin cage to transmembrane receptor molecules (Pearse and Robinson, 1990). A set of polypeptides of 28, 48, 70 and 110 kDa have been identified from Tris–HCl-extractable coat proteins of CCVs from pea cotyledons (Butler and Beevers, 1993) which, due to their molecular masses and capacity for autophosphorylation, have been suggested to be putative plant homologues of the Golgi-derived adaptor complexes of mammalian cells. Proteins from CCV preparations of *Zucchini* hypocotyls and pea cotyledons have also shown cross-reaction with antibodies to mammalian β-type adaptins (one component of the adaptor complex), and again the authors suggested that the immunoreactive polypeptide corresponded to the Golgi adaptor complexes from bovine brain (Holstein *et al.*, 1994). This result was supported by the hybridization of a β-adaptin cDNA clone to genomic DNA from *Zucchini* hypocotyls and pea cotyledons. However, there is some recent evidence that this adaptin may also be targeted to the plasma membrane, which would support the case for a receptor-based endocytic pathway in plants (Drucker *et al.*, 1995; see also Section 6).

Non-clathrin-coated vesicles. A population of coated vesicles which are morphologically distinct from clathrin-coated vesicles are also associated with the Golgi apparatus (*Figure 2c*; Staehelin *et al.*, 1990). However, in plants, their functions and the biochemical analyses of their coat components have yet to be established. It is possible that these are the plant equivalent of the well-known coatomer-coated vesicles in mammalian cells which are responsible for ER to Golgi and intercisternal transport in the Golgi stack (Kuge *et al.*, 1993). Purification of these non-clathrin-coated vesicles has resulted in the identification of seven coat proteins (COPs) which reside in the cytosol and which, in combination with a small GTP-binding protein (ADP-ribosylation factor, ARF), assemble on the donor membrane to form vesicle coats (Elazar *et al.*, 1994). The GTP form of ARF binds to a receptor in the donor membrane to establish a binding site for the coatomer complex. This ARF/membrane binding is sensitive to the drug brefeldin A (BFA), which inhibits coat formation, resulting in a retrograde reabsorption of Golgi membranes and processing enzymes into the ER (Helms and Rothman, 1992; Klausner *et al.*, 1992; Lippincott-Schwartz *et al.*, 1989). Whether these COP-coated vesicles are responsible for anterograde or retrograde transport through the Golgi stack is open to question. In yeasts it has been suggested that a third type of coated vesicle, termed COPII vesicles, are in fact responsible for anterograde transport (Letourneur *et al.*, 1994; Pelham, 1994).

Plant Golgi stacks are also sensitive to BFA, where treatment with the drug results in a gathering and vesiculation of cisternal stacks which is reversible on removal of the drug (Satiat-Jeunemaitre and Hawes, 1992, 1994). It has yet to be established whether the prime molecular target of the drug in plant cells is ARF and the coat of non-clathrin-coated vesicles. However, a low-molecular-mass (21 kDa) G-protein has been identified in peas which cross-reacts with an ARF antibody raised against mammalian ARF (Memon *et al.*, 1993), and cDNA clones encoding ARF have been isolated from both *Arabidopsis* (Regad *et al.*, 1993) and rice (Higo *et al.*, 1994). Perhaps these early data point towards some homology in the underlying mechanisms of vesicle production within the Golgi stack of plant and animal cells.

Vesicle uncoating and targeting

Once a vesicle has pinched off from its donor membrane, it uncoats prior to attachment to a suitable acceptor membrane. Whether such vesicles are directly transported to acceptor membranes, or whether the meeting of vesicle and acceptor membrane is a random event, is not clear. In plants, the main evidence for the targeting of vesicles comes from work on vacuolar targeted storage proteins (see Chapter 23 by Reynolds and Raikhel). If, as suggested for legumes, these proteins are initially transported in Golgi-derived CCVs (Harley and Beevers, 1989; Hoh *et al.,* 1991), this would be the only example of targeted coated vesicles in plants. However, these proteins are targeted through the action of signal peptides, and the involvement of molecules in the vesicle membranes in this targeting system has yet to be determined.

The uncoating of CCVs is mediated by an ATPase that is immunologically related to members of the 70-kDa heat-shock protein family (Kirsch and Beevers, 1993). The ATPase isolated from developing pea cotyledons also released clathrin from bovine brain CCVs. Limited proteolysis of the clathrin light chains inhibited vesicle uncoating, indicating a role for plant clathrin light chains in the maintenance of coat integrity and as a target for the uncoating ATPase. These results indicate a highly conserved homology in the role of light chains and clathrin release throughout the kingdoms (Brodsky *et al.,* 1991).

Vesicle fusion

The process of vesicle fusion in mammalian and yeast cells has also been shown to involve the interaction of a multiplicity of proteins prior to the formation of a fusion pore between the opposing lipid bilayers. The fact that the various proteins identified in this process from mammalian cells have been shown to have identical yeast homologues has resulted in the proposition of a unified concept of vesicle fusion in eukaryotic cells (Bennett and Scheller, 1993; Wilson *et al.,* 1992). A brief summary of the vesicle docking and fusion processes is as follows. Cytoplasmic proteins (*N*-ethylmaleimide-sensitive fusion protein, NSF, and soluble NSF attachment proteins, SNAPs) are required as docking proteins to bring together the vesicle-specific SNAP receptors (v-SNAREs) and the target membrane SNAP receptors (t-SNAREs) prior to the fusion event (Söllner *et al.,* 1993). The whole event is most probably regulated by a rab protein (Pfeffer, 1994). In *Arabidopsis,* a 31-kDa protein has been identified from a cDNA clone which functionally complements a yeast *pep12* mutant. This yeast protein is a homologue of syntaxin, a well-characterized vesicle fusion protein from nerves and the homologue of t-SNARE, thus hinting that plants, yeast and mammals may have similar vesicle fusion machineries (Bassham *et al.,* 1995).

Another family of proteins involved in the exocytotic machinery is the annexins. These calcium-dependent phospholipid-binding proteins are found in secretory cells, and have been implicated in the control of a variety of cellular processes including exocytosis. It has been postulated that they mediate the vesicle fusion process itself, perhaps by inducing the formation of the initial fusion pore between the two opposing lipid bilayers (Battey and Blackbourn, 1993; Blackbourn and Battey, 1993; Creutz, 1992). Plant homologues of annexins of molecular weights between 23 and 35 kDa have been described by several groups, and have been shown to have some sequence homology with mammalian annexins. There is also immunological cross-reactivity between these proteins from the two kingdoms (Andrawis *et al.,* 1993; Blackbourn and Battey, 1993; Blackbourn *et al.,* 1992; Clark *et al.,* 1992; Hobbs *et al.,* 1991; Smallwood *et al.,* 1990). Immunofluorescence microscopy has shown these plant annexins to be located at the tips of growing pollen tubes (Blackbourn *et al.,* 1992), and immunogold labelling has demonstrated peripheral Golgi and plasma membrane localization in secretory cells of pea root tips (Clark *et al.,* 1992). These putative plant annexins also induced the aggregation of liposomes and plant secretory vesicles at levels of free Ca^{2+} similar to that required for the binding of bovine annexins (Battey and Blackbourn, 1993; Blackbourn and Battey, 1993). Thus at least one piece of the molecular jigsaw of the plant secretory process may be falling into place.

Small GTP-binding proteins may regulate vesicle trafficking

It is highly likely that another level of regulation of the trafficking of vesicles exists in plants, as in both animal and yeast cells small GTPases related to the ras subfamily of GTP-binding proteins (rabs in mammalian cells, YPT/SEC in yeasts) have been shown to regulate vesicle docking/fusion steps (Hong, 1994; Pfeffer, 1994). What makes this family of GTP-binding proteins interesting is that, despite a high level of sequence homology (Moore *et al.*, 1995) different rabs appear to associate with different compartments of the secretory pathway. Thus each rab regulates a specific step in the transfer of vesicles through the endomembrane system, including the endocytic pathway. Newly synthesized rab proteins are soluble in the cytoplasm and attach to a donor membrane in their GDP-bound form, and membrane association is achieved by the attachment of one or more 20-carbon geranylgeranyl groups on to a C-terminal cysteine (Zerial and Stenmark, 1993). It is also the C-terminal hypervariable portion of the protein that contains targeting information for subcellular location (Chavrier *et al.*, 1991). Exactly how rabs mediate the vesicle docking/fusion process is not yet known, but they must somehow control the operation of the SNARE/SNAP/NSF complex during vesicle fusion (Staehelin and Moore, 1995; Zerial and Stenmark, 1993).

At least 13 classes of rabs and various isoforms have been identified in plants by molecular cloning techniques (identified from database searches), compared to a total of around 30 classes for yeast and mammalian cells (Simmons and Zerial, 1993). However, to date there is little information about the function of plant rabs in vesicle transport. Various plant *rab1* genes (implicated in ER to Golgi transport) have been shown to complement the growth of yeast *ypt1* mutants (Hong, 1994), and antisense constructs against legume rab1 and rab7 showed, respectively, lack of expansion of root nodule cells and an accumulation of multivesicular bodies in the cytoplasm (Cheon *et al.*, 1993). This latter study has provided the only data indicating a putative role for rabs in the growth of cells and the formation of membrane-bounded compartments (the biogenesis of the peribacteroid membrane).

3. Driving forces for vesicle transport

As well as the complex machinery that must exist to permit the budding, targeting and fusion of vesicles from donor to acceptor membranes, it is reasonable to assume that in many instances there must be a motor-driven mechanism to facilitate vesicular transport. The cytoskeleton and its array of accessory proteins are most likely to fulfil this role, and it is here that there may be a major divergence in mechanisms between animal and plant cells.

It is generally accepted that in animal cells microtubules play a major role in maintaining the spatial organization of the endomembrane system, being responsible for the positioning of both the Golgi apparatus (Kreis, 1990) and the endoplasmic reticulum (Terasaki *et al.*, 1986; Vale and Hontani, 1988). They also mediate the transport of membrane and vesicles between the different compartments of the endomembrane system. This can include retrograde transport of membrane from the Golgi intermediate compartment to the ER (Lippincott-Schwartz *et al.*, 1990), transport along the endocytic pathway to the lysosome (Kreis *et al.*, 1989), and Golgi vesicle transport such as that so well characterized in fast axonal transport (Okabe and Hirokawa, 1989). Both the positioning and movement of these organelles are dependent on motor proteins such as cytoplasmic dynein (Corthésy-Theulaz *et al.*, 1992) and kinesin (Okabe and Hirokawa, 1989).

In higher plant cells, recent data point to the actin cytoskeleton as being the most likely candidate for maintaining the spatial organization of the endomembrane system (see Chapter 22 by Lichtscheidl and Hepler). In onion epidermal cells, actin appears to be involved in the reorganization of the ER after disruption by cold treatment, whereas herbicides which disrupt microtubules have no effect on the ER (Quader, 1990; Quader *et al.*, 1989). Similarly, in our laboratory we have found that, in root cells, microtubule-disrupting drugs have no effect on the spatial localization of the Golgi stacks, while cytochalasin treatment, which disrupts the actin network, results in a redistribution and clumping of the Golgi stacks (Satiat-Jeunemaitre *et al.*, 1996a).

Evidence is also accumulating for the involvement of the actin cytoskeleton in vesicle transport. For instance, tip-growing cells such as pollen tubes have a well-documented system of actin filaments responsible for cytoplasmic streaming toward the tip. In these cells, myosin, located biochemically and by immunofluorescence staining in the tube cytoplasm, is often found to be concentrated at the tip, where the highest density of vesicles occurs (Miller *et al.*, 1995; Yokota *et al.*, 1995). These results are corroborated by the fact that characean actin bundles will support the movement of pollen tube organelles in an *in-vitro* system, and this movement is similar to that of myosin-coated latex beads (Kohno *et al.*, 1990). However, the pollen tube tip story is further complicated by a report of a plant homologue of the microtubule anterograde transporter kinesin in pollen tube extracts, and its localization at the tube tip, which is known to have few microtubules, most of these being restricted to the cortex of the tube itself (Tiezzi *et al.*, 1992).

Circumstantial evidence also indicates the involvement of actin in the transport of secretory vesicles from the *trans* Golgi. Cytochalasins have been used to block the movement of secretory vesicles to the cell surface in a variety of systems, in order to calculate Golgi vesicle production rates. A relatively short period of exposure to the drugs results in the accumulation of Golgi-derived vesicles in the cytoplasm, presumably due to the disruption of the actin cytoskeleton (Picton and Steer, 1981; Steer and O'Driscoll, 1991).

Cell division is one event where microtubules have traditionally been implicated in the directed flow of Golgi-derived vesicles to the phragmoplast (Baskin and Cande, 1990; Kakimoto and Shibaoka, 1988). However, actin filaments also represent a major structural element of the phragmoplast (Hepler *et al.*, 1993; Kakimoto and Shibaoka, 1987, 1988). It is possible that they may interact with the network of tubular ER at the plate, with the recently described tubulo-vesicular network of the plate (Samuels *et al.*, 1995) or with the plate vesicles, thus aiding in the control of the three-dimensional organization and two-dimensional growth of the phragmoplast.

4. Through the Golgi stack

In the centre of the vesicle-mediated transport pathways lies the Golgi apparatus, a unique organelle. Not only does it play a major role in the biosynthesis and sequential modification (glycosylation) of both extracellular and storage macromolecules, but also it is responsible for sorting and packaging these products prior to targeting them to the correct destination, such as the cell surface for secretion (exocytosis), or intracellular compartments (protein storage vacuoles). In addition to this well-characterized role in the secretory pathway, it is likely that the Golgi apparatus is one of the final destinations of molecules and membrane internalized via the endocytic pathway.

Golgi structure

The plant Golgi apparatus has been described, on the basis of ultrastructural investigations, as the compilation of a number of distinct units, the Golgi stacks (often termed dictyosomes), where each stack is composed of a pile of membrane-bounded discs known as the cisternae (Staehelin and Moore, 1995). The number of stacks forming the Golgi apparatus shows considerable variation, depending on the cell type (Melkonian *et al.*, 1991), from one (in some algal cells) to several hundreds (in higher plant cells; Staehelin and Moore, 1995), as has been confirmed by immunofluorescence of Golgi-associated glycoproteins (Horsley *et al.*, 1993). The number of cisternae per stack also varies, from between four and eight in higher plants to as many as 30 in some algal systems (Melkonian *et al.*, 1991). It has also been suggested that the number of cisternae per stack decreases with increasing secretory activity of the Golgi apparatus (Schnepf, 1963, 1969).

Fibrillar material lying between the cisternae, but restricted to the central areas of each plate, is a common feature of the Golgi stack (Kristen, 1978). The application of ultra-rapid freeze-fixation techniques combined with freeze-substitution (Staehelin *et al.*, 1990) or deep-etch replication (see Figure 2 in Satiat-Jeunemaitre and Hawes, 1993b) has shown that these intercisternal elements are not fixation artefacts, and that they can appear as distinct cross-bridges or as filaments lying parallel with the cisternae. Their function is as yet unknown, but they may be involved in bonding of the cisternae to preserve the integrity of the stack or to confine swelling and vesiculation of the cisternae to the stack margins. It has been suggested that in animal Golgi such elements represent cytoplasmic proteins involved in the retention of specific transferases within the correct cisternae (Machamer, 1993; Nilsson *et al.*, 1993; Slusarewicz *et al.*, 1994), a hypothesis that has yet to be tested in plant cells. It has also been suggested that the Golgi stacks themselves are embedded in some form of structured matrix which may function to help maintain the integrity of the stack, retain intracisternal transport vesicles and capture incoming vesicles (Staehelin and Moore, 1995). The presence of this matrix is supported by the fact that, upon treatment with brefeldin A, the resulting Golgi-derived vesicles stay clumped together in discreet BFA compartments as if glued together by a matrix (Satiat-Jeunemaitre and Hawes, 1992; Satiat-Jeunemaitre *et al.*, 1996b).

The Golgi stack is a polarized structure. Polarity of the Golgi stack has been established from ultrastructural, cytochemical and immunochemical data, all of which confirm gradients in structure and activity from the *cis* to the *trans* face (Driouich *et al.*, 1993; Robinson and Kristen, 1982). Structurally, these include concave curvature of the *cis* face and convex curvature of the *trans* face, a decrease in cisternal diameter, a decrease in intercisternal space, an increase in the number of vesicles at cisternal margins and a decrease in the osmiophilic nature of the cisternal membranes down the *cis* to *trans* gradient (*Figures 2* and *3*). Each cisternum is also heterogeneous in structure, as the margins often appear swollen. In face view, cisternal margins are fenestrated and are often associated with tubular membrane elements as revealed by osmium impregnation or freeze-fracture techniques (*Figures 2* and *3*). The degree of such fenestration/reticulation of the cisternal margins also increases along the *cis* to *trans* gradient.

Is there a post-Golgi compartment? A *trans* Golgi network (TGN) has been described in some plant cells, associated with the *trans*-face of the Golgi stack and presenting branched configurations of vesicle-budding membrane, which may be clathrin-coated (Zhang and Staehelin, 1992). This plant TGN-like structure has by implication been assumed to be homologous with the animal TGN. This view would assume that the plant TGN possesses both membranes and soluble proteins distinct from the rest of the Golgi stack. In fact, there is no evidence that the TGN-like structure in plants is functionally distinct from a mature *trans* cisternum. In sycamore cells, only 52% of the Golgi population was shown to have a TGN-like structure (Zhang and Staehelin, 1992). This structure could simply represent the ultimate maturation of the *trans* face before total vesiculation occurs. One of the key characteristics of the TGN in mammalian cells is its close relationship with the lysosome and its ability to fuse with the lysosome, in response to treatment with the drug BFA, when the rest of the Golgi stacks are reabsorbed into the ER (Lippincott-Schwartz *et al.*, 1991; Wong and Brodsky, 1992; Wood *et al.*, 1991). In plant cells, such differential effects of BFA on the Golgi stack/TGN have not been observed to date (Satiat-Jeunemaitre and Hawes, 1993a, 1994; Satiat-Jeunemaitre *et al.*, 1996a).

Therefore, the existence of a post-Golgi compartment as part of the secretory machinery but functionally distinct from the rest of the cisternae in the stack is still an open question in plant cell biology. This debate can be considered alongside the fidelity of the 'partially coated reticulum' (PCR), originally described by Pesacreta and Lucas in characean algae (Pesacreta and Lucas, 1984), but also considered to be part of the endocytic machinery in higher plants (see Section 6). The PCR, consisting of short ramifying tubules of membrane which are often clathrin-coated, shows structural similarities to the *trans*-Golgi network described in the plant literature. Therefore it is also possible that the PCR simply represents a TGN which has spatially separated from the rest of the Golgi stack as suggested by Hillmer *et al.* (1988), although it is one of the first compartments to receive endocytosed marker molecules (Fowke *et al.,* 1991).

Enzyme distribution, function and retention within the Golgi

In addition to its sorting and packaging function (Bednareck and Raikhel, 1992; Chrispeels, 1991; Vitale and Chrispeels, 1992; see also Section 5), the plant Golgi apparatus also has the task of maturing asparagine-linked oligosaccharides, and synthesizing *O*-linked oligosaccharides and complex cell wall polysaccharides (Driouich *et al.,* 1993). Glycosyltransferases that catalyse the stepwise addition of sugar residues to form oligo- and polysaccharides are generally regarded as working in a co-ordinated and highly ordered fashion. However, in contrast with the general acceptance of this assembly-line concept, there is little evidence for a spatially separate distribution of glycosyltransferases between the different cisternae of the Golgi apparatus.

Processing of glycoproteins. Plant glycoproteins fall into two general categories: the *N*-linked glycans (high-mannose and complex glycans, described for soybean lectins and many other

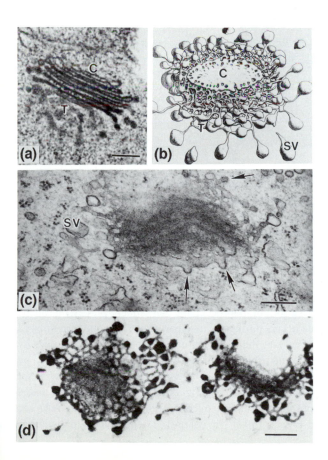

Figure 2. Different aspects of the Golgi apparatus in maize root tip cells. (a) Conventional thin section. (b) Reconstruction from transverse and *'en face'* views depicting the fenestrated/tubular nature of the cisternal margins. (c) Oblique view showing non-clathrin-coated vesicles budding from cisternal margins (arrows). (d) Osmium-impregnated stacks clearly showing the tubular nature of the cisternal margins towards the *trans* face. C, *cis* face; T, *trans* face; SV, secretory vesicles. Scale bar = 200 nm.

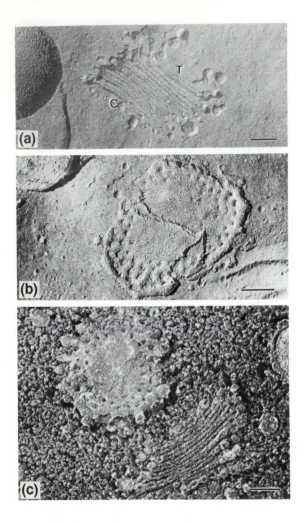

Figure 3. Freeze-fracture analysis of Golgi structure in unfixed, ultra-rapidly frozen carrot suspension culture cells. (a) Cross-section of Golgi stack showing vesicles budding from all cisternal margins. (b) Cisternum towards the *cis* face showing the swollen and fenestrated nature of the cisternal margins. (c) Deep-etch rotary-shadowed replica of stacks in face view and cross-section, again showing vesicles budding from all cisternal margins. C, *cis* face; T, *trans* face. Scale bars = 200 nm.

mature vacuolar or extracellular plant lectins and enzymes), and *O*-linked glycans (represented, for example, by the families of hydroxyproline-rich glycoproteins, or the arabinogalactan-proteins (Faye *et al.,* 1993; see also Chapter 6 by Knox).

The synthesis of *N*-linked oligosaccharides starts in the ER with the formation of a lipid-linked precursor oligosaccharide that has the composition $Glc_3Man_9(GlcNAc)_2$. The first seven sugar residues of this precursor oligosaccharide are assembled on the cytosolic face of the ER, and the biosynthesis of the molecule is finished on the lumenal face of this organelle. The precursor oligosaccharide is then transferred from the dolichol lipid carrier on which it was built to specific asparagine residues of the nascent polypeptide chain. Subsequent processing of the protein-linked precursor oligosaccharide occurs during the transport of the glycoprotein through the secretory pathway. This processing involves the trimming of a number of sugar residues by ER and Golgi-bound glycosidases, and the addition of new sugar residues by glycosyltransferases located in the Golgi apparatus. The maturation of the ER precursor oligosaccharide from $Glc_3Man_9(GlcNAc)_2$ to produce a Golgi $Man_5(GlcNAc)_2$ protein-linked oligosaccharide follows the same pattern in plant and mammalian cells (for recent reviews see Abeijon and Hirschberg, 1992; Faye *et al.,* 1992). In contrast, Golgi glycosyltransferases responsible for further processing of this $Man_5(GlcNAc)_2$ oligosaccharide into complex glycans generate structures which differ in plants and mammals. For instance, plant complex N-glycans differ from their mammalian counterparts in that they lack sialic acid, they have a β-1,2 xylose residue linked to the β-mannose of the glycan core, and they possess a fucose residue α-1,3-linked to the proximal GlcNAc of the chitobiose unit.

Few data are available regarding the subcellular localization of glycan-processing events in the Golgi apparatus of plant cells. Glycosyltransferases, such as xylosyl- and fucosyltransferases, have been shown to be located within the Golgi apparatus of cucumber (Sturm and Kindl, 1983) and bean (Johnson and Chrispeels, 1987). Recent attempts to describe more precisely the distribution of glycosyltransferases within the plant Golgi apparatus have taken advantage of the preparation, purification and characterization of polyclonal antibodies directed against the β-1,2 xylose or β-1,3 fucose residues which are specific components of plant complex N-glycans (Faye et al., 1993; Fitchette-Lainé et al., 1994). Using these antibodies it was shown that plant complex glycan β-1,2 xylosylation occurs mainly in the medial Golgi cisternae (Lainé et al., 1991), while α-1,3 fucosylation reaches its optimum level in the trans-Golgi cisternae and the TGN of suspension-cultured sycamore cells (Fitchette-Lainé et al., 1994). Consequently, these results are consistent with the non-even distribution of Golgi enzymes which has previously been illustrated by cytochemical staining techniques (Domozych, 1989; Goff, 1973). However, immunolabelling for both α-1,3 fucosylation and β-1,2 xylosylation of complex glycans is not strictly compartmentalized within the Golgi stacks, as both events start in the cis-Golgi cisternae. Therefore a strict compartmentation of glycosyltransferases responsible for complex N-glycan biosynthesis has not yet been demonstrated in plant cells using either immunolocalization with antibodies specific for glycosyltransferase products (Faye et al., 1993; Fitchette-Lainé et al., 1994) or Golgi fractionation techniques on sucrose gradients (Ali et al., 1986; Sturm et al., 1987).

Even though O-glycosylation patterns appear to be less complex than those of N-linked glycan biosynthesis, little information is currently available about the assembly pathway of O-linked glycans. O-linked glycans are important components of the hydroxyproline (Hyp)-rich glycoprotein extensin (Lamport, 1986). Extensin consists of a repeated pentapeptide sequence, Ser-(Hyp)$_4$, and Tyr-Lys-Tyr sequences (for review see Cassab and Varner, 1988; Showalter, 1993; Showalter and Varner, 1989). In the Ser-(Hyp)$_4$ repeated motif, most Hyp residues are O-glycosylated with one to four arabinose residues, and some Ser residues are O-substituted with one galactose (Showalter, 1993). These O-linked glycans probably reinforce the rod-like structure of extensin (Stafstrom and Staehelin, 1986). There is no information available about the compartment where Ser residues are O-glycosylated with galactose, but arabinosyltransferases acting on Hyp residues are probably located within the Golgi apparatus, and Hyp arabinosylation could progress in a cis to trans direction (Gardiner and Chrispeels, 1975; Kawasaki, 1981; Owens and Northcote, 1981; Swords and Staehelin, 1993). However, some authors have shown that the arabinosylation could start in the ER (Andreae et al., 1988; Robinson et al., 1985). There is also some ambiguity about prolyl-hydroxylation, which is described as occurring in plants not only in the ER, as in animal cells, but also in the Golgi apparatus (Gardiner and Chrispeels, 1975; Sauer and Robinson, 1985).

Polysaccharide biosynthesis. The major products synthesized in the plant Golgi apparatus are, however, not O- or N-linked glycans but complex polysaccharides constitutive of the cell wall (Levy and Staehelin, 1992; Roberts, 1990). Indeed, with the exception of cellulose and callose, which are in most cases assembled from UDP-glucose by multi-subunit complexes located in the plasma membrane (see Chapter 4 by Blanton and Haigler and Chapter 5 by Kauss), the other cell wall polymers such as hemicelluloses and pectic polysaccharides are assembled in the Golgi apparatus. Different enzymes involved in polysaccharide synthesis have been localized in isolated Golgi membranes by biochemical methods, and some studies suggest that polysaccharide assembly could be controlled by a compartmentalized set of glycan transferases within the Golgi stacks (Brummell *et al.*, 1990; Faye *et al.*, 1993; Hobbs *et al.*, 1991; Levy and Staehelin, 1992). However, it is only with the recent development of immunocytochemical studies using specific antibodies against sugar epitopes that models of synthesis and secretion of complex plant polysaccharides throughout the Golgi stack have been analysed further (Staehelin *et al.*, 1991). These studies mainly concern the assembly of polygalacturonan/rhamno-galaturonan-I (PGA/RG-I, a class of acid pectic polysaccharides) and xyloglucan (XG, a neutral hemicellulose particularly abundant in the cell walls of the dicotyledons). PGA/RG-I consists of two covalently bound domains, namely the PGA domain, which carries different amounts of negative charge as a function of the degree of methyl esterification, and the RG-I domain, which contains alternating galactosyl and rhamnosyl residues with approximately half of the rhamnosyl residues carrying oligosaccharide side-chains (Bacic *et al.*, 1988; Lau *et al.*, 1985).

The biosynthesis of the pectic polysaccharide PGA/RG-I has been studied by immuno-cytochemical techniques in sycamore suspension-cultured cells, and appears to involve all types of Golgi cisternae. Using four anti-sugar epitope antibodies, it has been shown that the unesterified backbone of PGA/RG-I is most probably assembled in *cis* and medial cisternae, methyl esterification of the carboxyl groups of the galacturonic acid residues in the PGA domains occurs in medial cisternae, and the arabinose-containing side-chains of the RG-1 domains are added in the *trans* cisternae (Zhang and Staehelin, 1992). Interestingly, when clover root-tip cells are labelled with the anti-unesterified PGA/RG-I backbone antibodies, the labelling is seen almost exclusively over *cis* and medial Golgi cisternae in cortical parenchyma cells (Moore *et al.*, 1991), but mostly over *trans* Golgi cisternae and the TGN in epidermal and peripheral root cells secreting mucilage (Lynch and Staehelin, 1992). In *Vicia* root hairs, there is an apparent absence of unesterified PGA in the Golgi complex, and methyl-esterified PGA are found in the medial and *trans* Golgi cisternae (Sherrier and VandenBosch, 1990). The variation between these studies may be related to differences in the probes used to trace the PGA (for discussion see Sherrier and VandenBosch, 1990), or alternatively they may reflect the fact that different secreted forms of PGA may occur among plant cells. Such differences may also be due to a differential regulation of polysaccharide synthesis according to the cell's needs, as demonstrated in clover root tissues (Lynch and Staehelin, 1992), or they could be a result of species-specific variations.

Xyloglucan (XG), the most abundant hemicellulose, has the same β-1,4 glucan backbone as cellulose, but many glucose residues are substituted with xylose. In dicots, some of these xylose residues are further substituted with galactose and fucose. The enzymes or the multi-enzyme complex responsible for the synthesis of xylosylated β-glucan backbone and the fucosyl- and galactosyltransferases that modify the glucose–xylose backbone have been identified as being located in the Golgi apparatus using either membrane fractionation on density gradients (Brummell *et al.*, 1990; Farkas and Maclachlan, 1988; Maclachlan *et al.*, 1992) or immunolocalization of their products (Levy and Staehelin, 1992; Moore *et al.*, 1991; Sherrier and VandenBosch, 1990; Zhang and Staehelin, 1992).

Renographin density gradient studies of a Golgi-enriched membrane fraction suggest that the XG backbone begins its assembly in the *cis* elements of the Golgi, and continues its synthesis in medial and *trans* cisternae (Brummell *et al.*, 1990). In contrast with these results, immunocytochemical localization of the xyloglucan backbone suggests that the xylosyl transferase activity occurs exclusively in the *trans* elements of the Golgi (Lynch and Staehelin, 1992; Staehelin *et al.*, 1991; Zhang and Staehelin, 1992). Thus these preliminary attempts to localize the different stages of XG biosynthesis within the plant Golgi stack have so far led to results that could be considered to be contradictory so long as a highly ordered organization of glycosyltransferases among different Golgi cisternae is the prevailing dogma. However, both biochemical and immunocytochemical studies indicate that the fucosylation of the xyloglucan side-chain (which is supposed to be the last step in the maturation of the xyloglucan molecule) probably takes place in the most *trans* part of the Golgi (Brummell *et al.,* 1990; Sherrier and VandenBosch, 1990; Staehelin *et al.,* 1991).

Such a description of a fucosyltransferase activity mainly localized in the *trans* side of the Golgi and involved in polysaccharide processing is reminiscent of the localization of fucosyltransferase activity described for glycoprotein processing, which is also localized at the *trans* side of the Golgi (Fitchette-Lainé *et al.,* 1994). Whether the fucosyl transferase enzymes involved in glycoprotein processing are the same enzymes as those involved in polysaccharide biosynthesis is, however, unclear.

Taken together, the immunocytochemical and fractionation results have shown that most glycosyltransferases involved in complex *N*-glycan processing and *O*-glycan and complex cell wall polysaccharide biosynthesis are probably Golgi membrane proteins. The enzyme distribution is not homogenous through the stack, and as such probably reflects the functional polarity of the latter. However, there is no longer a consensus on strict compartmentation of these enzymes between the different cisternae of a Golgi stack. This has already been illustrated in mammalian systems. For example, α-2,6 sialyltransferase is generally described as a *trans*-cisternae-resident membrane protein, but was immunodetected in almost all Golgi cisternae in absorptive cells of the gut (Roth *et al.,* 1988).

Retention of Golgi enzymes. At present, as discussed above, the extent to which glycosyl-transferases are compartmentalized within plant Golgi stacks is unknown. However, current progress in the isolation and molecular cloning of plant glycosyltransferases will help to answer a preliminary question of considerable interest in defining the signal(s) responsible for their targeting and retention in the Golgi complex. Glycosyltransferases located in the Golgi apparatus would be expected to contain retention signals to restrict their movement from this organelle and thus to prevent their being swept to the cell surface in the default bulk flow pathway. Several mammalian Golgi-specific glycosyltransferases have been cloned (for recent reviews see Kleene and Berger, 1993; Van den Eijnden and Joziasse, 1993), and preliminary findings suggest that sequences in and/or around the membrane spanning region of these type-II transmembrane proteins are important for their Golgi retention (Machamer *et al.,* 1990; Shaper and Shaper, 1992). Recently, an *Arabidopsis* mutant deficient in *N*-acetylglucosaminyltransferase I (GnT-I) (Von Schaewen *et al.,* 1993) has been successfully complemented by transformation with a cDNA encoding human GnT-I (Gomez and Chrispeels, 1994). Tang *et al.* (1992), using fusion proteins expressed from chimeric constructs, have shown that the transmembrane domain of GnT-I most probably contains the Golgi retention signal. Whether this apparently highly conserved signal is sufficient for retention of a chimeric protein in the Golgi apparatus of transgenic plants has not yet been demonstrated. Further investigations on the underlying mechanism of retention specificity will provide insights into the requirement for physical segregation of glycosyltransferases for an ordered and precise assembly of complex glycans and polysaccharides.

Transfer down the Golgi stack — cisternal maturation or vesicle shuttles?

Newly synthesized material (secretory products, processing enzymes and membrane) must move sequentially from the *cis* to the *trans* face of the stack, where it will then be targeted to the vacuole, carried by bulk flow to the cell surface, or retained within the stack. A retrograde pathway back through the stack has not been firmly established, although membrane return to different Golgi cisternae via the endocytic pathway may well occur (see Section 6). In actively secreting cells there is a rapid flow of membrane through the Golgi stack, with cisternal turnover times estimated to vary between 1 and 20 min (Steer, 1985), which is an important factor to be taken into account when constructing a hypothesis on the mechanics of Golgi function.

Currently there are two models which attempt to explain this forward transport through the stack. In the 'cisternal progression model', proposed some years ago (Morré and Mollenhauer, 1974), cisternae and their contents move as units from the *cis* to the *trans* face through the stack, and it is this progressive maturation of the cisternae which mediates the migration of secretory products from *cis* to *trans*. The *trans*-most cisternum vesiculates, and this loss of membrane is compensated for by the addition of new cisternae at the *cis* face. Such a model assumes that each cisternum possesses the capacity to process macromolecules sequentially as it matures, and it must therefore contain a complete set of processing enzymes. This model is still popular for many algae, especially those that produce extracellular scales within individual Golgi cisternae (Becker and Melkonian, 1995; Becker *et al.*, 1995).

The second model of Golgi function in higher plants, which is currently increasing in popularity, is based on a vesicle-mediated transport between the cisternae, each of which would have a specific processing task(s). This model is currently in vogue to explain the mechanics of the Golgi apparatus in animal cells (Rothman and Orci, 1992; Rothman, 1994; see Section 2), although such vesicle carriers have yet to be isolated from plant cells. However, as discussed in Section 2, coated vesicles that are morphologically similar to COP-coated vesicles (which are known to be involved in intercisternal transport in animal cells) have been reported to be associated with plant Golgi stacks. Thus this shuttle model may also apply to plants and be ubiquitous amongst eukaryotic cells. However, in view of the fact that exit from the plant Golgi is by no means restricted to the *trans* face (Staehelin *et al.*, 1991; see also Section 5), it is possible that the two theories are not mutually exclusive.

5. Transport to and from the Golgi apparatus

It is accepted that the vesicular transport of glycoproteins and polysaccharides from their sites of synthesis (the ER and Golgi apparatus, respectively) to their final destination also involves a flow of membrane beginning at the ER, passing through the Golgi apparatus and ending at the plasma membrane or tonoplast (Morré and Mollenhauer, 1974). The Golgi apparatus is the central structure in such a flow, not only because any secretory product as well as membrane has to pass through it, but also because, in terms of vesicle trafficking, the Golgi apparatus is the only cytoplasmic organelle known to function as a multidirectional membrane donor and membrane acceptor compartment.

ER/Golgi communication

The plant endoplasmic reticulum is a complex and dynamic compartment of the endomembrane system, and as such is reviewed elsewhere in this volume (see Chapter 21 by Denecke and Chapter 22 by Lichtscheidl and Hepler). In plants, it is the only pre-Golgi compartment identified, whereas in various animal cell types intermediate compartments between the ER and Golgi have been reported (Hauri and Schweizer, 1992; Orci *et al.*, 1993). Proteins destined to pass through the secretory pathway for delivery to the plasma membrane, cell wall or vacuole are firstly co-translationally targeted into the ER lumen, where they undergo their first post-translational modifications (see Section 4). Soluble resident ER proteins, including those involved in protein processing, are retained in the ER lumen, whilst those destined for secretion are routed to the Golgi stacks. The sorting/retention mechanisms which permit such discrimination between resident and secretory proteins are discussed elsewhere in this volume (see Chapter 21 by Denecke).

Is the ER–Golgi pathway unidirectional? To date, all data from mammalian, yeast and plant cells suggest that the proteins destined for secretion pass to the Golgi via a bulk flow mechanism. In mammalian and yeast systems, any ER-resident proteins (tagged with an H/KDEL C-terminus retention signal) caught up in this flow would be returned to the ER via a salvage pathway mediated by a receptor located in a post-ER intermediate compartment and in the Golgi apparatus itself (Hauri and Schweizer, 1992; Munro and Pelham, 1987; Pelham, 1991). This hypothesis suggests the existence of a bidirectional vesicle shuttle or membrane flow between the ER and the Golgi complex. In plant cells such ER retention signals have been identified (Denecke *et al.,* 1992; Napier *et al.,* 1992; see also Chapter 21 by Denecke), and a homologue of the sequence receptor (the *ERD2* protein of yeast; Semenza *et al.,* 1990) has now been identified (Bar-Peled *et al.,* 1995; D'Enfert *et al.,* 1992; Lee *et al.,* 1993; Vitale *et al.,* 1994). However, its subcellular location is unknown, and as yet we have no evidence for a retrograde Golgi–ER pathway in plant cells. In fact, based on the response of plant Golgi to the drug BFA, which appears not to result in a redistribution of Golgi membrane into the ER system as is observed in animal cells (Satiat-Jeunemaitre and Hawes, 1992, 1993a), the presence of such a retrograde pathway in plants has been questioned (Satiat-Jeunemaitre and Hawes, 1994; Satiat-Jeunemaitre *et al.,* 1996a).

How is ER–Golgi transport achieved? The diversity of the structural interrelationships between the ER and the Golgi in plant cells has often been debated (Harris and Oparka, 1983; Juniper *et al.,* 1982; Marty, 1978; Robinson, 1980; Robinson and Kristen, 1982). On the basis of electron microscopic studies of various cell types, it has been concluded that the structural relationship between the ER and the Golgi could very well depend on the nature of the predominant product secreted. It has been traditionally accepted that ER–Golgi communication is mediated by transition vesicles into which secretory proteins are packaged as soon as they acquire transport competence. Such vesicles, 50–65 nm in diameter, have been reported in various algal cells (Domozych, 1991; Domozych *et al.,* 1992; Melkonian *et al.,* 1991; Nogushi and Morré, 1991). Careful study of such electron micrographs suggests that these vesicles may be coated and appear morphologically similar to the coatomer-coated vesicles described in animal cell systems (see Section 2 above). In higher plants, such transition vesicles are rarely observed *in situ* (Dauwalder and Whaley, 1982) although the formation of similar vesicles 50–60 nm in diameter has been reported during initial attempts to reconstitute vesicle transport in cell-free systems from plants (Hellgren *et al.,* 1993; Morré *et al.,* 1989; Nogushi and Morré, 1991).

Ultrastructural observations on protein-secreting plant cells could in fact militate against identification of transition vesicles. For instance, in glands of insectivorous plants which are specialized with regard to enzyme secretion, the Golgi stacks appear tightly enmeshed in the ER network, with ER/cisternal connections being apparent throughout the stack (Juniper *et al.,* 1982). Similarly, in developing legume cotyledons, direct tubular connections between the ER and the Golgi have been reported (Harris and Oparka, 1983). Yet in carbohydrate-secreting cells such as those of the maize root cap, the ER and Golgi can appear spatially separate, with no evidence of direct contacts or vesicular intermediates, a situation which is perhaps the norm in most plant cells (Juniper *et al.,* 1982; Robinson, 1980; Robinson and Kristen, 1982; Satiat-Jeunemaitre *et al.,* 1996c). Moreover, if phospholipid insertion could take place in Golgi compartments as in other organelles (Van Meer, 1993; see also Chapter 10 by Kader *et al.*), this could potentially reduce the requirement for ER–Golgi transfer of membrane to service the packaging of carbohydrates into secretory vesicles.

Golgi–vacuole pathways

Most of our knowledge of trafficking between the Golgi and vacuoles comes from studies on the accumulation of a range of soluble storage proteins in developing cotyledons and other storage tissues (Chrispeels, 1983; Herman *et al.,* 1990; Höfte *et al.,* 1991; Matsuoka and Nakamura, 1992; Von Schaewen and Chrispeels, 1993; see also Chapter 23 by Reynolds and Raikhel). The application of molecular techniques in combination with the expression of these proteins in transgenic plants has resulted in spectacular advances in our understanding of the processing and targeting of such proteins (Höfte and Chrispeels, 1992; Vitale and Chrispeels, 1992; see also Chapter 23 by Reynolds and Raikhel). The transport of these storage proteins is a result of an active sorting process which takes place in the *trans*-most elements of the GA (Chrispeels, 1983, 1991; Craig and Goodchild, 1984a,b), and as such is most probably based on a receptor-mediated recognition process. Although various peptide-targeting sequences have been identified, to date there is only one tentative report that an 80-kDa protein from pea cotyledon CCVs recognizes the N-terminal targeting sequence of barley pro-aleuraine and may thus be a candidate for a *trans*-Golgi-based receptor (Kirsch *et al.,* 1994). However, the authors reported no affinity of the putative receptor for pea prolectin, which is also targeted to the vacuole, and they presented no data for the major pea storage proteins, such as legumin and vicilin, which have also been reported to be transported to the vacuole via *trans*-Golgi-derived coated vesicles (Harley and Beevers, 1989; Hoh *et al.,* 1991). This peptide-based signalling is more akin to the situation in yeast cells (Harris and Watson, 1991; Schekman, 1992) than to the carbohydrate residue-based system of targeting from the *trans* Golgi to the lysosome in animal cells (Kornfeld and Mellman, 1989).

The Golgi–vacuole pathway can be manipulated pharmacologically. For example, monensin, which modifies the Golgi-based processing and sorting function (Satiat-Jeunemaitre *et al.,* 1994), has been shown to induce a redirection of various vacuolar proteins, such as concanavalin A (Bowles *et al.,* 1986), vicilin (Craig and Goochild, 1984b) and phytohaemagglutinin (PHA) (Gomez and Chrispeels, 1993), to the cell surface. More specifically, deletion of the targeting sequence on these proteins results in the protein being secreted to the cell surface, thus identifying the Golgi–plasma membrane pathway as the default pathway (Denecke *et al.,* 1990).

The mechanisms underlying the transport of intrinsic tonoplast proteins from the Golgi to the vacuole are less clear. From the limited data on plant systems based on the targeting of α-TIP (a major tonoplast intrinsic protein from bean seeds) and PHA in transgenic tobacco plants, it appears that both monensin and the secretion inhibitor brefeldin A block the transport of PHA to the vacuole, but not that of α-TIP. This suggests that there may be at least two distinct vesicle-trafficking pathways from the Golgi to the vacuole (Gomez and Chrispeels, 1993).

No retrograde vacuole–Golgi pathway has yet been described in plant cells. However, endosome/lysosome to Golgi transport is well characterized in animal cell systems, and the possibility of vesicle-based recycling of receptors between the tonoplast and the Golgi should not be ignored.

Golgi–cell surface transport

The plant cell surface is a heterogeneous and complex structure composed of polysaccharides and glycoproteins which often interact with extrinsic and intrinsic proteins of the plasma membrane (Roberts, 1990). In higher plants the majority of the non-cellulosic macromolecules of the cell surface are processed in and packaged by the Golgi apparatus, and transferred to the cell surface in secretory vesicles. Similarly, the cellulose synthases active at the plasma membrane are transported to the cell surface in Golgi-derived vesicles (Haigler and Brown, 1986; see also Chapter 4 by Blanton and Haigler).

The sequence of events which occurs after processing in the lumen of the Golgi cisternae and before final fusion with the plasma membrane has yet to be fully elucidated. However, the production of banks of antibodies to cell-surface epitopes (Herman and Lamb, 1992; Knox, 1990; Knox *et al.*, 1991; Levy and Sataehelin, 1992; Pennel *et al.*, 1989; see also Chapter 6 by Knox), combined with a range of other affinity-labelling methods (Vian and Roland, 1991) to probe for vesicle cargoes has resulted in some progress in this difficult area of Golgi function.

What is the site of origin of secretory vesicles? Secretory molecules destined for the cell surface are packaged in smooth vesicles, and such products have never been located *in situ* in *trans*-Golgi-derived CCVs (Sherrier and VandenBosch, 1990). The commonly accepted view is that exocytotic vesicles are generated by scission of the swollen margin of the Golgi apparatus to produce smooth vesicles considerably larger (but variable in size) than the 50–70 nm coated vesicles associated with the stack (Morré *et al.*, 1989; Nogushi and Morré, 1991). The *trans*-most cisternum (or *trans*-Golgi network, depending on the existence of the latter as an autonomous compartment) appears to be the main producer of secretory vesicles, although it is obvious from the study of numerous electron micrographs and from affinity-labelling techniques that secretory vesicles can originate from the medial Golgi cisternae (*Figure 2a*; see also Staehelin *et al.*, 1991).

Heterogeneity in the stacks and secretory vesicles. Immunolabelling of polygalacturonic acids in mung bean cells has demonstrated that there may be a population of stacks which do not process the polysaccharide, thus indicating heterogeneity within the Golgi population (Vian and Roland, 1991). It has also been demonstrated in suspension-cultured cells that the stack can process glycoproteins (extensin) and polysaccharides (xyloglucans) simultaneously (Staehelin *et al.*, 1991), or two types of polysaccharides (Zhang and Staehelin, 1992), indicating functional heterogeneity within the stack. If this is the case, then is it possible that individual vesicles can transport more than one category of secretory macromolecule? In the root cap cells of clover, Moore *et al.* (1991) found that pectins (rhamnogalacturonans) and hemicelluloses (xyloglucans) resided in distinct secretory vesicles, suggesting a differential packaging of the two polysaccharides, although 9% of the vesicles did show mixed labelling (Staehelin *et al.*, 1991). More recently, in *Vicia* root hairs it has been shown that xyloglucan and pectin (methyl-esterified polygalacturonans) are packaged and transported in the same Golgi vesicles (Sherrier and VandenBosch, 1990). As different sources of antibodies recognizing pectins were used in the studies, it may be that the epitopes recognized differed from one study to the other, and as such gave rise to these apparently conflicting observations. Alternatively, they may be related to a difference in regulation of the Golgi function (and glycan transferase repartition within the Golgi stack) between species or tissues. It would then have to be assumed that, as in the case of *Vicia* root hairs, the final processing of methyl-esterified polygalacturonans would take place alongside that of xyloglucans, resulting in vesicles containing a mixture of the two polysaccharide products.

Differential packaging of secretory glycoproteins and polysaccharides has yet to be demonstrated, although, as the former are thought to be sorted in the *trans* Golgi and to occur in polysaccharide-synthesizing stacks (Levy and Staehelin, 1992; Staehelin *et al.*, 1991), it is probable that they can be packaged in the same vesicle.

It is also important to note that the release of a vesicle from the Golgi stack does not signal the end of the processing event. Vesicles themselves are not biochemically passive, and can be regarded as processing compartments in their own right. For example, fucosyl transferase activity involved in the maturation of the xyloglucan has been detected in the secretory vesicles (Brummell *et al.,* 1990). In mung bean cells, pectin methylesterase apparently starts the maturation (methyl esterification) of polygalacturans once the secretory vesicle has parted company with its parent cisternum (Vian and Roland, 1991).

The cell surface — a default pathway? Although a variety of experiments concerned with the deletion of vacuolar targeting sequences from storage peptides have demonstrated that the Golgi–cell surface route may simply be a default pathway (Denecke *et al.,* 1990), consideration of the polarized nature of many plant cell types suggests that some form of control of vesicle trafficking must take place (Battey and Blackbourn, 1993). Many tissues exhibit differential thickening of cell walls, some of the most obvious being the endodermal cell walls in monocotyledons and the construction of the cell plate, where it must be assumed that some form of vesicle targeting takes place. Even in sites of increased cellulose synthesis, such as wall thickenings in tracheary elements, it is possible that the Golgi-derived synthetic complexes are targeted in vesicles rather than simply being clustered by rearrangement of the pre-existing population in the plasma membrane. However, other than suggesting a cytoskeleton-mediated targeting system (Emons *et al.,* 1992; Satiat-Jeunemaitre, 1992; Schmid and Meindl, 1992), there is no information about the way in which secretory vesicles are targeted and fused to the correct plasma membrane domains at the correct time.

Bypassing the Golgi

Although in many cases the formation of protein storage bodies involves transport through the Golgi and vesicle-based transport, there are also many instances where the bodies correspond to a protein accumulation in the ER lumen without any Golgi-based processing (Galili *et al.,* 1993). In developing wheat endosperm, protein bodies forming in the ER appear to be internalized directly into a vacuole by the apparently autophagic action of small electron-lucent vesicles surrounding the protein aggregate (Levanony *et al.,* 1992). In castor bean seeds (Hiraiwa *et al.,* 1993) and pumpkin seeds (Hara-Hishimura *et al.,* 1993), storage proteins appear to be transported from the ER in vesicles and fuse to form a storage vacuole without any involvement of the Golgi.

A direct ER to cell surface secretion has been described in the extrafloral nectary of *Abutilon* where the prenectar is loaded into a specialized form of ER (secretory reticulum, SER), which fuses directly with the plasma membrane to expel the nectar on to the gland surface (Robards and Stark, 1988). This would overcome the problem of the cell having to recycle large areas of plasma membrane rapidly after a huge secretion event (Kronestedt-Robards and Robards, 1991).

6. Endocytosis — the way in

Mapping the pathway

The endocytic pathway is perhaps one of the best characterized vesicle-mediated pathways in animal cell systems. Unfortunately, it is well beyond the scope of this chapter to describe the outstanding work carried out in the last decade or so on the receptor-mediated uptake of physiologically significant molecules into CCVs, and the delivery of such molecules via the endosome to either the lysosomal system or the *trans* Golgi (Brodsky *et al.,* 1991; Goldstein *et al.,* 1985; Keen, 1990; Pearse and Robinson, 1990; Van Deurs *et al.,* 1989).

However, it has now been conclusively demonstrated that plant cells have the ability to undergo endocytic internalization of the plasma membrane and associated marker molecules

(Hawes *et al.*, 1995; Satiat-Jeunemaitre and Hawes, 1993b). Material can either be endocytosed in the fluid phase or bound to the plasma membrane non-specifically by charge or specifically via receptor/ligand interactions. Although the occurrence of the latter in plant cells is still a matter for contention, studies on the structure of plant CCVs have shown them to be morphologically and biochemically homologous with their mammalian counterparts. Thus it does appear to be acceptable to speculate about the existence of a receptor-mediated endocytic pathway in plants.

Various attempts have been made to study the kinetics of endocytosis and to map the pathway using fluorescent probes and, more successfully, electron-opaque plasma membrane markers (Hawes *et al.*, 1991, 1995). The obvious probes to use to study fluid-phase endocytosis, such as high-molecular-weight fluorescent dextrans and other membrane-impermeant dyes (e.g. Lucifer Yellow CH) have proved to be far from ideal. However, some uptake of contaminant-free fluoroscein isothiocyanate (FITC)-conjugated dextran has been demonstrated in carrot cells (Cole *et al.*, 1990) and pollen tubes (O'Driscoll *et al.*, 1993), but the possibility of spurious data arising as a result of enzymic cleavage of the dextran at the cell wall has to be considered.

Lucifer Yellow, one of the standard markers for fluid-phase endocytosis in animal cells, has been shown to be rapidly internalized into the vacuole of various plant cell types. However, it has now been demonstrated that this uptake does not take place by endocytosis but that it is mediated by inorganic anion transporters (Cole *et al.*, 1991; Oparka *et al.*, 1991, 1994; Oparka and Hawes, 1992). For example, when Lucifer Yellow was coupled to high-molecular-weight dextrans, the dye was excluded from oat aleurone protoplasts that had the ability to sequester unconjugated Lucifer Yellow (Wright *et al.*, 1992).

Ultrastructural studies have, through the use of adsorptive markers such as cationized ferritin and gold-conjugated lectins (Galway *et al.*, 1993; Hillmer *et al.*, 1986; Tanchak *et al.*, 1984) and fluid-phase heavy-metal salts (Hübner *et al.*, 1985; Samuels and Bisalputra, 1990), demonstrated the presence of a CCV-mediated endocytic pathway. Perhaps the studies on the uptake of cationized ferritin by soybean protoplasts (Fowke *et al.*, 1991) have provided the clearest picture of the various routes within the endocytic pathway. Uptake of cationized ferritin bound to the plasma membrane into coated pits occurred seconds after attachment at room temperature. After 2 min, some vesicles in the cytoplasm had lost their coats, but some CCVs retained ferritin for 5 min. Two minutes after treatment with ferritin, the first label appeared in the membranes of the PCR. After 4 min, labelling was observed at the periphery of one or two Golgi cisternae, with further labelling of the periphery of cisternae occurring with longer exposure to ferritin. Light-labelling of multivesicular bodies was observed after 6 min, with ferritin accumulating in the organelle with time. After 1 h of exposure to ferritin, the label was detected in clusters in the central vacuole and in peripheral vacuoles in the protoplasts. This time-course of events has also been confirmed in protoplasts from suspension-cultured cells of white spruce, using both conventional fixation and ultra-rapid freezing techniques (Galway *et al.*, 1993).

On the basis of these experiments, it has been suggested that plasma membrane-derived coated vesicles deliver their membrane first to the partially coated reticulum. Whether this organelle is the plant equivalent of the animal early endosome, or whether it is directly associated with the *trans*-Golgi network (see Section 4) is a question that will remain unanswered until suitable marker molecules are identified for these components of the system. The PCR buds off membrane as CCVs, which are sent to the Golgi or to multivesicular bodies. In the latter, ferritin accumulates before being sent to the central vacuole. It is significant that ferritin is observed not only in the *trans*-Golgi cisternae but also throughout the stack, indicating either the existence of a retrograde vesicle-mediated pathway through the stack or the fact that various cisternae in the stack have the ability to dock incoming vesicles (see Section 5 for further discussion).

Do smooth endocytic vesicles help to balance the membrane economy of the cell?

Although there is little evidence of naturally occurring, non-clathrin-coated, endocytic vesicles in plants (as discussed in Chapter 3 by Oparka *et al.*), it has been demonstrated that osmotic contraction of the plasma membrane can result in the formation of endocytic vesicles (Oparka *et al.*, 1990). However, the membranes of such vesicles do not appear to possess the necessary molecular organization and/or signals to permit fusion with other compartments within the endomembrane system. Thus they cannot be considered to be a genuine component of the endocytic machinery, and are not involved in membrane recycling. Under naturally imposed conditions of osmotic stress it is likely that other mechanisms have also evolved for contending with a dramatic reduction in the surface area of the plasma membrane. These may include the formation of Hechtian strands (Oparka *et al.*, 1994) and exocytotic extrusions from the plasma membrane (Gordon-Kamm and Steponkus, 1984; Steponkus, 1991).

Receptor-mediated endocytosis

The molecular machinery to permit receptor-mediated endocytosis undoubtedly exists in plant cells. However, there is very little evidence of the internalization of physiologically significant molecules. Just one report has proposed that the internalization of FITC-labelled poly-galacturonic acid and FITC-labelled elicitor from the fungus *Verticillium dahliae* into vacuoles of cultured soybean cells took place by receptor-mediated endocytosis (Horn *et al.*, 1989). As the investigation was solely by light microscopy, no evidence of endocytic vesicles or intermediate compartments in the uptake pathway was presented. It has also been claimed that soybean cells possess receptors which permit the internalization of biotin and biotin-conjugated macromolecules (Horn *et al.*, 1990; Low *et al.*, 1993), but again compartmentalization of the probes in the cytoplasm was not demonstrated (Hawes *et al.*, 1995).

Therefore, the following questions still remain unanswered: what are the functions of plant plasma membrane-coated vesicles, and what are their cargoes? An analysis of coated-vesicle fractions may appear to be the obvious method for elucidating their contents. However, as CCVs can bud from various organelles in the cell, pure fractions from the plasma membrane are difficult to obtain. For example, it has been shown that CCV fractions from various plant sources possess V-type H^+-ATPases, which are presumably incorporated into Golgi-derived CCVs targeted to the vacuole (Drucker *et al.*, 1993). The same authors have also suggested the presence of a plasma membrane E_1/E_2 ATPase. Similarly, it has been shown that CCV preparations from developing pea cotyledons contain storage proteins (Hoh *et al.*, 1991) and storage protein precursors (Harley and Beevers, 1989), which are presumably *trans*-Golgi-derived, and destined for storage vacuoles. However, to date, other than the simple cytochemical localization of polysaccharides in plasma membrane CCVs and isolated CCV fractions (Hawes *et al.*, 1989, 1995), no obvious cargo has been identified in plasma membrane-derived vesicles.

The simplest explanation is that plasma membrane CCVs are only involved in plasma membrane retrieval and recycling (Coleman *et al.*, 1988). However, given the established homology of clathrin genes between animals and yeast, and the biochemical and immunological similarity between plant and mammalian clathrin and its associated proteins (see Section 2), it may be safe to assume an equivalent homology of function across the kingdoms. Whilst plant cells may not have the same requirement as mammalian cells for the receptor-mediated extraction of signal molecules and various macromolecular assemblages from the surrounding medium, a receptor-mediated recycling of important plasma membrane proteins could be envisaged. Thus, for instance, plasma membrane pumps such as the H^+-ATPase could be recycled after auxin-stimulated exocytosis and insertion into the membrane (Hager *et al.*, 1991).

7. Conclusions and prospects

In many respects the study of vesicle-trafficking pathways in plant cells is still in its infancy. The basic mapping of the pathways is complete, and the functions of the major organelles within the endomembrane system are well established. Steady progress can be expected to be made at the biochemical level with continued improvements in techniques for the isolation of vesicles and their donor and docking compartments. Furthermore, our bank of affinity markers (antibodies, lectins, enzymes, etc.) and drugs with which to probe the various components of the system is growing rapidly. However, it is the use of these technologies in combination with the increasing armoury of molecular techniques that will undoubtedly result in rapid progress in this area of research in the next few years. Already, as we have seen in this chapter, plant homologues of several of the proteins involved in the secretory pathway of mammalian and yeast cells have been identified. It should soon be possible to assess their function in transgenic systems by the utilization of antisense technology, or by the overexpression of mutated inactive protein. The targeting of heterologous proteins to different compartments by expressing proteins such as mammalian Golgi transferases should provide valuable information about compartmentalization within the Golgi stack. Finally, the search for secretory and/or endocytic mutants should open an avenue towards the identification of plant-specific proteins involved in vesicle shuttling and Golgi function.

8. Acknowledgements

We are grateful to Clare Steele and Dr D.E. Evans (both of Oxford Brookes University) for critically reading this manuscript.

References

Abeijon C, Hirschberg CB. (1992) Topography of glycosylation reactions in endoplasmic reticulum. *Trends Biochem. Sci.* **17:** 32–36.

Ali MS, Mitsui T, Akazawa T. (1986) Golgi-specific localization of transglycosylases engaged in glycoprotein biosynthesis in suspension-cultured cells of sycamore (*Acer pseudoplatanus*). *Arch. Biochem. Biophys.* **251:** 421–431.

Andrawis A, Solomon M, Delmer DP. (1993) Cotton fiber annexins: a potential role in the regulation of callose synthase. *Plant J.* **3:** 763–772.

Andreae M, Lang WC, Barg C, Robinson DG. (1988) Hydroxyproline-arabinosylation in the endoplasmic reticulum of maize roots. *Plant Sci.* **56:** 205–212.

Bacic A, Harris PJ, Stone BA. (1988) Structure and function of plant cell walls. *Biochem. Plants* **14:** 297–371.

Bar-Peled M, da Silva Conceiçao A, Frigerio L, Raikhel NV. (1995) Expression and regulation of aERD2, a gene encoding the KDEL receptor homolog in plants, and other genes encoding proteins involved in ER–Golgi vesicular trafficking. *Plant Cell* **7:** 667–676.

Baskin TI, Cande WZ. (1990) The structure and function of the mitotic spindle in flowering plants. *Annu. Rev. Plant Physiol. Plant Mol. Biol.* **41:** 277–315.

Bassham DC, Gal S, da Silva Conceiçao A, Raikhel NV. (1995) An *Arabidopsis* syntaxin homologue isolated by functional complementation of a yeast PEP12 mutant. *Proc. Natl Acad. Sci. USA* **92:** 7262–7266.

Battey NH, Blackbourn HD. (1993) The control of exocytosis in plant cells. *New Phytol.* **125:** 307–338.

Becker B, Melkonian M. (1995) Intra-Golgi transport mediated by vesicles? *Bot. Acta* **108:** 172–173.

Becker B, Bölinger B, Melkonian M. (1995) Anterograde transport of algal scales through the Golgi complex is not mediated by vesicles. *Trends Cell Biol.* **5:** 305–307.

Bednareck SY, Raikhel NV. (1992) Intracellular trafficking of secretory proteins. *Plant Mol. Biol.* **20:** 133–150.

Bennett MK, Scheller RH. (1993) The molecular machinery for secretion is conserved from yeast to neurons. *Proc. Natl Acad. Sci. USA* **77:** 1496–1500.

Blackbourn HD, Battey NH. (1993) Annexin-mediated secretory vesicle aggregation in plants. *Physiol. Plant.* **89:** 27–32.

Blackbourn HD, Barker PJ, Huskisson NS, Battey NH. (1992) Properties and partial protein sequence of plant annexins. *Plant Physiol.* **99:** 864–871.

Bowles DJ, Marcus SE, Pappin DJC, Findlay JBC, Eliopoulos E, Maycox PR, Burgess J. (1986) Post-translational processing of concanavalin A precursors in jackbean cotyledons. *J. Cell Biol.* **102:** 1284–1297.

Brodsky FM, Hill BL, Acton SL, Näthke L, Wong DH, Ponnambalam S, Parham P. (1991) Clathrin light chains: arrays of protein motifs that regulate coated vesicle dynamics. *Trends Biochem. Sci.* **16:** 208–213.

Brummell DA, Camirand A, Maclachlan GA. (1990) Differential distribution of xyloglucan glycosyltransferases in pea Golgi dictyosomes and secretory vesicles. *J. Cell Sci.* **96:** 705–710.

Butler JM, Beevers L. (1993) Putative adaptor proteins of clathrin-coated vesicles from developing pea. In: *Molecular Mechanisms of Membrane Traffic* (Morré DJ, Howell KE, Bergeron JJM, ed.). NATO ASI Series, Vol. H74. Berlin: Springer-Verlag, p. 31.

Cassab GI, Varner JE. (1988) Cell wall proteins. *Annu. Rev. Plant Physiol. Plant Mol. Biol.* **39:** 321–353.

Chavrier P, Stelzer J-P, Simons K, Gruenberg J, Zerial M. (1991) Hypervariable C-terminal domain of Rab proteins acts as a targeting signal. *Nature* **353:** 769–772.

Cheon C III, Lee N-G, Siddique A-BM, Bal AK, Verma DPS. (1993) Roles of plant homologs of Rab 1p and Rab 7p in the biogenesis of the peribacteriod membrane, a subcellular compartment formed *de novo* during root nodule symbiosis. *EMBO J.* **12:** 4125–4135.

Chrispeels MJ. (1983) The Golgi apparatus mediates the transport of phytohemagglutinin to the protein bodies in bean cotyledons. *Planta* **158:** 140–151.

Chrispeels MJ. (1991) Sorting of proteins in the secretory system. *Annu. Rev. Plant Physiol. Plant Mol. Biol.* **42:** 21–53.

Clark GB, Dauwalder M, Roux SJ. (1992) Purification and immunolocalisation of an annexin–like protein in pea seedlings. *Planta* **187:** 1–9.

Cole L, Coleman J, Evans D, Hawes C. (1990) Internalisation of fluorescein isothiocyanate and fluorescein isothiocyanate–dextran by suspension cultured carrot cells. *J. Cell Sci.* **96:** 721–730.

Cole L, Coleman J, Kearns A, Morgan G, Hawes C. (1991) The organic anion transport inhibitor, probenecid, inhibits the transport of Lucifer Yellow at the plasma membrane and the tonoplast in suspension cultured plant cells. *J. Cell Sci.* **99:** 545–555.

Coleman J, Evans D, Hawes C. (1988) Plant coated vesicles. *Plant Cell Environ.* **11:** 669–684.

Coleman JOD, Evans, DE, Horsley D, Hawes CR. (1991) The molecular structure of plant clathrin and coated vesicles. In: *Endocytosis, Exocytosis and Vesicle Traffic in Plants* (Hawes CR, Coleman JOD, Evans DE, ed.). SEB Seminar Series 45. Cambridge: Cambridge University Press, pp. 41–63.

Corthésy-Theulaz I, Pauloin A, Pfeffer SR. (1992) Cytoplasmic dynein participates in the centrosomal localization of the Golgi complex. *J. Cell Biol.* **118:** 1333–1345.

Craig S, Goodchild DJ. (1984a) Periodate-acid treatment of sections permits on-grid immunogold localization of pea seed vicilin in ER and Golgi. *Protoplasma* **122:** 35–44.

Craig S, Goodchild DJ. (1984b) Golgi-mediated vicilin accumulation in pea cotyledons is redirected by monensin and nigericin. *Protoplasma* **122:** 91–97.

Creutz CE. (1992) The annexins and exocytosis. *Science* **258:** 924–931.

Dauwalder M, Whaley WG. (1982) Membrane assembly and secretion in higher plants. *J. Ultrastruct. Res.* **78:** 302–320.

Demmer A, Holstein SEH, Hinz G, Schauermann G, Robinson DG. (1993) Improved coated vesicle isolation allows better characterization of clathrin polypeptides. *J. Exp. Bot.* **44:** 23–33.

Denecke J, Botterman J, Deblaere R. (1990) Protein secretion in plant cells can occur via a default pathway. *Plant Cell* **2:** 51–59.

Denecke J, De Rycke R, Botterman J. (1992) Plant and mammalian sorting signals for protein retention in the endoplasmic reticulum contain a conserved epitope. *EMBO J.* **11:** 2345–2355.

D'Enfert C, Geusse M, Gaillardin C. (1992) Fission yeast and a plant have functional homologues of the Sar1 and Sec12 proteins involved in ER to Golgi traffic in budding yeast. *EMBO J.* **11:** 4205–4211.

Domozych DS. (1989) The endomembrane system and mechanism of membrane flow in the green alga *Gloeomonas kupfferi* (Volvocales, Chlorophyta) II. A cytochemical analysis. *Protoplasma* **149:** 108–119.

Domozych DS. (1991) The Golgi apparatus and membrane trafficking in green algae. *Int. Rev. Cytol.* **131:** 213–253.

Domozych DS, Wells B, Shaw PJ. (1992) Scale biogenesis in the green alga, *Megostigma viride*. *Protoplasma* **167:** 19–32.

Driouich A, Faye L, Staehelin LA. (1993) The plant Golgi apparatus: a factory for complex polysaccharides and glycoproteins. *Trends Biochem. Sci.* **18**: 210–214.

Drucker M, Hinz G, Robinson DG. (1993) ATPases in plant coated vesicles. *J. Exp. Bot.* **44**: 283–291.

Drucker M, Herkt B, Robinson DG. (1995) Demonstration of a β-type adaptin at the plasma membrane. *Cell Biol. Int. Rep.* **19**: 191–201.

Elazar Z, Orci L, Ostermann J, Amherdt M, Tanigawa G, Rothman JE. (1994) ADP-ribosylation factor and coatomer couple fusion to vesicle budding. *J. Cell Biol.* **124**: 415–424.

Emons AMC, Derksen J, Sassen MMA. (1992) Do microtubules orient plant cell wall microfibrils? *Physiol. Plant.* **84**: 486–493.

Farkas V, Maclachlan G. (1988) Fucosylation of exogenous xyloglucans by pea microsomal membranes. *Arch. Biochem. Biophys.* **264**: 48–53.

Faye L, Fitchette-Lainé AC, Gomord V, Chekkafi A, Delaunay AM, Driouich A. (1992) Detection, biosynthesis and some functions of glycans *N*-linked to plant secreted proteins. In: *Post-Translational Modifications in Plants* (Battey NH, Dickinson HG, Hetherington AM, ed.). SEB Seminar Series 53. Cambridge: Cambridge University Press, pp. 213–242.

Faye L, Gomord V, Fitchette-Lainé AC, Chrispeels MJ. (1993) Affinity purification of antibodies specific for Asn-linked glycans containing α-1,3 fucose or β-1,2 xylose. *Anal. Biochem.* **209**: 104–108.

Fitchette-Lainé AC, Gomord V, Chekkafi A, Faye L. (1994) Localization of xylosylation and fucosylation in the plant Golgi apparatus. *Plant J.* **5**: 673–682.

Fowke LC, Tanchak MA, Galway ME. (1991) Ultrastructural cytology of the endocytotic pathway in plants. In: *Endocytosis, Exocytosis and Vesicle Traffic in Plants* (Hawes CR, Coleman JOD, Evans DE, ed.). SEB Seminar Series 45. Cambridge: Cambridge University Press, pp. 15–40.

Galili G, Altschuler Y, Levanony H. (1993) Assembly and transport of seed storage proteins. *Trends Cell Biol.* **3**: 437–442.

Galway ME, Rennie PJ, Fowke LC. (1993) Ultrastructure of the endocytotic pathway in glutaraldehyde-fixed and high-pressure frozen/freeze-substituted protoplasts of white spruce (*Picea glauca*). *J. Cell Sci.* **106**: 847–858.

Gardiner M, Chrispeels MJ. (1975) Involvement of the Golgi apparatus in the synthesis and secretion of hydroxyproline-rich cell wall glycoproteins. *Plant Physiol.* **55**: 536–541.

Goff CW. (1973) Localization of nucleoside diphosphatase in the onion root tip. *Protoplasma* **78**: 397–416.

Goldstein JL, Brown MS, Anderson RGW, Russel DW, Schneider WJ. (1985) Receptor-mediated endocytosis: concepts emerging from the LDL receptor system. *Annu. Rev. Cell Biol.* **1**: 1–39.

Gomez L, Chrispeels MJ. (1993) Tonoplast and soluble vacuolar proteins are targeted by different mechanisms. *Plant Cell* **5**: 1113–1124.

Gomez L, Chrispeels MJ. (1994) Complementation of an *Arabidopsis thaliana* mutant that lacks complex asparagine-linked glycans with the human cDNA encoding *N*-acetylglucosaminyltransferase I. *Proc. Natl Acad. Sci. USA* **91**: 1829–1833.

Gordon-Kamm WJ, Steponkus PL. (1984) The influence of cold acclimation on the behaviour of the plasma membrane following osmotic contraction of isolated protoplasts. *Protoplasma* **123**: 161–173.

Goud B, McCaffrey M. (1991) Small GTP-binding proteins and their role in transport. *Curr. Opin. Cell Biol.* **3**: 626–633.

Gruenberg J, Clague M. (1992) Regulation of intracellular membrane transport. *Curr. Opin. Cell Biol.* **4**: 593–599.

Hager A, Debus G, Edel HG, Stransky H, Serrano R. (1991) Auxin induces exocytosis and the rapid synthesis of a high-turnover pool of plasma-membrane H+-ATPase. *Planta* **185**: 527–537.

Haigler CH, Brown RM Jr. (1986) Transport of rosettes from the Golgi apparatus to the plasma membrane in isolated mesophyll cells of *Zinnia elegans* during differentiation to tracheary elements in suspension culture. *Protoplasma* **134**: 111–120.

Hara-Hishimura I, Takeuchi Y, Inoue K, Nishimura M. (1993) Vesicle transport and processing of the precursor to 2S albumin in pumpkin. *Plant J.* **4**: 793–800.

Harley SM, Beevers L. (1989) Coated vesicles are involved in the transport of storage proteins during seed development in *Pisum sativum* L. *Plant Physiol.* **91**: 674–678.

Harris N, Oparka K. (1983) Connections between dictyosomes, ER and GERL in cotyledons of mung bean (*Vigna radiata* L.). *Protoplasma* **114**: 93–102.

Harris N, Watson MD. (1991) Vesicle transport to the vacuole and the central role of the Golgi apparatus. In: *Endocytosis, Exocytosis and Vesicle Traffic in Plants* (Hawes CR, Coleman JOD, Evans DE, ed.). SEB Seminar Series 45. Cambridge: Cambridge University Press, pp. 143–164.

Hauri HP, Schweizer A. (1992) The endoplasmic reticulum–Golgi intermediate compartment. *Curr. Opin. Cell Biol.* **4**: 600–608.

Hawes C, Coleman J, Evans D, Cole L. (1989) Recent advances in the study of plant coated vesicles. *Cell Biol. Int. Rep.* **13:** 119–128.

Hawes C, Evans D, Coleman JOD. (1991) An introduction to vesicle traffic in eukaryotic cells. In: *Endocytosis, Exocytosis and Vesicle Traffic in Plants* (Hawes CR, Coleman JOD, Evans DE, ed.). SEB Seminar Series 45. Cambridge: Cambridge University Press, pp. 1–15.

Hawes C, Crooks K, Coleman JOD, Satiat-Jeunemaitre B. (1995) Endocytosis in plants: fact or artefact? *Plant Cell Environ.* **18:** 1245–1252.

Hellgren L, Morré DJ, Selldén G, Sandelius AS. (1993) Isolation of a putative vesicular intermediate in the cell-free transfer of membrane from transitional endoplasmic reticulum to the Golgi apparatus of etiolated seedlings of garden pea. *J. Exp. Bot.* **44:** 197–207.

Helms JB, Rothman JE. (1992) Inhibition by brefeldin A of a Golgi membrane enzyme that catalyses exchange of guanine nucleotide bound to ARF. *Nature* **360:** 352–354.

Hepler PK, Cleary AL, Gunning BES, Wadsworth P, Wasteneys GO, Zhang DH. (1993) Cytoskeletal dynamics in living plant cells. *Cell Biol. Int. Rep.* **17:** 127–142.

Herman EM, Lam CJ. (1992) Arabinogalactan-rich glycoproteins are localized on the cell surface and in intravacuolar multivesicular bodies. *Plant Physiol.* **98:** 264–272.

Herman EM, Tague B, Hoffman LM, Kjemtrup SE, Chrispeels MJ. (1990) Retention of phytohemagglutinin with carboxyterminal tetrapeptide KDEL in the nuclear envelope and endoplasmic reticulum. *Planta* **182:** 305–312.

Higo H, Kishimoto N, Saito A, Higo K. (1994) Molecular cloning and characterization of a c-DNA encoding a small GTP-binding protein related to mammalian ADP-ribosylation factor from rice. *Plant Sci.* **100:** 41–49.

Hillmer S, Depta H, Robinson DG. (1986) Confirmation of endocytosis in higher plant protoplasts using lectin–gold conjugates. *Eur. J. Cell Biol.* **41:** 142–149.

Hillmer S, Freundt H, Robinson DG. (1988) The partially coated reticulum and its relationship to the Golgi apparatus in higher plant cells. *Eur. J. Cell Biol.* **47:** 206–212.

Hiraiwa N, Takeuchi Y, Nishimura M, Hara-Nishimura I. (1993) A vacuolar processing enzyme in maturing and germinating seeds: its distribution and associated changes during development. *Plant Cell Physiol.* **34:** 1197–1204.

Hobbs MC, Delagre MPH, Baydoun EAH, Brett CT. (1991) Differential distribution of glucuronyltransferase involved in glucuronoxylan synthesis within the Golgi apparatus of pea (*Pisum sativum* var. Alaska). *Biochem. J.* **277:** 653–658.

Höfte H, Chrispeels MJ. (1992) Protein sorting to the vacuolar membrane. *Plant Cell* **4:** 995–1004.

Höfte H, Faye L, Dickinson C, Herman EM, Chrispeels MJ. (1991) The protein-body proteins phytohemagglutinin and tonoplast intrinsic protein are targeted to vacuoles in leaves of transgenic tobacco. *Planta* **184:** 431–437.

Hoh B, Schauermann G, Robinson DG. (1991) Storage protein polypeptides in clathrin-coated vesicle fractions from developing pea cotyledons are not due to endomembrane contamination. *J. Plant Physiol.* **138:** 309–316.

Holstein SE, Drucker M, Robinson DG. (1994) Identification of a β-type adaptin in plant clathrin-coated vesicles. *J. Cell Sci.* **107:** 945–953.

Hong M. (1994) GTP-binding proteins in plants: new members of an old family. *Plant Mol. Biol.* **26:** 1611–1636.

Horn MA, Heinstein PF, Low PS. (1989) Receptor-mediated endocytosis in plant cells. *Plant Cell* **1:** 1003–1009.

Horn MA, Heinstein PF, Low PS. (1990) Biotin-mediated delivery of exogenous macromolecules into soybean cells. *Plant Physiol.* **3:** 1492–1496.

Horsley D, Coleman J, Evans D, Crooks K, Peart J, Satiat-Jeunemaitre B, Hawes C. (1993) A monoclonal antibody, JIM 84, recognizes the Golgi apparatus and plasma membrane in plant cells. *J. Exp. Bot.* **44:** 223–229.

Hübner R, Depta H, Robinson DG. (1985) Endocytosis in maize root cap cells; evidence obtained using heavy metal salt solutions. *Protoplasma* **129:** 214–222.

Johnson KD, Chrispeels MJ. (1987) Substrate specificities of *N*-acetylglucosaminyl-, fucosyl-, and xylosyltransferases that modify glycoproteins in the Golgi apparatus of bean cotyledons. *Plant Physiol.* **84:** 1301–1308.

Juniper B, Hawes CR, Horne JC. (1982) The relationship between the dictyosomes and the forms of the endoplasmic reticulum in plant cells with different export programs. *Bot. Gaz.* **143:** 135–145.

Kakimoto T, Shibaoka H. (1987) Actin filaments and microtubules in the preprophase band and phragmoplast of tobacco cells. *Protoplasma* **140:** 151–156.

Kakimoto T, Shibaoka H. (1988) Cytoskeletal ultrastructure of phragmoplast-nuclei complexes isolated from cultured tobacco cells. *Protoplasma* Suppl. 2: 95–103.

Kawasaki K. (1981) Synthesis of arabinose-containing cell wall precursors in suspension-cultured tobacco cells. I. Intracellular site of synthesis and transport. *Plant Cell Physiol.* **22:** 431–442.

Keen JH. (1990) Clathrin and its associated assembly and disassembly proteins. *Annu. Rev. Biochem.* **59:** 415–438.

Kirsch T, Beevers L. (1993) Uncoating of clathrin-coated vesicles by uncoating ATPase from developing peas. *Plant Physiol.* **103:** 205–212.

Kirsch T, Paris N, Butler JM, Beevers L, Rogers JC. (1994) Purification and initial characterization of a potential plant vacuolar targeting receptor. *Proc. Natl Acad. Sci. USA* **91:** 3403–3407.

Klausner RD, Donaldson JG, Lippincott-Schwartz J. (1992) Brefeldin A: insights into the control of membrane traffic and organelle structure. *J. Cell Biol.* **116:** 1071–1080.

Kleene R, Berger EG. (1993) The molecular and cell biology of glycosyltranferases. *Biochim. Biophys. Acta* **1154:** 283–325.

Knox P. (1990) Emerging patterns of organization at the plant cell surface. *J. Cell Sci.* **96:** 557–561.

Knox P, Linstead PJ, Peart J, Cooper C, Roberts K. (1991) Developmentally regulated epitopes of cell surface arabinogalactan proteins and their relation to root tissue pattern formation. *Plant J.* **1:** 317–326.

Kohno T, Chaen S, Shimmen T. (1990) Characterization of the translocator associated with pollen tube organelles. *Protoplasma* **154:** 179–183.

Kornfeld S, Mellman I. (1989) The biogenesis of lysosomes. *Annu. Rev. Cell Biol.* **5:** 483–525.

Kreis TE. (1990) Role of microtubules in the organisation of the Golgi apparatus. *Cell Motil. Cytoskel.* **15:** 67–70.

Kreis TE. (1992) Regulation of vesicular and tubular membrane traffic of the Golgi complex by coat proteins. *Curr. Opin. Cell Biol.* **4:** 609–615.

Kreis TE, Matteoni R, Hollinshead M, Tooze J. (1989) Secretory granules and endosomes show saltatory movements biased to the anterograde and retrograde directions, respectively, along microtubules in ATT20 cells. *Eur. J. Cell Biol.* **49:** 128–139.

Kristen U. (1978) Ultrastructure and a possible function of the intercisternal elements in dictyosomes. *Planta* **138:** 29–33.

Kronestedt-Robards E, Robards AW. (1991) Exocytosis in plant cells. In: *Endocytosis, Exocytosis and Membrane Traffic in Plant Cells* (Hawes CR, Coleman JOD, Evans DE, ed.). SEB Seminar Series 45. Cambridge: Cambridge University Press, pp. 199–233.

Kuge O, Hara-Kuge S, Orci L, Ravazzola M, Amherdt M, Tanigawa G, Wieland FT, Rothman JE. (1993) β-COP, a subunit of coatomer, is required for COP-coated vesicle assembly. *J. Cell Biol.* **123:** 1727–1734.

Lainé AC, Gomord V, Faye L. (1991) Xylose-specific antibodies as markers of subcompartmentation of terminal glycosylation in the Golgi apparatus of sycamore cells. *FEBS Lett.* **295:** 179–184.

Lamport DTA. (1986) The primary cell wall: a new model. In: *Cellulose: Structure, Modification and Hydrolysis* (Young RA, Rowell RM, ed.). New York: John Wiley, pp. 77–90.

Lau JM, McNeil M, Darvill AG, Albersheim P. (1985) Structure of the backbone of rhamnogalacturonan I, a pectic polysaccharide in the primary walls of plants. *Carbohydr. Res.* **137:** 111–125.

Lee HI, Gal S, Nezma TC, Raikhel NV. (1993) The *Arabidopsis* endoplasmic reticulum retention receptor functions in yeast. *Proc. Natl Acad. Sci. USA* **90:** 11433–11437.

Letourneur F, Gaynor EC, Hennecke S, Demouliere C, Duden R, Emr SD, Riezman H, Cosson P. (1994) Coatomer is essential for retrieval of dilysine-tagged proteins to the ER. *Cell* **79:** 1199–1207.

Levanony H, Rubin R, Altschuler Y, Galili G. (1992) Evidence for a novel route of wheat storage proteins to vacuoles. *J. Cell Biol.* **119:** 1117–1128.

Levy S, Staehelin LA. (1992) Synthesis, assembly and function of plant cell wall macromolecules. *Curr. Opin. Cell Biol.* **4:** 856–862.

Lin H-B, Harley SM, Butler JM, Beevers L. (1992) Multiplicity of clathrin light-chain-like polypeptides from developing pea (*Pisum sativum* L.) cotyledons. *J. Cell Sci.* **103:** 1127–1137.

Lippincott-Schwartz J, Yuan LC, Bonifacino JS, Klausner RD. (1989) Rapid redistribution of Golgi proteins into the ER in cells treated with brefeldin A: evidence for membrane recycling from Golgi to ER. *Cell* **56:** 801–813.

Lippincott-Schwartz J, Donaldson JG, Schweizer A, Berger E.G, Hauri H-P, Yuan L, Klausner RD. (1990) Microtubule-dependent retrograde transport of proteins into the ER in the presence of brefeldin A suggests an ER recycling pathway. *Cell* **60:** 821–836.

Lippincott-Schwartz J, Yuan L, Tipper C, Amherdt M, Orci L, Klausner R. (1991) Brefeldin A's effects on endosomes, lysosomes, and the TGN suggest a general mechanism for regulating organelle structure and membrane traffic. *Cell* **67**: 601–616.

Low PS, Legendre L, Heinstein PF, Horn MA. (1993) Comparison of elicitor and vitamin receptor-mediated endocytosis in cultured soybean cells. *J. Exp. Bot.* **44**: 269–275.

Lynch M, Staehelin LA. (1992) Domain-specific and cell type-specific localization of two types of cell wall matrix polysaccharides in the clover root tip. *J. Cell Biol.* **118**: 467–479.

Machamer C. (1993) Targeting and retention of Golgi membrane proteins. *Curr. Opin. Cell Biol.* **5**: 606–612.

Machamer CE, Mentone SA, Rose JK, Farquhar MG. (1990) The E1 glycoprotein of an avian coronavirus is targeted to the *cis* Golgi complex. *Proc. Natl Acad. Sci. USA* **87**: 6944–6948.

Maclachlan G, Levy B, Farkas V. (1992) Acceptor requirements for GDP-fucose: xyloglucan 1,2-a-1-fucosyltransferase activity solubilized from pea epicotyl membranes. *Arch. Biochem. Biophys.* **294**: 200–205.

Marty F. (1978) Cytochemical studies on GERL, provacuoles and vacuoles in root meristematic cells of *Euphorbia*. *Proc. Natl Acad. Sci. USA* **75**: 852–856.

Matsuoka K, Nakamura K. (1992) Transport of a sweet potato storage protein, sporamin, to the vacuole in yeast cells. *Plant Cell Physiol.* **33**: 453–462.

Melkonian M, Becker B, Becker D. (1991) Scale formation in algae. *J. Electron Microsc. Tech.* **17**: 165–178.

Memon AR, Clark GB, Thompson GA. (1993) Identification of an ARF-type low molecular mass GTP-binding protein in pea (*Pisum sativum*). *Biochem. Biophys. Res. Commun.* **193**: 809–813.

Miller DD, Scordilis SP, Hepler PK. (1995) Identification and localization of three classes of myosins in pollen tubes of *Lilium longiflorum* and *Nicotiana alata*. *J. Cell Sci.* **108**: 2549–2653.

Moore PJ, Swords KMM, Lynch MA, Staehelin LA. (1991) Spatial organization of the assembly pathways of glycoproteins and complex polysaccharides in the Golgi apparatus of plants. *J. Cell Biol.* **112**: 589–602.

Moore I, Schell J, Palme K. (1995) Subclass-specific sequence motifs identified in Rab GTPases. *Trends Biochem. Sci.* **20**: 10–12.

Morré DJ, Mollenhauer HH. (1974) The endomembrane concept: a functional integration of endoplasmic reticulum and Golgi apparatus. In: *Dynamic Aspects of Plant Ultrastructure* (Robards AW, ed.). London: McGraw-Hill, pp. 84–137.

Morré DJ, Nowack DD, Paulik M, Brightman AO, Thornborough K, Yim J, Auderset G. (1989) Transitional endoplasmic reticulum membranes and vesicles isolated from animals and plants. *Protoplasma* **153**: 1–13.

Munro S, Pelham HRB. (1987) A C-terminal signal prevents secretion of luminal ER proteins. *Cell* **48**: 899–907.

Napier RM, Fowke LC, Hawes C, Lewis M, Pelham H. (1992) Immunological evidence that plants use both HDEL and KDEL for targeting proteins to the endoplasmic reticulum. *J. Cell Sci.* **102**: 261–271.

Nilsson T, Slusarewicz P, Hoe MH, Warren G. (1993) Kin recognition. A model for the retention of Golgi enzymes. *FEBS Lett.* **330**: 1–14.

Nogushi T, Morré DJ. (1991) Vesicular membrane transfer between endoplasmic reticulum and Golgi apparatus of a green alga *Micrasterias americana*: a 16° block and reconstitution in a cell free system. *Protoplasma* **162**: 128–139.

O'Driscoll D, Hann C, Read SM, Steer MW. (1993) Endocytotic uptake of fluorescent dextrans by pollen tubes grown *in vitro*. *Protoplasma* **175**: 126–130.

Okabe S, Hirokawa N. (1989) Axonal transport. *Curr. Opin. Cell Biol.* **1**: 91–97.

Oparka KJ, Hawes CH. (1992) Vacuolar sequestration of fluorescent probes in plant cells: a review. *J. Microscopy* **166**: 15–27.

Oparka KJ, Prior DAM, Harris N. (1990) Osmotic induction of fluid phase endocytosis in onion epidermal cells. *Planta* **180**: 555–561.

Oparka KJ, Murant EA, Wright KM, Prior DAM, Harris N. (1991) The drug probenecid inhibits the transport of fluorescent anions across the tonoplast of onion epidermal cells. *J. Cell Sci.* **99**: 557–563.

Oparka KJ, Prior DAM, Crawford JW. (1994) Behaviour of plasma membrane, cortical ER and plasmodesmata during plasmolysis of onion epidermal cells. *Plant Cell Environ.* **17**: 163–171.

Orci L, Perrelet A, Ravazzola M, Wieland FT, Schekman R, Rothman JE. (1993) 'BFA bodies': a subcompartment of the endoplasmic reticulum. *Proc. Natl Acad. Sci. USA* **90**: 11089–11093.

Owens RJ, Northcote DH. (1981) The location of arabinosyl:hydroxyproline transferase in the membrane system potato of tissue culture cells. *Biochem. J.* **195**: 661–667.

Palme K, Diefenthal T, Vingron M, Sander C, Schell J. (1992) Molecular cloning and structural analysis of genes from *Zea mays* (L.) coding for members of the ras-related ypt gene family. *Proc. Natl Acad. Sci. USA* **89:** 787–791.

Pearse BMF, Robinson MS. (1990) Clathrin, adaptors, and sorting. *Annu. Rev. Cell Biol.* **6:** 151–171.

Pelham HR. (1991) Recycling of proteins between the endoplasmic reticulum and Golgi complex. *Curr. Opin. Cell Biol.* **3:** 585–591.

Pelham HR. (1994) About turn for the COPs? *Cell* **79:** 1125–1127.

Pennel RI, Knox JP, Scofield GN, Selvendran RR, Roberts K. (1989) A family of abundant plasma membrane-associated glycoproteins related to the arabinogalactan proteins is unique to flowering plants. *J. Cell Biol.* **198:** 1967–1977.

Pesacreta TC, Lucas WJ. (1984) Plasma membrane coat and a coated vesicle associated reticulum on membranes: their structure and possible interrelationship in *Chara corallina. J. Cell Biol.* **98:** 1537–1545.

Pfeffer S. (1994) Rab GTPases: master regulators of membrane trafficking. *Curr. Opin. Cell Biol.* **6:** 522–526.

Picton JM, Steer MW. (1981) Determination of secretory vesicle production rates by dictyosomes in pollen tubes of *Tradescantia* using cytochalasin D. *J. Cell Sci.* **49:** 261–272.

Quader H. (1990) Formation and disintegration of cisternae of the endoplasmic reticulum visualized in live cells by conventional fluorescence and confocal laser scanning microscopy: evidence for the involvement of calcium and the cytoskeleton. *Protoplasma* **155:** 166–175.

Quader H, Hofman A, Schnepf E. (1989) Reorganization of the endoplasmic reticulum in epidermal cells of onion bulb scales after cold stress: involvement of cytoskeletal elements. *Planta* **177:** 273–280.

Regad F, Bardet C, Tremousaygue D, Moisan A, Lescure B, Axelos M. (1993) cDNA cloning and expression of an *Arabidopsis* GTP-binding protein of the ARF family. *FEBS Lett.* **316:** 133–136.

Robards AW, Stark M. (1988) Nectar secretion in *Abutilon*: a new model. *Protoplasma* **142:** 79–91.

Roberts K. (1990) Structures at the plant cell surface. *Curr. Opin. Cell Biol.* **2:** 920–928.

Robinson DG. (1980) Dictyosome–endoplasmic reticulum associations in higher plant cells? A serial-section analysis. *Eur. J. Cell Biol.* **23:** 22–36.

Robinson DG, Kristen U. (1982) Membrane flow via the Golgi apparatus in higher plant cells. *Int. Rev. Cytol.* **77:** 89–127.

Robinson DG, Andreae M, Sauer A. (1985) Hydroxyproline-rich glycoprotein biosynthesis: a comparison with that of collagen. In: *Biochemistry of Plant Cell Walls* (Brett CT, Hillman JR, ed.). Cambridge: Cambridge University Press, pp. 155–176.

Robinson DG, Balusek K, Freundt H. (1989) Legumin antibodies recognize polypeptides in coated vesicles from developing pea cotyledons. *Protoplasma* **150:** 79–82.

Roth J, Taatjes DJ, Lucocq JM, Charest PM. (1988) Light and electron microscopical detection of sugar residues in tissue secretions by gold labeled lectins and glycoproteins. II. Applications in the study of the topology of Golgi apparatus glycosylation steps and the regional distribution of lectin binding sites in the plasma membrane. *Acta Histochem.* **36S:** 125–140.

Rothman JE. (1994) Mechanisms of intracellular protein transport. *Nature* **372:** 55–63.

Rothman JE, Orci L. (1992) Molecular dissection of the secretory pathway. *Nature* **355:** 409–415.

Sabatini DD, Louvard D, Adesnik M. (1993) Membranes. *Curr. Opin. Cell Biol.* **4:** 573–581.

Samuels AL, Bisalputra T. (1990) Endocytosis in elongating root cells of *Lobelia erinus. J. Cell Sci.* **97:** 157–165.

Samuels AL, Giddings TH Jr, Staehelin LA. (1995) Cytokinesis in tobacco BY-2 and root tip cells: a new model of cell plate formation in higher plants. *J. Cell Biol.* **130:** 1345–1357.

Satiat-Jeunemaitre B. (1992) Spatial and temporal regulations in helicoidal extracellular matrices: comparison between plant and animal systems. *Tissue Cell* **24:** 315–334.

Satiat-Jeunemaitre B, Hawes C. (1992) Redistribution of a Golgi glycoprotein in plant cells treated with brefeldin A. *J. Cell Sci.* **103:** 1153–1166.

Satiat-Jeunemaitre B, Hawes C. (1993a) The distribution of secretory products in plant cells is affected by brefeldin A. *Cell Biol. Int. Rep.* **17:** 183–193.

Satiat-Jeunemaitre B, Hawes C. (1993b) Insights into the secretory pathway and vesicular transport in plant cells. *Biol. Cell* **79:** 7–15.

Satiat-Jeunemaitre B, Hawes C. (1994) GATT (A general agreement on traffic and transport) and brefeldin A in plant cells. *Plant Cell* **6:** 463–467.

Satiat-Jeunemaitre B, Fitchette-Lainé AC, Alabouvette J, Marty-Mazars D, Hawes C, Faye L, Marty F. (1994) Differential effects of monensin on the plant secretory pathway. *J. Exp. Bot.* **45:** 685–698.

Satiat-Jeunemaitre B, Cole L, Bourett T, Howard R, Hawes CR. (1996a) Brefeldin A effects in plant and fungal cells: something new about vesicle trafficking? *J. Microscopy* **181:** 162–177.

Satiat-Jeunemaitre B, Steele C, Hawes C. (1996b) Golgi-membrane dynamics are cytoskeleton dependent. A study on Golgi movement induced by brefeldin A. *Protoplasma* (in press).

Satiat-Jeunemaitre B, Stele C, Hawes C. (1996c) Maintenance of the exocytotic and endocytic apparatus involved in protein targeting in plant cells. *Plant Physiol. Biochem.* **34:** 183–195.

Sauer A, Robinson DG. (1985) Intracellular localization of post-translational modifications in the synthesis of hydroxyproline-rich glycoproteins. Peptidyl proline hydroxylation in maize roots. *Planta* **164:** 287–294.

Schekman R. (1992) Genetic and biochemical analysis of vesicular traffic in yeast. *Curr. Opin. Cell Biol.* **4:** 587–592.

Schmid VHR, Meindl U. (1992) Microtubules do not control orientation of secondary cell wall microfibril deposition in *Micrasterias. Protoplasma* **169:** 148–154.

Schnepf E. (1963) Zur cytologie und physiologie pflanzlicher Drüsen. 2. Teil. Über die wirkung von Sauerstoffentzug und von Atmungsinhibitoren auf die Sekretion des Fangschleimes von *Drosophyllum* und auf die Feinstruktur der Drüsenzellen. *Flora* **153:** 23–48.

Schnepf E. (1969) Sekretion und Exkretion bei Pflanzen. *Protoplasmatologia* **VIII/8:** 1–181.

Semenza JC, Hardwick KG, Dean N, Pelham HRB. (1990) ERD2, a yeast gene required for the receptor mediated retrieval of luminal ER proteins from the secretory pathway. *Cell* **61:** 1349–1357.

Shaper JH, Shaper NL. (1992) Enzymes associated with glycosylation. *Curr. Opin. Struct. Biol.* **2:** 701–709.

Sherrier DJ, VandenBosch KA. (1990) Secretion of cell wall polysaccharides in *Vicia* root hairs. *Plant J.* **5:** 185–195.

Showalter AM. (1993) Structure and function of plant cell wall proteins. *Plant Cell* **5:** 9–23.

Showalter AM, Varner JE. (1989) Plant hydroxyproline-rich glycoproteins. In: *The Biochemistry of Plants* Vol. 15 (Stumpf PK, Conn EE, ed.). New York: Academic Press, pp. 485–520.

Simmons K, Zerial M. (1993) Rab proteins and road maps for intracellular transport. *Neuron* **11:** 789–799.

Slusarewicz P, Nilsson T, Hui N, Watson R, Warren G. (1994) Isolation of a matrix that binds *medial* Golgi enzymes. *J. Cell Biol.* **124:** 405–413.

Smallwood M, Keen JN, Bowles DJ. (1990) Purification and partial sequence analysis of plant annexins. *Biochemical J.* **270:** 157–161.

Söllner T, Whiteheart SW, Brunner M., Erdjument-Bromage H, Geromanos S, Tempst T, Rothman JE. (1993) SNAP receptors implicated in vesicle targeting and fusion. *Nature* **362:** 318–324.

Staehelin L, Moore I. (1995) The plant Golgi apparatus: structure, functional organization and trafficking mechanisms. *Annu. Rev. Plant Physiol. Plant Mol. Biol.* **46:** 261–288.

Staehelin LA, Giddings T Jr, Kiss JZ, Sack FD. (1990) Macromolecular differentiation of Golgi stacks in root tips of *Arabidopsis* and *Nicotiana* seedlings as visualized in high pressure frozen and freeze-substituted samples. *Protoplasma* **157:** 75–91.

Staehelin LA, Giddings TH, Levy S, Lynch MA, Moore PJ, Swords KMM. (1991) Organisation of the secretory pathway of cell wall glycoproteins and complex polysaccharides in plant cells. In: *Endocytosis, Exocytosis and Vesicle Traffic in Plants* (Hawes CR, Coleman JOD, Evans DE, ed.). SEB Seminar Series 45. Cambridge: Cambridge University Press, pp. 183–198.

Stafstrom JP, Staehelin LA. (1986) Cross-linking patterns in salt-extractable extensin from carrot cell walls. *Plant Physiol.* **81:** 234–241.

Stammes MA, Rothman JE. (1993) The binding of AP-1 clathrin adaptor particles to Golgi membranes requires ADP-ribosylation factor, a small GTP-binding protein. *Cell* **73:** 999–1005.

Steer MW. (1985) Vesicle dynamics. In: *Botanical Microscopy 1985* (Robards AW, ed.). Oxford: Oxford University Press, pp. 129–157.

Steer MW, O'Driscoll D. (1991) Vesicle dynamics and membrane turnover in plant cells. In: *Endocytosis, Exocytosis and Vesicle Traffic in Plants* (Hawes CR, Coleman JOD, Evans DE, ed.). SEB Seminar Series 45. Cambridge: Cambridge University Press, pp. 128–142.

Steponkus PL. (1991) Behaviour of the plasma membrane during osmotic excursions. In: *Endocytosis, Exocytosis and Vesicle Traffic in Plants* (Hawes CR, Coleman JOD, Evans DE, ed.). SEB Seminar Series 45. Cambridge: Cambridge University Press, pp. 103–128.

Sturm A, Kindl H. (1983) Fucosyl transferase activity and fucose incorporation *in vivo* as markers for subfractionating cucumber microsomes. *FEBS Lett.* **160:** 165–168.

Sturm A, Johnson KD, Szumilo T, Elbein AD, Chrispeels MJ. (1987) Subcellular localization of glycosidases and glycosyltranferases involved in the processing of *N*-linked oligosaccharides. *Plant Physiol.* **85:** 741–745.

Swords KMM, Staehelin A. (1993) Complementary immunolocalization patterns of cell wall hydroxy-proline-rich glycoproteins studied with the use of antibodies directed against different carbohydrate epitopes. *Plant Physiol.* **102:** 891–901.

Tanchak MA, Griffing LR, Mersey BG, Fowke LC. (1984) Endocytosis of cationized ferritin by coated vesicles of soybean protoplasts. *Planta* **162:** 481–486.

Tang BL, Wong SH, Low SH, Hong W. (1992) The transmembrane domain of *N*-glucosaminyltransferase I contains a Golgi retention signal. *J. Biol. Chem.* **267:** 10122–10126.

Terasaki M, Chen B, Fujiwara K. (1986) Microtubules and the endoplasmic reticulum are highly interdependent structures. *J. Cell Biol.* **103:** 1557–1568.

Tiezzi A, Moscatelli A, Cai G, Bartalasi A, Cresti M. (1992) An immunoreactive homolog of mammalian kinesin in *Nicotiana tabacum* pollen tubes. *Cell Motil. Cytoskel.* **21:** 132–137.

Vale RD, Hontani H. (1988) Formation of membrane networks *in vitro* by kinesin-driven microtubule movement. *J. Cell Biol.* **107:** 2233–2242.

Van den Eijnden DH, Joziasse DH. (1993) Enzymes associated with glycosylation. *Curr. Opin. Struct. Biol.* **3:** 711–721.

Van Deurs B, Petersen OW, Olsnes S, Sandvig K. (1989) The ways of endocytosis. *Int. Rev. Cytol.* **117:** 131–177.

Van Meer G. (1993) Transport and sorting of membrane lipids. *Curr. Opin. Cell Biol.* **4:** 661–675.

Vian B, Roland JC. (1991) Affinodetection of the sites of formation and further distribution of polygalacturonans and native cellulose in growing plant cells. *Biol. Cell* **71:** 43–55.

Vitale A, Chrispeels MJ. (1992) Sorting of proteins to the vacuoles of plant cells. *BioEssays* **14:** 151–160.

Vitale A, Ceriotti A, Denecke J. (1994) The role of the endoplasmic reticulum in protein synthesis, modification and intracellular transport. *J. Exp. Bot.* **44:** 1417–1444.

Von Schaewen A, Chrispeels MJ. (1993) Identification of vacuolar sorting information in phyto-hemagglutinin, an unprocessed vacuolar protein. *J. Exp. Bot.* **44:** 339–343.

Von Schaewen A, Sturm A, O'Neill J, Chrispeels MJ. (1993) Isolation of a mutant *Arabidopsis* plant that lacks *N*-acetyl glucosamyl transferase I and is unable to synthesize Golgi-modified complex *N*-linked glycans. *Plant Physiol.* **102:** 1109–1118.

Wilson DW, Whiteheart SW, Wiedmann M, Brunner M, Rothman JE. (1992) A multisubunit particle implicated in membrane fusion. *J. Cell Biol.* **117:** 531–538.

Wong DH, Brodsky FM. (1992) 100-kDa proteins of Golgi- and trans-Golgi network-associated coated vesicles have related but distinct membrane binding properties. *J. Cell Biol.* **117:** 1171–1179.

Wood SA, Park JE, Brown WJ. (1991) Brefeldin A causes a microtubule-mediated fusion of the trans-Golgi network and early endosomes. *Cell* **67:** 591–600.

Wright KM, Davies TGW, Steele SH, Leigh RA, Oparka KJ. (1992) Development of a probenecid-sensitive Lucifer Yellow transport system in vacuolating oat aleurone protoplasts. *J. Cell Sci.* **102:** 133–139.

Yokota E, McDonald AR, Liu B, Shimmen T, Palevitz BA. (1995) Localization of a 170-kDa myosin heavy chain in plant cells. *Protoplasma* **185:** 178–187.

Zerial M, Stenmark H. (1993) Rab GTPases in vesicular transport. *Curr. Opin. Cell Biol.* **5:** 613–620.

Zhang GF, Staehelin LA. (1992) Functional compartmentalization of the Golgi apparatus of plant cells. An immunochemical analysis of high pressure frozen and freeze-substituted sycamore maple suspension-cultured cells. *Plant Physiol.* **99:** 1070–1083.

Chapter 21

Mechanisms controlling function, identity and integrity of the plant endoplasmic reticulum

Jürgen Denecke

1. Introduction

The endoplasmic reticulum (ER) is the largest organelle of the endomembrane system with respect to the amount of membranes (see also Chapter 22 by Lichtscheidl and Hepler). Its major function is to synthesize proteins and to transport correctly folded proteins to the Golgi apparatus. Protein synthesis by the ER occurs on the cytosolic face of the rough ER membrane, and is accompanied by a simultaneous translocation of the nascent polypeptide to the lumen of the ER. The lumen differs from the cytosol in that it contains a specialized subset of molecular chaperones and protein-modifying enzymes to assist protein folding and maturation. Another unique feature is the more oxidizing environment compared to that of the cytosol, which allows the formation of disulphide bridges. Therefore, the ER lumen can also be regarded as a specialized environment which supports the biosynthesis of proteins which cannot be produced in the cytosol. The proteins synthesized by the ER can have biosynthetic functions in protein translocation and folding, protein modification/processing or carbohydrate synthesis, and may be located in the ER, the Golgi or the plasma membrane (cellulose synthase). A large group of vacuolar or secreted proteins have a lytic role (proteases, hydrolases), some of which may be defence related. Many membrane-spanning proteins have signalling functions, such as ion channels or receptors involved in signal transduction and/or protein transport. The ER also synthesizes structural proteins localized in the extracellular matrix. A continuous bulk flow of membranes delivers proteins to the Golgi (anterograde transport) if they are released by the quality control system formed by the set of chaperones present in the ER lumen. Membrane flow from the Golgi to the ER (retrograde transport) guarantees the maintenance of ER integrity and function. Very little is known about the regulatory mechanisms which ensure an equilibrium between anterograde and retrograde membrane traffic or correct function of the ER under different physiological conditions. During the last 5 years it has become clear that fundamental processes associated with the endomembrane system are conserved between plants, mammals and yeasts. For many biological questions it is therefore possible to switch from one model system to another. Rather than repeating experiments performed in other model organisms, it is now possible to conduct complementary experiments. The cloning of genes corresponding to key components of the protein synthesis and transport machinery and comparison of the primary sequence from organisms as different as plants and yeasts can give important information about functional protein domains. This chapter is written in the light of this new philosophy, with the emphasis on emerging questions concerning the function of the plant ER in a vast number of biological processes. The reader is referred to recent reviews about protein translocation (High and Stirling, 1993), protein folding within the lumen (Helenius *et al.*, 1992), transport of proteins through the endomembrane system of plants (Bednarek and Raikhel, 1992) and the diverse functions of the plant ER (Vitale *et al.*, 1993).

2. Polypeptide synthesis and translocation, a multi-step process

The targeting of mRNA to the ER membrane

Ribosome–mRNA complexes which synthesize proteins to be translocated across the ER membrane are retrieved from the cytosolic pool by a ribonucleoprotein complex, commonly known as the signal recognition particle (SRP). The signal peptide contains the information that specifies synthesis on the ER membrane (for reviews see Nunnari and Walter, 1992; Rapoport, 1992; Sanders and Schekman, 1992; Walter and Lingappa, 1986), and is located at the N-terminus of soluble ER-synthesized proteins and membrane proteins that contain only one transmembrane domain and a cytosolic tail (type-I membrane proteins). This signal is cleavable and is removed from the protein during the translocation process. Membrane proteins with a lumenal tail (type II) do not have a signal sequence, but possess an internal (non-cleavable) hydrophobic segment which causes membrane insertion and anchoring by an unusual mechanism (Kutay *et al.*, 1993), which will not be discussed here. SRP binds to the signal peptide as soon as it emerges from the ribosome and becomes accessible to the cytosolic components (Walter *et al.*, 1981). In mammalian cells, the signal peptide interacts directly with the 54-kDa protein subunit of SRP (Krieg *et al.*, 1986; Kurzchalia *et al.*, 1986). As a result of this interaction, translation is transiently arrested or slowed down (Walter and Blobel, 1981). The complex formed between SRP, the nascent protein, the ribosome and the mRNA will interact with the SRP receptor or docking protein on the surface of the ER, which results in binding of the ribosome to the ER membrane and release of the SRP/docking protein complex. Translation resumes and translocation of the nascent polypeptide is initiated. The rough membranes of the ER which contain attached ribosomes (Palade, 1975) constitute the protein-synthesizing region, and the sites of synthesis and translocation have been termed 'translocons' (Walter and Lingappa, 1986).

The role of the signal peptide in targeting ribosomes to the ER surface is well established. However, recent evidence suggests that mRNA targeting to the ER membrane may be more complex than was previously assumed. Developing rice endosperm cells display two distinct rough endoplasmic reticula, namely the cisternal ER and protein-body ER. Two types of storage protein, glutelin and prolamin, accumulate in the rice endosperm. Glutelins are transported to the vacuoles via the Golgi, whereas prolamins are retained in the ER, and prolamin-containing protein bodies are directly derived from the ER (Krishnan *et al.*, 1986; Yamagata *et al.*, 1986). Subcellular fractionation of membrane-bound polysomal mRNA of cisternal and protein-body ER as well as *in-situ* hybridization experiments indicate that protein-body ER membranes are enriched with prolamin mRNA, whereas cisternal ER contains more glutelin mRNA (Li *et al.*, 1993; Yamagata and Tanaka, 1986; Yamagata *et al.*, 1986). These observations suggest that the targeting of two different storage proteins to their final destinations may begin by differential segregation of SRP–nascent chain–ribosome–mRNA complexes on the rough ER membrane. The possibility that protein synthesis may be restricted to ER subdomains has been suggested before (Rose and Doms, 1988), and it has been observed that the two different ER types in rat liver cells possess distinct mRNA populations (Shore and Tata, 1977). Differential activity in protein synthesis between the rough ER and the nuclear envelope has also been reported (Pathak *et al.*, 1986). One possible explanation for mRNA segregation to different subdomains of the rough ER and nuclear envelope could be the presence of classes of signal peptides, SRPs and docking proteins. This view is supported by the fact that maize endosperm cells contain a heterologous population of 7SL RNAs (Campos *et al.,* 1989), an SRP component. However, it is also possible that segregation occurs during polypeptide synthesis and depends on lumenal proteins or membrane proteins that interact with the nascent chain (see below). It has also been suggested that mRNA molecules themselves contain targeting information in the 3′ untranslated end (Okita *et al.,* 1994). A clear mechanism for differential mRNA segregation on the ER surface has yet to be established.

Requirement for chaperones on either side of the membrane

In addition to SRP and the docking protein, cytosolic and lumenal components appear to be necessary to initiate and complete translocation (Chirico *et al.,* 1988; Deshaies *et al.,* 1988; Nicchitta and Blobel, 1993; Vogel *et al.,* 1990; Zimmermann *et al.,* 1988). One recognized function of such components is that of molecular chaperone, defined as a molecule which mediates translocation, folding and assembly without being part of the final structures (Ellis *et al.,* 1989). Chaperones are present on both sides of the ER membrane, either to maintain precursors in an unfolded, translocation-competent form at the cytosolic face of the membrane or to assist the protein-folding process on the lumenal side (Rothman, 1989). It is well established that creating or maintaining an unfolded translocation-competent configuration is crucial for the transport of precursors to mitochondria and chloroplasts or for the secretion of proteins in bacteria (Kumamoto, 1991; Lubben *et al.,* 1989; Phillips and Silhavy, 1990). Compared to prokaryotic cells, eukaryotes have solved a significant part of the problem imposed by the need to unfold precursor proteins simply by translocating the major portion of the polypeptide co-translationally. Nevertheless, Hsp70 facilitates the translocation of precursor proteins across the ER membrane of yeasts (Chirico *et al.,* 1988; Deshaies *et al.,* 1988) and mammalian cells (Zimmermann *et al.,* 1988). GroEL-related molecular chaperones have been found in the cytosol of oat cells (Grimm *et al.,* 1991). Direct evidence that cytosolic Hsc70 assists membrane translocation of plant secretory protein precursors was provided by *in-vitro* reconstitution experiments using wheat-germ extracts depleted for Hsc70 and maize endosperm microsomes (Miernyk *et al.,* 1992). ER lumenal proteins such as the lumenal binding protein (BiP) have been shown to be required for protein translocation in yeasts (Vogel *et al.,* 1990) and mammalian cells (Nicchitta and Blobel, 1993), but evidence for a similar requirement in plants has not been obtained. However, it is predictable that the identified plant homologues of protein disulphide isomerase (PDI) (Shorrosh and Dixon, 1991), endoplasmin (Walter-Larsen *et al.,* 1993) and BiP, the latter of which functionally complements yeast BiP mutants (Denecke *et al.,* 1991), are carrying out similar functions (see below). The availability of plant-derived *in-vitro* translation systems (Miernyk *et al.,* 1992; Osteryoung *et al.,* 1992) opens up the possibility of studying these functions in more detail in plants.

Protein modification during translocation

Very little is known about the role of the signal peptide during the initiation of the protein translocation process. Recent evidence suggests that N-terminal signal peptides are in an aqueous environment during the initial stages of the translocation process, and that the binding of the ribosome to the ER membrane results in the formation of an aqueous compartment that is sealed off from the cytosol and inaccessible to cytosolic ions (Crowley *et al.,* 1993; High, 1992). This aqueous channel contains the signal peptide and the nascent chain, which passes through the channel towards the lumen. The signal peptide is removed from the mature part of the protein by signal peptidase, a membrane-associated multi-subunit complex (Evans *et al.,* 1986; Shelness and Blobel, 1990). Removal of the signal peptide occurs prior to the completion of translation (Blobel and Dobberstein, 1975; Bollini *et al.,* 1983; Miyata and Akazawa, 1982; Sengupta *et al.,* 1981), but it is not yet understood how and where this takes place. Since the signal peptide is located in an aqueous compartment that is sealed off from the cytosol, and since it is prevented from partitioning into the non-polar core of the membrane early in translocation (Crowley *et al.,* 1993), it is possible that the signal peptide can emerge at the surface of the lumenal side of the ER membrane before translocation is completed. There would be no advantage of remaining in the aqueous compartment within the membrane, and due to its hydrophobic properties, the signal peptide would form a target for lumenal chaperones if it were exposed at the lumenal side of the ER membrane. The evidence that BiP depletion leads to the accumulation of secretory protein precursors (Nguyen *et al.,* 1991; Vogel *et al.,* 1990) could be explained by a direct involvement of this chaperone in signal peptidase action, probably by binding to the signal peptide directly, and thereby displacing the equilibrium towards the lumenal aqueous environment. Clearly, removal from the aqueous compartment which is sealed off from the cytosol by the ribosome is necessary in order for it to be accessible to signal peptidase.

Recently it has been shown that ribophorin I and II and an additional polypeptide of 48 kDa form a membrane-bound protein complex that co-purifies with oligosaccharyl transferase activity (Kelleher *et al.,* 1992). Since ribophorins are present in 1:1 stoichiometry with ribosomes on the rough ER, it is tempting to assume that glycosylation is a 'translocon'-associated process. The fact that PDI, a protein required for the formation of disulphide bridges, was accidentally identified as the glycosylation-site binding protein (Geetha-Habib *et al.,* 1988; Kelleher *et al.,* 1992; Noiva *et al.,* 1991a,b) could reflect a tight association of PDI with the oligosaccharyl transferase complex. A tight association of PDI with translocons is consistent with the view that protein folding occurs co-translationally (Helenius *et al.,* 1992). Support for co-localization of chaperones with the protein translocation machinery is also provided by evidence that calnexin, a newly identified membrane-spanning chaperone of the ER (David *et al.,* 1993; Jackson *et al.,* 1994; Margolese *et al.,* 1993; Ou *et al.,* 1993; Rajagopalan *et al.,* 1994; Wada *et al.,* 1991), is tightly associated with SSRα, a component of the translocation machinery. In conclusion, protein synthesis on the ER membrane is a process which requires the simultaneous action of a vast variety of membrane-bound and soluble proteins, the latter of which act on either side of the membrane. Indirect evidence suggests that the machinery for synthesis and translocation, signal peptide processing, glycosylation and folding is present as a tight complex.

3. ER proteins and their role in protein biosynthesis and folding

The ER lumen is a unique environment for specialized protein synthesis steps such as *N*-linked glycosylation and disulphide bridge formation, most of which occur during the synthesis and translocation of proteins across the ER membrane. *N*-linked glycosylation is dependent on the consensus site Asn-X-Ser on the primary structure of a polypeptide, where X is any amino acid other than proline. In addition, glycosylation is dependent on the protein conformation, as the presence of a consensus glycosylation site does not always lead to glycosylation (Vitale *et al.,* 1993). Until now, plant homologues of signal peptidase and the components of the oligosaccharyl transferase complex have not been identified. Information regarding the conformational dependence of protein glycosylation and ER-localized glycan modification in plants as well as other modifications, such as prolyl hydroxylation, have been reviewed recently (Vitale *et al.,* 1993), and will not be discussed in detail here. Protein folding and assembly in the lumen are assisted by chaperones as in the cytosol, and the ER lumen therefore contains a specialized subset of these proteins in order to accomplish its function. Many of the ER chaperones are homologues of chaperones in other compartments, but due to the differences in certain aspects of protein synthesis between the ER lumen and other compartments it is expected that compartment-specific chaperones exist as well. Several ER-resident proteins have been identified in plants, and some of the corresponding genes have been cloned.

The lumenal binding protein (BiP), a lumenal Hsp70 homologue

The glucose-regulated protein 78 (GRP78), also known as the lumenal binding protein (BiP), is perhaps the best characterized chaperone of the ER lumen. It is structurally and functionally related to cytosolic Hsp70, except for the presence of a signal peptide to target it for translocation across the ER membrane, an ER localization signal to prevent it from being secreted, and a more acidic C-terminus than that of Hsp70. BiP is required for cell viability in *Saccharomyces cerevisiae*, appears to be involved in protein translocation (Nguyen *et al.*, 1991; Vogel *et al.*, 1990), and plays a role in the folding and oligomerization of proteins in the ER lumen (Helenius *et al.*, 1992). BiP is likely to bind transiently to all nascent protein chains, and associates permanently with malfolded proteins. The binding appears to occur on exposed hydrophobic regions, which is thought to prevent excessive aggregation in the ER, but the mechanism whereby BiP assists in the folding process is unknown. The only information available is that BiP binds to ATP and has ATPase activity, and that binding to malfolded proteins can be disrupted by the addition of ATP.

Several BiP plant homologues have been identified to date (Vitale *et al.*, 1993), but few functional studies have been undertaken. Tunicamycin-mediated increases in BiP mRNA levels (Denecke *et al.*, 1991; Fontes *et al.*, 1991) and the BiP synthesis rate (D'Amico *et al.*, 1992) have been clearly demonstrated. Recent cloning and expression analysis of a BiP homologue of spinach demonstrated a modest induction during cold acclimation (Anderson *et al.*, 1994a). Interestingly, no BiP mRNA was detected during water stress and heat shock treatment, which suggests that BiP mRNA is unstable under these conditions. Protein levels remained constant under all conditions. In addition, it could be shown that BiP binds to ATP *in vitro* (Anderson *et al.*, 1994b). A well-established model system for assessing BiP function in folding and oligomerization is that of phaseolin biosynthesis in developing bean cotyledons. Experiments with tunicamycin-stressed cotyledons (D'Amico *et al.*, 1992) or assembly-defective phaseolin mutants synthesized in transfected tobacco protoplasts (Pedrazzini *et al.*, 1994) have shown preferential binding of BiP to malfolded or unassembled phaseolin. In these experiments, BiP could also be released in the presence of ATP. At present, the interaction of plant BiP with potentially malfolded proteins can be monitored using well-established routine techniques (Denecke and Vitale, 1995). However, the role of BiP in quality control by the ER and in ER retention of malfolded proteins is not clearly established. In mammalian cells, it has been observed that BiP overexpression leads to a reduction in the transport rates of secretory proteins (Dorner *et al.*, 1992), but the mechanism has yet to be demonstrated (Vitale *et al.*, 1993).

Endoplasmin, the lumenal Hsp90 homologue

Endoplasmin is related to the cytosolic Hsp90 protein. As in the case of BiP, the main protein backbone is very similar to the cytosolic counterpart, except for the presence of a signal peptide and a more acidic C-terminus, which also contains the ER localization motif. Endoplasmin has recently been shown to bind to unassembled immunoglobulin chains *in vivo* (Melnick *et al.*, 1992), providing the first evidence for a possible chaperone function. Endoplasmin homologues have been identified in tobacco (Denecke *et al.*, 1993, 1995) and barley (Walter-Larsen *et al.*, 1993) based on protein sequence homology. The corresponding transcripts increase in abundance when barley leaves are infected with fungi (Walter-Larsen *et al.*, 1993), and during the gibberellic acid-mediated onset of hydrolase secretion by barley aleurone layers (Denecke *et al.*, 1995). These data suggest that endoplasmin is needed in higher amounts when the endomembrane system is actively synthesizing and transporting proteins. Fungal cell wall-degrading enzymes induce the secretion or vacuolar transport of a variety of pathogenesis-related proteins in cells that do not normally secrete large amounts of protein. Functional complementation experiments with the plant homologue of endoplasmin have been hampered by the fact that no yeast endoplasmin homologue has yet been identified.

Protein disulphide isomerase (PDI)

The ER lumen is an oxidizing protein-folding compartment (Freedman, 1989; Hwang *et al.*, 1992) which allows the formation of disulphide bridges. Disulphides form cross-links between different regions of the same protein or between subunits of separate polypeptides. In mammalian cells, influenza haemagglutinin (HAO) is synthesized and co-translationally translocated into the ER lumen in 2 min, and folding and disulphide bond formation occur co- and post-translationally (Braakman *et al.*, 1991). Plant storage proteins are typical examples of proteins with disulphide bonds, the synthesis and transport of which have been extensively studied in recent years (Vitale *et al.*, 1993). Many secretory proteins and membrane glycoproteins contain disulphide bonds, and the formation of these bonds is a process catalysed by PDI.

In mammalian cells, PDI is a soluble homodimer which contains two thioredoxin-like sequence motifs on each subunit (Noiva and Lennarz, 1992). The cysteines of these active sites are thought to form reversible intramolecular disulphide bonds. Plant homologues of PDI have been identified from alfalfa (Shorrosh and Dixon, 1991) and tobacco (Denecke, unpublished results), based on sequence conservation. The protein is moderately conserved between mammals, yeasts and plants (35% sequence identity), but PDI activity and functional identity have been shown by the production of recombinant plant and mammalian proteins in *E. coli* (Shorrosh and Dixon, 1991).

The role of PDI in disulphide bond formation was assessed by *in-vitro* translation–translocation using PDI-depleted dog pancreas microsomes and wheat storage protein transcripts. Reconstitution with purified PDI led to increased disulphide bond formation (Bulleid and Freedman, 1988). ATP-depleted mammalian cells synthesize incorrectly folded HAO aggregates with incorrect disulphide bridges (Braakman *et al.*, 1992a). The addition of ATP can dissolve aggregates, suggesting that disulphide bridge formation is a reversible process which requires energy. It is possible that the formation of aggregates is due to lack of BiP function under conditions of ATP starvation. Interestingly, disulphide bond formation in the ER lumen can be reversibly manipulated *in vivo* without affecting cellular functions by the external addition of reducing agents such as diothiothreitol (DTT) to the growth medium and short incubations (Braakman *et al.*, 1992b). It has yet to be shown that incorrect disulphide bridges occur during the folding process under normal conditions, or that PDI can disrupt such disulphide bridges.

Calreticulin and calnexin

Calreticulin and calnexin are two highly conserved proteins which have been found in a variety of organisms, including plants (Denecke *et al.*, 1995; Huang *et al.*, 1993). They are calcium-binding proteins which are very abundant in the ER. Calreticulin is a soluble protein, which is a homo-multimer in its native form in tobacco (Denecke *et al.*, 1995). Calnexin is a type-I membrane protein with a small cytosolic tail, but the main part of the protein is present in the lumen. The primary structure of calreticulin is almost identical to that of calnexin, except for the replacement of the stop-transfer sequence and the cytosolic tail by an ER localization signal for soluble ER proteins. Calnexin has been suggested to be a novel type of chaperone, and there is an abundance of experimental data from mammalian cells to support this view (David et al., 1993; Jackson et al., 1994; Margolese et al., 1993; Ou et al., 1993; Rajagopalan et al., 1994). The most direct evidence for an in-vivo chaperone function is the ability of engineered calnexin lacking the cytosolic ER retention motif to re-direct normally ER-retained CD3 ε-chains to the Golgi in co-expression experiments. To date, no evidence has been obtained to support the view that calreticulin is a chaperone as well, but its primary structure and strong similarity to calnexin strongly suggest such a function (Denecke et al., 1995). The availability of antibodies to tobacco calreticulin (Denecke et al., 1995) opens up the possibility of assessing this hypothesis. An alternative function in regulating nuclear hormone receptors has also been proposed (Burns et al., 1994; Dedhar et al., 1994), but the biological relevance of these observations has yet to be demonstrated.

Novel components of the ER lumen and membrane with a suggested chaperone function

Recently, a new type of low-molecular-weight heat shock protein was discovered, which appears to be localized in the ER lumen (Helm et al., 1993). This is suggested by the presence of signal peptides and C-terminal sequences which might constitute ER localization signals. The presence of this additional class of heat shock-related lumenal proteins demonstrates the complexity of the system. A method based on immunological recognition of ER localization signals was recently used to identify a vast variety of ER resident proteins (Denecke et al., 1995). This methodology will lead to a rapid characterization of the lumenal content of the plant ER.

A secretory polypeptide, as deduced from cDNA sequences, which contains sequence similarity to PDI but which does not contain an ER retention motif, has been identified from alfalfa (Shorrosh and Dixon, 1992). The function of this protein is not clear, but it is interesting to note that a similar truncated version also exists for BiP (Otterson et al., 1994). This polypeptide was identified in S. cerevisiae and contains a signal peptide and only the ATPase domain. The simultaneous presence of secreted and ER-accumulated chaperones may ensure protein folding and maturation steps at multiple sites in the endomembrane system.

4. Autoregulatory mechanisms to guarantee ER integrity

The correct functioning of the ER as protein synthesis and delivery machinery depends on regulatory mechanisms which control the maintenance of proteins involved in translocation and modification, as well as chaperones and processing enzymes within the ER. This also involves a co-ordinated synthesis of components which act in concert, such as the different components of the protein translocation machinery. The functional dissection of isolated ER components has led to the development of a number of interesting working models of protein synthesis by the ER, but very little is known about the co-ordination of the various steps from protein translocation to the delivery of the products to the Golgi.

How do ER-resident proteins accumulate in the ER?

The recycling of membranes between the ER and the Golgi apparatus is a highly regulated process which is responsible for the delivery of membranes and lipids to the distribution centre of the endomembrane system, namely the Golgi apparatus. The ER and the Golgi apparatus can thus be regarded as a dynamic membrane system, and the identity of either of these organelles is dependent on an equilibrium between anterograde (from the ER to the Golgi) and retrograde (from the Golgi to the ER) vesicle transport (Lippincott-Schwartz, 1993). It should be noted that, to date, no clear information has been obtained as to whether the 'vesicular transport' model (Rothman and Orci, 1992) or the 'cisternal progression' model (Saraste and Kuismanen, 1984) provides the best description of the ER–Golgi system. The latter assumes dynamic progression of compartments, which mature (change in composition) by selective retrograde retrieval of components. The (more popular) vesicular transport model assumes that the ER and the Golgi are static compartments which are transiently connected via vesicular anterograde and retrograde transport. In mammalian cells, the so-called intermediate compartment (IC) is a compartment located between the ER and the Golgi which becomes visible when cells are incubated at 15°C. Both models can account for the existence and the role of the IC (Lippincott-Schwartz, 1993). The remainder of this section is written using the nomenclature employed by the vesicular transport model, but the reader is recommended to interpret open questions and final conclusions in the light of the cisternal progression model as well.

The transport of soluble proteins to the Golgi apparatus is generally believed to occur by default without sorting information on the protein backbone (Bednarek and Raikhel, 1992; Denecke et al., 1990). This implies that all molecules can diffuse freely into and out of nascent vesicles formed by the ER. Over time periods which exceed the half-life of proteins in the endomembrane system, the rate of protein synthesis on the rough ER membrane must be the limiting factor in both vacuolar and secretory protein transport, since otherwise ER and Golgi integrity would not be maintained. ER-resident proteins such as BiP, endoplasmin, PDI and calreticulin are soluble and accomplish their function in the ER lumen. They can thus 'escape' from the ER by diffusion into anterograde transport vesicles. Eukaryotic cells possess a receptor-mediated retrieval system which recognizes a conserved C-terminal motif in proteins which must be maintained in the ER lumen (for review see Vitale et al., 1993). This motif is often an acidic region followed by the C-terminal tetrapeptide HDEL or sometimes KDEL in plants, and can be recognized in most cases simply by analysing the primary structure of a protein. The analysis of different ER retention motifs with respect to plant proteins as well as experiments on the context dependence of the tetrapeptide motif have been reviewed extensively (Vitale et al., 1993).

ER accumulation of membrane proteins in mammalian cells is dependent on positively charged residues near the end of the cytosolic tail of the protein, which is either the C-terminus for type-I membrane proteins (Jackson et al., 1990) or the N-terminus for type-II membrane proteins (Schutze et al., 1994). From the sequence of Arabidopsis calnexin (Huang et al., 1993), a type-I membrane protein, it can be deduced that multiple positively charged residues at the cytosolic tail might also function as ER localization signals in plants. The recycling of ER-localized membrane proteins from a post-ER compartment has only been demonstrated in mammalian cells to date (Jackson et al., 1993).

The maintenance of ER components is not only important in the case of proteins. The intense bulk flow towards the Golgi requires a similarly intense retrograde transport of membranes to maintain the membranes of the ER and the Golgi, and thus the integrity of both organelles. The recycling principle for sorting receptors and pathways was first identified for mannose-6-phosphate receptor-mediated lysosomal targeting (von Figura and Hasilik, 1986). It is now recognized as a basic principle for membrane and protein transport between compartments of the endomembrane system. Sorting receptors and other components required for the formation, budding and fusion of vesicles have been identified using genetic approaches in S. cerevisiae or in-vitro reconstituted transport systems based on endomembrane components isolated from mammalian cells. The ERD2 gene product has been shown to act as the sorting receptor for soluble ER-resident proteins by demonstrating its ability to affect the specificity and capacity of the yeast lumenal retention system (Lewis *et al.*, 1990; Semenza and Pelham, 1992). ERD2p is a membrane-spanning protein which most probably binds to its ligands in a pH-dependent manner, and with high affinity in the Golgi/IC and low affinity in the ER (Wilson *et al.*, 1993). Overexpression of ERD2 homologues in mammalian cells leads to enhanced retrograde transport (Hsu *et al.*, 1992), which is consistent with its role in regulating the retrieval of proteins from post-ER compartments. Retrieval appears to be possible even from the *trans*-Golgi network (Pelham *et al.*, 1992; Peter *et al.*, 1992).

ER retention in plants has been shown to be difficult to saturate with higher amounts of ligands (Denecke *et al.*, 1992). There are two possible mechanisms which can account for this lack of saturability, both of which involve signal transduction across the membrane of the endomembrane system (see *Figure 1*).

(1) The number of ligands is monitored in the lumen and regulates the number of receptors accordingly, whereas the rate of vesicle traffic remains constant. If the number of ligands increases, more receptors will be synthesized, which increases the probability of interaction in the salvage compartment and thus the retention efficiency in the ER–Golgi system.
(2) The number of receptors remains constant but the recycling rate increases when the number of ligands increases.

Both models require an efficient anterograde delivery of empty receptors to the salvage compartment, since the number of receptors is likely to be lower than the number of ligands to be recycled. Model 1 would require an active mechanism to concentrate empty receptors relative to the number of leaking ER residents. Model 2 would require the existence of an active anterograde transport route of receptors to the retrieval compartment which is different from the default route. Interestingly, anterograde transport of membrane proteins was recently shown to include a concentration step in the ER (Balch *et al.*, 1994). The latter suggests the existence of such positive sorting information for membrane proteins, and it is possible that membrane-bound sorting receptors in particular possess such signals. Evidence for two independent transport pathways from the ER to the Golgi has also been suggested by recent work on plant cells (Gomez and Chrispeels, 1993). The finding that ERD2 redistributes to the ER if ligands are overexpressed (Lewis and Pelham, 1992) is inconsistent with the lack of saturation, since an increased retention capacity would require higher receptor concentrations in the Golgi/IC. Clearly, more work needs to be done in order to define the role and the mode of action of ERD2. The recent cloning of an *Arabidopsis* ERD2 homologue as well as continuous identification of other components of the vesicle transport machinery will provide the necessary tools to analyse this aspect of membrane transport in more detail.

How is the nucleus informed about the need for more chaperones?

A stress situation which requires increased recycling of ligands by ERD2 is that of enhanced synthesis of lumenal chaperones. It is well established that a variety of lumenal ER-resident proteins are induced by stresses that cause the accumulation of malfolded proteins in the ER lumen. Treatment of cells with tunicamycin is the stimulus most frequently used to induce the 'malfolded protein response'. Tunicamycin inhibits *N*-linked glycosylation, which leads to the accumulation of *de novo*-synthesized proteins lacking *N*-linked glycans. A subset of these unglycosylated proteins will be malfolded and lead to an increased rate of BiP mRNA synthesis (Kozutsumi *et al.*, 1988). It has been shown that overexpression of BiP diminishes or even abolishes the tunicamycin-mediated malfolded protein response in mammalian cells and yeasts (Dorner *et al.*, 1992; Kohno *et al.*, 1993). On the other hand, ERD2 mutants that are defective in

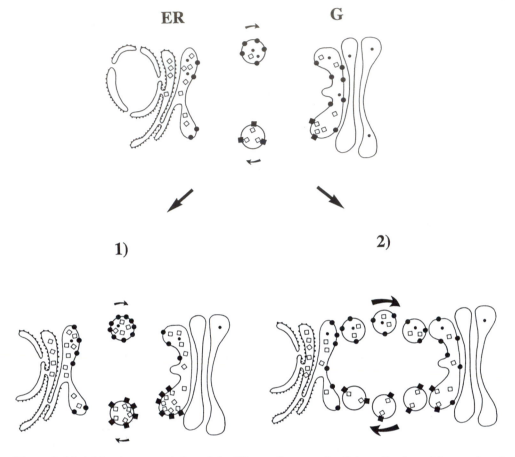

Figure 1. Model for the up-regulation of the ER retention capacity. Schematic view of the rough and smooth endoplasmic reticulum (ER) and the Golgi aparatus (G), and transport of vesicles from the ER to the Golgi (anterograde transport, arrow pointing to the right) and from the Golgi to the ER (retrograde transport, arrow pointing to the left). Reticuloplasmins (open squares) can escape from the ER via the anterograde transport route and reach the Golgi together with secretory proteins (small dots). The receptor (solid circles) binds exclusively to reticuloplasmins. The binding to its ligands is thought to cause a change in the conformation (solid squares) which leads to formation of a vesicle. In the ER, the receptor releases its ligands and returns to the Golgi. Up-regulation of the retention capacity can be achieved (1) by increasing the number of receptors or (2) by increasing the rate of vesicle transport (thick arrows).

ER retention of soluble ER proteins and which thus exhibit a lower steady-state concentration of BiP in the ER lumen appear to synthesize increased amounts of BiP (Hardwick *et al.,* 1990). These data suggest that the free BiP concentration in the ER lumen is monitored by the cell, and that low concentrations of free BiP lead to increased BiP transcription via a feedback mechanism. Such a mechanism would allow the cell to cope with changes in ER secretory activity or stress situations that require increased synthesis of chaperones. Obviously, increased BiP synthesis would also lead to a stimulation of the ER retention machinery to maintain these molecules intracellularly.

A genetic approach using yeast cells has led to the identification of a receptor kinase (IRE1p/ERN1p) which is required for the signal transduction process (Cox *et al.,* 1993; Mori *et al.*, 1993). This receptor kinase could be located in the ER, the nuclear envelope or the salvage compartment. The signal could be transduced via the cytosol and the nuclear pores, which would be comparable to plasma membrane-located receptor kinases. Another possibility would be direct signalling from the inner nuclear envelope membrane to nuclear transcription factors (Shamu *et al.*, 1994). Indirect evidence suggests that dimerization activates the malfolded protein response (Mori *et al.*, 1993). During the accumulation of malfolded proteins in the ER lumen, the free BiP concentration would decrease as BiP would bind to the unfolded protein aggregates. Two possible regulatory mechanisms could be used to increase BiP synthesis subsequently:

(1) if the IRE1p sends a negative signal in its active state (repression of BiP transcription), dimerization would require high lumenal BiP concentrations;
(2) if the active form of IRE1p sends a signal which stimulates BiP transcription, dimerization would be prevented by high BiP concentrations.

It remains to be shown where the IRE1 gene product is located and whether BiP directly interacts with the lumenal receptor domain.

The IRE1 gene product is not the only receptor kinase that could be involved in signal transduction from the lumen of the endomembrane system to the nucleus. The mouse Ltk transmembrane protein tyrosine kinase was recently found to be localized to the endoplasmic reticulum (Bauskin *et al.*, 1991). Its *in-vivo* catalytic activity is ligand-independent and appears to be regulated by changes in the lumenal redox potential, with the active form of the protein being a disulphide-linked multimer. It is tempting to assume that this transmembrane tyrosine kinase is involved in regulating processes related to protein disulphide bridge formation and the regulation of PDI transcription, but this has yet to be demonstrated.

5. External control of ER activity

The biosynthetic activity of the ER is not constant, and depends on the cell type and the physiology of the cell. During cell division, the ER synthesizes larger amounts of extracellular matrix proteins and enzymes which are required for carbohydrate synthesis in the Golgi apparatus and on the plasma membrane. The interaction of plant cells with pathogenic micro-organisms can result in the synthesis of defence-related proteins by the ER. Secretory plant tissues contain large amounts of ER membranes, and can regulate the ER activity via hormones. The ER–Golgi system contains efficient feedback mechanisms to respond to such changes, and can synthesize more lumenal chaperones or increase the rate of anterograde and retrograde protein transport if necessary. However, recent experiments have suggested the possibility that chaperone synthesis might not be exclusively regulated by feedback mechanisms.

The barley aleurone layer is a well-characterized secretory plant tissue and a suitable model system for assessing the external control of ER activity in plants. The aleurone layer secretes large quantities of hydrolases to the endosperm of germinating seeds, and the ER undergoes a proliferation or reorganization during the gibberellic acid-stimulated onset of secretory activity (Evins and Varner, 1971; Jacobsen and Beach, 1985; Jones, 1969a,b, 1980, 1985; Jones and Jacobsen, 1991). It was shown that gibberellic acid-induced mRNA accumulation corresponding to the ER-resident proteins BiP, endoplasmin and calreticulin was temporally co-ordinated with the onset of α-amylase mRNA accumulation (Denecke *et al.*, 1995). If chaperone synthesis is purely a feedback-controlled process, increased chaperone mRNA concentrations should become apparent with a certain delay compared to the onset of α-amylase mRNA synthesis. More dramatically, increased chaperone mRNA concentrations appeared with lower hormone dosage than required for α-amylase transcripts (Denecke *et al.*, 1995). This suggests that barley aleurone cells can anticipate the need for increased ER chaperone synthesis, and first induce the secretory machinery before the synthesis of hydrolases is initiated. Although this observation has yet to be confirmed using several hydrolases as markers for the secretory activity of the aleurone endomembrane system, it is tempting to assume that external factors can affect ER activity. Since chaperones such as BiP are already required during protein translocation, it would indeed be of advantage to anticipate the need for enhanced chaperone synthesis. In the future, the barley aleurone layer will be a useful model system for assessing the regulatory mechanisms which control co-ordinated events associated with secretory protein transport in the plant endomembrane system.

6. Future prospects

A combination of biochemical approaches using *in-vitro* reconstitution systems in combination with genetic and morphological approaches has led to significant progress in the elucidation of ER functions in protein synthesis and transport. These methodologies will continue to result in the isolation of novel ER components and provide new insights into their mode of action. Plants are ideal model systems for combining fundamental cell biology with the analysis of processes in entire organisms, due to the availability of transformation and regeneration systems. Several model systems, such as the synthesis and transport of storage proteins or hydrolase secretion, can be explored further. Rather than establishing similar *in-vitro* membrane-traffic assays to those in mammalian cells, it will be possible to use the available information and working models to advance the field and to test the available models *in vivo*.

Acknowledgements

This work was supported by the Swedish Natural Science Research Council (grant number B-BU 06863-301) and the Human Capital and Mobility Programme of the European Union (contract CHRX-CT94-0590).

References

Anderson JV, Li Q-B, Haskell DW, Guy CL. (1994a) Structural organization of the spinach endoplasmic reticulum luminal 70-kilodalton heat shock cognate gene and expression of 70-kilodalton heat-shock genes during cold acclimation. *Plant Physiol.* **104:** 1359–1370.

Anderson JV, Haskell DW, Guy CL. (1994b) Differential influence of ATP on native spinach 70-kilodalton heat-shock cognates. *Plant Physiol.* **104:** 1371–1380.

Balch WE, McCaffrey JM, Plutner H, Farquhar MG. (1994) Vesicular stomatitis virus glycoprotein is sorted and concentrated during export from the endoplasmic reticulum. *Cell* **76:** 841–852.

Bauskin AR, Alkalay I, Ben-Neriah Y. (1991) Redox regulation of a protein tyrosine kinase in the endoplasmic reticulum. *Cell* **66:** 685–696.

Bednarek SY, Raikhel NV. (1992) Intracellular trafficking of secretory proteins. *Plant Mol. Biol.* **20:** 133–150.

Blobel G, Dobberstein B. (1975) Transfer of proteins across the membrane. I. Presence of proteolytically processed and unprocessed nascent immunoglobulin light chains on membrane-bound ribosomes of murine myeloma. *J. Cell Biol.* **67:** 835–851.

Bollini R, Vitale A, Chrispeels MJ. (1983). *In vivo* and *in vitro* processing of seed reserve protein in the endoplasmic reticulum: evidence for two glycosylation steps. *J. Cell Biol.* **96:** 999–1007.

Braakman I, Hoover-Litty H, Wagner KR, Helenius A. (1991) Folding of influenza hemagglutinin in the endoplasmic reticulum. *J. Cell Biol.* **114:** 401–411.

Braakman I, Helenius J, Helenius A. (1992a) Role of ATP and disulphide bonds during protein folding in the endoplasmic reticulum. *Nature* **356:** 260–262.

Braakman I, Helenius J, Helenius A. (1992b) Manipulating disulphide bond formation and protein folding in the endoplasmic reticulum. *EMBO J.* **11:** 1717–1722.

Bulleid NJ, Freedman RB. (1988) Defective co-translational formation of disulphide bonds in protein disulphide-isomerase-deficient microsomes. *Nature* **335:** 649–651.

Burns K, Duggan B, Atkinson EA, Famulski KS, Nemer M, Bleakley RC, Michalak M. (1994) Modulation of gene expression by calreticulin binding to the glucocorticoid receptor. *Nature* **367:** 476–480.

Campos N, Palau J, Zwieb C. (1989) Diversity of 7 SL RNA from the signal recognition particle of maize endosperm. *Nucl. Acids Res.* **17:** 1573–1588.

Chirico WJ, Waters MG, Blobel G. (1988) 70K heat shock related proteins stimulate protein translocation into membranes. *Nature* **332:** 805–810.

Cox JS, Shamu CE, Walter P. (1993) Transcriptional induction of genes encoding endoplasmic reticulum resident proteins requires a transmembrane protein kinase. *Cell* **73:** 1197–1206.

Crowley KS, Reinhart GD, Johnson AE. (1993) The signal sequence moves through a ribosomal tunnel into a noncytoplasmic aqueous environment at the ER membrane early in translocation. *Cell* **73:** 1101–1115.

D'Amico L, Valsasina B, Daminati MG, Fabbrini MS, Nitti G, Bollini R, Ceriotti A, Vitale A. (1992) Bean homologues of the mammalian glucose-regulated proteins: induction by tunicamycin and interaction with newly synthesized storage proteins in the endoplasmic reticulum. *Plant J.* **2:** 443–455.

David V, Hochstenbach F, Rajagopalan S, Brenner MB. (1993) Interaction with newly synthesised and retained proteins in the endoplasmic reticulum suggests a chaperone function for human integral membrane protein IP90 (calnexin). *J. Biol. Chem.* **268:** 9585–9592.

Dedhar S, Rennie PS, Shago M et al. (1994) Inhibition of nuclear hormone receptor activity by calreticulin. *Nature* **367:** 480–483.

Denecke J, Vitale A. (1995) The use of plant protoplasts to study protein synthesis, quality control, protein modification and transport through the plant endomembrane system. *Meth. Cell Biol.* **50:** 335–348.

Denecke J, Botterman J, Deblaere R. (1990) Protein secretion in plant cells can occur via a default pathway. *Plant Cell* **2:** 51–59.

Denecke J, Souza Goldman MH, Demolder J, Seurinck J, Botterman J. (1991) The tobacco lumenal binding protein is encoded by a multigene family. *Plant Cell* **3:** 1025–1035.

Denecke J, De Rycke R, Botterman J. (1992) Plant and mammalian sorting signals for protein retention in the endoplasmic reticulum contain a conserved epitope. *EMBO J.* **11:** 2345–2355.

Denecke J, Ek B, Caspers M, Sinjorgo KMC, Palva ET. (1993) Analysis of sorting signals responsible for the accumulation of soluble reticuloplasmins in the plant endoplasmic reticulum. *J. Exp. Bot.* **44S:** 213–221.

Denecke J, Carlsson L, Vidal S, Ek B, Höglund A-S, van Zeijl M, Sinjorgo K, Palva ET. (1995) The tobacco homolog of mammalian calreticulin is present in protein complexes *in vivo. Plant Cell* **7:** 391–406.

Deshaies RJ, Kock BD, Werner-Washburne M, Craig EA, Scheckman R. (1988) A subfamily of stress proteins facilitates translocation of secretory and mitochondrial precursor polypeptides. *Nature* **332:** 800–805.

Dorner AJ, Wasley LC, Kaufman RJ. (1992) Overexpression of GRP78 mitigates stress induction of glucose regulated proteins and blocks secretion of selective proteins in Chinese hamster ovary cells. *EMBO J.* **11:** 1563–1571.

Ellis RJ, van der Vries SM, Hemmingsen SM. (1989) The molecular chaperone concept. *Biochem. Soc. Symp.* **55:** 145–153.

Evans EA, Gilmore R, Blobel G. (1986) Purification of microsomal signal peptidase as a complex. *Proc. Natl Acad. Sci. USA* **83:** 581–585.

Evins WH, Varner JE. (1971) Hormone-controlled synthesis of endoplasmic reticulum in barley aleurone cells. *Proc. Natl Acad. Sci. USA* **68:** 1631–1633.

Fontes EBP, Shank BB, Wrobel RL, Moose SP, O'Brian GR, Wurtzel ET, Boston RS. (1991) Characterisation of an immunoglobulin binding protein homolog in the maize floury-2 endosperm mutant. *Plant Cell* **3:** 483–496.

Freedman RB. (1989) Protein disulfide isomerase: multiple roles in the modification of nascent secretory proteins. *Cell* **57:** 1069–1072.

Geetha-Habib M, Noiva R, Kaplan HA, Lennarz WJ. (1988) Glycosylation site-binding protein, a component of oligosaccharyl transferase, is highly similar to three other 57 kDa luminal proteins in the ER. *Cell* **54:** 1053–1060.

Gomez L, Chrispeels MJ. (1993) Tonoplast and soluble vacuolar proteins are targeted by different mechanisms. *Plant Cell* **5:** 1113–1124.

Grimm R, Speth V, Gatenby AA, Schafer E. (1991) GroEL-related molecular chaperones are present in the cytosol of oat cells. *FEBS Lett.* **286:** 155–158.

Hardwick KG, Lewis MJ, Semenza J, Dean N, Pelham HRB. (1990) ERD1, a yeast gene required for the retention of luminal endoplasmic reticulum proteins, affects glycoprotein processing in the Golgi apparatus. *EMBO J.* **9**: 623–630.

Helenius A, Marquardt T, Braakman I. (1992) The endoplasmic reticulum as a protein folding compartment. *Trends Cell Biol.* **2**: 227–231.

Helm KW, LaFayette PR, Nagao RT, Key JL, Vierling E. (1993) Localization of small heat shock proteins to the higher plant endomembrane system. *Mol. Cell. Biol.* **13**: 238–247.

High S. (1992) Membrane protein insertion into the endoplasmic reticulum — another channel tunnel? *BioEssays* **14**: 535–540.

High S, Stirling CJ. (1993) Protein translocation across membranes: common themes in divergent organisms. *Trends Cell Biol.* **3**: 335–339.

Hsu VW, Shah N, Klausner RD. (1992) A brefeldin A-like phenotype is induced by the overexpression of a human ERD2-like protein, ELP-1. *Cell* **69**: 625–635.

Huang L, Franklin AE, Hoffman NE. (1993) Primary structure and characterisation of an *Arabidopsis thaliana* calnexin-like protein. *J. Biol. Chem* **268**: 6560–6566.

Hwang C, Sinskey AJ, Lodish HF. (1992) Oxidized redox state of glutathione in the endoplasmic reticulum. *Science* **257**: 1496–1502.

Jackson MR, Nilsson T, Peterson PA. (1990) Identification of a consensus motif for retention of transmembrane proteins in the endoplasmic reticulum. *EMBO J.* **9**: 3153–3162.

Jackson MR, Nilsson T, Peterson PA. (1993) Retrieval of transmembrane proteins to the endoplasmic reticulum. *J. Cell Biol.* **121**: 317–333.

Jackson MR, Cohen-Doyle MF, Peterson PA, Williams DB. (1994) Regulation of MHC Class I transport by the molecular chaperone, calnexin (p88, IP90). *Science* **263**: 384–387.

Jacobsen JV, Beach LR. (1985) Control of transcription of α-amylase and rRNA genes in barley aleurone protoplasts by gibberellin and abscisic acid. *Nature* **316**: 275–277.

Jones RL. (1969a) Gibberellic acid and the fine structure of barley aleurone cells. I. Changes during the lag phase of α-amylase synthesis. *Planta* **87**: 119–133.

Jones RL. (1969b) Gibberellic acid and the fine structure of barley aleurone cells. II. Changes during the synthesis and secretion of α-amylase. *Planta* **88**: 73–86.

Jones RL. (1980) Quantitative and qualitative changes in the endoplasmic reticulum of barley aleurone layers. *Planta* **150**: 70–81.

Jones RL. (1985) Protein synthesis and secretion by the barley aleurone, a perspective. *Israel. J. Bot.* **34**: 377–395.

Jones RL, Jacobsen JV. (1991) Regulation of synthesis and transport of secreted proteins in cereal aleurone. *Int. Rev. Cytol.* **126**: 49–88.

Kelleher DJ, Kreibich G, Gilmore R. (1992) Oligosaccharyl transferase activity is associated with a protein complex composed of ribophorins I and II and a 48 kDa protein. *Cell* **69**: 55–65.

Kohno K, Normington K, Sambrook J, Gething MJ, Mori K. (1993) The promoter region of the yeast KAR2 (BiP) gene contains a regulatory domain that responds to the presence of unfolded proteins in the endoplasmic reticulum. *Mol. Cell. Biol.* **13**: 877–890.

Kozutsumi Y, Segal M, Normington K, Slaughter C, Gething M-J, Sambrook J. (1988) The presence of malfolded proteins in the endoplasmic reticulum signals the induction of glucose-regulated proteins. *Nature* **332**: 462–464.

Krieg UC, Walter P, Johnson AE. (1986) Photocrosslinking of the signal sequence of nascent preprolactin to the 54-kilodalton polypeptide of the signal recognition particle. *Proc. Natl Acad. Sci. USA* **83**: 8604–8608.

Krishnan HB, Franceschi VR, Okita TW. (1986) Immunochemical studies on the role of the Golgi complex in protein-body formation in rice seeds. *Planta* **169**: 471–480.

Kumamoto CA. (1991) Molecular chaperones and protein translocation across the *Escherichia coli* inner membrane. *Mol. Microbiol.* **5**: 19–22.

Kurzchalia TV, Wiedmann M, Girshovich AS, Bochkareva ES, Bielka H, Rapoport TA. (1986) The signal sequence of nascent preprolactin interacts with the 54K polypeptide of the signal recognition particle. *Nature* **320**: 634–636.

Kutay U, Hartmann E, Rapoport TA. (1993) A class of membrane proteins with a C-terminal anchor. *Trends Cell Biol.* **3**: 72–75.

Lewis MJ, Pelham HRB. (1992) Ligand induced redistribution of a human KDEL receptor from the Golgi complex to the ER. *Cell* **68**: 353–364.

Lewis MJ, Sweet DJ, Pelham HRB. (1990) The ERD2 gene determines the specificity of the luminal ER protein retention system. *Cell* **61**: 1359–1363.

Li X, Franceschi VR, Okita TW. (1993) Segregation of storage protein mRNAs on the rough endoplasmic reticulum membranes of rice endosperm cells. *Cell* **72:** 869–879.

Lippincott-Schwartz J. (1993) Bidirectional membrane traffic between the endoplasmic reticulum and Golgi apparatus. *Trends Cell Biol.* **3:** 81–88.

Lubben TH, Donaldson GK, Viitanen PV, Gatenby AA. (1989) Several proteins imported into chloroplasts form stable complexes with the GroEL-related chloroplast molecular chaperone. *Plant Cell* **1:** 1223–1230.

Margolese L, Waneck GL, Suzuki CK, Degen E, Flavell RA, Williams DB. (1993) Identification of the region on the class I histocompatibility molecule that interacts with the molecular chaperone, p88 (calnexin, IP90). *J. Biol. Chem.* **268:** 17959–17966.

Melnick J, Aviel S, Argon Y. (1992) The endoplasmic reticulum stress protein GRP94, in addition to BiP, associates with unassembled immunoglobulin chains. *J. Biol. Chem.* **267:** 21303–21306.

Miernyk JA, Duck NB, Shatters RG, Folk WR. (1992) The 70-kilodalton heat shock cognate can act as a molecular chaperone during the membrane translocation of a plant secretory protein precursor. *Plant Cell* **4:** 821–829.

Miyata S, Akazawa T. (1982) α-amylase biosynthesis: evidence for temporal sequence of NH_2-terminal peptide cleavage and protein glycosylation. *Proc. Natl Acad. Sci. USA* **79:** 6566–6568.

Mori K, Ma W, Gething M-J, Sambrook J. (1993) A transmembrane protein with a cdc2+/cdc28-related kinase activity is required for signalling from the ER to the nucleus. *Cell* **74:** 743–756.

Nguyen TH, Law DTS, Williams DB. (1991) Binding protein BiP is required for translocation of secretory proteins into the endoplasmic reticulum in *Saccharomyces cerevisiae*. *Proc. Natl Acad. Sci. USA* **88:** 1565–1569.

Nicchitta CV, Blobel G. (1993) Lumenal proteins of the mammalian endoplasmic reticulum are required to complete protein translocation. *Cell* **73:** 989–998.

Noiva R, Lennarz WJ. (1992) Protein disulphide isomerase. A multifunctional protein resident in the lumen of the endoplasmic reticulum. *J. Biol. Chem.* **267:** 3553–3556.

Noiva R, Kaplan HA, Lennarz WJ. (1991a) Glycosylation site-binding protein is not required for *N*-linked glycoprotein synthesis. *Proc. Natl Acad. Sci. USA* **88:** 1986–1990.

Noiva R, Kimura H, Roos J, Lennarz WJ. (1991b) Peptide binding by protein disulphide isomerase, a resident protein of the endoplasmic reticulum lumen. *J. Biol. Chem.* **266:** 19645–19649.

Nunnari J, Walter P. (1992) Protein targeting to and translocation across the membrane of the endoplasmic reticulum. *Curr. Opin. Cell Biol.* **4:** 573–580.

Okita TW, Li X, Roberts MW. (1994) Targeting of mRNAs to domains of the endoplasmic reticulum. *Trends Cell Biol.* **4:** 91–96.

Osteryoung KW, Sticher L, Jones RL, Bennett AB. (1992) *In vitro* processing of tomato proteinase inhibitor I by barley microsomal membranes. A system for analysis of cotranslational processing of plant endomembrane proteins. *Plant Physiol.* **99:** 378–382.

Otterson GA, Flynn GC, Kratzke RA, Coxon A, Johnston PG, Kaye FJ. (1994) Stch encodes the ATPase core of a microsomal stress70 protein. *EMBO J.* **13:** 1216–1225.

Ou W-J, Cameron P, Thomas DY, Bergeron JJM. (1993) Association of folding intermediates of glycoproteins with calnexin during protein maturation. *Nature* **364:** 771–776.

Palade G. (1975) Intracellular aspects of the process of protein synthesis. *Science* **189:** 347–358.

Pathak RK, Luskey KL, Anderson RG. (1986) Biogenesis of crystalloid endoplasmic reticulum in UT-1 cells: evidence that newly formed endoplasmic reticulum emerges from the nuclear envelope. *J. Cell Biol.* **102:** 2156–2168.

Pedrazzini E, Giovinazzo G, Bollini R, Ceriotti A, Vitale A. (1994) Binding of BiP to an assembly-defective protein in plant cells. *Plant J.* **5:** 103–110.

Pelham HRB, Roberts LM, Lord JM. (1992) Toxin entry. How reversible is the secretory pathway? *Trends Cell Biol.* **2:** 183–185.

Peter F, Van Nguyen P, Söling H-D. (1992) Different sorting of Lys-Asp-Glu-Leu proteins in rat liver. *J. Biol. Chem* **267:** 10631–10637.

Phillips GJ, Silhavy TJ. (1990) Heat-shock proteins DnaK and GroEL facilitate export of lacZ hybrid proteins in *E. coli*. *Nature* **344:** 882–884.

Rajagopalan S, Xu Y, Brenner MB. (1994) Retention of unassembled components of integral membrane proteins by calnexin. *Science* **263:** 387–390

Rapoport TA. (1992) Transport of proteins across the endoplasmic reticulum membrane. *Science* **258:** 931–936.

Rose JK, Doms RW. (1988) Regulation of protein export from the endoplasmic reticulum. *Annu. Rev. Cell Biol.* **4:** 257–288.

Rothman JE. (1989) Polypeptide chain binding proteins: catalysts of protein folding and related processes in cells. *Cell* **59:** 591–601.

Rothman JE, Orci L. (1992) Molecular dissection of the secretory pathway. *Nature* **355:** 409–415.

Sanders SL, Schekman R. (1992) Polypeptide translocation across the endoplasmic reticulum membrane. *J. Biol. Chem.* **267:** 13791–13794.

Saraste J, Kuismanen E. (1984) Pre-Golgi and Post-Golgi vacuoles operate in the transport of Semliki Forest virus membrane-glycoproteins to the cell surface. *Cell* **38:** 535–549.

Schutze MP, Peterson PA, Jackson MR. (1994) An N-terminal double-arginine motif maintains type II membrane proteins in the endoplasmic reticulum. *EMBO J.* **7:** 1696–1705.

Semenza JC, Pelham HRB. (1992) Changing the specificity of the sorting receptor for luminal endoplasmic reticulum proteins. *J. Mol. Biol.* **224:** 1–5.

Sengupta C, DeLuca V, Bailey DS, Verma DPS. (1981) Post-translational processing of 7S and 11S components of soybean storage proteins. *Plant Mol. Biol.* **1:** 19–34.

Shamu CE, Cox JS, Walter P. (1994) The unfolded protein response pathway in yeast. *Trends Cell Biol.* **4:** 56–60.

Shelness GS, Blobel G. (1990) Two subunits of the canine signal peptidase complex are homologous to yeast SEC11 protein. *J. Biol. Chem.* **265:** 9512–9519.

Shore GC, Tata JR. (1977) Two fractions of rough endoplasmic reticulum from rat liver. *J. Cell Biol.* **72:** 726–743.

Shorrosh BS, Dixon RA. (1991) Molecular cloning of a plant putative endomembrane protein resembling vertebrate protein disulphide-isomerase and a phosphatidylinositol-specific phospholipase C. *Proc. Natl Acad. Sci. USA* **88:** 10941–10945.

Shorrosh BS, Dixon RA. (1992) Molecular characterization and expression of an alfalfa protein with sequence similarity to mammalian ERp72, a glucose regulated endoplasmic reticulum protein containing active site sequences of protein disulphide isomerase. *Plant J.* **2:** 51–58.

Vitale A, Ceriotti A, Denecke J. (1993) The role of the endoplasmic reticulum in protein synthesis, modification and intracellular transport. *J. Exp. Bot.* **44:** 1417–1444.

Vogel JP, Misra LM, Rose MD. (1990) Loss of BiP/GRP78 function blocks translocation of secretory proteins in yeast. *J. Cell Biol.* **110:** 1885–1895.

von Figura K, Hasilik A. (1986) Lysosomal enzymes and their receptors. *Annu. Rev. Biochem.* **55:** 167–193.

Wada I, Rindress D, Cameron P *et al.* (1991) SSRα and associated calnexin are major calcium binding proteins of the endoplasmic reticulum membrane. *J. Biol. Chem.* **266:** 19599–19610.

Walter P, Blobel G. (1981) Translocation of proteins across the endoplasmic reticulum. III. Signal recognition protein (SRP) causes signal sequence-dependent and site-specific arrest of chain elongation that is released by microsomal membranes. *J. Cell Biol.* **91:** 557–561.

Walter P, Lingappa VR. (1986) Mechanism of protein translocation across the endoplasmic reticulum membrane. *Annu. Rev. Cell Biol.* **2:** 499–516.

Walter P, Ibrahimi I, Blobel G. (1981) Translocation of proteins across the endoplasmic reticulum. I. Signal recognition protein (SRP) binds to *in-vitro*-assembled polysomes synthesizing secretory protein. *J. Cell. Biol.* **91:** 545–550.

Walter-Larsen H, Brandt J, Collinge DB, Thordal-Christensen H. (1993) A pathogen-induced-gene of barley encodes a HSP90 homologueue showing striking similarity to vertebrate forms resident in the endoplasmic reticulum. *Plant Mol. Biol.* **21:** 1097–1108.

Wilson DW, Lewis MJ, Pelham HRB. (1993) pH-dependent binding of KDEL to its receptor *in vitro*. *J. Biol. Chem.* **268:** 7465–7468.

Yamagata H, Tanaka K. (1986) The site of synthesis and accumulation of rice storage proteins. *Plant Cell Physiol.* **27:** 135–145.

Yamagata H, Tamura K, Tanaka K, Kasai Z. (1986) Cell-free synthesis of rice prolamin. *Plant Cell Physiol.* **27:** 1419–1422.

Zimmermann T, Sagstetter M, Lewis MJ, Pelham HRB. (1988) Seventy-kilodalton heat shock proteins and an additional component from reticulocyte lysate stimulate import of M13 procoat protein into microsomes. *EMBO J.* **7:** 2875–2880.

Chapter 22

Endoplasmic reticulum in the cortex of plant cells

Irene K. Lichtscheidl and Peter K. Hepler

1. Introduction

Membranes of the endoplasmic reticulum (ER) form highly pleiomorphic compartments in the cytoplasm of plant and animal cells. The quantity and conformation of ER often match the prevailing physiological situation of the cell and correspond to its many functions, including synthesis of proteins and lipids, inter- and intracellular transport of proteins and membrane material, and regulation of ionic conditions, especially levels of calcium and protons. ER membranes may form a network of tubules, stacks of more or less highly fenestrated lamellae, or disintegrate to form isolated vesicles, and may display many transitions between these states. In plant cells, cisternae may also vary greatly in diameter; tubules and vesicles with a diameter of 0.2–4 μm resembling small vacuoles may be continuous with thin tubules and lamellae with a diameter of 70–100 nm. They traverse the cytoplasm in three dimensions in an apparently haphazard distribution and occasionally exhibit close associations with other organelles (e.g. nucleus, plastids, aleurone grains), or are concentrated in certain regions of the cell. In this chapter, we shall focus on that subset of the ER that occurs in the periphery of the cytoplasm in close proximity to the plasma membrane (PM), and we shall discuss its spatial arrangements in relation to its possible functions.

2. Role of cortical cytoplasm

The PM and adjacent peripheral cytoplasm play a pivotal role in the response of a cell to its environment. The reception of and response to signals, the communication between cells, and the protection of the cell against pathogens and harmful conditions all involve the PM and its immediately subjacent cytoplasmic layer. These many and divergent capabilities require special mechanical and physiological properties of the cortical cytoplasm, and at least some of them seem to involve the ER. In the large coenocytic characean algal cells, for example, there is a rather rigid ectoplasm girdling the more fluid endoplasm (Palevitz and Hepler, 1975; Williamson, 1985), but in higher plant cells too, the ectoplasm is highly structured and tightly associated with the PM, as indicated by its resistance to centrifugal sedimentation (Quader et al., 1987) and to plasmolysis (Oparka et al., 1994). In addition, the organelles in the ectoplasm display greater stability than the same inclusions when observed in the more interior regions of the cell (Allen and Brown, 1988; Knebel et al., 1990; Lichtscheidl and Url, 1987; Masuda et al., 1991). Finally, the structural properties of the cortical cytoplasm appear to contribute to the generation of pattern in dividing cells (Hepler, 1981), the organization of growth in polar cells (Hepler et al., 1990; Schnepf, 1986), and the maintenance of pattern in mature cells without cell walls (Hess and Hesse, 1994; Luegmayr, 1993). These higher-order phenomena may owe their basis to the combined mechanical and physiological activities of the cortical ER, including in particular the ability of this membrane system to bind auxin and to generate and maintain specific ionic conditions. Through the local control of calcium and possibly other ions, the cortical ER can profoundly affect the structure, organization and motile processes of the cortical cytoskeleton (microtubules and microfilaments, MTs and MFs), facilitate the aggregation and exocytosis of vesicles, and in general regulate the transmission of signals to and from the PM (Hepler *et al.,* 1990).

3.   Organization of cortical ER in mature cells

In mature plant cells, a picture of the three-dimensional organization of ER elements has been derived largely from observations with light and fluorescence microscopy. In the highly vacuolated cells of the epidermis of onion bulb scales (Url, 1964, *Figure 1a*), small membrane tubules and vesicles with a diameter of 0.5–3 μm have been observed that resemble the small vacuoles described in leaf and stamen hair cells (Mersey and McCully, 1978; Thaine, 1965) and in guard cells (Palevitz *et al.,* 1981). They are in continuous motion and their shape and distribution change constantly, especially when they occur in cytoplasmic strands that cross the

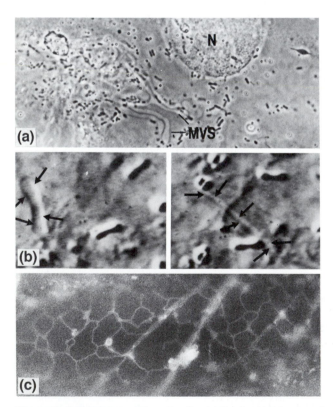

Figure 1. Onion inner epidermal cells. (a) In phase contrast, only membrane compartments with a diameter of more than 0.2 μm ('motile vacuolar system', MVS; Melillo *et al.,* 1990) can be resolved, together with the nucleus (N) and the organelles (magnification x 315). (b) Thinner tubules of ER (80 nm or more in diameter) become visible by video-enhanced light microscopy, which also allows observation of transitions from thick to thin ER compartments (arrows, magnification x 1400). (c) Staining with DiOC$_6$ (Terasaki, 1990) reveals thin tubules and sheets of ER under the fluorescence microscope (magnification x 210).

vacuole. In the peripheral cytoplasm they are more stationary and quiescent. Their velocity in both regions differs from that of most organelles, which also translocate mainly in the transvacuolar cytoplasm, with the changes in the network of small vacuoles being much slower than the movements of organelles (Palevitz *et al.,* 1981; Thaine, 1965).

The small vacuoles are continuous with thin tubules and lamellae of ER that have a diameter of *c.* 100 nm and become visible only by video and fluorescence microscopy and by electron microscopy (*Figure 1b* and *c*). On the basis of their size and staining characteristics, they correspond to the ER described in cultured animal cells (for review see Terasaki 1990). Based on studies in which several microscopic methods were applied to a wide variety of cell types from many different species, a remarkably consistent picture of the structure and pattern of these cortical membranes has emerged, namely that in the periphery of cells, immediately adjacent to the PM, there is a two-dimensional system of interconnected membrane tubules and lamellae representing a polygonal membrane net (Allen and Brown, 1988; Craig and Staehelin, 1988; Drawert and Rüffer-Bock, 1964; Hepler, 1981; Hepler *et al.,* 1990; Knebel *et al.,* 1990; Lichtscheidl *et al.,* 1990; Lichtscheidl and Url, 1987, 1990; Lichtscheidl and Weiss, 1988; McCauley and Hepler, 1990, 1992; Palevitz and Hodge, 1984; Quader, 1990; Quader and Fast, 1990; Quader and Schnepf, 1986; Quader *et al.,* 1987, 1989; Saunders and Hepler, 1981). The membranes are continuous with ER lying deeper within the cytoplasm and in cytoplasmic transvacuolar strands where the ER consists of parallel but interconnected extended membrane tubules and elongated lamellae. Although the ER within these transvacuolar strands appears to be highly dynamic, and its velocity and pattern of movement match those of the organelles, by contrast the ER net and organelles in the cortical cytoplasm appear to be rather stationary. Tubules of ER are combined to form polygons of variable mesh width, mainly by three-way junctions. They are continuous with flat lamellar sheets which are extended by the tubules to lobes and often occur at junction sites. At their borders, membrane sheets can disintegrate into small tubules and thus resemble fenestrated lamellae (Hepler, 1981; Hepler *et al.,* 1990; Knebel *et al.,* 1990), which may also occur when the tubular net tightens and the membranes subsequently fuse to form a homogenous membrane sheet (Knebel *et al.,* 1990; McCauley and Hepler, 1990, 1992; Quader, 1990; Quader and Fast, 1990). Tubular elements of ER may also radiate from a central core or knot and form asters, suggesting that the ER grows and proliferates from focal arrays (Allen and Brown, 1988; Hepler, 1981; *Figure 2*). The density and configuration of these basic components of cortical ER may vary between the different cell types and depend on physiological, developmental and ecological factors (e.g. *Figure 3*). The relatively stable appearance of the peripheral ER net may derive from fixed, immobile sites (Knebel *et al.,*

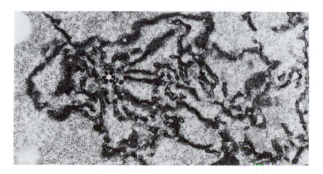

Figure 2. A focal aggregate of ER with tubular membranes radiating from a central point (*). Reproduced from Hepler (1981) with permission from Wissenschaftliche Verlagsgesellschaft. Magnification x 22 400.

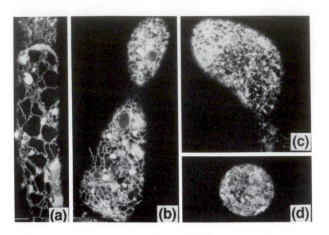

Figure 3. DiOC$_6$(3)-labelled cortical ER of living *Funaria* cells (scale bar = 5 µm). In caulonemal cells (a), the cortical ER forms a polygonal network (arrows) interspersed by lamellar cisternae (stars). The ER extends into the tip of a branch (b) and of a stalked bud (c), where it becomes very tight, particularly at the bud's apex (d). Reproduced from McCauley and Hepler (1992) with permission from Springer-Verlag.

1990). They are either flat lamellar cisternae, or blind ends and knots of tubules that remain anchored to the PM, even if the surrounding membranes and associated organelles move or cause modifications in the shape and size of such fixed lamellae. Through the storage of images of ER in a video computer and their comparison with subsequent images, it has been possible to distinguish fixed spots from mobile arrays of ER (Lichtscheidl and Url, 1990).

The occurrence of fixed spots in the periphery next to the PM suggests anchorage of at least part of the cortical ER to the PM. Experimental evidence for the binding of ER to the PM comes from various different approaches. Centrifugation of onion epidermal cells reveals that the cortical net of ER and MTs remains fixed, while the bulk ER and organelles are displaced to the centrifugal end. Recent studies employing plasmolysis show that elements of the cortical ER remained tightly adhered to the wall despite the extensive retraction of the bulk of the cytoplast (Oparka *et al.,* 1994). Interestingly, these islands of tightly adhered cortical ER are surrounded by PM drawn out to form Hechtian strands, and thus also remain connected to the plamolysed cytoplast. Further experiments using inhibitors against MTs and MFs have attributed the attachment of the ER to the PM to actin MFs, although the co-operation of MTs in anchoring the ER could not be completely excluded (Knebel *et al.,* 1990). A similar suggestion has emerged from studies of protoplast footprints of guard cells (Doohan and Palevitz, 1980; Wiedenhoeft, 1985), which also showed that a peripheral net of ER together with MTs and small vesicles remained attached to the PM when protoplasts were lysed and the bulk of the cytoplasm was washed away. Finally, dry-cleave preparations of root hairs (Traas *et al.,* 1985) and growing pollen tubes (Derksen *et al.,* 1985) provide additional evidence for an ER/MT/vesicle complex that is tightly linked to the PM.

Attempts to resolve the actual linkage between ER and PM using rapidly frozen tissues analysed by freeze-fracture or freeze-substitution have revealed that there are regions where the adjacent membrane leaflets of the ER and PM are extremely close to one another, although they do not fuse (Craig and Staehelin, 1988; Lancelle and Hepler, 1992; Lichtscheidl *et al.,* 1990; McCauley and Hepler, 1990, 1992, reviewed by Hepler *et al.,* 1990). Small crossbridging structures are observed between the ER and the PM (*Figure 4*); ribosomes are excluded from these purported 'attachment' domains and are found on the membrane regions pointing towards the cytoplasm (Fleurat-Lessard, 1986).

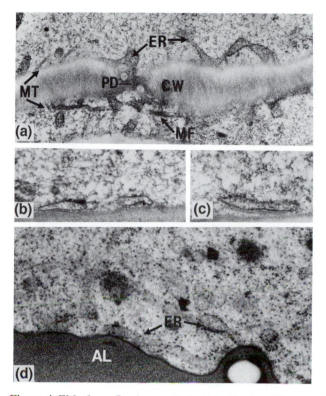

Figure 4. EM of cryofixed-cryosubstituted cells. (a–c) *Drosera* gland stalk epidermis. (a) A complex of cortical ER, MTs and MFs is associated with plasmodesmata (Lichtscheidl, Lancelle and Hepler, unpublished results; magnification x 18 900). (b and c) Cross-sections show intimate juxtaposition of ER and PM. (b) Reproduced from Lichtscheidl *et al.* (1990) with permission from Springer-Verlag; magnification x 21 000. (c) Lichtscheidl, Lancelle and Hepler, unpublished results; magnification x 21 000. (d) Tapetum cell of *Ledebouria,* which has no cell wall; the close association of ER and PM is suggested to contribute to the maintenance of the cell form (Hess, 1993; from Hess and Hesse, unpublished results). AL, polysaccharide fluid within the anther locule; magnification x 25 200).

4. ER dynamics and corresponding organelle movement

Despite the relatively stationary impression of the cortical ER, individual membrane tubules and lamellae exhibit constant motion; rapid Brownian oscillations are displayed by most of the tubules between the junctions, and rearrangements of the membranes cause permanent transformations of the polygonal net and gradually lead to completely different patterns (Knebel *et al.,* 1990; Lichtscheidl and Url, 1990). New elongating tubules grow out from already existing tubules or lamellae, and new flat lamellae form by extension of membranes from the existing net, thus increasing the amount of ER. Their path is either straight or curved, and on making contact with another membrane of the net they either pass by and continue on their path, or they fuse with the other membrane, thus forming an additional junction and an additional polygon of the net (*Figure 5a–d*). The tubule junctions are often not fixed but show lateral movements, in which tubules glide along one another. In combination with shortening and retraction of tubules and lamellae, existing meshes become drawn together and thus reduced, and may finally disappear altogether, eliminating the polygon from the net. Organelles, which lie mainly beneath the cytoplasmic face of the ER (Allen and Brown, 1988; Drawert and Rüffer-Bock, 1964), frequently move along the precise route of the ER tubules, as if using them as tracks (Allen and Schumm, 1990; Lichtscheidl and Weiss, 1988). Organelles often also closely adhere to the membranes and may drag ER tubules and lamellae in their wake (Drawert and Rüffer-Bock, 1964; Lichtscheidl and Url, 1990). The complex of organelles and adhering membranes moves in a very similar pattern to that for the ER described above; newly formed membranes may, on contact with other membranes, fuse and form new junctions and thus new polygons of the net (*Figure 5e* and *f*). The velocities of organelles and ER (8–10 µm sec^{-1}) correspond as well (Allen and Brown, 1988; Lichtscheidl and Url, 1990). Organelles may, however, be hindered in their movement by ER membranes and then either become repulsed or slowed down, or merge with the obstructing membrane and on continuing on their path pull out another membrane tubule fom the net (Drawert and Rüffer-Bock, 1964; Lichtscheidl and Url, 1990).

Given the dynamic, motile nature of the cortical ER, it becomes a matter of great interest how this motion is generated. In animal cells, where the motion of the ER is very similar to that described in onion inner epidermal cells (Dabora and Sheetz, 1988; Lee and Chen, 1988; Sanger *et al.,* 1989), the evidence points to a MT-based motile system, using either kinesin or dynein (Buckley and Porter, 1975; Dailey and Bridgman, 1988; Franke, 1971; Fuhrmann *et al.,* 1990; Terasaki *et al.,* 1986). In cell-free systems, observations that the ER moves along MTs and that these motile events are prevented by MT inhibitors support this assumption (Dabora and Sheetz,

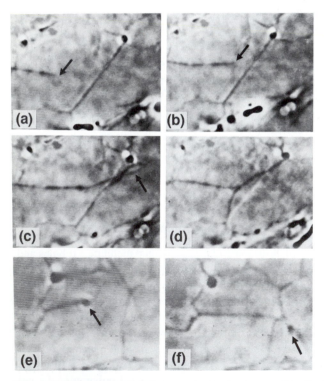

Figure 5. Video-enhanced phase contrast microscopy of cortical ER in living onion epidermal cells. (a–d) show an elongating ER tubule (a, b), which on contact with another membrane (c) fuses and forms a new mesh of the net (d) (magnification x 2520). In (e and f), the elongating tubule is associated with an organelle at its leading edge (arrow, magnification x 2800). Reproduced from Lichtscheidl and Url (1990) with permission from Springer-Verlag.

1988; for review see Terasaki, 1990). Despite the weight of evidence favouring a MT-based motile system, actin MFs cannot be discounted, since in some cultured animal cells co-alignment of ER with actin stress fibres has been demonstrated (Sanger *et al.,* 1989).

In plant cells, similar contacts of ER and MTs are documented. ER in contact with MTs was observed in the cortex of developing guard cells (Palevitz and Hodge, 1984), and MTs together with ER and small vesicles remain associated with the PM in membrane footprints of guard cell protoplasts (Wiedenhoeft, 1985) and in dry-cleave preparations of pollen tubes (Derksen *et al.,* 1985) and root hairs (Traas *et al.,* 1985). In tangential sections of pollen tubes prepared for electron microscopy by cryotechniques, Lancelle *et al.* (1987) observed peripheral MTs which were each accompanied by a thin tube of rough ER over long distances (*Figure 6a*) in the interior cytoplasm of pollen tubes (Lancelle and Hepler, 1992). Usually, however, MTs and ER, despite making contact at some points, are not co-aligned, and the significance of ER–MT contacts for the observed ER motility and stability is therefore still under discussion. In onion cells, inhibitor studies also yielded controversial results. Allen and Brown (1988) reported that high concentrations of colchicine, applied for several hours to destroy the MTs, disrupted the ER in the cortical and internal cytoplasm, although the movement of organelles remained unaffected, while Knebel *et al.* (1990) observed no effect of colchicine on either the structure or the motion of the ER. The more specific plant tubulin inhibitors, oryzalin and trifluralin, also had no effect on the pattern of the cortical ER (Quader *et al.,* 1989), or on the motility of ER elements and the site and behaviour of fixed ER domains (Knebel *et al.,* 1990). In marked contrast, the application of cytochalasins produced both a considerable alteration of ER pattern and inhibition of motion (Knebel *et al.,* 1990; Quader *et al.,* 1987), suggesting that MFs control ER motion (Hensel, 1987), as they do in the case of cytoplasmic streaming (Hepler and Palevitz, 1974; Williamson, 1986).

The MF/ER association is further supported by electron microscopy and video microscope observations, which revealed a close structural association between these two elements (Higashi-Fujime, 1988; Kachar and Reese, 1988; Lichtscheidl *et al.,* 1990; Nagai and Hayama, 1979; Williamson, 1979). The correlation between their physical interaction and the generation of

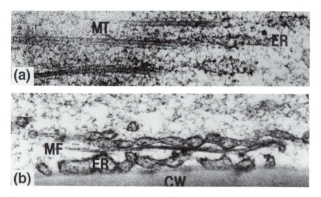

Figure 6. Electron micrographs of colocalizations of the ER with the cytoskeleton in plant cells prepared by freeze-fixation and freeze-substitution. (a) Close alignment of tubular elements of rough ER with cortical MTs in pollen tubes of *Nicotiana*. Magnification x 42 500. Reproduced from Lancelle *et al.* (1987) with permission from Springer-Verlag. (b) Structural association of cortical ER adjacent to the PM with bundles of MFS in *Drosera* tentacle cells. Additional ER elements contact these complexes. CW, cell wall. Magnification x 35 300. Reproduced from Lichtscheidl et al. (1990) with permission from Springer-Verlag.

movement was confirmed by visualization of ER tubules gliding along MF bundles together with organelles in extruded characean cytoplasm (Kachar and Reese, 1988), and by demonstration of the involvement of actin MFs in the transport of ER and vesicles towards the PM during plug formation (Foissner, 1991). Similar observations have been made in higher plant cells as well, where a close spatial proximity of the ER to bundles of MFs has been reported (Goosen-deRoo *et al.,* 1983; Lancelle and Hepler, 1989; Lichtscheidl *et al.,* 1990; Quader *et al.,* 1987). In addition to the interaction between cortical ER and MFs, it is also important to consider the structural association of this complex with the PM. In *Drosera* tentacle epidermis cells, electron micrographs of freeze-substituted material reveal an intimate attachment between cortical ER, the PM and MFs (Lichtscheidl *et al.,* 1990; *Figure 6b*). This evidence supports the view that the cortical ER, through its fixed union with the PM, might serve as an anchor for actin MFs which thus gain enough stability to serve as stable tracks along which force-generating processes may be applied for organelle motion (Kachar and Reese, 1988; Masuada *et al.,* 1991; Nagai and Hayama, 1979; Williamson, 1979). This stabilization of MFs by cortical ER membranes would not only explain the participation of the cortical ER, together with MFs, in moving the endoplasmic ER and organelles back to their former location after centrifugation (Quader *et al.,* 1987) or plasmolysis (Oparka *et al.,* 1994), but would also account for the observed tracks of moving organelles and ER elements, which often follow precisely the pattern of cortical ER elements (Allen and Brown, 1988; Lichtscheidl and Url, 1990; Lichtscheidl and Weiss, 1988). As a result of these structural properties of the ER, organelles together with ER and Golgi vesicles could thus be transported and also oriented in the cell cortex. Similarly, the cortical ER may contribute to the stabilizing of MTs. In the spindle apparatus, for example, MTs are closely associated with ER at the spindle poles. A clear example of an anaphase-I pollen mother cell reveals MTs of the meiotic spindle apparatus interdigitating with elements of the ER, which are in turn joined to the PM (Hepler *et al.,* 1990). The molecules responsible for these various associations have not yet been identified, and this will be an important area of future research.

Finally, it is important to mention that the close structural association of the ER with the cytoskeleton could have important implications with regard to the formation of new cytoskeletal elements. Recent studies on animal systems have revealed that the mRNAs for different actin isoforms are targeted to specific localities within the cell, and that poly(A) mRNAs in general display a strong structural association with ribosomes and with the F-actin cytoskeleton (Taneja *et al.,* 1992). It seems likely that these same mechanisms for spatially compartmentalizing translation in animal cells also apply to plants. Furthermore, to the extent that membrane-bound polysomes might be involved, this could provide a functional basis for the close attachment of MFs to the rough ER of the cortex, most clearly observed in *Drosera* (Lichtscheidl *et al.,* 1990).

5. Physiological properties of the cortical ER

With regard to the function of the cortical ER, it is difficult to make specific assignments and to distinguish it clearly from the bulk ER. To the extent that physiological/functional studies depend on the biochemical isolation of an ER membrane preparation, the findings will of necessity reflect the bulk ER, since the cortical ER, which is tightly associated with the PM, would be largely eliminated by most preparative procedures. Nevertheless, since the cortical ER may be continuous with the bulk ER, it seems reasonable to apply the data obtained from biochemical analysis of the latter in order to increase our understanding of the former.

One of the best and longest known characteristics of the ER is its role in synthesizing (glyco)proteins and (phospho)lipids. In specialized cells, for example, the synthesis of proteins and enzymes is always at least temporarily accompanied by an extensive proliferation of rough ER cisternae. After their synthesis by membrane-bound ribosomes, polypeptides destined for the secretory pathway enter the lumen of the ER, where they acquire their final structural and functional properties and are then delivered via secretory pathways to their final destination (for reviews see Chrispeels, 1980; Chrispeels and Tague, 1991; Vitale *et al.*, 1993). In addition, proteins may also be retained and even accumulated within the ER or ER-derived storage organelles such as protein bodies, vacuoles and microbodies (Chrispeels and Tague, 1991; Vitale *et al.*, 1993).

The ER is also a major site for the production and secretion of lipids in plant cells (eukaryotic pathway of lipid synthesis), including both storage lipids (triacylglycerols) and membrane lipids (phospholipids, glycolipids and sterols). Lipid synthesis occurs in smooth regions of the ER which are often continuous with sections of the rough ER (Morré, 1990), and it is often accompanied by extensive proliferation of smooth tubular ER, as was for example described during the secretion of a lipophilic stigmatic exudate in stigma tissue of *Petunia* (Konar and Linskens, 1966). The synthesis of membrane compounds such as phosphatidylcholine, phosphatidylethanolamine and phosphatidylinositol (Donaldson and Beevers, 1977) is accomplished by enzymes that are concentrated in the ER (Jones, 1983; Moore, 1982), for which reason the ER is held to be responsible for the biosynthesis of membranes in general. The transport of phospholipids from the ER to other organelles may occur via lipid exchange pathways by intimate contact between the different membrane systems (i.e. ER and organelles) (Chrispeels, 1980). A similar exchange of membrane material may take place during the frequently observed close juxtaposition of the ER to the PM. In addition, the synthesis of new ER membranes may account for the observed formation of new ER tubules and lamellae from an existing network of ER (Drawert and Rüffer-Bock, 1964; Lichtscheidl and Url, 1990); it may advance during the above-mentioned 'spinning out' of new ER elements.

In addition to these biosynthetic roles, it seems likely that the ER, especially perhaps those elements in the cortical cyoplasm, participates in regulation of the ionic calcium and proton milieu of the cytosol (Hepler and Wayne, 1985; Moore and Akerman, 1984). The ER, perhaps in combination with calciosomes (Allen and Schumm, 1990; Hashimoto *et al.*, 1988; Volpe *et al.*, 1988), is now generally accepted as one of the main intracellular storage compartments for calcium (Wick and Hepler, 1980), a feature which, among others, allows its visualization by the fluorescent dye chlortetracycline. In its membranes the ER contains three major components which allow the regulation of cytosolic calcium concentration: (i) ATPases to pump cytosolic calcium into its lumen (Schweitzer and Blaustein, 1980); (ii) proteins with a high capacity for binding calcium, such as calsequestrin and calreticulin, which store the calcium within the ER (Chou *et al.*, 1989); and (iii) ion channels, which may be activated by a chemical messenger (calcium, inositol (1, 4, 5)-triphosphate (IP_3), cyclic ADP-ribose, etc.) (Berridge, 1993; Berridge and Irvine, 1989; Putney, 1993).

The possible specific role of the cortical ER in calcium regulation draws its inspiration from that of the sarcoplasmic reticulum of muscle, wherein signals transmitted along the sarcolemma induce calcium release from the closely apposed terminal cisternae of the sarcoplasmic reticulum, and thus stimulate muscle contraction (Franzini-Armstrong, 1970; Haynes and Mandveno, 1987; Tada *et al.*, 1978). Muscle, of course, is a specialized cell type designed to respond with speed. However, in a variety of non-muscle systems there are also examples of close apposition between cortical ER and the PM that are thought to participate in the regulation of calcium levels. Cortical elements of the ER have been described in nerve axons, and, based on the presence of calcium, were postulated to participate in ion control. Another example occurs in oocytes, in which a well-developed system of cortical ER is found in close apposition to the PM and cortical granules. The calcium spike at fertilization induces cortical granule release, and also a modification of the ER. By comparison, it is tempting to speculate that ER elements, which are closely apposed to the PM in plant cells, could respond to signals received at the cell surface and cause the local release of calcium. The change in intracellular concentration of the ion in restricted regions could thus modulate a variety of processes, including fusion of vesicles and restructuring of the cytoskeleton, or act as an intermediate step in a larger signal transduction chain involving G-proteins and/or inositol polyphosphate metabolism. The central importance of calcium regulation in plant cells is demonstrated by the fact that cytoplasmic streaming is exquisitely sensitive to the intracellular calcium concentration, being inhibited when the concentration is increased from a basal level of 0.1–0.2 μM to 1.0 μM. The elevated levels of this ion appear to disrupt the structure of the actin MFs (Kohno and Shimmen, 1987, 1989), and may also down-regulate the myosin motor function. Elevated calcium concentrations also depolymerize MTs (Zhang *et al.,* 1990, 1992), again at levels of about 1.0 μM. As well as destroying these specific cytoskeletal elements, the elevated calcium may markedly decrease cytoplasmic viscosity (Russ *et al.,* 1991), thereby changing the structure and mobility of the cytoplasm. At the tip of a growing pollen tube, the naturally elevated calcium concentration, by down-regulating cytoplasmic streaming, may create conditions that favour the docking and fusion of vesicles necessary for tip growth (Kohno and Shimmen, 1987, 1989; Lancelle and Hepler, 1992; Malhó *et al.,* 1993; Miller *et al.,* 1992).

Besides possibly releasing and thereby generating domains of elevated calcium, the cortical ER may play an important role in responding to local increases resulting from wounding, exposure to cold temperature and other forms of stress, and may sequester these ions, thus protecting the cell against possible harmful effects. In this context the cortical ER becomes a second line of defence, next to the PM, which deals with activities that for some reason breach the PM. Furthermore, the elevated calcium concentration might modulate the structure and function of the ER. In onion cells, elevated cytoplasmic calcium levels cause the fusion of thin cortical ER tubules to form large flat membranous lamellae which cover most of the PM (Quader, 1990). Similar observations have been made in oocytes of *Xenopus,* where the cortical ER becomes disorganized when, on fertilization, a transient spike of calcium runs through the cortical cytoplasm (Gardiner and Grey, 1983).

The mechanism or mechanisms that control the release and sequestration of calcium by the ER are still not understood, particularly in plants. The release of calcium is often attributed to phosphoinositides as signalling agents (Berridge, 1986; Putney *et al.,* 1989) in animal systems (Streb *et al.,* 1983) and in plant cells (Alexandre *et al.,* 1990; Droback and Ferguson, 1985; Schumacher and Sze, 1987). Increased production of IP_3 and a subsequent increase in the levels of cytoplasmic ions may be initiated by the phytohormone auxin on its binding to the PM and ER (Ettlinger and Lehle, 1988), which is a prerequisite for processes related to cell growth and development (Hepler and Wayne, 1985). However, recent studies on plants indicate that the tonoplast/vacuole, rather than the ER, is the target for IP_3-induced calcium release (Schumacher and Sze, 1987). In animals cells, considerable interest has recently been focused on cyclic ADP-ribose as a system (an alternative to IP_3) that releases calcium from ER stores. Recently, evidence for the activity of cyclic ADP-ribose was also obtained in plant systems (Allen *et al.,* 1995). To the extent that this agonist works through the ryanodine channels, which are also sensitive to caffeine, it may thus be reasonable to suspect its activity in plants. Caffeine, for example, is well known for its ability to block cell-plate formation (Bonsignore and Hepler, 1985) and pollen tube-tip extension (Pierson *et al.,* unpublished results). This raises the possibility that caffeine modulates the activity of ER membranes and/or the nearby PM, which control the calcium concentration in the region of the cell plate or the pollen tube tip.

In addition to controlling calcium, the cortical ER could also regulate proton levels and thus possibly create local pH domains in the region of the PM. It is well established that acidification of the cell wall is necessary for cell wall expansion and growth of the cell. In this context it is important to note that auxin-binding proteins are located at the PM and also in the cortical ER;

such proteins would be expected to participate in the acidification process. Recent work has shown that H+-ATPases in the PM multiply and become activated, resulting in an increase in acidification of the apoplast and hence an increase in the plastic extensibility of the cell wall (Venis and Napier, 1991). Finally, it is pertinent to note that transient ionic changes have been suggested to activate clusters of genes which are then responsible for the synthesis of PM H+-ATPases and for the acceleration of exocytotic processes, although the mechanism underlying this function is still unclear (Hager *et al.,* 1991).

In summary, it seems likely that the cortical ER, together with the PM, regulates the ionic conditions within the local cytoplasmic regions of the PM, and that these local gradients and transients are part of a transduction system that is able to control a variety of cell processes involved in growth and development. An attractive feature of such a system is that the extensive structural ramifications of the cortical ER system may make it possible for signals received at the PM to be delivered to the cytoplast rapidly and under conditions in which the spatial information is retained. Important aspects of cytomorphogenesis, in particular those involving asymmetrical growth, might utilize the existence of local ion gradients, as established and maintained by the cortical ER/PM complex, as an underlying developmental mechanism.

6. ER and the formation of the cell plate and cell wall

A role for the ER in the formation of new cell wall has been observed in several different situations. We shall briefly discuss the following three examples: (i) the formation of the cell plate during cytokinesis; (ii) the deposition of local wall thickenings; and (iii) the expansion and development of primary walls during growth and differentiation.

Formation of the cell plate during cytokinesis

During cytokinesis the ER shows a close structural association with the vesicles that aggregate and fuse to form the new cell plate (Gunning, 1982; Hepler, 1982). Although initiating within the middle of the cell, the developing cell plate is an island of extracellular space, and thus the ER that associates with it is cortical. The ER migrates into the interzone between the two newly formed daughter nuclei during telophase, where it becomes distributed between MTs, MFs and Golgi vesicles (Goosen-deRoo *et al.,* 1983; Gunning, 1982; Hepler, 1982; Hepler and Bonsignore, 1990; Hepler and Jackson, 1968; Hepler and Newcomb, 1967; Jones and Payne, 1977; Kakimoto and Shibaoka, 1987; Palevitz, 1987; Schopfer and Hepler, 1991; Traas *et al.,* 1987). The ER, in the form of an extensive tubular reticulum, creates a planar web and appears to entrap the incoming Golgi vesicles (Hepler, 1982). By being held mechanically in the ER net, vesicles may be brought into and kept in close proximity with one another in the interzone, and thus caused to fuse in the correct position. In this instance it is tempting to speculate that the ER provides both the structural framework which traps the incoming vesicles, and the appropriate local ionic conditions to promote their fusion (Hepler *et al.,* 1990).

Deposition of local wall thickenings

A close structural relationship between the ER and the cell wall has also been observed in developing cells during the formation of cell wall thickenings. In both stomatal guard cells (Palevitz and Hodge, 1984) and phloem sieve plates (Esau and Thorsch, 1985), the ER is intimately associated with the PM during callose deposition. The regulation of cell wall deposition is suggested to occur by a dual action of auxin binding to the PM and to ER (Bret-Harte *et al.,* 1991). By acidifying the apoplast, auxin induces loosening of the wall. Within the cytoplasm auxin may, at the same time, promote the right cytoplasmic ionic conditions to allow fusion of vesicles and hence the deposition of additional wall material. So long as sufficient substrate is available, a net synthesis of additional cell wall is therefore postulated to take place.

The supply of the cell wall with proteins is also attributed to the ER. In developing pollen grains, an intimate juxtaposition of the ER with the PM is observed in different species (Van-Went and Gori, 1989), and has been assumed to be responsible for the production and secretion of proteins to the cell wall during intine formation (Rodríguez-García and Carmen Fernández, 1990). A similar apposition of the ER to the newly formed undulating cell wall of the generative cell (GC) was observed in pollen grains of *Tradescantia* (Noguchi and Ueda, 1990) and *Cyrtandra* (Luegmayr, 1993), and in germinating pollen tubes of *Lilium longiflorum* (Lancelle and Hepler, 1992). In these cases, however, tubular ER concentrates at the 'inner PM' of the vegetative cytoplasm and contacts the outside of the generative cell wall, where it lies in every invagination formed by the lobes of the GC (*Figure 7*). Here the ER is thought to direct nutrients to the GC rather than participating in the formation of the cell wall (Luegmayr, 1993), apart from contributing to the formation of the GC.

Expansion and development of primary walls

In addition to the formation of specific wall thickenings, noted above, cortical ER may also be associated with the process of cell expansion and shape formation. A particularly clear example derives from studies of the ER during bud formation in the moss *Funaria* (Conrad *et al.,* 1986; McCauley and Hepler, 1990, 1992; Saunders and Hepler, 1981), where a variety of microscopic methods indicate that cortical ER increases in the bud initials, and persists during differentiation. As a result of analyses under the confocal microscope of cells stained with the fluorescent dye, $DiOC_6$ (Terasaki *et al.*, 1986), it has been possible to document the changes in cortical ER in living cells (McCauley and Hepler, 1990, 1992; *Figure 5*). The findings demonstrate that, prior to bud formation, the ER is present in a loose net. However, following induction with benzyl adenine, there is an increase in the amount of ER and a dramatic change in its morphology, the

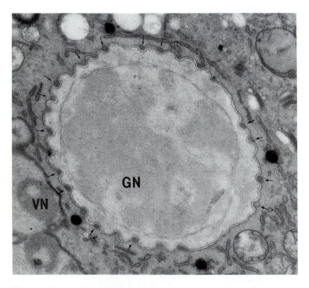

Figure 7. An electron micrograph of a mature pollen grain of *Cyrtandra* showing an undulating generative cell (GN, generative nucleus), which comprises a tubule of ER in each concavity (arrows). VN, vegetative nucleus; magnification x 21 700). Reproduced from Luegmayr (1993) with permission from Springer-Verlag.

loose net of cortical ER becoming progressively more tightly configured, until its reticulate nature is barely discernible. These observations thus provide direct evidence for profound changes in the cortical cytoplasm, especially the ER, that anticipate the subsequent events of cell growth and differentiation. Unfortunately, we do not know the function of the ER in these instances, but again ionic regulation appears to be important, since the marked cell expansion that will occur requires the aggregation and fusion of Golgi vesicles, and would be expected to utilize calcium-dependent processes.

7. Intracellular communication via plasmodesmata

Plasmodesmata are the cytoplasmic channels that connect adjacent cells and facilitate their chemical and electrical communication (see Chapter 27 by Cooke *et al.*). Briefly, plasmodesmata are characterized by a close, fixed physical relationship between the cortical ER and the PM. The PM forms the lining of the channel, and in higher plants there is a tightly furled tube of ER, the desmotubule, which extends through the plasmodesmata and is continuous with the cortical ER at both ends (López-Sáez *et al.*, 1966; for review see Robards and Lucas, 1990). The formation of primary plasmodesmata occurs during development of the cell plate. In the network of ER concentrated in the phragmoplast, membrane tubules become trapped during the fusion of vesicles and extend across the newly formed cell plate, each surrounded by a small layer of cytoplasm and PM (Hepler, 1982; Overall *et al.*, 1982). The participation of the ER in the formation of secondary plasmodesmata has been observed between cells of different plant origin (heterograft unions between *Vicia* and *Helianthus;* Kollmann and Glockmann, 1991), during normal plant development (e.g. between sieve elements and neighbouring companion cells; Fisher *et al.*, 1992) and in *Chara* algal cells (Franceschi *et al.*, 1994; for review see Lucas and Wolf, 1993). The ER closely connects to the PM and may participate in the thinning of the cell wall, which becomes invaginated and subsequently traversed by PM together with still-attached ER.

During desmotubule differentiation, transmembrane proteins in the ER bind to each other, thus contracting the desmotubule. Transmembrane proteins may also bind to the PM sleeve (Tilney *et al.*, 1991) and create a structurally stable unit (Ding *et al.*, 1992) that may be prevented from vesiculation (Tilney *et al.*, 1991). While earlier studies suggested that transport between cells might occur through the desmotubule, it now seems clear that intercellular transport of soluble molecules occurs in the cytoplasmic annulus between the PM and the desmotubule. With regard to the intercellular transport of lipids, however, Tilney *et al.* (1991) suggested that these bind to the PM and move to the neighbouring cell by lateral diffusion in the PM. On the other hand, Baron-Epel *et al.* (1988) and Grabski *et al.* (1993) infer from redistribution of fluorescence after photo-bleaching of fluorescent phospholipid analogues that lipid-soluble molecules are transported through the membrane of the desmotubules and connecting ER, and that the ER thus forms a dynamic communication pathway across the cell wall for lipids. The same route has been suggested for the propagation of electrical potentials. A 'symreticulum' connects the ER of statocytes via plasmodesmata with neighbouring cells in the zone of differential flank growth (Hejnowiez *et al.*, 1991; Zieschang and Sievers, 1991), and it has been argued that it thus transfers the gravity stimulus of statocytes to neighbouring cells, which subsequently respond to the stimulus by polar cell growth.

8. Role of cortical ER in response to endosymbionts and parasites

ER may not only be associated with the PM in the cortex, but may also contact PMs surrounding cells or organisms within the cytoplasm of the (host) cell. An example of this has been cited in pollen grains, where the ER of the vegetative cytoplasm is in close contact with the internal PM and generative cell (GC) through all stages of maturation (Hess, 1993; Lancelle and Hepler, 1992; Luegmayr, 1993; Noguchi and Ueda, 1990, *Figure 7*). This intimate association of the different membrane systems is especially striking in places where it appears to be mediated by strip-shaped projections of the PM pointing towards the ER (Hess, 1993). Another example occurs in endosymbiotic bacteria within the cells of *Chrysophyta* and *Pyrrhophyta,* which may also be closely associated with the ER of their host cell (Chesnick and Cox, 1986). In the dinoflagellate *Peridinium balticum*, symbiotic bacteria are delimited from the host cytoplasm solely by their own single PM and thin cell wall, the membranes of additional food vacuoles of the host cytoplasm being absent. The ER of the host cell surrounds each individual symbiont and makes contact with its cell surface by invaginations of the ER, which resemble tight junctions of animal cells. The dinoflagellate ER also plays a structural role by holding the bacteria together as a unit within the cell, possibly creating a micro-environment for them by compartmentalization produced by the infoldings. Similar membrane interactions between the macro- and microsymbiont are also essential for the functioning of symbiosis between bacteria and root nodules in legumes. The formation and maintenance of a peribacteroidal membrane involves cisternae of rough ER and dictyosomes (Andreeva *et al.,* 1989; Bassarab *et al.,* 1989).

Similar accumulations of ER in the periphery of host cells are observed on contact with multicellular endo/ectosymbionts and parasites, of which we shall give only a few examples. *Eucalyptus* roots, following invasion by mycorrhiza, show well-developed ER which frequently contacts the host PM (Boudarga and Dexheimer, 1988). On invasion by the parasite *Phytophthora,* rough ER accumulates in the parietal cytoplasm of *Capsicum annuum* (Saimmaime *et al.,* 1991). On infection of roots by nematodes, a stylet penetrates the host cell wall to induce a syncytium. Syncytial cells show considerable proliferation of ER, which accumulates in the host cell periphery around the secretions emanating from the tip of the stylet during feeding ('feeding tube') and may show differentiation into smooth and rough ER (Endo, 1991; Wyss *et al.,* 1984). While smooth ER is found close to the feeding tube, rough ER predominates towards the outer margin of the membrane system surrounding the feeding tube (Hussey and Mims, 1991), suggesting that rough ER, by partial digestion through the secretions of the stylet, has been modified to form smooth ER (Endo, 1991; Wyss *et al.,* 1984). The differences in ER proliferation may indicate variation in the susceptibility of the plants to nematode infection (Endo, 1991; Goinowski and Magnusson, 1991; Melillo *et al.,* 1990), for reasons which again appear to involve calcium (Renelt *et al.,* 1993).

The significance of the ER in contacts with hosts and parasites has been suggested to include the transfer of material needed for the formation of an isolating layer around these cells within the cytoplasm (Boudarga and Dexheimer, 1988), and several authors have discussed a connection between the amount and morphology of such ER and the resistance or susceptibility of the host cells (Endo, 1991; Melillo *et al.,* 1990; Saimmaime *et al.,* 1991). A similar protective function of ER against virus particles in fungal hosts has been suggested as well (Newhouse *et al.,* 1990).

9. Conclusions

From this brief review it can be concluded that cortical ER contributes to many fundamental processes in the plant cell. As a structural factor, cortical ER participates in orienting cytoskeletal elements and in trapping vesicles, in addition to anchoring desmotubules in plasmodesmata. ER as a transport system regulates ions, especially calcium ions and protons, and may create local gradients from which spatially defined activities can be generated. In this context the cortical ER may be a crucial link in the process of signal transduction. In addition, it may take part in the intra- and intercellular transport of lipophilic molecules along its membranes. The cortical elements also provide a site for auxin-binding proteins, and may thus contribute fundamentally to the mechanism of auxin action. When these activities are considered together with the well-known function of ER in protein and lipid synthesis, it becomes apparent that this cellular organelle is equipped to participate in a myriad of basic cellular activities that are indispensable for plant cell growth and development.

Acknowledgements

We thank Dr Hess for allowing us to reproduce *Figure 4c*. He, together with Professor Walter G. Url and Dr Ilse Foissner, assisted us with valuable discussions. This work was supported by grants from the US National Science Foundation (No. MCB-9304953) and the US Department of Agriculture (No. 94-37304-1180) to P.K.H., and from the Austrian Science Foundation (FWF; No. P6462) to W.G. Url.

References

Alexandre J, Lassalles JP, Kado RT. (1990) Opening of Ca^{2+} channels in isolated red beet root vacuole membrane by inositol 1,4,5-triphosphate. *Nature* **343:** 567–570.

Allen GJ, Muir SR, Sander D. (1995) Release of Ca^{2+} from individual plant vacuoles by both $InsP_3$ and cyclic ADP-ribose. *Science* **268:** 735–737.

Allen NS, Brown T. (1988) Dynamics of the endoplasmic reticulum in living onion epidermal cells in relation to microtubules, microfilaments, and intracellular particle movement. *Cell Motil. Cytoskel.* **10:** 153–163.

Allen NS, Schumm JH. (1990) Endoplasmic reticulum, calciosomes and their possible roles in signal transduction. *Protoplasma* **154:** 172–178.

Andreeva IN, Kozlova GI, Livanova GI, Zhiznevskaya GYA, Izmailov SF. (1989) Peribacteroid space in legume root nodules: an electron microscopic study. *Fiziol. Rastenii (Moscow)* **36:** 551–560.

Baron-Epel O, Hernandez D, Jiang LW, Meiners S, Schindler M. (1988) Dynamic continuity of cytoplasmic and membrane compartments between plant cells. *J. Cell Biol.* **106:** 715–721.

Bassarab S, Schenk SU, Werner D. (1989) Fatty acid composition of the peribacteroid membrane and the ER in nodules of *Glycine max* varies after infection by different strains of the microsymbiont *Bradyrhizobium japonicum*. *Bot. Acta* **102:** 196–201.

Berridge MJ. (1986) Cell signalling through phospholipid metabolism. *J. Cell Sci.* Suppl. 4: 137–153.

Berridge MJ. (1993) Inositol triphosphate and calcium signalling. *Nature* **361:** 315–325.

Berridge MJ, Irvine RF. (1989) Inositol phosphates and cell signalling. *Nature* **341:** 197–205.

Bonsignore CL, Hepler PK. (1985) Caffeine inhibition of cytokinesis: dynamics of cell plate formation — deformation *in vivo*. *Protoplasma* **129:** 28–35.

Boudarga K, Dexheimer J. (1988) Ultrastructural study of the VA endomycorrhiza of young plants of *Eucalyptus camaldulensis* (Dehnardt) (Myrtaceae). *Bull. Soc. Bot. France Lett. Bot.* **135:** 111–122.

Bret-Harte MS, Baskin TI, Green PB. (1991) Auxin stimulates both deposition and breakdown of material in the pea outer epidermal cell wall, as measured interferometrically. *Planta* **185:** 462–471.

Buckley I, Porter KR. (1975) Electron microscopy of critical point dried whole cultured cells. *J. Microscopy* **104:** 107–120.

Chesnick JM, Cox ER. (1986) Specialization of endoplasmic reticulum architecture in response to a bacterial symbiosis in *Peridinium balticum* (Pyrrhophyta). *J. Phycology* **22:** 291–298.

Chou MK, Krause K, Campbell K, Jensen K, Sjolund R. (1989) Calsequestrin: a novel calcium binding protein in plants. *Plant Physiol.* **89** (Suppl.): 149.

Chrispeels MJ. (1980) The endoplasmic reticulum. In: *The Biochemistry of Plants, Vol. 1* (Tolbert NE, ed.). New York: Academic Press, pp. 389–411.

Chrispeels MJ, Tague BW. (1991) Protein sorting in the secretory system of plant cells. *Int. Rev. Cytol.* **125:** 1–45.

Conrad PA, Steucek GL, Hepler PK. (1986) Bud formation in *Funaria*: organelle redistribution following cytokinin treatment. *Protoplasma* **131:** 211–223.

Craig S, Staehelin LA. (1988) High pressure freezing of intact plant tissues. Evaluation and characterization of novel features of the endoplasmic reticulum and associated membrane systems. *Eur. J. Cell Biol.* **46:** 80–93.

Dabora SL, Sheetz MP. (1988) The microtubule-dependent formation of a tubulovesicular network with characteristics of the ER from cultured cell extracts. *Cell* **54:** 27–35.

Dailey ME, Bridgman PC. (1988) Relationship between the endoplasmic reticulum and microtubules in cultured nerve growth cones. *J. Cell Biol.* **107:** 39a.

Derksen J, Pierson ES, Traas JA. (1985) Microtubules in vegetative and generative cells of pollen tubes. *Eur. J. Cell Biol.* **38:** 142–148.

Ding B, Turgeon T, Parthasarathy MV. (1992) Substructure of freeze-substituted plasmodesmata. *Protoplasma* **169:** 28–41.

Donaldson RP, Beevers H. (1977) Lipid composition of organelles from germinating castor bean endosperm. *Plant Physiol.* **59:** 259–263.

Doohan ME, Palevitz BA. (1980) Microtubules and coated vesicles in guard cell protoplasts of *Allium cepa* L. *Planta* **149:** 389–401.

Drawert H, Rüffer-Bock U. (1964) Fluorochromierung von Endoplasmatischem Reticulum, Dictyosomen und Chondriosomen mit Tetracyclin. *Ber. Dt. Bot. Ges.* **77:** 440–449.

Droback BK, Ferguson IB. (1985) Release of Ca^{2+} from plant hypocotyl microsomes by inositol 1,4,5,-triphosphate. *Biochem. Biophys. Res. Commun.* **145:** 1043–1047.

Endo BY. (1991) Ultrastructure of initial responses of susceptible and resistant soybean roots to infection by *Heterodera glycine. Revue Nématol.* **14:** 73–94.

Esau K, Thorsch J. (1985) Sieve plate pores and plasmodesmata, the communication channels of the symplast: ultrastructural aspects and developmental relations. *Am. J. Bot.* **72:** 1641–1653.

Ettlinger C, Lehle L. (1988) Auxin induces rapid changes in phosphatidylinositol metabolites. *Nature* **331:** 176–178.

Fisher DB, Wu Y, Ku MSB. (1992) Turnover of soluble proteins in the wheat sieve tube. *Plant Physiol.* **100:** 1433–1441.

Fleurat-Lessard P. (1986) Specialized formations of endoplasmic reticulum in parenchyma cells of *Mimosa pudica* L. *Protoplasma* **130:** 1–4.

Foissner I. (1991) Induction of exocytosis in characean internodal cells by locally restricted application of chlortetracycline and the effect of cytochalasin B, depolarizing and hyperpolarizing agents. *Plant Cell Environ.* **14:** 907–915.

Franceschi VR, Ding B, Lucas WJ. (1994) Mechanism of plasmodesmata formation in characean algae in relation to evolution of intercellular communication in higher plants. *Planta* **192:** 347–358.

Franke WW. (1971) Cytoplasmic microtubules linked to endoplasmic reticulum with cross bridges. *Exp. Cell Res.* **66:** 486–489.

Franzini-Armstrong C. (1970) Studies on the triad. I. Structure of the junction in frog twitch fibers. *J. Cell Biol.* **47:** 488–499.

Fuhrmann C, Bereiter-Hahn J, Brändle K. (1990) Influence of the cytoskeleton, energy supply, and protein synthesis on the structure of the endoplasmic reticulum. *Protoplasma* **158:** 53–65.

Gardiner DM, Grey RD. (1983) Membrane junctions in *Xenopus* eggs: their distribution suggests a role in calcium regulation. *J. Cell Biol.* **96:** 1159–1163.

Goinowski W, Magnusson C. (1991) Tissue response induced by *Heterodera schachtii* (Nematoda) in susceptible and resistant white mustard cultivars. *Can. J. Bot.* **69:** 53–62.

Goosen-deRoo L, Burggraaf PD, Libbenga KR. (1983) Microfilament bundles associated with tubular endoplasmic reticulum in fusiform cells in the active cambial zone of *Fraxinus excelsior* L. *Protoplasma* **116:** 204–208.

Grabski S, Feijter AW, Schindler M. (1993) Endoplasmic reticulum forms a dynamic continuum for lipid diffusion between contiguous soybean root cells. *Plant Cell* **5:** 25–38.

Gunning BES. (1982) The cytokinetic apparatus. In: *The Cytoskeleton in Plant Growth and Development* (Lloyd CW, ed.). New York: Academic Press, pp. 229–292.

Hager A, Debus G, Edel H-G, Stransky H, Serrano R. (1991) Auxin induces exocytosis and the rapid synthesis of a high-turnover pool of plasma-membrane H^+-ATPase. *Planta* **185:** 527–537.

Hashimoto S, Bruno B, Lew DP, Pozzan T, Volpe P, Meldolesi J. (1988) Immunocytochemistry of calciosomes in liver and pancreas. *J. Cell Biol.* **107:** 2523–2531.

Haynes DH, Mandveno A. (1987) Computer modeling of Ca^{2+} pump function of Ca^{2+}-Mg^{2+} ATPase of sarcoplasmic reticulum. *Physiol. Rev.* **67:** 244–286.

Hejnowicz Z, Krause E, Glebicki K, Sievers A. (1991) Propagated fluctuations in the electric potential in the apoplasm of *Lepidium sativum* L. roots. *Planta* **186:** 127–134.

Hensel W. (1987) Cytodifferentiation of polar plant cells: formation and turnover of endoplasmic reticulum in root statocytes. *Exp. Cell Res.* **172:** 377–384.

Hepler PK. (1981) The structure of the endoplasmic reticulum revealed by osmium tetroxide–potassium ferricyanide staining. *Eur. J. Cell Biol.* **26:** 102–110.

Hepler PK. (1982) Endoplasmic reticulum in the formation of cell plate and plasmodesmata. *Protoplasma* **111:** 121–133.

Hepler PK, Bonsignore CL. (1990) Caffeine inhibition of cytokinesis: ultrastructure of cell plate formation/degradation. *Protoplasma* **157:** 182–192.

Hepler PK, Jackson WT. (1968) Microtubules and early stages of cell-plate formation in the endosperm of *Haemanthus katherinae* Baker. *J. Cell Biol.* **38:** 437–446.

Hepler PK, Newcomb EH. (1967) Fine structure of cell plate formation in the apical meristem of *Phaseolus* roots. *J. Ultrastruct. Res.* **19:** 498–513.

Hepler PK, Palevitz BA. (1974) Microtubules and microfilaments. *Annu. Rev. Plant Physiol.* **25:** 209–362.

Hepler PK, Wayne O. (1985) Calcium and plant development. *Annu. Rev. Plant Physiol.* **36:** 397–439.

Hepler PK, Palevitz BA, Lancelle SA, McCauley MM, Lichtscheidl IK. (1990) Cortical endoplasmic reticulum in plants. *J. Cell Sci.* **96:** 355–373.

Hess MW. (1993) Membrane coatings on the generative cell surface of freeze-substituted monocotyledon pollen. *Protoplasma* **176:** 84–88.

Hess MW, Hesse M. (1994) Ultrastructural observations on anther tapetum development of freeze-fixed *Ledebouria socialis* Roth (Hyacinthaceae). *Planta* **192:** 421–430.

Higashi-Fujime S. (1988) Actin-induced elongation of fibers composed of cytoplasmic membranes from *Nitella. Protoplasma* Suppl. 2: 27–36.

Hussey RS, Mims CW. (1991) Ultrastructure of feeding tubes formed in giant cells induced in plants by the root-knot nematode *Meloidogyne incognita. Protoplasma* **162:** 99–107.

Jones MGK, Payne HL. (1977) Cytokinesis in *Impatiens balsamina* and the effect of caffeine. *Cytobios* **20:** 79–91.

Jones RL. (1983) Endoplasmic reticulum. In: *Cell Components* (Linskens HF, Jackson JF, ed.). Berlin: Springer-Verlag, pp. 304–330.

Kachar B, Reese TS. (1988) The mechanism of cytoplasmic streaming in Characean algal cells: sliding of endoplasmic reticulum along actin filaments. *J. Cell Biol.* **106:** 1545–1552.

Kakimoto T, Shibaoka H. (1987) Actin filaments and microtubules in the preprophase band and phragmoplast of tobacco cells. *Protoplasma* **140:** 151–156.

Knebel W, Quader H, Schnepf E. (1990) Mobile and immobile endoplasmic reticulum in onion bulb epidermis cells: short- and long-term observations with a confocal laser scanning microscope. *Eur. J. Cell Biol.* **52:** 328–340.

Kohno T, Shimmen T. (1987) Ca^{2+} induced F-actin fragmentation in pollen tubes. *Protoplasma* **141:** 177–179.

Kohno T, Shimmen T. (1989) Mechanism of Ca^{2+} inhibition of cytoplasmic streaming in lily pollen tubes. *J. Cell Sci.* **91:** 501–509.

Kollmann R, Glockmann C. (1991) Studies on graft unions. III. On the mechanism of secondary formation of plasmodesmata at the graft interface. *Protoplasma* **165:** 71–85.

Konar RN, Linskens HF. (1966) The morphology and anatomy of the stigma of *Petunia hybrida. Planta* **71:** 356–371.

Lancelle SA, Hepler PK. (1989) Immunogold labelling of actin on sections of freeze-substituted plant cells. *Protoplasma* **150:** 72–74.

Lancelle SA, Hepler PK. (1992) Ultrastructure of freeze-substituted pollen tubes of *Lilium longiflorum. Protoplasma* **167:** 215–230.

Lancelle SA, Cresti M, Hepler PK. (1987) Ultrastructure of the cytoskeleton in freeze-substituted pollen tubes of *Nicotiana alata. Protoplasma* **140:** 141–150.

Lee C, Chen LB. (1988) Dynamic behavior of endoplasmic reticulum in living cells. *Cell* **54:** 37–46.

Lichtscheidl IK, Url WG. (1987) Investigation of the protoplasm of *Allium cepa* inner epidermal cells using ultraviolet microscopy. *Eur. J. Cell Biol.* **43:** 93–97.

Lichtscheidl IK, Url WG. (1990) Organization and dynamics of cortical endoplasmic reticulum in inner epidermal cells of onion bulb scales. *Protoplasma* **157:** 203–215.

Lichtscheidl IK, Weiss DG. (1988) Visualization of submicroscopic structures in the cytoplasm of *Allium cepa* inner epidermal cells by video-enhanced contrast light microscopy. *Eur. J. Cell Biol.* **46:** 376–382.

Lichtscheidl IK, Lancelle SA, Hepler PK. (1990) Actin–endoplasmic reticulum complexes in *Drosera*: their structural relationship with the plasmalemma, nucleus and organelles in cells prepared by high pressure freezing. *Protoplasma* **155:** 116–126.

López-Sáez JF, Giménez-Martin G, Risueno MC. (1966) Fine structure of the plasmodesma. *Protoplasma* **61:** 81–84.

Lucas WJ, Wolf S. (1993) Plasmodesmata: the intercellular organelles of green plants. *Trends Cell Biol.* **3:** 308–315.

Luegmayr E. (1993) The generative cell and its close association with the endoplasmic reticulum of the vegetative cell in pollen of *Cyrtandra pendula* (Gesneriaceae). *Protoplasma* **177:** 73–81.

McCauley MM, Hepler PK. (1990) Visualization of the endoplasmic reticulum in living buds and branches of the moss *Funaria hygrometrica* by confocal laser scanning microscopy. *Development* **109:** 753–764.

McCauley MM, Hepler PK. (1992) Cortical ultrastructure of freeze-substituted protonemata of the moss *Funaria hygrometrica. Protoplasma* **169:** 215–230.

Malhó R, Read ND, Pais MS, Trewavas AJ. (1993) Role of cytosolic free calcium in the reorientation of pollen tube growth. *Plant J.* **5:** 331–341.

Masuda Y, Takagi S, Nagai R. (1991) Protease-sensitive anchoring of microfilament bundles provides tracks for cytoplasmic streaming in *Vallisneria. Protoplasma* **162:** 151–159.

Melillo MT, Bleve-Zacheo T, Zacheo G. (1990) Ultrastructural response of potato roots susceptible to cyst nematode *Globdera pallida* pathotype Pa3. *Revue Nématol.* **13:** 17–28.

Mersey B, McCully ME. (1978) Monitoring the course of fixation of plant cells. *J. Microscopy* **114:** 49–76.

Miller DD, Callaham DA, Gross DJ, Hepler PK. (1992) Free Ca^{++} gradient in growing pollen tubes of *Lilium. J. Cell Sci.* **101:** 7–12.

Moore AL, Akerman KEO. (1984) Calcium and plant organelles. *Plant Cell Environ.* **7:** 423–429.

Moore TS. (1982) Phospholipid biosynthesis. *Annu. Rev. Plant Physiol.* **33:** 235–259.

Morré DJ. (1990) Endomembrane system of plants and fungi. In: *Tip Growth in Plant and Fungal Cells* (Heath IB, ed.). San Diego: Academic Press, pp. 183–210.

Nagai R, Hayama T. (1979) Ultrastructure of the endoplasmic factor responsible for cytoplasmic streaming in *Chara* internodal cells. *J. Cell Sci.* **36:** 121–136.

Newhouse JR, MacDonald WL, Hoch HC. (1990) Virus-like particles in hyphae and conidia of European hypovirulent (ds-RNA-containing) strains of *Cryphonectria parasitica. Can. J. Bot.* **68:** 90–101.

Noguchi T, Ueda K. (1990) Structure of pollen grains of *Tradescantia reflexa* with special reference to the generative cell and the ER around it. *Cell Struct. Funct.* **15:** 379–384.

Oparka KJ, Prior DAM, Crawford JW. (1994) Behaviour of plasma membrane, cortical ER and plasmodesmata during plasmolysis of onion epidermal cells. *Plant Cell Environ.* **17:** 163–171.

Overall RL, Wolfe J, Gunning BES. (1982) Intercellular communication in *Azolla* roots. I. Ultrastructure of plasmodesmata. *Protoplasma* **111:** 134–150.

Palevitz BA. (1987) Accumulation of F-actin during cytokinesis in *Allium*. Correlation with microtubule distribution and the effects of drugs. *Protoplasma* **141:** 24–32.

Palevitz BA, Hepler PK. (1975) Identification of actin *in situ* at the ectoplasm–endoplasm interface of *Nitella*. Microfilament–chloroplast association. *J. Cell Biol.* **65:** 29–38.

Palevitz BA, Hodge LD. (1984) The endoplasmic reticulum in the cortex of developing guard cells: coordinate studies with chlorotetracycline and osmium ferricyanide. *Dev. Biol.* **101:** 147–159.

Palevitz BA, O'Kane DJ, Raikhel NV. (1981) The vacuole system in stomatal cells of *Allium*. Vacuole movements and changes in morphology in differentiating cells as revealed by epifluorescence, video and electron microscopy. *Protoplasma* **109:** 23–55.

Putney JW Jr. (1993) Excitement about calcium signalling in inexcitable cells. *Science* **262:** 676–678.

Putney JW Jr, Takemura H, Hughes AR, Horstman DA, Thorstrup O. (1989) How do inositol phosphates regulate calcium signalling? *Fed. Am. Soc. Exp. Biol. J.* **3:** 1899–1905.

Quader H. (1990) Formation and disintegration of cisternae of the endoplasmic reticulum visualized in live cells by conventional fluorescence and confocal laser scanning microscopy: evidence for the involvement of calcium and the cytoskeleton. *Protoplasma* **155:** 166–175.

Quader H, Fast H. (1990) Influence of cytosolic pH changes on the organization of the endoplasmic reticulum in epidermal cells of onion bulb scales: acidification by loading with weak organic acids. *Protoplasma* **157:** 216–224.

Quader H, Schnepf E. (1986) Endoplasmic reticulum and cytoplasmic streaming: fluorescence microscopical observations in adaxial cells of onion bulb scales. *Protoplasma* **131:** 250–252.

Quader H, Hofmann A, Schnepf E. (1987) Shape and movement of the endoplasmic reticulum in onion bulb epidermis cells: possible involvement of actin. *Eur. J. Cell Biol.* **44:** 17–26.

Quader H, Hofmann A, Schnepf E. (1989) Reorganization of the endoplasmic reticulum in epidermal cells of onion bulb scales after cold stress: involvement of cytoskeletal elements. *Planta* **177:** 273–280.

Renelt A, Colling C, Hahlbrock K, Nürnberger T, Parker JE, Sacks WR, Scheel D. (1993) Studies on elicitor recognition and signal transduction in plant defence. *J. Exp. Bot.* **44** (Suppl.): 257–268.

Robards AW, Lucas WJ. (1990) Plasmodesmata. *Annu. Rev. Plant Physiol. Plant Mol. Biol.* **41:** 369–419.

Rodríguez-García MI, Carmen Fernández M. (1990) Ultrastructural evidence of endoplasmic reticulum changes during maturation of the olive pollen grain (*Olea europaea* L., Oleaceae). *Plant Syst. Evol.* **171:** 221–231.

Russ U, Grolig F, Wagner G. (1991) Changes of cytoplasmic free Ca^{2+} in the green alga *Mougeotia scalaris* as monitored with indo-1, and their effect on the velocity of chloroplast movements. *Planta* **184:** 105–112.

Saimmaime I, Coulomb C, Coulomb PJ. (1991) *Trans*-cinnamate 4-hydroxylase activity in host–parasite interaction: *Capsicum annuum* and *Phytophthora capsici*. *Plant Physiol. Biochem. (Paris)* **29:** 481–488.

Sanger JM, Dome JS, Mittal B, Somlyo AV, Sanger JW. (1989) Dynamics of the endoplasmic reticulum in living non-muscle and muscle cells. *Cell Motil. Cytoskel.* **13:** 301–319.

Saunders MJ, Hepler PK. (1981) Localization of membrane-associated calcium following cytokinin treatment in *Funaria* using chlorotetracycline. *Planta* **152:** 272–281.

Schnepf E. (1986) Cellular polarity. *Annu. Rev. Plant Physiol.* **37:** 23–47.

Schopfer CR, Hepler PK. (1991) Distribution of membranes and the cytoskeleton during cell plate formation in pollen mother cells of *Tradescantia*. *J. Cell Sci.* **100:** 717–728.

Schumacher KS, Sze H. (1987) Inositol 1,4,5- triphosphate releases Ca^{2+} from vacuolar membrane vesicles of oat roots. *J. Biol. Chem.* **262:** 3944–3946.

Schweitzer ES, Blaustein MP. (1980) Calcium buffering in presynaptic nerve terminals. Free calcium levels measured with arsenazo III. *Biochim. Biophys. Acta* **600:** 912–921.

Streb H, Irvine RF, Berridge MJ, Schulz I. (1983) Release of Ca^{2+} from mitochondrial intracellular store in pancreatic acinar cells by inositol-1,4,5- triphosphate. *Nature* **306:** 67–69.

Tada M, Yamamoto T, Tonomura Y. (1978) Molecular mechanism of active calcium transport by sarcoplasmic reticulum. *Physiol. Rev.* **58:** 1–79.

Taneja KL, Lifshitz LM, Fay FS, Singer RH. (1992) Poly(A) RNA codistribution with microfilaments: evaluation by *in situ* hybridization and quantitative digital imaging microscopy. *J. Cell Biol.* **119:** 1245–1260.

Terasaki M. (1990) Recent progress on structural interactions of the endoplasmic reticulum. *Cell Motil. Cytoskel.* **15:** 71–75.

Terasaki M, Chen LB, Fujiwara K. (1986) Microtubules and the endoplasmic reticulum are highly interdependent structures. *J. Cell Biol.* **103:** 1557–1568.

Thaine R. (1965) Surface associations between particles and the endoplasmic reticulum in protoplasmic streaming. *New Phytol.* **64:** 118–130.

Tilney LG, Cooke TJ, Connelly PS, Tilney MS. (1991) The structure of plasmodesmata as revealed by plasmolysis, detergent extraction, and protease digestion. *J. Cell Biol.* **112:** 739–747.

Traas JA, Braat P, Emons AMC, Meekes H, Derksen J. (1985) Microtubules in root hairs. *J. Cell Sci.* **76:** 303–320.

Traas JA, Doonan JH, Rawlins DJ, Shaw PJ, Watts J, Lloyd CW. (1987) An actin network is present in the cytoplasm throughout the cell cycle of carrot cells and associates with the dividing nucleus. *J. Cell Biol.* **105:** 387–395.

Url W. (1964) Phasenoptische Untersuchungen an Innenepidermen der Zwiebelschuppe von *Allium cepa* L. *Protoplasma* **58:** 294–311.

Van-Went J, Gori P. (1989) The ultrastructure of *Capparis spinosa* pollen grains. *J. Submicrosc. Cytol. Pathol.* **21:** 149–156.

Venis MA, Napier RM. (1991) Auxin receptors: recent developments. *Plant Growth Regul.* **10:** 329–340.

Vitale A, Ceriotti A, Denecke J. (1993) The role of the endoplasmic reticulum in protein synthesis, modification and intracellular transport. *J. Exp. Bot.* **44:** 1417–1444.

Volpe P, Krause KH, Hashimoto S, Zorgato F, Pozzan T, Meldolesi J, Lew DP. (1988) 'Calciosome', a cytoplasmic organelle: the inositol 1,4,5-triphosphate sensitive Ca^{2+} store of non-muscle cells. *Proc. Natl Acad. Sci. USA* **85:** 1091–1095.

Wick SM, Hepler PK. (1980) Localization of Ca^{++} containing antimonate precipitates during mitosis. *J. Cell Biol.* **86:** 500–513.

Wiedenhoeft RE. (1985) *Comparative aspects of plant and animal coated vesicles.* MS Thesis. University of Georgia, Athens, GA.

Williamson RE. (1979) Filaments associated with the endoplasmic reticulum in the streaming cytoplasm of *Chara corallina*. *Eur. J. Cell Biol.* **20:** 177–183.

Williamson RE. (1985) Immobilisation of organelles and actin bundles in the cortical cytoplasm of the alga *Chara corallina* Klein ex. Wild. *Planta* **163:** 1–8.

Williamson RE. (1986) Organelle movement along actin filaments and microtubules. *Plant Physiol.* **82:** 631–634.

Wyss U, Stender C, Lehmann H. (1984) Ultrastructure of feeding sites of the cyst nematode *Heterodera schachtii* Schmidt in roots of susceptible and resistant *Raphanus sativus* L. var. *oleiformis* Pers. cultivars. *Physiol. Plant. Pathol.* **25:** 21–37.

Zhang DH, Callaham DA, Hepler PK. (1990) Regulation of anaphase chromosome motion in *Tradescantia* stamen hair cells by calcium and related signaling agents. *J. Cell Biol.* **111:** 171–182.

Zhang DH, Wadsworth P, Hepler PK. (1992) Modulation of anaphase spindle microtubule structure in stamen hair cells of *Tradescantia* by calcium and related agents. *J. Cell Sci.* **102:** 79–89.

Zieschang HE, Sievers A. (1991) Graviresponse and the localization of its initiating cells in roots of *Phleum pratense* L. *Planta* **184:** 468–477.

Chapter 23

Targeting and trafficking of vacuolar proteins

Tracey L. Reynolds and Natasha V. Raikhel

1. Introduction

One of the distinct features of plant cells is the presence of a large central vacuole characterized by low pH and the presence of acidic hydrolases. The plant vacuole is a multifunctional compartment involved in regulation of turgor and the storage of reserve proteins, inorganic ions, organic acids, sugars, lectins, secondary metabolites and plant defense proteins (Wink, 1993). The plant vacuole is also important for the degradation and hydrolysis of numerous macromolecules, making it analogous to the mammalian lysosome (Matile, 1985). Accurate and efficient delivery of resident proteins to this organelle is critical for its role in cellular metabolism. Therefore, it is important to understand the manner in which such proteins are directed to and accumulated in the vacuole.

2. Secretory system

The secretory system of plants transports soluble proteins to many different locations, such as the endoplasmic reticulum (ER), Golgi apparatus, vacuole, tonoplast, plasma membrane, cell wall matrix and extracellular space. Proteins traversing the secretory system of all eukaryotic cells must possess signals within the polypeptide backbone for entry into the ER. Additional signalling information must also be present that will result in their retention in, or targeting to, a specific organelle (Blobel, 1980).

Nearly all soluble vacuolar and secretory proteins are initially targeted to the secretory pathway by an N-terminal hydrophobic signal peptide that mediates transport into the lumen of the ER (Chrispeels, 1991). The signal peptide contains a stretch of apolar hydrophobic amino acids with little homology present between signal peptides from different proteins (von Heijne, 1986). The signal peptide allows co-translational insertion of the growing polypeptide into the ER, and thus permits access to the secretory system. During chain elongation, the signal polypeptide is removed; it can then be *N*-glycosylated and begins to fold. Completion of correct folding, formation of disulfide bonds, and subunit assembly in the case of multimeric proteins make the protein competent for transport from the ER to the Golgi complex (Vitale *et al.*, 1993). ER-resident proteins are retained in this compartment because they contain a specific C-terminal retention signal constituted by the sequence KDEL or other related tetrapeptides (Pelham, 1990). A receptor for KDEL-containing proteins, the ERD2 gene product, was first identified in *Saccharomyces cerevisiae* (Semenza *et al.*, 1990). This protein retrieves 'escaped' ER-resident proteins from the *cis* Golgi and returns them to the ER. Homologous genes have been isolated from *Kluyveromyces lactis* (Lewis *et al.*, 1990), bovine (Tang *et al.*, 1993), human (Lewis and Pelham, 1990) and *Plasmodium falciparum* (Elmendorf and Haldar, 1993) systems. Recently, a plant *ERD2* homolog was cloned during random sequencing of an *Arabidopsis thaliana* cDNA library (Lee *et al.*, 1993). Unlike its human counterpart, this homolog functionally complements the *erd2* deletion mutant of *S. cerevisiae*. Furthermore, the *Arabidopsis* ERD2 (*a*ERD2) protein shares 52% and 49% amino acid identity with the human and yeast proteins, respectively. The expression of *aERD2* was investigated by studying transcript levels under various developmental and environmental conditions (Bar-Peled *et al.*, 1995). Varying levels of expression of the *ERD2* gene are observed in different tissue and cell types. The highest level of expression of the *ERD2* mRNA is found in roots. Very low transcript levels are observed in leaves throughout development, but with an increased accumulation of transcript in trichomes in older leaves. This could reflect differing levels of secretory pathway activity in roots and leaves. Stress conditions affecting the secretory pathway (tunicamycin treatment and cold shock) cause an increase in expression of *aERD2*. This suggests that the secretory pathway can be regulated in response to environmental conditions (Bar-Peled *et al.*, 1995).

From the ER, proteins progress to the Golgi apparatus where glycans can be modified by enzymes such as glycosidases and glycosyltransferases. At this point, subsequent sorting must occur in the *trans*-Golgi network (TGN), where soluble proteins containing positive sorting information are transported to the vacuole, and proteins lacking such information are transported to the cell surface by the default or 'bulk flow' pathway (Kornfeld and Mellman, 1989). To determine whether or not secretion is the default pathway in plant systems, the signal peptide from the vacuolar seed protein phytohemaglutinin (PHA) was fused to the coding sequence of a cytosolic seed albumin (Wieland *et al.*, 1987). This fusion protein entered the secretory pathway but did not accumulate in the vacuole, indicating that positive sorting information, in addition to that provided by the signal peptide, is required for targeting to the vacuole. Additional experiments have shown that the signal peptide from PHA is sufficient for secretion of a cytosolic protein (Hunt and Chrispeels, 1991). Furthermore, the nonsecretory bacterial enzymes phosphinothricin acetyl transferase (PAT), β-glucuronidase (GUS), and neomycin phosphotransferase II (NPTII) were targeted to the lumen of the ER by the addition of signal sequences and subsequently secreted to the extracellular space of tobacco leaf protoplasts (Denecke *et al.*, 1990). This indicates that protein secretion in plant cells occurs via a default pathway.

This review will compare the targeting of soluble and membrane vacuolar proteins in plant systems. A much larger body of research has been completed in mammalian and yeast systems, and will be discussed in this review. Of great interest are *trans*-acting factors that recognize the targeting signals and direct transport of the proteins to the appropriate target organelle or membrane. Recently isolated proteins involved in neuronal secretion, together with homologous polypeptides in yeast systems, have been found to be necessary for vesicle formation, vesicle targeting and receptor membrane recognition. It is apparent from these studies that the recognition of soluble vacuolar proteins in the *trans*-Golgi network, vesicle budding, coating and release at the appropriate donor membrane require the co-operative effort of many specific proteins. It will be important to determine whether or not the mechanisms and components of the endomembrane system are conserved between distinct eukaryotic systems.

3. Soluble protein sorting to the vacuole

Cis-*acting factors*

Soluble vacuolar proteins contain distinct *cis*-acting sequences that determine their targeting to the vacuole. In general, plants seem to have a more diverse system of vacuolar protein transport than do mammals or yeast, as the plant vacuolar targeting signal can reside within amino-terminal propeptides (NTPP), carboxy-terminal propeptides (CTPP), or within the sequence of the mature protein. No common sequence motif necessary for targeting has been found among the various sequences (for review see Bednarek and Raikhel, 1992; Chrispeels and Raikhel, 1992; Vitale and Chrispeels, 1992). However, within the N-terminal targeting sequences studied to date, small patches of sequence similarity have been found (see Table 2 in Bednarek and Raikhel, 1992).

N-terminal signals. Sporamin is an abundant storage protein in sweet potato (*Ipomoea batatas*) tubers, that accumulates as a soluble monomer in the vacuole (Hattori *et al.,* 1985). The precursor contains a 16-amino acid NTPP that is removed post-translationally (Hattori *et al.,* 1987). The wild-type precursor is correctly targeted to the vacuole in transgenic tobacco cells, but a mutated sporamin without an NTPP is secreted into the medium. Mutational analysis of the NTPP demonstrated the requirement for a sequence of four amino acids: Asn-Pro-Ile-Arg. The asparagine and isoleucine residues seem to be most critical, as substitution of these residues with glycine resulted in secretion of the protein. In yeast, sporamin with or without the NTPP is directed to the vacuole, suggesting that a different signal operates in yeast (Matsuoka and Nakamura, 1992).

Aleurain is a barley thiol protease related to cathepsin H, a lysosomal enzyme (Holwerda *et al.,* 1990). Specific amino acids were identified in aleurain that target a normally secreted protein to the vacuole. The NTPP of aleurain contains the Asn-Pro-Ile-Arg sequence previously shown to be critical for vacuolar targeting of sporamin. The aleurain-targeting determinant (SSSFADSNPIR) can be further divided into two shorter segments (SSSFADS and SNPIR), each directing the protein to the vacuole but with less efficiency than the full-length sequence. Furthermore, mutations or disruptions of the smaller signals resulted in decreased targeting efficiency of the protein. Another sequence (VTDRAAST), adjacent to the above signal, was also necessary for efficient targeting (Holwerda *et al.,* 1992). In general, it was concluded that the efficient targeting of proaleurain is mediated by the co-operative action of several small N-terminal motifs.

Sequence comparison of several N-terminal propeptides (Chrispeels and Raikhel, 1992), such as a potato 22-kDa protein and the potato cathepsin D inhibitor, reveal a common hydrophobic amino acid sequence with a central arginine: Asn-Pro-Ile-Arg-Leu-Pro. The asparagine and isoleucine residues are conserved, while the other amino acids may vary. Therefore, N-terminal vacuolar targeting sequences contain a consensus motif.

C-terminal signals. The targeting signal of barley lectin (BL) has been studied extensively. This protein is initially synthesized as a glycosylated 23-kDa polypeptide that must dimerize in the ER to form the active proprotein. In transgenic tobacco plants, BL has been shown to be correctly synthesized, proteolytically processed, and targeted to the vacuole, indicating that the sorting signals and machinery are conserved between dicots and monocots (Wilkins *et al.,* 1990). Using a transgenic plant system, the 15-amino-acid CTPP of probarley lectin has been shown to be necessary (Bednarek *et al.,* 1990) for sorting BL to the vacuole, because mutant forms lacking the CTPP are secreted. Furthermore, proteins containing the CTPP fused to a normally extracellular cucumber chitinase (cuc-chit) caused this fusion protein to be redirected to the plant vacuole (Bednarek and Raikhel, 1991). Redirection was 100% complete only when the entire proBL was fused to cuc-chit, whereas the CTPP alone redirected only 70–75% of the cuc-chit/CTPP fusion protein. Thus, fully efficient vacuolar sorting may require the mature BL for correct three-dimensional presentation of the CTPP to the sorting machinery.

Extensive mutational analysis of the CTPP suggested that the presence of specific amino acids did not confer vacuolar sorting capability (Dombrowski *et al.*, 1993). A large number of deletions, substitutions and alternate sequences of the CTPP were tolerated by the vacuolar sorting machinery resulting in correct delivery of these mutants to the vacuole. The addition of two glycines to the C-terminal glutamic acid or the relocation of the glycosylation site close to the C-terminal end of the CTPP resulted in secretion of BL from the cell. This indicates that components of the sorting machinery must interact at the C-terminus of the peptide, and the presence of glycines or carbohydrates may interfere with recognition of the targeting determinant sequence, resulting in secretion.

The addition of the BL CTPP to the secreted yeast protein invertase did not cause redirection of this fusion protein to the vacuole in yeast cells (Gal and Raikhel, 1993). Therefore, yeast cells do not recognize a C-terminal vacuolar targeting sequence.

Many other vacuolar proteins also contain C-terminal vacuolar targeting sequences. The vacuolar form of tobacco chitinase contains a C-terminal 7-amino-acid sequence (Glu-Leu-Leu-Val-Asp-Thr-Met) that is absent from a homologous cell wall chitinase (Shinshi *et al.*, 1990). This hexapeptide was shown to be necessary and sufficient for vacuolar targeting (Neuhaus *et al.*, 1991). Furthermore, the addition of a glycine residue to the CTPP resulted in secretion (Neuhaus *et al.*, 1994) analogous to the mutational analysis of the BL CTPP (Dombrowski *et al.*, 1993). Class I β-1,3-glucanases are basic vacuolar proteins implicated in plant pathogen defense (Mauch and Staehelin, 1989). These proteins contain an *N*-glycosylated C-terminal extension that is lost during maturation of the protein. Similarly, two other vacuolar pathogenesis-related proteins, AP24 and a basic chitinase, were found to contain targeting information in a short C-terminal propeptide that is removed during or after transport to the vacuole (Melchers *et al.*, 1993).

Comparison of the C-terminal signal sequences of lectins and vacuolar hydrolases (see Table 1 in Chrispeels, 1992) reveals no consensus sequence; however, a large number of hydrophobic residues are present. Thus, a recognition signal may be present that relies on a specific secondary structure formed by various combinations of amino acids.

Although there are no similarities between the sequences of the NTPP and the CTPP, sporamin and BL have been shown to be transported to the same vacuoles when they are co-expressed in tobacco plants (Schroeder *et al.*, 1993). It has also been demonstrated that the CTPP and NTPP are functionally interchangeable in their ability to direct each of these proteins to the vacuole (Matsuoka *et al.*, 1995).

Internal signals. Many vacuolar proteins are transported to the vacuole without subsequent proteolytic processing of the propeptide, indicating that targeting information resides within the sequence of the mature protein. The lectin of the common bean *Phaseolis vulgaris*, phytohemagglutinin (PHA), accumulates in cotyledon storage vacuoles during seed development. Except for the removal of the signal sequence during translocation into the ER, PHA does not undergo further proteolytic processing upon delivery to the vacuole (Vitale *et al.*, 1984). PHA, expressed in yeast, was retained intracellularly and correctly targeted to the vacuole (Tague and Chrispeels, 1987). PHA/invertase fusions were made to identify the targeting sequence in yeast, and a portion of the N-terminus was sufficient for sorting the fusion protein to the yeast vacuole (Tague *et al.*, 1990). Deletions and mutations made within this sequence demonstrated that the sorting information was contained within a tetrapeptide, Leu-Gln-Arg-Asp (Tague *et al.*, 1990). This sequence is similar to the vacuolar sorting sequence of carboxypeptidase Y from yeast (Gln-Arg-Pro-Leu).

Using a transient expression system in plants (von Schaewen and Chrispeels, 1993), invertase fusions were tested to confirm the results of targeting in the yeast expression system. Neither glycosylation of the domain recognized by yeast nor deletion of a large N-terminal portion caused secretion of the PHA/invertase fusion protein. However, an internal portion of PHA (amino acids 84–113) caused retention of 50% of the invertase activity within the cell. The crystal structure of related lectins has shown that amino acids 91–117 are present on the surface of the molecule, making this sequence accessible to the sorting machinery. Therefore, the yeast system is recognizing signals that are not utilized by the plant's sorting machinery.

Targeting information in legumin, the major storage protein in bean seeds, was analyzed using gene fusions in yeast (Saalbach *et al.,* 1991). Short N-terminal segments of legumin did not cause targeting of the fusion proteins to the vacuole. The entire α chain (281 amino acids) was necessary to target more than 90% of the invertase fusion protein to the vacuole. A shorter segment of 76 amino acids at the C-terminal end was also found to be sufficient for vacuolar targeting. Therefore, legumin contains long multiple targeting segments that may act to establish a 'signal patch' on the surface of the protein for vacuolar targeting in yeast. However, the identity of the plant signal in legumin has not been established.

Glycan modifications. In mammalian systems, lysosomal protein-sorting information resides in a glycan modification, namely the selective phosphorylation of mannose residues. This signal is recognized by a mannose 6-phosphate (M6P) receptor that is responsible for targeting newly synthesized enzymes to the lysosome. This was determined first by the secretion of lysosomal enzymes in mucolipidosis type-II patients who fail to phosphorylate mannose residues. Furthermore, phosphorylated exogenous lysosomal enzymes were endocytosed, a process that could be inhibited by phosphorylated sugars (Kaplan *et al.,* 1977).

The M6P system of vacuolar protein sorting does not operate in yeast cells. After applying tunicamycin (a glycosylation inhibitor), two yeast enzymes, carboxypeptidase Y (CPY) (Schwaiger *et al.,* 1982) and alkaline phosphatase (Clark *et al.,* 1982), were correctly targeted to the yeast vacuole.

In plants, M6P residues were not detected on glycoproteins from pea (Gaudreault and Beevers, 1984). To investigate the role of other glycan modifications in processing and targeting of plant vacuolar proteins, the glycan attachment sites on PHA were deleted by site-directed mutagenesis. This mutation did not alter the delivery of this protein to the vacuole in transgenic plants (Voelker *et al.,* 1989). Similarly, the *N*-linked glycosylation site residing within the CTPP of BL was deleted by site-directed mutagenesis. Transgenic plants expressing the *gly⁻* construct correctly sorted BL to the vacuole (Wilkins *et al.,* 1990). Similar experiments have been performed on patatin (Sonnewald *et al.,* 1990) and β-1,3-glucanases (Sticher *et al.,* 1992). Therefore, in plants, as in yeast, glycans are not necessary for vacuolar protein sorting.

Yeast vacuolar targeting signals. Yeast vacuoles store inorganic ions, metabolites and amino acids in a manner that is more similar to the plant vacuole than to the mammalian lysosome (Klionsky *et al.,* 1990). Although many yeast proteins contain M6P moieties, they do not utilize this modification for vacuolar targeting (Schwaiger *et al.,* 1982). They contain a sorting signal within the N-terminal propeptide (Johnson *et al.,* 1987; Valls *et al.,* 1987). Two soluble yeast vacuolar hydrolases, CPY and proteinase A (PrA), have been the most extensively characterized. Both contain N-terminal propeptides that are necessary and sufficient for targeting of these proteins to the yeast vacuole (Johnson *et al.,* 1987; Klionsky *et al.,* 1988). Site-directed mutagenesis of the CPY targeting sequence defined four amino acids (Gln-Arg-Pro-Leu) that are critical for targeting (Valls *et al.,* 1990). The targeting signal of PrA (amino acids 61–76) differs from that found in CPY (Klionsky *et al.,* 1988). Therefore, yeast vacuolar proteins contain N-terminal targeting sequences, and the two proteins studied are targeted by distinct recognition systems or form a common structural feature that has yet to be identified.

Yeast and plant vacuolar targeting appear to have a similar mechanism: both systems rely on signal peptides within the polypeptide sequence of the protein; both lack a consensus targeting sequence; both are independent of M6P-based targeting; and plant proteins, expressed in yeast cells, are targeted to the yeast vacuole. Nevertheless, yeast cells are unable to utilize plant vacuolar targeting signals from the C-terminus, N-terminus or internal to the polypeptide chain (Gal and Raikhel, 1993; Matsuoka and Nakamure, 1992; von Schaewen and Chrispeels, 1993). Despite this, plant vacuolar proteins are correctly targeted to the yeast vacuole, indicating that other determinants present in plant proteins are recognized by the yeast vacuolar transport machinery. Because consensus targeting sequences have not been ascertained for yeast or plant vacuolar targeting sequences, structural determinants formed by various combinations of amino acids may be critical to the recognition system for vacuolar targeting in yeast and plant cells.

Trans-*acting factors*

Cis-acting factors such as the sorting signals for vacuolar targeting have been extensively characterized in mammalian, yeast and plant systems. Now, extensive research efforts are being focused on *trans*-acting factors that recognize these sequences and are responsible for their transport to the vacuole. It has been shown that overproduction of CPY and PrA in yeast cells results in secretion, possibly due to saturation of a receptor (Rothman *et al.,* 1986; Stevens *et al.,* 1986). However, overproduction of CPY does not affect vacuolar localization of PrA, or vice versa, so different receptors or sorting pathways must be involved. Similarly, the overproduction of a specific resident lysosomal enzyme in a malignantly transformed mammalian cell line results in its secretion, without affecting the localization of other lysosomal proteins (Gal *et al.,* 1985). Therefore, mammalian and yeast systems possess saturable components that recognize *cis*-acting elements and thus allow delivery of soluble proteins to lysosomes and vacuoles.

Potato sporamin, with an N-terminal signal sequence, and BL, containing a C-terminal signal sequence, were found to co-localize to the same vacuole in transgenic tobacco plants with little or no secretion (Schroeder *et al.,* 1993). It can be concluded that these proteins are not competing for the same receptor, or that this system is not saturable. Recently, it has been shown that the vacuolar retention of a tobacco chitinase is not affected by expression of up to four times the endogenous level. However, when greater amounts of chitinase are produced, as much as 50% of the chitinase activity is recovered in the medium (Neuhaus *et al.,* 1994) during transient assay analysis.

In plant systems, increased flux through the secretory system may result in the synthesis of vacuolar-sorting components to accommodate the elevated level of proteins traversing this pathway. Increased secretory pathway activity may be caused by the synthesis of plant defense proteins during pathogen attack or of cell wall components during development. Furthermore, plant cells can contain as many as 400 individual stacks dispersed throughout the cytoplasm, while animal cells contain only 10 stacks clustered around the nucleus (Driouich *et al.,* 1993). Saturability of vacuolar targeting in transient assays reflects the system's inability to tolerate an overload through the secretory pathway over a short period of time. Therefore, plant systems may contain a saturable receptor responsible for vacuolar targeting, but components of this system may be able to adapt to secretory pathway flux, thereby preventing mis-sorting of vacuolar proteins.

Mammalian and yeast vacuolar targeting receptors. Soluble proteins targeted to the mammalian lysosome contain a M6P moiety that facilitates lysosomal targeting in a receptor-dependent manner (for review, see Kornfeld and Mellman, 1989). The asparagine-linked, high-mannose sugars of lysosomal enzymes are modified in the *cis* Golgi to a M6P group. This M6P residue is recognized by specific receptors in the TGN, and this complex is sorted into clathrin-coated vesicles that bind and fuse with endosomes. The acidity of the endosomes causes release of the ligand from the M6P receptor, and the receptor returns to the TGN to reinitiate the cycle. Two M6P receptors have been described (for review, see Dahms *et al.,* 1989), and are differentiated by their dependence on cations, intracellular functions and molecular mass. Thus, M6P receptors can divert soluble phosphorylated enzymes from the default secretory pathway to the lysosome.

In patients with I-cell disease, who lack the *N*-acetylglucosamine-1-phosphotransferase enzyme responsible for the synthesis of M6P residues, lysosomal enzyme activities were normal in liver, spleen and β-lymphocyte cells (Owada and Neufeld, 1982; Waheed *et al.,* 1982). Therefore, in certain organs, the transport of lysosomal enzymes such as cathepsin D must be accomplished by a process independent of the receptor-mediated M6P pathway. In HepG2 cells, cathepsin D was efficiently targeted to the lysosome independent of binding to the M6P receptor (Rijnboutt *et al.,* 1991). Further studies of this phenomenon have used cathepsin D/pepsinogen (lysosomal/secretory) protein fusions to determine the sequences responsible for M6P-independent targeting. The carboxy lobe of cathepsin D (amino acids 188–265) was found to contain sufficient sorting information to confer lysosomal targeting when fused to pepsinogen (Glickman and Kornfeld, 1993).

Recent genetic analysis of yeast has demonstrated that the VPS10 gene product is the putative vacuolar-sorting receptor for CPY (Marcusson *et al.,* 1994). The 110-kDa protein is a type-I integral-membrane protein that cross-links to CPY but not to mutants in the CPY sorting signal. This protein is localized to the late Golgi compartment where vacuolar protein sorting occurs. Specific receptors for other soluble vacuolar proteins such as PrA have yet to be isolated.

Targeting receptors in plant systems. Several laboratories are currently trying to isolate specific receptors that recognize plant vacuolar targeting sequences. Recently, Kirsch *et al.* (1993) have identified a protein from clathrin-coated vesicles of developing pea cotyledons that will bind to the N-terminal targeting sequence of barley proaleurain. This glycosylated 80-kDa protein (BP-80) was present in a less dense membrane fraction containing high activity of inosine diphosphatase, a Golgi marker enzyme. It possesses a cytosolic C-terminal tail and an N-terminal region within the lumen of the vesicle. Interestingly, this putative receptor binds specifically to the NTPP of aleurain and sporamin and the C-terminal targeting sequences of 2S albumin from Brazil nut; however, BP-80 does not bind to the CTPP of barley lectin (Kirsch *et al.,* 1995). The NTPP targeting motif does not share any amino acid similarity with the C-terminal sequence of 2S albumin. These findings suggest that, although the putative vacuolar targeting receptor has a broad binding specificity for various motifs, distinct receptor proteins and possibly targeting mechanisms are involved in the transport of different vacuolar proteins.

4. Vacuolar membrane (tonoplast) protein sorting

Plants

The tonoplast membrane acts to separate physically the neutral cytoplasmic environment from the acidic hydrolytic constituents present in the vacuolar sap. One of the few plant tonoplast proteins that has been characterized is the tonoplast intrinsic protein (TIP) family. The common bean α-TIP protein has been shown to be targeted correctly to the tonoplast in mesophyll cells of transgenic tobacco (Höfte *et al.,* 1991). Initially, it was determined that the C-terminal 48 amino acids (the 18-amino-acid tail plus the sixth membrane-spanning domain) from α-TIP are sufficient to direct a reporter protein to the tonoplast. However, deletion of the 18-amino-acid tail from α-TIP did not deter its accumulation in the tonoplast (Höfte and Chrispeels, 1992).

Therefore, targeting information may be localized to the sixth transmembrane domain. Alternatively, the tonoplast may be the default destination for membrane proteins.

A more recent study of vacuolar membrane protein transport (Gomez and Chrispeels, 1993) involved a comparison between the tonoplast membrane protein α-TIP and PHA-L, a soluble vacuolar protein. The fungal toxin monensin and brefeldin A (BFA) were used to block vesicle transport. Monensin is a monovalent ionophore that may affect acidification across the TGN membrane, and BFA inhibits anterograde vesicle transport between the ER and the Golgi. Both of these drugs blocked the vacuolar transport of PHA-L and slowed the processing of glycans; however, neither drug inhibited the transport of α-TIP to the tonoplast. This confirms the hypothesis that soluble and tonoplast proteins have different mechanisms for transport to the vacuole. A more extensive study of other plant tonoplast and Golgi membrane proteins is required to determine whether the tonoplast membrane is the default pathway for vacuolar membrane proteins, as in yeast, or whether specific targeting sequences are present, as in mammalian systems.

Acidification of the vacuolar contents is generated by an H^+-ATPase located on the tonoplast membrane (Sze, 1985). A purified oat H^+-ATPase consisted of 10 polypeptides containing a large peripheral sector of six polypeptides (Ward and Sze, 1992). Specific vacuolar targeting sequences have not yet been identified for these proteins, and it will be of interest to determine how targeting and co-ordination of these polypeptides to form a multi-subunit complex is regulated.

Mammalian and yeast systems

Integral membrane proteins of the mammalian lysosome do not contain M6P moieties and must therefore reach the lysosomal membrane independently of the M6P receptor-mediated system (Fukuda, 1991). Two possible avenues may be taken by lysosomal membrane proteins; either they are directly targeted to the lysosomal membrane, or they are initially delivered to the plasma membrane followed by selective retrieval to the endosomal/lysosomal system. Experimental evidence indicates that both methods are used, with preference for a particular path determined by the protein in question (Baun et al., 1989; Harter and Mellman, 1992; Nabi et al., 1991). For example, Lamp-1, a lysosomal-associated membrane glycoprotein, can be transported to the lysosome by both pathways (Carlsson and Fukuda, 1992).

A tyrosine motif within the C-terminal portion of lysosomal membrane proteins has been shown to be important for localization. This sequence, when fused to a secreted protein, results in delivery of the fusion protein to the lysosomal membrane (Fukuda, 1991; Williams and Fukuda, 1990). A recently discovered dileucine motif present near the tyrosine-containing lysosomal signal may enhance sorting efficiency (Letourneur and Klausner, 1992). Chimeras containing the tyrosine and dileucine motifs were targeted predominantly to the lysosomal membrane without initial transport to the cell surface. However, a tyrosine-lacking cytoplasmic tail of a lysosomal integral membrane protein (LIMP) is necessary and sufficient to target this protein, as well as two plasma membrane reporter proteins, to the lysosomal membrane (Vega et al., 1991). Therefore, mammalian lysosomal membrane targeting may use different signal sequences depending on the protein or its pathway to the lysosome.

No specific sorting signals have been found for several vacuolar membrane proteins in yeast cells. Alkaline phosphatase (ALP), an integral membrane protein of the vacuolar membrane, is anchored in the membrane by an N-terminal hydrophobic domain that may function as an internal signal sequence (Klionsky and Emr, 1989). Dipeptidyl aminopeptidase B (DPAP B), a vacuolar membrane protein, and dipeptidyl aminopeptidase A (DPAP A), a Golgi membrane protein, were analyzed in yeast cells to determine whether targeting sequences were present.

Removal and replacement of the cytoplasmic, transmembrane or lumenal domain of DPAP B with that from DPAP A did not affect localization of DPAP B to the vacuolar membrane (Roberts *et al.*, 1992). Therefore, a specific vacuolar membrane targeting sequence has not been identified. Mutation or overproduction of several Golgi membrane proteins such as DPAP A (Roberts *et al.*, 1992), Kex2p (Wilcox *et al.*, 1992) and Kex1p (Cooper and Bussey, 1992) resulted in localization of these proteins to the vacuolar membrane rather than to the cell surface. Thus, in yeast, vacuolar membrane proteins may not require sorting information, and the vacuolar membrane may be the default destination for membrane proteins of the secretory pathway.

The targeting and assembly of multi-subunit complexes destined for the vacuolar membrane involves the co-ordination of integral and peripheral proteins synthesized in the cytoplasm or from the secretory pathway. Vacuolar proton-translocating ATPases are present in all eukaryotic cells and contain multiple subunits of 70, 60 and 17 kDa, as well as additional subunits. In yeast, studies of the subunits of the vacuolar H^+-ATPase (three peripheral subunits of 69, 60 and 42 kDa and two integral membrane proteins of 100 and 17 kDa) indicate that the peripheral subunits are synthesized in the cytoplasm and the integral membrane components enter the secretory system through the ER (Kane *et al.*, 1992).

How are the subunits of this complex, originally synthesized in two locations within the cell, assembled and targeted to the same membrane? The assembly and vacuolar targeting of some of the subunits are very interdependent (Kane *et al.*, 1992); for instance, the 100-kDa integral membrane subunit is transported to the vacuole independently of the peripheral membrane subunits, so long as the 17-kDa integral membrane subunit is present. In the absence of the 17-kDa subunit, the 100-kDa subunit is not competent for transport to the vacuole. Furthermore, in yeast cells lacking the 69- or 60-kDa subunit peripheral protein, the 42-kDa subunit is not transported to the vacuolar membrane. Thus, targeting information may reside in only one component of a multi-subunit complex. Studies of this type will aid our understanding of the regulation and control of vacuolar targeting in multi-subunit complexes.

5. Components of the vacuolar sorting machinery

The secretory system is divided into distinct membrane-bound organelles. The process of protein trafficking through the secretory pathway is mediated by the specific budding and targeting of vesicles from one compartment and their fusion to a subsequent one. The efficiency and regulation of this system require controlled interactions between organelles, and are critical for the orderly flow of proteins through the secretory system. Several laboratories have recently isolated proteins that appear to function in the initiation, targeting or docking of vesicles. These data indicate that the fundamental mechanisms of vesicle initiation and fusion are conserved between mammalian neurones and yeast cells (Bennett and Scheller, 1993).

Vesicle-mediated transport of proteins has two distinct stages: (1) vesicle formation and budding from the donor membrane, and (2) vesicle docking and fusion with the acceptor membrane. There appear to be many classes of proteins involved in the regulation of vesicular trafficking. Vesicle fusion processes require soluble cytoplasmic proteins such as the *N*-ethylmaleimide-sensitive fusion protein (NSF) and soluble NSF attachment proteins (SNAPs). The search for receptors of SNAP proteins (SNAREs) has yielded specific integral membrane proteins associated with the donor and target membranes. Vesicle-associated membrane proteins (VAMPs)/synaptobrevins are associated with the transport vesicle (v-SNARE), while syntaxins (t-SNARE) are found localized to the target membrane. It is hypothesized that a specific v-SNARE will interact with its corresponding t-SNARE through the NSF–SNAP complex to mediate vesicle targeting and fusion. In principle, regulation of the targeting reaction is thought to be mediated by small GTP-binding proteins specific for each compartment of the secretory pathway (for reviews, see Bennett and Sheller, 1993; Novick and Brennwald, 1993; Zerial and Stenmark, 1993).

Attempts have been made to identify cellular factors involved in vesicular trafficking to the plant vacuole. Biochemical analyses have identified small GTP-binding proteins which might function in the vesicle-mediated transport of plant vacuolar proteins (Shimada *et al.*, 1994). Genes encoding GTPases (possibly involved in the vesicular transport system of plants; Anai *et al.*, 1994; Bednarek *et al.*, 1994; Dombrowski *et al.*, 1995) as well as syntaxin homologs (Bassham *et al.*, 1995; Bar-Peled *et al.*, 1995) (proteins involved in the docking and fusion of vesicles in mammalian and yeast cells) have been isolated and their expression patterns characterized. Furthermore, the *Arabidopsis* PEP12 gene complements a yeast mutation and encodes a protein homologous to Pep12 and other members of the syntaxin family (Bassham *et al.*, 1995). This protein may thus be a component of the recognition and/or fusion apparatus for vesicles targeted to the plant vacuole.

vps *mutants*

Extensive genetic analysis in *Saccharomyces cerevisiae* has been undertaken to improve our understanding of *trans*-acting factors involved in targeting proteins to the vacuole. The *vps* (vacuolar-protein-sorting defective) mutants mis-sort these enzymes to the cell surface instead of to the vacuole. The selection criteria are mislocalization of CPY to the cell surface of mutated yeast cells (Rothman *et al.*, 1986) and mutant sucΔ yeast cells that are unable to utilize sucrose as a carbon source (Bankaitis *et al.*, 1986). In the latter, selection of mutants was based on the ability of cells expressing a vacuole-localized CPY/invertase fusion protein to mis-sort this protein to the cell surface, where it survives via invertase activity on the external sucrose medium.

Many *vps* mutants display morphologically defective vacuoles, temperature-sensitive growth, and mis-sorting of vacuolar membrane proteins. At least 50 genes may be involved in some manner in vacuolar sorting (Raymond *et al.*, 1992). Of the many *vps* mutants analyzed, VPS15 and VPS34 gene products have been characterized and recent experiments indicate that they are associated in a heteroligomeric protein complex. The VPS15 gene encodes a 1455-amino-acid protein with significant homology to the serine/threonine family of protein kinases (Herman *et al.*, 1991). This kinase seems to play a direct role in regulating protein sorting to the vacuole, and is associated with the cytoplasmic face of the late Golgi. The VPS34 gene product is homologous to the catalytic subunit of mammalian phosphatidylinositol 3-kinase (PI 3-kinase) (Hiles *et al.*, 1992). In mammalian cells, PI 3-kinase functions in secondary-messenger signalling involved in cell growth and proliferation.

Yeast strains that contain a deleted or point-mutated *VPS4* gene lack PI 3-kinase activity and are severely defective in vacuolar protein sorting. Overexpression of the Vps34 protein results in an increase in PI 3-kinase activity (Schu *et al.*, 1993) and suppression of defects associated with *Vps15* kinase domain mutants. Immunoprecipitation and cross-linking experiments indicate that Vps15 and Vps34 proteins are physically associated, and that Vps15 can facilitate the association of Vps34 with a membrane fraction (Stack *et al.*, 1993). Furthermore, Vps34 lipid kinase activity is believed to be regulated by Vps15-mediated phosphorylation. Therefore, it has been proposed that the Vps15/Vps34 complex may interact with the cytoplasmic tails of soluble vacuolar protein receptors, promoting a conformational change and activating the Vps15 protein kinase. A Vps15 phosphorylation event would activate the PI 3-kinase activity of Vps34, triggering a cascade of events resulting in vesicle formation and ultimate delivery to the vacuole.

In yeast, the VPS34 gene is essential for vacuolar protein sorting and encodes a protein with PI 3-kinase and protein kinase activities (Stack *et al.*, 1993; Stack and Emr, 1994). PI 3-kinases have recently been cloned from *Arabidopsis thaliana* and soybean which show homology to VPS34 (Hong and Verma, 1994; Welters *et al.,* 1994), although the *Arabidopsis* gene is unable to complement the yeast *VPS34* mutant (Welters *et al.,* 1994). Wortmannin is a specific inhibitor of PI 3-kinase that is involved in vesicle transport and membrane structure in various cell types (Matsuoka *et al.,* 1995, and references therein). The effect of wortmannin on vacuolar delivery of NTPP- and CTPP-containing proteins, and on the synthesis of phospholipids, has been examined in tobacco cells. The results obtained suggest that at least two different mechanisms exist in plants for the transport of soluble proteins to the vacuole. Pulse-chase analysis indicated that 33 μM wortmannin caused almost complete inhibition of CTPP-mediated transport to the vacuoles, while NTPP-mediated transport displays almost no sensitivity to wortmannin at this concentration. The dose-dependency of wortmannin on the inhibition of CTPP-mediated vacuolar delivery of proteins and on the inhibition of the synthesis of phospholipids in tobacco cells have been compared. Mis-sorting caused by wortmannin displays a dose-dependency similar to that which inhibits the synthesis of PI 4-phosphate and major phospholipids. Thus, the synthesis of phospholipids may well be involved in CTPP-mediated vacuolar transport (Matsuoka *et al.,* 1995).

Small GTP-binding proteins

Small GTP-binding proteins (of approximately 22 kDa) function in a variety of cellular processes and are characterized by four highly conserved GTP-binding domains for hydrolysis and exchange of GTP and GDP (see black boxes in *Figure 1*) (Bourne *et al.,* 1990). These proteins exist in two interconvertible conformational states: active (GTP-bound) and inactive (GDP-bound). Another more divergent domain, the 'effector loop', is where the GTPase-activating protein (GAP) interacts to increase the intrinsic GTPase activity (Pai *et al.,* 1989). Finally, GTP-binding proteins contain one or two conserved cysteine residues at the C-terminus that are necessary for attachment of a modified lipid group, resulting in membrane association (Balch, 1990).

A large number of small GTPases of the Rab family have been localized to various sites along the secretory pathway. By alternating between the two conformations, these proteins may function as molecular switches involved in docking and/or fusion of vesicles with their appropriate target membrane. Rab1a, Rab1b and Rab2 are involved in transport from the ER to the Golgi (Pultner *et al.,* 1991; Tisdale *et al.,* 1992). Rab3a is involved in regulated secretion via synaptic vesicles (Oberhauser *et al.,* 1992), Rab4 is associated with early endosomes (Van der Sluijs *et al.,* 1991), and Rab5 with early endosomes and the plasma membrane (Bucci *et al.,* 1992). Rab7 and Rab9 (Chauvier *et al.,* 1990; Wichmann *et al.,* 1992) are found in late endosomes, and Rab6 in the TGN (Antony *et al.,* 1992). Several homologs of Rab proteins have recently been isolated from plants (Terryn *et al.,* 1993).

It can be argued that Rab6 functions in the regulation of vesicle transport from the TGN, and yeast mutants have been reported that mis-sort vacuolar proteins to the medium (Hengst and Gallwitz, personal communication). Recently, however, several reports have shown that this protein may be involved in the transport of secretory vesicles to the plasma membrane in mammalian systems (Karniguian *et al.,* 1993; Tixier-Vidal *et al.,* 1993). To determine the function of this protein in plant systems, our laboratory used PCR primers from conserved regions of mammalian RAB6 to isolate an *Arabidopsis* homolog of this gene (Bednarek *et al.,* 1994). A.t.Rab6 has approximately 79% sequence identity to the mammalian (Rab6) and yeast (Ryh1 and Ypt6) protein counterparts. In addition, A.t.RAB6 functionally complemented the *YPT6* mutant from yeast.

Two mutations were made to the wild-type gene. First, a specific amino acid change was made to the third GTP-binding domain (Asn^{122} → Ile^{122}) (*Figure 2*). Previous work on

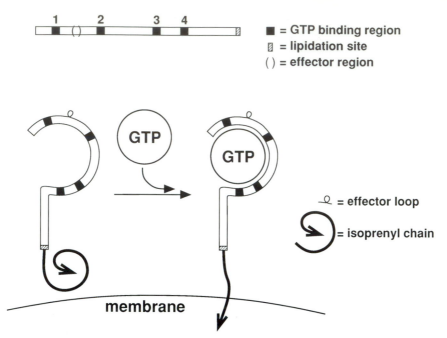

Figure 1. Diagrammatic representation of a GTP-binding protein. The four highly conserved GTP-binding domains (black boxes) co-operate to bind GTP after displacement of GDP. The GTPase-activating protein (GAP) interacts at the effector region or loop to increase the intrinsic GTPase activity of the protein. Finally, post-translational addition of an isoprenyl chain at the C-terminal lipidation site results in membrane attachment.

mammalian systems indicated that this mutation obliterated GTP binding and resulted in a dominant mutation (Wagner *et al.,* 1987). Secondly, the C-terminal conserved cysteine residues were mutated so that isoprenylation and subsequent membrane attachment could not occur. This mutation should not alter the ability of the protein to bind GTP. Each mutant construct was expressed in *E. coli,* and the ability of the resultant proteins to bind GTP was altered as expected. That is, the dominant mutant construct could not bind GTP, while the C-terminal mutant retained GTP-binding capability. To assess the function of A.t.Rab6 in vesicular transport, the dominant mutant construct as well as an antisense construct of the wild-type *A.t.RAB6* gene will be transformed into tobacco and *Arabidopsis* plants to determine the effects on protein trafficking.

6. Conclusions

Comparison of sorting signals for soluble and membrane-associated vacuolar proteins between mammalian, yeast and plant cells has only emphasized the diverse nature of these systems. No consensus sequence could be determined for *cis*-acting vacuolar sorting sequences from comparable regions of their respective proteins, although small stretches of homology have been found within N-terminal signal sequences in plants. What does this tell us about receptors that recognize these signals? First, different receptors may recognize specific sequences. Secondly, a smaller number of receptors may identify secondary or tertiary structural sections on folded proteins. Various biochemical and genetic approaches could be used to identify vacuolar sorting receptors. The recent isolation of specific point mutations that cause secretion of vacuolar proteins (Dombrowski *et al.,* 1993) could serve as a control in these experiments.

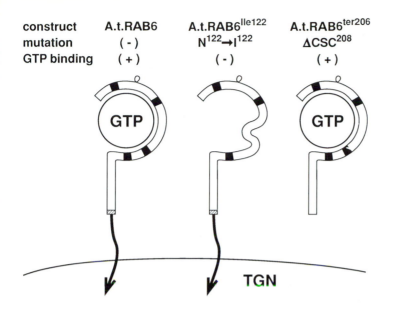

Figure 2. Diagrammatic representation of the A.t.RAB6 gene and two mutant constructs. The wild-type gene (A.t.RAB6) was able to bind GTP and associate with its target membrane, the TGN. The first mutant construct has a single amino acid change in the third GTP-binding domain (*A.t.RAB6^{Ile122}*), which obliterates GTP-binding yet allows membrane association. The second mutant has a mutated C-terminal lipidation site (*A.t.RAB6^{ter206}*) that renders the protein incapable of post-translational lipidation and subsequent membrane association. This mutant is still capable of binding GTP.

In conclusion, much remains to be understood, particularly in view of the fact that no fully conserved mechanism appears to operate in the different systems analyzed. Identification of the key components of vacuolar sorting, namely *cis*-acting vacuolar targeting sequences, is the initial stage in identification of the components involved in the sorting mechanism. In plants, we are now in a position to make rapid progress, especially given our ability to produce transgenic plants routinely. This will permit us to examine the fate of endomembrane components in a complex multicellular organism.

Acknowledgments

The authors would like to thank Dr Alessandro Vitale, Dr Glenn Hicks, Dr Susannah Gal, Dr Jim Dombrowski and Antje Heese-Peck for critically reviewing this manuscript. This research was supported by Grant DCB-9002652 from the National Science Foundation and Grant DE-FG02-91ER20021 from the United States Department of Energy to N.V.R.

References

Anai T, Matsui M, Nomura N, Ishizaki R, Uchimiya H. (1994) In vitro mutation analysis of *Arabidopsis thaliana* small GTP-binding proteins and detection of GAP-like activities in plant cells. *FEBS Lett.* **346:** 175–180.

Antony C, Cibert C, Geraud G, Santa Maria A, Maro B, Mayau V, Goud B. (1992) The small GTP-binding protein rab6p is distributed from medial Golgi to the *trans* Golgi network as determined by a confocal microscopic approach. *J. Cell Sci.* **103:** 785–796.

Balch WE. (1990) Small GTP-binding proteins in vesicular transport. *Trends Biochem. Sci.* **15:** 473–477.

Bankaitis VA, Johnson LN, Emr SD. (1986) Isolating yeast mutants defective in protein targeting to the vacuole. *Proc. Natl Acad. Sci. USA* **83**: 9075–9079.

Bar-Peled M, da Silva Conceição A, Frigerio L, Raikhel NV. (1995) Expression and regulation of aERD2, a gene encoding the KDEL receptor homolog in plants, and other genes encoding proteins involved in ER–Golgi vesicular traffic. *Plant Cell* **7**: 667–676.

Bassham DC, Gal S, da Silva Conceição A, Raikhel NV. (1995) An *Arabidopsis* syntaxin homologue isolated by functional complementation of a yeast *pep*12 mutant. *Proc. Natl Acad. Sci. USA* **92**: 7262–7266.

Bednarek SY, Raikhel NV. (1991) The barley lectin carboxyl-terminal propeptide is a vacuolar protein sorting determinant in plants. *Plant Cell* **3**: 1195–1206.

Bednarek SY, Raikhel NV. (1992) Intracellular trafficking of secretory proteins. *Plant Mol. Biol.* **20**: 133–150.

Bednarek SY, Wilkins TA, Dombrowski JE, Raikhel NV. (1990) A carboxy-terminal propeptide is necessary for proper sorting of barley lectin to vacuoles of tobacco. *Plant Cell* **2**: 1145–1155.

Bednarek SY, Reynolds TL, Schroeder M, Grabowski L, Gallwitz D, Raikhel NV. (1994) A small GTP-binding protein from *Arabidopsis thaliana* functionally complements the yeast *YPT6* null mutant. *Plant Physiol.* **104**: 591–596.

Bennett MK, Scheller RH. (1993) The molecular machinery for secretion is conserved from yeast to neurons. *Proc. Natl Acad. Sci. USA* **90**: 2559–2563.

Blobel G. (1980) Intracellular protein topogenesis. *Proc. Natl Acad. Sci. USA* **77**: 1496–1500.

Bourne HR, Sanders DA, McCormick F. (1990) The GTPase superfamily: a conserved switch for diverse cell functions. *Nature* **348**: 125–131.

Braun M, Waheed A, von Figura K. (1989) Lysosomal acid phosphatase is transported to lysosomes via the cell surface. *EMBO J.* **8**: 3633–3640.

Bucci C, Parton RG, Mather IM, Stunnenberg H, Simons K, Hoflack B, Zerial M. (1992) The small GTPase Rab5 functions as a regulatory factor in the early endocytic pathway. *Cell* **70**: 715–728.

Carlsson SR, Fukuda M. (1992) The lysosomal membrane glycoprotein Lamp-1 is transported to lysosomes by two alternative pathways. *Arch. Biochem. Biophys.* **296**: 630–639.

Chauvier P, Parton RG, Hauri HP, Simons K, Zerial M. (1990) Localization of low molecular weight GTP-binding proteins to exocytic and endocytic compartments. *Cell* **62**: 317–329.

Chrispeels MJ. (1991) Sorting of proteins in the secretory system. *Annu. Rev. Plant Physiol. Plant Mol. Biol.* **42**: 21–53.

Chrispeels MJ, Raikhel NV. (1992) Short peptide domains target proteins to plant vacuoles. *Cell* **68**: 613–616.

Clark DW, Tkacz JS, Lampen JO. (1982) Asparagine-linked carbohydrate does not determine the cellular location of yeast vacuolar nonspecific alkaline phosphatase. *J. Bacteriol.* **152**: 865–873.

Cooper A, Bussey H. (1992) Yeast Kex1p is a Golgi-associated membrane protein: deletions in a cytoplasmic targeting domain result in mislocalization to the vacuolar membrane. *J. Cell Biol.* **119**: 1459–1468.

Dahms NM, Lokel P, Kornfeld S. (1989) Mannose 6-phosphate receptors and lysosomal enzyme targeting. *J. Biol. Chem.* **264**: 12115–12118.

Denecke J, Botterman J, Deblaere R. (1990) Protein secretion in plant cells can occur via a default pathway. *Plant Cell* **2**: 51–59.

Dombrowski JE, Raikhel NV. (1995) Isolation of a cDNA encoding a novel GTP-binding protein of *Arabidopsis thaliana. Plant Mol. Biol.* **28**: 1121–1126.

Dombrowski JE, Schroeder MR, Bednarek S, Raikhel NV. (1993) Determination of the functional elements within the vacuolar targeting signal of barley lectin. *Plant Cell* **5**: 587–596.

Driouich A, Faye L, Staehelin LA. (1993) The plant Golgi apparatus: a factory for complex polysaccharides and glycoproteins. *Trends Biochem. Sci.* **18**: 210–214.

Elmendorf HG, Haldar K. (1993) Identification and localization of ERD2 in the malaria parasite *Plasmodium falciparum*: separation from sites of sphingomyelin synthesis and implications for organization of the Golgi. *EMBO J.* **12**: 4763–4773.

Fukuda M. (1991) Lysosomal membrane glycoproteins. *J. Biol. Chem.* **266**: 21327–21330.

Gal S, Raikhel NV. (1993) Protein sorting in the endomembrane system of plant cells. *Curr. Opin. Cell Biol.* **5**: 636–640.

Gal S, Willingham MC, Gottesman MM. (1985) Processing and lysosomal localization of a glycoprotein whose secretion is transformation stimulated. *J. Cell Biol.* **100**: 535–544.

Gaudreault PR, Beevers L. (1984) Protein bodies and vacuoles as lysosomes: investigation into the role of mannose-6-phosphate in intracellular transport of glycosidases in pea cotyledons. *Plant Physiol.* **76:** 228–232.

Glickman JN, Kornfeld S. (1993) Mannose 6-phosphate-independent targeting of lysosomal enzymes in I-cell disease β lymphoblasts. *J. Cell Biol.* **123:** 99–108.

Gomez L, Chrispeels MJ. (1993) Tonoplast and soluble vacuolar proteins are targeted by different mechanisms. *Plant Cell* **5:** 1113–1124.

Harter C, Mellman I. (1992) Transport of the lysosomal membrane glycoprotein lgp120 (lgp-A) to lysosomes does not require appearance on the plasma membrane. *J. Cell Biol.* **117:** 311–325.

Hattori T, Nakagawa T, Maeshima M, Nakamura K, Asahi T. (1985) Molecular cloning and nucleotide sequence of cDNA for sporamin, the major soluble protein of sweet potato tuberous roots. *Plant Mol. Biol.* **5:** 313–320.

Hattori T, Ichihara S, Nakamura K. (1987) Processing of a plant vacuolar protein precursor *in vitro. Eur. J. Biochem.* **166:** 533–538.

Herman PK, Stack JH, Emr SD. (1991) A genetic and structural analysis of the yeast Vps15 protein kinase: evidence for a direct role of Vps15p in vacuolar protein delivery. *EMBO J.* **10:** 4049–4060.

Hiles ID, Otsu M, Volinia S *et al.* (1992) Phosphatidylinositol 3-kinase: structure and expression of the 110 kD catalytic subunit. *Cell* **70:** 419–429.

Höfte H, Chrispeels MJ. (1992) Protein sorting to the vacuolar membrane. *Plant Cell* **4:** 995–1004.

Höfte H, Faye L, Dickinson C, Herman EM, Chrispeels MJ. (1991) The protein-body proteins phytohemagglutinin and tonoplast intrinsic protein are targeted to vacuoles in leaves of transgenic tobacco. *Planta* **184:** 431–437.

Holwerda BC, Galvin NJ, Baranski TJ, Rogers JC. (1990) In vitro processing of aleurain, a barley vacuolar thiol protease. *Plant Cell* **2:** 1091–1106.

Holwerda BC, Padgett HS, Rogers JC. (1992) Proaleurain vacuolar targeting is mediated by short contiguous peptide interactions. *Plant Cell* **4:** 307–318.

Hong A, Verma DPS. (1994) A phosphatidylinositol 3-kinase is induced during soybean module organogenesis and is associated with membrane proliferation. *Proc. Natl Acad. Sci. USA* **91:** 9617–9621.

Hunt DC, Chrispeels MJ. (1991) The signal peptide of a vacuolar protein is necessary and sufficient for the efficient secretion of a cytosolic protein. *Plant Physiol.* **96:** 18–25.

Johnson LM, Bankaitis VA, Emr SD. (1987) Distinct sequence determinants direct intracellular sorting and modification of a yeast vacuolar protease. *Cell* **48:** 875–885.

Kane PN, Kuehn NC, Howald-Stevenson J, Stevens TH. (1992) Assembly and targeting of peripheral and integral membrane subunits of the yeast vacuolar H+-ATPase. *J. Biol. Chem.* **267:** 447–454.

Kaplan A, Achord DT, Sly WS. (1977) Phosphohexosyl components of a lysosomal enzyme are recognized by pinocytosis receptors on human fibroblasts. *Proc. Natl Acad. Sci. USA* **74:** 2026–2030.

Karniguian A, Zahraoui A, Tavitian A. (1993) Identification of small GTP-binding rab proteins in human platelets: thrombin-induced phosphorylation of rab3B, rab6, and rab8 proteins. *Proc. Natl Acad. Sci. USA* **90:** 7647–7651.

Kirsch T, Rogers JC, Beevers L. (1993) Clathrin coated vesicles contain a binding protein for the N-terminal vacuolar targeting sequence of barley pro-aleurain. *Plant Physiol.* **102** (Suppl.): 149.

Kirsch T, Paris N, Butler M, Beevers L, Rogers JC. (1995) Purification and initial characterization of a potential plant vacuolar targeting receptor. *Proc. Natl Acad. Sci. USA* **91:** 3403–3407.

Klionsky DJ, Emr SD. (1989) Membrane protein sorting: biosynthesis, transport and processing of yeast vacuolar alkaline phosphatase. *EMBO J.* **8:** 2241–2250.

Klionsky DJ, Banta LM, Emr SD. (1988) Intracellular sorting and processing of a yeast vacuolar hydrolase: proteinase A propeptide contains vacuolar targeting information. *Mol. Cell. Biol.* **8:** 2105–2116.

Klionsky DJ, Herman PK, Emr SD. (1990) The fungal vacuole: composition, function and biogenesis. *Microbiol. Rev.* **54:** 266–292.

Kornfeld S, Mellman I. (1989) The biogenesis of lysosomes. *Annu. Rev. Cell Biol.* **5:** 483–525.

Lee H, Gal S, Newman TC, Raikhel NV. (1993) The *Arabidopsis* endoplasmic reticulum retention receptor functions in yeast. *Proc. Natl Acad. Sci. USA* **90:** 11433–11437.

Letourneur F, Klausner RD. (1992) A novel di-leucine motif and a tyrosine-based motif independently mediate lysosomal targeting and endocytosis of CD3 chains. *Cell* **69:** 1143–1157.

Lewis MJ, Pelham HRB. (1990) A human homolog of the yeast HDEL receptor. *Nature* **348:** 162–163.

Lewis MJ, Sweet DJ, Pelham HRB. (1990) The *ERD2* gene determines the specificity of the luminal ER protein retention system. *Cell* **61:** 1359–1363.

Marcusson EG, Horazdovsky BF, Cereghino JL, Gharakhanian E, Emr SD. (1994) The sorting receptor for yeast vacuolar carboxypeptidase Y is encoded by the VPS10 gene. *Cell* **77:** 579–586.

Matile P. (1975) *The Lytic Compartment of Plant Cells.* Berlin: Springer-Verlag.

Matsuoka K, Nakamura K. (1992) Transport of a sweet potato storage protein, sporamin, to the vacuole in yeast cells. *Plant Cell Physiol.* **33:** 453–462.

Matsuoka K, Bassham DC, Raikhel NV, Nakamura K. (1995) Different sensitivity to wortmannin of 2 vacuolar sorting signals indicates the presence of distinct sorting machineries in tobacco cells. *J. Cell Biol.* **130:** 1307–1318.

Mauch F, Staehelin LA. (1989) Functional implications of the sub-cellular localization of ethylene-induced chitinase and β-1,3-glucanase in bean leaves. *Plant Cell* **1:** 447–457.

Melchers LS, Sela-Buurlage MB, Vloemans SA, Woloshuk CP, Van Roekel JSC, Pen J, van der Elzen PJM, Cornelissen BJC. (1993) Extracellular targeting of the vacuolar tobacco proteins AP24, chitinase and β-1,3-glucanase in transgenic plants. *Plant Mol. Biol.* **21:** 583–593.

Nabi IR, Le Bivic A, Fambrough D, Rodriguez-Boulan E. (1991) An endogenous MDCK lysosomal membrane glycoprotein is targeted basolaterally before delivery to lysosomes. *J. Cell Biol.* **115:** 1573–1585.

Neuhaus JM, Sticher L, Meins F Jr, Boller T. (1991) A short C-terminal sequence is necessary and sufficient for the targeting of chitinases to the plant vacuole. *Proc. Natl Acad. Sci. USA* **88:** 10362–10366.

Neuhaus JM, Pietrzak M, Boller T. (1994) Mutation analysis of the C-terminal vacuolar targeting peptide of tobacco chitinase: low specificity of the sorting system, and gradual transition between intracellular retention and secretion into the extracellular space. *Plant J.* **5:** 45–54.

Novick P, Brennwald P. (1993) Friends and family: the role of the Rab GTPases in vesicular traffic. *Cell* **75:** 597–601.

Oberhauser AF, Monck JR, Balch WE, Fernandez JM. (1992) Exocytotic fusion is activated by Rab3a peptides. *Nature* **360:** 270–273.

Owada M, Neufeld EF. (1982) Is there a mechanism for introducing acid hydrolases into liver lysosomes that is independent of mannose 6-phosphate recognition? Evidence from I-cell disease. *Biochem. Biophys. Res. Commun.* **105:** 814–820.

Pai EF, Kabsch W, Kiengel U, Holmes KC, John J, Wittinghofer A. (1989) Structure of the guanine-nucleotide binding domain of the Ha-ras oncogene product p21 in the triphosphate conformation. *Nature* **341:** 209–214.

Pelham HRB. (1990) The retention signal for soluble proteins of the endoplasmic reticulum. *Trends Biochem. Sci.* **15:** 483–486.

Pultner H, Cox A, Pind S, Khosravi-Far R, Birne J, Schwaninger R, Der C, Balch W. (1991) Rab1b regulates vesicular transport between the endoplasmic reticulum and successive Golgi compartments. *J. Cell Biol.* **115:** 31–43.

Raymond CK, Howald-Stevenson I, Vater CA, Stevens TH. (1992) Morphological classification of the yeast vacuolar protein sorting mutants: evidence for a prevacuolar compartment in Class E *vps* mutants. *Mol. Biol. Cell* **3:** 1389–1402.

Rijnboutt S, Kal AJ, Geuze HJ, Aerts H, Strous GJ. (1991) Mannose 6-phosphate-independent targeting of cathepsin D to lysosomes in HepG2 cells. *J. Biol. Chem.* **266:** 23586–23592.

Roberts CJ, Nothwehr SF, Stevens TH. (1992) Membrane protein sorting in the yeast secretory pathway: evidence that the vacuole may be the default compartment. *J. Cell Biol.* **119:** 69–83.

Rothman JH, Hunter CP, Valls LA, Stevens TH. (1986) Overproduction-induced mislocalization of a yeast vacuolar protein allows isolation of its structural gene. *Proc. Natl Acad. Sci. USA* **83:** 3248–3252.

Saalbach G, Jung R, Kunze G, Saalbach I, Adler K, Muntz K. (1991) Different legumin protein domains act as vacuolar targeting signals. *Plant Cell* **3:** 695–708.

Schroeder MR, Borkhsenious ON, Matsuoka K, Nakamura K, Raikhel NV. (1993) Co-localization of barley lectin and sporamin in vacuoles of transgenic tobacco plants. *Plant Physiol.* **101:** 451–458.

Schu PV, Takegawa K, Fry MJ, Stack JH, Waterfield MD, Emr SD. (1993) Phosphatidylinositol 3-kinase encoded by yeast *VPS34* gene essential for protein sorting. *Science* **260:** 88–91.

Schwaiger H, Hasilik A, von Figura K, Wiemken A, Tanner W. (1982) Carbohydrate-free carboxy-peptidase Y is transferred into the lysosome-like yeast vacuole. *Biochem. Biophys. Res. Commun.* **104:** 950–956.

Semenza JC, Hardwick KG, Dean N, Pelham HRB. (1990) *ERD2,* a yeast gene required for the receptor-mediated retrieval of luminal ER proteins from the secretory pathway. *Cell* **61:** 1349–1357.

Shimada T, Nishimura M, Hara-Nishimura I. (1994) Small GTP-binding proteins are associated with the vesicles that are targeted to vacuoles in developing pumpkin cotyledons. *Plant Cell Physiol.* **35**: 995–1001.

Shinshi H, Neuhaus JM, Ryals J, Meins F Jr. (1990) Structure of a tobacco endochitinase gene: evidence that different chitinase genes can arise by transposition of sequences encoding a cysteine-rich domain. *Plant Mol. Biol.* **14**: 357–368.

Sonnewald U, van Schaewen A, Willmitzer L. (1990) Expression of mutant patatin protein in transgenic tobacco plants: role of glycans and intracellular location. *Plant Cell* **2**: 345–355.

Stack JH, Emr SD. (1994) Vps34p required for yeast vacuolar protein sorting is a multiple specificity kinase that exhibits both protein kinase and phosphatidylinositol-specific PI 3-kinase activities. *J. Biol. Chem.* **269**: 31552–31562.

Stack JH, Herman PK, Schu PV, Emr SD. (1993) A membrane-associated complex containing the Vps15 protein kinase and Vps 34 PI 3-kinase is essential for protein sorting to the yeast lysosome-like vacuole. *EMBO J.* **12**: 2195–2204.

Stevens TH, Rothman JH, Payne GS, Schekman R. (1986) Gene dosage-dependent secretion of yeast vacuolar carboxypeptidase. *J. Cell Biol.* **102**: 1551–1557.

Sticher L, Hinz U, Meyer AD, Meins F Jr. (1992) Intracellular transport and processing of a tobacco vacuolar β-1,3-glucanase. *Planta* **188**: 559–565.

Sze H. (1985) H+-translocating ATPases: advances using membrane vesicles. *Annu. Rev. Plant Physiol.* **36**: 175–208.

Tague BW, Chrispeels MJ. (1987) The plant vacuolar protein, phytohemagglutinin, is transported to the vacuole of transgenic yeast. *J. Cell Biol.* **105**: 1971–1979.

Tague BW, Dickinson CD, Chrispeels MJ. (1990) A short domain of the plant vacuolar protein phytohemagglutinin targets invertase to the yeast vacuole. *Plant Cell* **2**: 533–546.

Tang BL, Wong SH, Qi XL, Low SH, Hong W. (1993) Molecular cloning, characterization, subcellular localization and dynamics of p23, the mammalian KDEL receptor. *J. Cell Biol.* **120**: 325–338.

Terryn N, Von Montagu M, Inze D. (1993) GTP-binding proteins in plants. *Plant Mol. Biol.* **22**: 143–152.

Tisdale E, Bourne JR, Khosravi-Far R, Der CJ, Balch WE. (1992) GTP-binding mutants of Rab1 and Rab2 are potent inhibitors of vesicular transport from the endoplasmic reticulum to the Golgi complex. *J. Cell Biol.* **119**: 749–761.

Tixier-Vidal A, Barret A, Picart R, Mayau V, Vogt D, Wiedenmann B, Goud B. (1993) The small GTP-binding protein, Rab6p, is associated with both Golgi and post-Golgi synaptophysin-containing membranes during synaptogenesis of hypothalamic neurons in culture. *J. Cell Sci.* **105**: 935–947.

Valls LA, Hunter CP, Rothman JH, Stevens TH. (1987) Protein sorting in yeast: the localization determinant of yeast vacuolar carboxypeptidase Y resides in the propeptide. *Cell* **48**: 887–897.

Valls LA, Winther JR, Stevens TH. (1990) Yeast carboxypeptidase Y vacuolar targeting signal is defined by four propeptide amino acids. *J. Cell Biol.* **111**: 361–368.

Van der Sluijs P, Hull M, Zahraoui A, Tavitian A, Goud B, Mellman I. (1991) The small GTP binding protein Rab4 is associated with early endosomes. *Proc. Natl Acad. Sci. USA* **88**: 6313–6317.

Vega MA, Rodriguez F, Segui B, Cales C, Alcalde J, Sandoval IV. (1991) Targeting of lysosomal integral membrane protein LIMPII. The tyrosine-lacking carboxyl cytoplasmic tail of LIMPII is sufficient for direct targeting to the lysosome. *J. Biol. Chem.* **266**: 16269–16272.

Vitale A, Chrispeels MJ. (1992) Sorting of proteins to the vacuoles of plant cells. *BioEssays* **14**: 151–160.

Vitale A, Ceriotti A, Bollini R, Chrispeels MJ. (1984) Biosynthesis and processing of phytohemagglutinin in developing bean cotyledons. *Eur. J. Biochem.* **141**: 97–104.

Vitale A, Ceriotti A, Denecke J. (1993) The role of the endoplasmic reticulum in protein synthesis, modification and intracellular transport. *J. Exp. Bot.* **44**: 1417–1444.

Voelker TA, Herman EM, Chrispeels MJ. (1989) *In vitro* mutated phytohemagglutinin genes expressed in tobacco seeds: role of glycans in protein targeting and stability. *Plant Cell* **1**: 95–104.

von Heijne H. (1986) Towards a comparative anatomy of N-terminal topogenic protein sequences. *J. Mol. Biol.* **189**: 239–242.

von Schaewen A, Chrispeels MJ. (1993) Identification of vacuolar sorting information in phytohemaggutinin, an unprocessed vacuolar protein. *J. Exp. Bot.* **44**: 339–342.

Wagner P, Molenaar CMI, Rauh AJG, Brokel R, Schmitt HD, Gallwitz D. (1987) Biochemical properties of the *ras*-related YPT proteins in yeast: a mutational analysis. *EMBO J.* **6**: 2373–2379.

Waheed A, Pohlmann R, Hasilik A, von Figura K, van Eisen A, Leroy JG. (1982) Deficiency of uridine diphosphate-N-acetylglucosamine, 1-phosphotransferase in organs of I-cell patients. *Biochem. Biophys. Res. Commun.* **105**: 1052–1058.

Ward JM, Sze H. (1992) Subunit composition and organization of the vacuolar H⁺-ATPase from oat roots. *Plant Physiol.* **99:** 170–179.

Welters P, Takegawa K, Emr SD, Chrispeels MJ. (1994) *AtVPS34,* a phosphatidylinositol 3-kinase of *Arabidopsis thaliana,* is an essential protein with homology to a calcium-dependent lipid binding domain. *Proc. Natl Acad. Sci. USA* **91:** 11398–11412.

Wichmann H, Hengst I, Gallwitz D. (1992) Endocytosis in yeast: evidence for the involvement of a small GTP-binding protein (Ypt7p) *Cell* **71:** 1131–1142.

Wieland FT, Gleason ML, Serafina TA, Rothman JE. (1987) The rate of bulk flow from the endoplasmic reticulum to the cell surface. *Cell* **50:** 289–300.

Wilcox CA, Redding K, Wright R, Fuller RS. (1992) Mutation of a tyrosine localization signal in the cytosolic tail of yeast Kex2 protease disrupts Golgi retention and results in default transport to the vacuole. *Mol. Biol. Cell* **3:** 1353–1371.

Wilkins TA, Bednarek SY, Raikhel NV. (1990) Role of propeptide glycan in post-translational processing and transport of barley lectin to vacuoles in transgenic tobacco. *Plant Cell* **2:** 301–313.

Williams MA, Fukuda M. (1990) Accumulation of membrane glycoproteins in lysosomes requires a tyrosine residue at a particular position in the cytoplasmic tail. *J. Cell Biol.* **111:** 955–966.

Wink M. (1993) The plant vacuole: a multifunctional compartment. *J. Exp. Bot.* **44** (Suppl.): 231–246.

Zerial M, Stenmark H. (1993) Rab GTPases in vesicular transport. *Curr. Opin. Cell Biol.* **5:** 613–620.

Chapter 24

Biogenesis of plant peroxisomes

Alison Baker

1. Introduction

The past decade has seen considerable progress in the understanding of peroxisome biogenesis, but, with one exception (Olsen and Harada, 1995), it has been a number of years since the publication of a review which focuses primarily on the biogenesis of plant peroxisomes (Beevers, 1979; Huang *et al.,* 1983; Trelease, 1984). In contrast, a number of more recent general reviews on peroxisomes have appeared (Borst, 1986, 1989; deHoop and Ab, 1992; Lazarow and Fujiki, 1985; Subramani, 1993). It therefore seems timely to review this progress from the plant perspective, and to try to place this knowledge in the context of peroxisome biogenesis in general.

Peroxisomes are organelles that are delimited by a single membrane and lack DNA; consequently, all their proteins are nuclear encoded and imported from the cytosol. The metabolic functions of peroxisomes are reviewed in Tolbert (1981).

In plants, peroxisomes can be subdivided into the following four categories (Huang *et al.,* 1983).

(1) Glyoxysomes (specialized peroxisomes which contain the glyoxylate cycle) are primarily involved in the mobilization of stored fats and their conversion into precursors for gluconeogenesis.
(2) Leaf peroxisomes, in co-operation with mitochondria, salvage phosphoglycolate produced in chloroplasts as a by-product of photorespiration.
(3) Peroxisomes for ureide metabolism are found in the root nodules of those nitrogen-fixing legumes, such as soybean, which export ureides as a source of nitrogen in organic combination. The peroxisomes of the uninfected cells of the nodule are particularly prominent and contain uricase as a major protein (van den Bosch and Newcomb, 1986).
(4) Unspecialized peroxisomes are found in other tissues, but their function is uncertain.

These definitions are based on function, but it appears that different peroxisome types can be interconverted by the synthesis and import of new enzyme activities. Because peroxisomes have no DNA, genetic control resides with the nucleus but may be exercised at various levels; the control of catalase expression, discussed later, is an extreme example. Once synthesized, peroxisomal proteins must be correctly targeted to the organelle, imported into the matrix and assembled, or inserted into the membrane. While the mechanisms which control these processes are beginning to be understood, many important questions concerning peroxisome biogenesis remain unanswered.

2. Ontogeny and development

Germination

Glyoxysomes are commonly found in the cotyledons and/or endosperm of oil-storing plant species, but there is also evidence for their presence in developing and mature seeds in some species (Miernyk et al., 1979). Numerous studies have investigated the behaviour of characteristic glyoxysomal enzymes during germination and post-germinative growth in a variety of oil seeds, including cucumber (Becker et al., 1978; Smith and Leaver, 1986; Weir et al., 1980), cotton (Turley and Trelease, 1990), sunflower (Allen et al., 1988; Schnarrenberger et al., 1971), Brassica (Comai et al., 1989) and castor bean (Martin et al., 1984; Rodriguez et al., 1987). The appearance of enzyme activity, and accumulation of proteins and mRNAs have been documented. The activities of enzymes such as isocitrate lyase (ICL) and malate synthase increase rapidly upon imbibition from a low or undetectable level in dry seeds. Maximum activities are present 3–4 days after imbibition, which correlates with the maximal rate of lipid mobilization, after which the activities rapidly decline in the light as the seedling becomes photosynthetically competent. In the dark, the activities persist for somewhat longer. The appearance of specific proteins and their mRNAs mirrors the appearance of their corresponding enzyme activities. Thus the synthesis of these enzymes is regulated by the availability of the cognate RNA, primarily but not exclusively through control of transcription (Comai et al., 1989). The isolation of genomic clones of ICL and malate synthase (Comai et al., 1992; Graham et al., 1989) and the demonstration that the promoter regions of these clones can direct faithful expression of attached coding regions in transgenic plants (Graham et al., 1990; Zhang et al., 1993) have opened up the way for detailed analysis of the regulation of these genes.

Greening

In the light, cotyledons of epigeous species green and become the first photosynthetic organs of the developing seedling. As this occurs, the activities of the glyoxylate cycle enzymes decrease and become undetectable in fully green cotyledons. Concomitantly, enzymes characteristic of glycolate metabolism (photorespiration), such as glycolate oxidase (GO), hydroxypyruvate reductase (HPR) and serine:glyoxylate amino transferase (SGAT), increase from low or undetectable levels to become major components of leaf peroxisomes. Enzyme activities and the corresponding mRNAs appear around 2–3 days post-imbibition, and their accumulation is strongly light dependent (Hondred et al., 1987; Greenler et al., 1989). HPR gene expression is induced by red light and the induction is reversed by far red light, indicating the involvement of phytochrome, although there is evidence for the involvement of other photoreceptors as well (Bertoni and Becker, 1993). Interestingly, the expression of HPR is dependent on the presence of functional chloroplasts. Similar observations have been made for several nuclear-encoded chloroplast genes (Taylor, 1989). In the case of HPR this may reflect the origin of phosphoglycolate, the substrate for the photorespiratory pathway, in the chloroplast (Schwartz et al., 1992). The mechanism by which chloroplasts affect nuclear gene expression is unknown. The hydroxypyruvate reductase gene has been cloned and sequenced (Schwartz et al., 1991), and an analysis of cis-acting promoter elements has been undertaken in transgenic plants (Sloan et al., 1993). Just over 1 kb of 5' sequence was able to direct faithful expression in transgenic plants, and it appears that critical elements for light inducibility reside between –299 and –218 (defined relative to the start of transcription). Within this region are sequences similar to the 'I box' and 'G box' motifs found in other light-regulated genes (Guiliano et al., 1988; Manzara et al., 1991).

Catalase is present in both glyoxysomes and leaf-type peroxisomes, but its activity is markedly lower in the latter. The control of catalase biosynthesis is complex, with post-transcriptional processes playing an important role. Catalase is a tetramer, and a number of isoenzymes can be produced through the assembly of different subunits in varying proportions (Eising *et al.*, 1990; Ni *et al.*, 1990) The subunits may be the product of different genes, as in cotton (Ni and Trelease, 1991a), or they may be generated by differential proteolytic processing of a single precursor, as in pumpkin (Yamaguchi *et al.*, 1984, 1986), or probably by a combination of the two in sunflower (Eising *et al.*, 1990).

The two cotton catalase genes, S1 and S2, are transcribed to a similar extent throughout post-germinative growth, but the steady-state levels of their respective mRNAs differ. S1 mRNA predominates in the early stages, leading to the formation of isoenzymes which contain mostly S1 subunits and which are more active. Later, the steady-state level of S1 mRNA declines while that of S2 increases, leading to a switch in isoenzyme type (Ni and Trelease, 1991b).

In pumpkin, glyoxysomes contain 55-kDa catalase subunits generated by proteolytic processing of a 59-kDa precursor. Homotetrameric catalase containing the 55-kDa subunits has 10-fold greater specific activity than leaf peroxisomal catalase which contains unprocessed 59-kDa subunits (Yamaguchi *et al.*, 1984, 1986). Note that protein processing is not required for peroxisomal import of catalase because the unprocessed polypeptides are found, in peroxisomes, assembled into a tetrameric holoenzyme.

The glyoxysome–peroxisome transition

The change in function of peroxisomes during the transition from heterotrophy to autotrophy could be due to the destruction of one population of organelles and the synthesis of a new population, or to the conversion of the existing organelle to a new function (Beevers, 1979). It is now clear that the latter is the case. This was first convincingly demonstrated by immunogold double-labelling of ultra-thin sections of cotyledons at the transition stage (where both sets of enzymes can be detected). Both glyoxysome-specific and leaf-peroxisome-specific enzymes were detected within the same organelles (Nishimura *et al.*, 1986; Sautter, 1986; Titus and Becker, 1985).

The observation that different types of plant peroxisomes interconvert raises the following question. Does the organelle simply import the proteins which are being synthesized at that time, or is modification of the import machinery required in order to switch from one type of peroxisome to another? Two research groups expressed ICL in the green leaves of transgenic plants by placing it under the control of a constitutive (Olsen *et al.*, 1993) or light-induced (Onyeocha *et al.*, 1993) promoter. Both groups found that ICL was transported to leaf peroxisomes, and Olsen *et al.* (1993) also showed that ICL was imported by root peroxisomes. Double-labelling immunofluoresence demonstrated that all leaf peroxisomes in the transformed plants contained both ICL and GO (Marrison *et al.*, 1993). Therefore, all peroxisomes were capable of importing a glyoxysomal enzyme which would not normally be expressed in this tissue. It had previously been shown that malate synthase could be imported into leaf peroxisomes *in vitro* (Mori and Nishimura, 1989), and Onyeocha *et al.* (1993) extended this to show that the import *in vitro* of ICL into peroxisomes from green cotyledons was identical in all respects examined to the import of ICL into glyoxysomes (Behari and Baker, 1993). The conclusion drawn from these studies is that the enzymic content and therefore the function of peroxisomes is dictated by the pool of precursors available for import, and that this in turn reflects changes in the pattern of gene expression. Thus the interconversion of peroxisome types appears to be analogous to the interconversion of plastid types during plant development, and the control of gene expression is fundamental to the regulation of organelle biogenesis.

Senescence

Recently, evidence has been obtained for a reverse transition from leaf peroxisome to glyoxysome during the senescence of cotyledons and leaves. Green leaves and cotyledons do not contain detectable levels of glyoxylate cycle enzymes, but in naturally senescing organs or in detached organs maintained in the dark these activities and their corresponding mRNAs reappear, and the enzymes co-migrate upon density gradient centrifugation with accepted peroxisomal marker enzymes. This appears to be a general phenomenon during senescence in both monocotyledons and dicotyledons (De Bellis and Nishimura, 1991; De Bellis *et al.,* 1990; Graham *et al.,* 1992; Gut and Matile, 1988). Glyoxylate cycle enzymes and their mRNAs are also found in senescing petals (De Bellis *et al.,* 1991; Graham *et al.,* 1992). Recently, immunogold double labelling has been used to study this reverse transition. As was observed for the transition from glyoxysome to leaf peroxisome during greening, enzymes characteristic of both types of peroxisome occur within the same organelles during senescence (Nishimura *et al.,* 1993). This reverse transition may allow the salvage of phospholipids from the breakdown of organelles (Gut and Matile, 1988) and/or purine nucleotides from DNA and RNA degradation in senescing tissue (Rodriguez *et al.,* 1990).

The pattern of expression of ICL and malate synthase suggests that these enzymes, like their counterparts in lower eukaryotes (Einerhand *et al.,* 1992), may be under metabolic control. Experiments with anise tissue cultures (Kudielka *et al.,* 1981; Kudielka and Thiemer, 1983) demonstrated that the withdrawal of sucrose from the cultures resulted in the appearance of ICL and malate synthase activities in particulate fractions. Acetate enhanced the appearance of these enzyme activities, but had no effect when administered in the presence of sucrose. Using a cucumber cell culture system, Graham *et al.* (1994) have shown that transcription of the ICL and malate synthase genes is closely correlated with a decrease in intracellular levels of sucrose, glucose and fructose below a threshold value, whereas the addition of these sugars to the culture medium results in repression of transcription. Additionally, they showed that 2-deoxyglucose and mannose (which are taken up and phosphorylated but not further metabolized) cause repression, but 3-methylglucose (which is not phosphorylated) does not do so, suggesting an important role for phosphorylated sugars in sensing the metabolic status of the cells. This 'metabolic control' model could explain the observed tissue-specific and developmental pattern of malate synthase gene expression, as ICL and malate synthase are present in non-photosynthetically active tissues, such as cotyledons, prior to the development of photosynthetic capacity, and in senescent leaves and cotyledons, but not in photosynthetically active tissues such as green leaves, or in non-green tissues which import photosynthate, such as roots (Graham *et al.,* 1992). However, the possibility cannot be excluded that additional 'developmental' signals contribute to the overall pattern of expression.

3. Peroxisomal protein targeting

Protein import into animal and lower eukaryotic peroxisomes has recently been reviewed (de Hoop and Ab, 1992; Subramani, 1993). The peroxisomal proteins are synthesized on free polysomes and imported post-translationally (de Hoop and Ab, 1992; Lazarow and Fujiki, 1985). This is also the case for plant peroxisomes (Huang *et al.,* 1983; Kindl, 1982). Targeting signals have been identified at the carboxyl-terminus of some proteins (PTS-1) and at the amino-terminus of others (PTS-2). Some proteins appear to have internal or multiple peroxisomal targeting signals. General principles are only gradually emerging, and it seems likely that current ideas concerning peroxisomal protein targeting may be naive.

Carboxyl-terminal signals

Mammals and lower eukaryotes. The class-1 peroxisomal targeting signal (PTS-1) consists of the sequence serine-lysine-leucine or a relatively conserved variant of it at the carboxyl-terminus of the protein. Experiments *in vivo* (Gould *et al.,* 1987, 1988, 1989) and *in vitro* (Miura *et al.,* 1992) have shown that in mammalian cells the following consensus

$$-[S/A/C]-[R/H/K]-L-(COOH)$$

is both necessary and sufficient to function as a PTS if present at the extreme carboxyl-terminus of the protein. The addition of extra amino acids (Gould *et al.,* 1989) or amidation of the free α-carboxyl (Miura *et al.,* 1992) abolished the targeting function.

Many peroxisomal proteins end with a related tripeptide (see Table 2 in de Hoop and Ab, 1992). This class of PTS appears to be widespread; a polyclonal antiserum raised against the last nine amino acids of rat peroxisomal acyl CoA oxidase, which ends in SKL, specifically decorates peroxisomes from a number of different species, including castor bean (Keller *et al.,* 1991).

In lower eukaryotes the PTS-1 appears to be more degenerate. AKI, GKI and AQI were all functional in *Candida tropicalis*; only AKI functioned in *Saccharomyces cerevisiae* (Aitchison *et al.,* 1991), but a C-terminal sequence SKI is not a PTS in mammals (Gould *et al.,* 1989). SKL on its own was insufficient to target dihydrofolate reductase (DHFR) to peroxisomes in *S. cerevisiae*, although it was necessary for correct targeting of luciferase (Distel *et al.,* 1992). Similarly, AKI attached to chloramphenicol acetyl transferase failed to target this reporter to peroxisomes in *S. cerevisiae,* although it was necessary for targeting the trifunctional enzyme in both *C. tropicalis* and *S. cerevisiae* (Aitchison *et al.,* 1991). The observation that some sequences are necessary but not sufficient may stem from the need for the signal to be displayed in the right context. Linker insertion mutations in luciferase abolished peroxisomal import even though the carboxyl-terminal SKL was intact (Gould *et al.,* 1987). The requirement for targeting signals to be exposed is well documented in other systems (e.g. mitochondria; Hurt and Schatz, 1987).

Plants. A number of plant peroxisomal proteins end with tripeptides which in other systems have been shown to be functional PTSs, but given the apparent species (or maybe fusion protein)-dependent variability of the PTS-1, experimental verification is required. The finding by Keller *et al.* (1991) that an SKL-specific antibody decorated castor bean endosperm glyoxysomes suggests that plants use this type of signal. Also noteworthy is the observation that luciferase is peroxisomal when expressed in plants, although it was not established that the C-terminal SKL was responsible for peroxisomal targeting (Gould *et al.,* 1990). These observations all suggest that this peroxisomal import pathway may be conserved across the plant and animal kingdoms, but direct experimental evidence has only recently been obtained (Banjoko and Trelease, 1995; Volokita, 1991). Crucial to this has been the development of experimental systems to tackle the problem. Broadly speaking, two approaches have been used: the study of sorting *in vivo* in transgenic plants or transformed suspension cell cultures, and the complementary *in-vitro* approach.

The first clear demonstration of a PTS-1 in plants was by Volokita (1991), who showed that the last six amino acids of GO (RAVARL-COOH) could direct the passenger protein β-glucuronidase to peroxisomes in transgenic tobacco. Because the localization of the fusion protein was established by cell fractionation, it was difficult to assess the efficiency of targeting. Much of the fusion protein appeared to be cytosolic, but it may have leaked from damaged organelles, since peroxisomes are very fragile. Control experiments designed to examine leakage of other peroxisomal enzymes are of limited use as the physical properties of fusion proteins are generally unknown; it is difficult to be sure that they would behave similarly to native proteins. For this reason, immunofluoresence or immunoelectron microscopy are useful analytical techniques, but were not used in this study. In contrast, *in-vitro* import experiments with GO revealed that neither N- nor C-terminal regions of the protein were required for import, but the last 20 amino acids added to a passenger protein conferred the ability to bind to glyoxysomes (Horng *et al.*, 1995). However, in the light of recent results which indicate that oligomeric proteins are capable of import (see Section 5), while unlikely, the possibility cannot be excluded that these mutant proteins were imported in a 'piggy-back' manner together with native GO molecules which may have been present in the import assay.

Two recent reports suggest that the carboxyl-termini of two glyoxylate cycle enzymes, malate synthase and ICL, contain peroxisomal targeting signals (Banjoko and Trelease, 1995; Olsen *et al.*, 1993). A further interesting development is the finding that an SKL-containing peptide can bind specifically and with high affinity to carbonate-washed glyoxysomal membranes of castor bean (Wolins and Donaldson, 1993).

Although it is clear that carboxyl-terminal tripeptides related to SKL are an important component of one type of peroxisomal targeting signal, it is very difficult to see how such a tripeptide on its own can confer sufficient specificity to the targeting process. This problem appears even more acute when all the variations on SKL which have been reported to function in different systems are taken into consideration. There may be some organism-specific differences, but it seems likely that there is some other component to the signal, perhaps context or higher order structure, or perhaps some other element, which has so far eluded us.

Amino-terminal signals

Not all peroxisomal proteins contain a sequence related to SKL at the extreme carboxyl-terminus, and the existing evidence suggests that SKL cannot function at an internal location (Gould *et al.*, 1989; Miura *et al.*, 1992). Peroxisomal thiolases have long been recognized as unusual proteins in that they are synthesized as precursors and have amino-terminal sequences which are removed upon import. Most peroxisomal proteins do not undergo detectable processing. Thiolases also lack a C-terminal SKL-like tripeptide (Arakawa *et al.*, 1987). It has been demonstrated that the first 11 amino acids of rat peroxisomal thiolase can target a passenger protein to peroxisomes (Swinkels *et al.*, 1991). The thiolase signal has been dubbed a PTS-2, and mutational analysis has identified the consensus sequence $R.[I/L].X_5H.L$ as the targeting signal (Glover *et al.*, 1994b; Tsukamoto *et al.*, 1994). A plant peroxisomal thiolase has recently been cloned and sequenced, and is also synthesized as a precursor which contains the PTS-2 consensus sequence (Preisig-Mueller and Kindl, 1993).

Another protein which is synthesized as a precursor with an amino-terminal PTS is glyoxysomal malate dehydrogenase (gMDH) from water-melon (Gietl, 1990). Both the thiolases and gMDH share the PTS-2 consensus (de Hoop and Ab, 1992), and this conserved region is required for the import of gMDH into peroxisomes when expressed in the heterologous host *Hansenula polymorpha* (Gietl *et al.*, 1994).

Among yeast mutants that are defective in peroxisomal protein import and assembly, there are mutants which are defective in the import of SKL-containing proteins but not thiolase, and vice versa. This provides independent genetic evidence for at least two separate import pathways (Erdmann and Kunau, 1992).

Internal signals

Small *et al.* (1988) studied the import of *Candida tropicalis* acyl CoA oxidase in a *Candida in-vitro* import system. Unlike rat liver acyl CoA oxidase, the *Candida* enzyme does not end with

SKL. As previously discussed, *Candida* appears to have slightly different permissible PTS-1s. However, the C-terminus does not seem to play a role in this case. Two separate regions, one from the amino-terminus (amino acids 1–112) and an internal region (amino acids 309–464), were able to target DHFR to peroxisomes, although again, the possibility of 'piggy-back' import (see Section 5) needs to be considered.

Catalase from *Saccharomyces cerevisiae* also appears to have multiple topogenic sequences. The last six amino acids, SSNSKF, will target a reporter protein to peroxisomes *in vivo*, but deletion of the terminal tripeptide SKF still permits truncated catalase to enter peroxisomes. Deletion and fusion experiments pointed to a region between amino acids 126 and 143 as being important for targeting. However, this signal appears to be context dependent, as not all fusion proteins consisting of more than 140 amino acids of catalase were import-competent (Kragler *et al.*, 1993).

Isocitrate lyase

In my own laboratory we have recently developed an improved *in-vitro* import assay for plant peroxisomes and glyoxysomes (Behari and Baker, 1993). Most of our studies to date have focused on the import of isocitrate lyase. Castor bean isocitrate lyase ends with the C-terminal tripeptide ARM, a candidate for a PTS-1. A series of deletions was created, producing polypeptides which had the same amino-terminus as authentic isocitrate lyase, but which lacked varying proportions of the carboxyl-terminus. All of these truncated polypeptides were imported into glyoxysomes with identical characteristics to the full-length protein (Behari and Baker, 1993). On the basis of these experiments it was concluded that targeting information resides within the first 143 amino acids of castor bean ICL, and that the carboxyl-terminal tripeptide is not essential for import into glyoxysomes *in vitro*. It is not clear at this point whether the amino-terminal targeting information in isocitrate lyase is analogous to the thiolase signal. Furthermore, until more refined data are obtained, it is difficult to establish whether the ICL targeting signal actually resides at the amino-terminus, or whether it is really an internal signal. As a C-terminally truncated form of castor bean ICL is imported into leaf peroxisomes *in vivo* (X. Gao *et al.*, unpublished data), and green leaves do not contain ICL, the possibility of 'piggy-back' import can be excluded in this instance.

How can these results be reconciled with the finding of Olsen *et al.* (1993) that the C-terminus of ICL constitutes a PTS? In their hands, a truncated version of *Brassica napus* ICL lacking the last 37 amino acids failed to target to peroxisomes in transgenic plants, whereas the last 37 or even the last five amino acids caused a passenger protein to become associated with peroxisomes (Olsen *et al.*, 1993). Isocitrate lyase may be an example of a protein with multiple topogenic sequences, and perhaps aberrant folding of the *B. napus* mutant ICL masked an otherwise functional PTS. Alternatively, the castor bean enzyme may have a signal that is not present in the *Brassica* ICL.

Membrane proteins

At present our knowledge of peroxisomal membranes, their function and protein composition is incomplete. As a result, there is relatively little information available about the targeting of membrane proteins, although circumstantial evidence suggests that they may be inserted by yet another pathway. Peroxisomal membrane proteins are made on free polysomes. The integral membrane proteins which have been sequenced so far do not end with SKL, consistent with the observation that anti-SKL antibodies do not label membranes. In fibroblasts from some patients suffering from Zellweger syndrome, a generalized defect in the import of peroxisomal matrix proteins, at least one membrane protein is inserted into peroxisomal membrane ghosts (Lazarow *et al.*, 1986; Santos *et al.*, 1988). One integral peroxisomal membrane protein of *Candida boidinii*, PMP 47, has a carboxyl-terminal tripeptide AKE and an internal SKL motif. Mutation of the lysines to alanine did not abolish peroxisomal targeting either as single or double mutations, while transmembrane domains 4 and 5 were found to be necessary for correct localization (McCammon *et al.*, 1994).

In plants, there are few glyoxysomal membrane proteins with a known function. Evidence has been presented that electron transport components are present in glyoxysome membranes, including a cytochrome b_5, NAD(P)H cytochrome c reductase and NADH ferricyanide reductase (FCR) activity (Alani *et al.*, 1990). Cytochrome b_5 and its reductase insert post-translationally into a number of cell membranes (Okada *et al.*, 1982), so the presence of cytochrome b_5 in glyoxysomal membranes is perhaps not surprising. NADH:FCR activity has been partially purified from both endoplasmic reticulum (ER) and glyoxysomal membranes of castor bean. The ER activity appeared to be associated with a 32-kDa polypeptide and the glyoxysomal activity with a 33-kDa polypeptide (Luster *et al.*, 1988). However, since ferricyanide is a non-physiological and relatively non-specific electron acceptor, it is not clear how closely related these two activities really are. It will be interesting to determine the similarity of these proteins with regard to amino acid sequence, and to establish whether they are encoded by different genes. Alkaline lipase is an integral membrane protein of about 62 kDa in castor bean glyoxysomes (Maeshima and Beevers, 1985), but to date the gene has not been cloned.

4. Membrane lipids

The source of lipids for the expansion of the membrane in cotyledon glyoxysomes appears to be degenerating oil bodies (Chapman and Trelease, 1991), rather than the ER as was originally thought (Huang *et al.*, 1983). This conclusion was based on the direct transfer of [^{3}H]triolein and [^{3}H]phosphatidylcholine from lipid bodies to glyoxysomes in an *in-vitro* assay using highly purified organelles, and on the inability to chase [^{14}C]phosphatidylcholine or [^{14}C]-phospatidylethanolamine from ER to glyoxysomes *in vivo* (Chapman and Trelease, 1991). However, oil bodies are themselves formed from the ER during seed maturation (Huang, 1992). Therefore the phospholipids may well be synthesized in the ER but routed to glyoxysomes via oil bodies. The origin of the membrane lipids of other plant peroxisomes does not appear to have been investigated experimentally.

5. Import mechanisms

Energy requirement

Import *in vitro* into both rat liver (Imanaka *et al.*, 1987) and sunflower (Behari and Baker, 1993) peroxisomes has been shown to be ATP-dependent. In the case of rat liver peroxisomes, ATP hydrolysis was required, but no effect of uncouplers was observed. However, there is contradictory evidence concerning the permeability properties of the peroxisomal membrane. Some groups have presented evidence for a non-specific pore (Lemmens *et al.*, 1989), while others have provided evidence for a membrane-associated ATPase in rat liver (Del Valle *et al.*, 1988) and in *Hansenula polymorpha* (Douma *et al.*, 1987), and were able to detect a proton gradient (acidic inside) *in vivo* by ^{31}P nuclear magnetic resonance (NMR) (Nicolay *et al.*, 1987). Recent elegant genetic studies have demonstrated that *in vivo* the peroxisomal membrane of *S. cerevisiae* is impermeable to NAD(H) and acetyl CoA (van Roermund *et al.*, 1995). As *in-vivo* studies tend to support the notion of peroxisomes as closed compartments, failure to detect an effect of uncouplers *in vitro* may reflect leaky organelles, or the presence of a pore which might be gated *in vivo* but open *in vitro*. It is possible that, like the bacterial Sec A/Y/E secretion system, ATP alone can drive protein translocation, but that a proton-motive force increases efficiency (Wickner *et al.,* 1991). Alternatively, protein import may not be dependent on a membrane potential.

Recently, permeabilized mammalian cells have been established as a system to study peroxisomal import. In such systems, import is also temperature- and ATP-dependent, and requires cytosol. GTP could not substitute for ATP, and uncouplers apparently have no effect, but as import is determined by immunofluoresence, quantitative measurements are difficult to make (Rapp *et al.,* 1993; Wendland and Subramani, 1993). Evidence has been presented that even in osmotically shocked yeast spheroplasts, the interior of visibly intact peroxisomes is no longer acidic (Waterham *et al.,* 1990). Thus treatments which leave peroxisomes intact by most accepted criteria may still collapse the ΔpH.

It is not known whether plant peroxisomal membranes contain either pores or ATPases analogous to those found in peroxisomes from other species. Many plant peroxisomal enzymes show latency, which is generally assumed to indicate that the substrates are not freely permeable across the membrane. This view of latency has recently been challenged. Heupel *et al.* (1991) demonstrated that osmotically shocked leaf peroxisomes, which were shown by electron microscopy to lack an intact boundary membrane but to have retained a core structure of matrix material, still had latent enzyme activity which could be unmasked by detergent treatment. Their interpretation was that many of the matrix enzymes are highly organized into a proteinaceous core which facilitates metabolite channelling and reduces the release of noxious pathway intermediates. Disaggregation of the core by detergent treatment gives the appearance of latency regardless of the intactness of the boundary membrane. These observations may partially explain the paradoxical properties of isolated peroxisomes.

At present it is only possible to speculate as to how ATP is used in peroxisomal import and whether it is required inside or outside the organelle. PMP 70, a peroxisomal membrane protein which shows marked sequence similarity to the ABC cassette transport proteins (Kamijo *et al.,* 1990, 1992), could be an import ATPase, although there is no evidence for this. Indeed, PMP70 has only been found in mammalian peroxisomal membranes. *PAS 1,* a yeast gene essential for peroxisome assembly, is also predicted on the basis of its sequence to encode a matrix-located ATPase, but its function in peroxisome assembly is unknown (Erdmann *et al.,* 1991). Alternatively, the ATP requirement could reflect a need for chaperones, as in other systems. It seems likely that there are multiple ATP requirements.

Protein conformation during import

In most systems, proteins need to be in a loose conformation in order to cross a biological membrane (Verner and Schatz, 1988). Recent experimental evidence suggests that this is unlikely to be the case for peroxisomes. Two research groups have independently demonstrated that an epitope-tagged subunit of a peroxisomal protein lacking its own targeting signal can be imported *in vivo* if, and only if, it is co-expressed in the same cell with a wild-type subunit (Glover *et al.,* 1994a; McNew and Goodman, 1994). The interpretation of these results is that hetero-oligomers of the wild-type and mutant subunits can form in the cytosol, and are substrates for the peroxisome import machinery. While these results do not prove that it is the oligomeric form which actually traverses the membrane, a recent demonstration that peroxisomes are capable of importing gold particles up to 9 nm in size, if these are derivatized with a PTS, shows that the import machinery can accomodate large, bulky molecules (Walton *et al.,* 1995). How this can occur while denying free permeability to small metabolites remains a mystery.

The ability of peroxisomes to import mixed oligomers of mutant and wild-type proteins has important consequences for the design and interpretation of experiments to delimit targeting signals, since mutant proteins which have lost their targeting signals but not their ability to oligomerize will score as import-competent unless expressed in a background devoid of the wild-type subunit.

Protein assembly

Many peroxisomal proteins have co-factors or prosthetic groups and many are oligomeric. Our knowledge of the assembly processes inside peroxisomes is almost non-existent. Rat liver catalase is synthesized as an apomonomer, transported into peroxisomes and acquires a haem group before assembling into tetrameric enzymically active catalase (Lazarow and de Duve, 1973), but the mechanistic details remain unknown. Catalase is capable of assembling into an active holoenzyme in the cytosol in the absence of functional peroxisomes. Alcohol oxidase from *Hansenula polymorpha* is imported into peroxisomes when expressed in *Saccharomyces cerevisiae,* but fails to oligomerize and is not enzymically active, implying that some post-import modification fails to take place (Distel *et al.,* 1987).

Receptors

The best characterized components of the peroxisomal import machinery are the receptors for PTS-1 and PTS-2 proteins. Circumstantial evidence for the existence of a receptor for the SKL pathway came from observations that SKL-containing peptides can compete with the import of proteins which use this pathway both in *in-vitro* experiments (Miura *et al.,* 1992) and in a permeabilized cell system (Wendland and Subramani, 1993). The *Pichia pastoris PAS8* gene product which is essential for the import of peroxisomal proteins with a PTS-1 was shown to be an SKL-binding protein (McCollum *et al.,* 1993). Homologues of this protein are now known from several other yeast species and mammals (Brocard *et al.,* 1994; Dodt *et al.,* 1995; Fransen *et al.,* 1995; van der Klei *et al.,* 1995; van der Leij *et al.,* 1993). They all contain several tetratricopeptide (or TPR) repeats which are essential for binding the PTS-1 (Brocard *et al.,* 1994; Dodt *et al.,* 1995; Terlecky *et al.,* 1995). Surprisingly, there is disagreement between the different laboratories concerning the cellular location of the PTS-1 receptor. The *Pichia pastoris* Pas8p is reported to be predominantly associated with the peroxisome membrane (McCollum *et al,* 1993; Terlecky *et al.,* 1995), as is the human protein studied by Fransen *et al.* (1995). However, Dodt *et al.* (1995) isolated the same protein and found it to be predominantly cytosolic. A proportion of the *Hansenula* homologue is found inside peroxisomes (van der Klei *et al.,* 1995). How can these apparently contradictory results be reconciled? There is at present no direct evidence, but a model has been proposed in which the receptor is envisaged as shuttling between the cytosol, where it binds PTS-1-containing proteins, and the peroxisome membrane (or even the matrix, if the receptor can be co-imported with its cargo) (Dodt *et al.,* 1995). Thus the observed differences in distribution might arise from differences in growth conditions or metabolic status, as well as from differences in experimental techniques.

A similar situation pertains with regard to the receptor for PTS-2. To date this has only been identified in *S. cerevisiae,* although many other organisms, including plants, are known to use a PTS-2 import pathway (see Section 3). In *S. cerevisiae* the PTS-2 receptor is encoded by the *PAS7/PEB1* gene. Disruption of this gene leads to an inability to import only PTS-2-containing proteins (Marzioch *et al.,* 1994; Zhang and Lazarow, 1995). Marzioch *et al.* (1994) found that the Pas 7 protein is predominantly cytosolic, with a small amount being associated with peroxisomes, whereas Zhang and Lazarow (1995) found a substantial amount of the protein inside peroxisomes. Again a cycling model for the function of Pas7p has been proposed (Marzioch *et al.,* 1994).

6. Unanswered questions

Import mechanisms

The recent findings concerning the import of folded proteins have required a revaluation of ideas concerning peroxisomal protein import. Prior to this, most models were based on mitochondrial or ER protein import. It now appears that the peroxisome import machinery may be significantly different, but as yet there is no alternative model which has gained general acceptance amongst workers in the field. Many of the genes involved in peroxisome biogenesis have been cloned, largely as a result of yeast genetics, but, with the exception of the PTS-1 and PTS-2 receptors, their function has not been clearly established. Even in the case of the receptors, the important issue of cellular localization has yet to be resolved.

Evolutionary origins

The evolutionary origin of peroxisomes has long been debated (Borst, 1986, 1989; de Duve, 1969; de Hoop and Ab, 1992; Stabenau, 1991). Is the peroxisome a remnant of a former endosymbiont or is its origin similar to that of other compartments of the eukaryotic endomembrane system? Peroxisomes have a single membrane and lack DNA, and thus at first sight they are unlike chloroplasts and mitochondria, which are widely accepted to have an endosymbiotic origin. However, peroxisomes do not appear to be like other compartments of the endomembrane system which are maintained by a process of membrane flow and membrane fusion. With the exception of the ER (and one or two specialized cases of proteins being directly transported across the eukaryotic plasma membrane or into lysosomes), none of the other single membranes of the eukaryotic cell are capble of translocating proteins, whereas peroxisomes clearly acquire most if not all their proteins in this way. All the peroxisomal enzymes studied so far are synthesized on free polysomes, have targeting signals which are demonstrably different to those which direct proteins into the endomembrane system, and can be imported into peroxisomes post-translationally.

An endosymbiotic origin would explain the observation that core metabolic pathways (dealing with the β-oxidation of fatty acids and the metabolism of hydrogen peroxide and glyoxylate) are present in most peroxisomes, but secondary losses and acquisitions of enzymes would have to be postulated in order to account for the many differences between species. The lack of DNA means that phylogenies based on nucleic acid sequences cannot be constructed. Comparisons of proteins found in peroxisomes with those that catalyse similar or identical reactions in other eukaryotic compartments and in prokaryotes might shed some light on peroxisomal origins. However, even if the ancestral peroxisome was an endosymbiont, the secondary loss and acquisition of proteins during peroxisome evolution leads to difficulties in picking the 'right' proteins for comparison. The peroxisomal β-oxidation system is probably a good choice because all peroxisomes have this pathway (Borst, 1989) and it is thus likely to be an ancestral trait. Peroxisomal β-oxidation enzymes are more similar to prokaryotic enzymes than to mitochondrial ones in that peroxisomes and prokaryotes have multifunctional polypeptides, whereas mitochondria have separate polypeptides for each activity (Kunau *et al.,* 1988). However, this can hardly be regarded as definitive evidence.

Growth and division of peroxisomes

Originally it was thought that peroxisomes were derived from the endoplasmic reticulum by budding. This concept has now lost favour with workers on mammalian and yeast peroxisomes, as more information has been obtained and better techniques have been developed (for review see Lazarow and Fujiki, 1985), and at the very least it has had to be radically revised by workers on plant peroxisomes (see the extensive discussions in Huang *et al.*, 1983; Kindl, 1982; Trelease, 1984).

In animals and yeasts there is clear evidence that peroxisomes grow and divide, and can therefore be regarded as self-perpetuating organelles, probably independent of other membranes for their formation. In glucose-grown *Candida boidinii* cells, one or two very tiny peroxisomes can be detected which subsequently divide and enlarge upon transfer of the cells to methanol. Under these circumstances the induction of integral membrane proteins precedes the accumulation of matrix enzymes (Veenhuis and Goodman, 1990). In contrast, the peroxisomes of methanol-grown *Hansenula polymorpha* enlarge dramatically before budding off small peroxisomes which migrate into the developing bud as the mother cell divides (Veenhuis *et al.*, 1978). Treatment of rats with hypolipodaemic drugs, or with partial hepatectomy, induces peroxisome proliferation in the liver. Electron micrographs of dividing peroxisomes have been published, as have micrographs of structures which may represent the site of incorporation of newly synthesized membrane proteins (Fahimi *et al.*, 1993).

It has been suggested that newly synthesized proteins are incorporated preferentially into small peroxisomes in rat liver (Heinemann and Just, 1992; Luers *et al.*, 1993), and there are several reports of plant peroxisomes increasing in buoyant density during development, which suggests enlargement of smaller organelles (Choinski and Trelease, 1978; Feierabend and Beevers, 1972; Kudielka *et al.*, 1981). There is also evidence from studies of *Hansenula* that the peroxisome population is heterogeneous and that 'mature' peroxisomes may not take up newly synthesized enzymes (Veenhuis *et al.*, 1989). This is in apparent contrast to the situation in cotyledons of higher plants, where no evidence for heterogeneity was observed during the glyoxysome–peroxisome–glyoxysome transitions (Nishimura *et al.*, 1993; Titus and Becker, 1985), or when isocitrate lyase was synthesized in green leaves (Marrison *et al.*, 1993).

Numerous ultrastructural studies have addressed the question of peroxisome proliferation in plant cells. In cotton seeds and germinating seedlings, a careful morphometric study failed to find evidence for peroxisome division, but demonstrated a dramatic enlargement of existing organelles (Kunce *et al.*, 1984). Similar conclusions were drawn from developmental studies on cucumber cotyledons (Trelease *et al.*, 1971), greening bean leaves (Gruber *et al.*, 1973) and

---

Figure 1. Models of peroxisome proliferation. (a) Upon transfer of *Candida boidinii* from glucose to methanol, several peroxisomes develop from a pre-existing organelle. These peroxisomes undergo further division, forming one or two clusters which subsequently enlarge. Not all peroxisomes proliferate, as is indicated by the shaded circle. Peroxisomes are partitioned to daughter cells, resulting in a decrease in the number of peroxisomes per cell until a new steady state is reached. Based on the data and model of Veenhuis and Goodman (1990). (b) Glucose-grown *Hansenula polymorpha* has small peroxisomes. In the early log phase there is no crystal of alcohol oxidase, but crystalline inclusions are apparent in peroxisomes of stationary-phase cells (indicated by shading in the figure). Upon transfer from glucose to methanol, very large peroxisomes filled with crystalline alcohol oxidase are formed. These produce small crystal-free peroxisomes by fission, the smaller peroxisomes migrating into the bud of the developing daughter cell. Based on the data of Veenhuis *et al.*, (1978). (c) In rat liver regenerating after partial hepatectomy, some peroxisomes appear to grow membranous loops which expand, bud off and form new peroxisomes. Based on the data of Fahimi (1993). (d) In higher plant cotyledons, glyoxysomes were not observed to divide during germination and early post-germinative growth, but they enlarged dramatically (Kunce *et al.*, 1984). Glyoxysomes are converted to peroxisomes upon greening (Nishimura *et al.*, 1986; Sautter *et al.*, 1986; Titus and Becker, 1985) and revert back during senescence (Nishimura *et al.*, 1993). (e) Exposure of *Lemna minor* to high light intensity results in rapid growth of the organism, and the peroxisomes appear to divide (Ferreira *et al.*, 1989). The dotted lines joining (d) and (e) represent the speculation that peroxisome division in plants may only occur in actively dividing cells.

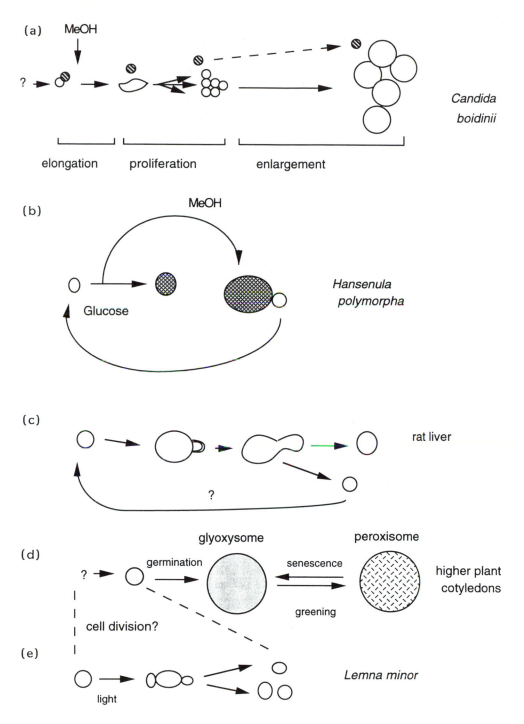

soybean root nodules (Newcomb *et al.,* 1985). Dividing peroxisomes were not observed in these studies, and the conclusion that proliferation was due to enlargement rather to an than increase in numbers was based on counts of peroxisome profiles in observed sections, corrected to account for the increased probability of sectioning a larger organelle. In striking contrast to these results is the effect of light on peroxisomes of *Lemna minor* (Ferreira *et al.,* 1989). Increased irradiance led to increased activity of catalase and GO, and the *Lemna* peroxisomes had highly irregular shapes and looked very much as if they were dividing or fusing, as occurs with fungal and animal peroxisomes. These various models of peroxisome proliferation are summarized in *Figure 1.*

What is to be made of the apparent lack of peroxisome division in many plant tissues? I would like to speculate that the answer lies in the relationship with cell division. In the plant examples referred to earlier, most (and in oilseed cotyledons all) of the growth is due to cell expansion, whereas the *Lemna* fronds were cultivated under conditions that gave a doubling time of less than 2 days at the higher irradiances. Perhaps there is no need for peroxisome division in plant cells which are not themselves dividing. The enlargement of existing organelles or the turnover of their contents may be sufficient to supply the needs of cells which are expanding and differentiating but no longer dividing.

If peroxisome division never occurs, even in meristematic cells, this would represent a fundamental difference in the biology of plant peroxisomes. One of the dogmas of cell biology is that membranes only form from pre-existing membranes. In some mutant yeast and mammalian cells which lack morphologically identifiable peroxisomes, careful investigation has revealed vestigial membrane structures which could act as a target for incorporation of the missing proteins upon complementation of the genetic defect (Kunau *et al.,* 1993; Santos *et al.,* 1988). Although the origin of these structures is unknown, it would seem likely that they are (like mitochondria and plastids) present in all cells throughout their life and that they are elaborated in response to the appropriate cellular signals, although it is conceiveable that they could be formed from some other membrane, such as the ER. In plants, lack of mutants and specific probes for membrane proteins has prevented definitive experiments in this area. Since the membrane must always be import-competent even at the earliest stages, identification of the components of the peroxisomal import machinery and study of their biogenesis might provide the answer. This in turn may shed some light on the evolutionary origin of peroxisomes.

While it is clear that plant peroxisomes have the capacity to enlarge dramatically, comparatively little is known about the chain of events which triggers proliferation, or what determines the final size of the organelle. There is evidence that peroxisomal enlargement is driven by the accumulation of matrix proteins, since the down-regulation of uricase, the main protein of nodule peroxisomes, led to a proportionate decrease in peroxisome size (Lee *et al.,* 1993). Therefore, unravelling the signal transduction pathways that regulate the expression of peroxisomal proteins may be an important step in understanding the proliferation of the organelles.

Acknowledgements

I would like to thank the following members of the plant peroxisome research community for sending me their preprints and reprints; W. Becker, L. de Bellis, L. del Rio, C. Gietl, J. Harada, H. Kindl, H. Stabenau, R. Trelease and D. Verma. I am also greatly indebted to the members of my laboratory and to Dr C.J. Baker and Dr I.A. Graham for critical reading of the manuscript.

References

Aitchison JD, Murray WW, Rachubinski RA. (1991) The carboxyl-terminal tripeptide ala-lys-ile is essential for targeting *Candida tropicalis* trifunctional enzyme to yeast peroxisomes. *J. Biol. Chem.* **266:** 23197–23203.

Alani AA, Luster DG, Donaldson RP. (1990) Development of endoplasmic reticulum and glyoxysomal membrane redox activities during castor bean germination. *Plant Physiol.* **94:** 1842–1848.

Allen RD, Trelease RN, Thomas TL. (1988) Regulation of isocitrate lyase gene expression in sunflower. *Plant Physiol.* **86:** 527–532.

Arakawa H, Takiguchi M, Amaya Y, Nagata S, Hayashi H, Mori M. (1987) cDNA derived amino acid sequence of rat mitochondrial 3-oxoacyl-CoA thiolase with no transient presequence: structural relationship with peroxisomal enzyme. *EMBO J.* **6:** 1361–1366.

Banjoko A, Trelease RN. (1995) Development and application of an *in vivo* plant peroxisome import system. *Plant Physiol.* **107:** 1201–1208.

Becker WM, Leaver CJ, Weir EM, Riezman H. (1978) Regulation of glyoxysomal enzymes during germination of cucumber. I. Developmental changes in cotyledonary protein, RNA and enzyme activities during germination. *Plant Physiol.* **62:** 542–549.

Beevers H. (1979) Microbodies in higher plants. *Annu. Rev. Plant Physiol.* **30:** 159–193.

Behari R, Baker A. (1993) The carboxyl terminus of isocitrate lyase is not essential for import into glyoxysomes in an *in vitro* system. *J. Biol. Chem.* **268:** 7315–7322.

Bertoni GP, Becker WM. (1993) Effects of light fluence and wavelength on expression of the gene encoding cucumber hydroxypyruvate reductase. *Plant Physiol.* **103:** 933–941.

Borst P. (1986) How proteins get into microbodies (peroxisomes, glyoxysomes, glycosomes). *Biochem. Biophys. Acta.* **866:** 179–203.

Borst P. (1989) Peroxisome biogenesis revisited. *Biochem. Biophys. Acta.* **1008:** 1–13.

Brocard C, Kragler F, Simon M, Schuster T, Hartig A. (1994) The tetratricopeptide repeat domain of the PAS 10 protein of *Saccharomyces cerevisiae* is essential for binding the peroxisomal targeting signal SKL. *Biochem. Biophys. Res. Commun.* **204:** 1016–1022.

Chapman KD, Trelease RN. (1991) Accquisition of membrane lipids by differentiating glyoxysomes: role of lipid bodies. *J. Cell Biol.* **115:** 995–1007.

Choinski JS, Trelease RN. (1978) Control of enzyme activities in cotton cotyledons during maturation and germination. II. Glyoxysomal enzyme development in embryos. *Plant Physiol.* **62:** 141–145.

Comai L, Dietrich RA, Maslyar DJ, Baden S, Harada JJ. (1989) Coordinate expression of transcriptionally regulated isocitrate lyase and malate synthase genes in *Brassica napus* L. *Plant Cell* **1:** 293–300.

Comai L, Matsudaira KL, Heupel RC, Dietrich RA, Harada JJ. (1992) Expression of a *Brassica napus* malate synthase gene in transgenic tomato plants during the transition from late embryogeny to germination. *Plant Physiol.* **98:** 53–61.

De Bellis L, Nishimura M. (1991) Development of enzymes of the glyoxylate cycle during senescence of pumpkin cotyledons. *Plant Cell Physiol.* **32:** 555–561.

De Bellis L, Picciarelli P, Pistelli L, Alpi A. (1990) Localisation of glyoxylate cycle marker enzymes in peroxisomes of senescent leaves and green cotyledons. *Planta* **180:** 435–439.

De Bellis L, Tsugeki R, Nishimura M. (1991) Glyoxylate cycle enzymes in peroxisomes isolated from petals of pumpkin (*Cucubita* sp.) during senescence. *Plant Cell Physiol.* **32:** 1227–1235.

De Duve C. (1969) Evolution of the peroxisome. *Ann. NY Acad. Sci.* **168:** 369–381.

De Hoop MJ, Ab G. (1992) Import of proteins into peroxisomes and other microbodies. *Biochem. J.* **286:** 657–669.

Del Valle R, Soto U, Necochea C, Leighton F. (1988) Detection of an ATPase in rat liver peroxisomes. *Biochem. Biophys. Res. Commun.* **156:** 1353–1359.

Distel B, Veenhuis M, Tabak HF. (1987) Import of alcohol oxidase into peroxisomes of *Saccharomyces cerevisiae*. *EMBO J.* **6:** 3111–3116.

Distel B, Gould SJ, Voorn-Brouwer T, van der Berg M, Tabak HF, Subramani S. (1992) The carboxyl-terminal tripeptide serine-lysine-leucine of firefly luciferase is necessary but not sufficient for peroxisomal import in yeast. *New Biol.* **4:** 157–165.

Dodt G, Braverman N, Wong C, Moser A, Moser HW, Watkins P, Valle D, Gould SJ. (1995) Mutations in the PTS-1 receptor gene PXR-1 define complementation group 2 of the peroxisome biogenesis disorders. *Nature Genet.* **9:** 115–125.

Douma AC, Veenhuis M, Sulter GJ, Harder W. (1987) A proton translocating ATPase is associated with the peroxisomal membrane of yeasts. *Arch. Microbiol.* **147:** 42–47.

Einerhand AWC, van der Leij I, Kos WT, Distel B, Tabak HF. (1992) Transcriptional regulation of genes encoding proteins involved in biogenesis of peroxisomes in *S. cerevisiae. Cell Biochem. Funct.* **10:** 185–191.

Eising R, Trelease RN, Ni W. (1990) Biogenesis of catalase in glyoxysomes and leaf-type peroxisomes of sunflower cotyledons. *Arch. Biochem. Biophys.* **278:** 258–264.

Erdmann R, Kunau WH. (1992) Genetic approach to the biogenesis of peroxisomes in the yeast *Saccharomyces cerevisiae. Cell Biochem. Funct.* **10:** 167–174.

Erdmann R, Wiebel FF, Flessau A, Rytka J, Beyer A, Froehlich KU, Kunau WH. (1991) PAS 1: a yeast gene required for peroxisome biogenesis encodes a member of a novel family of putative ATPases. *Cell* **64:** 499–510.

Fahimi HD, Baumgart E, Voekl A. (1993) Ultrastructural aspects of the biogenesis of peroxisomes in rat liver. *Biochemie* **75:** 201–209.

Feierabend J, Beevers H. (1972) Developmental studies on microbodies in wheat leaves. II. Ontogeny of particulate enzyme associations. *Plant Physiol.* **49:** 33–39.

Ferreira RMB, Bird B, Davies DD. (1989) The effect of light on the structure and organisation of *Lemna* peroxisomes. *J. Exp. Bot.* **40:** 1029–1035.

Fransen M, Brees C, Baumgart E, Vanhooren CT, Baes M, Mannaerts GP, van Veldhoven PP. (1995) Identification and characterisation of the putative human peroxisomal c terminal targeting signal import receptor. *J. Biol. Chem.* **270:** 7731–7736.

Gietl C. (1990) Glyoxysomal malate dehydrogenase from watermelon is synthesised with an amino terminal transit peptide. *Proc. Natl Acad. Sci. USA* **87:** 5773–5777.

Gietl C, Faber KN, van der Klei IJ, Veenhuis M. (1994) Mutational analysis of the N-terminal topogenic signal of watermelon glyoxysomal malate dehydrogenase using the heterologous host *Hansenula polymorpha. Proc. Natl Acad. Sci. USA* **91:** 3151–3155.

Glover JR, Andrews DW, Rachubinski RA. (1994a) *S. cerevisiae* peroxisomal thiolase is imported as a dimer. *Proc. Natl Acad. Sci. USA* **91:** 10541–10545.

Glover JR, Andrews DW, Subramani S, Rachubinski RA. (1994b) Mutagenesis of the amino terminal targeting signal of *Saccharomyces cerevisiae* 3-ketoacyl-CoA thiolase reveals conserved amino acids required for import into peroxisomes *in vivo. J. Biol. Chem.* **269:** 7558–7563.

Gould SJ, Keller GA, Subramani S. (1987) Identification of a peroxisomal targeting signal at the carboxy terminus of firefly luciferase. *J. Cell Biol.* **105:** 2923–2931.

Gould SJ, Keller GA, Subramani S. (1988) Identification of peroxisomal targeting signals located at the carboxy terminus of four peroxisomal proteins. *J. Cell Biol.* **107:** 897–905.

Gould SJ, Keller GA, Hosken N, Wilkinson J, Subramani S. (1989) A conserved tripeptide sorts proteins to peroxisomes. *J. Cell Biol.* **108:** 1657–1664.

Gould SJ, Keller GA, Schneider M, Howell SH, Garrard LJ, Goodman J, Distel B, Tabak H, Subramani S. (1990) Peroxisomal protein import is conserved between yeast, plants, insects and mammals. *EMBO J.* **9:** 85–90.

Graham IA, Smith LA, Brown JWS, Leaver CJ, Smith SM. (1989) The malate synthase gene of cucumber. *Plant Mol. Biol.* **13:** 673–684.

Graham IA, Smith LM, Leaver CJ, Smith SM. (1990) Developmental regulation of expression of the malate synthase gene in transgenic plants. *Plant Mol. Biol.* **15:** 539–549.

Graham IA, Leaver CJ, Smith SM. (1992) Induction of malate synthase gene expression in senescent and detached organs of cucumber. *Plant Cell* **4:** 349–357.

Graham IA, Denby KJ, Leaver CJ. (1994) Carbon catabolite repression regulates glyoxylate cycle gene expression in cucumber. *Plant Cell* **6:** 761–772.

Greenler JMcC, Sloan JS, Schwartz BW, Becker WM. (1989) Isolation, characterisation and sequence analysis of a full-length cDNA clone encoding NADH-dependent hydroxypyruvate reductase from cucumber. *Plant Mol. Biol.* **13:** 139–150.

Gruber PJ, Becker WM, Newcom EH. (1973) The development of microbodies and peroxisomal enzymes in greening bean leaves. *J. Cell Biol.* **56:** 500–518.

Guiliano G, Pichersky E, Malik VS, Timko MP, Scolnik PA, Cashmore AR. (1988) An evolutionarily conserved protein binding sequence upstream of a plant light regulated gene. *Proc. Natl Acad. Sci. USA* **85:** 7089–7093.

Gut H, Matile P. (1988) Apparent induction of key enzymes of the glyoxylic acid cycle in senescent barley leaves. *Planta* **176:** 548–550.

Heinemann P, Just WW. (1992) Peroxisomal protein import; *in vivo* evidence for a novel protein translocation competent compartment. *FEBS Lett.* **300:** 179–182.

Heupel R, Markgraf T, Robinson D, Heldt HW. (1991) Compartmentation studies on spinach leaf peroxisomes; evidence for channeling of photorespiratory metabolites in peroxisomes devoid of intact boundary membrane. *Plant Physiol.* **96:** 971–979.

Hondred D, Wadle DM, Titus DE, Becker WM. (1987) Light-stimulated accumulation of the peroxisomal enzymes hydroxypyruvate reductase and serine:glyoxylate aminotransferase and their translatable mRNAs in cotyledons of cucumber seedlings. *Plant Mol. Biol.* **9:** 259–275.

Horng J-T, Behari R, Burke LECA, Baker A. (1995) Investigation of the energy requirement and targeting signal for the import of glycolate oxidase into glyoxysomes. *Eur. J. Biochem.* **230:** 157–163.

Huang AC, Trelease RN, Moore TS. (ed.) (1983) *Plant Peroxisomes.* New York: Academic Press.

Huang AHC. (1992) Oil bodies and oleosins in seeds. *Annu. Rev. Plant Physiol.* **43:** 177–200.

Hurt E, Schatz G. (1987) A cytosolic protein contains a cryptic mitochondrial targeting signal. *Nature* **325:** 499–503.

Imanaka T, Small GM, Lazarow PB. (1987) Translocation of acyl-CoA oxidase into peroxisomes requires ATP hydrolysis but not a membrane potential. *J. Cell Biol.* **105:** 2915–2922.

Kamijo K, Taketani S, Yokota S, Osumi T, Hashimoto T. (1990) The 70-kDa peroxisomal membrane protein is a member of the Mdr (P-glycoprotein)-related ATP-binding superfamily. *J. Biol. Chem.* **265:** 4534–4540.

Kamijo K, Kamijo T, Ueno I, Osumi T, Hashimoto T. (1992) Nucleotide sequence of the human 70 kDa peroxisomal membrane protein: a member of the ATP-binding cassette transporters. *Biochem. Biophys. Acta* **1129:** 323–327.

Keller G-A, Krisans S, Gould SJ, Sommer JM, Wang CC, Schliebs W, Kunau W, Brody S, Subramani S. (1991) Evolutionary conservation of a microbody targeting signal that targets proteins to peroxisomes, glyoxysomes and glycosomes. *J. Cell Biol.* **114:** 893–904.

Kindl H. (1982) The biosynthesis of microbodies (peroxisomes, glyoxysomes). *Int. Rev. Cytol.* **80:** 193–229.

Kragler F, Langeder A, Raupachova J, Binder M, Hartig A. (1993) Two independent peroxisomal targeting signals in catalase A of *Saccharomyces cerevisiae*. *J. Cell Biol.* **120:** 665–673.

Kudielka RA, Kock H, Theimer RR. (1981) Substrate dependent formation of glyoxysomes in cell suspension cultures of anise (*Pimpinella anisum* L.). *FEBS Lett.* **136:** 8–12.

Kudielka RA, Thiemer RR. (1983) Derepression of glyoxylate cycle enzyme activities in anise suspension culture cells. *Plant Sci. Lett.* **31:** 237–244.

Kunau WH, Buehne S, de la Garza M, Kionka C, Mateblowski M, Schultz-Borchard U, Thieringer R. (1988) Comparative enzymology of beta oxidation. *Biochem. Soc. Trans.* **16:** 418–420.

Kunau WH, Beyer A, Franken T, Gotte K, Marzioch M, Saidowsky J, Skaletz-Rorowski A, Wiebel FF. (1993) Two complementary approaches to study peroxisome biogenesis in *S. cerevisiae*; forward and reverse genetics. *Biochemie* **75:** 209–224.

Kunce CM, Trelease RN, Doman C. (1984) Ontogeny of glyoxysomes in maturing and germinated cotton seeds — a morphometric analysis. *Planta* **161:** 156–164.

Lazarow P, De Duve C. (1973) The synthesis and turnover of rat liver peroxisomes; intracellular pathway of catalase synthesis. *J. Cell Biol.* **59:** 507–529.

Lazarow PB, Fujiki Y. (1985) Biogenesis of peroxisomes. *Annu. Rev. Cell Biol.* **1:** 489–530.

Lazarow PB, Fujiki Y, Small GM, Watkins P, Moser H. (1986) Presence of the peroxisomal 22 kDa integral membrane protein in the liver of a person lacking recognisable peroxisomes (Zellweger syndrome). *Proc. Natl Acad. Sci. USA* **83:** 9193–9196.

Lee NG, Stein B, Suzuki H, Verma DPS. (1993) Expression of antisense nodulin-35 RNA in *Vigna aconitifolia* transgenic root nodules retards peroxisome development and affects nitrogen availability to the plant. *Plant J.* **3:** 599–606.

Lemmens M, Verheyden K, van Veldhoven P, Vereecke J, Mannaerts GP, Carmeliet E. (1989) Single-channel analysis of a large conductance channel in peroxisomes of rat liver. *Biochem. Biophys. Acta* **984:** 351–359.

Luers G, Hashimoto T, Fahimi HD, Völkl A. (1993) Biogenesis of peroxisomes: isolation and characterisation of two distinct peroxisomal populations from normal and regenerating rat liver. *J. Cell Biol.* **121:** 1271–1280.

Luster DG, Bowditch MI, Eldridge KM, Donaldson RP. (1988) Characterisation of membrane-bound electron transport enzymes from castor bean glyoxysomes and endoplasmic reticulum. *Arch. Biochem. Biophys.* **265:** 50–51.

McCammon MT, McNew JA, Willy PJ, Goodman JM. (1994) An internal region of the peroxisomal membrane protein PMP47 is essential for sorting to peroxisomes. *J. Cell Biol.* **124:** 915–925.

McCollum D, Monosov E, Subramani S. (1993) The pas8 mutant of *Pichia pastoris* exhibits the peroxisomal protein import deficiencies of zellweger syndrome cells — the pas 8 protein binds to the COOH-terminal tripeptide peroxisomal targeting signal and is a member of the TPR protein family. *J. Cell Biol.* **121:** 761–774.

McNew JA, Goodman JM. (1994) An oligomeric protein is imported into peroxisomes *in vivo. J. Cell Biol.* **127:** 1245–1257.

Maeshima M, Beevers H. (1985) Purification and properties of glyoxysomal lipase from castor bean. *Plant Physiol.* **79:** 489–493.

Manzara T, Carrasco P, Gruissem W. (1991) Developmental and organ specific changes in promoter DNA–protein interactions in the tomato *rbcS* gene family. *Plant Cell* **3:** 1305–1316.

Marrison JL, Onyeocha I, Baker A, Leech RM. (1993) Recognition of peroxisomes by immunofluoresence in transformed and untransformed tobacco cells. *Plant Physiol.* **103:** 1055–1059.

Martin C, Beeching JR, Northcote DH. (1984) Changes in levels of transcripts in endosperms of castor beans treated with exogenous gibberellic acid. *Planta* **162:** 68–76.

Marzioch M, Erdman R, Veenhuis M, Kunau W-H. (1994) PAS 7 encodes a member of the WD40 protein family essential for the import of 3-oxo-acyl CoA thiolase as a PTS-2-containing protein into peroxisomes. *EMBO J.* **13:** 4908–4918.

Miernyk JA, Trelease RN, Choinski JS. (1979). Malate synthase activity in cotton and other ungerminated oilseeds. *Plant Physiol.* **63:** 1068–1071.

Miura S, Kasuya-Arai I, Mori H, Miyazawa S, Osumi T, Hashimoto T, Fujiki Y. (1992) Carboxyl-terminal consensus ser-lys-leu related tripeptide of peroxisomal proteins functions *in vitro* as a minimal peroxisome-targeting signal. *J. Biol. Chem.* **267:** 14405–14411.

Mori H, Nishimura M. (1989) Glyoxysomal malate synthetase is specifically degraded in microbodies during greening of pumpkin cotyledons. *FEBS Lett.* **244:** 163–166.

Newcomb EH, Tandon SHR, Kowal RR. (1985) Ultrastructural specialization for ureide production in uninfected cells of soybean root nodules. *Protoplasma* **125:** 1–12.

Nicolay K, Veenhuis M, Douma AC, Harder W. (1987) A ^{31}P NMR study of the internal pH of yeast peroxisomes. *Arch. Microbiol.* **147:** 37–41.

Ni W, Trelease RN. (1991a) Two genes encode the two subunits of cottonseed catalase. *Arch. Biochem. Biophys.* **289:** 237–243.

Ni W, Trelease RN. (1991b) Post-transcriptional regulation of catalase isozyme expression in cotton seeds. *Plant Cell* **3:** 737–744.

Ni W, Trelease RN, Eising R. (1990) Two temporally synthesised charge subunits interact to form the five isoforms of cottonseed (*Gossypium hirsutum*) catalase. *Biochem. J.* **269:** 233–238.

Nishimura M, Yamaguchi J, Mori H, Akazawa T, Yokota S. (1986) Immunocytochemical analysis shows that glyoxysomes are directly transformed to leaf peroxisomes during greening of pumpkin cotyledons. *Plant Physiol.* **81:** 313–316.

Nishimura M, Takeguchi Y, De Bellis L, Hara-Nishimura I. (1993) Leaf peroxisomes are directly transformed to glyoxysomes during senescence of pumpkin cotyledons. *Protoplasma* **175:** 131–137.

Okada Y, Frey AB, Günthner TM, Sabatini DD, Kreibich G. (1982) Studies on the biosynthesis of microsomal membrane proteins. Site of synthesis and mode of insertion of cytochrome b_5, cytochrome b_5 reductase, cytochrome P450 reductase and epoxide reductase. *Eur. J. Biochem.* **122:** 393–402

Olsen LJ, Harada JJ. (1995) Peroxisomes and their assembly in higher plants. *Annu. Rev. Plant Physiol. Plant Mol. Biol.* **46:** 123–146.

Olsen LJ, Ettinger WF, Damsz B, Matsudaira K, Webb A, Harada JJ. (1993) Targeting of glyoxysomal proteins in leaves and roots of a higher plant. *Plant Cell* **5:** 941–952.

Onyeocha I, Behari R, Hill D, Baker A. (1993) Targeting of castor bean glyoxysomal isocitrate lyase to tobacco leaf peroxisomes. *Plant Mol. Biol.* **22:** 385–396.

Preisig-Mueller R, Kindl H. (1993) Thiolase mRNA translated *in vitro* yields a peptide with a putative N-terminal presequence. *Plant Mol. Biol.* **22:** 59–66.

Rapp S, Soto U, Just WW. (1993) Import of firefly luciferase into peroxisomes of permeabilised Chinese hamster ovary cells: a model system to study peroxisomal protein import *in vivo. Exp. Cell Res.* **205:** 59–65.

Rodriguez D, Dommes J, Northcote DH. (1987) Effect of abscisic and gibberellic acids on malate synthase transcripts in germinating castor bean seeds. *Plant Mol. Biol.* **9:** 227–235.

Rodriguez D, Ginger RS, Baker A, Northcote DH. (1990) Nucleotide sequence analysis of a cDNA clone encoding malate synthase of castor bean *(Ricinus communis)* reveals homology to *DAL7,* a gene involved in allantoin degradation in *Saccharomyces cerevisiae. Plant Mol. Biol.* **15:** 501–504.

Santos MJ, Imanaka T, Shio H, Small GM, Lazarow B. (1988) Peroxisomal membrane ghosts in Zellweger syndrome-aberrant organelle assembly. *Science* **239:** 1536–1538.

Sautter C. (1986) Microbody transition in greening watermelon cotyledons: double immunocytochemical labeling of isocitrate lyase and hydroxypyruvate reductase. *Planta* **167:** 491–503.

Schnarrenberger C, Oeser A, Tolbert NE. (1971) Development of microbodies in sunflower cotyledons and castor bean endosperm during germination. *Plant Physiol.* **48:** 566–574.

Schwartz BW, Sloan JS, Becker WM. (1991) Characterisation of genes encoding hydroxypyruvate reductase in cucumber. *Plant Mol. Biol.* **7:** 941–947.

Schwartz BW, Daniel SG, Becker WM. (1992) Photooxidative destruction of chloroplasts leads to reduced expression of peroxisomal NADH-dependent hydroxypyruvate reductase in developing cucumber cotyledons. *Plant Physiol.* **99:** 681–685.

Sloan JS, Schwartz BW, Becker WM. (1993) Promoter analysis of a light regulated gene encoding hydroxypyruvate reductase, an enzyme of the photorespiratory glycolate pathway. *Plant J.* **3:** 867–874.

Small GM, Szabo LJ, Lazarow PB. (1988) Acyl-CoA oxidase contains two targeting sequences each of which can mediate protein import into peroxisomes. *EMBO J.* **7:** 1167–1173.

Smith SM, Leaver CJ. (1986) Glyoxysomal malate synthase of cucumber: molecular cloning of a cDNA and regulation of enzyme synthesis during germination. *Plant Physiol.* **81:** 762–767.

Stabenau H. (1991) Phylogeny of peroxisomes in algae and higher plants. In: *Phylogenetic Changes in Peroxisomes of Algae/Phlogeny of Plant Peroxisomes. Proceedings of an International Symposium on Phylogeny of Algal and Plant Peroxisomes* (Stabenau H, ed.). Oldenberg: University of Oldenberg, pp. 143–150.

Subramani S. (1993) Protein import into peroxisomes and biogenesis of the organelle. *Annu. Rev. Cell Biol.* **9:** 445–478.

Swinkels BW, Gould SJ, Bodnar AG, Rachubinski RA, Subramani S. (1991) A novel cleavable peroxisomal targeting signal at the amino-terminus of rat 3-ketoacyl-CoA thiolase. *EMBO J.* **10:** 3255–3262.

Taylor WS. (1989) Regulatory interactions between nuclear and plastid genomes. *Annu. Rev. Plant Physiol.* **40:** 211–233.

Terlecky SR, Nuttley WM, McCollum D, Sock E, Subramani S. (1995) The *Pichia pastoris* peroxisomal protein Pas8p is the receptor for the C-terminal tripeptide peroxisomal targeting signal. *EMBO J.* **14:** 3627–3634.

Titus DE, Becker WM. (1985) Investigation of the glyoxysome–peroxisome transition in germinating cucumber cotyledons using double label immunoelectron microscopy. *J. Cell Biol.* **101:** 1288–1299.

Tolbert NE. (1981) Metabolic pathways in glyoxysomes and peroxisomes. *Annu. Rev. Biochem.* **50:** 133–157.

Trelease RN. (1984) Biogenesis of glyoxysomes. *Annu. Rev. Plant Physiol.* **35:** 321–347.

Trelease RN, Becker WM, Gruber PJ, Newcomb EH. (1971) Microbodies (glyoxysomes and peroxisomes) in cucumber cotyledons: correlative biochemical and ultrastructural study in light and dark grown seedlings. *Plant Physiol.* **48:** 461–465.

Tsukamoto T, Hata S, Yokota S, Miura S, Fujiki Y, Hijkata M, Miyazawa S, Hashimoto T, Osmi T. (1994) Characterisation of the signal peptide at the amino terminus of the rat peroxisomal 3-ketoacyl-CoA thiolase precursor. *J. Biol. Chem.* **269:** 6001–6010.

Turley RB, Trelease RN. (1990) Development and regulation of three glyoxysomal enzymes during cotton seed maturation and growth. *Plant Mol. Biol.* **4:** 137–146.

van den Bosch KA, Newcomb EH. (1986) Immunogold localization of nodule specific uricase in developing soybean root nodules. *Planta* **167:** 425–436.

van der Klei IJ, Hilbrands RE, Swaving GJ, Waterham HR, Vrieling EG, Titorenko VI, Cregg JM, Harder W, Veenhuis M. (1995) The *Hansenula polymorpha PER 3* gene is essential for the import of PTS-1 proteins into the peroxisomal matrix. *J. Biol. Chem.* **270:** 17229–17236.

van der Leij I, Franse M, Elgersma Y, Distel B, Tabak HF. (1993) PAS 10 is a tetratricopeptide repeat protein that is essential for the import of most matrix proteins into peroxisomes of *S. cerevisiae. Proc. Natl Acad. Sci. USA* **90:** 11782–11786.

van Roermund CWT, Elgersma Y, Singh N, Wanders RJA, Tabak HF. (1995) The membrane of peroxisomes in *S. cerevisiae* is impermeable to NAD(H) and acetyl CoA under *in vivo* conditions. *EMBO J.* **14:** 3480–3486.

Veenhuis M, Goodman J. (1990) Peroxisomal assembly: membrane proliferation precedes the induction of the abundant matrix proteins in the methylotropic yeast *Candida boidinii. J. Cell Sci.* **96:** 583–590.

Veenhuis M, van Dijken JP, Pilon SAF, Harder W. (1978) Development of crystalline peroxisomes in methanol grown cells of the yeast *Hansenula polymorpha* and its relation to environmental conditions. *Arch. Microbiol.* **117:** 153–163.

Veenhuis M, Sulter G, van der Klei I, Harder W. (1989) Evidence for functional heterogeniety among microbodies in yeast. *Arch. Microbiol.* **151:** 105–110.

Verner K, Schatz G. (1988) Protein translocation across membranes. *Science* **241:** 1307–1313.

Volokita M. (1991). The carboxy-terminal end of glycolate oxidase directs a foreign protein into tobacco leaf peroxisomes. *Plant J.* **1:** 361–366.

Walton PA, Hill PE, Subramani S. (1995) Import of stably folded proteins into peroxisomes. *Mol. Biol. Cell* **6:** 675–683.

Waterham HR, Kiezer-Gunnick I, Goodman J, Harder W, Veenhuis M. (1990) Immunocytochemical evidence for the acidic nature of peroxisomes in methylotropic yeasts. *FEBS Lett.* **262:** 17–19.

Weir EM, Riezman H, Grienenberger JM, Becker WM, Leaver CJ. (1980) Regulation of glyoxysomal enzymes during germination of cucumber; temporal changes in translatable mRNA for isocitrate lyase and malate synthase. *Eur. J. Biochem.* **112:** 469–477.

Wendland M, Subramani S. (1993) Cytosol-dependent peroxisomal protein import in a permeabilised cell system. *J. Cell Biol.* **120:** 675–685.

Wickner W, Driessen AJM, Hartl FU. (1991) The enzymology of protein translocation across the *Escherichia coli* plasma membrane. *Annu. Rev. Biochem.* **60:** 101–124.

Wolins NE, Donaldson RP. (1993) Specific binding of the peroxisomal protein targeting sequence to glyoxysomal membranes. *J. Biol. Chem.* **269:** 1149–1153.

Yamaguchi Y, Nishimura M, Akazawa T. (1984) Maturation of catalase precursor proceeds to a different extent in glyoxysomes and leaf peroxisomes of pumpkin cotyledons. *Proc. Natl Acad. Sci. USA* **81:** 4809–4813.

Yamaguchi J, Nishimura M, Akazawa T. (1986) Purification and characterisation of heme containing low activity form of catalase from greening pumpkin cotyledons. *Eur. J. Biochem.* **159:** 315–322.

Zhang JW, Lazarow PB. (1995) *PEB1* (*PAS7*) in *S. cerevisiae* encodes a hydrophilic intra-peroxisomal protein that is a member of the WD40 repeat family and is essential for the import of thiolase into peroxisomes. *J. Cell Biol.* **129:** 65–80.

Zhang JZ, Gomez-Pedrozo M, Baden CS, Harada JJ. (1993) Two classes of isocitrate lyase are expressed during late embryogeny and postgermination in *Brassica napus* L. *Mol. Gen. Genet.* **238:** 177–184.

Chapter 25

Biogenesis of chloroplasts in higher plants

John C. Gray

1. Introduction

Chloroplasts are organelles specialized to carry out the reactions of photosynthesis. In higher plants, they are found primarily in leaves, but they provide the green coloration of stems, sepals, stamens, carpels and many fruits, and are often found in petals, where their presence is masked by other pigments. The distinctive feature of chloroplasts at the ultrastructural level is the extensive internal thylakoid membrane which houses the photosynthetic energy transduction system. The thylakoid membrane system may account for up to 90% of the membranes in a green leaf, and probably represents the most abundant membrane system in biology. Like all other plastids, chloroplasts are bounded by two envelope membranes which separate the plastid contents from the cytosol. The outer envelope membrane is a barrier to cytosolic proteins, but is freely permeable to many small molecules. The inner envelope membrane provides the permeability barrier to small molecules, and is the location of a wide variety of translocators or permeases. The three membrane systems of the chloroplasts enclose three soluble compartments: the intermembrane space of the envelope, the stroma and the thylakoid lumen. Each has a different function and distinctive enzyme and protein components. The main soluble phase of the chloroplasts is the stroma, which houses the enzymes of the CO_2-fixation cycle and of a large number of biosynthetic pathways. The thylakoid lumen is the location of some of the components of the photosynthetic electron transfer chain, but little is known about the components of the intermembrane space of the envelope.

Like all other plastids, chloroplasts contain their own genetic system which is responsible for the synthesis of a small number ($c.$ 100) of RNA and protein components. Chloroplasts contain numerous copies of a circular plastid DNA molecule, and the machinery for the replication, transcription and translation of the genetic material. However, the vast majority of chloroplast components are nuclear-encoded, synthesized in the cytosol and imported into the chloroplasts. One of the major problems in chloroplast biogenesis is understanding how the expression of the chloroplast and nuclear genes is co-ordinated to ensure the synthesis of a functional chloroplast. This brief review will consider some of the background information and recent advances in our understanding of chloroplast biogenesis in higher plants. The coverage is selective rather than comprehensive, and tends to concentrate on areas where genetic approaches appear to be likely to provide major insights into previously poorly understood areas.

2. Chloroplast development and division

Plastid differentiation

Chloroplasts normally develop from undifferentiated proplastids in meristematic cells of shoots and leaves, although occasionally they may also develop from differentiated plastids such as chromoplasts, amyloplasts and leucoplasts. In all instances the development of fully functional chloroplasts depends on the presence of light. In the absence of light in leaf and shoot tissues, distinctive plastids known as etioplasts develop in place of chloroplasts. Etioplasts do not contain chlorophyll, and possess a membranous prolamellar body in place of a thylakoid membrane system. On illumination, the individual lamellae of the thylakoid membrane system develop from the prolamellar body and associate to form the distinctive granal structure of the thylakoids. Illumination can also provide the signal for chloroplast development in tissues and organs, such as roots and tubers, which are normally below ground and contain other specialized plastids. For example, illumination of potato tubers and carrot roots results in the differentiation of amyloplasts and chromoplasts, respectively, into chloroplasts. Prolonged illumination of root tissue in many species leads to the development of chloroplasts from leucoplasts. These examples illustrate the developmental plasticity of plastids, and it is believed that all plastid types are interconvertible. Chloroplasts develop into chromoplasts in many ripening fruits (e.g. tomato and pepper), and may dedifferentiate into proplastids during callus formation after wounding or in tissue culture.

All plastids arise by growth and division of pre-existing plastids, and there is no evidence for the assembly of plastids *de novo*. Indeed, our current understanding of membrane biogenesis precludes the assembly of membranes *de novo* because of the requirement for receptors for targeting of proteins to the correct membrane system. All plastids are bounded by a double-membrane envelope which contains the receptors and translocation systems for the import of nuclear-encoded plastid proteins synthesized in the cytosol. The thylakoid membrane also contains receptors and translocation systems for the insertion and translocation of thylakoid membrane and lumen proteins. It is necessary that all these receptors and translocation systems are present in all plastid types, and this suggests that all plastid types must contain an internal membrane system, distinct from the inner envelope membrane, which can develop into the thylakoid system in chloroplasts. The distinction in function and composition of the inner envelope membrane and the thylakoid membrane suggests that separate protein insertion and translocation systems must be present in each membrane. Evidence for a functional internal membrane system, equivalent to the thylakoid membrane, in root leucoplasts is provided by transgenic tomato plants expressing a chimeric plastocyanin gene (de Boer *et al.,* 1988). In these plants the plastocyanin precursor is targeted and processed to the mature size in plastids in roots and petals (de Boer *et al.,* 1988); this requires translocation across both the envelope and the internal membranes. The presence of an internal membrane system in all plastid types appears to be a requirement for the interconversion of plastid types and for full plastid differentiation.

Chloroplast number and size

The number of chloroplasts in leaf mesophyll cells varies widely depending on the species. As few as three chloroplasts are found in the mesophyll cells of cocoa (*Cacao theobroma*) and *Peperomia metallica* leaves, whereas as many as 300 chloroplasts are found in the mesophyll cells of radish (*Raphanus sativus*) leaves (Boffey and Leech, 1982). The most thorough studies of chloroplast development have been carried out in spinach, pea, *Arabidopsis thaliana* and the cereals wheat, barley and oats. In most of these plants, mature mesophyll cells contain 50–150 chloroplasts. The number of chloroplasts per cell is closely correlated with the size of the mesophyll cell, such that a constant proportion of the mesophyll cell surface is covered with chloroplasts (Ellis and Leech, 1985; Pyke and Leech, 1987; Robertson and Laetsch, 1974). However, the proportion of the surface covered varies widely between different plants. In wheat, 70% of the mesophyll cell surface is covered with chloroplasts, although only 15% of the cell surface is covered in maize mesophyll cells (Ellis and Leech, 1985). In *Arabidopsis thaliana*, the proportion of the mesophyll cell surface covered with chloroplasts is constant, even in different mutants where the number of chloroplasts per cell varies enormously. In a series of *arc* (accumulation and replication of chloroplasts) mutants, decreased chloroplast number per cell is compensated for by increased chloroplast size (Pyke and Leech, 1992, 1994). Similarly, in other *Arabidopsis* mutants, increased numbers of chloroplasts per cell are compensated for by a decrease in chloroplast size (McCourt *et al.*, 1987; Pyke and Leech, 1991). This indicates that there must be efficient mechanisms for coordinately regulating the size and number of chloroplasts in mesophyll cells. The nature of these mechanisms is currently unknown, but the molecular characterization of the *arc* mutants should provide important insights into the processes involved.

Many studies have shown that the number of chloroplasts per mesophyll cell increases during leaf development (Boffey *et al.*, 1979; Hashimoto and Possingham, 1989a,b; Lamppa *et al.*, 1980; Mullett, 1988; Possingham, 1980; Scott and Possingham, 1980). Increases of 3- to 4-fold have been observed in developing leaves of cereals and pea (Baumgartner *et al.*, 1989; Boffey *et al.*, 1979; Boffey and Leech, 1982; Hashimoto and Possingham, 1989a; Lamppa *et al.*, 1980), whereas 10- to 15-fold increases in chloroplast number per cell have been observed in developing spinach leaves *in vivo* and *in vitro* (Hashimoto and Possingham, 1989b; Scott and Possingham, 1980). Relatively small numbers of undifferentiated proplastids are present in meristematic cells. Estimates of 10–15 proplastids per meristematic cell have been made for a number of species, including spinach, barley and *Arabidopsis* (Mullett, 1988; Possingham, 1980; Pyke and Leech, 1992, 1994). Microscopic examination has shown that these proplastids develop into chloroplasts, which in turn divide to produce more chloroplasts (Hashimoto and Possingham, 1989b; Leech *et al.*, 1981; Possingham, 1980). Hashimoto and Possingham (1989b) have distinguished four characteristic shapes of chloroplasts in developing leaves of spinach *in vitro*, and, from the proportion of chloroplasts of each type, they have attempted to define the duration of different stages of the chloroplast-division cycle. From cells with a chloroplast doubling time of 19.4 h, they estimate that chloroplasts are spherical for 13.4 h, oval for 2.8 h, dumb-bell-shaped for 3.1 h, and in the final stages of division, showing a thin neck separating two otherwise spherical chloroplasts, for 0.3 h. Similarly shaped chloroplasts have been observed in other plants and have been implicated in the division process (Leech *et al.*, 1981; Possingham, 1980). Only a few rounds (2–4) of chloroplast division are required to generate the numbers of chloroplasts found in mature mesophyll cells, provided that there is no further division of mesophyll cells. In wheat leaves, chloroplast division is temporally and spatially separated from cell division. Mesophyll cell division takes place exclusively in a region 1 cm above the leaf base in 7-day-old primary leaves, whereas chloroplast division occurs predominantly in the region 1.7–4.5 cm above the leaf base (Boffey *et al.*, 1979).

Chloroplast DNA replication and distribution

Replication of chloroplast DNA is not co-ordinated with chloroplast division, and the number of plastid DNA molecules per chloroplast changes during chloroplast development. Values in the range of 50–1000 plastid DNA molecules per chloroplast have been estimated for a number of higher plants (Baumgartner *et al.*, 1989; Boffey and Leech, 1982; Hashimoto and Possingham, 1989a; Lamppa *et al.*, 1980; Miyamura *et al.*, 1990; Scott and Possingham, 1980). In cereals, plastid DNA replication is initiated in the leaf base and precedes chloroplast division (Miyamura *et al.*, 1986). This results in an increase in the number of plastid DNA molecules per chloroplast in the lower part of the leaf, followed by decreased numbers of plastid DNA molecules per chloroplast as chloroplast division distributes the plastid DNA molecules between an increased number of chloroplasts (Baumgartner *et al.*, 1989; Boffey and Leech, 1982; Hashimoto and Possingham, 1989a; Miyamura *et al.*, 1990). Degradation of plastid DNA in the light has been suggested to contribute to the decline in plastid DNA copy number per chloroplast in barley (Baumgartner *et al.*, 1989) and oats (Hashimoto and Possingham, 1989a). A similar decline in the number of plastid DNA molecules per chloroplast during mesophyll cell development in spinach and pea can be accounted for solely by chloroplast division (Lamppa *et al.*, 1980; Scott and Possingham, 1980). Expanding spinach and pea leaves contain approximately constant amounts of plastid DNA per mesophyll cell, although the number of chloroplasts increases by chloroplast division (Lamppa *et al.*, 1980; Scott and Possingham, 1980).

The spatial distribution of plastid DNA, as detected by fluorescence microscopy with the fluorochrome 4,6-diamidino-2-phenylindole (DAPI), changes markedly during chloroplast development, although different patterns of staining are observed in different species (Kuroiwa, 1991). In wheat, plastid DNA is located in a continuous band around the chloroplast periphery in young pre-division chloroplasts and during chloroplast division (Marrison and Leech, 1992; Miyamura *et al.*, 1986). However, following chloroplast division, plastid DNA disperses and is observed as discrete nucleoids within mature chloroplasts (Marrison and Leech, 1992; Miyamura *et al.*, 1986; Selldén and Leech, 1981). In wheat, the nucleoids are located at the periphery of the chloroplasts, whereas in mature chloroplasts of tobacco, spinach and pea the nucleoids are more centrally located (Marrison and Leech, 1992; Miyamura *et al.*, 1986; Sato *et al.*, 1993).

The location of plastid DNA at the chloroplast periphery before and during chloroplast division may indicate that an association with the chloroplast envelope membrane is required for the replication and segregation of plastid DNA molecules. Sato *et al.* (1993) have detected an envelope membrane protein of 130 kDa which binds to specific AT-rich sequences within pea plastid DNA. The protein is present in the chloroplast envelope when plastid DNA is located at the chloroplast periphery, but is undetectable in mature chloroplasts when the plastid DNA forms nucleoids within the stroma (Sato *et al.*, 1993). Topoisomerase II, which is required for decatenating plastid DNA molecules after division, also shows a similar distribution to plastid DNA in pre-division and dividing wheat chloroplasts, and then disperses in mature chloroplasts (Marrison and Leech, 1992). An association of bacterial DNA with cell membranes is required for the initiation of replication (Jacob *et al.*, 1963; Landoulsi *et al.*, 1990). However, the binding sites recognized by the 130-kDa protein do not correspond to the origins of replication of the pea plastid genome identified from the D-loop regions (Meeker *et al.*, 1988).

3. The role of the plastid genome in chloroplast biogenesis

The plastid genome of green plants contains essential genetic information for chloroplast biogenesis and the assembly of the photosynthetic apparatus. In most higher plants the plastid genome consists of covalently closed circles of double-stranded DNA of 120–160 kbp, although the size ranges from 70 kbp in beechdrops (*Epifagus virginiana*) to 218 kbp in *Pelargonium* (Palmer, 1991). The complete nucleotide sequences of plastid DNA from the higher plants rice (134 525 bp), maize (140 387 bp), tobacco (155 844 bp), black pine (119 707 bp) and beechdrops (70 028 bp), have been determined (Hiratsuka *et al.*, 1989; Maier *et al.*, 1995; Shinozaki *et al.*, 1986; Wakasugi *et al.*, 1994; Wolfe *et al.*, 1992a). With the exception of the beechdrops plastid genome, which has large-scale deletions of photosynthesis genes, similar sets of genes are present in each of these plastid genomes. Genes for more than 110 different RNA or protein components of chloroplasts have been identified (Sugiura, 1992), and only a small number (<10) of conserved open reading frames (*ycfs*) have yet to be identified. The identification of the genes present suggests that the plastid genome is principally concerned with providing genetic information for the synthesis of photosynthesis components. Over 50 genes encode polypeptides of the photosynthetic apparatus, and an additional 60 components of the chloroplast transcription and translation apparatus are specified by the chloroplast genome to ensure the synthesis of the photosynthesis components (see *Table 1*).

Approximately 40 genes encode components of the photosynthetic apparatus, and an additional 11 genes encode components of a putative NAD(P)H-plastoquinone oxidoreductase (*Table 1*). The *rbcL* gene encodes the large subunit of ribulose-1,5-bisphosphate carboxylase (Rubisco), the CO_2-fixing enzyme of the reductive pentose phosphate pathway (Calvin cycle), and the most abundant protein on this planet. Approximately half of the known components of the thylakoid membrane complexes involved in photosynthetic electron transfer and ATP synthesis are encoded by plastid DNA (*Table 1*). Genes encoding components of a putative NAD(P)H-plastoquinone oxidoreductase were identified by homology to genes encoding components of the mitochondrial NADH-ubiquinone oxidoreductase (complex I of the mitochondrial electron transfer chain). They are present in the plastid genomes of tobacco, rice and maize, but are absent from the plastid genomes of black pine and beechdrops (Wakasugi *et al.*, 1994; Wolfe *et al.*, 1992a). The plastid *ndh* genes have been shown to be transcribed (Schantz and Bogorad, 1988; Steinmüller *et al.*, 1989; Matsubayashi *et al.*, 1987), and some of the polypeptides have been detected in thylakoid membranes (Berger *et al.*, 1993; Nixon *et al.*, 1989; Steinmetz *et al.*, 1986). The function of the putative NAD(P)H-plastoquinone oxidoreductase is

Table 1. Genes in chloroplast DNA[a]

Chloroplast complex or component	Gene designation
Ribulose 1,5-bisphosphate carboxylase	*rbcL*
Photosystem I	*psaA, B, C, I, J*
Photosystem II	*psbA, B, C, D, E, F, H, I, J, K, L, M, N, T*
Cytochrome *bf* complex	*petA, B, D, G, L*
ATP synthase	*atpA, B, E, F, H, I*
NAD(P)-plastoquinone oxidoreductase	*ndhA, B, C, D, E, F, G, H, I, J, K*
RNA polymerase	*rpoA, B, C1, C2*
Ribosomal RNA	*rrn23, 16, 5, 4.5*
Ribosomal proteins	*rpl2, 14, 16, 20, 22, 23, 32, 33, 36*
	rps2, 3, 4, 7, 8, 11, 12, 14, 15, 16, 18, 19
Translation initiation factors	*infA*
Transfer RNAs	*trnA, R(ACG), R(UCU), N, D(GUC), C, Q, E,*
	G(GCC), G(UCC), H, I(GAU), I(CAU), L(CAA),
	L(UAA), L(UAG), K, fM, M, F, P, S(GGA), S(UGA),
	S(GCU), T(GGU), T(UGU), W, Y, V(GAC), V(UAC)
RNA maturase	*matK*
ATP-dependent protease	*clpP*
Acetyl-CoA carboxylase	*accD*
Conserved open reading frames	*ycf1, 2, 3, 4, 5, 6, 9, 10*

[a] This table presents a compilation of genes present in plastid DNA from different species of higher plants. All plastid gene products identified to date are components of complexes which require additional nuclear-encoded subunits for assembly of a functional complex. The exact identity of the individual gene products can be obtained from Sugiura (1992) or the recommendations of the Commission on Plant Gene Nomenclature (1993).

not yet clear. It may operate in a chlororespiratory pathway from NAD(P)H to O_2 to reoxidize reducing equivalents generated by the oxidative pentose phosphate pathway in chloroplasts in the dark (Bennoun, 1982). However, it may also function in the light to maintain the correct redox poise of electron transfer components or to maintain the balance between NADPH and ATP synthesis required for biosynthetic reactions in the chloroplast (Berger *et al.*, 1993), perhaps by catalysing cyclic electron transfer around photosystem I.

Approximately 60 components of the chloroplast transcription and translation apparatus are plastid gene products. Four *rpo* genes encode subunits of RNA polymerase showing homology to subunits of RNA polymerase from *Escherichia coli*. The presence of these subunits in an active RNA polymerase preparation has been shown by N-terminal sequencing and immunochemical methods (Hu and Bogorad, 1990; Hu *et al.*, 1991; Purton and Gray, 1987). Four ribosomal RNAs and 21 ribosomal proteins are specified by plastid DNA, indicating a major contribution of the chloroplast genome to the structure of the chloroplast 70S ribosome. A region encoding a protein homologous to protein synthesis initiation factor 1 is present in several chloroplast genomes, although it is not yet clear whether this represents an active gene. Genes for approximately 30 transfer RNAs are present in plastid DNA (Sugiura, 1992), and this complement of transfer RNAs is apparently able to decode all codons used in chloroplast genes.

Several other open reading frames in plastid DNA of higher plants have been identified over the past few years. The gene *clpP* encoding the proteolytic subunit of an ATP-dependent protease was identified by homology of the putative chloroplast and *E. coli* proteins (Gray *et al.*, 1990; Maurizi *et al.*, 1990). The β-subunit of the carboxyltransferase complex of acetyl-CoA carboxylase from *E. coli* shows homology to the product of an open reading frame present in plastid DNA from tobacco and pea (Li and Cronan, 1992a,b; Smith *et al.*, 1991). A severely truncated, and probably non-functional, version of the gene *accD* is present in the rice and wheat genomes (Smith *et al.*, 1991). The association of the *accD* gene product with protein complexes showing acetyl-CoA carboxylase activity has been reported for pea chloroplasts (Sasaki *et al.*, 1993a). Both of these genes are retained in the plastid genome of the parasitic plant beechdrops (*Epifagus virginiana*), which has apparently lost all genes encoding components of the photosynthetic apparatus (Wolfe *et al.*, 1992a). The 70-kbp genome contains only 42 protein-coding genes, of which at least 38 genes specify components of the plastid protein synthesis system (Wolfe *et al.*, 1992b). These genes are actively transcribed (Morden *et al.*, 1991), suggesting that the genome is functional. This indicates that plastid genomes have not been maintained exclusively for the production of the photosynthetic apparatus.

4. The role of the nuclear genome in chloroplast biogenesis

The nuclear genome contributes to chloroplast biogenesis in two ways. First, it specifies all those chloroplast proteins that are not encoded by the plastid genome, and secondly, it encodes components that are involved in the regulation of chloroplast biogenesis but which may never enter the chloroplast. This latter category includes components involved in regulating nuclear gene expression and in co-ordinating chloroplast development with the development of the whole plant.

Nuclear genes encoding chloroplast proteins

The vast majority of the protein components of chloroplasts are encoded by the nuclear genome. It is difficult to make precise estimates of the numbers of different polypeptides in chloroplasts, but more than 1000 polypeptides may be a conservative estimate, in view of the complexity of the biochemical pathways known to be present in chloroplasts. This would suggest that the nuclear genome encodes more than 90% of the polypeptides in chloroplasts. Ever increasing numbers of cDNA and genomic clones encoding components of chloroplasts are being isolated and characterized. The first nuclear genes to be isolated encoded components of the photosynthetic apparatus (Cashmore, 1984; Coruzzi *et al.,* 1983), and there has been considerable effort in many laboratories to isolate and characterize the nuclear genes and cDNAs encoding components of photosystems I and II, the cytochrome *bf* and ATP synthase complexes and the soluble electron transfer components plastocyanin, ferredoxin and ferredoxin-NADP$^+$ reductase (Tittgen *et al.,* 1986). In addition, many of the nuclear genes encoding enzymes of the reductive pentose phosphate pathway (Calvin cycle) have now been isolated (Lloyd *et al.,* 1991; Longstaff *et al.,* 1989; Raines *et al.,* 1989). The widespread use of gene-cloning technology has led to the isolation of nuclear genes or cDNAs encoding components of many chloroplast-located biochemical pathways, including starch synthesis and degradation, fatty acid synthesis and desaturation, nitrogen and sulphur assimilation and amino acid biosynthesis, tetrapyrrole (including chlorophyll and haem) biosynthesis, carotenoid and isoprenoid biosynthesis and pyrimidine biosynthesis. Furthermore, many nuclear genes or cDNAs for components of the chloroplast protein-synthesizing system, the molecular chaperone systems and protein degradation systems have been isolated and characterized. cDNAs for many more chloroplast proteins have undoubtedly been isolated and characterized as part of the expressed sequence tag (EST) sequencing projects on *Arabidopsis* and rice (Höfte *et al.,* 1993; Uchimiya *et al.,* 1992), but the identity of the proteins has not yet been established. It is clear that the nuclear genome plays a vital role in encoding the vast majority of the proteins of chloroplasts.

Sequence analysis of cDNAs and of the isolated polypeptides that they encode has demonstrated that most nuclear-encoded chloroplast proteins are synthesized initially with an N-terminal presequence (von Heijne *et al.,* 1989), which is used to target the protein to the chloroplast envelope (Keegstra *et al.,* 1989). After translocation across the outer and inner envelope membranes, most probably at contact sites (Gray and Row, 1995), the presequences are removed in the stroma by a soluble stromal processing peptidase (Oblong and Lamppa, 1992; Robinson and Ellis, 1984). The processed polypeptides may associate with molecular chaperones, such as Hsp70 and Cpn60, to prevent unproductive folding during and after translocation across the envelope membrane (Lubben *et al.,* 1989; Madueño *et al.,* 1993; Tsugeki and Nishimura, 1993). The polypeptides may then fold to form monomeric stromal proteins, or they may assemble into multi-subunit complexes containing other nuclear- or chloroplast-encoded subunits. A role for molecular chaperones in preventing the improper interactions of unassembled subunits may account for the large amounts of Cpn60 and Hsp70 proteins present in the chloroplast stroma (Ellis and van der Vies, 1988; Marshall *et al.,* 1990).

Imported nuclear-encoded proteins must be targeted to the correct compartment within the chloroplasts. Chloroplasts are composed of six different compartments: the outer and inner envelope membranes and the space between them, the stroma, the thylakoid membrane and the thylakoid lumen. Several proteins destined for the outer envelope membrane are not synthesized with N-terminal presequences, and appear to insert into the membrane by a process different to that used by other imported chloroplast proteins (Li *et al.,* 1991; Salomon *et al.,* 1990). The pathways of import of proteins of the inner envelope membrane are not yet clear. Although several inner envelope proteins are synthesized with N-terminal presequences (Dreses-Werringloer *et al.,* 1991; Flügge *et al.,* 1989; Li *et al.,* 1992) which are removed by stromal processing peptidase (P. Balint-Kurti and J.C. Gray, unpublished results), it has not yet been established whether the processed proteins are fully imported into the stroma and then insert into the membrane from the stromal side, or whether the proteins insert directly into the inner envelope by a stop-transfer process (Knight *et al.,* 1993; Knight and Gray, 1995). A similar, apparently unresolved, problem exists with regard to the pathway of insertion of nuclear-encoded proteins into the mitochondrial inner membrane (Glick *et al.*, 1992; Hartl and Neupert, 1990). No information is available about the targeting of nuclear-encoded proteins to the intermembrane space of the chloroplast envelope. Nuclear-encoded proteins destined for the thylakoid membrane and the thylakoid lumen are first directed to the stroma and then targeted to the appropriate compartment using cleavable or non-cleavable targeting signals. Molecular chaperones, including Hsp70 and Cpn60 proteins, have been shown to associate with stromal forms of thylakoid membrane proteins (Lubben *et al.*, 1989; Madueño *et al.*, 1993; Yalovsky *et al.*, 1992), and may be required to prevent the formation of translocation-incompetent forms of the proteins. Fuller details of the translocation of proteins across chloroplast envelope membranes and the thylakoid membrane are given in Chapter 26 by Robinson.

Nuclear genes regulating chloroplast development

A second category of nuclear genes involved in chloroplast biogenesis includes those encoding transcription factors required for regulating nuclear gene expression and those required for co-ordinating chloroplast development with the development of the whole plant. Although there has been an enormous effort to characterize transcription factors involved in light-regulated and tissue-specific expression of nuclear genes encoding chloroplast proteins (Gilmartin *et al.*, 1990; Kuhlemeier *et al.*, 1987), there is as yet little real understanding of their role in chloroplast biogenesis. Details of the transcriptional regulation of nuclear genes encoding chloroplast proteins are beyond the scope of this brief review.

Chloroplast development is co-ordinated with plant development and is influenced by several environmental signals, the most important of which is light. Nuclear gene products play an important role in the perception of light and the integration of plant responses. Several genetic loci have been identified which are involved in co-ordinating chloroplast development within the framework of normal plant development. Disruption of these genetic loci in *Arabidopsis* has produced mutants which show de-etiolated (*det*) or constitutively photomorphogenic (*cop*) phenotypes (Chory *et al.*, 1989, 1991; Chory and Peto, 1990; Deng *et al.*, 1991; Deng and Quail, 1992; Wei and Deng, 1992). An apparently similar mutant, *lip*1, showing light-independent photomorphogenesis, has also been described for pea (Frances *et al.*, 1992). The *Arabidopsis det* and *cop* mutant seedlings when grown in the dark have a morphology similar to that of light-grown wild-type seedlings, with shortened hypocotyls and expanded cotyledons. The plastids in the *det*1, *cop*1 and *cop*9 mutants lack prolamellar bodies and develop an extended thylakoid membrane system, although they do not accumulate chlorophyll (Chory *et al.*, 1989; Deng *et al.*, 1991; Wei and Deng, 1992). In the *det*2 mutants, the plastids resemble the etioplasts found in dark-grown wild-type seedlings (Chory *et al.*, 1991). However, in all *det* and *cop* mutants, several nuclear (*RbcS*, *Cab* and *PetF*) and chloroplast (*rbcL*, *psbA* and *psaA*) genes which are normally light-regulated are expressed constitutively in the dark (Chory *et al.*, 1989, 1991; Chory and Peto, 1990; Deng *et al.*, 1991; Deng and Quail, 1992; Wei and Deng, 1992). This suggests that the wild-type DET and COP gene products are components of a signal transduction system coupling light perception to chloroplast and plant development (Duckett and Gray, 1995).

The *Arabidopsis DET*1 gene has been isolated by positional cloning, and encodes a novel, 62-kDa nuclear-localized protein (Pepper *et al.*, 1994). This is consistent with a putative role in gene regulation, although the DET1 protein does not appear to bind DNA directly (Pepper *et al.*, 1994). The *Arabidopsis COP*1 gene has been isolated following the characterization of a T-DNA-tagged *cop*1 mutant (Deng *et al.*, 1992). The gene encodes a predicted soluble protein of 75 kDa, containing a conserved zinc-binding domain in the N-terminal region and four consecutive copies of the WD-40 repeat found in β-subunits of trimeric G-proteins in the C-terminal region.

The presence of a putative DNA-binding motif and a G-protein-related domain suggests a possible regulatory role for the polypeptide. The recessive nature of the *det* and *cop* mutations implies that the wild-type gene products act to repress photomorphogenic chloroplast and plant development in the dark. This has been demonstrated directly in stable transgenic lines of *Arabidopsis* overexpressing *COP1*, which resulted in a partial suppression of light-mediated development in these plants (McNellis *et al.*, 1994). The novel structure of the COP1 protein may suggest a function as a negative transcriptional regulator capable of direct interaction with components of the G-protein signalling pathway. The involvement of heterotrimeric G-proteins in the light-regulated expression of several nuclear and chloroplast genes has been suggested by microinjection of GTPγS and cholera toxin A chain into single cells of hypocotyls of the *aurea* mutant of tomato (Neuhaus *et al.*, 1993). The *aurea* mutant is deficient in active phytochrome (PhyA), and microinjection of G-protein agonists or calcium and calmodulin is able to stimulate expression from the *Cab* promoter and to induce synthesis and assembly of some photosynthesis-related proteins. It seems highly likely that the DET and COP gene products are involved in signal transduction events downstream of phytochrome (Duckett and Gray, 1995). Further analysis of these gene products and the signal transduction system is likely to lead to greater understanding of the co-ordination of chloroplast biogenesis with plant development.

5. Co-ordination of chloroplast and nuclear gene expression

The nuclear genome plays a pre-eminent role in the control of chloroplast biogenesis. As outlined above, the nuclear genome encodes regulatory components to co-ordinate chloroplast development with plant development, as well as encoding the vast majority of proteins located in the chloroplasts. Mutations in many of these nuclear genes may be expected to result in defects in chloroplast development or in chloroplast dysfunction. This previously led to the view that control of chloroplast biogenesis is invested solely in the nuclear genome. Nuclear genes encode key components of the chloroplast systems for DNA replication, transcription and translation and their control. All of the identified chloroplast gene products require nuclear gene products for assembly into functional complexes. However, the enormous contribution of the nuclear genome and the energy required to synthesize and assemble the chloroplast would be wasted if photosynthetically active, functional chloroplasts were not produced. This has led to the suggestion of a feedback mechanism from the chloroplast to the nucleus, to regulate the expression of nuclear genes encoding chloroplast components (Oelmüller, 1989; Susek and Chory, 1992; Taylor, 1989).

Chloroplast control of nuclear gene expression

In many mutant plants with defective chloroplasts, the nuclear genes encoding photosynthesis-related proteins are not expressed (Taylor, 1989). This is most clearly seen in carotenoid-deficient mutants of maize, barley and tomato grown under high light intensities (Batschauer *et al.*, 1986; Guiliano and Scolnik, 1988; Harpster *et al.*, 1984; Taylor, 1989). The absence of carotenoids results in photobleaching of chlorophyll and the inactivation of the photosynthetic apparatus in high light, but not in low light. Under high-light conditions, transcripts of nuclear genes encoding the light-harvesting chlorophyll *a/b* protein (*Cab*) and the small subunit of Rubisco (*RbcS*) are severely depleted, whereas transcripts encoding the cytosolic enzyme phosphoenolpyruvate (PEP) carboxylase are unaffected (Mayfield and Taylor, 1984, 1987). A similar phenotype can be induced by treating wild-type plants with the carotenoid synthesis inhibitor norflurazon in the light (Batschauer *et al.*, 1986; Longstaff *et al.*, 1989; Oelmüller and Mohr, 1986; Reiss *et al.*, 1983). Norflurazon-treated plants showed decreased levels of transcripts of a large number of nuclear genes for photosynthesis proteins, including *Cab*, *RbcS*, plastocyanin, ferredoxin, Rieske

FeS protein and the 23-kDa and 33-kDa extrinsic polypeptides of photosystem II, but transcripts for most cytosolic proteins were unaffected (Burgess and Taylor, 1987, 1988; Sagar *et al.,* 1988). Experiments with nuclei isolated from norflurazon-treated plants demonstrated that the effect was on the rate of transcription of the nuclear genes (Ernst and Schefbeck, 1988; Sagar *et al.,* 1988). This led to the suggestion that functional chloroplasts are required for the expression of these nuclear genes (Oelmüller, 1989; Taylor, 1989). Studies with promoters of photosynthesis genes fused to reporter genes, such as those encoding neomycin phosphotransferase and β-glucuronidase (GUS), in transgenic plants has established the requirement for functional chloroplasts for the expression of these genes (Lübberstedt *et al.,* 1994; Oelmüller *et al.,* 1993; Simpson *et al.,* 1986; Stockhaus *et al.,* 1989; Vorst *et al.,* 1993). The expression of these chimeric genes containing the *Cab, RbcS, PsbR, PetE* and *PetH* promoters was inhibited by norflurazon in high light, even though the reporter proteins remained in the cytosol and were not targeted to the chloroplasts. These experiments provide additional evidence for an effect on transcription. The most widely accepted suggestion is that functional chloroplasts produce a positive signal, the 'plastidic factor', required for the expression of nuclear genes for photosynthesis proteins, and that the production of this factor is prevented by photo-oxidation of chloroplasts (Oelmüller, 1989; Taylor, 1989). An alternative explanation, that photo-oxidation of chloroplasts prevents the import of precursor proteins from the cytosol, and that this leads to feedback inhibition of nuclear gene expression, has been suggested (Oelmüller, 1989; Rajasekhar, 1991), but has not received much support.

The 'plastidic factor'

The identity of the 'plastidic factor' is currently unknown, but experiments with the inhibitor tagetitoxin have indicated that chloroplast transcription is necessary for the expression of photosynthesis-related nuclear genes (Rapp and Mullett, 1991). Tagetitoxin is a specific inhibitor of chloroplast RNA polymerase in higher plants (Mathews and Durbin, 1990), and its application to wheat and barley seedlings results in inhibition of the accumulation of *Cab* and *RbcS* transcripts (Rapp and Mullett, 1991). There is conflicting evidence for a role for chloroplast translation in the formation of the 'plastidic factor'. Nuclear-encoded photosynthesis proteins are present in several plants with defects in chloroplast ribosomes, due either to mutations or to growth at elevated temperatures (Bradbeer *et al.,* 1979; Feierabend and Schrader-Reichhardt, 1976), and this was taken to suggest that chloroplast protein synthesis is not necessary for the production of the 'plastidic factor'. However, chloramphenicol, a chloroplast translation inhibitor, prevented the expression of *Cab* and *RbcS* genes when presented during the early stages, but not during the later stages, of seedling development in mustard (Oelmüller and Mohr, 1986). Similar results have also been obtained with lincomycin and erythromycin, inhibitors of chloroplast protein synthesis, in transgenic tobacco seedlings containing chimeric genes with the *RbcS, PetE* and *PetH* promoters (R. Sornarajah and J.C. Gray, unpublished results). Inhibition of chloroplast protein synthesis during the first 2–3 days of seedling growth, but not subsequently, prevented the normal light-induced expression from these promoters in transgenic plants. It has been suggested that this requirement for chloroplast protein synthesis at an early stage of development may be necessary for the formation of the chloroplast RNA polymerase, and that once the polymerase is present, further protein synthesis is not needed for production of the 'plastidic factor' (Rapp and Mullett, 1991). There is currently no evidence for the export of RNA from the chloroplast, suggesting that the transcripts which give rise to the 'plastidic factor' act within the chloroplast.

The involvement of the chloroplast glutamyl-tRNA in the synthesis of 5-aminolaevulinate, a key intermediate in chlorophyll synthesis (Schön *et al.,* 1986), led to suggestions that glutamyl-tRNA was the essential transcript produced by the chloroplasts, and that intermediates of the chlorophyll biosynthetic pathway may function as the 'plastidic factor' (Johanningmeier, 1988; Oelmüller, 1989; Susek and Chory, 1992). However, studies with the plastid-ribosome-deficient *albostrians* mutant of barley, which lacks the plastid glutamate tRNA, indicate that it would have to be the absence and not the accumulation of products of this pathway which would stimulate nuclear gene expression (Hess *et al.*, 1992). In addition, the results of recent studies with gabaculine, a potent inhibitor of 5-aminolaevulinate synthesis, appear to rule out the involvement of intermediates of the chlorophyll biosynthetic pathway (R. Sornarajah and J.C. Gray, unpublished results). Gabaculine had no effect on expression from the *PetE* and *PetH* promoters in transgenic tobacco seedlings, even though it completely inhibited the synthesis of chlorophyll in these plants.

The role of chloroplast protein synthesis in the production of the 'plastidic factor' requires further investigation. With transgenic tobacco seedlings, similar limited periods of both chloroplast transcription *and* translation are required for high-level light-induced expression from the *RbcS*, *PetE* and *PetH* promoters (R. Sornarajah and J.C. Gray, unpublished results). This may be taken to suggest that a chloroplast-encoded protein (or proteins) is required for the production of the 'plastidic factor' in normal plants. The absence of evidence for protein export by chloroplasts may be used to suggest a function within the chloroplast. The synthesis of many of the components of the photosynthetic apparatus in chloroplasts has led to suggestions that the 'plastidic factor' may be a product of photosynthesis (Tonkyn *et al.*, 1992). However, during early development, the expression of nuclear photosynthesis genes would have to precede the production of a functional photosynthetic apparatus. A role for photosynthesis in producing the 'plastidic factor' appears to be ruled out by experiments with 3-(3,4-dichlorophenyl)-1,1-dimethylurea (DCMU), an inhibitor of electron transfer through photosystem II (R. Sornarajah and J.C. Gray, unpublished results). DCMU had no effect on expression from the *RbcS*, *PetE* and *PetH* promoters in transgenic tobacco seedlings, even though it effectively inhibited photosystem II electron transfer in these plants.

The identification of the 'plastidic factor' remains a major challenge. Chloroplast transcription and translation may be required for the synthesis of a signal-generating machinery which, in the chloroplast, produces a mobile component of a signal transduction pathway leading to the nucleus. Alternatively, the 'plastidic factor' may be a chloroplast inner envelope membrane protein involved in signal transduction across the chloroplast envelope. A conserved open reading frame (*ycf10*) in plastid DNA from green plants, but not from *Epifagus virginiana*, has been shown to encode an inner envelope membrane protein (Sasaki *et al.*, 1993b; J. Craig, J.S. Knight and J.C. Gray, unpublished results). This putative membrane-spanning protein has been suggested to bind haem (Willey and Gray, 1990), which might conceivably be involved in redox signalling across the chloroplast envelope.

An alternative, genetic approach may provide important information for identification of the 'plastidic factor' or the signal transduction pathway from the chloroplast to the nucleus. Susek *et al.* (1993) have identified at least three *Arabidopsis* nuclear genes that are necessary for coupling the expression of nuclear genes encoding photosynthesis-related proteins to the functional state of the chloroplasts. Homozygous recessive *gun* (genomes uncoupled) mutations allow expression from the *Cab* and *RbcS* promoters in plants treated with norflurazon or chloramphenicol, or in white sectors of variegated leaves produced by the chloroplast mutator (*chm*) mutation (Susek *et al.*, 1993). The recessive nature of the *gun* mutations suggests that the signal transduction pathway normally functions to repress nuclear gene expression in the absence of the 'plastidic factor'.

Evidence for a possible repressor protein has been obtained by examining the DNA-binding activity of nuclear proteins extracted from green leaves or from white leaves of norflurazon-treated tobacco plants (R. Sornarajah and J.C. Gray, unpublished results). Nuclei from white leaves, but not from green leaves, contained a protein which bound to TATA-proximal regions of the pea *PetE* and *PetH* promoters (R. Sornarajah and J.C. Gray, unpublished results). This may be one of the targets influenced by the signal transduction pathway from the chloroplasts.

6. The role of mitochondria in chloroplast biogenesis

Chloroplast biogenesis requires large amounts of energy for the synthesis, translocation and assembly of the organelle components. However, during the early stages of chloroplast development the photosynthetic apparatus is not functional and is unable to provide the energy required. Oxidative phosphorylation in mitochondria is the most likely source of the necessary energy, and there is now considerable evidence that mitochondrial dysfunction may result in abnormal or prematurely arrested chloroplast development. Several non-chromosomal stripe (*ncs*) mutants of maize, displaying striped sectors of pale-green tissue in the leaves, have been analysed and shown to be caused by mutations and rearrangements of mitochondrial DNA (Hunt and Newton, 1991; Newton and Coe, 1988; Newton *et al.,* 1990; Roussell *et al.,* 1991). The development of chloroplasts in both bundle sheath and mesophyll cells is abnormal in the maternally inherited *ncs*2 mutant, and examination of the protein composition of the chloroplasts indicated a reduction in the reaction centre polypeptides of photosystem I (Roussell *et al.,* 1991). Rearrangements in mitochondrial DNA which affect the development of chloroplasts are also observed in mutations at the *chloroplast mutator* (*CHM*) locus in *Arabidopsis* (Martinez-Zapater *et al.,* 1992). These mutations generate a variegated phenotype which is inherited in a non-Mendelian fashion. Contrary to expectations, rearrangements of mitochondrial DNA co-segregated with the variegated phenotype in these mutants. Martinez-Zapater *et al.* (1992) speculated that *CHM* may encode a protein involved in the control of specific mitochondrial genome rearrangements. Uncontrolled rearrangements would lead to mitochondrial dysfunction and the inability to provide sufficient ATP for chloroplast development. The disruption of chloroplast biogenesis results in the variegated regions and stripes on the leaves. These genetic studies clearly indicate the importance of mitochondrial function for chloroplast development. However, it is not yet clear whether the co-ordination of mitochondrial function and chloroplast development occurs solely at the level of energy provision, or whether there is active signalling between the organelles.

7. Conclusions

Chloroplast biogenesis is regulated by an intricate set of interactions between the nuclear and chloroplast genomes which responds to extrinsic signals, such as light, or to intrinsic signals as part of normal plant development. Biochemical approaches have been successful in characterizing many of the functional components of chloroplasts and in identifying genes and cDNAs encoding these components. Such approaches have been less successful in characterizing the regulatory networks involved in chloroplast biogenesis. Genetic approaches, particularly with *Arabidopsis,* have recently identified novel classes of genetic loci whose mutation interferes with normal chloroplast development. Characterization of these genetic loci promises to provide important information about the regulatory pathways involved in chloroplast biogenesis. It is essential that the genetic and biochemical approaches are integrated to produce testable models of the pathways involved in co-ordinating nuclear and chloroplast gene expression.

Acknowledgements

I am extremely grateful to Renuka Sornarajah and Kate Duckett for discussion of chloroplast control of nuclear gene expression. The unpublished work from my laboratory cited in this article was supported by the Biotechnology and Biological Sciences Research Council.

References

Batschauer A, Mösinger E, Kreuz K, Dörr I, Apel K. (1986) The implication of a plastid-derived factor in the transcriptional control of nuclear genes encoding the light-harvesting chlorophyll *a/b* protein. *Eur. J. Biochem.* **154**: 625–634.

Baumgartner BD, Rapp JC, Mullet JE. (1989) Plastid transcription activity and DNA copy number increase early in barley chloroplast development. *Plant Physiol.* **89**: 1011–1018.

Bennoun P. (1982) Evidence for a respiratory chain in the chloroplast. *Proc. Natl Acad. Sci. USA* **79**: 4342–4356.

Berger S, Ellersiek U, Westhoff P, Steinmüller K. (1993) Studies on the expression of NDH-H, a subunit of the NAD(P)H-plastoquinone-oxidoreductase of higher plant chloroplasts. *Planta* **190**: 25–31.

Boffey SA, Leech RM. (1982) Chloroplast DNA levels and the control of chloroplast division in light grown wheat leaves. *Plant Physiol.* **69**: 1387–1391.

Boffey SA, Ellis JR, Selldén G, Leech RM. (1979) Chloroplast division and DNA synthesis in light grown wheat leaves. *Plant Physiol.* **64**: 502–505.

Bradbeer JW, Atkinson YE, Börner T, Hageman R. (1979) Cytoplasmic synthesis of plastid polypeptides may be controlled by plastid-synthesized RNA. *Nature* **279**: 816–817.

Burgess DG, Taylor WC. (1987) Chloroplast photo-oxidation affects the accumulation of cytosolic mRNAs encoding chloroplast proteins in maize. *Planta* **170**: 520–527.

Burgess DG, Taylor WC. (1988) The chloroplast affects the transcription of a nuclear gene family. *Mol. Gen. Genet.* **214**: 89–96.

Cashmore AR. (1984) Structure and expression of a pea nuclear gene encoding a chlorophyll *a/b*-binding polypeptide. *Proc. Natl Acad. Sci. USA* **81**: 2960–2964.

Chory J, Peto CA. (1990) Mutations in the *DET1* gene affect cell-type-specific expression of light-regulated genes and chloroplast development in *Arabidopsis. Proc. Natl Acad. Sci. USA* **87**: 8776–8780.

Chory J, Peto CA, Feinbaum R, Pratt L, Ausubel F. (1989) *Arabidopsis thaliana* mutant that develops as a light-grown plant in the absence of light. *Cell* **58**: 991–999.

Chory J, Nagpal P, Peto CA. (1991) Phenotypic and genetic analysis of *det2*, a new mutant that affects light-regulated seedling development in *Arabidopsis. Plant Cell* **3**: 445–459.

Commission on Plant Gene Nomenclature (1993) A nomenclature for sequenced plant genes. *Plant Mol. Biol. Rep.* **11**: 291–312.

Coruzzi G, Broglie R, Cashmore AR, Chua N-H. (1983) Nucleotide sequences of two pea cDNA clones encoding the small subunit of ribulose 1,5-bisphosphate carboxylase and the major chlorophyll *a/b*-binding thylakoid polypeptide. *J. Biol. Chem.* **258**: 1399–1402.

de Boer D, Cremers F, Teetstra R, Smits L, Hille J, Smeekens S, Weisbeek P. (1988) In vivo import of plastocyanin and a fusion protein into developmentally different plastids of transgenic plants. *EMBO J.* **7**: 2631–2635.

Deng X-W, Quail PH. (1992) Genetic and phenotypic characterization of *cop1* mutants of *Arabidopsis thaliana. Plant J.* **2**: 83–95.

Deng X-W, Caspar T, Quail PH. (1991) *cop1*: a regulatory locus involved in light-controlled development and gene expression in *Arabidopsis. Genes Dev.* **5**: 1172–1182.

Deng X-W, Matsui M, Wei N, Wagner D, Chu AM, Feldmann KA, Quail PH. (1992) *COP1*, an *Arabidopsis* regulatory gene, encodes a protein with both a zinc-binding motif and a G_β homologous domain. *Cell* **71**: 791–801.

Dreses-Werringloer U, Fischer K, Wachter E, Link TA, Flügge U-I. (1991) cDNA sequence and deduced amino acid sequence of the precursor of the 37-kDa inner envelope membrane polypeptide from spinach chloroplasts. Its transit peptide contains an amphiphilic α-helix as the only detectable structural element. *Eur. J. Biochem.* **195**: 361–368.

Duckett CM, Gray JC. (1995) Illuminating plant development. *BioEssays* **17**: 101–103.

Ellis JR, Leech RM. (1985) Cell size and chloroplast size in relation to chloroplast replication in light-grown wheat leaves. *Planta* **165**: 120–125.

Ellis RJ, van der Vies S. (1988) The Rubisco subunit binding protein. *Photosyn. Res.* **16:** 101–115.

Ernst D, Schefbeck K. (1988) Photo-oxidation of plastids inhibits transcription of nuclear encoded genes in rye (*Secale cereale*). *Plant Physiol.* **88:** 255–258.

Feierabend J, Schrader-Reichhardt U. (1976) Biochemical differentiation of plastids and other organelles in rye leaves with a high-temperature-induced deficiency of plastid ribosomes. *Planta* **129:** 133–145.

Flügge U-I, Fischer K, Gross A, Sebald W, Lottspeich F, Eckershorn C. (1989) The triose phosphate–3-phosphoglycerate-phosphate translocator from spinach chloroplasts. Nucleotide sequence of a full-length cDNA clone and import of the in vitro synthesized precursor protein into chloroplasts. *EMBO J.* **8:** 39–46.

Frances S, White MJ, Edgerton MD, Jones AL, Elliott RC, Thompson WF. (1992) Initial characterization of a pea mutant with light-independent photomorphogenesis. *Plant Cell* **4:** 1519–1530.

Gilmartin PM, Sarokin L, Memelink J, Chua N-H. (1990) Molecular light switches for plant genes. *Plant Cell* **2:** 369–378.

Glick BS, Beasley EM, Schatz G. (1992) Protein sorting in mitochondria. *Trends Biochem. Sci.* **17:** 453–459.

Gray JC, Row PE. (1995) Protein translocation across chloroplast envelope membranes. *Trends Cell Biol.* **5:** 243–247.

Gray JC, Hird SM, Dyer TA. (1990) Nucleotide sequence of a wheat chloroplast gene encoding the proteolytic subunit of an ATP-dependent protease. *Plant Mol. Biol.* **15:** 947–950.

Guiliano G, Scolnik PA. (1988) Transcription of two photosynthesis-associated nuclear gene families correlates with the presence of chloroplasts in leaves of the variegated tomato ghost mutant. *Plant Physiol.* **86:** 7–9.

Harpster MN, Mayfield SP, Taylor WC. (1984) Effects of pigment-deficient mutants on the accumulation of photosynthesis proteins in maize. *Plant Mol. Biol.* **3:** 59–71.

Hartl F-U, Neupert W. (1990) Protein sorting to mitochondria: evolutionary conservations of folding and assembly. *Science* **247:** 930–938.

Hashimoto H, Possingham JV. (1989a) DNA levels in dividing and developing plastids in expanding primary leaves of *Avena sativa*. *J. Exp. Bot.* **40:** 257–262.

Hashimoto H, Possingham JV. (1989b) Effect of light on the chloroplast division cycle and DNA synthesis in cultured leaf discs of spinach. *Plant Physiol.* **89:** 1178–1183.

Hess WR, Schendel R, Rüdiger W, Fieder B, Börner T. (1992) Components of chlorophyll biosynthesis in a barley albino mutant unable to synthesize δ-aminolevulinic acid by utilizing the transfer RNA for glutamic acid. *Planta* **188:** 19–27.

Hiratsuka J, Shimada H, Whittier RF, et al. (1989) The complete sequence of the rice (*Oryza sativa*) chloroplast genome: intermolecular recombination between distinct tRNA genes accounts for a major plastid DNA inversion during the evolution of the cereals. *Mol. Gen. Genet.* **217:** 185–194.

Höfte H, Desprez T, Amselem J, et al. (1993) An inventory of 1152 expressed sequence tags obtained by partial sequencing of cDNAs from *Arabidopsis thaliana*. *Plant J.* **4:** 1051–1061.

Hu J, Bogorad L. (1990) Maize chloroplast RNA polymerase: the 180-, 120-, and 38-kilodalton polypeptides are encoded in chloroplast genes. *Proc. Natl Acad. Sci. USA* **87:** 1531–1535.

Hu J, Troxler RF, Bogorad L. (1991) Maize chloroplast RNA polymerase: the 78-kilodalton polypeptide is encoded by the plastid rpoC1 gene. *Nucleic Acids Res.* **19:** 3431–3434.

Hunt MD, Newton KJ. (1991) The NCS3 mutation: genetic evidence for the expression of ribosomal protein genes in *Zea mays* mitochondria. *EMBO J.* **10:** 1045–1052.

Jacob F, Brenner S, Cuzin F. (1963) On the regulation of DNA replication in bacteria. *Cold Spring Harbor Symp. Quant. Biol.* **28:** 329–348.

Johanningmeier U. (1988) Possible control of transcript levels by chlorophyll precursors in *Chlamydomonas*. *Eur J. Biochem.* **177:** 417–424.

Keegstra K, Olsen LJ, Theg SM. (1989) Chloroplastic precursors and their transport across the envelope membranes. *Annu. Rev. Plant Physiol. Plant Mol. Biol.* **40:** 471–501.

Knight JS, Gray JC. (1995) The N-terminal hydrophobic region of the mature phosphate translocator is sufficient for targeting to the chloroplast inner envelope membrane. *Plant Cell* **7:** 1421–1432.

Knight JS, Madueño F, Gray JC. (1993) Import and sorting of proteins by chloroplasts. *Biochem. Soc. Trans.* **21:** 31–36.

Kuhlemeier C, Green PJ, Chua N-H. (1987) Regulation of gene expression in higher plants. *Annu. Rev. Plant Physiol.* **38:** 221–257.

Kuroiwa T. (1991) The replication, differentiation, and inheritance of plastids with emphasis on the concept of organelle nuclei. *Int. Rev. Cytol.* **128:** 1–62.

Lamppa GK, Elliot LV, Bendich AJ. (1980) Changes in chloroplast number during pea leaf development. *Planta* **148**: 437–443.

Landoulsi A, Malki A, Kern R, Kohiyama M, Hughes P. (1990) The *E. coli* cell surface specifically prevents the initiation of DNA replication at *oriC* on hemimethylated DNA templates. *Cell* **63**: 1053–1060.

Leech RM, Thomson WW, Platt-Aloia KA. (1981) Observations on the mechanism of chloroplast division in higher plants. *New Phytol.* **87**: 1–9.

Li H-M, Moore T, Keegstra K. (1991) Targeting of proteins to the outer envelope membrane uses a different pathway than transport into chloroplasts. *Plant Cell* **3**: 709–717.

Li H-M, Sullivan TD, Keegstra K. (1992) Information for targeting to the chloroplastic inner envelope membrane is contained in the mature region of the maize *Bt1*-encoded protein. *J. Biol. Chem.* **267**: 18999–19004.

Li S-J, Cronan JE. (1992a) The genes encoding the two carboxyltransferase subunits of *Escherichia coli* acetyl-CoA carboxylase. *J. Biol. Chem.* **267**: 16841–16847.

Li S-J, Cronan JE. (1992b) A putative zinc finger protein encoded by a conserved chloroplast gene is very likely a subunit of a biotin-dependent carboxylase. *Plant Mol. Biol.* **20**: 759–761.

Lloyd JC, Raines CA, John UP, Dyer TA. (1991) The chloroplast FBPase gene of wheat: structure and expression of the promoter in photosynthetic and meristematic cells of transgenic tobacco plants. *Mol. Gen. Genet.* **225**: 209–216.

Longstaff M, Raines CA, McMorrow EM, Bradbeer JW, Dyer TA. (1989) Wheat phosphoglycerate kinase: evidence for recombination between genes for the chloroplastic and cytosolic enzymes. *Nucleic Acids Res.* **16**: 6569–6580.

Lubben TH, Donaldson GK, Viitanen PV, Gatenby AA. (1989) Several proteins imported into chloroplasts form stable complexes with the GroEL-related chloroplast molecular chaperone. *Plant Cell* **1**: 1223–1230.

Lübberstedt T, Oelmüller R, Wanner G, Herrmann RG. (1994) Interacting *cis* elements in the plastocyanin promoter from spinach ensure regulated high-level expression. *Mol. Gen. Genet.* **242**: 602–613.

McCourt P, Kunst L, Browse J, Somerville CR. (1987) The effects of reduced amounts of lipid unsaturation on chloroplast ultrastructure and photosynthesis in a mutant of *Arabidopsis. Plant Physiol.* **84**: 353–360.

McNellis TW, von Arnim AG, Deng X-W. (1994) Overexpression of *Arabidopsis COP1* results in partial suppression of light mediated development: evidence for a light-inactivable repressor of photomorphogenesis. *Plant Cell* **6**: 1391–1400.

Madueño F, Napier JA, Gray JC. (1993) Newly imported Rieske iron-sulfur protein associates with both Cpn60 and Hsp70 in the chloroplast stroma. *Plant Cell* **5**: 1865–1976.

Maier RM, Neckermann K, Igloi GL, Kössel H. (1995) Complete sequence of the maize chloroplast genome: gene content, hotspots of divergence and fine tuning of genetic information by transcript editing. *J. Mol. Biol.* **251**: 614–628.

Marrison JL, Leech RM. (1992) Co-immunolocalization of topoisomerase II and chloroplast DNA in developing, dividing and mature wheat chloroplasts. *Plant J.* **2**: 783–790.

Marshall JS, DeRocher AE, Keegstra K, Vierling E. (1990) Identification of heat shock protein hsp70 homologues in chloroplasts. *Proc. Natl Acad. Sci. USA* **87**: 374–378.

Martinez-Zapater JM, Gil P, Capel J, Somerville CR. (1992) Mutations at the *Arabidopsis CHM* locus promote rearrangements of the mitochondrial genome. *Plant Cell* **4**: 889–899.

Mathews DE, Durbin RD. (1990) Tagetitoxin inhibits RNA synthesis directed by RNA polymerases from chloroplasts and *Escherichia coli. J. Biol. Chem.* **265**: 493–498.

Matsubayashi T, Wakasugi T, Shinozaki K, et al. (1987) Six chloroplast genes (*ndhA–F*) homologous to human mitochondrial genes encoding components of the respiratory chain NADH dehydrogenase are actively expressed: determination of the splice sites in *ndhA* and *ndhB* pre-mRNAs. *Mol. Gen. Genet.* **210**: 385–393.

Maurizi MR, Clark WP, Kim S-H, Gottesman S. (1990) ClpP represents a unique family of serine proteases. *J. Biol. Chem.* **265**: 12546–12552.

Mayfield SP, Taylor WC. (1984) Carotenoid-deficient maize seedlings fail to accumulate light-harvesting chlorophyll *a/b* binding protein (LHCP) mRNA. *Eur. J. Biochem.* **144**: 79–84.

Mayfield SP, Taylor WC. (1987) Chloroplast photo-oxidation inhibits the expression of a set of nuclear genes. *Mol. Gen. Genet.* **208**: 309–314.

Meeker R, Nielsen B, Tewari KK. (1988) Localization of replication origins in pea chloroplast DNA. *Mol. Cell. Biol.* **8**: 1216–1223.

Miyamura S, Nagata T, Kuroiwa T. (1986) Quantitative fluorescence microscopy of dynamic changes of plastid nucleoids during wheat development. *Protoplasma* **133**: 66–72.

Miyamura S, Kuroiwa T, Nagata T. (1990) Multiplication and differentiation of plastid nucleoids during development of chloroplasts and etioplasts from proplastids in *Triticum aestivum*. *Plant Cell Physiol.* **31**: 597–602.

Morden CW, Wolfe KH, dePamphilis CW, Palmer JD. (1991) Plastid translation and transcription in a non-photosynthetic plant: intact, missing and pseudogenes. *EMBO J.* **10**: 3281–3288.

Mullet JE. (1988) Chloroplast development and gene expression. *Annu. Rev. Plant Physiol. Plant Mol. Biol.* **39**: 475–502.

Neuhaus G, Bowler C, Kern R, Chua N-H. (1993) Calcium/calmodulin-dependent and -independent phytochrome signal transduction pathways. *Cell* **73**: 937–952.

Newton KJ, Coe EH. (1988) Mitochondrial DNA changes in abnormal growth (non-chromosomal stripe) mutants of maize. *Proc. Natl Acad. Sci. USA* **83**: 7363–7366.

Newton KJ, Knudsen C, Gabay-Laughnan S, Laughnan JR. (1990) An abnormal growth mutant in maize has a defective mitochondrial cytochrome oxidase gene. *Plant Cell* **2**: 107–113.

Nixon PJ, Gounaris K, Coomber SA, Hunter CN, Dyer TA, Barber J. (1989) *psbG* is not a photosystem two gene but may be an *ndh* gene. *J. Biol. Chem.* **264**: 14129–14135.

Oblong JE, Lamppa GK. (1992) Identification of two structurally related proteins involved in proteolytic processing of precursors targeted to the chloroplast. *EMBO J.* **11**: 4401–4409.

Oelmüller R. (1989) Photo-oxidative destruction of chloroplasts and its effects on nuclear gene expression and extraplastidic enzyme levels. *Photochem. Photobiol.* **49**: 229–239.

Oelmüller R, Mohr H. (1986) Photo-oxidative destruction of chloroplasts and its consequences for expression of nuclear genes. *Planta* **167**: 106–113.

Oelmüller R, Bolle C, Tyagi AK, Niekrawietz N, Breit S, Herrmann RG. (1993) Characterization of the promoter for the single-copy gene encoding ferredoxin-NADP$^+$-oxidoreductase from spinach. *Mol. Gen. Genet.* **237**: 261–272.

Palmer JD. (1991) Plastid chromosomes: structure and evolution. In: *Molecular Biology of Plastids and the Photosynthetic Apparatus, Vol. 7A, Cell Culture and Somatic Genetics of Plants* (Bogorad L, Vasil IK, ed). New York: Academic Press, pp. 5–53.

Pepper A, Delaney T, Washburn T, Poole D, Chory J. (1994) *DET1*, a negative regulator of light-mediated development and gene expression in *Arabidopsis,* encodes a novel nuclear-localised protein. *Cell* **78**: 109–116.

Possingham JV. (1980) Plastid replication and development in the life cycle of higher plants. *Annu. Rev. Plant Physiol.* **31**: 113–129.

Purton S, Gray JC. (1989) The plastid *rpo*A gene encoding a protein homologous to the bacterial RNA polymerase alpha subunit is expressed in pea chloroplasts. *Mol. Gen. Genet.* **217**: 77–84.

Pyke KA, Leech RM. (1987) The control of chloroplast number in mesophyll cells. *Planta* **170**: 416–420.

Pyke KA, Leech RM. (1991) Rapid image analysis screening procedure for identifying chloroplast number mutants in mesophyll cells of *Arabidopsis thaliana* (L.) Heynh. *Plant Physiol.* **96**: 1193–1195.

Pyke KA, Leech RM. (1992) Chloroplast division and expansion is radically altered by nuclear mutations in *Arabidopsis thaliana. Plant Physiol.* **99**: 1005–1008.

Pyke KA, Leech RM. (1994) A genetic analysis of chloroplast division and expansion in *Arabidopsis thaliana. Plant Physiol.* **104**: 201–207.

Raines CA, Longstaff M, Dyer TA. (1989) Complete coding sequence of wheat phosphoribulokinase: developmental and light-dependent expression of the mRNA. *Mol. Gen. Genet.* **220**: 43–48.

Rajasekhar VK. (1991) Regulation of nuclear gene expression for plastidogenesis affected by developmental stage of plastids. *Biochem. Physiol. Pflanzen* **187**: 257–271.

Rapp JC, Mullet JE. (1991) Chloroplast transcription is required to express the nuclear genes rbcS and cab. Plastid DNA copy number is regulated independently. *Plant Mol. Biol.* **17**: 813–823.

Reiss T, Bergfeld R, Link G, Thien W, Mohr H. (1983) Photo-oxidative destruction of chloroplasts and its consequences for cytosolic enzyme levels and plant development. *Planta* **159**: 518–528.

Robertson D, Laetsch WM. (1974) Structure and function of developing barley plastids. *Plant Physiol.* **54**: 148–159.

Robinson C, Ellis RJ. (1984) Transport of proteins into chloroplasts. Partial purification of a chloroplast protease involved in the processing of imported precursor polypeptides. *Eur. J. Biochem.* **142**: 337–342.

Roussell DL, Thompson DL, Pallardy SG, Miles D, Newton KJ. (1991) Chloroplast structure and function is altered in the NCS2 maize mitochondrial mutant. *Plant Physiol.* **96**: 232–238.

Sagar AD, Horwitz BA, Elliott RC, Thompson WF, Briggs WR. (1988) Light effects on several chloroplast components in norflurazon-treated pea seedlings. *Plant Physiol.* **88**: 340–347.

Salomon M, Fischer K, Flügge U-I, Soll J. (1990) Sequence analysis and protein import studies of an outer chloroplast envelope polypeptide. *Proc. Natl Acad. Sci. USA* **87:** 5578–5582.

Sasaki Y, Hakamada K, Suama Y, Nagano Y, Furusawa I, Matsuno R. (1993a) Chloroplast-encoded protein as a subunit of acetyl-CoA carboxylase in pea plant. *J. Biol. Chem.* **268:** 25118–25123.

Sasaki Y, Sekiguchi K, Nagano Y, Matsuno R. (1993b) Chloroplast envelope protein encoded by chloroplast genome. *FEBS Lett.* **316:** 93–98.

Sato N, Albrieux C, Joyard J, Douce R, Kuroiwa T. (1993) Detection and characterization of a plastid envelope DNA-binding protein which may anchor plastid nucleoids. *EMBO J.* **12:** 555–561.

Schantz R, Bogorad L. (1988) Maize chloroplast genes *ndhD, ndhE* and *psaC.* Sequences, transcripts and transcript pools. *Plant Mol. Biol.* **11:** 239–247.

Schön A, Krupp G, Berry-Lowe S, Kannangara G, Söll D. (1986) The RNA required in the first step of chlorophyll biosynthesis is a chloroplast glutamate tRNA. *Nature* **322:** 281–284.

Scott NS, Possingham JV. (1980) Chloroplast DNA in expanding spinach leaves. *J. Exp. Bot.* **31:** 1081–1092.

Selldén G, Leech RM. (1981) Localization of DNA in mature and young wheat chloroplasts using the fluorescent probe 4, 6-diaminidino-2-phenylindole. *Plant Physiol.* **68:** 731–734.

Shinozaki K, Ohme M, Tanaka M, *et al.* (1986) The complete nucleotide sequence of the tobacco chloroplast genome: its gene organization and expression. *EMBO J.* **5:** 2043–2049.

Simpson J, van Montagu M, Herrera-Estrella L. (1986) Photosynthesis-associated gene families: differences in response to tissue-specific and environmental factors. *Science* **233:** 34–38.

Smith AG, Wilson RM, Kaethner TM, Willey DL, Gray JC. (1991) Pea chloroplast genes encoding a 4kDa polypeptide of photosystem I and a putative enzyme of C_1 metabolism. *Curr. Genet.* **19:** 403–410.

Steinmetz AA, Castroviejo M, Sayre RT, Bogorad L. (1986) Protein PSII-G. An additional component of photosystem II identified through its plastid gene in maize. *J. Biol. Chem.* **261:** 2485–2488.

Steinmüller K, Ley AC, Steinmetz AA, Sayre RT, Bogorad L. (1989) Characterization of the *ndhC-psbG-ORF157/159* operon of maize plastid DNA and of the cyanobacterium *Synechocystis* sp. PCC6803. *Mol. Gen. Genet.* **216:** 60–69.

Stockhaus J, Schell J, Willmitzer L. (1989) Correlation of the expression of the nuclear photosynthesis gene ST-LS1 with the presence of chloroplasts. *EMBO J.* **8:** 2445–2451.

Sugiura M. (1992) The chloroplast genome. *Plant Mol. Biol.* **19:** 149–168.

Susek RE, Chory J. (1992) A tale of two genomes: role of a chloroplast signal in coordinating nuclear and plastid genome expression. *Aust. J. Plant Physiol.* **19:** 211–233.

Susek RE, Ausubel FM, Chory J. (1993) Signal transduction mutants of *Arabidopsis* uncouple nuclear *CAB* and *RBCS* gene expression from chloroplast development. *Cell* **74:** 787–799.

Taylor WC. (1989) Regulatory interactions between nuclear and chloroplast genomes. *Annu. Rev. Plant Physiol.* **40:** 211–233.

Tittgen J, Hermans J, Steppuhn J, Jansen T, Jansson C, Andersson B, Nechushtai R, Nelson N, Herrmann RG. (1986) Isolation of cDNA clones for fourteen nuclear-encoded thylakoid membrane proteins. *Mol. Gen. Genet.* **204:** 258–265.

Tonkyn JC, Deng X-W, Gruissem W. (1992) Regulation of plastid gene expression during photo-oxidative stress. *Plant Physiol.* **99:** 1406–1415.

Tsugeki R, Nishimura M. (1993) Interaction of homologues of Hsp70 and Cpn60 with ferredoxin-NADP$^+$ reductase upon its import into chloroplasts. *FEBS Lett.* **320:** 198–202.

Uchimiya H, Kidou S-I, Shimazaki T, *et al.* (1992) Random sequencing of cDNA libraries reveals a variety of expressed genes in cultured cells of rice (*Oryza sativa* L.). *Plant J.* **2:** 1005–1009.

von Heijne G, Steppuhn J, Herrmann RG. (1989) Domain structure of mitochondrial and chloroplast targeting peptides. *Eur. J. Biochem.* **180:** 535–545.

Vorst O, Kock P, Lever A, Wetering B, Weisbeek P, Smeekens S. (1993) The promoter of the *Arabidopsis thaliana* plastocyanin gene contains a far upstream enhancer-like element involved in chloroplast-dependent expression. *Plant J.* **4:** 933–945.

Wakasugi T, Tsudzuki J, Ito S, Nakashima K, Tsudzuki T, Sugiura M. (1994) Loss of all *ndh* genes as determined by sequencing the entire chloroplast genome of the black pine *Pinus thunbergii. Proc. Natl Acad. Sci. USA* **91:** 9794–9798.

Wei N, Deng X-W. (1992) *COP9*: a new genetic locus involved in light-regulated development and gene expression in *Arabidopsis. Plant Cell* **4:** 1507–1516.

Willey DL, Gray JC. (1990) An open reading frame encoding a putative haem-binding polypeptide is cotranscribed with the pea chloroplast gene for apocytochrome *f. Plant Mol. Biol.* **15:** 347–356.

Wolfe KH, Morden CW, Palmer JD. (1992a) Function and evolution of a minimal plastid genome from a nonphotosynthetic parasitic plant. *Proc. Natl Acad. Sci. USA* **89:** 10648–10652.

Wolfe KH, Morden CW, Ems SC, Palmer JD. (1992b) Rapid evolution of the plastid translational apparatus in a nonphotosynthetic plant: loss or accelerated sequence evolution of tRNA and ribosomal protein genes. *J. Mol. Evol.* **35:** 304–317.

Yalovsky S, Paulsen H, Michaeli D, Chitnis PR, Nechushtai R. (1992) Involvement of a chloroplast HSP70 heat shock protein in the integration of a protein (light-harvesting complex protein precursor) into the thylakoid membrane. *Proc. Natl Acad. Sci. USA* **89:** 5616–5619.

Chapter 26

Translocation of proteins across chloroplast membranes

Colin Robinson

1. Introduction

The sheer variety of organelles in plant cells ensures that protein translocation is a central feature of plant cell biology. Cytosolically synthesized proteins are targeted through the endomembrane system to the endoplasmic reticulum, Golgi body, protein bodies and other destinations, whereas other proteins are targeted directly to the chloroplasts and mitochondria and glyoxysomes. In all, a considerable proportion of newly synthesized proteins are rapidly and specifically translocated into one or other of these organelles, and a considerable effort has been made to understand the underlying translocation and sorting mechanisms. The protein traffic to the chloroplast is especially intensive, partly because this is the site of numerous metabolic functions, and partly because this organelle is the most complex of all in structural terms. The presence of three distinct membranes (outer and inner envelope and thylakoid membranes) and three soluble phases (intermembrane space, stroma and thylakoid lumen) requires that proteins must first be targeted to the chloroplast on a large scale, and subsequently 'sorted' within the organelle so that the proteins end up in the correct locations. The entire process is made even more complex by the fact that some *chloroplast*-encoded proteins have to be directed into or across membranes; many of the key photosynthetic proteins are encoded by the chloroplast genome and targeted into or across the thylakoid membrane, and there is now evidence for the targeting of chloroplast-encoded proteins into the envelope (Sasaki *et al., 1993*). It should also be borne in mind that the chloroplast is only one member of the large plastid family, and numerous proteins must also be transported into amyloplasts, leucoplasts, chromoplasts and other members of this family, probably by rather similar basic mechanisms.

A variety of *in-vitro* assays have been developed for the study of chloroplast protein import and sorting, and many of the individual translocation/integration processes can now be studied in detail. The purpose of this article is to review recent advances in studies of the translocation of proteins across both the envelope and thylakoid membranes, but particular emphasis will be placed on the biogenesis of thylakoid proteins. Recent studies in this area point to the operation of a completely unexpected variety of translocation and integration mechanisms, and there are signs that these findings have major implications for the evolution of the translocases themselves.

2. Translocation of proteins into and across the envelope membranes

Early steps in the import process

Several hundred different proteins are imported from the cytosol into chloroplasts, and most of these are destined for either the soluble stromal phase or the thylakoid network. The import process was first reconstituted *in vitro* in 1978 (Chua and Schmidt, 1978; Highfield and

Ellis, 1978) in studies of the import of the small subunit of ribulose bisphosphate carboxylase/oxygenase (Rubisco SSU), and these studies established two key tenets in this field which appear to apply to all imported stromal and thylakoidal proteins: the proteins are synthesized as larger precursors and they are imported *post-translationally*. Strictly speaking, these studies showed that protein translocation across the chloroplast envelope *can* take place post-translationally *in vitro* when the imported protein is synthesized in a cell-free system and incubated with isolated chloroplasts. However, the apparent absence of bound ribosomes in electron micrographs of chloroplasts suggests that co-translational protein import occurs at low levels, if at all, *in vivo*, and a recent study in *Chlamydomonas reinhardtii* (Howe and Merchant, 1993) has shown the transient appearance of precursor forms of chloroplast proteins *in vivo*. This is an interesting point because post-translational protein import into mitochondria is also known to proceed efficiently *in vitro*, but there is now evidence for co-translational import *in vivo* in yeast (Fujiki and Verner, 1993).

All imported proteins studied to date (with the exception of outer envelope membrane proteins, discussed later) are initially synthesized with amino-terminal presequences, sometimes termed transit peptides, which are typically about 50–80 residues in length, but can be much longer. Not surprisingly, these presequences have been found to contain essential targeting signals specifying targeting of the precursor protein into chloroplasts; a variety of studies have shown that chimeric proteins, consisting of a chloroplast protein presequence linked to 'foreign' mature proteins, can be imported by chloroplasts both *in vitro* and *in vivo* (Kavanagh *et al.,* 1988; Lubben *et al., 1989*; van den Broeck *et al.,* 1985), although with widely varying efficiencies. The targeting signals appear to be recognized very specifically by chloroplasts, since there is no evidence for the mis-targeting of chloroplast (or mitochondrial) proteins on a significant scale (Boutry *et al.,* 1987; de Boer *et al.,* 1988; Whelan *et al.,* 1990).

Although the role of the presequence in the import process is now fairly certain, the molecular details of the presequence–chloroplast recognition process are anything but clear. The presequences of different proteins are almost invariably basic and enriched in hydroxylated residues, but they otherwise share virtually no primary sequence similarity or identifiable common secondary structure (Pilon *et al.,* 1992; Theg and Geske, 1992; von Heijne *et al.,* 1989; von Heijne and Nishikawa, 1991). It is therefore extremely difficult to predict how they might interact in such a specific manner with the chloroplast import apparatus, and this crucial aspect of chloroplast biogenesis is therefore a virtual mystery.

Translocation of proteins across the envelope membranes

Although the key membrane-translocation events are undoubtedly mediated by chloroplast proteins, there is some evidence to suggest that the initial interaction may be between precursor protein and chloroplast membrane lipids. Chloroplast protein presequences possess high affinities for non-polar environments (Theg and Geske, 1992; van't Hof *et al.,* 1993), and it was suggested that the presequence–lipid interaction may be the one that confers organelle-specific recognition, since parts of the Rubisco SSU presequence interact specifically with lipids characteristic of the chloroplast outer envelope membrane. Horniak *et al.* (1993) also found that secondary structure is induced in the ferredoxin presequence by anionic lipids, and suggested that such lipid-induced structural motifs may serve as recognition signals for the import apparatus.

However, the first clearly identified step in the import process involves the binding of the precursor protein to proteinaceous receptors on the chloroplast surface. It has been shown (Cline *et al.,* 1985) that the treatment of chloroplasts with proteases leads to a drastic reduction in their ability to bind or import precursor molecules subsequently, and binding studies (Friedman and Keegstra, 1988) have provided some indication of the number of specific binding sites on the chloroplast surface (about 2000–3000 per chloroplast). Some precursor proteins have been shown to be import-competent after expression in *Escherichia coli* and subsequent purification (Pilon *et al.,* 1990; Cline *et al.,* 1993), suggesting that the precursors interact directly with the import apparatus in the chloroplast envelope. However, in one case (Waegemann *et al.,* 1990), plant extracts appeared to be required for import to take place, raising the possibility that cytosolic factors may be involved in the import mechanism. Even in those cases where purified precursors are shown to be imported in the absence of other factors, it remains a distinct possibility that cytosolic factors may assist import *in vivo,* in order to improve the efficiency of the import process. The binding of newly synthesized precursors to cytosolic 'chaperone' molecules would also explain why no co-translational import has been observed, despite the demonstrations that amino-terminal presequences have a high affinity for chloroplasts.

Identification and characterization of the import receptors has proved difficult and controversial. Early reports (Pain *et al.,* 1988; Schnell *et al.,* 1990) that the import receptor is a 36-kDa envelope protein were challenged (Flügge *et al.,* 1991), and more recent analyses of the import machinery have failed to detect this protein. A number of proteins have now been identified as belonging to the import machinery, and characterization of these proteins is now progressing rapidly. These studies have been aided by the finding that at least two steps in the overall import process require ATP. Moreover, the sites at which ATP is required have interesting implications for the import mechanism. ATP at low concentrations (100 μM) is needed for the *binding* of precursors to the chloroplast (Olsen *et al.,* 1989), suggesting that binding may be an active process, perhaps involving the unfolding of the precursor protein, and subsequent translocation across the envelope is dependent on *stromal* ATP at higher concentrations (Theg *et al.,* 1989). This may be an indication that stromal proteins are involved in 'pulling' the precursor through the envelope membrane, as has been proposed for the action of matrix-localized Hsp70 in mitochondrial protein import (for review see Glick *et al.,* 1992). Cross-linking studies by Perry and Keegstra (1994) at low ATP concentrations have implicated two outer membrane proteins, of molecular mass 86 kDa and 75 kDa, in the early binding events. The 86-kDa protein alone could be labelled even in the absence of ATP, suggesting that the 86-kDa protein may be the actual import receptor, which then perhaps passes the precursor to the 75-kDa protein in an ATP-dependent fashion. The 75-kDa protein was also found to be mainly located at contact sites between the outer and inner membranes, providing further evidence that it participates in the translocation process slightly later than the 86-kDa protein. The next steps appear to involve translocation into the stroma at 'contact sites' between the outer and inner envelope membranes; studies using fusion proteins have identified a translocation intermediate which spans both membranes (Schnell and Blobel, 1993).

Wu *et al.* (1994) used different constructs in another cross-linking study and identified several envelope proteins among the adducts. These included an outer envelope Hsp70-related protein, 44-kDa proteins localized in both the outer and inner membranes, and a 97-kDa protein in the inner membrane. Finally, two groups have attempted to obtain highly enriched fractions containing import components, with reassuringly similar results. Schnell *et al.* (1994) used pre-ferredoxin–protein A chimeras to 'tag' the import apparatus and identified proteins of 86 kDa, 75 kDa (two different ones) and 34 kDa when the ATP concentration was kept at 100 μM. If the import process was allowed to proceed further by the inclusion of 2 mM ATP, two other proteins of 100 kDa and 36 kDa were detected, both from the inner membrane. The 100-kDa protein almost certainly corresponds to the 97-kDa protein identified by Wu *et al.* (1994), while the 86-kDa and 75-kDa proteins correspond to those identified by Perry and Keegstra. The second 75-kDa protein was identified by Schnell *et al.* (1994) as an Hsp70-related protein, and also probably corresponds to one of the proteins identified by Wu *et al.* (1994). Further independent identification of the 86-kDa, 75-kDa, 34-kDa and Hsp70 proteins has emerged from fractionation studies by Waegemann and Soll (1991).

Figure 1, adapted from a diagram by Gray and Row (1995), summarizes a model for the overall chloroplast import process, although further work is required to verify some of these steps. After an initial interaction with membrane lipids, the precursor encounters the 86-kDa protein, which may be a GTP-hydrolysing protein (Kessler *et al.*, 1994). The component denoted 70? represents an outer envelope Hsp70 homologue identified by Ko *et al.* (1992), which may possibly be involved in presenting an unfolded precursor protein to the import apparatus at some stage. The complex then interacts with the 75-kDa protein and the 34-kDa protein, which is also predicted to hydrolyse GTP (Kessler *et al.*, 1994). Another Hsp70 homologue is present in the intermembrane space and forms part of the isolated translocation complex (Schnell *et al.*, 1994; Waegemann and Soll, 1991). Next, the precursor interacts with the import machinery in the inner envelope membrane (100- and 36-kDa proteins) before being pulled across the membranes by the action of a stromal Hsp70 molecule.

Events in the stroma

During or shortly after transport across the envelope membranes, precursors of stromal proteins are processed to the mature forms by a stromal processing peptidase (SPP), a metal-dependent enzyme which is highly specific for imported precursor proteins (Robinson and Ellis, 1984), and which in pea consists of one or two subunits of about 145 kDa (Oblong and Lamppa, 1992). As with the chloroplast-targeting signals, however, the basis for the high degree of reaction specificity of SPP has proved difficult to decipher, because the SPP cleavage sites of different stromal protein precursors display essentially no primary sequence similarity (von Heijne *et al.*, 1989).

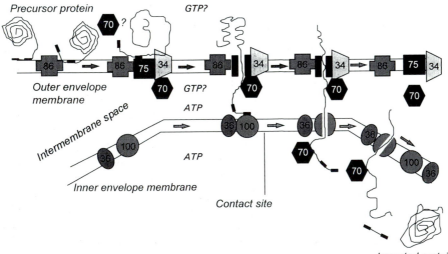

Figure 1. A model for the import of proteins across the envelope membranes of higher plant chloroplasts. Cytosolically synthesized stromal proteins, for example Rubisco SSU, are synthesized with amino-terminal presequences (oval) which specify ATP-dependent binding to import receptors on the chloroplast surface. After binding, precursors are imported across the two envelope membranes at contact sites, probably through a proteinaceous pore, after which they are processed to the mature size. The diagram illustrates a current model for the translocation process which is described in the text.

Although the details are still unclear, there is good evidence that further critical protein–protein interactions must take place before the import of stromal proteins is correctly completed. Newly imported proteins have been found to associate transiently with a stromal member of the Hsp60 family, termed chaperonin 60 (Cpn60), which is believed to be essential for the correct folding and assembly of chloroplast-encoded proteins (notably the large subunit of Rubisco) and the refolding of imported proteins, including the Rubisco SSU (Hemmingsen et al., 1988). There is also some evidence from studies of the import of ferredoxin-NADP reductase that at least some proteins may interact with a stromal Hsp70 protein (Tsugeki and Nishimura, 1993). The precise sequence of events has yet to be determined, but the overall message seems to be that the import of functional proteins into the stroma requires the participation of a variety of both membrane-bound and soluble chloroplast proteins.

A different import mechanism for outer envelope membrane proteins

As far as is known, all stromal and thylakoidal proteins enter the chloroplast by the same receptor-mediated, ATP-dependent mechanism; it is only after this stage that intra-organellar sorting takes place. It is also likely that inner envelope membrane proteins are imported by the same mechanism, since these are similarly synthesized with presequences and imported by an ATP-dependent process (Flügge et al., 1989), although it is not known whether these proteins integrate into the inner envelope membrane during the translocation process, or from the stromal phase after translocation has taken place. However, outer envelope membrane proteins are clearly imported by a completely different pathway. Cloning of two outer envelope membrane proteins has revealed that neither is synthesized with a cleavable presequence, and import studies have shown that each of these proteins integrates into the envelope in the absence of ATP, and even after protease treatment of the chloroplasts (Li et al., 1991; Salomon et al., 1990). It thus appears that the proteins integrate spontaneously into the outer membrane, perhaps by a mechanism involving specific interaction with the lipids in this membrane.

3. Integration of proteins into the thylakoid membrane

The thylakoid membrane contains a number of hydrophobic, integral membrane proteins, many of which are synthesized in the cytosol. The import of these proteins is of interest because two major problems have to be overcome. First, these proteins must avoid integrating into either of the envelope membranes, and secondly, the proteins must traverse the soluble stromal phase without aggregating. Strangely, in view of the large number of integral proteins which have been cloned, the vast majority of studies in this area have focused on a single protein, the abundant light-harvesting chlorophyll-binding protein of photosystem II (LHCP). The biogenesis of this protein has been examined using an in-vitro assay for the integration into isolated thylakoids, and a fairly detailed picture has emerged for the overall import pathway. Interestingly, LHCP is synthesized with a stroma-targeting presequence, which can be replaced by the Rubisco SSU presequence without affecting import or integration (Lamppa, 1988; Viitanen et al., 1988). Information in the mature protein sequence must therefore specify integration into the thylakoid membrane, although it is at present unclear how this is achieved (or how integration into the inner envelope membrane is avoided). Early studies showed that the integration process requires a stromal protein factor and ATP (Chitnis et al., 1987; Cline, 1986; Cline et al., 1992), and suggested that the stromal factor may act as a chaperone molecule to maintain the stromal form of LHCP in a soluble form by shielding the hydrophobic surfaces. More recently (Li et al., 1995), it has been found that an essential element of the stromal factor is a homologue of the 54-kDa polypeptides of signal recognition particles (SRPs), soluble components involved in targeting proteins to the endoplasmic reticulum and the bacterial plasma membrane. It is therefore highly likely that this targeting mechanism was inherited from the cyanobacterial ancestor of chloroplasts.

4. The biogenesis of nuclear-encoded thylakoid lumen proteins

A general two-stage import pathway for thylakoid lumen proteins

A number of hydrophilic proteins are imported from the cytosol across all three chloroplast membranes into the thylakoid lumen, and the biogenesis of these proteins has proved to be particularly interesting. The best-studied lumenal proteins are plastocyanin and the 33-, 23- and 16-kDa proteins (33K, 23K and 16K) of the photosynthetic oxygen-evolving complex. All four proteins are imported by a pathway which can be divided into two stages: (1) import of a cytosolically synthesized precursor protein and processing to an *intermediate* form by SPP; and (2) transport of the intermediate across the thylakoid membrane followed by processing to the mature size by a thylakoidal processing peptidase (TPP) (Hageman *et al.,* 1986; James *et al.,* 1989; Kirwin *et al.,* 1987, 1989). As might be expected from this type of import pathway, lumenal proteins are synthesized with bipartite presequences consisting of two signals in tandem. The first 'envelope transit' signals specify transport into the stroma, and are structurally and functionally equivalent to the presequences of stromal proteins (Hageman *et al.,* 1990; Ko and Cashmore, 1989). The second 'thylakoid transfer' signals, however, are notable in resembling bacterial export or 'signal' sequences in several respects, particularly in possessing hydrophobic core domains and small-chain residues, usually Ala, at the −3 and −1 positions relative to the terminal cleavage site (Halpin *et al.,* 1989; von Heijne *et al.,* 1989). These properties are summarized in *Figure 2.* Functional similarities between the two types of signal have been confirmed by the ability of the 33K thylakoid transfer domain to direct export in *E. coli* (Meadows and Robinson, 1991), and the finding that the reaction specificities of pea TPP and *E. coli* signal peptidase are virtually identical (Halpin *et al.,* 1989). These results have prompted speculation that the thylakoidal protein import system may have evolved from a prokaryotic export system, a theory which will be explored in greater detail later in this section.

Two distinct mechanisms for the translocation of proteins across the thylakoid membrane

The general two-stage import pathway for lumenal proteins emerged from studies of the import of precursor proteins into intact chloroplasts, and particularly from the reconstruction of the maturation sequence using partially purified SPP and TPP. The actual translocation of these proteins across the thylakoid membrane has been examined in greater detail using *in-vitro* assays for the import of proteins by isolated thylakoids. Unexpectedly, these studies have revealed that the four lumenal proteins described above are translocated by two distinct mechanisms with totally different requirements. The translocation of 33K and plastocyanin is strictly dependent on the presence of at least one stromal protein factor and ATP, but the thylakoidal ΔpH is not required (although it may stimulate translocation to some extent). In complete contrast, stromal factors and ATP are not required at all for the translocation of 23K or 16K, but the thylakoidal ΔpH is an absolute requirement (Cline *et al.,* 1992; Klösgen *et al.,* 1992; Mould and Robinson, 1991; Mould *et al.,* 1991).

At first it appeared fairly likely that the two groups of import requirements might reflect differences in the mature proteins being transported, but recent work suggests that this is probably not the case. In competition experiments using saturating concentrations of *E. coli*-expressed precursors, it has been found that pre-23K competes with pre-16K for transport across

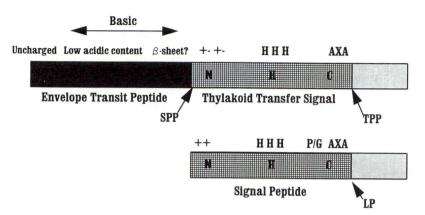

Figure 2. The bipartite presequences of thylakoid lumen proteins. Nuclear-encoded proteins such as plastocyanin and the components of the oxygen-evolving complex are synthesized with bipartite presequences consisting of two signals in tandem. The first 'envelope transfer' signal specifies transport into the stroma, and is usually uncharged in the N-terminal region and basic in the remainder. The signals are hydrophilic overall, and enriched in serine and threonine residues. This signal is removed by the stromal processing peptidase (SPP), and the 'thylakoid transfer' signal mediates translocation of the intermediate form across the thylakoid membrane. Thylakoid transfer signals resemble bacterial leader sequences in possessing hydrophobic core sections (H) and small-chain residues, usually Ala, at the −3 and −1 positions, relative to the terminal cleavage site. The only notable difference is that many transfer signals contain both acidic and basic residues in the N-terminal domain, whereas leader peptides usually contain only basic residues. Transfer signals are cleaved by thylakoidal processing peptidase (TPP), an enzyme with similar properties to bacterial leader peptidase (LP).

the thylakoid membrane, and that pre-33K competes with pre-plastocyanin, but that the two groups of proteins do not compete with each other (Cline *et al.*, 1993). In another study (Robinson *et al.*, 1994) it has been found that a chimeric protein, consisting of the 23K presequence linked to mature plastocyanin, is transported across the thylakoid membrane only by the pre-23K-type mechanism, ruling out the possibility that the mature protein dictates the translocation requirements, and instead suggesting (as do the competition data) that two different translocators operate in the thylakoid membrane, with the 23K/16K system recognizing specific signals in the presequences of these proteins. The two import pathways are depicted in *Figure 3*, together with the LHCP import/integration pathway described earlier.

The possible origins of parallel translocation pathways for lumenal proteins

Two obvious questions are raised by the above data. Why are there two distinct translocases (assuming that separate translocases operate — this has yet to be formally proved), and how did this arrangement arise? The first clues emerged in studies of the coding properties of the plastid genomes of red algae and chromophytes. In several cases (Scaramuzzi *et al.*, 1992; Valentin, 1993) it has been found that the plastid genomes of such algae contain homologues of *sec* genes that are known to encode elements of the protein export apparatus in bacteria. These genes are not present in the plastid genomes of higher plants, but may well have been transferred to the nucleus. It therefore appeared likely that a *sec*-type translocation system operates to transport proteins across the thylakoid membrane, and that this system was inherited from the cyanobacterial-type progenitor of higher plant chloroplasts. There is good evidence that this was indeed the case. A nuclear-encoded, stromal SecA homologue has been shown to be required for

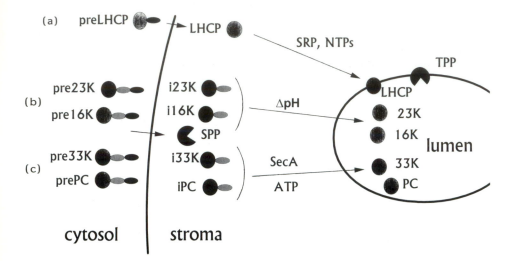

Figure 3. Multiple pathways for the import of thylakoid proteins. Thylakoid lumen proteins are synthesized with bipartite presequences as depicted in *Figure 2*, and then imported into the stroma where they are cleaved to intermediate forms by SPP. The intermediate forms are then translocated across the thylakoid membrane by two distinct mechanisms. One mechanism relies only on the thylakoidal ΔpH, whereas the other is strictly dependent on the presence of stromal protein factor(s), including SecA, and ATP. Complete maturation is carried out by a membrane-bound thylakoidal processing peptidase (TPP). LHCP is imported by a different pathway; the protein is synthesized with a stroma targeting signal only, and integration into the thylakoid membrane is mediated by information in the mature protein by a process requiring the presence of a stromal homologue of the 54-kDa protein of signal recognition particles (SRPs) and nucleoside triphosphates (NTPs). PC, plastocyanin.

the import of 33K into chloroplasts (Nakai *et al.*, 1994; Yuan *et al.*, 1994), and a SecY homologue has been shown to be present in the thylakoid membrane (Laidler *et al.*, 1995). Consistent with this theory, 33K and plastocyanin are present in the thylakoid lumen in cyanobacteria, and are synthesized in these organisms with presequences which resemble both bacterial signal sequences and the thylakoid transfer sequences of their higher plant counterparts (Kuwabara *et al.*, 1987). Finally, it is interesting that 23K and 16K are *not* present in cyanobacteria, and it is therefore possible that the evolution of the photosynthetic apparatus in higher plant chloroplasts has involved the acquisition both of new lumenal proteins and a new system for their translocation across the thylakoid membrane. This scheme would account for all of the available mechanistic data, although it would not answer one of the questions alluded to above, namely, why the new components were not simply transported by the pre-existing, prokaryotic system.

5. Concluding remarks

The last few years have witnessed a number of genuine breakthroughs in the field of chloroplast protein import, and several parts of the overall picture are now much clearer. The biogenesis of thylakoidal proteins in particular is now understood in some detail, and in the near future it should prove possible to rationalize to some extent the surprising variety of pathways for the import of integral and lumenal thylakoidal proteins. However, many aspects of chloroplast protein import remain poorly understood, and this is especially true of the critical early stages of the import process.We still have very little information about the means whereby the chloroplast recognizes precursor proteins, or how these proteins are translocated across the two envelope membranes. Nevertheless, the recent identification of envelope proteins involved in these processes should ensure rapid progress in the future, although it may be that new approaches, for example involving the isolation of import mutants, will be required to fill in some of the gaps.

References

Boutry M, Nagy F, Poulsen C, Aoyagi K, Chua, N-H. (1987) Targeting of bacterial chloramphenicol acetyltransferase to mitochondria in transgenic plants. *Nature* **328:** 340–342.

Chitnis PR, Nechushtai R, Thornber JP. (1987) Insertion of the precursor of the light-harvesting chlorophyll *a/b* protein into the thylakoids requires the presence of a developmentally regulated stromal factor. *Plant Mol. Biol.* **10:** 3–11.

Chua N-H, Schmidt GW. (1978) Post-translational transport into intact chloroplasts of a precursor to the small subunit of ribulose-1,5-bisphosphate carboxylase. *Proc. Natl Acad. Sci. USA* **75:** 6110–6114.

Cline K. (1986) Import of proteins into chloroplasts: membrane integration of a thylakoid precursor protein reconstituted in chloroplast lysates. *J. Biol. Chem.* **261:** 14804–14810.

Cline K, Werner-Washburne M, Lubben TH, Keegstra K. (1985) Precursors to two nuclear-encoded chloroplast proteins bind to the outer envelope membrane before being imported into chloroplasts. *J. Biol. Chem.* **260:** 3691–3696.

Cline K, Ettinger WF, Theg SM. (1992) Protein-specific energy requirements for protein transport across or into thylakoid membranes. Two lumenal proteins are transported in the absence of ATP. *J. Biol. Chem.* **267:** 2688–2696.

Cline K, Henry R, Li C, Yuan J. (1993) Multiple pathways for protein transport into or across the thylakoid membrane. *EMBO J.* **12:** 1405.

de Boer D, Cremers F, Teertstra R, Smits L, Hille J, Smeekens S, Weisbeek P. (1988) *In vivo* import of plastocyanin and a fusion protein into developmentally different plastids of transgenic plants. *EMBO J.* **7:** 2631–2635.

Flügge U-I, Fischer K, Gross A, Sebald W, Lottspeich F, Eckerskorn C. (1989) The triose phosphate–3-phosphoglycerate–phosphate translocator from spinach chloroplasts: nucleotide sequence of a full-length cDNA clone and import of the *in vitro* synthesised precursor protein into chloroplasts. *EMBO J.* **8:** 39–46.

Flügge U-I, Weber A, Fischer K, Lottspeich F, Eckerskorn C, Waegemann K, Soll J. (1991) The major chloroplast envelope polypeptide is the phosphate translocator and not the protein import receptor. *Nature* **353:** 364–367.

Friedman AL, Keegstra K. (1988) Chloroplast protein import: quantitative analysis of precursor binding. *Plant Physiol.* **89:** 993–999.

Fujiki M, Verner K. (1993) Coupling of cytosolic protein synthesis and mitochondrial protein import in yeast: evidence for cotranslational import *in vivo*. *J. Biol. Chem.* **268:** 1914–1920.

Glick BS, Beasley EM, Schatz G. (1992) Protein sorting in mitochondria. *Trends Biochem. Sci.* **17:** 453–459.

Gray JC, Row PE. (1995) Protein translocation across chloroplast envelope membranes. *Trends Cell Biol.* **5:** 243–247.

Hageman J, Robinson C, Smeekens S, Weisbeek P. (1986) A thylakoid processing protease is required for complete maturation of the lumen protein plastocyanin. *Nature* **324:** 567–569.

Hageman J, Baecke C, Ebskamp M, Pilon R, Smeekens S, Weisbeek P. (1990) Protein import into and sorting inside the chloroplast are independent processes. *Plant Cell* **2:** 479–494.

Halpin C, Elderfield PD, James HE, Zimmermann R, Dunbar B, Robinson C. (1989) The reaction specificities of the thylakoidal processing peptidase and *Escherichia coli* leader peptidase are identical. *EMBO J.* **8:** 3917–3921.

Hemmingsen SM, Woolford C, van der Vies SM, Tilly K, Dennis DT, Georgopoulos C, Hendrix RW, Ellis RJ. (1988) Homologous plant and bacterial proteins chaperone oligomeric protein assembly. *Nature* **339:** 483.

Highfield PE, Ellis RJ. (1978) Synthesis and transport of the small subunit of chloroplast ribulose bisphosphate carboxylase. *Nature* **271:** 420–424.

Horniak L, Pilon M, van't Hof R, de Kruijff B. (1993) The secondary structure of the ferredoxin transit sequence is modulated by its interaction with negatively charged lipids. *FEBS Lett.* **334:** 241–246.

Howe G, Merchant S. (1993) Maturation of thylakoid lumen proteins proceeds post-translationally through an intermediate *in vivo*. *Proc. Natl Acad. Sci. USA* **90:** 1862–1866.

James HE, Bartling D, Musgrove JE, Kirwin PM, Herrmann RG, Robinson C. (1989) Transport of proteins into chloroplasts: import and maturation of precursors to the 33-, 23-, and 16-kDa proteins of the photosynthetic oxygen-evolving complex. *J. Biol. Chem.* **264:** 19573–19576.

Kavanagh TA, Jefferson RA, Bevan MW. (1988) Targeting a foreign protein to chloroplasts using fusions to the transit peptide of a chlorophyll *a/b* protein. *Mol. Gen. Genet.* **215:** 38–45.

Kessler F, Blobel G, Patel HA, Schnell DJ. (1994) Identification of two GTP-binding proteins in the chloroplast import machinery. *Science* **266:** 1035–1039.

Kirwin PM, Elderfield PD, Robinson C. (1987) Transport of proteins into chloroplasts: partial purification of a thylakoidal processing peptidase involved in plastocyanin biogenesis. *J. Biol. Chem.* **262:** 16386–16390.

Kirwin PM, Meadows JW, Shackleton JB, Musgrove JE, Elderfield PD, Mould R, Hay NA, Robinson C. (1989) ATP-dependent import of a lumenal protein by isolated thylakoid vesicles. *EMBO J.* **8:** 2251–2255.

Klösgen RB, Brock IW, Herrmann RG, Robinson C. (1992) Proton gradient-driven import of the 16kDa oxygen-evolving complex protein as the full precursor protein by isolated thylakoids. *Plant Mol. Biol.* **18:** 1031–1034.

Ko K, Cashmore AR. (1989) Targeting of proteins to the thylakoid lumen by the bipartite transit peptide of the 33kDa oxygen-evolving protein. *EMBO J.* **8:** 3187–3194.

Ko K, Bornemisza O, Kourtz L, Ko ZW, Plaxton WC, Cashmore AR. (1992) Isolation and characterisation of a cDNA clone encoding a cognate 70-kDa heat shock protein of the chloroplast envelope. *J. Biol. Chem.* **267:** 2986–2993.

Kuwabara T, Reddy KJ, Sherman LA. (1987) Nucleotide sequence of the gene from the cyanobacterium *Anacystis nidulans* R2 encoding the Mn-stabilising 8234. *Proc. Natl Acad. Sci. USA* **84:** 8230–8234.

Laidler V, Chaddock AM, Knott TG, Walker D, Robinson C. (1995) A SecY homolog in *Arabidopsis thaliana*: sequence of a full-length cDNA clone and import of the precursor protein into chloroplasts. *J. Biol. Chem.* **270:** 17664–17667.

Lamppa GK. (1988) The chlorophyll *a/b*-binding protein inserts into the thylakoids independent of its cognate transit peptide. *J. Biol. Chem.* **263:** 14996.

Li H-M, Moore T, Keegstra K. (1991) Targeting of proteins to the outer envelope membrane uses a different pathway than transport into chloroplasts. *Plant Cell* **3:** 709–717.

Li X, Henry R, Yuan J, Cline K, Hoffman NE. (1995) A chloroplast homologue of the signal recognition particle subunit SRP54 is involved in the post-translational integration of a protein into thylakoid membranes. *Proc. Natl Acad. Sci. USA* **92:** 3789–3793.

Lubben TH, Gatenby AA, Ahlquist P, Keegstra K. (1989) Chloroplast import characteristics of chimeric proteins. *Plant Mol. Biol.* **12:** 13–18.

Meadows JW, Robinson C. (1991) The full precursor of the 33kDa oxygen-evolving complex protein of wheat is exported by *Escherichia coli* and processed to the mature size. *Plant Mol. Biol.* **17:** 1241–1243.

Mould RM, Robinson C. (1991) A proton gradient is required for the transport of two lumenal oxygen-evolving proteins across the thylakoid membrane. *J. Biol. Chem.* **266:** 12189–12193.

Mould RM, Shackleton JB, Robinson C. (1991) Transport of proteins into chloroplasts: requirements for the efficient import of two lumenal oxygen-evolving complex proteins into isolated thylakoids. *J. Biol. Chem.* **266:** 17286–17289.

Nakai M, Goto A, Nohara T, Sugita D, Endo T. (1994) Identification of the SecA protein homolog in pea chloroplasts and its possible involvement in thylakoidal protein transport. *J. Biol. Chem.* **269:** 31338–31341.

Oblong JE, Lamppa GK. (1992) Identification of two structurally related proteins involved in proteolytic processing of precursors targeted to the chloroplast. *EMBO J.* **11:** 4401–4409.

Olsen LJ, Theg SM, Selman BR, Keegstra K. (1989) ATP is required for the binding of precursor proteins to chloroplasts. *J. Biol. Chem.* **264:** 6724–6729.

Pain D, Kanwar YS, Blobel G. (1988) Identification of a receptor for protein import into chloroplasts and its localisation to envelope contact zones. *Nature* **331:** 232–237.

Perry SE, Keegstra K. (1994) Envelope membrane proteins that interact with chloroplastic precursor proteins. *Plant Cell* **6:** 93–105.

Pilon M, de Boer AD, Knols SL, Koppelman MHGM, van der Graaf RM, de Kruijff B, Weisbeek PJ. (1990) Expression in *E. coli* and purification of a translocation-competent precursor of the chloroplast protein ferredoxin. *J. Biol. Chem.* **265:** 3358–3361.

Pilon M, Rietveld AG, Weisbeek PJ, de Kruijff B. (1992) Secondary structure and folding of a functional chloroplast precursor protein. *J. Biol. Chem.* **267:** 19907–19913.

Robinson C, Ellis RJ. (1984) Transport of proteins into chloroplasts: partial purification of a chloroplast protease involved in the processing of imported precursor polypeptides. *Eur. J. Biochem.* **142:** 337–342.

Robinson C, Cai D, Hulford A, Hazell L, Michl D, Brock IW, Herrmann RG, Klösgen RB. (1994) The presequence of a chimeric construct dictates which of two mechanisms is utilised for translocation across the thylakoid membrane: evidence for the existence of two distinct translocation systems. *EMBO J.* **14:** 2715–2722.

Salomon M, Fischer K, Flügge U-I, Soll J. (1990) Sequence analysis and protein import studies of an outer chloroplast envelope polypeptide. *Proc. Natl Acad. Sci. USA* **87:** 5578–5782.

Sasaki Y, Sekiguchi K, Nagano Y, Matsuno R. (1993) Chloroplast envelope protein encoded by chloroplast genome. *FEBS Lett.* **316:** 93–98.

Scaramuzzi CD, Hiller RG, Stokes HW. (1992) Identification of a chloroplast-encoded secA gene homologue in a chromophytic alga: possible role in chloroplast protein translocation. *Curr. Genet.* **22:** 421.

Schnell DJ, Blobel G. (1993) Identification of intermediates in the pathway of protein import into chloroplasts and their localisation to envelope contact sites. *J. Cell Biol.* **120:** 103–115.

Schnell DJ, Blobel G, Pain D. (1990) The chloroplast import receptor is an integral membrane protein of chloroplast envelope contact sites. *J. Cell Biol.* **111:** 1825–1838.

Schnell DJ, Kessler F, Blobel G. (1994) Isolation of components of the chloroplast import machinery. *Science* **266:** 1007–1012.

Theg SM, Geske FJ. (1992) Biophysical characterisation of transit peptide directing chloroplast protein import. *Biochemistry* **31:** 5053–5060.

Theg SM, Bauerle C, Olsen LJ, Selman BR, Keegstra K. (1989) Internal ATP is the only energy requirement for the translocation of precursor proteins across chloroplastic membranes. *J. Biol. Chem.* **264:** 6730–6736.

Tsugeki R, Nishimura M. (1993) Interaction of homologues of Hsp70 and Cpn60 with ferredoxin-NADP⁺ reductase upon its import into chloroplasts. *FEBS Lett.* **320:** 198–202.

Valentin K. (1993) SecA is plastid-encoded in a red alga: implications for the evolution of plastid genomes and the thylakoid protein import apparatus. *Mol. Gen. Genet.* **236:** 245.

Van den Broeck G, Timko MP, Kausch AP, Cashmore AR, Montagu MV, Herrera-Estrella L. (1985) Targeting of a foreign protein to chloroplasts by fusion to the transit peptide from the small subunit of ribulose-1,5-bisphosphate carboxylase. *Nature* **313:** 358–363.

van't Hof R, van Klompenburg W, Pilon M, Kosubek A, de Korte-kool G, Demel RA, Wiesbeek P, de Kruijff B. (1993) The transit sequence mediates the specific interaction of the precursor of ferredoxin with chloroplast envelope membrane lipids. *J. Biol. Chem.* **266:** 4037–4042.

Viitanen PV, Doran ER, Dunsmuir P. (1988) What is the role of the transit peptide in thylakoid integration of the light-harvesting chlorophyll a/b protein? *J. Biol. Chem.* **263:** 15000–15007.

von Heijne G, Nishikawa K. (1991) Chloroplast transit peptides — the perfect random coil? *FEBS Lett.* **278:** 1–3.

von Heijne G, Steppuhn J, Herrmann RG. (1989) Domain structure of mitochondrial and chloroplast targeting peptides. *Eur. J. Biochem.* **180:** 535–545.

Waegemann K, Soll J. (1991) Characterisation of the protein import apparatus in isolated outer envelopes of chloroplasts. *Plant J.* **1:** 149–158.

Waegemann K, Paulsen H, Soll J. (1990) Translocation of proteins into isolated chloroplasts requires cytosolic factors to obtain import competence. *FEBS Lett.* **261:** 89–92.

Whelan J, Knorpp C, Glaser E. (1990) Sorting of precursor proteins between isolated spinach leaf mitochondria and chloroplasts. *Plant Mol. Biol.* **14:** 977–982.

Wu C, Seibert FS, Ko K. (1994) Identification of chloroplast envelope proteins in close proximity to a partially translocated chimeric precursor protein. *J. Biol. Chem.* **269:** 32264–32271.

Yuan J, Henry R, McCaffery M, Cline K. (1994) SecA homolog in protein transport within chloroplasts: evidence for endosymbiont-derived sorting. *Science* **266:** 796–798.

Chapter 27

Plasmodesmatal networks in apical meristems and mature structures: geometric evidence for both primary and secondary formation of plasmodesmata

Todd J. Cooke, Mary S. Tilney and Lewis G. Tilney

1. Introduction

Despite the widespread belief that plasmodesmata (cytoplasmic channels through the common walls between plant cells) play very important roles in the processes of transport, pathogenesis, communication and development in plants (Gunning and Robards, 1976; Robards and Lucas, 1990), little research has been done to investigate how plasmodesmatal networks are organized in mature plant organs. Even more surprising is the virtual absence of research effort being devoted to the distribution of plasmodesmata in apical meristems. Since growing plant cells expand as a co-ordinated unit with no apparent slippage, then it follows that cellular organization within each meristem is more or less preserved in the mature tissues derived from it. Thus, one would predict from first principles that the plasmodesmatal networks in apical meristems should profoundly affect the manner in which plasmodesmata are distributed in mature organs.

In this chapter, we shall first examine those situations in the lower vascular plants for which this prediction holds true, and then we shall address more interesting situations in the angiosperms, where the meristems and their derivative organs have different plasmodesmatal networks. These latter examples demonstrate that we must drastically revise our concepts about the relative timing of wall formation and plasmodesmatal insertion, at least in the angiosperms. In particular, it is concluded that the ferns develop lineage-specific networks of primary plasmodesmata which are formed in expanding cell plates during cytokinesis, whereas the angiosperms develop interface-specific networks of both primary and secondary plasmodesmata, the latter of which are inserted into pre-existing cell walls.

2. The definition of primary and secondary plasmodesmata

Before we describe plasmodesmatal networks in apical meristems and derived organs, it is first necessary to discuss the standard definitions of primary and secondary plasmodesmata (Jones, 1976). Primary plasmodesmata are defined as those cytoplasmic channels that develop from endoplasmic reticulum trapped in the coalescing cell plate during cytokinesis (Hepler, 1982). In contrast, secondary plasmodesmata are inserted into existing cell walls. Interestingly, the mechanism of secondary plasmodesmata formation is very similar to that reported for primary plasmodesmata (i.e. they originate at very thin regions of existing cell walls via the fusion of plasma membrane and endoplasmic reticulum across the middle lamella) (Kollmann and Glockmann, 1991; Monzer, 1991). Although early workers considered that almost all plasmodesmata across plant cell walls arose as primary structures in the cell plate, except in extraordinary circumstances such as graft unions and parasitic haustoria (Gunning and Steer, 1975; Jones, 1976; Ledbetter and Porter, 1970), more recent reviewers have mentioned the possibility that secondary plasmodesmata may commonly arise in existing walls under normal conditions of plant growth (Monzer, 1991; Robards and Lucas, 1990).

Clearly, the problem with these definitions is that primary and secondary plasmodesmata are distinguished on the basis of their origin, which is next to impossible to determine from single thin sections of individual cells. There are three conceivable approaches for circumventing this problem: (1) the structural approach — one could look for diagnostic structural features that distinguish between plasmodesmata known to originate via either primary or secondary mechanisms; (2) the developmental approach — one could look at plasmodesmatal structure in the same tissues at different developmental stages in an attempt to distinguish between those plasmodesmata that develop in young dividing cells and those that appear later in mature cells; or (3) the geometric approach — one could use the fixed geometry of cell walls in growing plant structures to characterize the manner in which the timing of the origin of specific walls relates to the nature of the plasmodesmata inserted in them.

Structural approach

In certain apical meristems described in the following sections, new cell walls from recent cell divisions are packed with unbranched plasmodesmata, each of which has a well-defined central desmotubule. Thus it seems reasonable to conclude that these unbranched plasmodesmata represent primary plasmodesmata, although it must be conceded that very few published micrographs depict endoplasmic reticulum being trapped in expanding cell plates in dividing cells (Hepler, 1982). In contrast, the plasmodesmata which have unequivocally developed as secondary structures across fused walls at graft unions (Kollmann and Glockmann, 1985; Kollmann et al., 1985) and in other situations without cell division (Jones, 1976) are often highly branched structures whose internal components may appear either as compact desmotubules or as conventional unit membranes which are typically associated with prominent median cavities at the middle lamella. However, since unbranched plasmodesmata are also seen in these so-called non-division walls (e.g. Kollmann and Glockmann, 1985; Kollmann et al., 1985), plasmodesmatal structure is not a sufficient criterion by itself for designating plasmodesmatal type in more ambiguous situations found in growing plant organs.

Developmental approach

Ding et al. (1992) confronted the problem of plasmodesmatal definition in virus-infected tobacco leaves by studying plasmodesmatal structure in successive leaves at different developmental stages. The plasmodesmata in the first leaf, which is designated in their study as the youngest leaf to have attained a length of 5 cm, are predominantly unbranched in structure, but each successive older leaf displays more branched plasmodesmata, until they eventually represent over 90% of the total plasmodesmata in the mature fifth and sixth leaves.

Thus the terms 'primary' and 'secondary' were assigned to all unbranched and branched plasmodesmata, respectively (Ding et al., 1992). Although plausible, these operational terms suffer from the limitation that, in the absence of exhaustive (and exhausting) studies on plasmodesmatal number at different stages of leaf development, it is unknown whether the branched plasmodesmata are formed as genuine secondary plasmodesmata in existing walls or as modified groups of adjacent primary plasmodesmata which are secondarily joined together at the middle lamella to form the median cavity.

Geometric approach

In this chapter, a geometric approach, which can partially circumvent the problems encountered with the other two approaches, is used to characterize the features of plasmodesmata in developing plant organs. Certain plant structures, such as the roots of the fern *Azolla,* exhibit almost invariable patterns of cell division, so that it is possible to follow individual cell walls from their inception in the apical meristem to their final mature state. Although most plant organs lack such precise cell arrangements, they do exhibit prominent cell files, where it is also possible to determine the origins of cell walls and the changes in their plasmodesmatal densities. Thus, by the judicious selection of plant organs with favorable geometry, we can determine how various characteristics of the plasmodesmata, such as number, density and structure, change with subsequent growth of the cell wall. Although this approach does not allow us to identify whether a given plasmodesma is primary or secondary in origin, it does permit informative estimates about the relative proportions of primary and secondary plasmodesmata in any given wall.

Plasmodesmatal networks in lower vascular plants

General aspects of apical meristem construction in ferns

The distribution of plasmodesmata in apical meristems has been investigated in a few lower vascular plants, namely the ferns. Each apical meristem in almost all the ferns studied to date is composed of a single apical cell whose immediate derivatives, known as merophytes, are ultimately responsible for generating all the cells in mature organs (Bierhorst, 1977; Gifford, 1983). Since the original walls of these merophytes can often be traced into more mature regions of the plant body, the term merophyte is applied to either the most recent unicellular derivative of the apical cell or the older multicellular unit contained within the original merophyte walls.

Several features of this type of meristem construction are relevant to the issue of plasmodesmatal distribution. First, all walls within the meristem are recently formed via cell division (i.e. there are no internal walls in these meristems which are positioned such that they persist in the apical region). (This situation is very different from the so-called stratified meristems often observed in the seed plants described below.) Secondly, the patterns of cell wall insertion are highly regular in the apical cells and adjacent merophytes, thereby permitting the investigator to determine the relative timing of the origin of each cell wall and its subsequent growth from fixed sections. *Figures 1, 2* and *3* illustrate these features of meristem construction in three types of fern apical meristem.

The typical prothallus (mature gametophyte) of many species, such as the sensitive fern (*Onoclea sensibilis*) (Tilney *et al.*, 1990), consists of a single cell layer with a prominent triangular apical cell located at the base of the apical notch (*Figure 1*). This apical cell divides in the two planes parallel to its two lateral sides. This division sequence, which alternates between the two planes with almost clock-like precision, is thus responsible for generating the merophytes of which the two symmetrical lobes of the prothallus are composed. For instance, in the prothallus depicted in *Figure 1*, all the walls and merophytes are lettered and numbered in reverse order of their origin. The most recent division in the apical cell occurred on its left-hand side by means of the wall AA, which cut off the youngest merophyte marked '1' in the prothallus. Its second most recent division occurred as wall BB on the right-hand side, cutting off the second youngest merophyte ('2'), and so forth. It is thus possible to proceed to older regions of the

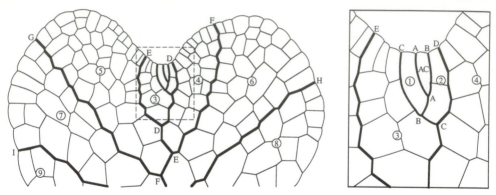

Figure 1. A frontal view of a prothallus of the sensitive fern *Onoclea sensibilis*, illustrating the sequence of apical cell divisions responsible for generating the merophytes of which this prothallus is comprised. The box on the right shows a magnified view of the apical meristem, with the triangular apical cell marked as 'AC'. The sequence of apical cell divisions and the resulting merophytes are lettered and numbered, respectively, in reverse order. The most recent division in the apical cell occurred on its left-hand side by means of the wall AA, which cut off the youngest merophyte marked '1' in the prothallus. The second most recent apical cell division occurred on the right-hand side by means of the wall BB, which cut off the second youngest merophyte ('2'). The third most recent division occurred by means of the wall CC on the same side as the most recent division to cut off the third youngest merophyte ('3'), and so forth. Redrawn from Cooke and Lu (1992) with permission. © 1992 by The University of Chicago Press.

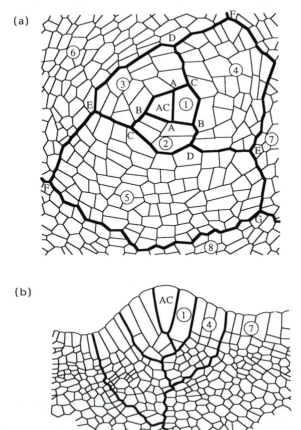

Figure 2. Two views of the typical shoot apical meristem of the ferns. (a) A surface view of the shoot apical meristem of the royal fern (*Osmunda regalis*). (b) A median longitudinal section of a comparable apex from the polypody fern (*Polypodium glaucum*). The reverse sequences of apical cell divisions (AA, BB, etc.) and the resulting merophytes (1, 2, etc.) are marked as indicated in the legend to *Figure 1*. The sites where the walls of successive merophytes contact the outer epidermal wall can be visualized in (a). The sequence of internal walls defining the merophytes that are generated from two lateral faces of the apical cell can be seen in (b). The third set of merophytes is out of the plane of this section. (a) Redrawn from Bierhorst (1977) with permission from the *American Journal of Botany* and the author; (b) redrawn from Wetter and Wetter (1954) with permission from Gustav Fisher Verlag Jena.

prothallus in order to identify all the original walls and merophyte boundaries generated from successive apical cell divisions. Therefore one can reconstruct the entire sequence of apical cell divisions from a single frontal section of the prothallus. It is obvious from such reconstructions that the lateral walls of the apical cell are continually being displaced to older prothallial regions following subsequent divisions. The upper wall of the apical cell is continuously being stretched such that only part of this non-division wall remains located within the apical cell following each division.

The division sequences in the apical meristems of fern sporophytes can also be determined from fixed material, except that such reconstructions require serial sections because the tetrahedral apical cells divide in more than two planes to generate three-dimensional structures. *Figure 2a* shows a surface view of a shoot apical meristem of the royal fern (*Osmunda regalis*) (Bierhorst, 1977), while *Figure 2b* shows a median longitudinal section of a comparable apex from the polypody fern (*Polypodium glaucum*) (Wetter and Wetter, 1954). Again, the wall from the most recent cell division of the apical cell is marked as AA and the resulting merophyte as '1', the wall from the second most recent division is marked as BB and the resulting merophyte as '2', etc. In the shoot apex shown in *Figure 2a*, the apical cell has divided in a clockwise pattern parallel to its three lateral faces, and therefore the sites where the walls of successive merophytes contact the outer epidermal wall can be visualized from this surface view. *Figure 2b* shows the sequence of internal walls defining the merophytes from two lateral faces of the apical cell as these merophytes are displaced from the apical cell following subsequent divisions. (The third set of merophytes is out of the plane of this section.) Just like the walls of the triangular apical cell in the prothallus, the three lateral walls of these tetrahedral apical cells are repeatedly

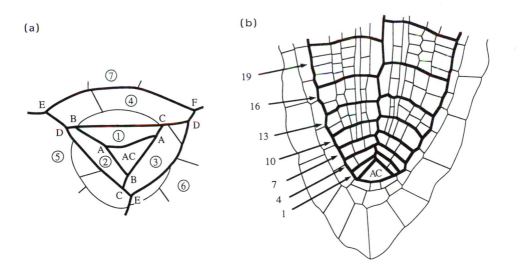

Figure 3. Two views of the typical root apical meristem of the water fern *Azolla filiculoides*. These views are approximately comparable to those of the shoot apical meristem presented in *Figure 2*. (a) A transverse section of the root apical meristem parallel to the distal face of the apical cell. (b) A median longitudinal section, with the tetrahedral apical cell and two out of three sets of the merophytes. The reverse sequences of apical cell divisions (AA, BB, etc.) and the resulting merophytes (1, 2, etc.) are marked as indicated in the legend to *Figure 1*. (a) Redrawn from Gunning *et al.* (1978) with permission from Springer-Verlag and the authors. © 1978 Springer-Verlag. (b) redrawn from Nitayangkura *et al.* (1980) with permission from the *American Journal of Botany* and the authors.

displaced toward more mature regions, and the fourth outer wall is continuously stretched with the formation of additional merophytes.

The tetrahedral apical cell in the root apical meristem operates in a very similar manner to the equivalent cell in the shoot apical meristem, except that it divides in the fourth plane to generate the cells comprising the root cap in some, but not all, species (Clowes, 1961; Gifford, 1983). *Figure 3a* shows a transverse view of the root apical meristem of the water fern *Azolla filiculoides,* which is sectioned parallel to the distal face of the apical cell opposite the root cap. In the meristem illustrated here, the apical cell has divided in a clockwise sequence along the three lateral faces. In this species, the apical cell in the early root primordium undergoes a single division parallel to the distal face to generate the root cap initial, after which all of its divisions are restricted to being parallel to the lateral faces (Gunning *et al.*, 1978). *Figure 3b* shows the median longitudinal section of the root apex of this species, with the tetrahedral apical cell and two out of three sets of its merophytes (Nitayangkura *et al.*, 1980).

Plasmodesmatal densities in fern apical meristems

Table 1 lists the available data on the density of plasmodesmata in the cell walls of fern apical meristems. The greatest values of these densities are much higher than those reported for any other cell wall in plant structures, including secretory hairs specialized for high-volume transport (for references, see Tilney *et al.*, 1990). Indeed, using the typical value of 25 nm as the plasmodesmatal radius, it can be calculated that the plasmodesmata in the *Dryopteris* root meristem and the *Osmunda* shoot meristem may cover up to 27 and 35% of the surface areas of their respective walls!

The plasmodesmata in these cell walls have several other noteworthy characteristics. Judging from serial sections (e.g. *Onoclea sensibilis*; Tilney *et al.*, 1990) or face views (e.g. *Dryopteris filix-mas*; Burgess, 1971), plasmodesmata are distributed rather uniformly throughout the internal walls of these cells, with no evidence for their localization in primary pit-fields. Secondly, all the plasmodesmata in the published micrographs appear as simple unbranched channels without swollen median cavities, thereby supporting the notion that branched plasmodesmata must be secondary in their origin.

Plasmodesmatal densities in mature structures of ferns

Why do the walls in fern apical meristems have such high plasmodesmatal frequencies? A partial answer to this question can be obtained from developmental studies on *Azolla* roots (Gunning, 1978) and *Onoclea* gametophytes (Tilney *et al.*, 1990). Both of these investigations exploited the very regular pattern of cell divisions in the two structures in order to monitor the change in plasmodesmatal frequency in the expanding walls as they were displaced from the meristem. The results obtained from such an analysis of *Onoclea* gametophytes are summarized in *Figure 4*. The walls originating from successive apical cell divisions maintain a constant number of plasmodesmata even though they undergo considerable expansion as they are displaced from the apical cell. These constant plasmodesmatal numbers imply that the fern gametophyte is unable to produce secondary plasmodesmata. Thus it follows that the density of plasmodesmata in these walls must decrease with wall expansion, and indeed *Figure 4* shows that wall GG averages a 10-fold reduction in plasmodesmatal density as compared to the value in the newly formed wall AA. Another intriguing observation is that new walls within the enlarging merophytes acquire primary plasmodesmata at the same density as that observed in the older boundary walls in that merophyte *(Figure 4)*. Thus the gametophyte can somehow specify the appropriate number of primary plasmodesmata that should be inserted in a new wall at a given position. Finally, we had

Table 1. Plasmodesmatal densities in the apical meristems of various ferns

Fern species	Structure	Meristem organization	Cell walls	Plasmodesmatal density (number μm^{-2})	Reference
Onoclea sensibilis	Prothallus	Single apical cell	Lateral walls of apical cell	15	Tilney et al. (1990)
Pteris cretica	Shoot	Single apical cell	Lateral walls of apical cell	32	Michaux (1970)[a]
Osmunda cinnamomea	Shoot	Single apical cell	Anticlinal walls of adjacent derivatives	c. 180	Hicks and Steeves (1970) as calculated in Juniper (1976)
			Periclinal walls of adjacent derivatives	c. 80	
Azolla pinnata	Root	Single apical cell	Proximal walls of apical cell	≤ 80	Gunning (1978)
			Distal wall of apical cell	Low	Gunning et al. (1978)
Dryopteris filix-mas	Root	Single apical cell	Unspecified wall	≤ 140	Burgess (1970)

[a] Plasmodesmatal densities were calculated from electron micrographs according to the procedures described by Tilney et al. (1990). Unless specified in the original article, the section thickness and plasmodesmatal diameter were assumed to be 80 nm (i.e. the thickness of silver sections) and 40 nm (i.e. the diameter of Onoclea plasmodesmata), respectively.

initially concluded that the walls of dividing apical cells have such high numbers of plasmodesmata in order to carry out efficient cell-to-cell communication (Tilney et al., 1990). However, it is equally probable that the high numbers of plasmodesmata in apical cell walls ensure that expanded walls of mature cells, which appear to lack the potential for secondary plasmodesmata, will have sufficient plasmodesmata to carry out their general physiological activities.

Comparable studies indicate that no secondary plasmodesmata arise in the cell walls of Azolla roots (Gunning, 1978). The total number of primary plasmodesmata which are originally inserted into nascent cell walls is never less than the numbers observed in mature expanded walls, even though the expansion process can reduce the plasmodesmatal densities in certain longitudinal walls from over 100 to less than 0.1 μm^{-2}. Indeed, the only changes in total plasmodesmatal numbers noted in these walls are the losses of plasmodesmata that accompany the formation of certain specialized cells. Again, the plasmodesmatal densities of new walls in expanding merophytes do not match the original densities in apical cell walls, but rather match the plasmodesmatal densities of the neighboring expanded walls. This regulation of plasmodesmatal density may be due to the gradual decrease in endoplasmic reticulum associated with cell differentiation (Gunning, 1978).

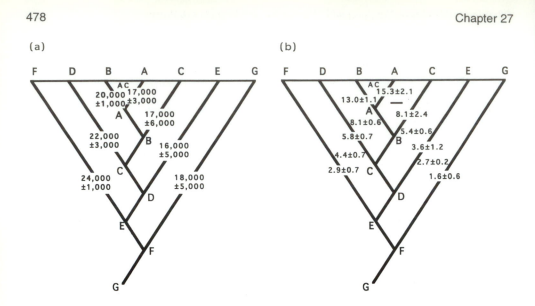

Figure 4. Schematic representation of *Onoclea* prothalli with summary data for the number and density of the plasmodesmata in their cell walls. The letters represent the walls derived from apical cell divisions as indicated in the legend to *Figure 1.* (a) Mean total plasmodesmatal number ± standard error in each expanded wall. (b) Mean plasmodesmatal density ± standard error in each expanded wall. The numbers between the walls denote the densities for the younger walls that arose between the merophyte boundaries. These numbers are calculated from the data in Tilney *et al.* (1990) and from unpublished data.

The most surprising outcome of both studies is the observation that the patterns of cell lineage in the apical meristem do not correlate with the patterns of specialized cells in mature tissues. In the fern gametophyte, specialized cells, including rhizoids, antheridia and archegonia, arise in the basal, lateral and subapical regions, respectively. Thus a single merophyte may produce two different specialized cells. The cellular construction of the fern root is based on three-part symmetry associated with the formation of three merophyte sectors from three lateral sides of the apical cell. Nevertheless, the mature regions of the root exhibit bilateral symmetry as manifested by two protoxylem poles. Since cell lineage determines the quantitative features of plasmodesmatal networks in these organs, it appears that plasmodesmatal frequency by itself does not determine the distribution of the diffusible substances postulated to control cell differentiation.

Conclusions

The results of research to date on fern plasmodesmata lead to the following conclusions:

(1) all plasmodesmata in fern cell walls are apparently primary in their origin;

(2) other features of these plasmodesmata that correlate with their primary origin are an unbranched structure and a random distribution in cell walls; and

(3) the plasmodesmatal density in the cell walls of mature structures depends on (a) the plasmodesmatal network in the apical cell, (b) the relative amount of wall expansion and (c) the selective loss, if any, of primary plasmodesmata during cell differentiation.

4. Plasmodesmatal networks in angiosperms

General aspects of apical meristem construction in angiosperms

In contrast to the single apical cells in fern meristems, the apical meristems of many angiosperms exhibit stratified cellular arrangements (i.e. the initiating cells within the meristem are organized into distinct layers). In this section, we shall establish that this layering has important implications for the predicted distribution of plasmodesmata within the meristem.

The shoot apical meristems of almost all angiosperms are said to have a tunica–corpus organization (Schmidt, 1924). The outer region, known as the tunica, consists of one or more peripheral cell layers where the divisions are generally restricted to the anticlinal plane perpendicular to the surface. The inner region, known as the corpus, consists of the central core where the cells are capable of dividing in all planes. The shoot apical meristem depicted in *Figure 5* has a well-defined tunica composed of two cell layers positioned on top of the unorganized corpus. Direct evidence that these layers are remarkably stable, at least in dicot meristems, comes from numerous studies on periclinal chimeras in which cell layers composed of different genotypes with visible markers show persistent arrangements over considerable periods of plant growth (Clowes, 1961; Poethig, 1989; Tilney-Bassett, 1986).

The anticlinal divisions in the tunica reflect the surface growth of this region, while the variable divisions in the corpus reflect its tendency to undergo volume growth (*Figure 5*). It is an unavoidable geometric consequence of these growth patterns that the walls within the meristem have different relative ages. In particular, the periclinal walls parallel to the surface between adjacent tunica layers and between the innermost tunica layer and the corpus are much older and have thus undergone much more expansion than the younger anticlinal walls in the tunica and all the walls in the corpus. In stable shoot apical meristems, the periclinal wall between the two outermost tunica layers may indeed be one of the oldest walls in the plant, because it arises as the protoderm is demarcated during the segmentation of the early embryo (Maheshwari, 1950).

In general, the apical meristems of angiosperm roots do not exhibit comparable layering. However, it turns out that the most reliable data available for the plasmodesmatal distribution in angiosperm root tips have been obtained from the *Zea mays* root, which happens to exhibit a

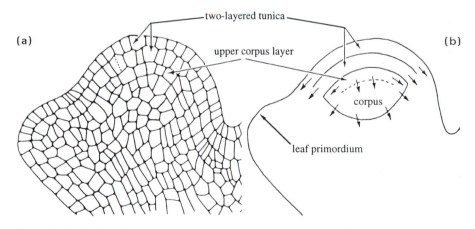

(a)   two-layered tunica   (b)

upper corpus layer

corpus

leaf primordium

Figure 5. Median longitudinal representations of the shoot apical meristem of garden pea (*Pisum sativum*), illustrating the tunica–corpus organization found in many angiosperms. (a) Cellular details. (b) Interpretative diagram with arrows depicting the directions of cell flow. Redrawn from Esau (1977) with permission from John Wiley & Sons, Inc.

stratified apical meristem (*Figure 6*). This root apical meristem exhibits three so-called histogens known as the calyptrogen, periblem and plerome, which are the ultimate sources of the cells for the root cap, cortex and vascular cylinder, respectively (Hanstein, 1868). The geometry here is somewhat misleading because it turns out that the periblem and the plerome comprise a quiescent center in which the cells divide on average once every 175 h (Esau, 1977). The proximal meristem, which consists of rapidly dividing cells located just proximal to the quiescent center, is directly responsible for generating the longitudinal cell files that extend toward the mature regions of the root body. The cells in the calyptrogen divide every 12 h to generate the longitudinal cell files that radiate toward the tip of the root cap. Although the root apical meristems of most other angiosperms, such as the onion (*Allium cepa*), do not exhibit a three-histogen organization, the growth patterns of these meristems, including marked differences in localized cell division rates and the longitudinal cell files, are very similar to those observed in the maize meristem (Esau, 1977).

As in the angiosperm shoot apical meristem, the arrangement of cell walls and resulting growth vectors shows that the walls within the root apical meristem must have different relative ages. It is a geometric imperative that, in order to maintain prominent longitudinal cell files in both the body and the cap, the longitudinal walls parallel to these files must be much older and more expanded than the transverse walls in the files. For instance, given the rapid rate of transverse cell division in the maize calyptrogen, the longitudinal walls in these cells must double their size every 12 h in order to maintain a constant cell size, which means that their walls must undergo 2^{20}-fold (or over 1 000 000-fold) expansion every 10 days. Furthermore, the walls that demarcate the three histogens in the maize meristem are established very early in root development, with the wall at the body–cap interface being one of the oldest in the entire root (Bell and McCully, 1970).

If we assume that all the plasmodesmata in angiosperm apical meristems are formed as primary plasmodesmata inserted into expanding cell plates like fern plasmodesmata, then the above geometric considerations lead to the following testable predictions:

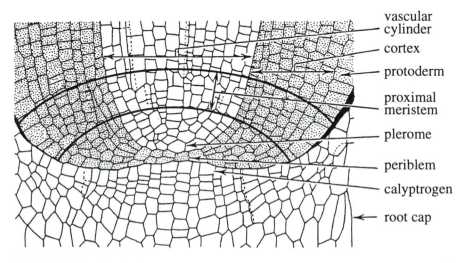

vascular cylinder
cortex
protoderm
proximal meristem
plerome
periblem
calyptrogen
root cap

Figure 6. Median longitudinal section of the root apical meristem of maize (*Zea mays*), illustrating the three-histogen organization common to the grass and several other families. The approximate position of the proximal meristem for the root body (Feldman, 1994) is indicated by solid lines. Redrawn from Esau (1965) with permission from the author.

(1) young walls (e.g. the anticlinal walls in the tunica layers of shoot apical meristems and the transverse walls in the cell files from root apical meristems) should have very high plasmodesmatal densities, comparable to those seen in fern apical cells;
(2) in contrast, the plasmodesmatal densities should be very diluted in older, much expanded walls (e.g. the periclinal walls between tunica layers, the longitudinal walls in the root cell files, and the cap–body interface).

Interestingly, there have been no previous attempts to test these predictions.

Plasmodesmatal densities in angiosperm apical meristems

Despite several comprehensive literature searches, we were unable to find any papers that cross-referenced the terms angiosperm, shoot apical meristem and plasmodesma, or their equivalents. However, the reason for this puzzling lack of pertinent papers became obvious once we started to examine published electron micrographs of meristematic cells in angiosperm shoot apices. The occasional presence of plasmodesmata is pointed out in these figures, but the plasmodesmatal densities are not high enough to become especially noteworthy in general ultrastructural papers. Therefore, we identified the few papers available on angiosperm shoot apical meristems in which both tunica and corpus cells are presented at a sufficiently high magnification to allow plasmodesmata to be easily visualized from the figures, and then we calculated the plasmodesmatal densities using conventional procedures (Tilney *et al.*, 1990). *Table 2* shows that plasmodesmatal densities in the shoot apical meristems of three dicot species, *Glechoma hederacea* (ground ivy), *Chenopodium album* (lambs quarters) and *Lycopersicon esculentum* (tomato), range from 5 to 11 plasmodesmata μm^{-2}, which is 10- to 20-fold lower than the densities reported for fern shoot apices. Even more remarkable is the observation that there are no significant differences between the plasmodesmatal densities of the cell walls in different orientations, which implies that plasmodesmatal density is completely independent of wall age or relative expansion in angiosperm shoot apical meristems. It appears that the angiosperms have evolved a hitherto unrecognized mechanism for maintaining a narrow range of plasmodesmatal densities in the cell walls in their apical meristems.

A comparable situation is seen in the meristematic cells of root apices (*Table 2*). Although we could find only two relevant papers on these apices (Juniper and Barlow, 1969; Strugger, 1957), the published values for plasmodesmatal densities in the root apices of two monocots, *Zea mays* (maize) and *Allium cepa* (onion), fall within the same range as the above values for angiosperm shoot apices. Again, these plasmodesmatal densities are approximately 10-fold lower than those in the comparable fern apices, and they are virtually independent of wall age and relative expansion. It is especially striking that the elongating older longitudinal walls in both the proximal meristem and the cap initials exhibit densities that are only slightly lower than those of the new transverse walls.

We can conceive of no plausible explanation for the constant plasmodesmatal densities in all cell walls of stratified angiosperm meristems other than that these meristems must routinely develop primary plasmodesmata in new walls and secondary plasmodesmata in existing walls. Consequently, the balance between primary and secondary plasmodesmata in any given meristematic cell wall depends on its developmental history, with secondary plasmodesmata predominating in the more expanded older walls. Although this conclusion contradicts the prevailing dogma in the literature, it does follow directly from well-established principles of cell geometry.

Other observations support the concept that some plasmodesmata in angiosperm meristems arise via secondary mechanisms. For example, these plasmodesmata are sometimes clustered into small groups (Bowes, 1965; Chandra Sehkar and Sawhney, 1985; Gifford and Stewart,

Table 2. Plasmodesmatal densities in the apical meristems of various angiosperms

Angiosperm species	Structure	Meristem organization	Cell walls	Plasmodesmatal density (number μm^{-2})	Reference
Glechoma hederacea	Shoot	Tunica–corpus (T–C)	T – anti-clinal walls	11	Bowes (1965)[a]
			T – peri-clinal walls	6	
			C – all walls	9	
Chenopodium album	Shoot	Tunica–corpus (T–C)	T – anti-clinal walls	5	Gifford and Stewart (1967)[a]
			T – peri-clinal walls	5	
			C – all walls	8	
Lycopersicon esculentum	Shoot	Tunica–corpus (T–C)	T – anti-clinal walls	8	Chandra Sehkar and Sawhney (1985)[a]
			T – peri-clinal walls	8	
			C – all walls	7	
Zea mays	Root	Plerome–periblem–calyptrogen (PL–PE–C)	Proximal meristem:		Juniper and Barlow (1969)
			transverse walls	14	
			longitudinal walls	10	
			PL and PE – all walls	10	
			PE/C inter-face	6	
			C – trans-verse walls	15	
			C – longi-tudinal walls	5	
Allium cepa	Root	Unorganized cells	Unspecified walls	6–7	Strugger (1957)

[a] Plasmodesmatal densities were calculated from electron micrographs according to the procedures described in the footnote to *Table 1*.

1967), as opposed to their more uniform distribution in fern apices. Moreover, the plasmodesmata in angiosperm apices are occasionally compound (Sawhney *et al.*, 1981), just like confirmed secondary plasmodesmata in non-division walls (Kollmann and Glockmann, 1985; Kollmann *et al.,* 1985). Finally, it should be relatively easy to characterize secondary plasmodesmata formation in angiosperm meristems by studying the maize calyptrogen, because geometric considerations predict that almost all of the plasmodesmata in its longitudinal walls must be secondary in origin.

Plasmodesmatal densities in mature structures of angiosperms

In general, angiosperm organs do not exhibit the organized cellular arrangements needed to demonstrate that secondary plasmodesmata contribute to the total plasmodesmatal numbers in expanded walls. Nevertheless, there is considerable evidence which suggests that the mature walls in angiosperm organs contain numerous secondary plasmodesmata. In a classic but often ignored paper, Seagull (1984) compared the characteristics of the plasmodesmata in the longitudinal walls of meristematic and elongated cortical cells in the roots of four different species. Even though this study did not adjust for the marked expansion of the longitudinal walls as they were displaced from the meristem, the plasmodesmata in the longitudinal walls of the maize root cortex were observed to undergo a fourfold increase in their density as the cells moved from the meristematic region to more mature regions, which provides unequivocal evidence that secondary plasmodesmata formation is a routine process in growing angiosperm organs. The other three roots investigated in this paper showed slight decreases in plasmodesmatal density, which must still represent large increases in the total number of plasmodesmata per original meristematic wall, since these walls become extremely expanded during root growth. Moreover, the plasmodesmata in all four roots shifted from being dispersed in meristematic cells to becoming clustered into primary pit-fields in elongate cells, which could only result from the secondary insertion of new plasmodesmata at specific sites in the expanding walls (Seagull, 1984).

In an attempt to understand how organic carbon is transported from photosynthetic cells to the phloem, considerable attention has been paid to the distribution of plasmodesmata in both dicot and monocot leaves (Robinson-Beers and Evert, 1991b; Warmbrodt and Vanderwoude, 1990; references cited therein). *Table 3* lists the plasmodesmatal densities in the cell walls between different cell types in two C_3 dicots, namely *Spinacia oleracea* (spinach) and *Populus deltoides* (cottonwood), one C_4 dicot, namely *Amaranthus retroflexus* (green amaranth), and one C_4 monocot, namely a *Saccharum* hybrid, whose differences are presumably related to their different photosynthetic carbon cycles, transport pathways and evolutionary origins. Nevertheless, these plasmodesmatal densities are often less than one order of magnitude lower than the values for the apical meristems of angiosperm shoots (*Table 2*). Since the cells in the leaf primordium must repeatedly divide and expand by several orders of magnitude in order to generate the mature leaf, the presence of such high densities in mature walls provides additional evidence for the insertion of secondary plasmodesmata in the original walls of the young leaf. Secondly, it is noteworthy that common walls between different cell types have characteristic plasmodesmatal densities with little variability within each species (e.g. Robinson-Beers and Evert, 1991b; Warmbrodt and Vanderwoude, 1990), thereby suggesting that these densities are determined by specific communication across the common walls between neighboring cells. Such cell interface-specific values are not observed in mature fern structures such as the *Azolla* root (Gunning, 1978), because cell lineage and relative wall expansion determine plasmodesmatal densities in the ferns.

Other features of the plasmodesmata are also consistent with their secondary origin in leaf cells:

(1) the plasmodesmata are typically positioned in clustered groups in both thick and thin regions of mature cell walls (Evert *et al.*, 1977; Fisher and Evert, 1982; Robinson-Beers and Evert, 1991b; Russin and Evert, 1985; Warmbrodt and Vanderwoude, 1990); and

Table 3. Plasmodesmatal densities in the common walls between different cell types in representative angiosperm leaves

Cell interface [a]	Plasmodesmatal density (number μm^{-2})			
	Saccharum hybrid [b]	*Amaranthus* retroflexus [c]	*Populus* deltoides [d]	*Spinacia* oleracea [e]
M/M	1.9	ND	1.1	0.7
M/BS	4.3	7.3	1.8	0.5
BS/BS	0.9	1.2	2.2	0.5
BS/VP	3.7	3.5	2.7	0.8
BS/CC	ND	0.7	4.8	0.4
VP/VP	4.3	3.9	4.6	0.9
VP/CC	0.3	1.7	2.7	0.9
VP/ST	0.3	0.3	0.6	0.2
CC/CC	ND	0.1	0.0	1.0
CC/ST	0.4	0.9	1.7	1.0

[a] M, mesophyll; BS, bundle sheath; VP, vascular parenchyma; CC, companion cell; ST, sieve tube. Some categories of cell interfaces were combined and averaged to make the observations from all species compatible. The data from the cell walls of the mestome sheath in Saccharum leaves and the bundle-sheath extension of *Populus* leaves are not included here because the other leaves do not have comparable structures.
[b] Reference: Figure 13a in Robinson-Beers and Evert (1991b). The frequency values (number μm^{-1}) for the cells associated with all sizes of vascular bundles were averaged and then converted to density values (number μm^{-2}) according to the procedures described by Tilney *et al.* (1990), using 90 nm as the section thickness (Robinson-Beers and Evert, 1991a) and 80 nm as the plasmodesmatal diameter (Figures 1 and 2 in Robinson-Beers and Evert, 1991a).
[c] Reference: Table 1 in Fisher and Evert (1982). Density values were calculated as described in footnote b.
[d] Reference: Figure 21 in Russin and Evert (1985). Density values were calculated as described in footnote b.
[e] Reference: Table 1 in Warmbrodt and Vanderwoude (1990).

(2) they exhibit cell interface-specific differences in fine structure, including the position of localized constrictions or swollen membranes in the desmotubule, the position of electron-opaque structures called sphincters between the desmotubule and the plasma membrane, and the amount of plasmodesmatal branching (Robinson-Beers and Evert, 1991a; Warmbrodt and Vanderwoude, 1990).

It was proposed that these structural differences result from additional modifications of the plasmodesmata after their initial formation during cytokinesis (Robinson-Beers and Evert, 1991a). However, it is equally likely, and indeed easier to envisage, that such differences arise from specific interactions between differentiating cells that regulate the location and structure of interconnecting plasmodesmata formed via secondary mechanisms.

Conclusions

The work on angiosperm plasmodesmata discussed in the above sections leads to the following new and provocative conclusions:

(1) the plasmodesmata in the apical meristems of angiosperms develop via both primary and secondary mechanisms;

(2) the relative proportions of primary and secondary plasmodesmata in the different cell walls in angiosperm meristems should largely depend on their developmental history, with older, more expanded walls having a higher proportion of secondary plasmodesmata; and

(3)   the plasmodesmata in mature organs of the angiosperms are predominantly secondary in their origin, and their distribution depends on specific intercellular interactions across common walls, rather than being determined by cell lineage and wall expansion like fern plasmodesmata.

5. Discussion

General features of plasmodesmatal networks

It is clear from our geometric analysis that different groups of land plants use different mechanisms to construct their plasmodesmatal networks. In particular, the ferns can only insert primary plasmodesmata into cell plates in dividing cells in their apical meristems. Consequently, the mature structures of the ferns develop *lineage-specific* plasmodesmatal networks, with the observed plasmodesmatal density in any given wall being dependent on its relative elongation. In contrast, the geometric evidence presented here argues very cogently that the angiosperms are able to insert both primary and secondary plasmodesmata in the cell walls of apical meristems, as well as secondary plasmodesmata into the walls of more mature organs. This routine insertion of secondary plasmodesmata signifies that the mature structures of the angiosperms develop *interface-specific* plasmodesmatal networks, where two differentiating cells across each common wall specify the appropriate number and fine structure of their plasmodesmata in order to carry the appropriate intercellular communication. The rest of this section will briefly address several profound evolutionary changes in angiosperm structure, development and physiology that may be mediated by these interface-specific plasmodesmatal networks.

Evolutionary consequences of plasmodesmatal networks

Most lower plants, including bryophytes (mosses and liverworts) and pteridophytes (ferns, horsetails and certain clubmosses), exhibit single apical initials in their apical meristems (e.g. *Figures 1–3*), whereas many seed plants (gymnosperms and angiosperms) have numerous apical initials in stratified arrangements in their apical meristems (e.g. *Figures 5* and *6*). One geometric consequence of a single apical cell is that all new walls with primary plasmodesmata are displaced via subsequent growth to mature regions, thereby allowing the plant to maintain a certain minimum plasmodesmatal destiny in all its walls. In contrast, the walls of which the layers within stratified meristems and subapical regions are composed maintain fixed positions in the apex even though they are also constantly being diluted by cell expansion. The density of plasmodesmata in these walls will necessarily approach zero, unless the cells develop a sec-ondary mechanism for plasmodesmatal formation. Thus the need for all neighboring cells to engage in intercellular communication via their plasmodesmata seems to have provided a strong selection pressure for maintaining single apical cells in those lower plants which have only primary plasmodesmata. Since the seed plants evolved the ability to insert secondary plasmo-desmata into their walls, this selection pressure was apparently relaxed so that their meristems were free to evolve a wide range of different stratified and unstratified structures, as illustrated in Clowes (1961) and Popham (1951).

   One fundamental difference between plants and animals is that the cells in developing animals are routinely capable of morphogenetic movements, whereas typical plant cells, which are held within rigid cellulose walls, maintain fixed positions relative to adjoining cells. Since the position of mature animal cells is not dictated by their original lineage, animals have evolved transmembrane structures known as gap junctions which connect adjacent cells in a manner analogous to secondary plasmodesmata. Gap junctions are organized into a cell-specific network for intercellular communication where their number and relative permeability are tightly regulated between different cells (Finbow, 1982; Larsen, 1989; Warner, 1992). The inherent flexibility associated with secondary formation and regulated transmission allows the gap junctions to mediate the sophisticated intercellular signalling necessary to construct complex higher animals (Warner, 1992). It seems reasonable to argue that a communication network based on primary plasmodesmata is intrinsically inferior to a gap junction-based network, because the distribution of primary plasmodesmata depends on cell lineage rather than on cell function.

Therefore, the ability to develop secondary plasmodesmata has permitted the angiosperms to evolve a more elaborate communication network which is truly comparable to the network in animals. To mention just one likely consequence of secondary plasmodesmatal formation, it seems quite conceivable that the evolution of C_4 photosynthesis has been restricted to the angiosperms partly because only these plants have the potential to use secondary plasmodesmata to construct the high-capacity cytoplasmic pathway needed for rapid metabolic exchange between the mesophyll and bundle-sheath cells. Our future research will address the evolutionary consequences of secondary plasmodesmatal formation in greater detail.

Acknowledgements

We thank Dana Emery and Bin Lu for their assistance with the figures.

References

Bell JK, McCully ME. (1970) A histological study of lateral root initiation and development in *Zea mays*. *Protoplasma* **70:** 179–205.

Bierhorst DW. (1977) On the stem apex, leaf initiation, and early leaf ontogeny. *Am. J. Bot.* **64:** 125–152.

Bowes BG. (1965) The ultrastructure of the shoot apex and young shoot of *Glechoma hederacea. Cellule* **65:** 349–356.

Burgess J. (1971) Observations on structure and differentiation in plasmodesmata. *Protoplasma* **73:** 83–95.

Chandra Sehkar KN, Sawhney VK. (1985) Ultrastructure of the shoot apex of tomato *(Lycopersicon esculentum). Am. J. Bot.* **72:** 1813–1822.

Clowes FAL. (1961) *Apical Meristems.* Oxford: Blackwell Science Ltd.

Cooke TJ, Lu B. (1992) The independence of cell shape and overall form in multicellular algae and land plants: cells do not act as building blocks for constructing plant organs. *Int. J. Plant Sci.* **153:** S7–S27.

Ding B, Haudenshield JS, Hull RJ, Wolf S, Beachy RN, Lucas WJ. (1992) Secondary plasmodesmata are specific sites of localization of the tobacco mosaic virus movement protein in transgenic tobacco plants. *Plant Cell* **4:** 915–928.

Esau K. (1965) *Plant Anatomy,* 2nd edn. New York: John Wiley & Sons.

Esau K. (1977) *Anatomy of Seed Plants*, 2nd edn. New York: John Wiley & Sons.

Evert RF, Eschrich W, Heyser W. (1977) Distribution and structure of the plasmodesmata in mesophyll and bundle-sheath cells of *Zea mays* L. *Planta* **136:** 77–89.

Feldman L. (1994) The maize root. In: *The Maize Handbook* (Freeling M, Walbot V, ed.). New York: Springer-Verlag, pp. 29–37.

Finbow ME. (1982) A review of junctional mediated intercellular communication. In: *The Functional Integration of Cells in Animal Tissue* (Pitts JD, Finbow ME, ed.). Cambridge: Cambridge University Press, pp. 1–37.

Fisher DG, Evert RF. (1982) Studies on the leaf of *Amaranthus retroflexus* (Amaranthaceae): ultrastructure, plasmodesmatal frequency, and solute concentration in relation to phloem loading. *Planta* **155:** 377–387.

Gifford EM Jr. (1983) Concept of apical cells in bryophytes and pteridophytes. *Annu. Rev. Plant Physiol.* **34:** 419–440.

Gifford EM Jr, Stewart KD. (1967) Ultrastructure of the shoot apex of *Chenopodium album* and certain other seed plants. *J. Cell Biol.* **33:** 131–142.

Gunning BES. (1978) Age-related and origin-related control of the numbers of plasmodesmata in cell walls of developing *Azolla* roots. *Planta* **143:** 181–190.

Gunning BES, Steer MW. (1975) *Ultrastructure and the Biology of Plant Cells.* London: Edward Arnold Publishers.

Gunning BES, Robards AW, eds. (1976) *Intercellular Communication in Plants: Studies on Plasmodesmata.* Heidelberg: Springer-Verlag.

Gunning BES, Hughes JE, Hardham AR. (1978) Formative and proliferative cell divisions, cell differentiation, and developmental changes in the meristem of *Azolla* roots. *Planta* **143:** 121–144.

Hanstein J. (1868) Die Scheitelzellgruppe im Vegetationspunkt der Phanerogramen. *Festschr. Niederrhein. Gesell. Natur. Heilkunde* 109–134.

Hepler PK. (1982) Endoplasmic reticulum in the formation of the cell plate and plasmodesmata. *Protoplasma* **111:** 121–133.

Hicks GS, Steeves TA. (1973) Plasmodesmata in the shoot apex of *Osmunda cinnamomea*. *Cytologia* **38:** 449–453.

Jones MGK. (1976) The origin and development of plasmodesmata. In: *Intercellular Communication in Plants: Studies on Plasmodesmata* (Gunning BES, Robards AW, ed.). Heidelberg: Springer-Verlag, pp. 81–102.

Juniper BA. (1976) Junctions between plant cells. In: *The Developmental Biology of Plants and Animals* (Graham CF, Wareing PF, ed.). Philadelphia: W.B. Saunders Company, pp. 111–126.

Juniper BA, Barlow PW. (1969) The distribution of plasmodesmata in the root tip of maize. *Planta* **89:** 352–360.

Kollmann R, Glockmann C. (1985) Studies on graft unions. I. Plasmodesmata between cells of plants belonging to different unrelated taxa. *Protoplasma* **124:** 224–235.

Kollmann R, Glockmann C. (1991) Studies on graft unions. III. On the mechanism of secondary formation of plasmodesmata at the graft interface. *Protoplasma* **165:** 71–85.

Kollmann R, Yang S, Glockmann C. (1985) Studies on graft unions. II. Continuous and half plasmodesmata in different regions of the graft interface. *Protoplasma* **126:** 19–29.

Larsen WJ. (1989) Mechanisms of gap junction modulation. In: *Cell Interactions and Gap Junctions, Vol. 1* (Sperelakis N, Cole WC, ed.). Boca Raton: CRC Press, pp. 3–27.

Ledbetter MC, Porter KR. (1970) *Introduction to the Fine Structure of Plant Cells.* New York: Springer-Verlag.

Maheshwari P. (1950) *An Introduction to the Embryology of Angiosperms.* New York: McGraw-Hill Book Company.

Michaux N. (1970) Etude comparee de la structure et du fonctionnement du meristeme apical adulte de l'*Isoetes setacea* Lam. et du *Pteris cretica* L. *Bull Soc. Bot. Fr. Mem.* **117:** 83–101.

Monzer J. (1991) Ultrastructure of secondary plasmodesmata formation in regenerating *Solanum nigrum*-protoplast cultures. *Protoplasma* **165:** 86–95.

Nitayangkura S, Gifford EM Jr, Rost TL. (1980) Mitotic activity in the root apical meristem of *Azolla filiculoides* Lam., with special reference to the apical cell. *Am. J. Bot.* **67:** 1484–1492.

Poethig RS. (1989) Genetic mosaics and cell lineage analysis in plants. *Trends Genet.* **5:** 274–278.

Popham RA. (1951) Principal types of vegetative shoot apex organization in vascular plants. *Ohio J. Sci.* **51:** 249–270.

Robards AW, Lucas WJ. (1990). Plasmodesmata. *Annu. Rev. Plant Physiol. Plant Mol. Biol.* **41:** 369–419.

Robinson-Beers K, Evert RF. (1991a) Fine structure of plasmodesmata in mature leaves of sugarcane. *Planta* **184:** 307–314.

Robinson-Beers K, Evert RF. (1991b) Ultrastructure of and plasmodesmatal frequency in mature leaves of sugarcane. *Planta* **184:** 291–306.

Russin WA, Evert RF. (1985) Studies on the leaf of *Populus deltoides* (Salicaceae): ultrastructure, plasmodesmatal frequency, and solute concentration. *Am. J. Bot.* **72:** 1232–1247.

Sawhney VK, Rennie PJ, Steeves TA. (1981) The ultrastructure of the central zone cells of the shoot apex of *Helianthus annuus*. *Can. J. Bot.* **59:** 2009–2015.

Schmidt A. (1924) Histologische Studien an phanerogramen Vegetationspunkten. *Bot. Arch.* **8:** 345–404.

Seagull RW. (1984) Differences in the frequency and disposition of plasmodesmata resulting from root cell elongation. *Planta* **159:** 497–504.

Strugger S. (1957) Die elektronenmikroskopische Nachweis von Plasmodesmen mit Hilfe der Uranyl-imprägnierung an Wurzelmeristemen. *Protoplasma* **48:** 231–236.

Tilney LG, Cooke TJ, Connelly PS, Tilney MS. (1990) The distribution of plasmodesmata and its relationship to morphogenesis in fern gametophytes. *Development* **110:** 1209–1221.

Tilney-Bassett RAE. (1986) *Plant Chimeras.* London: Edward Arnold Publishers.

Warmbrodt RD, Vanderwoude WJ. (1990) Leaf of *Spinacea oleracea* (spinach): ultrastructure, and plasmodesmatal distribution and frequency, in relation to sieve-tube loading. *Am. J. Bot.* **77:** 1361–1377.

Warner A. (1992) Gap junctions in development — a perspective. *Semin. Cell Biol.* **3:** 81–91.

Wetter R, Wetter C. (1954) Studien über das Erstarkungswachstum und das primare Dickenwachstum bei leptosporangiaten Farnen. *Flora* **141:** 598–631.

Chapter 28

Secondary plasmodesmata: biogenesis, special functions and evolution

Biao Ding and William J. Lucas

1. Introduction

During the course of evolution, green plants advanced in terms of structural and functional complexity. This complexity is well manifested by the process of morphogenesis, in which groups of cells become differentiated to form specialized tissues possessing distinct structures and functions. As the fate of a plant cell is position-dependent, that cell must know its particular position within the plant body in order to express relevant genes during differentiation. Co-ordination, at the supracellular level, of differential gene expression is made possible only by means of effective and specific cell-to-cell communication. Plasmodesmata, the intercellular organelles of higher plants, presumably provide the pathway(s) for direct cell-to-cell communication (Beebe and Turgeon, 1991; Gunning and Robards, 1976; Gunning and Overall, 1983; Lucas *et al.*, 1993; Lucas and Wolf, 1993; Oparka, 1993; Robards and Lucas, 1990).

Higher plant plasmodesmata can be formed in two basic ways. During cytokinesis, pro-grammed positioning of the endoplasmic reticulum (ER) between fusing vesicles of the deve-loping cell plate and subsequent transformation of the ER into an appressed cylinder lead to the formation of the so-called 'primary plasmodesmata' (Hepler, 1982; Lucas *et al.*, 1993). In general appearance, primary plasmodesmata are single-stranded, linear entities, and their substructure has been subjected to numerous electron microscopic studies (for reviews see Ding *et al.*, 1992b; Lucas *et al.*, 1993; Robards and Lucas, 1990). A current model suggests that proteinaceous particles of approximately 3 nm in diameter are embedded in the plasma membrane and appressed ER membrane of the plasmodesma. Spaces between these protein particles are thought to form micro-channels, 2.5 nm in diameter, that function as the potential sites for intercellular transport of small molecules and ions (*Figure 1*) (Ding *et al.*, 1992b). Primary plasmodesmata are discussed in Chapter 27 by Cooke *et al.*

Ample evidence exists to support the hypothesis that plasmodesmata can be generated across established cell walls (Ding *et al.*, 1992a; Jones, 1976; Kollmann and Glockmann, 1985, 1991; Kollmann *et al.*, 1985; Lucas *et al.*, 1993). As these plasmodesmata are formed post-cytokinetically, they are referred to as 'secondary plasmodesmata' (Jones, 1976; Lucas *et al.*, 1993; Robards and Lucas, 1990). Recent studies indicate that a number of mechanisms exist for the biogenesis of secondary plasmodesmata, and that these plasmodesmata are structurally polymorphic. Moreover, higher plant secondary plasmodesmata appear to have evolved the ability to perform certain special function(s), in addition to participating in general intercellular

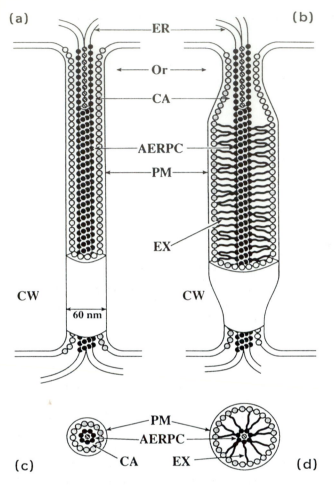

Figure 1. Substructural model of primary plasmodesmata of the tobacco leaf (Ding *et al.,* 1992b). (a) Longitudinal view of a newly formed plasmodesma. (b) Longitudinal view of a primary plasmodesma that has undergone structural changes by the formation of a central cavity during leaf development. (c) Transverse view of any given region along the length of a newly formed plasmodesma as shown in (a), or the orifice region of a developing plasmodesma as shown in (b). (d) Transverse view of the middle portion of a developing plasmodesma as shown in (b). AERPC, appressed ER–protein complex; CA, cytoplasmic annulus containing micro-channels; CW, cell wall; ER, endoplasmic reticulum; EX, spoke-like extension; Or, orifice; PM, plasma membrane. Reproduced from Lucas *et al.* (1993) with permission from *The New Phytologist.*

transport. This chapter will focus on recent advances in our understanding of the biogenesis, structure and special functions of secondary plasmodesmata, as well as discussing future research prospects in these areas. For previous reviews on secondary plasmodesmata, readers are referred to Jones (1976), Robards and Lucas (1990), and Lucas *et al.* (1993). Comprehensive treatises on various forms of cytoplasmic bridges involved in intercellular transport in higher plants, algae and fungi can be found in Gunning and Robards (1976) and Lucas *et al.* (1993).

2. Biogenesis and structure of higher plant secondary plasmodesmata

Mechanisms of biogenesis

As the formation of secondary plasmodesmata creates new cytoplasmic bridges through established primary cell walls, this process must be mechanistically coupled to the wall dynamics. Hence, knowledge of the molecular architecture and metabolism of the primary cell walls is essential for studies on the biogenesis of secondary plasmodesmata. Broadly speaking, the primary plant cell wall of dicots and non-graminaceous monocots is composed of a skeletal framework of cellulose microfibrils that is intertwined with matrix polymers including hemicelluloses, pectins and proteins. Based on current interpretations (Carpita and Gibeaut, 1993; McCann and Roberts, 1991; McCann *et al.*, 1990, 1992; Talbot and Ray, 1992), the microfibrils are cross-linked non-covalently by hemicellulosic polymers, particularly xyloglucans. Pectins in the primary walls appear to form an independent network that is as extensive as the cellulose/hemicellulose network; removal of this pectin network does not disrupt the structural integrity of the cellulose/hemicellulose network. Pectin molecules of the middle lamella are cross-linked by Ca^{2+}. These pectins presumably cement together the walls of adjacent cells by cross-linking the pectins of both cell walls.

Probably the most important aspect of cell wall dynamics related to secondary plasmodesmal formation is the control of wall porosity. Normally, the diameter of the pores within the primary cell wall is less than 10 nm (Baron-Epel *et al.*, 1988; McCann *et al.*, 1990), which is much smaller than the diameter of a plasmodesma (50–60 nm). Thus the cell wall porosity must be increased locally for plasmodesmata to be formed across the wall. In this context, it is important to note that the removal of pectins has been shown to increase the pore size of the onion primary wall from approximately 10 nm to up to 40 nm (McCann *et al.*, 1990). These results, together with the findings of other studies (Baron-Epel *et al.*, 1988; Shedletzky *et al.*, 1990), indicate that pectins most probably play an important role in controlling wall porosity. Further removal of xyloglucans, *in vitro*, leads to the collapse of the microfibrillar network, creating spaces as large as 100 nm in diameter (McCann *et al.*, 1990). Thus, these polymers also appear to play important roles in regulating the spacing between microfibrils, and hence the porosity of the cell wall. These data suggest that the metabolism of pectins and xyloglucans may be critical to the formation of secondary plasmodesmata in dicots and non-graminaceous monocots.

These basic features of primary cell walls provide the basis for the following discussions on the likely mechanisms underlying secondary plasmodesmal formation. However, it should be borne in mind that as certain biochemical constituents, the matrix polymers in particular, vary qualitatively as well as quantitatively between different plants (e.g., graminaceous monocots versus dicots and non-graminaceous monocots) (Carpita and Gilbeaut, 1993; McCann and Roberts, 1991), the mechanics of secondary plasmodesmal formation may vary from plant to plant.

De novo *formation.* Kollmann and Glockmann (1991) have recently provided a detailed description of how secondary plasmodesmata are formed *de novo* between heterografts of *Vicia faba* and *Helianthus annuus*. As is shown in *Figure 2*, one of the earliest events during secondary plasmodesmal formation in this system involves the thinning and loosening of localized regions of the contact wall between neighbouring cells. Meanwhile, the ER becomes attached to the plasma membrane on both sides of the thinning regions of the adjoining walls (*Figure 2c*). When the cell wall material is completely removed the neighbouring plasma membranes become fused, a process that is presumably driven by turgor pressure (*Figure 2d, e*). Coalescence of wall material-containing Golgi vesicles, either among themselves or with the plasma membrane,

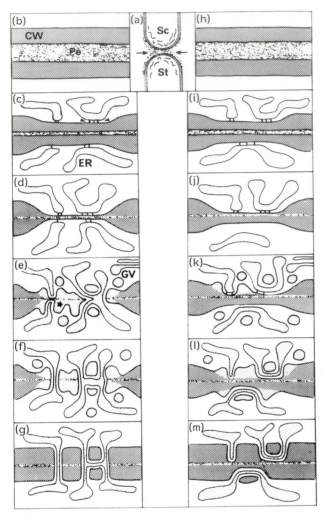

Figure 2. Diagram summarizing the cellular processes underlying the biogenesis of secondary plasmodesmata at a graft interface. (a) Approaching callus cells of scion (Sc) and stock (St) Pectic substance (Pe) is present between the contact regions of these cells. Arrows indicate the wall region where secondary plasmodesmata will be formed. (b–g) Formation of continuous secondary plasmodesmata by fusion of the ER and the plasma membrane of neighbouring cells. In (c), arrowheads indicate the 5 nm particles that connect the ER to the plasma membrane. In (e), the star indicates newly deposited wall material. (h–m) Formation of discontinuous secondary plasmodesmata (see text for details). CW, cell wall; GV, Golgi vesicles. Reproduced from Kollmann and Glockmann (1991) with permission from Springer-Verlag and the authors.

results in the reconstruction of the cell wall (*Figure 2f, g*). Along with these events, the ER is positioned and then transformed between the coalescing Golgi vesicles to generate secondary plasmodesmata. *Figure 2h–m* illustrates the events that are often associated with the formation of incomplete secondary plasmodesmata at the graft union interface. A similar mechanism has been proposed by Monzer (1991) for secondary plasmodesmal formation between regenerating protoplasts of *Solanum nigrum* L.

The thinning and removal of the cell wall at the site of the graft union apparently involves the activities of cellulases, hemicellulases and pectinases. How the activities of these enzymes are regulated has yet to be resolved. It is possible that the necessary enzymes are secreted into the cell wall either by the ER which becomes attached to the plasma membrane, or by specific enzyme-carrying vesicles that fuse with the plasma membrane. Alternatively, the enzymes may reside in the cell wall in an inactive conformation and can be activated by signal molecules that are targeted to this specific region of the plasma membrane. In this context, it is interesting to note that Kollmann and Glockmann (1991) reported that electron-dense particles (proteins) of diameter approximately 5 nm appear to link the ER to the plasma membrane at the beginning of the wall-thinning process (see *Figure 2c*). Perhaps these particles serve as initial signals that trigger the cascade of events leading to secondary plasmodesmal formation. Characterization of these particles in terms of their biochemical nature and function will provide important information about their role in secondary plasmodesmal formation.

At present there is a complete lack of information about the biochemical and molecular nature of the cellular agents involved in secondary plasmodesmal formation across the graft interface. Until such information is available, it will be impossible to ascertain whether this mechanism shares any common features with the *de novo* formation of secondary plasmodesmata that takes place during normal plant development.

Modification of primary plasmodesmata. Modification of primary plasmodesmata represents an alternative mechanism for generating secondary plasmodesmata during plant development (Ding *et al.*, 1992a, 1993; Jones, 1976; Lucas *et al.*, 1993). Based on the available data, it is possible to reconstruct the major cellular events involved in primary-to-secondary transformation of plasmodesmata (*Figure 3*) (Lucas *et al.*, 1993). In some cases, as shown in *Figure 3a–c*, lateral fusion of neighbouring primary plasmodesmata appears to initiate the modification process, as occurs between tobacco mesophyll cells. Such lateral fusion produces an intermediate form, the so-called 'H-shaped' modified primary plasmodesma (Ding *et al.*, 1992a, 1993). The addition of new cytoplasmic bridges to these modified primary plasmodesmata converts the latter into bona fide complex secondary plasmodesmata. In other cases, however, lateral fusion of primary plasmodesmata does not occur. As illustrated in *Figure 3d–f*, for the development of secondary plasmodesmata between the sieve element and the companion cell, new protoplasmic bridges can be added asymmetrically to a primary plasmodesma from one cell side (from the companion cell side in this case), leading to the formation of a 'deltoid-shaped' secondary plasmodesma (Esau and Thorsch, 1985). It should be noted that such asymmetrical addition of new cytoplasmic bridges is not restricted to secondary plasmodesmal development between the sieve element and the companion cell. It is also common between other cell types, but with the number of branches being smaller (more variable).

Details of the cellular processes underlying such primary-to-secondary plasmodesmal transformation have yet to be further elucidated. A notable feature is that fusion of neighbouring primary plasmodesmata starts exclusively at the middle lamella region, and all of the branches of a 'fully developed' secondary plasmodesma are united via a central cavity created by local removal of the middle lamella and adjacent cell wall material. In 'deltoid-shaped' secondary plasmodesmata, all of the branches from one side of the cell also meet at the middle lamella. These features imply that special signals exist either in the middle lamella, or at the portion of the plasmodesmal plasma membrane that interfaces with the middle lamella, and wall degradation is initiated at this site.

As illustrated in *Figure 4*, there may be several mechanisms by which localized degradation of the middle lamella and adjacent cell wall material could be achieved. First, active cell wall-

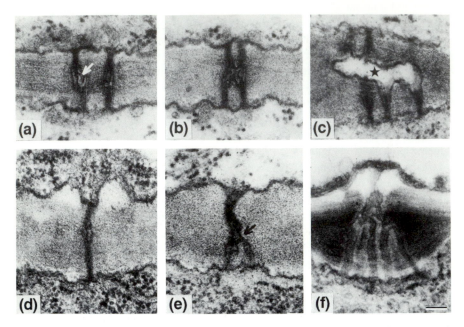

Figure 3. (a–c) Development of morphologically complex secondary plasmodesmata via lateral fusion of primary plasmodesmata between mesophyll cells of tobacco. (a) At an early stage, the plasma membranes of neighbouring plasmodesmata extend out into the region of the middle lamella that has presumably been enzymatically digested. Note the separation of the appressed ER membranes (arrow). (b) At a later stage, the neighbouring primary plasmodesmata fuse to form an 'H-shaped' modified primary plasmodesma. (c) Addition of new cytoplasmic bridges to the modified primary plasmodesma results in the formation of a morphologically complex secondary plasmodesma. The star indicates the location of the central cavity that develops within the region of the middle lamella. Scale bar = 0.1 μm. (d–f) Formation of the 'deltoid-shaped' secondary plasmodesma, via asymmetrical addition of new cytoplasmic bridges to a primary plasmodesma, between the sieve element (upper) and the companion cell (lower). Primary plasmodesma are formed between the sieve element and the companion cell in the early stages of phloem differentiation (cytokinetic event) (d). During the course of differentiation, new cytoplasmic bridges (arrow in e) are gradually added to the primary plasmodesmata, on the companion cell side, to give rise to the secondary plasmodesma (e and f). Scale bar = 0.1 μm. (b) and (f) are reproduced from Ding *et al.* (1993) with permission from Blackwell Science Ltd. (c) is reproduced from Ding *et al.* (1992a) with permission from the American Society of Plant Physiologists.

degrading enzymes may be directly transported, via the primary plasmodesma, into the middle lamella (*Figure 4a*). Secondly, such (inactive?) enzymes may be secreted into the wall via vesicle fusion with the plasma membrane, and triggering molecules may be transported, via the primary plasmodesma, into the middle lamella to achieve a local activation of these enzymes (*Figure 4b*). Alternatively, the triggering molecules may be located in the plasma membrane interfacing with the middle lamella. Signalling molecules are transported, via the primary plasmodesmata, to this region of the plasma membrane to interact with the triggering molecules, which then move into the middle lamella to activate the enzymes (*Figure 4c*). Thirdly, as previously suggested (Lucas *et al.*, 1993), the inactive enzymes may have been deposited in the walls by fusing Golgi vesicles during cell plate formation, and the triggering molecules may enter the middle lamella through plasmodesmata, by means illustrated in *Figure 4b* or *Figure 4c*, to activate the enzymes at a later stage.

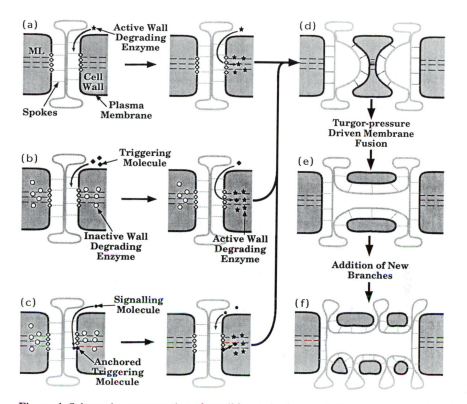

Figure 4. Schematic representation of possible mechanisms underlying primary-to-secondary plasmodesmal transformation. (a), (b) and (c) illustrate various possible modes of enzymatic digestion of the middle lamella and adjacent primary cell wall material involved in the modification of a single primary plasmodesma. At a certain stage of such modification, neighbouring primary plasmodesmata become fused (d and e) to give rise to an 'H-shaped' modified primary plasmodesma. Subsequent addition of new cytoplasmic branches leads to the formation of a morphologically complex secondary plasmodesma (e) (see text for a more detailed discussion). ML, middle lamella.

Initial removal of the middle lamella and nearby wall material results in an extension of the plasma membrane out into the primary wall. In concert, the appressed ER membranes separate in the middle lamella region, and the cisterna facing the plasma membrane is pulled along with the plasma membrane, presumably due to specific interactions between the two membrane systems (*Figure 4d*). When the middle lamella between the two (or more) neighbouring plasmodesmata is completely removed, the plasma membranes of these plasmodesmata coalesce with each other via a turgor pressure-driven process, and the ER membranes unite into a continuum, forming a lumen in the central cavity where the middle lamella once existed (*Figure 4e*). Subsequent addition of new cytoplasmic bridges results in the formation of a complex secondary plasmodesma (*Figure 4f*).

While the middle lamella pectins apparently need to be removed, presumably by a pectinase or by chelation of Ca^{2+} (McCann *et al.*, 1990), it would appear that cellulose microfibrils do not have to be enzymatically digested; rather, they may be mechanically rearranged to create spaces to accommodate penetration of the plasma membrane–ER network. This speculation is based on the finding that extraction of xyloglucans *in vitro* leads to lateral association of microfibrils and the formation of spaces as large as 100 nm in diameter that are encircled by microfibrils (McCann *et al.*, 1990). A space of this dimension would easily accommodate a plasmodesma. Interestingly, some of the elliptic porous structures revealed in the published micrographs of onion primary cell wall, after sequential extraction of pectin and hemicellulosic polymers (McCann *et al.*, 1990, 1992), look remarkably like transverse views of complex secondary plasmodesmata near the middle lamella.

Enzymatic removal or rearrangement of hemicelluloses, mainly xyloglucans, from microfibrils *in vivo* may be achieved via hydrolysis by endo-1,4-β-glucanase activity (Hayashi, 1991; Hayashi *et al.*, 1984), or via non-hydrolytic cleavage by xyloglucan endotransglycosylase (XET) (Fry *et al.*, 1992; Nishitani and Tominaga, 1992; Smith and Fry, 1991). While the activity of the endo-1,4-β-glucanase has been shown to be stimulated by auxin (Hayashi, 1991), little information is available about regulation of the XET activity. Clearly, a combination of structural, biochemical and molecular approaches is required to identify the set of enzymes involved in the co-ordinated degradation of the wall matrix material to permit secondary plasmodesmal formation.

Regulation

How the biogenesis of secondary plasmodesmata is regulated, in terms of when, where and how many secondary plasmodesmata are to be laid down, is currently unknown. A related issue is whether structural and/or biochemical turnover of secondary plasmodesmata, either individually or as a population, occurs as a function of cell or tissue development.

Regulation of the biogenesis and potential turnover of secondary plasmodesmata is conceivably an important means of exerting control over the morphogenesis, growth and development of the plant. As such, the underlying molecular and cellular mechanisms warrant further study.

Structure

In contrast with the situation regarding primary plasmodesmata, there have been no specific studies on the substructure of secondary plasmodesmata, particularly that of complex secondary plasmodesmata. While the substructure of a simple, linear secondary plasmodesma may not differ from that of a primary plasmodesma (see *Figure 1*), the substructure of a morphologically 'advanced' secondary plasmodesma is much more complicated. Examination of many published and unpublished micrographs (see also *Figure 3a*) has enabled us to develop a structural model for complex secondary plasmodesmata (*Figure 5*). A notable feature of this model is that the ER membranes within all of the branches are united into a functional continuum. In this model we have omitted details of the membrane structure. However, as primary and secondary plasmodesmata exhibit an equivalent size exclusion limit (SEL), globular proteins in the appressed ER of each secondary plasmodesmal branch are likely to be organized in a manner similar to that observed in primary plasmodesmata (see *Figure 1*).

To understand fully how individual plasmodesmata function mechanistically in cell-to-cell transport, we need to know the specific molecular components that combine together to form the structural framework of the plasmodesma that potentiates intercellular trafficking of macromolecules. Numerous studies have shown that plasmodesmata exhibit activities of several enzymes, including ATPases (Didehvar and Baker, 1986; Gilder and Cronshaw, 1973a, b; Hall, 1969; Zhen *et al.*, 1985) and phosphatases (Robards and Kidwai, 1969). Recently, the tobacco mosaic virus movement protein (TMV MP)-phosphorylating activity of a putative protein kinase was detected in the cell wall fraction of tobacco leaves, and the presence of this kinase activity was found to be developmentally correlated with primary-to-secondary transformation of

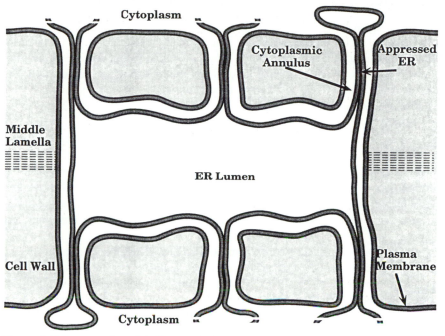

Figure 5. Substructural model of a morphologically complex secondary plasmodesma. A notable feature of this model is that the ER membranes within all of the cytoplasmic branches are united into a functional continuum. Hence an ER lumen exists in a central cavity created by the removal of the middle lamella and adjacent cell wall material. Although not shown here, the globular proteinaceous particles of the appressed ER of each cytoplasmic bridge are assumed to be organized in a pattern similar to that found in a primary plasmodesma (see *Figure 1*).

mesophyll plasmodesmata (Citovsky *et al.*, 1993). Thus, it is possible that this particular kinase activity is localized within secondary plasmodesmata. These studies provide some insights into possible mechanisms of plasmodesmal transport. However, real progress must await the development of a protocol that enables the isolation of intact plasmodesmata. Although Epel and co-workers (Kotlizky *et al.*, 1992; Yahalom *et al.*, 1991) have made some advances in this area, progress in general has been disappointingly slow.

3. Special functions of higher plant secondary plasmodesmata

Secondary plasmodesmata occur commonly in higher plants. They are formed during normal plant development and in special systems such as plant chimeras, graft unions, protoplast fusion, and parasite–host interaction (for reviews see Jones, 1976; Lucas *et al.*, 1993; Robards and Lucas, 1990). Recent studies indicate that, in addition to establishing new cytoplasmic continuity between neighbouring cells in order to mediate general intercellular solute transport, secondary plasmodesmata may have evolved to fulfil special functions in plant growth and development. However, it should be emphasized that there is no definitive means presently available to distinguish, in a functional sense, between primary and secondary plasmodesmata. Therefore, there is no experimental evidence which demonstrates that the functions that are known to be performed by secondary plasmodesmata, as discussed below, are not also performed by primary plasmodesmata.

Co-ordination of growth and development

Secondary plasmodesmata between L1 and L2 layers. The shoot apical meristem of most dicots and some monocots is a highly organized structure composed of three distinct layers of cells: the outermost L1, the subsurface L2, and the inner L3 (Medford, 1992; Satina *et al.*, 1940; Smith *et al.*, 1992). While L1 cells divide anticlinally to generate the epidermis of various parts of a differentiated shoot, L2 and L3 cells divide both anticlinally and periclinally to produce inner tissues such as the shoot cortex, leaf mesophyll and vasculature (Medford, 1992; Satina *et al.*, 1940; Sussex, 1989).

Although cell division is lacking between L1 and L2, plasmodesmata are frequently observed between these cells. *Figure 6* illustrates such secondary plasmodesmata which develop between the epidermal and mesophyll cells in tobacco leaves. The fact that these plasmodesmata are formed in the absence of cell division clearly indicates that their formation involves *de novo* processes. The simple, linear secondary plasmodesmata that are formed initially can be further modified to form morphologically complex secondary plasmodesmata during organ development.

Circumstantial evidence exists for the involvement of secondary plasmodesmata between L1 and L2 in the communication of developmental signals. The dominant mutation *Knotted-1 (Kn1)* of maize alters the normal pattern of leaf development, resulting in periclinal cell divisions in the epidermis and hence extra growth in localized regions of maize leaves expressing this gene.

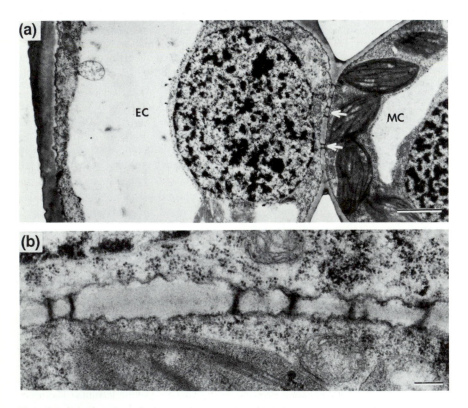

Figure 6. Simple secondary plasmodesmata between neighbouring epidermal and mesophyll cells of a young tobacco leaf. (a) Low-magnification view of an epidermal cell (EC) and a mesophyll cell (MC). Note the plasmodesmata (arrows) across the cell wall between these two cell types. (b) High-magnification view of plasmodesmata shown in (a). Scale bars = 0.2 μm.

Analysis of genetic mosaics shows that the unusual cell divisions in the epidermis are induced by inner tissue layers (Hake and Freeling, 1986; Sinha and Hake, 1990). Immunocytochemical studies have consistently demonstrated that ectopic expression of the *Kn1* gene is associated with leaf veins. Thus epidermal cells must respond to certain signals from the inner tissues to initiate periclinal division in the mutant leaves (Smith *et al.*, 1992). It is possible that one such signal is the Kn1 protein that traffics via plasmodesmata. A recent study has provided direct evidence that Kn1 is able to move from cell to cell via the plasmodesmata (Lucas *et al.*, 1995).

For study of the biogenesis and function of secondary plasmodesmata between cells of the L1 and inner layers, it would appear that periclinal chimeras, in which L1 and inner layers of cells derive from different plants with distinct genetic, morphological and/or functional markers (Heichel and Anagnostakis, 1978; Sussex, 1989), offer an excellent experimental approach.

Mesophyll secondary plasmodesmata are indispensable for normal leaf development. In tobacco, primary plasmodesmata predominate at an early leaf developmental stage. As the leaf progresses towards maturation, most of these primary plasmodesmata are gradually converted into highly branched secondary plasmodesmata (*Figure 7a, b*; see also *Figure 3a*) (Ding *et al.*, 1992a). Studies on transgenic tobacco plants expressing the gene for the 30-kDa tobacco mosaic virus (TMV) movement protein (MP) provided initial evidence that these complex secondary plasmodesmata differ from primary plasmodesmata in certain functional aspects (Ding *et al.*, 1992a). In leaves of such transgenic plants, the TMV MP is localized exclusively in secondary, but not primary, plasmodesmata of non-vascular cells during the later stages of leaf development. Moreover, this localization pattern is developmentally correlated with a TMV MP-induced increase in the plasmodesmal SEL (Deom *et al.*, 1990; Wolf *et al.*, 1989).

Analysis of a transgenic tobacco plant (*Nicotiana tabacum* L. var. Samsun, line A41-10) that expresses a yeast acid invertase gene (Ding *et al.*, 1993; see also Von Schaewen *et al.*, 1990) further indicates that secondary and primary plasmodesmata have different functional properties, and that secondary plasmodesmata appear to be indispensable for normal leaf development/ function. In this transgenic tobacco line, primary-to-secondary transformation of plasmodesmata between mesophyll cells is arrested in leaf sectors which express high levels of invertase activity (*Figure 7c*). This arrest precedes, and is correlated with, the development of symptoms such as interveinal chlorosis and necrosis, as well as early leaf senescence. Importantly, the development of complex secondary plasmodesmata within the vasculature and between bundle sheath and phloem parenchyma does not appear to be affected by the high levels of invertase activity. Thus it is evident that the arrest of the primary-to-secondary transformation in the mesophyll sectors exhibiting high invertase activity is not simply a pathological consequence; rather, the complex secondary plasmodesmata between mesophyll cells may be required to transport special informational molecules to co-ordinate various cellular processes underlying normal tobacco leaf development. The interruption of this communication pathway would therefore prevent such cell-to-cell information exchange, leading to symptom development and early leaf senescence (Ding *et al.*, 1993).

Secondary plasmodesmata between bundle sheath and vascular parenchyma cells may govern unique signal exchange. Vascular parenchyma cells derive from the cambium, whereas bundle sheath cells, in general, originate from the ground tissue (Esau, 1960). A known exception occurs in some C4 plants, where the bundle sheath appears to originate from procambium (Nelson and Langdale, 1989). Thus, in most cases, the plasmodesmata between bundle sheath and vascular parenchyma cells must be formed via secondary processes. Both simple and branched secondary plasmodesmata are found at this interface, at early and late leaf developmental stages. The ontogenetic relationship between these two forms of secondary plasmodesmata (i.e. whether the

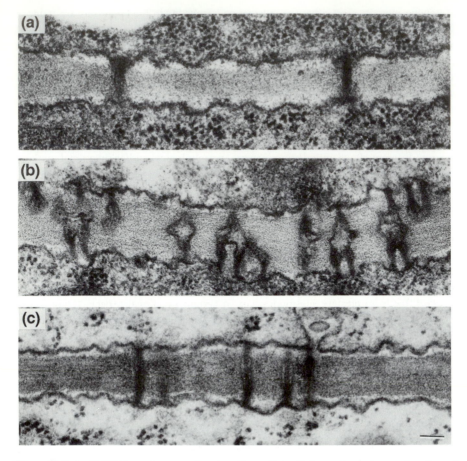

Figure 7. (a and b) Primary-to-secondary transformation of plasmodesmata as a function of mesophyll development in a tobacco leaf. (a) Primary plasmodesmata between mesophyll cells in a young expanding leaf. (b) Secondary plasmodesmata between mesophyll cells in a mature leaf. (c) In a transgenic tobacco plant expressing a yeast acid invertase gene, accumulation of high levels of the acid invertase is correlated with both the arrest of primary-to-secondary transformation of plasmodesmata between mesophyll cells and the development of chlorotic symptoms and early leaf senescence. Scale bar = 0.1 μm. (a) is reproduced from Ding *et al.* (1992a) with permission from the American Society of Plant Physiologists. (b) and (c) are reproduced from Ding *et al.* (1993) with permission from Blackwell Science Ltd.

branched secondary plasmodesmata are formed *de novo*, or via modification of simple secondary plasmodesmata) has yet to be established.

Secondary plasmodesmata between bundle sheath and vascular parenchyma cells appear to possess unique properties that distinguish them functionally from secondary plasmodesmata between other cell or tissue types. In transgenic tobacco plants that express the 30-kDa TMV MP, the TMV MP is localized to secondary plasmodesmata between bundle sheath and vascular parenchyma cells, as well as between various non-vascular cells. However, whereas the TMV MP induces a 10-fold increase in the SEL of secondary plasmodesmata between mesophyll cells, as well as between other non-vascular cells, it does not increase the SEL of secondary plasmodesmata between cells of the bundle sheath and vascular parenchyma (Ding *et al.*, 1992a). Furthermore, as previously mentioned, while the expression of a yeast acid invertase gene in the transgenic tobacco plant line A41-10 causes an arrest in the primary-to-complex secondary transformation of plasmodesmata between leaf mesophyll cellls, it does not affect the development of complex secondary plasmodesmata between bundle sheath and vascular parenchyma cells (Ding *et al.*, 1993).

Virological as well as pathological studies have also provided data consistent with the above observations. For TMV and several other viruses, when the coat protein is mutated or removed, the mutant virus is able to move from cell to cell for short distances, but is unable to move over a long distance to initiate systemic infection (Culver and Dawson, 1989; Dalmay *et al.*, 1992; Petty and Jackson, 1990; Xiong *et al.*, 1993; Ziegler-Graff *et al.*, 1991). The simplest explanation for this finding is that mutation or removal of the coat protein impairs the ability of these viruses to breach the bundle sheath–vascular parenchyma boundary in order to enter the phloem tissue for long-distance transport (Ding *et al.*, 1992a). Thus, although the viruses with coat protein that has been mutated or removed are able to interact with secondary plasmodesmata between mesophyll and bundle sheath cells, they are unable to interact with secondary plasmodesmata between bundle sheath and vascular parenchyma to potentiate trafficking of their infectious material.

Other plant viruses, such as geminiviruses and luteoviruses, are transmitted to plants via specific insect vectors (e.g. whiteflies and aphids) that probe into the phloem during feeding. The distribution of these viruses is often phloem-limited during infection (Barker and Harrison, 1986; D'Arcy and de Zoeten, 1979; Esau *et al.*, 1967; Esau, 1977; Kim *et al.*, 1978; Russo *et al.*, 1980; Shepardson *et al.*, 1980; Thongmeearkom *et al.*, 1981), although isolated mesophyll protoplasts can be infected by the viruses (Kubo and Takanami, 1979). This suggests that the inability of these phloem-limited viruses to invade non-vascular cells may not be due to their inability to replicate in the mesophyll and other non-vascular cells, but rather to their inability to escape from the vasculature. This hypothesis is supported by recent immunogold labelling studies in which it was established that the bundle sheath and vascular parenchyma boundary is the site of a barrier to the egress of beet western yellows virus (a member of the luteovirus group) from the vasculature (Sanger *et al.*, 1994).

Collectively, these data support the hypothesis that secondary plasmodesmata at the bundle sheath–vascular parenchyma interface possess distinct functional properties. This could be due to unique substructural features, biochemical composition or unique aspects of functional regulation. Obviously, cell-to-cell communication across this intercellular boundary may play a key role in co-ordinating various cellular and physiological processes between the vasculature and non-vascular tissues.

Secondary plasmodesmata between sieve element and companion cell in relation to long-distance signal transduction . Using [^{35}S]methionine labelling and gel electrophoresis, Fisher *et al.* (1992) identified over 200 soluble proteins which were collected, via severed aphid stylets, from wheat sieve elements. The majority of these proteins were smaller than 36 kDa, with a few being as large as 70 kDa. Autoradiographic studies performed on these wheat plants indicated that the ^{35}S-labelling was predominantly localized to sieve elements and companion cells. Moreover, proteins were constantly exchanged between sieve elements and companion cells along the translocation path from source to sink. Similar results have been obtained from studies on phloem transport in *Ricinus communis* (Sakuth *et al.*, 1993) and rice (Nakamura *et al.*, 1993). Some of these proteins have been identified, namely thioredoxin h in rice (Ishiwatari *et al.*, 1995) and ubiquitin and chaperones in *Ricinus communis* (Schobert *et al.*, 1995). Since mature sieve elements of higher plants in general are enucleate and probably lack ribosomes (Parthasarathy, 1975; Raven, 1991), these data indicate that proteins are synthesized in the companion cells and then transported into the sieve elements. Such protein transport presumably occurs through secondary plasmodesmata between these two cell types (see *Figure 3b*) (Fisher *et al.*, 1992; Schobert *et al.*, 1995).

The functions of these proteins that shuttle between the sieve element and the companion cell while moving along the long-distance pathway of the phloem have yet to be established. While some of them may be involved in maintaining the proper functioning of the phloem (Fisher *et al.*, 1992), others may well carry information that is used to co-ordinate cellular processes over long distances (Lucas *et al.*, 1993; Lucas and Wolf, 1993).

Interactions with viral infection

Secondary plasmodesmata appear to play important roles in viral infection. Some viruses, such as TMV, encode MPs that appear to interact with existing secondary plasmodesmata to permit cell-to-cell transfer of the viral genetic material (Ding *et al.*, 1992a; Lucas *et al.*, 1993; Wolf *et al.*, 1989). However, certain plant viruses appear to utilize a mechanism similar to the *de novo* formation of secondary plasmodesmata to create new passages through plant cell walls to permit cell-to-cell movement of virions. Typical examples are the trans-wall tubular structures formed by the comoviruses (Van Lent *et al.*, 1991) and caulimoviruses (Perbal *et al.*, 1993). The 58/48-kDa movement protein(s) encoded by the genome of cowpea mosaic virus (CPMV) have been shown to be involved in the formation of these tubular structures (Van Lent *et al.*, 1991). The gene 1 product (putative movement protein) of cauliflower mosaic virus (CaMV) has also been suggested to play a similar role in viral cell-to-cell movement (Perbal *et al.*, 1993). It will be interesting to ascertain whether the viral genes encoding the ability to produce such simple secondary plasmodesmata bear any homology to the plant genes involved in such a process.

4. Evolutionary aspects of secondary plasmodesmata

Given their potentially pivotal roles in regulating cell-to-cell communications during plant morphogenesis, plasmodesmata have conceivably played an important role in shaping the evolutionary pattern of green plants. How plasmodesmata have evolved to perform their various functions is therefore one of the most interesting issues in plant evolution.

In terms of the evolutionary origin of plasmodesmata, perhaps the earliest detected forms of intercellular connections that are structurally similar to higher plant plasmodesmata are found between various cell types of brown algae (Bisalputra, 1966; Oliveira and Bisalputra, 1973; Schmitz and Srivastava, 1974, 1975; Schmitz and Kühn, 1982). However, there is a notable exception in that the brown algal plasmodesmata lack the presence of the appressed ER. As brown algae produce septal walls by annular furrowing, it is almost impossible that the plasma membrane becomes trapped in the forming walls (Lucas *et al.*, 1993). Thus the most logical explanation is that these brown algal species evolved a cellular mechanism for forming secondary plasmodesmata across both septal walls and sites of cell-to-cell contact (Lucas *et al.*, 1993).

Ancestors of characean algae are generally recognized as the most likely candidates to have given rise to modern higher plants (Graham *et al.*, 1991; Graham and Kaneko, 1991). Given such an evolutionary position, the characean algae appear to be a logical place to search for ancestral forms of higher plant plasmodesmata and mechanisms involved in their formation. A recent detailed electron microscopic study indicated that plasmodesmata in characean algae are formed post-cytokinetically, or secondarily (Franceschi *et al.*, 1994). Interestingly, the secondary plasmodesmata which are initially formed are simple, linear structures that are similar to higher plant simple plasmodesmata in general appearance, with the exception that they, like brown algal plasmodesmata, also lack the presence of the appressed ER. During algal growth, these simple secondary plasmodesmata can undergo further structural modification to become highly branched secondary plasmodesmata across mature cell walls (Franceschi *et al.*, 1994).

These results raise several fundamental questions that need to be addressed in future research.

(1) When, where and how did the mechanism for the inclusion of the ER in plasmodesmata evolve?
(2) Did certain aspects of the mechanism(s) for secondary plasmodesma formation in higher plants evolve directly from the characean algae?
(3) Does the mechanism for primary formation of plasmodesmata exist in an unidentified algal species, or did it evolve later in a species intermediate between the characean algae and the primitive land plants?

Molecular approaches in combination with electron microscopic techniques will prove to be powerful tools for tracing the origin and evolutionary pattern of plasmodesmata. Once plasmodesmal proteins and the corresponding genes have been isolated from either higher plants or characean algae, it will be possible to use appropriate antibodies or RNA/DNA probes to screen the plant kingdom in order to establish possible evolutionary trees for plasmodesmata.

4. Future prospects

This brief account of the current status of research on secondary plasmodesmata highlights some of the important roles that these intercellular organelles appear to play in plant growth and development. We emphasize that most of these roles are deduced on the basis of circumstantial evidence. It is imperative that future studies employ experimental approaches to address these and potentially many other functions of secondary plasmodesmata.

To appreciate fully how plasmodesmata (primary as well as secondary) perform various functions, it is necessary to know how they are formed and how their formation is regulated. Thus the mechanism and regulation of the biogenesis of secondary as well as primary plasmodesmata are among the most important subjects for future research. In this context, substantial progress can only be made by the integration of structural, biochemical, biophysical, molecular and genetic approaches.

Once proteinaceous components and the corresponding genes of plasmodesmata have been isolated and the signals regulating the biogenesis of secondary plasmodesmata have been identified, it will be possible to manipulate the biogenesis of these plasmodesmata experimentally. This will not only allow experimental studies of the function of plasmodesmata, but also provide opportunities to engineer plants, with modified patterns of biogenesis and/or functions of plasmodesmata, that possess desirable traits such as higher yield, resistance to pathogens, and novel developmental patterns. Clearly much remains to be done, but the results obtained from future studies on plasmodesmata will surely provide critical information about the evolutionary pathway that gave rise to the supracellular nature of higher plants.

Acknowledgements

Research on plasmodesmata in our laboratory has been supported by National Science Foundation Grant DCB-90-05722 and a University of California Biotechnology Award to W.J.L. Our special thanks go to Debby Delmer for valuable discussions on the intricacies of the primary cell wall.

References

Barker H, Harrison BD. (1986) Restricted distribution of potato leafroll virus antigen in resistant potato genotypes and its effect on transmission of the virus by aphids. *Ann. Appl. Biol.* **109:** 595–604.
Baron-Epel O, Gharyal PK, Schindler M. (1988) Pectins as mediators of wall porosity in soybean cells. *Planta* **175:** 389–395.

Beebe DU, Turgeon R. (1991) Current perspectives on plasmodesmata: structure and function. *Physiol. Plant.* **83:** 194–199.

Bisalputra T. (1966) Electron microscopic study of the protoplasmic continuity in certain brown algae. *Can. J. Bot.* **44:** 89–93.

Carpita NC, Gibeaut DM (1993) Structural models of primary cell walls in flowering plants: consistency of modular structure with the physical properties of the walls during growth. *Plant J.* **3:** 1–30.

Citovsky V, McLean BG, Zupan JR, Zambryski P. (1993) Phosphorylation of tobacco mosaic virus cell-to-cell movement protein by a developmentally regulated plant cell wall-associated protein kinase. *Genes Dev.* **7:** 904–910.

Culver JN, Dawson WO. (1989) Tobacco mosaic virus coat protein: an elicitor of the hypersensitive reaction but not required for the development of mosaic symptoms in *Nicotiana sylvestris*. *Virology* **173:** 755–758.

Dalmay T, Rubino L, Burgyan J, Russo M. (1992) Replication and movement of a coat protein mutant of cymbidium ringspot tombusvirus. *Mol. Plant–Microbe Interact.* **5:** 379–383.

D'Arcy CJ, de Zoeten GA. (1979) Beet western yellows virus in phloem tissue of *Thlaspi arvense*. *Phytopathology* **69:** 1194–1198.

Deom CM, Schubert K, Wolf S, Holt C, Lucas WJ, Beachy RN. (1990) Molecular characterization and biological function of the movement protein of tobacco mosaic virus in transgenic plants. *Proc. Natl Acad. Sci. USA* **87:** 3284–3288.

Didehvar F, Baker DA. (1986) Localization of ATPase in sink tissues of *Ricinus*. *Ann. Bot.* **57:** 823–828.

Ding B, Haudenshield JS, Hull RJ, Wolf S, Beachy RN, Lucas WJ. (1992a) Secondary plasmodesmata are specific sites of localization of the tobacco mosaic virus movement protein in transgenic tobacco plants. *Plant Cell* **4:** 915–928.

Ding B, Turgeon R, Parthasarathy MV. (1992b) Substructure of freeze substituted plasmodesmata. *Protoplasma* **169:** 28–41.

Ding B, Haudenshield JS, Willmitzer L, Lucas WJ. (1993) Correlation between arrested secondary plasmodesmal development and onset of accelerated leaf senescence in yeast acid invertase transgenic tobacco plants. *Plant J.* **4:** 179–189.

Esau K. (1960) *Plant Anatomy.* New York: John Wiley & Sons, Inc.

Esau K. (1977) Virus-like particles in nuclei of phloem cells in spinach leaves infected with the curly top virus. *J. Ultrastruct. Res.* **61:** 78–88.

Esau K, Thorsch J. (1985) Sieve plate pores and plasmodesmata, the communication channels of the symplast: ultrastructural aspects and developmental relations. *Am. J. Bot.* **72:** 1641–1653.

Esau K, Cronshaw J, Hoefert LL. (1967) Relation of beet yellows virus to the phloem and to movement in the sieve tube. *J. Cell Biol.* **32:** 71–87.

Fisher DB, Wu Y, Ku MSB. (1992) Turnover of soluble proteins in the wheat sieve tube. *Plant Physiol.* **100:** 1433–1441.

Franceschi V, Ding B, Lucas WJ. (1994) Mechanism of plasmodesmata formation in characean algae in relation to evolution of intercellular communication in higher plants. *Planta* **192:** 347–358.

Fry SC, Smith RC, Renwick KF, Martin DJ, Hodge SK, Matthews KJ. (1992) Xyloglucan endotransglycosylase, a new wall-loosening enzyme activity from plants. *Biochem. J.* **282:** 821–828.

Gilder J, Cronshaw J. (1973a) Adenosine triphosphatase in the phloem of *Cucurbita. Planta* **110:** 189–204.

Gilder J, Cronshaw J. (1973b) The distribution of adenosine triphosphatase activity in differentiating and mature phloem cells of *Nicotiana tabacum* and its relationship to phloem transport. *J. Ultrastruct. Res.* **44:** 388–404.

Graham LE, Kaneko Y. (1991) Subcellular structure of relevance to the origin of land plants (Embryophytes) from green algae. *Crit. Rev. Plant Sci.* **10:** 323–342.

Graham LE, Delwiche CF, Mishler BD. (1991) Phylogenetic connections between the 'green algae' and the 'bryophytes'. *Adv. Bryol.* **4:** 213–244.

Gunning BES, Robards AW. (1976) *Intercellular Communication in Plants: Studies on Plasmodesmata.* Berlin: Springer-Verlag.

Gunning BES, Overall RL. (1983) Plasmodesmata and cell-to-cell transport in plants. *BioScience* **33:** 260–265.

Hake S, Freeling M. (1986) Analysis of genetic mosaics shows that the extra epidermal cell divisions in *Knotted* mutant maize plants are induced by adjacent mesophyll cells. *Nature* **320:** 621–623.

Hall JL. (1969) Localization of cell surface adenosine triphosphatase activity in maize roots. *Planta* **85:** 105–107.

Hayashi T. (1991) Biochemistry of xyloglucans in regulating cell elongation and expansion. In: *The Cytoskeletal Basis of Plant Growth and Form* (Lloyd CW, ed.). London: Academic Press, pp. 131–144.

Hayashi T, Wong YS, Maclachlan G. (1984) Pea xyloglucan and cellulose. II. Hydrolysis by pea endo-1,4-β-glucanases. *Plant Physiol.* **75:** 605–610.

Heichel GH, Anagnostakis SL. (1978) Stomatal response to light of *Solanum pennellii, Lycopersicon esculentum,* and a graft-induced chimera. *Plant Physiol.* **62:** 387–390.

Hepler PK. (1982) Endoplasmic reticulum in the formation of the cell plate and plasmodesmata. *Protoplasma* **111:** 121–133.

Ishiwatari Y, Honda C, Kawashima I, Nakamura S-I, Hirano H, Mori S, Fujiwara T, Hayashi H, Chino M. (1995) Thioredoxin h is one of the major proteins in rice phloem sap. *Planta* **195:** 456–463.

Jones MGK. (1976) The origin and development of plasmodesmata. In: *Intercellular Communication in Plants: Studies on Plasmodesmata* (Gunning BES, Robards AW, ed.). Berlin: Springer-Verlag, pp. 81–105.

Kim KS, Shock TL, Goodman RM. (1978) Infection of *Phaseolus vulgaris* by bean golden mosaic virus: ultrastructural aspects. *Virology* **89:** 22–33.

Kollmann R, Glockmann C. (1985) Studies on graft unions. I. Plasmodesmata between cells of plants belonging to different unrelated taxa. *Protoplasma* **124:** 224–235.

Kollmann R, Glockmann C. (1991) Studies on graft unions. III. On the mechanism of secondary formation of plasmodesmata at the graft interface. *Protoplasma* **165:** 71–85.

Kollmann R, Yang S, Glockmann C. (1985) Studies on graft unions. II. Continuous and half plasmodesmata in different regions of the graft interface. *Protoplasma* **126:** 19–29.

Kotlizky G, Shurtz S, Yahalom A, Malik Z, Traub O, Epel BL. (1992) An improved procedure for the isolation of plasmodesmata embedded in clean maize cell walls. *Plant J.* **2:** 623–630.

Kubo S, Takanami Y. (1979) Infection of tobacco mesophyll protoplasts with tobacco necrotic dwarf virus, a phloem-limited virus. *J. Gen. Virol.* **42:** 387–398.

Lucas WJ, Wolf S. (1993) Plasmodesmata: the intercellular organelles of green plants. *Trends Cell Biol.* **3:** 308–315.

Lucas WJ, Ding B, van der Schoot C. (1993) Plasmodesmata and the supracellular nature of plants. *New Phytol.* **125:** 435–476.

Lucas WJ, Bouché-Pillon S, Jackson DP, Nguyen L, Baker L, Ding B, Hake S. (1995) Selective plasmodesmal trafficking of KNOTTED1 and its mRNA demonstrates supracellular control over plant development. *Science* **270:** 1980–1983.

McCann MC, Roberts K. (1991) Architecture of the primary cell wall. In: *The Cytoskeletal Basis of Plant Growth and Form* (Lloyd CW, ed.). London: Academic Press, pp. 109–129.

McCann MC, Wells B, Roberts K. (1990) Direct visualization of cross-links in the primary cell wall. *J. Cell Sci.* **96:** 323–334.

McCann MC, Wells B, Roberts K. (1992) Complexity in the spatial localization and length distribution of plant cell-wall matrix polysaccharides. *J. Microscopy* **166:** 123–136.

Medford JI. (1992) Vegetative apical meristems. *Plant Cell* **4:** 1029–1039.

Monzer J. (1991) Ultrastructure of secondary plasmodesmata formation in regenerating *Solanum nigrum*-protoplast cultures. *Protoplasma* **165:** 86–95.

Nakamura S, Hayashi H, Mori S, Chino M. (1993) Protein phosphorylation in the sieve tubes of rice plants. *Plant Cell Physiol.* **34:** 927–933.

Nelson T, Langdale JA. (1989) Patterns of leaf development in C4 plants. *Plant Cell* **1:** 3–13.

Nishitani K, Tominaga R. (1992) Endo-xyloglucan transferase, a novel class of glycosyltransferase that catalyzes transfer of a segment of xyloglucan molecule to another xyloglucan molecule. *J. Biol. Chem.* **267:** 21058–21064.

Oliveira L, Bisalputra T. (1973) Studies in the brown alga *Ectocarpus* in culture. I. General ultrastructure of the sporophytic vegetative cells. *J. Submicrosc. Cytol.* **5:** 107–120.

Oparka KJ. (1993) Signalling via plasmodesmata — the neglected pathway. *Sem. Cell Biol.* **4:** 131–138.

Parthasarathy MV. (1975) Sieve-element structure. In: *Encyclopedia of Plant Physiology, New Series: Transport in Plants. I. Phloem Transport* (Zimmermann MH, Milburn JA, ed.). Berlin: Springer-Verlag, pp. 3–38.

Perbal MC, Thomas CL, Maule AJ. (1993) Cauliflower mosaic virus gene-1 product (P1) forms tubular structures which extend from the surface of infected protoplasts. *Virology* **195:** 281–285.

Petty ITD, Jackson AO. (1990) Mutational analysis of barley stripe mosaic virus RNA beta. *Virology* **179:** 712–718.

Raven JA. (1991) Long-term functioning of enucleate sieve elements: possible mechanisms of damage avoidance and damage repair. *Plant Cell Environ.* **14:** 139–146.

Robards AW, Kidwai P. (1969) Cytochemical localization of phosphatase in differentiating secondary vascular cells. *Planta* **87:** 227–238.

Robards AW, Lucas WJ. (1990) Plasmodesmata. *Annu. Rev. Plant Physiol. Plant Mol. Biol.* **41**: 369–419.

Russo M, Cohen S, Martelli GP. (1980) Virus-like particles in tomato plants affected by the yellow leaf curl disease. *J. Gen. Virol.* **49**: 209–213.

Sakuth T, Schobert C, Pecsvaradi A, Eichholz A, Komor E, Orlich G. (1993) Specific proteins in the sieve-tube exudate of *Ricinus communis* L. seedlings: separation, characterization and in-vivo labelling. *Planta* **191**: 207–213.

Sanger M, Passmore B, Falk B, Bruening G, Ding B, Lucas WJ. (1994) Symptom severity of beet western yellows virus strain ST9 is conferred by the ST9-associated RNA and is not associated with virus release from the phloem. *Virology* **200**: 48–55.

Satina S, Blakeslee AF, Avery AG. (1940) Demonstration of the three germ layers in the shoot apex of *Datura* by means of induced polyploidy in periclinal chimeras. *Am. J. Bot.* **27**: 895–905.

Schmitz K, Kühn R. (1982) Fine structure, distribution and frequency of plasmodesmata and pits in the cortex of *Laminaria hyperborea* and *L. saccharina. Planta* **154**: 385–392.

Schmitz K, Srivastava LM. (1974) Fine structure and development of sieve tubes in *Laminaria groenlandica* Rosenv. *Cytobiologie* **10**: 66–87.

Schmitz K, Srivastava LM. (1975) On the fine structure of sieve tubes and the physiology of assimilate transport in *Alaria marginata. Can. J. Bot.* **53**: 861–876.

Schobert C, Großmann P, Gottschalk M, Komor E, Pecsvaradi A, Nieden Uz. (1995) Sieve-tube exudate from *Ricinus communis* L. seedlings contains ubiquitin and chaperones. *Planta* **196**: 205–210.

Shedletzky E, Shmuel M, Delmer DP, Lamport DTA. (1990) Adaptation and growth of tomato cells on the herbicide 2,6-dichlorobenzonitrile leads to production of unique cell walls virtually lacking a cellulose–xyloglucan network. *Plant Physiol.* **94**: 980–987.

Shepardson S, Esau K, McCrum R. (1980) Ultrastructure of potato leaf phloem infected with potato leafroll virus. *Virology* **105**: 379–392.

Sinha NR, Hake S. (1990) Mutant characters of *Knotted* maize leaves are determined in the innermost tissue layers. *Dev. Biol.* **141**: 203–210.

Smith LG, Greene B, Veit B, Hake S. (1992) A dominant mutation in the maize homeobox gene, *Knotted-1*, causes its ectopic expression in leaf cells with altered fates. *Development* **116**: 21–30.

Smith RC, Fry SC. (1991) Endotransglycosylation of xyloglucans in plant suspension cultures. *Biochem. J.* **279**: 529–535.

Sussex IM. (1989) Developmental programming of the shoot meristem. *Cell* **56**: 225–229.

Talbot LD, Ray PM. (1992) Molecular size and separability of pea cell wall polysaccharides. Implications for models of primary wall structure. *Plant Physiol.* **98**: 357–368.

Thongmeearkom P, Honda Y, Saito Y, Syamananda R. (1981) Nuclear ultrastructural changes and aggregates of viruslike particles in mungbean cells affected by mungbean yellow mosaic disease. *Phytopathology* **71**: 41–44.

Van Lent J, Storms M, Van der Meer F, Wellink J, Goldbach R. (1991) Tubular structures involved in movement of cowpea mosaic virus are also formed in infected cowpea protoplasts. *J. Gen. Virol.* **72**: 2615–2623.

Von Schaewen A, Stitt M, Schmidt R, Sonnewald U, Willmitzer L. (1990) Expression of a yeast-derived invertase in the cell wall of tobacco and *Arabidopsis* plants leads to accumulation of carbohydrate and inhibition of photosynthesis and strongly influences growth and phenotype of transgenic tobacco plants. *EMBO J.* **9**: 3033–3044.

Wolf S, Deom CM, Beachy RN, Lucas WJ. (1989) Movement protein of tobacco mosaic virus modifies plasmodesmatal size exclusion limit. *Science* **246**: 377–379.

Xiong Z, Kim KH, Giesman-Cookmeyer D, Lommel SA. (1993) The roles of the red clover necrotic mosaic virus capsid and cell-to-cell movement proteins in systemic infection. *Virology* **192**: 27–32.

Yahalom A, Warmbrodt RD, Laird DW, Traub O, Revel J-P, Willecke K, Epel BL. (1991) Maize mesocotyl plasmodesmata proteins cross-react with connexin gap junction protein antibodies. *Plant Cell* **3**: 407–417.

Zhen G-C, Nie X-W, Wang Y-X, Jian L-C, Sun L-H, Sun D-L. (1985) Cytochemical localization of adenosine triphosphatase activity during cytomixis in pollen mother cells of David lily and its relation to the intercellular migrating chromatin substance. *Act. Bot. Sinica* **27**: 26–32.

Ziegler-Graff V, Guilford PJ, Baulcombe DC. (1991) Tobacco rattle virus RNA-1 29K gene product potentiates viral movement and also affects symptom induction in tobacco. *Virology* **182**: 145–155.

Chapter 29

Plant membrane structure and function in the *Rhizobium*–legume symbiosis

Nicholas J. Brewin

1. Endosymbiosis and biological nitrogen fixation

Question: when is a plasma membrane not a plasma membrane? Answer: when it is a peribacteroid membrane. To understand the importance of the question (and the subtlety of the answer) it is necessary to consider the unique role of *Rhizobium* spp. as nitrogen-fixing endo-symbionts of legume root nodules (Brewin, 1990, 1991, 1993; Brewin *et al.*, 1992). Following a long and complicated series of specific host–microbe interactions (*Figure 1*), *Rhizobium* bacteria (or 'rhizobia') come to occupy much of the cytoplasmic space of host plant cells (*Figure 2*). The intracellular bacteria (now termed bacteroids) are differentiated with respect to morphology, biochemistry and gene expression. Within the central tissues of the legume nodule, bacteroids occupy a microaerobic niche in which the concentration of free oxygen is maintained in the range 3–10 nM (Appleby, 1984), enabling them to express and operate the oxygen-sensitive nitrogenase enzyme system. As a result of biological nitrogen fixation, they convert nitrogen gas (N_2) into

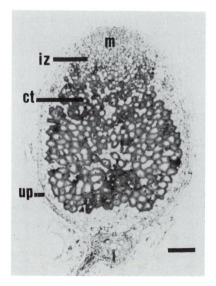

Figure 1. Median longitudinal section through a pea root nodule showing apical meristem (m), invasion zone (iz), central infected tissue (ct) with host cells containing nitrogen-fixing *Rhizobium* bacteroids in the cytoplasmic space, and uninfected parenchyma (up). Scale bar represents 0.5 mm. Reproduced from Perotto *et al.* (1994) with permission from The American Phytopathological Society.

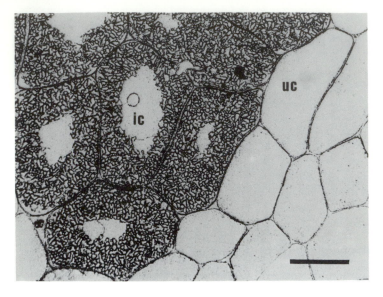

Figure 2. Light micrograph showing infected cells (ic) and uninfected cells (uc) taken from the central tissue of a pea nodule. The section has been treated with MAC 255, a monoclonal antibody which recognizes a glycolipid component present in the PBM of infected cells and the plasma membrane of all cells (Perotto *et al.,* 1991). The section was stained by immunogold labelling and silver enhancement; there was no counterstaining, and hence the only image visible is that of the membrane-associated MAC 255 antigen. This 'glycocalyx' antigen is not present on the tonoplast or nuclear membranes. The outline of the individual endosymbiotic *Rhizobium* bacteroids is clearly visible in the infected host cells. As host cells develop further, the central vacuole becomes completely occluded. Scale bar represents 50 μm. Reproduced from Perotto *et al.* (1991) with permission from The Company of Biologists Limited.

ammonia (NH_3), which is excreted by the bacterium and assimilated by enzymes in the host cell cytoplasm or its associated organelles (Stacey *et al.*, 1992). Bacteroids derive their energy from oxidative respiration of organic acids (principally malate) absorbed from the host cell cytoplasm. Because bacteroids are obligate aerobes that occupy a microaerobic niche, their respiration depends on receiving a sufficient supply of oxygen. This is provided as a result of facilitated diffusion involving leghaemoglobin, a carrier protein present in very high concentrations in the host cell cytoplasm.

2. Peribacteroid membrane: the symbiotic interface

The above description of the specialized biochemistry of bacteroids omits the single most salient fact, namely that every bacteroid is separated from the host cytoplasm by a plant-derived membrane, termed the peribacteroid membrane (PBM) (*Figure 3*). In many legumes (e.g. pea, alfalfa), bacteroids are enclosed individually by PBM so that, when the intracellular bacteroids divide, the peribacteroid membrane sac divides concomitantly. In other legumes (e.g. soybean, *Phaseolus*), this synchrony breaks down in the later stages of nodule development, so that groups of 6–12 bacteroids are enclosed within a single peribacteroid membrane envelope. In a mature host cell containing several thousand bacteroids, it has been estimated that the total surface area of PBM exceeds that of the plasma membrane by 30- to 100-fold (Cheon *et al.*, 1993; Robertson *et al.*, 1985), depending on the legume species. The integrity of the PBM is of paramount

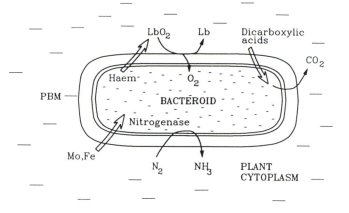

Figure 3. Metabolic exchanges between nitrogen-fixing bacteroids and the host cell cytoplasm that are mediated through the PBM. Lb represents cytoplasmic leghaemoglobin, a facilitated oxygen carrier. Reproduced from Brewin *et al.* (1992) with permission from Academic Press, Inc.

importance to the success of the symbiotic interaction (Werner *et al.*, 1985). Its role can be rationalized from two points of view, depending on whether the intracellular bacteria are perceived as potential friend or as potential foe (Long and Staskawicz, 1993).

Regarding rhizobia as potential friends, the bacteroids and their enclosing plant-derived PBM are almost equivalent to a prokaryotic organelle. The endosymbiotic bacteroid is the under-evolved equivalent of a mitochondrion or a chloroplast, a specialist cytoplasmic organelle concerned with the metabolism of dinitrogen gas. Following this line of argument, the term 'symbiosome' has been coined to describe the functional unit comprising the plant-derived PBM and the enclosed bacteroid or bacteroids (Mellor and Werner, 1987; Roth and Stacey, 1989a). One role of the PBM is therefore to facilitate the exchange of metabolites between the bacteroid and the host cell cytoplasm in a way that will optimize the overall efficiency of biological nitrogen fixation. Thus, as it differentiates, the PBM loses some functions characteristic of the plasma membrane. For example, it is not associated with the ability to synthesize a cell wall. At the same time, the PBM acquires other specialist functions that are more reminiscent of the tonoplast membrane or endoplasmic reticulum than of the plasma membrane (Day and Copeland, 1991; Verma *et al.*, 1994).

Regarding rhizobia as potential foes can also be justified from the point of view of the host legume, because not all rhizobia that are capable of invading nodule tissue will necessarily be capable of fixing nitrogen. For example, the bacteria may be better adapted to function in an alternative legume host, or they may simply be mutant in some aspect of the nitrogen fixation process (Huang *et al.*, 1993; Perotto *et al.*, 1994). In such circumstances, intracellular rhizobia become pathogenic rather than symbiotic, because they create growth deformities (nodules) and consume plant photosynthate without contributing to the survival of the host plant. Thus, another role of the PBM is to control the progress of cell invasion by *Rhizobium* and to maintain the bacteria in an extracytoplasmic (but intracellular) compartment. The PBM therefore appears to retain those properties of the plasma membrane that are concerned with the capacity to respond to microbial attack and invasion by a potentially pathogenic endophyte.

Components of the peribacteroid membrane

Glycocalyx antigens. Symbiosomes with intact peribacteroid membranes can be isolated from nodule homogenates by density gradient centrifugation under conditions which protect against physical or osmotic rupture (Blumwald *et al.*, 1985; Day *et al.*, 1988; Garbers *et al.*, 1988; Herrada *et al.*, 1989; Price *et al.*, 1987). It is interesting to note that these preparations expose the cytoplasmic face of the PBM to the exterior, and are thus equivalent to an inside-out version of the plasma membrane. The antigenic characteristics of the glycocalyx associated with the PBM have been investigated by the isolation and use of a wide range of monoclonal antibodies as immunological probes (Perotto *et al.*, 1991). Three major groups of carbohydrate epitopes were identified that were associated with glycoprotein or glycolipid components of the PBM (*Figures 2* and *4a*). These antigens were also present on plasma membranes of infected or uninfected host cells and on Golgi membranes, but not on endoplasmic reticulum. The epitopes recognized are probably components of membrane arabinogalactan proteins and glycolipids (Baldwin *et al.*, 1993). Immunolocalization studies of longitudinal sections of pea nodules revealed enhanced expression of these antigens in nodule tissue, with the three classes of antigen showing different degrees of expression in different developmental zones of the tissue (Perotto *et al.*, 1991). The reason for this tissue-specific distribution is still unclear.

Membrane lipids. The lipid composition of the mature PBM appears to be more similar to that of the microsomal membranes than to that of the plasma membrane. Compared to plasma membrane, PBM contains high levels of phosphatidylcholine and low levels of phosphatidylethanolamine. Choline kinase II has been shown to be very active in infected nodule tissue (Hernandez and Cooke, 1996; Mellor *et al.*, 1986). Using monoclonal antibody JIM 18 (*Figure 4b*), a novel

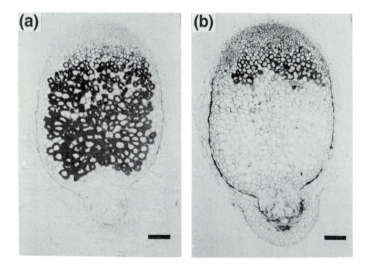

Figure 4. Median longitudinal sections of a pea nodule following immunostaining with monoclonal antibodies which react with plasma membrane and PBM. (a) Staining with MAC 206, a monoclonal antibody reacting with a plant membrane glycolipid present on plasma membranes but expressed strongly on PBM in the central infected tissue of the nodule. (b) Staining with JIM 18, a monoclonal antibody recognizing plasma membranes and juvenile PBM but not 'mature' PBM present in differentiated host cells in the tissue region capable of nitrogen fixation. Scale bar represents 0.5 mm. Reproduced from Perotto *et al.* (1995) with permission from The American Phytopathological Society.

complex inositol-containing glycolipid has recently been identified and tentatively characterized as a glycophosphosphingolipid (Brewin, 1993; Cross, 1990; Laine and Hsieh, 1987; Perotto *et al.*, 1995) (*Figure 5*). Immunolocalization studies with JIM 18 on pea nodule sections revealed that the antigen is present on plasma membranes and juvenile (undifferentiated) peribacteroid membranes, but that it is lost from the PBM (but not the plasma membrane) of differentiating host cells (*Figure 4b*). Loss of this antigen is the earliest known cytological marker for the differentiation of the PBM, and perhaps signals the advent of a specific targeting pathway for PBM biogenesis (Brewin *et al.*, 1993; Perotto *et al.*, 1995; Verma *et al.*, 1994).

Analysis of acetylated sterols showed them to be five times less abundant in peribacteroid membranes from pea nodules than in plasma membranes from pea roots (Hernandez and Cooke, 1996). Moreover, peribacteroid membranes contained β-amyrin (*Figure 6*), a novel plant triterpenoid that is not found in root plasma membranes. Because this sterol is a smaller molecule than cholesterol or stigmasterol, and because the total quantity of sterols is lower in the PBM than in plasma membranes, it was suggested that the PBM might have enhanced membrane fluidity. It was also noted that the bound fatty acids from the PBM (i.e. those esterified as components of fatty acids) showed a relatively high degree of unsaturation compared to plasma membrane fractions, which again suggests a higher degree of membrane fluidity (Bassarab *et al.*, 1989; Hernandez and Cooke, 1996).

Callose synthase. 1,3-β-glucan synthase, previously recognized as a plasma membrane marker (see also Chapter 5 by Kauss), has been identified in purified preparations of PBM from soybean (Ahlborn and Werner, 1992). This enzyme is often considered to be an important component of

INOSITOL - GLYCAN

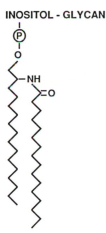

Figure 5. General structure of a glycophosphosphingolipid, a possible structure for the JIM 18 antigen illustrated in *Figure 4* (Cross, 1990; Laine, 1987).

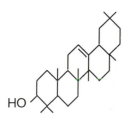

Figure 6. Chemical structure of β-amyrin, a plant sterol that accumulates in the PBM membrane of pea nodules. (Hernandez and Cook, 1996.)

the plant defence system responding to pathogen attack, but, in the context of root nodules, it seems that callose synthase is only activated when the normal balance of symbiosis is disturbed. Callose deposits could only be visualized in thin sections of clover or soybean root nodules when these harboured symbiotically defective or incompatible strains of rhizobia (Ahlborn and Werner, 1992; Kumarasinghe and Nutman, 1977).

Vectorial (electrogenic) ATPase. Using isolated symbiosomes from nodules of various legumes (e.g. siratro – *Macroptilium atropurpureum*), evidence was obtained for a H⁺-translocating ATPase in the peribacteroid membrane (Ouyang and Day, 1992; Udvardi and Day, 1989). Inhibitor studies and the pH response of ATPase activity indicated that only a single plasma-membrane-type ATPase is present on the PBM of both siratro and soybean.

Transport proteins. Unlike the plasma membrane, a primary function of the PBM is the exchange of metabolites and hence the control of bacteroid growth and metabolism (Day and Udvardi, 1993). Its transport properties have been investigated by comparing the uptake of compounds by intact and ruptured symbiosomes isolated from nodules of siratro (Ouyang and Day, 1992), soybean (Udvardi *et al.*, 1990) and pea (Rosendahl *et al.*, 1992). Dicarboxylate uptake into the symbiosomes showed saturation kinetics, suggesting that uptake was carrier-mediated (Yang *et al.*, 1990). In contrast, no evidence was obtained for transport of sugars (e.g. sucrose, glucose or fructose). The rapid conversion of ¹⁴C-labelled glutamate into the amino-acid fraction of pea symbiosomes indicated the presence of an associated transaminase activity (Rosendahl *et al.*, 1992), and the existence of a malate/aspartate shuttle between the plant cytoplasm and the bacteroid has been proposed (Appels and Haaker, 1991).

Nodule-specific proteins. Because the PBM has a specialized metabolic role and represents the major membrane component of infected plant cells, it is not surprising that the products of several nodule-specific genes (nodulins) are targeted to the PBM (Verma 1992; Verma *et al.*, 1992). The best studied component of a peribacteroid membrane is nodulin-26 (N-26) from soybean (Miao *et al.*, 1992). N-26 is specifically expressed in root nodules of soybean, although a nodule-specific counterpart has not yet been identified in any other host legume. The protein is encoded by a member of an ancient gene family, conserved from bacteria to humans. A plant protein with homology to N-26 has also been identified as the tonoplast intrinsic protein of vacuolar membranes (Höfte *et al.*, 1992; Maurel *et al.*, 1993), and at least two vegetative homologues of N-26 are expressed in vegetative tissues of soybean, with their maximum level of expression being found in the root elongation zone (Miao and Verma, 1993). During the evolution of legume symbiosis, it therefore seems probable that a pre-existing gene expressed in root tissues was recruited for symbiotic function and brought under nodule-specific developmental control.

Soybean N-26 is an integral membrane protein of the PBM (*Figure 7*). It comprises 271 amino acids with six potential transmembrane domains. *In-vitro* transcription and translation of N-26 sequences in a rabbit reticulocyte system suggested that the protein was co-translationally inserted into microsomal membranes when these were supplied (Miao *et al.*, 1992). Membrane insertion does not involve a cleavable N-terminal transit peptide, and sequence deletion studies showed that the first two transmembrane domains were necessary and sufficient for microsomal insertion. The *in-vitro* synthesized protein was capable of being glycosylated following insertion into canine microsomal preparations. Similarly, the native N-26 isolated from root nodule fractions is glycosylated and binds to the lectin concanavalin A. Moreover, *in-vitro* phosphorylation experiments showed that N-26 is a major phosphorylated protein of the PBM

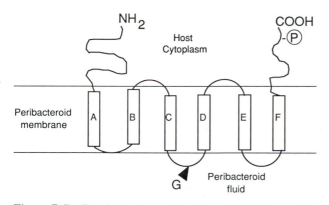

Figure 7. Predicted structure and topology of N-26 comprising 271 amino acids, an integral membrane channel protein from soybean PBM (Miao *et al.,* 1992; Miao and Verma, 1993). Boxes A to F represent six hydorphobic membrane-spanning domains determined by computer analysis; G represents the position of a glycosylation site determined by binding to the lectin concanavalin A; P represents the approximate position of several phosphorylation sites near the C-terminus.

(Miao *et al.,* 1992; Weaver and Roberts, 1992). Phosphorylation occurs at the C-terminus of the protein and is mediated by a Ca^{2+}-dependent calmodulin-independent protein kinase located in the PBM. Based on its homology with several eukaryotic and prokaryotic channel-type membrane proteins, it is suggested that N-26 functions as a channel protein that facilitates the transport of small molecules or ions across the PBM. An alternative suggestion, that it functions as a dicarboxylate transport protein (Ouyang and Day, 1992), is based on observed correlations between malate uptake and the phosphorylation status of N-26 in intact symbiosomes, but this hypothesis lacks any direct evidence at the present time.

Several other classes of nodulin gene have been identified in which the protein sequence carries an N-terminal leader sequence (Verma, 1992). Some of these proteins, such as nodulin-24, may be integral membrane proteins or proteins associated with the PBM (de Blank *et al.,* 1993; Richter *et al.,* 1991), while others are more probably released through the PBM into the peribacteroid fluid that occupies the lumen of the symbiosome compartment (Kardailsky *et al.,* 1993). Unfortunately, the biochemical functions of most of these nodulin proteins are still unknown. Most nodulins identified by cDNA cloning and DNA sequencing have not yet been isolated as proteins from nodule tissue, nor have they been localized to a particular subcellular compartment.

Properties of the peribacteroid membrane

Biogenesis. The PBM has properties of both tonoplast and plasma membranes, and its biogenesis and differentiation are not well understood (Mellor and Werner, 1987; Verma *et al.,* 1994). Initially, bacteria are taken into the host cytoplasm by endocytosis (i.e. by involution of the plasma membrane). Subsequent growth and division of intracellular bacteroids necessitates further extension of the PBM by inclusion of additional vesicles derived from the Golgi or endoplasmic reticulum (Roth and Stacey, 1989a, b). Host cells containing differentiating bacteroids usually become enlarged; the cytoplasm becomes tightly packed with bacteroids and the tonoplast is partially or completely occluded. Thus the mosaic nature of the PBM is commensurate with its dual functions in host defence (plasma membrane) and metabolite exchange (tonoplast).

Successive stages in the differentiation of the PBM have been investigated in several ways. The use of cytochemical stains or immunological markers (e.g. JIM 18; *Figure 4b*) has helped to define the point at which the PBM becomes distinct from the plasma membrane. Symbiotically defective bacterial mutants have been identified that are blocked at the point of bacteroid release (Morrison and Verma, 1987). These mutants should help to identify the signals necessary to stimulate the production and differentiation of PBM material. In addition, a symbiotically defective pea mutant has recently been described in which the normal synchrony of bacteroid division and PBM division has broken down and the bacteroids fail to develop the capacity for nitrogen fixation (Borisov *et al.*, 1993).

A novel approach to the study of PBM biogenesis has recently been adopted by Verma and co-workers (Cheon *et al.*, 1993), who suggested that small GTP-binding proteins might be involved in endocytosis of bacteria and development of the symbiosome compartment. (Although such proteins had been described previously in plants (Drøbak *et al.*, 1988), a functional role had not been established.) Three cDNA clones were isolated from legumes by virtue of their homology with mammalian *Rab1p* and *Rab7p* sequences. Two of the corresponding genes, *sRab1p* from soybean and *vRab7p* from *Vigna*, were shown to be expressed at high levels in developing nodules. Expression of these genes was experimentally reduced in nodules of transgenic plants by the use of antisense constructs driven by nodule-specific promoters. This was observed to reduce the size of the nodules, affecting the compartment-alization of bacteria and the efficiency of nitrogen fixation.

Physical interaction with the bacterial cell surface. Using isolated fragments of PBM and of the bacteroid membrane, a physical interaction between these two membranes has been demonstrated *in vitro* (Bradley *et al.*, 1986). Such a direct interaction between plant and bacterial membrane surfaces could be involved in the uptake of rhizobial cells (endocytosis) into the host cytoplasm, which normally proceeds from structures termed infection droplets (unwalled intrusions of the plasma membrane). Similarly, surface interactions between these two membranes could explain how subsequent division of intracellular bacteroids is accompanied by concomitant division of the PBM. It is interesting to note that not only is the PBM devoid of an associated cell wall, but also the bacteroid membrane is not encapsulated by acidic extracellular polysaccharides, as it is in extracellular rhizobia (Latchford *et al.*, 1991). Thus the close association of these two surfaces is physically possible in the context of endosymbiotic interaction.

The major component of the bacterial outer membrane is lipopolysaccharide (LPS). Pea nodules containing mutant forms of *Rhizobium* defective in LPS structure show a reduced capacity for host cell invasion, a lack of synchrony between bacteroid cell division and segregation of the PBM, and a very abnormal morphology of bacteroids (Perotto *et al.*, 1994). Thus the structure of LPS is apparently important for tissue and cell invasion by rhizobia. Its possible functions are either to promote physical association with plant membranes or, alternatively, to suppress the normal host defence responses during this invasion process (Carlson *et al.*, 1992). The lipopolysaccharide macromolecule is conceptually divided into several structural domains (O-antigen, core polysaccharide and lipid A), which may have different and discrete functions in plant–microbe interactions (Kannenberg and Brewin, 1994).

Functions in metabolite exchange. The exchange of metabolites across the PBM has been recently reviewed (Brewin, 1990, 1991; Udvardi and Kahn, 1993). The exact role of the PBM in initiating and regulating the process of biological nitrogen fixation by bacteroids is still a matter for conjecture. The symbiosome system appears to be maintained in a state of dynamic equilibrium, and if the process of nitrogen fixation does not take place because of a bacterial mutation, a plant mutation, or some other physiological constraint, the bacteroids rapidly become senescent and the host cell subsequently dies (Grosskopf *et al.*, 1993; Huang *et al.*, 1993; Kneen *et al.*, 1990; Perotto *et al.*, 1994). One of the curious and little understood features of the PBM is the mechanism whereby the enclosed bacteroids are supplied with respiratory substrates and other materials necessary for nitrogen fixation, but the further growth and division of bacteroids is inhibited. Several models have been proposed based on the concept of coupled metabolic exchanges (Udvardi and Kahn, 1993) or of a dynamic equilibrium regulating pH in the peribacteroid compartment (Brewin, 1990). Components of the peribacteroid fluid that have been identified include α-mannosidases (Kinnbach *et al.*, 1987), proteases and protease inhibitors (Manen *et al.*, 1991; Mellor *et al.*, 1984), and a protein with strong homology to a vegetative lectin from pea (Kardailsky *et al.*, 1996).

Host defence responses. In addition to callose synthase, a number of inducible defence-related responses have been shown to be induced in incompatible *Rhizobium*–legume associations. These include the synthesis of antimicrobial phytoalexins (Niehaus *et al.*, 1993; Werner *et al.*, 1985) (*Figure 8*), the secretion and oxidation of lignins and phenolics (Perotto *et al.*, 1994), and the accumulation of hydroxyproline-rich glycoproteins (Benhamou *et al.*, 1991). It is probable that these functions, which are all standard features of plasma membranes, are all retained on the PBM.

Membrane lipid peroxidation. The intactness of the PBM appears to be a precondition for bacteroid function. Conversely, degradation of the PBM appears to be a very early consequence of bacteroid malfunction. One of the major processes involved in the degradation of biological membranes is lipid peroxidation (Richter *et al.*, 1991), which causes loss of fluidity, loss of membrane potential and eventual rupture of the membrane. Although the chemistry of the initiation of membrane lipid peroxidation is not understood, it is believed that iron ions may play an important role (Herrada *et al.*, 1993; Minotti and Aust, 1987). This is particularly relevant to the PBM because, on its cytoplasmic face, it is surrounded by a very high concentration (approximately 3 mM) of the haem-carrying protein leghaemoglobin (Appleby, 1984). In a functional nodule, the active respiration of bacteroids involved in the process of biological nitrogen fixation is normally sufficient to maintain leghaemoglobin in the reduced state. However, if bacteroids fail to fix nitrogen, either because of genetic mutations or as a consequence of physiological impairment, oxidized leghaemoglobin (met-leghaemoglobin) will accumulate, which could conceivably be active in lipid peroxidation. Using enriched PBM preparations from *Phaseolus vulgaris* nodules, lipid peroxidation and the production of malondialdehyde was observed in the presence of met-leghaemoglobin and H_2O_2, but not with reduced (ferrous) leghaemoglobin (Puppo *et al.*, 1991). Lipid peroxidation of PBM was also observed with iron ions, with a mixture of iron(III) and iron(II) producing maximal peroxidation. In addition, it was shown that senescing nodules were able to provoke lipid peroxidation, and, moreover, that they

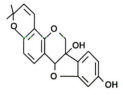

Figure 8. Chemical structure of glyceollin I, a phytoalexin that accumulates in an ineffective type of soybean nodule with early loss of the PBM (Werner *et al.*, 1985).

contained free iron ions. Soybean nodules also contain enzymes of the ascorbate–glutathione pathway, presumably to minimize the risk of oxidative damage (Dalton *et al.*, 1993). Significantly, the distribution of ascorbate peroxidase closely follows the distribution of leg-haemoglobin itself.

3. Plasma membrane

The peribacteroid membrane is an example of a differentiated form of plant membrane that arises during the course of legume nodule development, but, in the earlier stages of nodule initiation, plant–microbe interactions involve the plasma membrane of relatively unspecialized root epidermal and cortical cells. Because of the intense interest in the *Rhizobium*–legume symbiosis that has been sustained over many years, the biochemistry, molecular genetics and cytology of these processes are probably better understood than those of any other plant–microbe interaction. The study of this system provides interesting insights into the functioning of plasma membranes in general, particularly with regard to the coupling between plasma membranes and cell wall biogenesis, and also with regard to the issue of cell-to-cell signalling and signal transduction.

Tissue and cell invasion by Rhizobium

Tissue and cell invasion by *Rhizobium* involves the reorganization of plant cell wall growth, which must in some way be driven by the underlying plasma membrane. In many legumes the invasion route involves a specialized intracellular or transcellular structure termed an infection thread (Rae *et al.*, 1992). This takes the form of a tunnel of plant cell wall and plant cell membrane, with bacteria contained in the central lumen and embedded in a matrix of secreted plant glycoprotein (VandenBosch *et al.*, 1989). Infection threads are normally initiated in root hair cells that have undergone cell wall deformation following infection of seedling roots with *Rhizobium* (Smit *et al.*, 1992; van Spronsen *et al.*, 1994). Bacterial attachment and the initiation of infection threads may be enhanced by the presence of plant lectin on the root hair surface (Diaz *et al.*, 1989).

Secreted plant matrix glycoprotein. The form of plasma membrane that encircles the infection thread is normally termed the infection thread membrane. It secretes unusually large quantities of an intercellular plant glycoprotein (*Figure 9a*) that has been identified using monoclonal antibody MAC 265 (Rae *et al.*, 1992; VandenBosch *et al.*, 1989). This protein is also found to accumulate in the three-way junctions associated with intercellular spaces in uninfected tissue (*Figure 9b*), but it appears to be specifically targeted into the lumen of infection threads (Rae *et al.*, 1991). Its rate of secretion is even further enhanced following infection with a mutant strain of *Rhizobium* that has a defective lipopolysaccharide structure and is impaired in its capacity for host tissue and cell invasion (Perotto *et al.*, 1994). This glycoprotein material can become cross-linked under oxidative conditions (Bradley *et al.*, 1992; Gardner and Brewin, unpublished observations). Thus under some circumstances it seems possible that it could serve as an anti-microbial plug, preventing access to plant tissues by forms of bacteria that evoke a host defence response.

Association with the plant cytoskeleton. The initiation of infection thread development in a curled root hair cell involves the cessation of apical cell wall growth and the establishment of a new internally oriented system of inverted tip growth (Kijne, 1992). This creates the inwardly growing tunnel of the infection thread. Cytoskeletal connections maintain the nucleus at a distance of approximately 5 µm from the growing point of the infection thread, in just the same

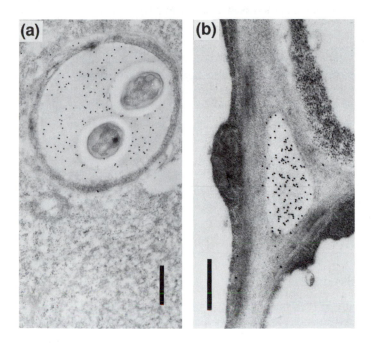

Figure 9. Localization of a plant matrix glycoprotein by immunogold staining following treatment of pea nodule sections with monoclonal antibody MAC 265 (Rae *et al.,* 1992). (a) *Rhizobium*-induced infection thread, seen here in transverse section as it traverses the cytoplasm of the host cell. The infection thread is a transcellular tunnel sheathed by plant cell wall and plant cell membrane with *Rhizobium* bacteria embedded in a matrix material containing plant glycoprotein (MAC 265 antigen). (b) Three-way junction between host plant cell walls, showing that the glycoprotein identified by MAC 265 is also present in intercellular spaces. Scale bars represent 5 µm.

way as the nucleus is maintained at a constant distance of 5 µm behind the tip of a root hair cell undergoing normal tip growth (Lloyd *et al.,* 1987).

The transcellular orientation of infection thread growth also appears to be imposed by the cytoskeleton (Rae *et al.,* 1992; Van Brussel *et al.,* 1992). Prior to the initiation of an infection thread, host cells in the root cortex become internally reorganized to form transcellular cytoplasmic strands which establish the pathway for future infection thread growth. This process somewhat resembles a dummy cell division (Yang *et al.,* 1994), although there is no nuclear division and instead of a cell plate there is merely the development of a transcellular tunnel. The importance of this comparison is that it helps to explain the orientation of transcellular infection thread growth. It also provides a partial explanation of how an infection thread can exit from a host cell — by fusion with the mother cell wall in a manner analogous to the way in which a cell plate fuses with the mother cell wall after a cycle of cell division (Brewin, 1990; Goodbody *et al.,* 1990; Venverloo, 1990).

Association with early nodulins. In-situ localization of transcripts for certain early nodulin genes provides circumstantial evidence that some of them, for example *ENOD12* and *ENOD5,* are involved in infection thread formation (Vijn *et al.,* 1993). Moreover, the fact that some of these genes have N-terminal leader sequences and proline-rich sequences suggests that they may associate with the plant cell membrane or wall in such a way as to adapt it to the constraints of infection thread growth.

Suppression of the host defence response. It is not known why invading rhizobia of the appropriate strain do not provoke a host defence response that prevents infection thread formation and consequent tissue invasion. However, it seems that the response of the host is always very delicately balanced. Even in a successful infection of alfalfa roots by *R. meliloti* it is observed that the majority of infection threads actually abort as a result of the induction of a localized host defence response (Vasse *et al.*, 1993), which may or may not be equivalent to a classical hypersensitive response (Jakobek and Lindgren, 1993).

An interesting observation that is perhaps relevant to the suppression of host defence responses is that about half of the non-nodulating pea mutants that have been isolated are also unable to establish a symbiosis with a mycorrhizal fungus (Duc *et al.*, 1989). This suggests a common mechanism for the suppression of the host defence response, or perhaps the involvement of a common component in the infection mechanisms for these two very different symbionts.

Signal transduction in nodule initiation

As a result of the intensive study of *Rhizobium* nodulation (*nod*) genes during the last decade, the basic components of a plant–microbial signalling system have become firmly established. This subject has been extensively reviewed (Dénarié *et al.*, 1992; Fisher and Long, 1992; Spaink *et al.*, 1993; Vijn *et al.*, 1993). Clearly, the whole system has major implications for intercellular signalling and signal transduction involving plant membranes. This has provoked a great deal of excited conjecture, but has so far yielded very little hard experimental data relevant to plant membrane biology.

Flavonoids. Flavonoids and chalcones secreted by legume roots act as host-specific inducers of *nod*-gene transcription (Djordjevic *et al.*, 1987). Paradoxically, roots secrete not only flavonoid inducers but also anti-inducers (Firmin *et al.*, 1986; Van Brussel *et al.,* 1990), and the relative proportions of these different products vary during the course of the infection process (*Figure 10*). The mechanisms which regulate the synthesis and secretion of flavonoids are still completely obscure (Recourt *et al.*, 1992), but it is worth noting that flavonoid compounds are also active as host-defence compounds; in legumes, they serve as biochemical precursors of the group of antimicrobial compounds known as the phytoalexins (Peters and Verma, 1990). Hence, from the very outset of the symbiotic interaction there is an undertone of the involvement of host defence systems that are mediated through the plasma membrane.

Lipochitin-oligosaccharides. The *Rhizobium* nodulation genes are responsible for the synthesis and secretion of a novel class of plant growth regulators, which are active at extremely low concentrations (10^{-10}–10^{-12} M), inducing root hair curling and initiating cortical cell division on the appropriate host legume (Dénarié *et al.*, 1992). These Nod factors are all lipophilic chitin oligomers (*Figure 11*) consisting of 3–5 residues of *N*-acetyl glucosamine linked 1,4-β with a long-chain unsaturated fatty acid substituted on the sugar moiety at the non-reducing end. In addition, there are several other forms of chemical substitution at the reducing or non-reducing end which confer host specificity on the system.

Circumstantial evidence for a possible signal transduction pathway comes from the observation that the application of lipochitin-oligosaccharide to alfalfa root hairs causes a membrane depolarization within 20 min (Ehrhardt *et al.*, 1992). No receptor molecule for lipochitin-

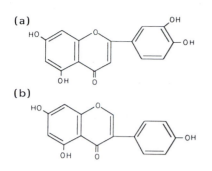

Figure 10. Chemical structures of flavonoid compounds secreted by legume roots which act as chemical signals from host legume to *Rhizobium*. (a) Luteolin (a flavone) acts as an inducer for transcription of nodulation genes in *Rhizobium leguminosarum* biovar *viciae* (which nodulates peas). (b) Genistein (an isoflavone) acts as an anti-inducer (competitive inhibitor) of *nod* gene induction in *Rhizobium leguminosarum* biovar *viciae*, but acts positively as an inducer for *Bradyrhizobium japonicum* (which nodulates soybeans). (Recourt *et al.*, 1992.)

oligosaccharides has been identified either biochemically or genetically, but plant chitinases have been implicated in the degradation of these signal molecules within host plant tissues (Collinge *et al.*, 1993).

Nod-O protein — a secreted channel-forming protein. In *R. leguminosarum* biovar *viciae* (which nodulates peas and *Vicia* spp.), a nodulation gene has been found that is not directly involved in the synthesis or secretion of lipochitin-oligosaccharide. The *nodO* gene encodes a secreted protein with a repeated Ca^{2+}-binding domain. This protein has homology to the bacterial haemolysin group, and functions as an ion channel-forming protein in artificial membrane systems (Sutton *et al.*, 1994). Its function in nodule initiation is still unclear. In *R. leguminosarum* biovar *viciae* it appears to potentiate the action of lipochitin-oligosaccharide, particularly in the case of a *nodE* mutant (which is unable to incorporate the correct fatty acid moiety on to the *N*-acetylglucosamine backbone). Furthermore, when introduced into a *nodE* mutant of *R. leguminosarum* biovar *trifolii* (which normally nodulates clover but not *Vicia* or pea), the *nodO* gene confers nodulation ability for *Vicia* on this strain.

4. Conclusions

At all stages of symbiotic interaction, a plant membrane separates the bacterial cell from the host plant cytoplasm. The specialized functions of this membrane differ at successive stages during

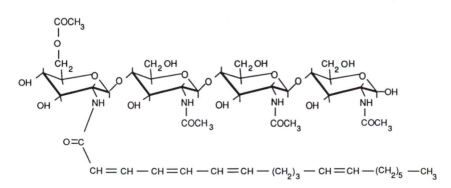

Figure 11. Chemical structure of a lipophilic chitin oligomer (lipochitin-oligosaccharide, NodRlv – IV (C18:4, Ac)) synthesized by *Rhizobium leguminosarum* biovar *viciae*. Such compounds act as chemical signals from *Rhizobium*, inducing root hair deformation and cell division in the root cortex of the host legume. (Dénarié *et al.*, 1992.)

the initiation and development of the legume root nodule. Initially, there is an exchange of chemical signals between root epidermal cells and soil rhizobia. Signal transduction at the plant cell surface results either positively in reorganized root hair cell wall growth and the initiation of cortical cell wall growth, or negatively in the induction of host defence responses. Tissue and cell invasion by *Rhizobium* involves host cell membranes in the construction of a specialized trans-cellular tunnel termed an infection thread, the orientation of which is controlled by the host cell cytoskeleton. Following the release of bacteria into the intracellular compartment, the rhizobia (now termed bacteroids) are enveloped by a plant-derived peribacteroid membrane which retains many properties of the plasma membrane, but lacks an associated sheath of plant cell wall material. The peribacteroid membrane is capable of very close association with the bacteroid outer membrane because of the absence of the capsular sheath normally synthesized by free-living bacteria. Bacteroid metabolism is regulated by the transport properties of the peribacteroid membrane, which are somewhat similar to those of the tonoplast membrane. The peribacteroid membrane therefore seems to be of hybrid origin, being derived partly from plasma membrane precursors and partly from endoplasmic reticulum and tonoplast precursors. Many of the antimicrobial host defence functions of the plasma membrane appear to be retained on the peribacteroid membrane, although they are normally suppressed in a successful symbiosis.

Acknowledgements

I am grateful for the contributions of many co-workers, and in particular I wish to thank Silvia Perotto, Elmar Kannenberg, Anne Rae, Igor Kardailsky, Chris Gardner, Allan Downie and Bjorn Drøbak for allowing me to present some of their unpublished results, and Janine Sherrier for her comments on the manuscript. I also thank L.E. Hernandez and D.T. Cooke for disclosing their results on the lipid analysis of peribacteroid membranes.

References

Ahlborn B, Werner D. (1992) 1,3-β-glucan synthase on the peribacteroid membrane (symbiosome membrane) from soybean root nodules. *Physiol. Mol. Plant Pathol.* **40**: 299–314.

Appels MA, Haaker H. (1991) Glutamate oxaloacetate transaminase in pea root nodules. *Plant Physiol.* **95**: 740–747.

Appleby CA. (1984) Leghaemoglobin and *Rhizobium* respiration. *Annu. Rev. Plant Physiol.* **35**: 443–478.

Baldwin TC, McCann MC, Roberts K. (1993) A novel hydroxyproline-deficient arabinogalactan protein secreted by suspension-cultured cells of *Daucus carota*. *Plant Physiol.* **103**: 115–123.

Bassarab S, Scheuk SU, Werner D. (1989) Fatty acid composition of the peribacteroid membrane and the endoplasmic reticulum in nodules of *Glycine max* varies after infection with different strains of the microsymbiont *Bradyrhizobium japonicum*. *Bot. Acta* **102**: 196–201.

Benhamou N, Lafontaine PJ, Mazau D, Esquerré-Tugayé M-T. (1991) Differential accumulation of hydroxyproline-rich glycoproteins in bean root nodules infected with a wild-type strain or a C_4-dicarboxylic acid mutant of *Rhizobium leguminosarum* bv. *phaseoli*. *Planta* **184**: 457–467.

Blumwald E, Fortin MG, Rea PA, Verma DPS. (1985) Presence of host plasma membrane type H⁺-ATP-ase in the membrane envelope enclosing the bacteroids in soybean root nodules. *Plant Physiol.* **78**: 665–672.

Borisov AY, Morzina EV, Kulikova OA, Tchetkova SA, Lebsky VK, Tikhonovich IA. (1993) New symbiotic mutants of pea (*Pisum sativum* L.) affecting either nodule initiation or symbiosome development. *Symbiosis* **14**: 297–313.

Bradley DJ, Butcher GW, Galfrè G, Wood EA, Brewin NJ. (1986) Physical association between the peribacteroid membrane and lipopolysaccharide from the bacteroid outer membrane in *Rhizobium*-infected pea root nodule cells. *J. Cell Sci.* **85**: 47–61.

Bradley DJ, Kjellbom P, Lamb CJ. (1992) Elicitor- and wound-induced oxidative cross-linking of a proline-rich plant cell wall protein: a novel, rapid defense response. *Cell* **70**: 21–30.

Brewin NJ. (1990) The role of the plant plasma membrane in symbiosis. In: *The Plant Plasma Membrane: Structure, Function and Molecular Biology* (Larsson C, Moller IM, ed.). Berlin: Springer-Verlag, pp. 351–375.

Brewin NJ. (1991) Development of the legume root nodule. *Annu. Rev. Cell Biol.* **7**: 191–226.

Brewin NJ. (1993) The *Rhizobium*–legume symbiosis: plant morphogenesis in a nodule. *Sem. Cell Biol.* **4**: 149–156.

Brewin NJ, Downie JA, Young JPW. (1992) Nodule formation in legumes. In: *Encyclopedia of Microbiology, Vol. 3* (Lederberg J, ed.). San Diego: Academic Press Inc., pp. 239–248.

Brewin NJ, Perotto S, Kannenberg EL, Rae AL, Rathbun EA, Lucas MM, Kardailsky I, Donovan N, Drobak BK. (1993) Mechanism of cell and tissue invasion by *Rhizobium leguminosarum*: the role of cell surface interactions. In: *Advances in Molecular Genetics of Plant–Microbe Interactions* (Nester EW, Verma DPS, ed.). Dordrecht: Kluwer Academic Publishers, pp. 369–380.

Carlson RW, Bhat UR, Reuhs B. (1992) *Rhizobium* lipopolysaccharides: their structure and evidence for their importance in the nitrogen-fixing symbiotic infection of their host legumes. In: *Plant Biotechnology and Development* (Gresshoff PM, ed.). Boca Raton: CRC Press, pp. 33–44.

Cheon C III, Lee N-G, Siddique A-BM, Bal AK, Verma DPS. (1993) Roles of plant homologs of Rab1p and Rab7p in the biogenesis of the peribacteroid membrane, a subcellular compartment formed *de novo* during root nodule symbiosis. *EMBO J.* **12**: 4125–4135.

Collinge DB, Kragh KM, Mikkelson JD, Nielsen KK, Rasmussen U, Vad K. (1993) Plant chitinases. *Plant J.* **3**: 31–40.

Cross GAM. (1990) Glycolipid anchoring of plasma membrane proteins. *Annu. Rev. Cell Biol.* **6**: 1–39.

Dalton DA, Baird LM, Langeberg L, Taugher CY, Anyan WR, Vance CP, Sarath G. (1993) Subcellular localisation of oxygen defense enzymes in soybean (*Glycine max* [L.] Merr.) root nodules. *Plant Physiol.* **102**: 481–489.

Day DA, Copeland L. (1991) Carbon metabolism and compartmentation in nitrogen-fixing legume nodules. *Plant Physiol. Biochem.* **29**: 185–201.

Day DA, Udvardi MK. (1993) Metabolite exchange across symbiosome membranes. *Symbiosis* **14**: 175–189.

Day DA, Price GD, Udvardi MK. (1988) Membrane interface of the *Bradyrhizobium japonicum*–*Glycine max* symbiosis: peribacteroid units from soybean nodules. *Aust. J. Plant Physiol.* **16**: 69–84.

de Blank C, Mylona P, Yang WC, Katinakis P, Bisseling T, Franssen HJ. (1993) Characterization of the soybean early nodulin cDNA clone GmENOD55. *Plant Mol. Biol.* **22**: 1167–1171.

Dénarié J, Debelle F, Rosenberg C. (1992) Signalling and host range variation in nodulation. *Annu. Rev. Microbiol.* **46**: 497–531.

Diaz CL, Melchers LS, Hooykas PJJ, Lugtenberg BJJ, Kijne JW. (1989) Root lectin as a determinant of host-plant specificity in the *Rhizobium*–legume symbiosis. *Nature* **338**: 579–581.

Djordjevic MA, Redmond JW, Batley M, Rolfe BG. (1987) Clovers secrete specific phenolic compounds which either stimulate or repress *nod* gene expression in *Rhizobium trifolii*. *EMBO J.* **6**: 1173–1179.

Drøbak BK, Allan EF, Comerford JG, Roberts K, Dawson AP. (1988) Presence of guanine nucleotide-binding proteins in a plant hypocotyl microsomal fraction. *Biochem. Biophys. Res. Commun.* **150**: 899–903.

Duc G, Trouvelot A, Gianinazzi-Pearson V, Gianinazzi S. (1989) First report of non-mycorrhizal plant mutants (Myc⁻) obtained in pea (*Pisum sativum* L.) and faba bean (*Vicia faba* L.). *Plant Sci.* **60**: 215–222.

Ehrhardt DW, Atkinson EM, Long SR. (1992) Depolarization of alfalfa root hair membrane potential by *Rhizobium meliloti* Nod factors. *Science* **256**: 998–1000.

Firmin JL, Wilson KE, Rossen L, Johnston AWB. (1986) Flavonoid activation of nodulation genes in *Rhizobium* reversed by other compounds present in plants. *Nature* **324**: 90–92.

Fisher RF, Long SR. (1992) *Rhizobium*–plant signal exchange. *Nature* **357**: 655–660.

Garbers C, Meckbach R, Mellor RB, Werner D. (1988) Protease (thermolysin) inhibition activity in the peribacteroid space of *Glycine max* root nodules. *J. Plant Physiol.* **132**: 442–445.

Goodbody KC, Lloyd CW. (1990) Actin filaments line up across *Tradescantia* epidermal cells, anticipating wound-induced division planes. *Protoplasma* **157**: 92–101.

Grosskopf E, Ha DTC, Wingender R, Rohrig H, Szecsi J, Kondorosi E, Schell J, Kondorosi A. (1993) Enhanced levels of chalcone synthase in alfalfa nodules induced by a Fix⁻ mutant of *Rhizobium meliloti*. *Mol. Plant–Microbe Interact.* **6**: 173–181.

Hernandez LE, Cooke DT. (1996) Lipid composition of symbiosomes from pea root nodules. *Phytochemistry* **42**: 341–346.

Herrada G, Puppo A, Rigaud J. (1989) Uptake of metabolites by bacteroid-containing vesicles and by free bacteroids from French bean nodules. *J. Gen. Microbiol.* **135**: 3165–3171.

Herrada G, Puppo A, Moreau S, Day DA, Rigaud J. (1993) How is leghaemoglobin involved in peribacteroid membrane degradation during nodule senescence? *FEBS Lett.* **326**: 33–38.

Höfte H, Hubbard L, Reizer J, Ludevid D, Herman EM, Chrispeels MJ. (1992) Vegetative and seed-specific forms of tonoplast intrinsic protein in the vacuolar membrane of *Arabidopsis thaliana*. *Plant Physiol.* **99:** 561–570.

Huang SS, Djordjevic MA, Rolfe BG. (1993) Microscopic analysis of the effect of *Rhizobium leguminosarum* biovar *trifolii* host specific nodulation genes in the infection of white clover. *Protoplasma* **172:** 180–190.

Jakobek JL, Lindgren PB. (1993) Generalized induction of defense responses in bean is not correlated with the induction of the hypersensitive response. *Plant Cell* **5:** 49–56.

Kannenberg EL, Brewin NJ. (1994) Molecular mechanisms of host plant invasion by *Rhizobium*. *Trends Microbiol.* **2:** 277–283

Kardailsky I, Yang W-C, Zalensky A, Van Kammen A, Bisseling T. (1993) The pea late nodulin gene PsNOD6 is homologous to the early nodulin genes PsENOD3/14 and is expressed after the leghaemoglobin genes. *Plant Mol. Biol.* **23:** 1029–1037.

Kardailsky I, Sherrier DJ, Brewin NJ. (1996) Identification of a new pea gene, *PsNlec1*, encoding a lectin-like glycoprotein isolated from the symbiosomes of pea root nodules. *Plant Physiol.* (in press).

Kijne JW. (1992) The *Rhizobium* infection process. In: *Biological Nitrogen Fixation* (Stacey G, Burris RH, Evans HJ, ed.). New York: Chapman & Hall, pp. 349–398.

Kinnbach A, Mellor RB, Werner D. (1987) Alpha-mannosidase II isoenzyme in the peribacteroid space of *Glycine max* root nodules. *J. Exp. Bot.* **38:** 1373–1377.

Kneen BE, LaRue TA, Hirsch AM, Smith CA, Weeden NF. (1990) *sym*13 — a gene conditioning ineffective nodulation in *Pisum sativum*. *Plant Physiol.* **94:** 899–905.

Kumarasinghe RMK, Nutman PS. (1977) *Rhizobium*-stimulated callose formation in clover root hairs and its relation to infection. *J. Exp. Bot.* **28:** 961–976.

Laine RA, Hsieh TC-Y. (1987) Inositol-containing sphingolipids. *Methods Enzymol.* **138:** 186–195.

Latchford JW, Borthakur D, Johnston AWB. (1991) The products of *Rhizobium* genes *psi* and *pss*, which affect exopolysaccharide production, are associated with the bacterial cell surface. *Mol. Microbiol.* **5:** 2107–2114.

Lloyd CW, Pearce KJ, Rawlins DJ, Ridge RW, Shaw PJ. (1987) Endoplasmatic microtubules connect the advancing nucleus to the tip of legume root hairs, but F-actin is involved in basipetal migration. *Cell Motil. Cytoskel.* **8:** 27–36.

Long SR, Staskawicz BJ. (1993) Prokaryotic plant parasites. *Cell* **73:** 921–935.

Manen JF, Simon P, Slooten JC, Osteras M, Frutiger S, Hughes GJ. (1991) A nodulin specifically expressed in senescent nodules of winged bean is a protease inhibitor. *Plant Cell* **3:** 259–270.

Maurel C, Reizer J, Schroeder JI, Chrispeels MJ. (1993) The vacuolar membrane protein g-TIP creates water specific channels in *Xenopus* oocytes. *EMBO J.* **12:** 2241–2247.

Mellor RB, Werner D. (1987) Peribacteroid membrane biogenesis in mature legume root nodules. *Symbiosis* **3:** 75–100.

Mellor RB, Morschel E, Werner D. (1984) Proteases and protease inhibitors present in the peribacteroid space. *Z. Naturforsch.* **39c:** 123–125.

Mellor RB, Christensen TMIE, Werner D. (1986) Choline kinase II is present only in nodules that synthesise stable peribacteroid membranes. *Proc. Natl Acad. Sci. USA* **83:** 659–663.

Miao G-H, Verma DPS. (1993) Soybean nodulin-26 gene encoding a channel protein is expressed only in the infected cells of nodules and is regulated differently in roots of homologous and heterologous plants. *Plant Cell* **5:** 781–794.

Miao G-H, Hong Z, Verma DPS. (1992) Topology and phosphorylation of soybean nodulin-26, an intrinsic protein of the peribacteroid membrane. *J. Cell Biol.* **118:** 481–490.

Minotti G, Aust SD. (1987) The requirement for iron (III) in the initiation of lipid peroxidation by iron (II) and hydrogen peroxide. *J. Biol. Chem.* **262:** 1098–1104.

Morrison N, Verma DPS. (1987) A block in the endocytosis of *Rhizobium* allows cellular differentiation in nodules but affects the expression of some peribacteroid membrane nodulins. *Plant Mol. Biol.* **9:** 185–196.

Niehaus K, Kapp D, Puhler A. (1993) Plant defence and delayed infection of alfalfa pseudonodules induced by an exopolysaccharide (EPS I)-deficient *Rhizobium meliloti* mutant. *Planta* **190:** 415–425.

Ouyang L-J, Day DA. (1992) Transport properties of symbiosomes isolated from siratro nodules. *Plant Physiol. Biochem.* **30:** 613–623.

Perotto S, VandenBosch KA, Butcher GW, Brewin NJ. (1991) Molecular composition and development of the plant glycocalyx associated with the peribacteroid membrane of pea root nodules. *Development* **112:** 763–773.

Perotto S, Brewin NJ, Kannenberg EL. (1994) Cytological evidence for a host defense response that reduces cell and tissue invasion in pea nodules by lipopolysaccharide-defective mutants of *Rhizobium leguminosarum* strain 3841. *Mol. Plant–Microbe Interact.* **7:** 99–112.

Perotto S, Donovan N, Drøbak BK, Brewin NJ. (1995) Differential expression of a glycosyl inositol phospholipid antigen on the peribacteroid membrane during pea nodule development. *Mol. Plant–Microbe Interact.* **8:** 560–568.

Peters NK, Verma DPS. (1990) Phenolic compounds as regulators of gene expression in plant–microbe interactions. *Mol. Plant–Microbe Interact.* **3:** 4–8.

Price GD, Day DA, Gresshoff PM. (1987) Rapid isolation of intact peribacteroid envelopes from soybean nodules and demonstration of selective permeability to metabolites. *J. Plant Physiol.* **130:** 157–164.

Puppo A, Herrada G, Rigaud J. (1991) Lipid peroxidation in peribacteroid membranes from French-bean nodules. *Plant Physiol.* **96:** 826–830.

Rae AE, Perotto S, Knox JP, Kannenberg EL, Brewin NJ. (1991) Expression of extracellular glycoproteins in the uninfected cells of developing pea nodule tissue. *Mol. Plant–Microbe Interact.* **4:** 563–570.

Rae AL, Bonfante-Fasolo P, Brewin NJ. (1992) Structure and growth of infection threads in the legume symbiosis with *Rhizobium leguminosarum*. *Plant J.* **2:** 385–395.

Recourt K, Verkerke M, Schripsema J, Van Brussel AAN, Lugtenberg BJJ, Kijne JW. (1992) Major flavonoids in uninoculated and inoculated roots of *Vicia sativa* subsp. *nigra* are four conjugates of the nodulation gene-inhibitor kaempferol. *Plant Mol. Biol.* **18:** 505–513.

Richter AE, Sandal NN, Marcker KA, Sengupta-Gopalan C. (1991) Characterization and genomic organisation of a highly expressed late nodulin gene subfamily in soybeans. *Mol. Gen. Genet.* **229:** 445–452.

Robertson JG, Wells B, Brewin NJ, Wood EA, Knight CD, Downie JA. (1985) The legume–*Rhizobium* symbiosis: a cell surface interaction. *J. Cell Sci.* Suppl. 2: 317–331.

Rosendahl L, Dilworth MJ, Glenn AR. (1992) Exchange of metabolites across the peribacteroid membrane in pea root nodules. *J. Plant Physiol.* **139:** 635–638.

Roth LE, Stacey G. (1989a) Cytoplasmic membrane systems involved in bacterium release into soybean nodule cells as studied with two *Bradyrhizobium japonicum* mutant strains. *Eur. J. Cell Biol.* **49:** 24–32.

Roth LE, Stacey G. (1989b) Bacterium release into host cells of nitrogen-fixing soybean nodules: the symbiosome membrane comes from three sources. *Eur. J. Cell Biol.* **49:** 13–23.

Smit G, Swart S, Lugtenberg BJJ, Kijne JW. (1992) Molecular mechanisms of attachment of *Rhizobium* bacteria to plant roots. *Mol. Microbiol.* **6:** 2897–2903.

Spaink HP, Wijfjes AHM, van Vilet TB, Kijne JW, Lugtenberg BJJ. (1993) Rhizobial lipo-oligosaccharide signals and their role in plant morphogenesis; are analogous lipophilic chitin derivatives produced by the plant? *Aust. J. Plant Physiol.* **20:** 381–392.

Stacey G, Burris RH, Evans HJ. (1992) *Biological Nitrogen Fixation.* New York: Chapman & Hall.

Sutton MJ, Lea EJA, Downie JA. (1994) The nodulation signalling protein NodO from *Rhizobium leguminosarum* biovar *viciae* forms ion channels in membranes. *Proc. Natl Acad. Sci. USA* **91:** 9990–9994.

Udvardi MK, Day DA. (1989) Electrogenic ATPase activity on the peribacteroid membrane of soybean (*Glycine max* L.) root nodules. *Plant Physiol.* **90:** 982–987.

Udvardi MK, Kahn ML. (1993) Evolution of the (*Brady*)*Rhizobium*–legume symbiosis; why do bacteroids fix nitrogen? *Symbiosis* **14:** 87–101.

Udvardi MK, Yang L-JO, Young S, Day DA. (1990) Sugar and amino acid transport across symbiotic membranes from soybean nodules. *Mol. Plant–Microbe Interact.* **3:** 334–340.

Van Brussel AAN, Recourt K, Pees E, Spaink HP, Tak T, Wijffelman CA, Kijne JW, Lugtenberg BJJ. (1990) A biovar-specific signal of *Rhizobium leguminosarum* bv. *viciae* induces increased nodulation gene-inducing activity in root exudate of *Vicia sativa* subsp. *nigra*. *J. Bacteriol.* **172:** 5394–5401.

Van Brussel AAN, Bakhuizen R, van Spronsen PC, Spaink HP, Tak T, Lugtenberg BJJ, Kijne JW. (1992) Induction of pre-infection thread structures in the leguminous host plant by mitogenic lipo-oligosaccharides of *Rhizobium*. *Science* **257:** 70–72.

VandenBosch KA, Bradley DJ, Knox JP, Perotto S, Butcher GW, Brewin NJ. (1989) Common components of the infection thread matrix and the intercellular space identified by immunocytochemical analysis of pea nodules and uninfected roots. *EMBO J.* **8:** 335–342.

van Spronsen PC, Bakhuizen R, van Brussel AAN, Kijne JW. (1994) Cell wall degradation during infection thread formation by the root nodule bacterium *Rhizobium leguminosarum* is a two-step process. *Eur. J. Cell Biol.* **64:** 88–94

Vasse J, de Billy F, Truchet G. (1993) Abortion of infection during the *Rhizobium meliloti*–alfalfa symbiotic interaction is accompanied by a hypersensitive reaction. *Plant J.* **4:** 555–566.

Venverloo CJ. (1990) Regulation of the plane of cell division in vacuolated cells. II. Wound induced changes. *Protoplasma* **155:** 85–94.

Verma DPS. (1992) Signals in root nodule organogenesis and endocytosis of *Rhizobium*. *Plant Cell* **4:** 373–382.

Verma DPS, Hu C-A, Zhang M. (1992) Root nodule development: origin, function and regulation of nodulin genes. *Physiol. Plant.* **85:** 253–265.

Verma DPS, Cheon C-I, Hong Z. (1994) Small GTP-binding proteins and membrane biogenesis in plants. *Plant Physiol.* **106:** 1–6.

Vijn I, das Neves L, Van Kammen A, Franssen HJ, Bisseling T. (1993) Nod factors and nodulation in plants. *Science* **260:** 1764–1765.

Weaver CD, Roberts DM. (1992) Determination of the site of phosphorylation of nodulin 26 by the calcium-dependent protein kinase from soybean nodules. *Biochemistry* **31:** 8954–8959.

Werner D, Mellor RB, Hahn MG, Grisebach H. (1985) Glyceollin I accumulation in an ineffective type of soybean nodule with an early loss of peribacteroid membrane. *Z. Naturforsch.* **40c:** 179–181.

Yang L-JO, Udvardi MK, Day DA. (1990) Specificity and regulation of the dicarboxylate carrier on the peribacteroid membrane of soybean nodules. *Planta* **182:** 437–444.

Yang W-C, de Blank C, Meskiene I, Hirt H, Bakker J, van Kammen A, Franssen HJ, Bisseling T. (1994) *Rhizobium* Nod factors reactivate the cell cycle during infection and nodule primordium formation, but the cycle is only completed in primordium formation. *Plant Cell* **6:** 1415–1426.

Chapter 30

Membranes in mycorrhizal interfaces: specialized functions in symbiosis

Sally E. Smith and F. Andrew Smith

1. Introduction

The symbioses between mycorrhizal fungi and host plants are biotrophic, and, with the exception of orchid mycorrhizas, they are also mutualistic. Mycorrhizas are one of the most common examples of mutualism, potentially involving 80–90% of the species of vascular plants and having a fossil history (at least for vesicular–arbuscular (VA) mycorrhizas) extending from the time when plants first colonized land (Nicolson, 1975; Raven *et al.*, 1978; Remy *et al.,* 1994; Simon *et al.,* 1993). This means that, as we and others have emphasized, mycorrhizas are important organs for the absorption of nutrients by the majority of plants, and it is at least as important to understand the way in which they function as it is to understand how non-mycorrhizal roots behave.

Key features which distinguish mycorrhizas from parasitic symbioses involving biotrophic fungi include low levels of specificity (Harley and Smith, 1983; Smith and Douglas, 1987), very prolonged duration of the compatible interaction between the symbionts, and bidirectional (rather than unidirectional) nutrient transfer (Harley and Smith, 1983; Smith and Smith, 1989, 1990a,b).

Membrane structure and function in mycorrhizas are of interest because they are involved in the development and function of the specialized interfaces between plant and fungal cells, across which all the interactions between the heterologous organisms take place. Membranes and their products play key roles in recognition between the symbionts, and in cellular modifications which underlie the maintenance of the compatible interactions. Furthermore, nutrient transfer between the symbionts is one of the most important functional attributes of the mutualistic relationship, and it is controlled by membrane transport processes.

2. Membranes in the interfaces of mycorrhizal associations

Types of interface

The adjoining plasma membranes in mycorrhizal associations must be considered within the context of the symbiotic interfaces in which they are found. In all cases the protoplasts of the symbionts remain separate, with the fungal symbiont 'outside' the plant protoplast in an apoplastic compartment, even when the plant cell wall has been penetrated during the intracellular stages of colonization (Smith and Smith, 1990b; *Table 1*). Cellular independence of the symbionts is retained and the way in which the symbiosis functions is the result of modifications to either or both of the organisms. Membrane transport processes have the potential to modify

525

Table 1. Characteristics of the interfaces between the plant (P) and fungal (F) cells in different types of mycorrhizal association

Attribute	Mycorrhizal type			
	Ecto	VA	Ericoid	Orchid
Symbosis [a]	mut	mut	mut	par
Interface				
Intercellular	+	+	–	–
Wall origin	F/P	F/P	NA	NA
Matrix	+	–	NA	NA
ATPase (P)	+	(+)	NA	NA
ATPase (F)	+	+	NA	NA
Intracellular	–	+	+	+
	NA	arb/coil	coil	coil
Wall origin	NA	F	F	F
Matrix	NA	+(P)	?	P
ATPase(P)	NA	++	?	+
ATPase (F)	NA	(+)	?	+
Nutrition [b]	B	B	B	B/N
Nutrient transport [c]	bi	bi	bi	uni

NA, not applicable; ?, data not available; +, reaction observed; –, reaction absent; (+), reaction not always observed.
[a] mut, mutualistic; par, parasitic.
[b] B, biotrophic; N, necrotrophic.
[c] bi, bidirectional; uni, unidirectional.

the delivery of solutes to the apoplast and in some cases their polymerization, with consequences for wall deposition, recognition, resistance to pathogens and nutrient movements.

Broadly, two types of interface, intercellular and intracellular, are involved (*Table 1*). In ectomycorrhizas, the developing and mature interfaces are entirely intercellular (in the Hartig net region) or involve superficial contact between fungal sheath cells and the epidermal cells of the root. Although there are extensive modifications of wall structure (e.g. Duddridge and Read, 1984; Massicotte *et al.*, 1986, 1989), the plasma membranes of the symbionts remain separated by recognizable walls of both symbionts, which form an essential feature of the interfacial apoplast. The area of contact is increased by branching of the Hartig net hyphae (Massicotte *et al.*, 1986) and by the development of transfer cells in the plant (Allaway *et al.*, 1985; Massicotte *et al.*, 1986, 1989).

In endomycorrhizas in the Orchidaceae and Ericales, the fungi penetrate cells directly from the soil and form intracellular hyphal coils (*Figure 1a*). The interfaces are almost entirely intracellular and involve the fungal and plant plasma membranes and an interfacial apoplast which contains recognizable fungal wall and a matrix of plant-derived material (Bonfante-Fasolo *et al.*, 1981; Gollotte *et al.*, 1993; Hadley *et al.*, 1971; Peterson and Currah, 1990). As the fungal coils in orchid mycorrhizas are lysed, they remain surrounded by the plant plasma membrane and by electron-lucent material which is probably callose (Peterson and Currah, 1990). This kind of reaction does not occur in other mycorrhizal types, and provides further evidence that orchid mycorrhizas are more similar to plant pathogen interactions than they are to other mycorrhizas.

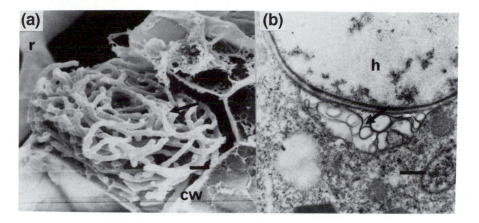

Figure 1. Orchid mycorrhiza. (a) Scanning electron micrograph of intracellular coils (arrowed) of *Rhizoctonia* with the cells of a protocorm of *Orchis morio*. cw, cell wall; r, epidermal rhizoid of the protocorm. Scale bar = 10 μm. (b) Proliferations (arrowed) of the plant plasma membrane (*Orchis morio*) adjacent to an intracellular fungal hypha (h). Scale bar = 0.5 μm. From H. Beyrle and R. L. Peterson, unpublished results.

In VA mycorrhizas, both intercellular and intracellular interfaces are important. The fungus first penetrates the epidermal and outer cortical cells, and at this stage the plant may react by synthesizing additional wall material (apposition layers) akin to the papillae which are produced during infection by pathogenic fungi. However, once it has penetrated these cell layers, the fungus spreads in the root cortex either by longitudinal growth of hyphae in the intercellular spaces between the cortical parenchyma cells (*Figure 2b*) or directly from cell to cell, frequently forming both intracellular coils and arbuscules (Smith and Smith, 1996). The hyphae retain their characteristic structure and their walls are little modified (Bonfante-Fasolo and Grippiolo, 1982). The same is true of the cortical cells themselves, although there are indications that the fungus may produce pectinases which depolymerize the middle lamella (Bonfante-Fasolo *et al.*, 1992). Thus the interface formed between the intercellular hyphae and the cortical cells is comparable with the Hartig net interface, and it includes recognizable walls of both symbionts. The interface between intracellular hyphal coils and cortical cells has not been studied extensively, but is presumably composed of the membranes of both symbionts, plus the fungal wall.

The main intracellular interfaces of VA mycorrhizas are the highly specialized arbuscules. These arise as short branches from the intercellular hyphae which penetrate the walls of the cortical cells, and, in the course of repeated dichotomous branching, they invaginate the plasma membrane of the host plant extensively (*Figure 2c,d*). The plant membrane continues to produce polysaccharide material in the interface, but this is only organized into a thick apposition layer at the point of penetration, and tapers gradually along the main arbuscular trunk hypha (Dexheimer *et al.*, 1979). At the same time, the walls of the hyphal branches become progressively thinner and the chitin content is markedly reduced (Bonfante-Fasolo and Grippiolo, 1982). The consequence is an interface where the plasma membranes are in close proximity, the material deposited in the interface is quite unlike the walls of the symbionts growing separately, and the area of contact between the symbionts is increased 3- to 7-fold (Alexander *et al.*, 1988, 1989; Cox and Tinker, 1976).

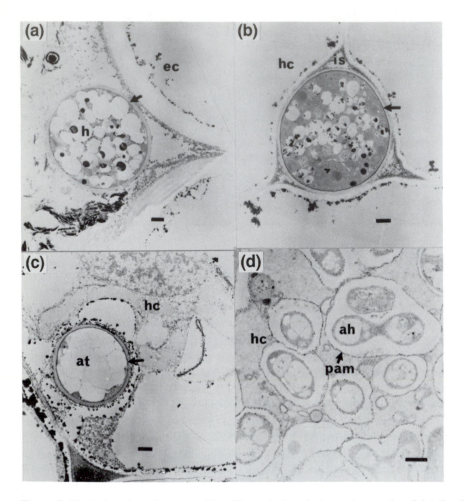

Figure 2. Vesicular arbuscular mycorrhiza. Transmission electron micrographs of details of the mycorrhizal interaction between *Allium cepa* and *Glomus intraradices*. (a–d) Segments of root fixed and stained for ATPase activity as described by Gianinazzi-Pearson *et al.* (1991). (a,b,c) No inhibitor; (d) effect of molybdate. Membrane-associated ATPase activity is indicated by single solid arrows. ec, epidermal cell; hc, host cortical cell; is, intercellular space. (a) Extramatrical hypha (h) at the root surface. (b) Intercellular hypha. (c) Arbuscular trunk hypha (at) in a cortical cell. (d) Fine arbuscular branches; ah, arbuscular hypha; pam, periarbuscular membrane. (a–c) From Gianinazzi-Pearson, unpublished results. (d) Reproduced from Gianinazzi-Pearson *et al.* (1991) with permission from *The New Phytologist*.

Plasma membranes of plant and fungus

The plant membranes. The periarbuscular membrane (PAM) is derived from the plant plasma membrane, presumably as a result of synthesis of new membrane, giving the opportunity for alterations in specific membrane components. The PAM retains continuity with the peripheral plasma membrane (PPM) of the cell, and reacts in the same way with phosphotungstic acid, which is assumed to be specific for the plasma membrane (Dexheimer *et al.*, 1979, 1985). Striking formations of membranous vesicles and tubules continuous with the plant membrane adjacent to the fungus have been observed in all stages of development. These membranes show the same reactions to cytochemical stains, including that for ATPase and the silver proteinate reaction for polysaccharides (Dexheimer *et al.*, 1985; Marx *et al.*, 1982). No function for these formations has been established, although their occurrence in the interfacial regions may indicate involvement in symbiotic processes, including an increase in plasma membrane surface area and the deposition of polysaccharides. Similar formations occur on the fungal membranes (see below), and on the plant membranes in orchid mycorrhizal interfaces (see *Figure 1b*).

The activities of the invaginated membrane are clearly modified. Immunolabelling with monoclonal antibodies, originally raised to components of the peribacteroid membrane (PBM) in peas, has shown that there is considerable similarity between the PBM, the PAM and the PPM (Perotto, unpublished results; see Bonfante-Fasolo *et al.*, 1992). However, the monoclonal anti-body (mAb) MAC 64, which binds to the PBM, PPM and Golgi apparatus in peas, only labelled the PPM in cells containing arbuscules, and did not label the PAM (Gianinazzi-Pearson *et al.*, 1990). No tonoplast-like features of the PAM have been observed (e.g. Day *et al.*, 1989; Kinnbach *et al.*, 1987), so that, unlike the situation for the PBM, we do not yet need to conclude that the PAM is a mixed membrane with both plasma membrane and tonoplast components (Smith and Smith, 1990b).

Enzyme activities, as determined by cytochemical methods, are also changed. The PAM shows increased ATP-hydrolysing ability compared with the PPM while the arbuscules are active (see *Figure 2*), but this activity declines when they collapse (Marx *et al.*, 1982). Such activity is almost certainly due to an increase in the activity of H^+-ATPase on the PAM which, together with fungal H^+-ATPases, is likely to play an important role in nutrient transfer between the symbionts (see below) (Gianinazzi-Pearson *et al.*, 1991; Smith and Smith, 1990b). The plant membranes in the interfaces of ecto- (Lei and Dexheimer, 1988) and orchid mycorrhizas (Serrigny and Dexheimer, 1985) also exhibit ATPase activity. As with VA mycorrhizas, the ATPase activity in plant membranes in ectomycorrhizas disappears when the opposing fungal cell is dead, but in orchids the activity is retained even when the fungus has completely collapsed, reflecting the different physiological relationships between the symbionts.

Neutral phosphatase activity, which is thought to be involved in polysaccharide synthesis, is normally localized on the plasma membrane, tonoplast and endoplasmic reticulum (ER) in non-colonized root cells. In colonized cells the activity on the tonoplast and ER is unchanged, but on the PPM it may be reduced, and around the fine arbuscular branches it is considerably increased (Jeanmaire *et al.*, 1985). Similar activity is found on the plasma membrane of colonized orchid cells as well as in the interface (Serrigny and Dexheimer, 1985), and in this case it may be associated with callose synthesis (see above).

In orchid mycorrhizas, the plant membranes surrounding the coils and the interfacial apoplast are associated with high acid phosphatase activity (Dexheimer and Serrigny, 1983), which is again likely to be involved in the control of fungal invasion.

Doubts have been cast on the techniques used to localize these phosphatase activities, but a recent reinvestigation of ATPase activity in VA mycorrhizas using a range of inhibitors (Gianinazzi-Pearson *et al.*, 1991) has gone some way towards confirming the above conclusions which, in the absence of investigations using other techniques, are the only data available. As we have said before (Gianinazzi-Pearson *et al.*, 1991; Smith and Smith, 1990b), the use of monoclonal antibodies would confirm the presence of these important enzymes, but they have proved difficult to apply (V. Gianinazzi-Pearson, personal communication). A recent finding (Murphy, Smith and Langridge, unpublished results; Murphy *et al.*, 1995), that the expression of a plant gene with considerable homology to H^+-ATPase from *Arabidopsis thaliana* is increased during VA mycorrhizal colonization, provides some independent confirmation of the cytochemical data and indicates one direction for future investigations.

If we accept that there is increased activity of H^+-ATPase on the PAM and on the orchid membranes surrounding the coils, it is tempting to speculate that the increase is due to a stimulus, as yet unknown, from the infection process, as well as to an increase in the surface area of membrane as a consequence of invagination. Identification of the stimulus is an important goal for future research.

Activities of plant membranes deduced from components of the interface. Further information about the modification and activity of the plant membranes can be deduced from the molecules found in the interfacial apoplast. The invaginated PAM rapidly loses any ability to produce recognizable walls in the periarbuscular apoplast. However, it clearly retains the ability to produce polysaccharide and pectin materials (Dexheimer *et al.*, 1979; Scannerini and Bonfante-Fasolo, 1979) that have been identified respectively as $\beta(1\text{-}4)$-glucan using cellobiohydrolase as a gold probe, and non-esterified and methyl-esterified polygalacturonans using monoclonal antibodies (Bonfante-Fasolo *et al.*, 1990). The deposition of these molecules is consistent with the observations of neutral phosphatase activity (see above), suggesting that the PAM continues to deliver polysaccharide to the apoplast.

Hydroxyproline-rich glycoproteins (HRGPs) are present in the arbuscular interface, and also in the cell walls of epidermal and cortical cells and in the stele of the root (Bonfante-Fasolo *et al.*, 1991). Their role has not been determined, but speculative ideas include cell wall extension and involvement in defence against pathogens.

The fungal membranes. There is even less information available about fungal plasma membranes than there is for the plant membranes. The external hyphae have an H^+- ATPase on their plasma membranes, consistent with their role in nutrient uptake (see *Figure 2a*). Furthermore, Harrison and van Buuren (1996) have identified a cDNA clone that represents a phosphate transporter from *Glomus versiforme* (*GvPT*). The gene appears to code for the high-affinity phosphate transporter, and it is expressed in external mycelium, but not in the fungus colonizing the roots of *Medicago truncatula*. Intercellular hyphae are relatively unmodified compared with the structure of external hyphae. They remain active (with respect to succinate dehydrogenase activity), exhibit membrane-bound ATPase activity for many weeks, and essentially provide the 'skeleton' of the fungus in VA mycorrhizal roots. Wall synthesis in the intercellular apoplast shows little change.

Inside the cells the surface area of the interface is increased either by the formation of intracellular coils, or by the dichotomous branching of the arbuscules, or both (Smith and Smith, 1996). In the fine arbuscular branches, plasma membrane formations of the same type as those on the plant membrane have been observed, but they do not necessarily occur opposite the plant formations, and are not associated with polysaccharides which give a positive result with the silver proteinate test (Dexheimer *et al.*, 1985). In VA mycorrhizas, plasma membrane ATPase activity is present not only in external hyphae (see *Figure 2a*), but also in appressoria attached to the epidermal cell walls, in intercellular hyphae (see *Figure 2b*), and sometimes on the membranes of the arbuscular branches (Gianinazzi-Pearson *et al.*, 1991). In ectomycorrhizas the activity is present in the external hyphae, those of the sheath and also those of the Hartig net, except where the latter are adjacent to apparently dead cortical cells of the plant (Lei, 1988). In orchids, the fungus also has ATPase activity on the plasma membrane of the fungal coils, which disappears as the hyphae collapse.

Different interfaces have different periods of activity. In many VA mycorrhizas the intercellular hyphae apparently remain alive for many weeks (Smith and Dickson, 1991), while the life span of individual arbuscules arising from them is short (4–14 days; Alexander *et al.*, 1988, 1989; Cox and Sanders, 1974), so that the relative importance of the two types of interface changes during the development of an infected root system, with consequences for symbiotic function if the interfaces carry out different processes (see below). In those VA mycorrhizas that are characterized by abundant intracellular coils, it is thought that the coils and arbuscules may be relatively long-lived (Smith and Smith, 1996). In orchid mycorrhizas, the intracellular fungal coils are also transitory, but in this case turnover has been associated with the expression of host defence against invasion.

The apoplast

The organization and composition of the apoplastic compartments in mycorrhizas have important consequences for the function of the symbioses, particularly for nutrient transfer mediated by membrane transport processes. The arrangement of apoplastic compartments needs to be considered at the tissue level as well as the cellular level. Ashford and co-workers (Ashford et al., 1988, 1989) have highlighted the importance of apoplastic barriers in ectomycorrhizal roots. In some associations it appears that the walls of the sheath hyphae are modified to provide an impermeable barrier in the apoplast between the root cortex (including hyphae in the Hartig net) and the soil, with the consequence that all nutrient uptake and loss must be under the control of the fungal symplast. Furthermore, the apoplastic compartment in the root cortex is 'insulated' by the development of apolastic barriers in the walls of the sheath, as well as in the endodermis of the root, and the apoplastic environment might be modified to facilitate inter-symbiont nutrient transport.

Endomycorrhizal roots do not develop similar apoplastic barriers at the outer surface of the root as a consequence of fungal colonization. However, the roots of many plant species normally have an exodermis (hypodermis) with suberized or lignified walls which, together with the endodermis, may perform the same 'insulating' functions and place the cortical apoplast under the control of the membrane transport processes of the two symbionts (Smith et al., 1989; Smith and Smith, 1990b). In contrast to many biotrophic pathogens where the haustorium is delimited by an impermeable neck-band, the intracellular phase of most mycorrhizal fungi is not isolated within a sealed, intracellular apoplastic compartment. As far as we can determine, the cortical apoplast is continuous in both intra- and intercellular phases, and would provide an exchange compartment similar to that proposed for ectomycorrhizal roots.

As yet we know very little about the composition of the different apoplasts, apart from the wall modifications described above. In the interfaces where transfer of nutrients occurs between the symbionts, we need to know the pH and concentration of key molecules and ions which are either transferred from one symbiont to another or which may influence transport processes. At this stage we can only guess that the interfacial apoplast is likely to be acid as a consequence of the activity of H^+-ATPases on either or both of the membranes. We have no idea of the concentrations of sugars, inorganic phosphate (P_i) or amino compounds, or indeed of Ca^{2+} or other cations that might influence transport (e.g. by changing the membrane potential or opening channels).

Intracellular membranes

The intracellular membranes associated with organelles of plant and fungal cells have not been the focus of much research. However, the fungal vacuoles, delimited by the tonoplast, do contain alkaline phosphatase and there has been considerable speculation about the possible importance of this enzyme in symbiotic phosphate metabolism, because of the increased activities localized in the fine branches of the arbuscules (Gianinazzi-Pearson and Gianinazzi, 1976, 1978; Gianinazzi et al., 1979). Another important feature of the vacuole is its role in phosphate storage. It was long believed that *granules* of polyphosphate were stored in fungal vacuoles (see *Figure*

2) and that, particularly in VA mycorrhizas, the actual transport of these vacuoles by cytoplasmic streaming was the mechanism for long-distance translocation of phosphate in hyphae. However, recent work has shown that the deposition of calcium polyphosphate in granular form is an artefact of the method of preparation. If freeze substitution is used, it is clear that the polyphosphate is stabilized by monovalent cations and that it is soluble (Orlovich and Ashford, 1993). Shepherd *et al.* (1993) have shown that the vacuoles (at least in ectomycorrhizal fungi) are not discrete, but form part of a continuum of pleiomorphic tubules which pulsate, transferring globules of material by peristalsis-like movements. Different tubules appear to operate in different directions within a single hypha, and the movements are independent of cytoplasmic streaming. These observations open up a new area with regard to mechanisms of bidirectional translocation in fungi, including mycorrhizal fungi. As Shepherd *et al.* (1993) point out, one of the 'attractions' of the system is that translocation of the contents of the tubules by peristalsis would not result in translocation of large amounts of membrane. However, if there is mass flow of solution in the tubules, then a mechanism for the 'disposal' of the solvent must be envisaged.

3. Transport processes

Extramatrical mycelium and phosphate uptake from soil

In all mycorrhizas the fungal symbiont grows extensively outside the root in the soil. Nutrient uptake by the hyphae and translocation of the nutrients for many centimetres along them results in the delivery of nutrients to the symbiotic interfaces within the roots. There is no suggestion that the membranes or membrane transport and translocation processes in the extramatrical phase of mycorrhizal fungi differ markedly from those in free-living fungi (Beever and Burns, 1980; Cairney *et al.*, 1988; Harley and Smith, 1983; Straker and Mitchell, 1987; Thomson *et al.*, 1990). Two phosphate uptake systems with different kinetic characteristics which operate simultaneously have been identified in ecto- (Cairney *et al.*, 1988), ericoid (Straker and Mitchell, 1987) and VA mycorrhizal fungi (Thomson *et al.,* 1990) grown in the absence of hosts, and in the mycorrhizal roots themselves (Cress *et al.*, 1979; Hampp *et al.*, 1993; Harley and McCready, 1952). Whether the fungus alters the kinetic characteristics of the uptake systems in the plant cells in colonized roots is not very clear, because of experimental difficulties in controlling the phosphate (P) status of the mycorrhizal and non-mycorrhizal roots, and the effects of this and of solution P concentration on the operation of the transport systems (Pearson and Jakobsen, 1993). Whatever the mechanisms involved, the fungus makes an important contribution to increasing the efficiency with which roots absorb P and also zinc (Zn) and copper (Cu).

There are some conflicting data with regard to uptake of P by germ tubes of *Gigaspora margarita*. Thomson *et al.* (1990) showed that germlings took up P in the absence of roots, and that there were (as indicated above) two transport systems for P with different K_m and V_{max} values and different sensitivities to phosphate concentration. Lei *et al.* (1991), using autoradiography, failed to show ^{32}P uptake by very young germ tubes of the same fungus, unless they were 'stimulated' by the presence of uncolonized roots (Bécard and Piché, 1989). The 'stimulated' germ tubes had membrane-bound, diethylstilbesterol (DES)-sensitive ATPases localized about 70 µm behind their apices, while 'non-stimulated' hyphae did not. In addition, while the growth of 'non-stimulated' hyphae was unaffected by DES, the growth of 'stimulated' hyphae was reduced to that of the 'non-stimulated' ones. The reasons for the different findings have not been established, but might be related to the different ages of the germlings used, as well as to some volatile factor in the 'stimulating' atmosphere. Use of the GvPT clone as a probe for the high-affinity transporters could be very useful for resolving this problem.

Transport at the interface

Nutrient transfer between plant and fungal symbionts at the interfaces is fundamental to the prolonged compatible interaction between them. We have reviewed what is known of the mechanisms of transfer (Smith and Smith, 1986, 1989, 1990a,b, 1995, 1996; Smith et al., 1994; see also Gianinazzi-Pearson et al., 1991). A brief summary of the important features is all that is appropriate here.

The information required for a clear understanding of transport processes at the interface includes: (i) the arrangement of membranes in the interface, (ii) the composition of the interfacial apoplast, (iii) the direction of transfer of nutrients in the symbioses as a whole (normally bidirectional), (iv) the identity of the molecules or ions delivered to the interfacial apoplast or transferred between symbionts (where known), (v) the concentrations of these molecules and ions in the apoplast, (vi) other ions that may need to be transferred to maintain the charge balance or pH, (vii) the distribution of H^+-ATPase activity on the membranes in the interface, as potential markers for energized membranes and the development of proton motive force, (viii) the pH of the interfacial apoplast, (ix) the permeability of the membranes, (x) the membrane potentials of cells of both symbionts and the membrane potential across the whole interface, (xi) the activity of other enzymes in the interface as potential modifiers of molecules delivered to the interface (e.g. phosphatases and invertases) and (xii) the rate of transfer or flux of different molecules or ions across the interface, and the identity and activity of the individual membrane transport proteins.

It is evident from the previous sections that most of this information is lacking, but methods for investigating some of these questions are becoming available. We do know that transfer of any nutrient from one symbiont to the other requires efflux or loss from the source organism and uptake into the sink organism. For soil-derived nutrients (e.g. P, Zn, N) the fungus is the source and the plant is the sink; for carbohydrates, the converse is true, except for the rather special case of orchid mycorrhizas, in which transfer is unidirectional from fungus to plant.

It is usually assumed that this opposed transfer of nutrients is *bidirectional across the same interface*. This must be true of the Hartig net in ectomycorrhizas and probably for the coils of ericoid mycorrhizas, and it may also be the case for VA mycorrhizas (but see below). The distribution of the ATPases in the interfacial membranes indicates that the plant membranes are probably energized in the Hartig net regions of ectomycorrhizas, the PAM of VA mycorrhizas and the plant membrane surrounding the coils of orchid mycorrhizas (*Table 1*), and that 'normal' uptake of P_i (for example) from the interfacial apoplast is likely. The fluxes of P across the VA mycorrhizal interface as a whole (Cox and Tinker, 1976; Smith et al., 1994a, 1995) are of the same order as the measured rates of uptake of P by giant algal cells or roots (2–12 nmol m^{-2} sec^{-1}). When soil P is relatively high, the fluxes may be higher (Dickson, unpublished results), and appear to exceed uptake by external hyphae perhaps by a factor of 10 (Sukarno et al., 1996). Values are not available for other mycorrhizal types, and for the purposes of this discussion we shall assume that they are similar. In symbiosis such rates must be supported by equivalent rates of efflux from the source organism, and in the case of P this suggests that the fungus must preferentially lose P at the interface faster than would be expected in a free-living organism (Smith and Smith, 1990b; Smith et al., 1995; Tester et al., 1993). We believe that this loss is specific and not due to generally increased permeability or 'leakiness', otherwise, the fungus itself would be incapable of controlling its own uptake from the apoplast (carbohydrate) or maintenance of turgor. The efflux of P in the cultured ericoid mycorrhizal fungus, *Hymenoscyphus ericae*, is around 13 pmol m^{-2} sec^{-1}, 100- to 1000-fold less than the measured overall flux accross the interface (Smith et al., 1995). Efflux from ectomycorrhizal fungi is also low, although it has not been calculated on a surface area basis (Cairney and Smith, 1993a,b). At this stage it seems necessary to postulate special mechanisms which promote efflux in the symbiosis.

Efflux in ectomycorrhizal fungi is stimulated by high P concentrations in the mycelium and high concentrations of monovalent cations in the bathing medium (Cairney and Smith, 1993a,b), leading to the suggestion that the concentrations of these ions in the interfacial apoplast could be important, and could lead to membrane depolarization and consequent loss of through anion channels (Tester et al., 1993). This effect of monovalent cations was not observed in the fungus *H. ericae* (Tester, Cherry and Reid, personal communication), but intracellular acidification, induced by the addition of butyrate, was followed by increased efflux of both P and chloride. It is likely that the increased efflux was a consequence of an increased cytoplasmic concentration of $H_2PO_4^-$, rather than loss of P through a channel, as is also suggested by parallel work with *Chara corallina* and excised cereal roots (Smith et al., 1995).

The few available data on membrane potentials in mycorrhizal roots are not particularly helpful in elucidating transport processes at the interface. The fungus appears to exert a long-distance influence in hyperpolarizing (rather than depolarizing, see above) the plant cells (Fieschi et al., 1992). An increase in membrane potential difference (p.d.) in infected cells might be associated with the increase in H^+-ATPase activity and might influence the cortical apoplast at a tissue level, rather than at the level of the apoplast in individual cells. We do not know what happens to the fungal membrane potential. The relative values for plant and fungus might be important, because changes in the *net* p.d. between the two organisms could influence the driving forces for transport.

The only other information about biophysical aspects of membrane function has been obtained using a vibrating electrode (Berbara et al., 1995). These workers observed that the positive current flow behind the root tips changes from outward to inward at VA mycorrhizal entry points. It has been postulated that currents act as signals which influence the colonization of roots by micro-organisms (Morris et al., 1992), but the reasons for changes in current during infection are not clear. Speculative ideas include reorientation of the cytoskeleton during invagination, like the changes stimulated by wounding (Hush and Overall, 1991). In VA mycorrhizal spores there is a marked polarity of inward and outward current flow, with germination always occurring at the inward positive pole.

In VA mycorrhizas, the existence of three types of interface (intercellular hyphae, intracellular coils and arbuscules) means that the transfer between symbionts, which is bidirectional in the root system as a whole, might be polarized in one direction or the other at these different cellular locations (Gianinazzi-Pearson et al., 1991; Smith and Smith, 1995). It is possible to envisage a scenario in which the transfer of P from fungus to plant occurs across the arbuscular interface, with highly energized PAM and possibly fungal membranes which are not energized and/or have the phosphate transporter down-regulated (see below). This would provide a mechanism whereby reabsorption of P by the fungus from the arbuscular apoplast might be prevented (see below). The distribution of H^+-ATPases (*Table 1*) is certainly consistent with this view. Carbon transfer might occur at the hyphal interfaces, with the long-lived intercellular hyphae or intracellular coils playing an important role in fungal carbohydrate acquisition from the intercellular spaces or cortical cells (Gianinazzi-Pearson et al., 1991; Smith and Smith, 1996), in a way that might be similar to transport in the Hartig net region of ectomycorrhizal roots.

Patrick (1989) has discussed the efflux of reduced carbon from cells supporting micro-organisms in the context of what is known of import–export characteristics of cells in different plant tissues. He concluded that (i) plant genomes must carry information for the specialized efflux that occurs in some tissues, (ii) elevated cytoplasmic concentrations of sugars may be required in tissues committed to efflux, and (iii) transport of reduced carbon to symbionts may need to involve enhanced efflux and hence elicitors from the organisms may be involved in causing the expression of the genetic information in tissues that would not otherwise export reduced carbon. He made another important point, that in most plant tissues the cells are specialized to retrieve solutes from the apoplast, so that net transfer to sinks, including invading microorganisms, may require that the uptake (retrieval) mechanism is switched off. This would also apply to phloem unloading, and again suggests that the fungi may exploit mechanisms which occur normally in some plant tissues. Similar arguments need to be applied to the transfer of P from fungus to plant where the fungal retrieval of P from the apoplast must be reduced (Smith et al., 1994b). The apparent lack of expression of the fungal phosphate transporter (GvPT) in intraradical fungus (Harrison and Van Burren, 1996) would certainly fit well within this picture.

Patrick's (1989) discussion is consistent with what is known about the factors influencing carbon transfer to the fungus in beech mycorrhizas (Lewis and Harley, 1965; Smith et al., 1969), where the continued unloading of the plant cell is envisaged as occurring down a concentration gradient that is maintained by the hydrolysis of sucrose to glucose and fructose in the apoplast, their uptake by the fungus and their conversion to fungal carbohydrates, such as trehalose and mannitol, that cannot be re-utilized by the plant. As has been suspected for some time (Harley and Smith, 1983), plant-derived acid invertases in the Hartig net interface of ectomycorrhizas may be implicated in facilitating the directional transport of sugars (Salzer and Hager, 1993). The ectomycorrhizal fungi Amanita muscaria and Hebeloma crustuliniforme have no wall-bound invertase and are incapable of growing on sucrose, although they grow well on both glucose and fructose. Both form mycorrhizas with Picea abies, which does have acid wall invertase isoforms that were studied in suspension-cutured cells. The key features are that they have relatively high K_m values of 16 mM and 8.6 mM, very narrow pH optima of around 4.5, with some activity between 4.5 and 6.0, and are both competitively inhibited by fructose but not by glucose. The fungi might have the potential to regulate the invertase activity via changes in the apoplastic pH and utilization of fructose, hence manipulating the concentration gradient of sucrose between root cells and apoplast and controlling the efflux from the plant (Salzer and Hager, 1993). Protoplasts of A. muscaria have a higher affinity for glucose than for fructose (K_m values of 1.25 and 11.3 mM, respectively), and while the uptake of fructose is inhibited by glucose, the opposite is not true (Hampp et al., 1993). Whether these apparently complex regulatory mechanisms operate in intact mycorrhizas is not yet clear. New techniques for measurement of ion concentrations and pH in vivo, such as the use of ion-sensitive ratio fluorochromes in confocal microscopy, will be useful for determining the conditions in the apoplast. Using ^{13}C nuclear magnetic resonance (NMR) spectroscopy, Shachar-Hill et al. (1995) have shown that glucose supplied to mycorrhizal leek roots can be utilized by the fungal symbiont. It will be most interesting to determine whether fructose and sucrose can also pass to the fungus, and whether the latter is hydrolysed before uptake.

A second hypothetical mechanism involving alterations in wall metabolism which might support carbohydrate transfer in VA mycorrhizas has also been proposed (Harley and Smith, 1983). It is clear that, although walls are not polymerized in the arbuscular apoplast, the PAM continues to deliver both glucans and pectins (see above). These could be used by the fungus so long as it had the appropriate uptake mechanisms in place, or maintained a concentration gradient by rapid utilization in the arbuscular branches.

4. Recognition and colonization: speculation about the involvement of plasma membrane receptors

Development of a symbiosis involves a regulated sequence of events leading to establishment of the characteristic long-term compatible interaction (Smith, 1995). One of the earliest stages must be the mutual recognition of the symbionts and the initial attachment of the cells. In mycorrhizas, this initial stage, operating at the root surface, is followed by a number of more or less well-defined steps in which changes in the development of the fungus or host indicate that 'recognition' of a developmental branch point has again occurred, followed by changes in gene expression which underlie the structural and functional alterations. In some cases cytological changes indicating that the plant has 'recognized' the presence of the fungus occur before tissue penetration. Examples include an increase in nuclear size and activity (Münzenberger *et al.*, 1992) and membrane hyperpolarization (Fieschi *et al.*, 1992). As we have seen, many of these alterations involve membranes and membrane processes, and the recognition processes are likely to do so as well.

Recognition in mycorrhizas is not associated with phenomena conferring a high degree of specificity. The mechanisms which in most encounters between micro-organisms and plants lead to restriction of the activities of the micro-organism and failure of tissue penetration (basic incompatibility mechanisms; Ride, 1992) are switched off (e.g. Harrison and Dixon, 1994). Cell wall thickenings at the point of initial contact are slight, phenolic deposition is minimal, and the delivery of pathogenesis-related proteins (e.g. hydrolases and chitinases) and peroxidases to the interfacial apoplasts may occur initially, but is not maintained and is much less important than in parasitic interactions (Dumas Gaudot *et al.*, 1992a, b; Dumas *et al.*, 1989). In peas carrying a recessive allele (*myc⁻*) which results in failure of mycorrhizal development, the recognition step and formation of appressoria are not followed by tissue penetration; the plants appear to have regained their ability to restrict access to the fungi by synthesizing wall thickenings and phenolic compounds (Gollotte *et al.*, 1993). Gollotte *et al.* (1993) propose that full symbiotic development depends on the activation of 'sensor' genes in the plant by an elicitor from the fungus. They suggest that the sensor genes code for a specific binding site (receptor), which is absent in the Myc⁻ plants.

This idea can be extended, using information about diseases caused by *Alternaria alternata* on a range of host species (for review see Kohmoto and Otani, 1991). As with mycorrhizas, resistance to fungal development is determined by a recessive allele and it is probable that the prevalence of the disease is the result of horticultural selection of these genotypes in desirable varieties. Susceptibility is determined by suppression of the basic resistance mechanisms by host-specific toxins (HST) produced by particular pathotypes of the fungus. The HST molecules bind to receptors in the plant cells, often on the plasma membrane, and a cascade of events follows which permits fungal colonization of the tissues. Thus HST molecules modify the plant response and confer both compatibility and specificity on the fungus–plant interaction. Although specificity is linked with compatibility in the *Alternaria* symbioses, there is no *a priori* reason why this should be so. It may be that in mycorrhizas the fungus produces a signal molecule which binds to a receptor site on the plant membrane and hence switches off the defence mechanisms. Different groups of mycorrhizal fungi might have different signal–receptor systems, and plants that act as hosts to more than one type of mycorrhizal fungus might have multiple systems of signals and receptors. The limited levels of specificity observed in mycorrhizas could easily be built into the system by small changes in the design of signals and receptors. Non-mycorrhizal plants (e.g. brassicas and chenopods) could have evolved via loss of receptors.

It might be argued that drawing parallels between the necrotrophic *A. alternata* and biotrophic mycorrhizal fungi is inappropriate. However, here we are proposing a model by which the mechanisms underlying basic incompatibility may be switched off to permit fungal colonization of the tissues, and the nutritional mode of the fungus during colonization is irrelevant. Some of the HSTs do induce changes in the plant plasma membrane which are supposed to assist colonization by necrotrophs, such as invagination, depolarization and increased permeability.

However, the linkage between these effects and fungal invasion can be uncoupled, and, furthermore, the HSTs do not appear to inhibit H^+-ATPases (Khomoto *et al.*, 1989; Shimomura *et al.*, 1991).

The model for plant–fungus recognition (*Figure 3*) promoting the development of a compatible mycorrhizal interaction requires that the plant carry a dominant gene coding for a receptor which, if triggered, switches off the basic defence mechanisms. We have proposed that this is followed by modifications of plant membrane transport processes which switch on carbohydrate efflux and switch off carbohydrate retrieval. Similar changes in the fungus are also required to support bidirectional nutrient movements, and in both organisms the changes must be such that membrane integrity and capacity for active uptake are maintained. It is worth noting that specificity is not a prerequisite for biotrophy, as is sometimes implied in the literature on plant–pathogen interactions (e.g. Pryor, 1987). Race–cultivar specificity is superimposed on basic compatibility and has evolved where the compatible biotrophic interaction has highly negative results for the plant and in consequence there is a strong selection pressure against it. A second set of signals and receptors (the products of avirulence genes and resistance genes) may be involved here, as has been suggested by Gabriel *et al.* (1988). Their model proposes that channel proteins may double as specific receptors for signals from plant pathogens carrying

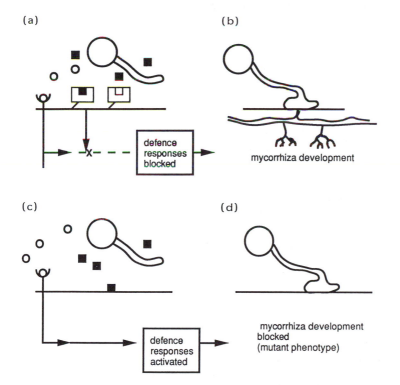

Figure 3. Hypothetical scheme for the operation of a signal–receptor system in the interactions between mycorrhizal fungi and host or mutant (Myc^-) plants. (a and b) Mycorrhizal interaction, receptors present (dominant condition) and defence responses blocked. (c and d) Non-mycorrhizal interaction, mycorrhizal receptors absent, defence responses activated. Developed from the suggestions of Gollotte *et al.* (1993).

⬚, mycorrhiza receptor; ■, mycorrhizal signal molecule;
◡, elicitor receptor; ○, elicitor.

avirulence genes, resulting in the opening of the channel(s) with consequent membrane leakage, depolarization and hypersensitive cell death. The receptors are envisaged as being products of dominant resistance genes, and the variation in the receptors (generated by random duplication, recombination and mutation) as providing a basis for the high degree of race–cultivar specificity observed in interactions between biotrophic pathogens and their hosts.

In compatible parasitic interactions, either the receptors would be absent (susceptible cultivars) or the fungus would not produce the signal (virulent parasites). A compatible interaction would develop, including the development of characteristic interfaces involved in unidirectional transport of nutrients across the interface from plant to fungus. As in mycorrhizas, this transport requires modification of the export and retrieval mechanisms of the cells. Interestingly, the extrahaustorial membrane (EHM) in the interface between pathogenic fungi and their hosts appears to be much more strongly modified than that in mycorrhizas (Smith and Smith, 1990b). The continuity between the PPM and the EHM can be masked by the development of a neckband, and the EHM reacts quite differently to cytochemical stains. The neckband also acts as an apoplastic seal between the interfacial apoplast and the general leaf apoplast, a situation contrasting with that in most mycorrhizal roots. Furthermore, in some associations, such as that between *Erisyphe pisi* and *Pisum*, the plant membrane lacks ATPase, which may indicate a lack of energization associated with 'leakiness' and polarized net flux of nutrients in favour of the fungus.

5. Summary and conclusions

The arrangement and function of membranes are of key importance in mycorrhizal symbioses, and much remains to be discovered about them, particularly with respect to modifications which permit controlled net export and conditions in the interfacial apoplast which may affect transport mechanisms.

We have discussed the role of the plasma membrane in a signal–receptor system permitting the extreme form of compatible fungus–plant interaction seen in mycorrhizal symbioses, because this is an important new area of research, linking membrane function to gene expression and the integrated development of two quite dissimilar organisms in symbiosis.

The inclusion of this chapter in a book on the functions of plant membranes is important. It means that it will be read alongside detailed treatments of a number of different aspects of membrane physiology. These will not only provide essential background for the discussions in this chapter, but should also help to point the way for ongoing investigations into membrane biology in mycorrhizas.

Acknowledgements

We would like to thank friends and colleagues who have helped us in discussions and who have allowed us to use unpublished results. Sandy Dickson deserves special thanks for her technical support. Our research is supported by the Australian Research Council (F.A.S. and S.E.S.) and The Cooperative Research Centre for Soil and Land Management (S.E.S.).

References

Alexander T, Meier R, Toth R, Weber HC. (1988) Dynamics of arbuscule development and degeneration in mycorrhizas of *Triticum aestivum* L. and *Avena sativa* L. with reference to *Zea mays* L. *New Phytol.* **110**: 363–370.

Alexander T, Toth R, Meier R, Weber HC. (1989) Dynamics of arbuscule development and degeneration in onion, bean and tomato with reference to vesicular–arbuscular mycorrhizae with grasses. *Can. J. Bot.* **67**: 2505–2513.

Allaway WG, Carpenter JL, Ashford AE. (1985) Amplification of the inter-symbiont surface by root epidermal transfer cells in the *Pisonia* mycorrhiza. *Protoplasma* **128:** 227–231.

Ashford AE, Peterson CA, Carpenter JL, Cairney JWG, Allaway WG. (1988) Structure and permeability of the fungal sheath in *Pisonia* mycorrhiza. *Protoplasma* **147:** 149–161.

Ashford AE, Allaway WG, Peterson CA, Cairney JWG. (1989) Nutrient transfer and the fungus–root interface. *Aust. J. Plant Physiol.* **16:** 85–97.

Bécard G, Piché Y. (1989) Fungal growth stimulation by CO_2 and root exudates in vesicular–arbuscular mycorrhizal symbiosis. *Appl. Environ. Microbiol.* **55:** 2320–2325.

Beever RE, Burns DJW. (1980) Phosphorus uptake, storage and utilisation by fungi. *Adv. Bot. Res.* **8:** 128–219.

Berbara RLL, Morris BM, Fonseca HMAC, Reid B, Gow NAR, Daft MJ. (1995) Electrical currents associated with arbuscular mycorrhizal interactions. *New Phytol.* **129:** 433–438.

Bonfante-Fasolo P, Grippiolo R. (1982) Ultrastructural and cytochemical changes in the wall of a vesicular–arbuscular mycorrhizal fungus during symbiosis. *Can. J. Bot.* **60:** 2303–2312.

Bonfante-Fasolo P, Berta G, Gianinazzi-Pearson V. (1981) Ultrastructural aspects of endomycorrhiza in the Ericaceae. II. Host endophyte relationships in *Vaccinium myrtillus. New Phytol.* **89:** 219–224.

Bonfante-Fasolo P, Vian B, Perotto S, Faccio A, Knox JP. (1990) Cellulose and pectin localization in roots of mycorrhizal *Allium porrum*: labelling continuity between host cell wall and interfacial material. *Planta* **180:** 537–547.

Bonfante-Fasolo P, Tamagnone L, Peretto R, Esquerre-Tugaye MT, Mazau D.Mosiniak M, Vian B. (1991) Immunocytochemical location of hydroxyproline-rich glycoproteins at the interface between a mycorrhizal fungus and its host plants. *Protoplasma* **165:** 127–138.

Bonfante-Fasolo P, Peretto R, Perotto S. (1992) Cell surface interactions in endomycorrhizal symbiosis. In: *Perspectives in Plant Cell Recognition* (Callow JA, Green JR, ed.). Cambridge: Cambridge University Press, pp. 239–255.

Cairney JWG, Smith SE. (1993a) Efflux of phosphate from the ectomycorrhizal basidiomycete *Pisolithus tinctorius*: general characteristics and the influence of intracellular phosphorus concentration. *Mycol. Res.* **97:** 1262–1266.

Cairney JWG, Smith SE. (1993b) The influence of monovalent cations on efflux of phosphate from the ectomycorrhizal basidiomycete *Pisolithus tinctorius. Mycol. Res.* **97:** 1267–1271.

Cairney JWG, Jennings DH, Ratcliffe RG, Southon TE. (1988) The physiology of basidiomycete linear organs. II. Phosphate uptake by rhizomorphs of *Armillaria mellea. New Phytol.* **109:** 327–333.

Cox G, Sanders FE. (1974) Ultrastructure of the host–fungus interface in a vesicular–arbuscular mycorrhiza. *New Phytol.* **73:** 901–912.

Cox G, Tinker PB. (1976) Translocation and transfer of nutrients in vesicular–arbuscular mycorrhizas. I. The arbuscule and phosphorus transfer: a quantitative ultrastructural study. *New Phytol.* **77:** 371–378.

Cress WA, Throneberry GO, Lindsey DL. (1979) Kinetics of phosphorus absorption by mycorrhizal and non-mycorrhizal tomato roots. *Plant Physiol.* **64:** 484–487.

Day DA, Price GD, Udvardi MK. (1989) The membrane interface of the *Bradyrhizobium japonicum–Glycine max* symbiosis: peribacteroid units from soybean nodules. *Aust. J. Plant Physiol.* **16:** 69–84.

Dexheimer J, Serrigny J. (1983) Étude ultrastructurale des endomycorhizes d'une orchidée tropicale: *Epidendrum ibaguense* H.B.K.I. Localisation des activité phosphatasiques acides et alkalines. *Bull. Soc. Bot. France* **130:** 187–194.

Dexheimer J, Gianinazzi S, Gianinazzi-Pearson V. (1979) Ultrastructural cytochemistry of the host–fungus interface in the endomycorrhizal association *Glomus mosseae/Allium cepa. Z. Pflanzenphysiol.* **92:** 191–206.

Dexheimer J, Marx C, Gianinazzi-Pearson V, Gianinazzi S. (1985) Ultracytological studies on plasmalemma formations produced by host and fungus in vesicular–arbuscular mycorrhizae. *Cytologia* **50:** 461–471.

Duddridge JA, Read DJ. (1984) Modification of the host–fungus interface in mycorrhizas synthesised between *Suillus bovinus* (Fr.) O. Kuntz and *Pinus sylvestris* L. *New Phytol.* **96:** 583–588.

Dumas E, Gianinazzi-Pearson V, Gianinazzi S. (1989) Production of new soluble proteins during VA endomycorrhiza formation. *Agric. Ecosyst. Environ.* **29:** 111–114.

Dumas-Gaudot E, Grenier J, Furlin V, Asselin A. (1992a) Chitinase, chitosanase and β-1,3 glucanase activities in *Allium* and *Pisum* roots colonized by *Glomus* species. *Plant Sci.* **84:** 17–24.

Dumas-Gaudot E, Furlan V, Grenier J, Asselin A. (1992b) New acidic chitinase isoforms induced in tobacco roots by vesicular–arbuscular mycorrhizal fungi. *Mycorrhiza* **1:** 133–136.

Fieschi M, Alloatti G, Sacco S, Berta G. (1992) Membrane potential hyperpolarisation in vesicular–arbuscular mycorrhizae of *Allium porrum* L.: a non-nutritional long-distance effect of the fungus. *Protoplasma* **168:** 136–140.

Gabriel DA, Loschke DC, Rolfe BG. (1988) Gene for gene recognition: the ion channel defense model. In: *Molecular Genetics of Plant–Microbe Interactions* (Palacios R, Varma DPS, ed.). St Paul, MN: PS Press, pp. 3–14.

Gianinazzi S, Gianinazzi-Pearson V, Dexheimer J. (1979) Enzymatic studies on the metabolism of vesicular–arbuscular mycorrhiza. III. Ultrastructural localisation of acid and alkaline phosphatase in onion roots infected with *Glomus mosseae* (Nicol. and Gerd.) Gerdemann and Trappe. *New Phytol.* **82:** 127–132.

Gianinazzi-Pearson V, Gianinazzi S. (1976) Studies on the metabolism of vesicular–arbuscular mycorrhiza. I. Effect of mycorrhiza formation and phosphorus nutrition on soluble phosphatase activities in onion roots. *Physiol. Vég.* **14:** 833–841.

Gianinazzi-Pearson V, Gianinazzi S. (1978) Enzymatic studies on the metabolism of vesicular–arbuscular mycorrhiza. II. Soluble alkaline phosphatase specific to mycorrhizal infection in onion roots. *Physiol. Plant Pathol.* **12:** 45–53.

Gianinazzi-Pearson V, Gianinazzi S, Brewin NJ. (1990) Immunocytochemical localisation of antigenic sites in the perisymbiotic membrane of vesicular–arbuscular endomycorrhiza using monoclonal antibodies reacting against the peribacteroid membrane of nodules In: *Endocytobiology IV* (Nardon P, Gianinazzi-Pearson V, Grenier AH, Margulis L, Smith DC, ed.). Paris: INRA Press, pp. 127–131.

Gianinazzi-Pearson V, Smith SE, Gianinazzi S, Smith FA. (1991) Enzymatic studies on the metabolism of vesicular–arbuscular mycorrhizas. V. Is H⁺ATPase a component of ATP-hydrolysing enzyme activities in plant–fungus interfaces? *New Phytol.* **117:** 61–74.

Gollotte A, Gianinazzi-Pearson V, Giovanetti M, Sbrana C, Avio L, Gianinazzi S. (1993) Cellular localisation and cytochemical probing of resistance reactions to arbuscular mycorrhizal fungi in a 'locus a' myc⁻ mutant of *Pisum sativum* L. *Planta* **119:** 112–122.

Hadley G, Johnson RPC, John DA. (1971) Fine structure of the host fungus interface in orchid mycorrhiza. *Planta* **100:** 191–199.

Hampp R, Chen X-Y, Stulten C. (1993) Sugar uptake by protoplasts of ectomycorrhizal fungi. In: *Abstracts of the SEB Symposium on Membrane Transport in Plants and Fungi: Molecular Mechanisms and Control.* Wye: Wye College, p 9.

Harley JL, McCready CC. (1952) The uptake of phosphate by excised mycorrhizal roots of the beech. IV. The effect of the fungal sheath on the availability of phosphate to the core. *New Phytol.* **51:** 342–348.

Harley JL, Smith SE. (1983) *Mycorrhizal Symbiosis.* London: Academic Press.

Harrison MJ, Dixon RA. (1994) Spatial patterns of expression of flavonoid/isoflavonoid pathway genes during interactions between roots of *Medicago truncatula* and the mycorrhizal fungus *Glomus versiform.* *Plant J.* **6:** 9–20.

Harrison MJ, Van Buuren ML. (1996) A phosphate transporter from the mycorrhizal fungus *Glomus versiforme. Nature* **373:** 626–629.

Hush J, Overall R. (1991) Electrical and mechanical fields orient cortical microtubules in higher plant tissues. *Cell Biol. Int. Rep.* **15:** 50–60.

Jeanmaire C, Dexheimer J, Marx C, Gianinazzi S, Gianinazzi-Pearson V. (1985) Effect of vesicular–arbuscular mycorrhizal infection on the distribution of neutral phosphatase activities in root cortical cells. *J. Plant Physiol.* **119:** 285–293.

Kinnbach A, Mellor RB, Werner D. (1987) Alpha mannosidase II isoenzyme in the peribacteroid space of *Glycine max* root nodules. *J. Exp. Bot.* **38:** 1373–1377.

Kohmoto K, Otani H. (1991) Host recognition by toxigenic plant pathogens. *Experientia* **47:** 755–764.

Kohmoto K, Otani H, Kodama M, Nishimura S. (1989) Host recognition: can accessibility to fungal invasion be induced by host specific toxins without necessitating necrotic cell death? In: *Phytotoxins and Plant Pathogenesis* (Graniti A, ed.). Nato ASI series Vol. H27. Berlin: Springer-Verlag, pp. 249–265.

Lei J. (1988) *Étude experimentale des systems symbiotique mycorhiziens de quelques essences ligneuses. Application pratique à la mycorhization de vitroplants.* PhD Thesis, University of Nancy, France.

Lei J, Dexheimer J. (1988) Ultrastructural localisation of ATPase activity in *Pinus sylvestris/Laccaria laccata* mycorrhizal association. *New Phytol.* **108:** 329–334.

Lei J, Bécard G, Catford JG, Piché Y. (1991) Root factors stimulate ³²P uptake and plasmalemma ATPase activity in vesicular–arbuscular mycorrhizal fungus, *Gigaspora margarita. New Phytol.* **118:** 289–294.

Lewis DH, Harley JL. (1965) Carbohydrate physiology of mycorrhizal roots of beech. III. Movement of sugars between host and fungus. *New Phytol.* **64:** 256–269.

Marx C, Dexheimer J, Gianinazzi-PearsonV, Gianinazzi S. (1982) Enzymatic studies on the metabolism of vesicular–arbuscular mycorrhiza. IV. Ultracytoenzymological evidence (ATPase) for active transfer processes in the host–arbuscular interface. *New Phytol.* **90**: 37–43.

Massicotte HB, Peterson RL, Ackerley CA, Piché Y. (1986) Structure and ontogeny of *Alnus crispa–Alpova diplophloeus* ectomycorrhizae. *Can. J. Bot.* **64**: 177–192.

Massicotte HB, Ackerley CA, Peterson RL. (1989) Ontogeny of *Alnus rubra–Alpova diplophloeus* ectomycorrhizae. II. Transmission electron microscopy. *Can. J. Bot.* **67**: 201–210.

Morris BM, Reid B, Gow NAR. (1992) Electrotaxis of zoospores of *Phytophthora palmivora* at physiologically relevant field strengths. *Plant Cell Environ.* **15**: 645–653.

Münzenberger B, Kottke I, Oberwinkler F. (1992) Ultrastructural investigations of *Arbutus unedo–Laccaria amethystea* mycorrhiza synthesised *in vitro*. *Trees* **7**: 40–47.

Murphy PJ, Karakousis A, Smith SE, Langridge P. (1995) Cloning functional endomycorrhiza genes: potential for use in plant breeding. In: *Biotechnology of Ectomycorrhizas* (Stocchi V, Bonfante P, Nuti M, ed.). New York: Plenum Press, pp. 77–83.

Nicolson TH. (1975) Evolution of vesicular–arbuscular mycorrhizas. In: *Endomycorrhizas* (Sanders FE, Mosse B, Tinker PB, ed.). London: Academic Press, pp. 25–34.

Orlovich DA, Ashford AE. (1993) Polyphosphate granules are an artefact of specimen preparation in the ectomycorrhizal fungus *Pisolithus tinctorius*. *Protoplasma* **173**: 91–102.

Patrick JW. (1989) Solute efflux to the apoplast at plant/microorganism interfaces. *Aust. J. Plant Physiol.* **16**: 53–67.

Pearson JN, Jakobsen I. (1993) The relative contribution of hyphae and roots to phosphorus uptake by arbuscular mycorrhizal plants measured by dual labelling with ^{32}P and ^{33}P. *New Phytol.* **124**: 489–494.

Peterson RL, Currah RS. (1990) Synthesis of mycorrhizae between protocorms of *Goodyera repens* (Orchidaceae) and *Ceratobasidium cereale*. *Can. J. Bot.* **68**: 1117–1125.

Pryor T. (1987) The origin and structure of fungal disease resistance genes in plants. *Trends Genetics* **3**: 157–161.

Raven JA, Smith SE, Smith FA. (1978) Ammonium assimilation and the role of mycorrhizas in climax communities in Scotland. *Trans. Bot. Soc. Edin.* **43**: 27–35.

Remy W, Taylor TN, Haas H, Kerp H. (1994) Four hundred-million-year-old vesicular–arbuscular mycorrhizae. *Proc. Natl Acad. Sci. USA* **91**: 11841–11843.

Ride JP. (1992) Recognition signals and initiation of host responses controlling basic incompatibility between fungi and plants. In: *Perspectives in Plant Cell Recognition* (Callow JA, Green JR, ed.). Cambridge: Cambridge University Press, pp. 213–237.

Salzer P, Hager A. (1993) Characterisation of wall-bound invertase isoforms of *Picea abies* cells and regulation by ectomycorrhizal fungi. *Physiol. Plant.* **88**: 52–59.

Scannerini S, Bonfante-Fasolo P. (1979) Ultrastructural cytochemical demonstration of polysaccharides and proteins within the host–arbuscule interfacial matrix in an endomycorrhiza. *New Phytol.* **83**: 87–94.

Serrigny J, Dexheimer J. (1985) Étude ultrastructurale des endomycorhizes d'une orchidée tropicale: *Epidendrum ibaguense*. II. Localisation des ATPases et des nucleosides diphosphatases. *Cytologia* **50**: 779–788.

Shachar-Hill Y, Pfeffer PE, Douds D, Osman SF, Doner LW, Ratcliffe RG. (1995) Partitioning of intermediary carbon metabolism in VAM colonised leek. *Plant Physiol.* **108**: 7–15.

Shepherd VA, Orlovich DA, Ashford AE. (1993) A dynamic continuum of pleiomorphic tubules and vacuoles in growing hyphae of a fungus. *J. Cell Sci.* **104**: 495–507.

Shimomura N, Otani H, Tabira H, Kodama M, Kohmoto K. (1991) Two primary action sites for AM toxin produced by *Alternaria alternata* apple pathotype and their pathological significance. *Ann. Phytopath. Soc. Jpn.* **57**: 247–255.

Simon L, Bousquet J, Levesque RC, Lalonde M. (1993) Origin and diversification of endomycorrhizal fungi and coincidence with vascular land plants. *Nature* **363**: 67–69.

Smith DC, Douglas AE. (1987) *The Biology of Symbiosis*. London: Edward Arnold.

Smith DC, Muscatine L, Lewis DH. (1969) Carbohydrate movement from autotrophs to heterotrophs in parasitic and mutualistic symbioses. *Biol. Rev.* **44**: 17–90.

Smith FA, Smith SE. (1986) Movement across membranes: physiology and biochemistry. In: *Physiological and Genetical Aspects of Mycorrhizae* (Gianinazzi-Pearson V, Gianinazzi S, ed.). Paris: INRA Press, pp. 75–84.

Smith FA, Smith SE. (1989) Membrane transport at the biotrophic interface: an overview. *Aust. J. Plant Physiol.* **16**: 33–43.

Smith FA, Smith SE. (1990a) Solute transport at the interface: ecological implications. *Agric. Ecosyst. Environ.* **28**: 475–478.

Smith FA, Smith SE. (1995) Nutrient transfer in vesicular–arbuscular mycorrhizas: a new model based on the distribution of ATPases on fungal and plant membranes. *Biotropia* **8:** 1–10.

Smith FA, Smith SE. (1996) Mutualism and parasitism: biodiversity in function and structure in the 'arbuscular' (VA) mycorrhizal symbiosis. *Adv. Bot. Res.* **22:** 1–43.

Smith FA, Dickson S, Morris C, Reid RJ, Tester M, Smith SE. (1995) Phosphate transfer in VA mycorrhizas — special mechanisms or not? In: *Structure and Function of Roots* (Baluska F, Ciamporova M, Gasparikova O, Barlow PW, ed.). Dordrecht: Kluwer Academic Publishers, pp. 155–161.

Smith SE. (1995) Discoveries, discussion and directions in mycorrhizal research. In: *Mycorrhiza: Function, Molecular Biology and Biotechnology* (Varma A, Hock B, ed.). Berlin: Springer-Verlag, pp. 5–24.

Smith SE, Dickson S. (1991) Quantification of active vesicular–arbuscular mycorrhizal infection using image analysis and other techniques. *Aust. J. Plant Physiol.* **18:** 637–648.

Smith SE, Smith FA. (1990b) Structure and function of the interfaces in biotrophic symbioses as they relate to nutrient transport. *New Phytol.* **114:** 1–38.

Smith SE, Long CM, Smith FA. (1989) Infection of roots with a dimorphic hypodermis: possible effects on solute uptake. *Agric. Ecosyst. Environ.* **29:** 403–407.

Smith SE, Dickson S, Morris C, Smith FA. (1994a) Transport of phosphate from fungus to plant in VA mycorrhizas: calculations of the area of symbiotic interface and of fluxes from two different fungi to *Allium porrum* L. *New Phytol.* **127:** 93–99.

Smith SE, Gianinazzi-Pearson V, Koide R, Cairney JWG. (1994b) Nutrient transport in mycorrhizas: structure, physiology and consequences for the symbiosis. *Plant Soil* **159:** 103–113.

Straker CJ, Mitchell DT. (1987) Kinetic characterisation of a dual phosphate uptake system in the endomycorrhizal fungus of *Erica hispidula. New Phytol.* **106:** 129–137.

Sukarno N, Smith FA, Smith SE, Scott ES. (1996) The effect of fungicides on vesicular–arbuscular mycorrhizal symbiosis. II. The effects on area of interface and efficiency of P uptake and transfer to plant. *New Phytol.* (in press).

Tester M, Smith FA, Smith SE. (1993) The role of ion channels in controlling solute exchange in mycorrhizal associations. In: *Mycorrhizas in Ecosystems* (Alexander IJ, Fitter AH, Lewis DH, Read DJ, ed.). London: CAB International, pp. 348–351.

Thomson BD, Clarkson DT, Brain P. (1990) Kinetics of phosphorus uptake by the germ-tubes of the vesicular–arbuscular mycorrhizal fungus, *Gigaspora margarita. New Phytol.* **116:** 647–653.

Chapter 31

The plant plasma membrane in fungal disease

J.A. Callow and J.R. Green

1. Introduction

The plasma membrane plays a leading role in providing a stable milieu within the cell through the regulated exchange of metabolites with the environment and the perception of a range of external stimuli that may herald the switch to new developmental pathways or signal particular responses, including those concerned with pathogenic stress. The two aims of this article are (i) to review the major functions performed by the plant plasma membrane and to consider how interactions with pathogens affect these, and (ii) to consider the contribution made by the plant plasma membrane to the formation and function of an effective interface between the host plant cell and the specialized infection structures used by biotrophic pathogens in the form of haustoria. Such interfaces are involved in nutrient transport and other forms of molecular communication, recognition and signalling, and the analysis of their structure and function presents a significant challenge to molecular plant pathologists.

2. Consequences of disease for specific membrane functions

There is now a fairly detailed understanding at the molecular level of the role of the plasma membrane in regulating disease in animal cells. For example, ion channels and transporters are essential for a wide range of cellular functions, and in the past few years there has been a dramatic expansion in the number and type of such membrane proteins that have been cloned and sequenced from animal cells, and in some cases disease has been associated with malfunctions in such proteins (Ashcroft and Roper, 1993).

In contrast, the level of understanding of perturbed plasma membrane functions in diseased plants is minimal. One problem is that the plant plasma membrane is an excitable structure with transporters and channels that are capable of responding to a wide range of specific and non-specific stimuli, the latter including surface charge, pH and mechanical stretching. Such non-specific stimuli may all be presented by fungal invasion, and it then becomes difficult to distinguish primary causes and effects from these less specific events.

In addition, whilst the basic physiological principles governing solute transfer across plant plasma membranes have been understood for some time, only recently has it become possible to describe such functions in terms of precisely characterized molecular entities, to clone the genes for such molecules, and to study their expression. Correspondingly, with such advances it has now become possible to explore the influence of pathogens on plant membrane functions in specific molecular terms. In general, we shall try to cite examples of perturbations that involve well-characterized membrane functions and which appear to be the consequence of direct effects of fungal pathogens on integral membrane molecules, but inevitably a degree of speculation is necessary in order to explore areas in which this increasing understanding might most profitably be applied.

The H+-ATPase

The H+-ATPase of the plant plasma membrane, consisting of a single 100-kDa polypeptide with eight transmembrane domains and a regulatory cytoplasmic C-terminal domain (Maathuis and Sanders, 1992), pumps protons from the cytoplasm to the apoplastic space, thus creating an electrochemical gradient across the plasma membrane, which is then used to drive uptake and extrusion of solutes by uniport, symport and antiport mechanisms. In addition, proton extrusion and the resulting acidification of the apoplast regulates hydrogen bonding of cell wall polymers and therefore the plasticity of the cell wall. Its activity is thus central to the nutrition and growth of the cell.

Given this pivotal role for H+-ATPase in the regulation of cell metabolism, it is not surprising that its activity is regulated by a number of growth regulators and other stimuli, including microbial metabolites. A more general survey of such effects has been published by Novacky (1983). It is also not surprising that H+-ATPase activity is implicated in both compatible and incompatible associations with pathogens. However, detailed understanding of the molecular mechanisms involved in either case is slight.

H+-ATPase in compatible associations. A classical effect of infection of plants by micro-organisms is enhanced host cell permeability with leakage of metabolites into the apoplast. The metabolites may then support microbial growth and multiplication and constitute a particularly significant source of assimilates for biotrophic pathogens with an intracellular growth habit. Such effects may be direct, involving a primary action of some fungal toxin, for example, specifically on the H+-ATPase, or they may be indirect, due to effects on ATP-generating systems or some secondary effect on other components of the plasma membrane.

The clearest, most rapid effect of a fungal toxin on ATPase activity is that caused by the host non-selective fungal toxin fusicoccin of *Fusicoccum amygdali*. This toxin elicits a range of physiological effects in higher plants, including stimulation of cell growth, seed germination, stomatal opening and ion uptake, through stimulation of the plasma membrane H+-ATPase (Marré, 1979). A high affinity 30- to 34-kDa receptor with an apparent dissociation constant in the nanomolar range has been identified from plasma membrane fractions (Meyer *et al.*, 1989), but the precise signal transduction pathway between this receptor and H+-ATPase has yet to be firmly established. Proteolytic removal of a 7- to 10-kDa fragment from the C-terminus of the H+-ATPase strongly activates both ATP hydrolysis and H+ pumping suggesting that this region constitutes an autoregulatory domain (Johansson *et al.*, 1993). It is possible that removal or displacement of this region by some as yet unknown mechanism forms the ultimate basis of fusicoccin action (and that of other agonists). The finding that a protein kinase inhibitor K-252a and fusicoccin induce similar initial changes in ion transport in parsley cells (Kauss *et al.*, 1992) opens up the possibility that dephosphorylation of this domain may be involved.

The plasma membrane H+-ATPase of plants is encoded by a multi-gene family, and there is evidence that individual isoforms are tissue-specific in their expression (Harper *et al.*, 1990). A fruitful line of investigation might be to explore the possibility that expression of specific isoforms may occur in infected tissues when source–sink relationships and translocation of assimilates are altered.

H+-ATPase in incompatible associations. Physiological changes which can be detected in plant cells undergoing the hypersensitive response (HR) to incompatible strains of bacterial or fungal pathogens, such as depolarization (Tomiyama *et al.*, 1983), electrolyte leakage, or failure of cells to deplasmolyse after plasmolysis (Woods *et al.,* 1988a) are generally consistent with a mechanism involving activation of plasma membrane H^+-ATPase, although other factors such as lipid peroxidation and active oxygen production should also be considered (see Section 3). More illuminating studies of the mechanism of HR have been conducted on bacterial systems. For example, a variety of ATPase and other metabolic inhibitors were used to demonstrate that activation of the plasma membrane H^+-ATPase is required for the rapid loss of intracellular K^+ and disruption of H^+ gradients that accompany host cell death in tobacco cell suspensions treated with incompatible races of *Pseudomonas syringae* pv. *pisi* (Atkinson and Baker, 1989).

The effects of resistance-inducing elicitors and suppressors on H^+-ATPase are discussed in more detail in a later section, but De Wit (1995) has pointed out an interesting difference between specific- and non-specific elicitors in terms of their respective effects on H^+-ATPase, the former causing extracellular acidification, and the latter alkalinization. This may suggest that similar defence responses induced by the two classes of elicitor may proceed through very different primary biochemical mechanisms.

Specific carriers and ion channels

Infection of plants, particularly by biotrophic pathogens, has long been known to modify source–sink relationships and the kinetics of sugar transport, with infected source leaves, for example, becoming sinks in the vicinity of lesions. These changes in carbon partitioning within the host must presumably be carefully regulated and are likely to involve some form of modulation of the expression and/or activity of the various carriers which mediate the movement of small solutes, such as ions or sugars, into and out of the cell, although another consideration must be the potential, in infected tissues, for increased leakage of solutes into apoplastic spaces following general increases in membrane permeability. Transport systems for sugars play an important role in facilitating the movement of assimilated carbon from source to sink, and there has been good progress recently in the molecular analysis of such channels with the cloning of several gene families of H^+/monosaccharide co-transporters (Sauer and Stadler, 1993) and their immunolocalization to the plasma membrane. One of the H^+/monosaccharide co-transporters of tobacco is sink-specific (Sauer and Stadler, 1993). Plasma membrane-localized sucrose–H^+ symporter proteins and encoding cDNAs have been studied in spinach (Riesmeier *et al.*, 1992), potato (Riesmeier *et al.*, 1993), *Arabidopsis* (Sauer and Stoltz, 1994) and *Plantago major* (Gahrtz *et al.,* 1994). Spinach and potato appear to have only one sucrose transporter (*SUC*) gene, but two genes have been detected in *Arabidopsis* and *P. major*, and expression studies in yeast indicate two classes of transporter. 'Acid' sucrose transporters, such as the PmSUC2 transporter of *P. major* (Stadler *et al.*, 1995), are present in all four species and are localized in the phloem, where they appear to be involved in sucrose loading from apoplast to phloem companion cells. In contrast, the 'neutral' sucrose transporter of *P. major* (PmSUC1) has a different expression pattern to that of PmSUC2, showing vascular expression in leaves, but inducible non-vascular expression in developing ovules and seeds (Gahrtz *et al.*, 1996). Given this rapid improvement in our understanding of the molecular basis of transport systems, a detailed study of the expression and localization of sugar transporter gene families in situations where infection alters the dynamics of carbon partitioning is likely to prove fruitful.

At the actual host–pathogen interface there is a net transfer of solutes from autotrophic host to heterotrophic pathogen, and the presumption is that this involves enhanced efflux of solutes across the host plasma membrane. As yet there is no direct evidence concerning the role of specific plasma membrane transporters or permeases responsible for this, and physiological mechanisms involved are also far from clear. As more transporter genes are cloned we are likely to see increased efforts devoted to exploring their expression at plant–fungal interfaces. A transmembrane phosphate transporter (GvPT) was recently cloned from the vesicular–arbuscular mycorrhizal fungus *Glomus versiforme* (Harrison and van Buuren, 1995). Whilst this particular gene appears to play a role in initial phosphate uptake from the soil, rather than in promoting phosphate transfer across plant–fungal interfaces, this demonstration is an important first step towards understanding solute transport in plant–fungal interactions in general. Solute transport at haustorial interfaces will be dealt with in more detail in Section 4.

Cell wall synthesis and repair

Although direct evidence for an appropriate enzyme activity is still lacking, there is a wealth of indirect evidence that plants synthesize cellulose via a cellulose synthase embedded within the plasma membrane as organized, transmembrane protein complexes (Delmer, 1990). In addition, another membrane glycosyltransferase, callose synthase, is responsible for synthesis of the polysaccharide callose, which is not a normal cell wall constituent, but is synthesized rapidly in response to mechanical wounding, stress and pathogenic infection. It is possible that there is in fact only one enzyme, which has alternating activities, switching between cellulose and callose synthesis when cells are perturbed (Delmer, 1990). The callose synthase activity is stimulated by Ca^{2+}, and there has therefore been considerable interest in exploration of the signalling responses associated with the activation of this enzyme during infection.

Rapid callose synthesis, often localized to papillae, is a common (but not invariable) resistant response to pathogenic invasion of plants. A wide range of treatments which lead directly or indirectly to increased permeability of the plasma membrane to Ca^{2+} (including toxins, saponins, certain elicitors such as chitosan or just general permeability changes caused by lipid bilayer degradation) appear to activate latent callose synthase. The reader is referred to a recent comprehensive review of this area (Kauss, 1990).

Redox functions

The plant plasma membrane contains redox components in the form of *b*-type cytochromes, cytochromes P-420 and P-450, cytochrome *c* reductase and non-covalently bound flavins and peroxidases (Moller and Crane, 1990). These components transfer electrons from electron donors such as NAD(P)H to a range of electron acceptors, including oxygen, although the physiological significance of most of the activities that can be demonstrated has yet to be clearly established. Of most interest to the pathologist are the redox functions terminating in oxygen as the electron acceptor that may lead to the so-called 'oxidative burst' associated with early defence responses, in which case the relevant immediate products are 'active oxygen (AO) species' which include the superoxide radical ($\bullet O_2^-$) and hydrogen peroxide formed from superoxide either by dismutation or by a non-enzymic reaction with ferrous ions. There is now a plethora of reports implicating AO species in defence (for recent reviews see Dixon and Lamb, 1995; Smith, 1996). For example, within a few hours of application of an incompatible race of *Phytophthora infestans*, or substances contained in germination fluids, to potato leaves, a burst of superoxide production was observed (Chai and Doke, 1989). At penetration there was a second burst of superoxide production, which was suggested to act as a trigger to other biochemical events associated with the HR (see Section 3). Legrendre *et al.* (1993) obtained an oxidative burst within 90 sec of applying a purified oligogalacturonide elicitor to cultured soybean cells. Tomato cells treated with race-specific elicitors from *Cladosporium fulvum* showed a rapid accumulation of AO species, accompanied by lipid peroxidation and increased lipoxygenase activity (Vera-Estrella *et al.*, 1992). Further aspects of both of these reports that are relevant to underlying signalling mechanisms are discussed in Section 2. It is interesting that defence responses to microbes in certain types of animal cell, such as leukocytes, also appear to involve inducible electron transfer systems in the plasma membrane, which accept electrons from cytosolic NADPH and transfer them to O_2 (Karnovsky, 1980).

It is generally assumed that, since superoxide is highly toxic, it must be generated outside the cell, presumably via one or more of the redox components referred to above. However, with the exception of often-reported increases in peroxidase activity, some of which may represent plasma membrane-bound forms of the enzyme (possibly in transit between the cytoplasm and the cell wall, where it engages in cell wall synthesis (Askerlund et al., 1987)), there have been relatively few clear studies of the impact of infection or elicitors on the activity and expression of transmembrane redox components. It was recently reported that enriched plasma membrane preparations from tomato cell lines showed a 200% increase in NADH oxidase activity and a 150% increase in NADH-dependent cytochrome c reductase activity when treated with a race-specific elicitor from Cladosporium fulvum (Vera-Estrella et al., 1994b).

Extracellular $\cdot O_2^-$ or H_2O_2 produced in plant defence responses may contribute to resistance in a number of ways:

(i)   by the participation of H_2O_2 in lignification responses mediated via apoplastic peroxidase;
(ii) by direct antibiotic activity of superoxide radicals or the even more potent hydroxyl radicals ($\cdot$OH);
(iii) by membrane damage and the generation of further signals. $\cdot$OH may attack many molecules, but the acyl groups of polyunsaturated fatty acids in membranes are particularly vulnerable. $\cdot$OH can initiate an exothermic chain reaction leading to peroxidation of many fatty acid molecules (Epperlein et al., 1986);
(iv) by signalling to the nucleus for defence gene activation (Lamb and Dixon, 1994).

The consequences of plant membrane lipid peroxidation can be expected to be wide-ranging. Specific attack and loss of the unsaturated fatty acid acyl groups will lead to decreased membrane fluidity. If this occurs in discrete regions of membrane, it will lead to the formation of 'islands' of crystalline-phase lipid within the normal gel phase at physiological temperatures, which could produce 'cracks' between the two lipid phases, giving a dramatic change in membrane permeability such as that involved in the HR (Keppler and Novacky, 1989). Furthermore, fatty acid hydroperoxides could transport Ca^{2+} across membranes, allowing Ca^{2+} influx from the cell wall and associated electrolyte leakage and depolarization, which in turn could activate the H^+-ATPase, leading to further proton efflux and associated K^+ transport. Additionally, the function of membrane proteins such as the various transporters could be affected directly by an altered lipid environment.

Recognition and transduction of pathogen signals

Plasma membrane receptors allow cells to sense and respond to a variety of external stimuli. In most cases this is achieved by the binding of specific ligands by the receptors, which induces conformational changes that in turn influence the activity of downstream effectors. It is beyond the scope of this chapter to look at the general evidence that plasma membranes of plant cells bear receptors involved in specific recognition phenomena (see, for example, Callow and Green, 1992; Gilroy and Trewavas, 1990). However, it is germane to consider the evidence for surface receptors that are involved in sensing the presence of pathogens, and the associated coupling mechanisms linking these receptors to downstream effectors.

Cloned resistance (recognition) genes. The 'elicitor-receptor' model of race/cultivar specificity postulates that the products of pathogen avirulence genes are race-specific elicitors which serve as ligands for surface-localized receptors encoded by dominant *R* (resistance) genes (Callow, 1986; De Wit, 1995; Gabriel and Rolfe, 1990). Binding to the R gene product activates signal transduction cascades, leading to a physiological defence response. The recent cloning of a number of plant *R* genes (for reviews see De Wit, 1995; Hammond-Kosack and Jones, 1995) has permitted at least a partial evaluation of this model. The current situation is that there are five cloned and characterized *R* genes (excluding the rather different case of the *Hml* gene for HC-toxin resistance in maize (Johal and Briggs, 1992)), one encoding resistance to virus, two to bacteria and two to fungal pathogens. For only one of these *R* genes has the complementary race-specific elicitor (i.e. the avirulence gene product) been completely characterized, enabling evaluation of the elicitor–receptor model. The *Cf9* gene, which confers resistance in tomato to appropriate races of *Cladosporium fulvum*, encodes a putative membrane-anchored extracyto-plasmic glycoprotein with 28 imperfect repeats of a conserved 24-amino-acid leucine-rich repeat (LRR) (Hammond-Kosack and Jones, 1995). The LRR domain is present in a wide range of proteins involved in specific protein–protein interactions, including receptor-like protein kinases and antifungal polygalacturonase-inhibiting proteins, but CF9 does not itself contain an obvious signalling domain. The receptor-like structure of the CF9 protein is consistent with direct binding to the fungal AVR9 elicitor polypeptide elicitor, but this remains to be demonstrated, and an unanticipated result is that ^{125}I-labelled AVR9 appears to bind equally well to plasma membranes of both Cf9 and the susceptible Cf0 tomato genotype (De Wit, 1995). Southern hybridization showed that *Cf9* is a member of a clustered gene family, and Northern analysis indicated that susceptible Cf0 lines expressed mRNAs homologous to those of Cf9 lines. It has therefore been suggested that CF9 and CF0 proteins may represent two-component ligand-binding proteins with the extracellular LRR domains binding the AVR9 peptide (De Wit, 1995), and that specificity must reside in the activation of the signal transduction cascade rather than in binding *per se*. As yet there is no clear evidence as to how binding of elicitor to any component of the CF9/CF0 system leads through into signal transduction.

Three other cloned resistance genes, *L6* for flax rust resistance (Lawrence *et al.*, 1994), *RPS2* for bacterial blight resistance in *Arabidopsis* (Bent *et al.*, 1994; Mindrinos *et al.*, 1994) and the *N* gene for mosaic virus resistance in tobacco (Whitham *et al.*, 1994), also contain LRRs, as well as nucleotide binding sites (P-loops) that are potentially involved in signal transduction. However, *L6* and *N* are cytoplasmic, and the situation for *RPS2* is uncertain. The remaining cloned *R* gene, the *Pto* gene for bacterial blight resistance in tomato, is a cytoplasmic serine–threonine protein kinase with a putative membrane anchor (Martin *et al.*, 1993).

Biochemical studies on elicitor receptors. From this survey of cloned *R* genes, it appears that while they all share features which suggest that they function in receptor/signal transduction mechanisms, support for the concept that specificity-conferring resistance genes might form part of a surface-located 'surveillance' system for plant pathogens and their elicitors is essentially confined to CF9. On the other hand, there has been a plethora of biochemical studies on both specific and non-specific receptors for elicitors. A wide range of fungal polysaccharides, oligo-saccharides and some peptides can elicit AO species, phytoalexins, callose or hydrolytic enzyme synthesis (for a recent review see Smith, 1996). It would be anticipated that such external signals, based on high-molecular-weight hydrophilic molecules with poor ability to penetrate plasma membranes, would be perceived and transduced by specific receptor molecules located within the plasma membrane, and much effort has been devoted to the biochemical characterization of the receptors for such pathogen recognition cues. High-affinity binding sites for several pathogen-derived glycan and protein elicitors have been identified on the plasma membranes of soybean, tomato, parsley and rice cells (Basse *et al.*, 1993; Cheong and Hahn, 1991; Cosio *et al.*,

1990; Ebel and Cosio, 1994; Langen *et al.*, 1993; Nurnberger *et al.*, 1994; Shibuya *et al.*, 1993). Elicitor receptors are dealt with in more detail in Chapter 7 by Hahn. The remainder of this section will consider aspects of the stimulus–response system.

A putative fungal avirulence gene product, NIP1, has been purified and sequenced from *Rhynchosporium secalis* (Gierlich *et al.*, 1993; Wevelsiep *et al.*, 1991). Genomic and cDNA clones of NIP1 reveal a 61-amino-acid mature protein (including 10 cysteine residues) and a 22-amino-acid leader sequence. Whilst there is no evidence that the product of the host resistance gene is a plasma membrane-located receptor for NIP1, it is relevant that NIP1 has an additional cultivar-non-specific toxic activity involving stimulation of the activity of H^+-ATPase of the plasma membrane (Wevelsiep *et al.*, 1993). At this stage, the mechanism of ATPase stimulation and the relationship between these specific and non-specific activities of NIP1 is unclear.

Aspects of the plasma membrane-associated signalling pathways involved in the rapid stimulation of synthesis of AO species in soybean cells exposed to a purified polygalacturonic acid (PGA) elicitor have been examined by Legrendre *et al.* (1993). They demonstrated a desensitization effect — a diminished response of cells following a second exposure to the same or a different stimulus. This is a common property of signal transduction pathways in animal cells, and is believed to protect the cells against deleterious effects of over-stimulation. Prior treatment of soybean cells with mastoparan (the G-protein activator peptide, which also elicits the oxidative burst), but not its inactive analogue, mas 17, was able to desensitize cells to subsequent exposure to the PGA elicitor. This may suggest that the mechanism of elicitation involves G-protein signalling, and that desensitization involves temporary inactivation of this pathway, rather than a mechanism based on competition for available elicitor receptors by the primary and secondary exposure to PGA.

Increased H^+-ATPase activity, NADH oxidase activity and NADPH-cytochrome *c* reductase activity were observed in enriched preparations of plasma membranes isolated from tomato cell suspension cultures containing the *Cf5* resistance gene, when treated with a crude preparation of the race-specific AVR5 elicitor of *Cladosporium fulvum* (Vera-Estrella *et al.*, 1992, 1994a, b). The increases in H^+-ATPase and NADH oxidase activity were abolished by okadaic acid, an inhibitor of protein phosphatases, suggesting a role for dephosphorylation in the activation of these enzymes. ^{32}P-labelling experiments revealed an elicitor-induced dephosphorylation of the 100-kDa H^+-ATPase polypeptide, and a role for G-proteins in activating these protein phosphatases has been claimed since GTP(γ)S, a non-hydrolysable GTP analogue, inhibited the increased H^+-ATPase activity (Vera-Estrella *et al.*, 1994a). It is known that the C-terminus of plasma membrane H^+-ATPase is subject to phosphorylation/dephosphorylation by membrane-bound protein kinases and phosphatases (Serrano, 1989), suggesting that this region may be part of a domain that serves to regulate pump activity. It was therefore proposed that the signal initiated by recognition of the elicitor by the hypothetical tomato membrane receptor is amplified and transduced by G-protein-associated phosphatases which dephosphorylate the H^+-ATPase (Vera-Estrella *et al.*, 1994a). Important requirements for substantiation of this hypothesis include the use of purified, defined elicitor fractions, and the demonstration of the hypothetical elicitor ligand-binding plasma membrane receptor.

Evidence from a number of host–pathogen systems supports the general idea that many compatible pathogens are able to suppress the hypersensitive reactions or other plant defence responses induced by non-specific elicitor signals. Both glucan and glycopeptide suppressors have been characterized, but the precise molecular mechanisms by which they act to inhibit the initial signal transduction events triggered by the elicitors are poorly understood. However, in at least one system it appears that the mechanism of suppression involves direct interactions of 'suppressor' molecules with the plasma membrane H^+-ATPase (Kato *et al.*, 1993; Shiraishi *et al.*,

1992). The pea pathogen *Mycosphaerella pinodes* produces a polysaccharide elicitor of the phytoalexin pisatin, and the mode of action of this elicitor appears to involve transmembrane signalling and rapid activation of enzymes of phosphoinositide metabolism (Toyoda *et al.*, 1993). Two glycopeptide suppressors, Supprescins A and B, have been characterized as GalNAc-Ser-Ser-Gly and Gal-GalNAc-Ser-Ser-Gly-Asp-Glu-Thr, respectively (Shiraishi *et al.*, 1992). Both inhibit the activation of the phosphoinositide-metabolizing enzymes and accumulation of pisatin in pea leaves, but only Supprescin B inhibits ATPase *in vivo* and *in vitro*. Orthovanadate, an inhibitor of P-type (plasma membrane) ATPases and Supprescin B, suppresses the expression of genes associated with phytoalexin accumulation, suggesting that the primary site of Supprescin B activity is the plasma membrane H^+-ATPase. The activity of Supprescin B appears to reside in the hexapeptide, and studies with synthetic oligopeptides have revealed two functional elements (Kato *et al.*, 1993). The Ser-Ser-Gly moiety inhibits ATPase of pea plasma membranes competitively, probably through the ATP-binding site, whilst the acidic Asp-Glu moiety interacts with the ATPase non-competitively and also inhibits acid phosphatase in pea membranes, suggesting that this moiety is interacting with the phosphatase domain of the ATPase. The evidence suggests that the plasma membrane ATPase plays an important role in the signal transduction cascades controlling the defence response, although the nature of this link remains to be determined.

Vesicle transport across plasma membranes

The transport of macromolecules to the plasma membrane or extracellular compartment takes place by means of vesicles. Although extensive studies have been made of exocytotic secretion pathways in model systems such as barley aleurone, root cap cells and germinating pollen grains, there have been no comparable studies of vesicle dynamics in relation to the secretion of products from plant cells in response to pathogen invasion, although *a priori* the secretion of hydrolases in resistant responses and the changes in plasma membrane synthesis which occur in plants infected by biotrophs (see Section 4) would seem to implicate exocytotic pathways.

Plant cells also possess endocytotic systems for the internalization of high-molecular-weight compounds, and one of the lines of evidence for this derives from the treatment of cells with elicitors. A crude protein elicitor from *Verticillium dahliae* and a PGA elicitor from citrus pectin bound to their receptors on soybean cell surfaces, eliciting a burst of H_2O_2 synthesis within 5 min, followed by synthesis of the phytoalexin, glyceollin (Horn *et al.*, 1989; Low *et al.*, 1993). Elicitors were internalized after application, being quantitatively delivered to the vacuole within 2 h in the case of PGA or 5 h for *Verticillium* elicitor. Internalization was temperature-sensitive and competitively inhibited by unlabelled ligand, both of which are characteristics of specific processes of receptor-mediated endocytosis in animal cells. The kinetics of internalization and defence response induction suggest that gradual sequestration of elicitor within cells is not involved in signal transduction, and it was postulated that the endocytosis of elicitors clears signal-generating molecules from the surface, delivering them to the plant vacuole for degradation. Internalization of ligands by receptor-mediated endocytosis is a common feature of cell-signalling systems once the signal has been transmitted, in order to clear the cell surface of unwanted ligands and to regulate the number of receptors on the plasma membrane.

3. Hypersensitive resistance: the role of substrates provided by the lipid bilayer

The most comprehensive view of the importance of oxidative and peroxidative processes in disease resistance is associated with the form of resistance known as the 'hypersensitive response' (HR). Models for cultivar/race specificity and the associated HR generally invoke specific recognition events at the host–pathogen interface involving products of the host resistance (*R*) genes and pathogen avirulence (*Av*) genes. Either directly, or indirectly through non-specific 'stress' responses, the recognition may result in the production of signals in the form of 'elicitors' which then trigger the cascade of reactions which comprise the response, including the generation of AO species, synthesis of phytoalexins and hydrolases, lignification, production of callose, etc. (Kombrink and Somssich, 1995; Lamb and Dixon, 1994; Smith, 1996).

The HR is also often associated with the induction of an increased resistance of the plant to a second challenge by the same or a different pathogen, and this 'induced or acquired resistance' may be systemic. The role of the HR in producing systemic effects in uninfected tissues is thought to involve the production of mobile chemical messengers. However, although most attention has been devoted to carbohydrate-based signals, or elicitors derived from cell wall degradation, and to compounds such as salicylic acid, some signals may be generated from within the host plasma membrane, and it is therefore relevant to consider these.

Lipoxygenases (LOX) are dioxygenases which catalyse the oxidation of polyunsaturated fatty acids (PUFAs) such as linolenic acid (18:3) and arachidonic acid (20:4), which are released from membrane lipids by lipolytic acyl hydrolases. Recently, a number of studies have shown that increased LOX activity and PUFAs are associated with plant responses to fungal pathogens or elicitors. The low basal level of LOX activity in tobacco cells was strongly but transiently increased in response to infection and elicitor preparations from *Phytophthora parasitica* var. *nicotiniae* (Fournier *et al.*, 1993). The tobacco LOX enzyme preferentially reacts with arachidonic acid, which in animal cells is the first step in the synthesis of leukotrienes, which are important mediators of various pathophysiological processes in animals (Parker, 1987). It has been proposed that the elicitor activity of exogenous arachidonic acid in plants is due to the action of LOX (Preisig and Kuć, 1987) and it is well known that lipid hydroperoxides and associated free radicals can exert deleterious effects on cell membranes, leading to cell death. The major substrates for LOX in plants are linoleic and linolenic acids, and the tobacco LOX enzyme reacts with these C18 PUFAs to produce 13-hydroperoxides and 9-hydroperoxides. The former can be converted to jasmonic acid or traumatic acid, which are structurally similar to animal eicosanoids that act as lipid mediators of physiopathological processes. Both compounds are implicated in defence and wound-related responses in plants; jasmonate in particular induces gene expression, and its role in signalling has been discussed previously (Staswick, 1992). Other volatile compounds with roles in defence may be produced via the lipoxygenase pathway (Croft *et al.*, 1993).

4. Structure, function and assembly of modified plasma membranes associated with haustoria

In interactions with biotrophic plant pathogens, the infected plant cells and tissues remain alive and active for extensive periods. A specialized host–parasite interface, the haustorium, is formed during infection of plants by rust, powdery mildew and downy mildew fungi. This structure is surrounded by an invagination of a modified form of the host plant plasma membrane, termed the extrahaustorial membrane (EHM). The EHM is separated from the fungal structures in the haustorium by a gel-like substance rich in carbohydrate, the extrahaustorial matrix. Together, the EHM and the extrahaustorial matrix keep the pathogen compartmentalized from the cytoplasm of the host. The EHM thus provides the primary interface for the exchange of nutrients and signals, and the elucidation of its functional properties in molecular terms is therefore crucial to an understanding of biotrophic parasitism as a mode of pathogenesis. Such understanding can be expected to generate novel opportunities for the control of these important pathogens through the manipulation of key features of this subtle interaction.

The complete structural unit, consisting of the haustorium, extrahaustorial matrix and EHM, is termed the haustorial complex (HC; *Figure 1*). For many haustoria, annular structures termed neckbands develop in the neck region. In pea powdery mildew, two such structures have been identified as the A- and B-neckbands. The A-neckband joins the inner face of the host cell wall, whereas the B-neckband is separate and nearer to the haustorial body, and the EHM of *Erysiphe pisi* is firmly attached to it (Bracker, 1968; Gil and Gay, 1977). Similar neckbands have been described for *E. graminis* and several rusts (Allen *et al.*, 1979; Heath, 1976; Spencer-Phillips and Gay, 1981). In most powdery mildews the haustoria provide the only nutritional interface with the plant; however, infection of host tissues by rust and downy mildew fungi leads to extensive development of intercellular hyphae which then form haustoria in host cells.

Of all the biotrophic interactions, the powdery mildew systems (and in particular the pea mildew interaction caused by *E. pisi*) are the most amenable to study, since they occur in epidermal cells which can be removed in strips from the plant, and intact haustoria surrounded by the EHM can be isolated (Gil and Gay, 1987; Mackie *et al.*, 1991). The EHM and extrahaustorial matrix are retained in these HCs, due to the sealing neckbands and the resistance of the EHM to both physical and biochemical treatments (see below); the haustorial cytoplasm is retained by a septal plug.

Several methods have also been developed for the isolation of rust haustoria, probably the most successful involving the use of the lectin concanavalin A in affinity chromatography (Hahn and Mendgen, 1992), but the isolated structures in general do not have an EHM. Similarly, haustoria of the downy mildew, *Peronospora viciae*, have been isolated by enzymic maceration,

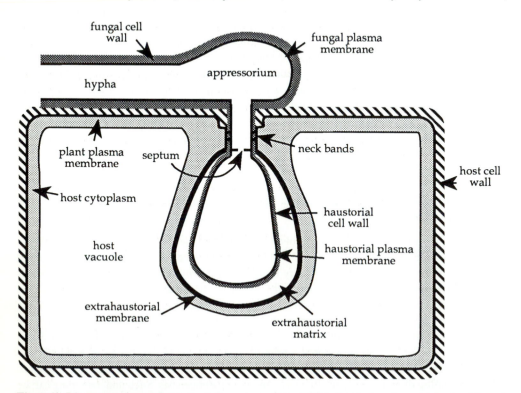

Figure 1. Diagrammatic representation of a pea powdery mildew haustorium in a pea host epidermal cell, illustrating the main structural features.

but these structures again lack surrounding membranes (Clarke and Spencer-Phillips, 1990). Such systems are therefore less useful for studies on the structure and transport properties of the invaginated host plasma membrane, and most of the available information about these membranes comes from ultrastructural and cytochemical studies of infected plant tissue.

Structure and composition of the EHM

The main methods used for examining the invaginated plasma membranes surrounding fungal infection structures have involved the use of electron microscopy (EM) (freeze-fracture and freeze-substitution) and probes for carbohydrate, polysaccharides and glycoproteins (lectins, antibodies) and enzymatic activity (e.g. ATPase). Since little is known about the detailed molecular composition of host plant plasma membranes, the findings have been fairly limited in scope. Several lines of evidence described below show that the EHM surrounding haustoria differs from the normal plant plasma membrane both structurally and functionally.

The EHM formed during infection of plants by rust, powdery mildew and downy mildew fungi differs in appearance from the normal host plant plasma membrane. In general, the EHM has a clear membrane bilayer, but is approximately 1.5–2.3 times thicker than the normal membrane (Bracker, 1968; Gil and Gay, 1977). Material which is carbohydrate in nature is present on both sides of the EHM, contributing to this extra thickness. The EHM is also often very convoluted, and this increases its effective surface area. Freeze-fracture has shown that the EHM formed in powdery mildews and uredinal-stage rust infections lacks the intramembranous particles (approximately 10 nm in diameter) characteristic of host plasma membranes (Bracker, 1968; Knauf et al., 1989; Mendgen et al., 1988). It has also been noted that, in rust infections, the host plasma membrane has fewer particles in infected cells than in non-infected cells. These particles probably represent integral membrane proteins, but although the freeze-fracture evidence suggests the absence of major membrane proteins in the EHM, protein has been detected cytochemically in the EHM of E. pisi using fluorescent reagents specific for protein or for constituent groups of proteins (Chard and Gay, 1984).

Lectin-binding studies have shown that the EHM of E. pisi contains sugars, including α-glucose, α-mannose, galactose and β-N-acetylglucosamine (Chard and Gay, 1984). In addition, the lectin concanavalin A (which recognizes glucose and mannose) bound to the cytoplasmic face of haustoria, and wheat germ agglutinin (which recognizes N-acetylglucosamine) bound to the haustorial face. However, although these studies indicated the occurrence of glycoconjugates on the EHM, their molecular nature has not been identified, and comparisons with host plasma membranes have not been made. Most of the extra thickness of the EHM in E. pisi can be removed by pectinase and cellulase treatments, and is therefore presumably composed of polysaccharide material (Bushnell and Gay, 1978). However, neither a lectin specific for galacturonic acid residues nor gold-labelled-cellobiohydrolase bound to the EHM of rose powdery mildew, suggesting the absence of pectins and cellulose (Hajlaoui et al., 1991). Yet other evidence suggests the presence of β,1–4-linked carbohydrate, since the EHM in E. pisi is stained by Tinopal BOPT and Calcofluor White (Chard and Gay, 1984) Since these histochemical treatments have been performed on different mildew systems, the overall results are difficult to compare, and the exact nature of the carbohydrate associated with the EHM remains uncertain.

The EHM of E. pisi is extremely resistant to some treatments which normally dissolve cell membranes, including detergent treatment or osmotic stress (Gil and Gay, 1977). The strength of the EHM allows the isolation of HCs, as they are not easily disrupted. Removal of the polysaccharide associated with the EHM using pectinase and cellulase renders the membrane much more permeable to water, and susceptible to osmotic or detergent lysis (Gil and Gay, 1977). Thus the associated carbohydrate would appear to give the EHM its characteristic strength and resistance to various chemical and mechanical stresses. Ca^{2+} has also been detected in the EHM of *E. pisi* by staining with the Ca^{2+}-specific compound chlorotetracycline, and has additionally been implicated in the stability of the EHM (Chard and Gay, 1989).

Development and application of monoclonal antibody probes to explore molecular differentiation in the EHM

None of the studies described above has yielded any detailed information on the molecular composition of the EHM surrounding haustoria. One problem is the small number of HCs that can be isolated from infected plant material for biochemical procedures such as protein purification and characterization. An alternative approach, pioneered in the authors' laboratory, is to produce monoclonal antibodies (mAbs) to HCs, since antibodies can be raised to previously unknown or undetected proteins or glycoproteins. The resultant mAbs have been used to identify components of the EHM formed in the pea powdery mildew system, and have been used to compare the EHM with the normal host plasma membrane (Callow *et al.*, 1992; Green *et al.*, 1992; Mackie *et al.*, 1991, 1993; Roberts *et al.*, 1993). After immunizing mice with isolated HCs from powdery mildew-infected pea leaves, mAb UB11 was obtained, which bound to the periphery of all isolated HCs as shown by indirect immunofluorescence, but did not bind to the host plasma membrane or any other plant or fungal structure (Roberts *et al.*, 1993). EM immunogold labelling suggested that it bound to the outer face of the EHM. UB11 recognized a carbohydrate epitope on a 250-kDa glycoprotein. The importance of these results is that they showed for the first time that there is a specific glycoprotein located in the EHM, and thus there is a clear difference between the EHM and the host plasma membrane in molecular terms.

The development of the HC in pea leaves can be observed using epidermal strips removed from the leaf, labelled with mAbs and other fluorescent probes (Chard and Gay, 1984; Mackie *et al.*, 1993; Roberts *et al.*, 1993). It has been possible to immunolabel the HC at different stages of development, and this system is particularly amenable to observations using confocal laser-scanning microscopy. Results with UB11 have shown that the specific glycoprotein recognized by this antibody is incorporated into the EHM at very early stages of HC development, suggesting that new gene expression for the differentiation of the EHM is switched on early in the plant–pathogen interaction. These results parallel the situation within the haustorium, since two mAbs (UB8 and UB10) that recognize glycoproteins (62 and 45 kDa, respectively) specific to the haustorial plasma membrane also label the HC at a very early stage in development (Mackie *et al.*, 1993).

Another mAb (UB9) was raised against enriched pea leaf plasma membrane fractions, and this antibody bound to the pea membrane and to the EHM of a proportion of HCs in infected leaves (Roberts *et al.*, 1993). UB9 binds to a carbohydrate epitope of a large glycoprotein ($M_r >$ 250 kDa). Immunofluorescence studies showed that UB9 did not bind to HCs until 5–7 days after infection, and at this stage it identified a sub-population of HCs, labelling up to approximately 20% of them. These results showed that at least one glycoprotein which is normally present in pea plasma membranes is *excluded* from the EHM at early stages of its development. The EHM of a sub-population of HCs is modified by the incorporation of this host plant glycoprotein at a late stage of infection. Thus the composition of the EHM in the pea powdery mildew interaction changes during development of the HC.

Localization of ATPase activity and the role of neckbands

Studies of the functional properties of the invaginated host plasma membranes surrounding fungal infection structures have mainly been concerned with the localization of ATPase activity.

The results have been correlated with the presence or absence of neckbands and incorporated into models for the mechanisms of nutrient transport from the host plant to the fungus (see below). Cytochemical detection of ATPase has mainly used ATP or β-glycerophosphate as substrate. However, the interpretation of these cytochemical studies should take into account the uncertainty that has been expressed concerning the validity of the stains for ATPase activity (Chauhan *et al.*, 1991). Observations of pea epidermal cells infected with powdery mildew showed that high levels of ATPase activity were present in the host plasma membranes, whereas there was no activity associated with the EHM, and the abrupt transition was found to occur at the point of invagination where the plant plasma membrane joins the A-neckband (Spencer-Phillips and Gay, 1981). Recent evidence from the authors' laboratory, using an antibody that recognizes the 100-kDa H$^+$-ATPase, suggests that the deficiency of ATPase activity in the EHM is due to the absence of the enzyme protein rather than to inactivation.

The distribution of ATPase activity and the occurrence of neckbands have been investigated in several different biotrophic interactions. The EHM formed by the dikaryon of *Uromyces appendiculatus* infecting *Phaseolus vulgaris* lacked ATPase activity in comparison with the host plasma membrane, and a neckband was observed to delimit these domains of ATPase activity (Spencer-Phillips and Gay, 1981). However, in the infection of *Tussilago farfara* by the monokaryon of *Puccinia poarum*, the haustoria were filamentous and lacked a visible structure resembling a neckband (Al-Khesraji and Losel, 1981). The EHM surrounding the distal part of the haustorium lacked ATPase activity, but activity was retained in the neck region. Unlike the situation in powdery mildews and rust dikaryons, a gradual transition from activity to inactivity occurred over a distance of 1.5 μm, instead of an abrupt change.

In comparison with powdery mildews and rusts, cytological studies of downy mildews have been limited. Observations of the haustorium formed by *Albugo candida* in *Cardamine hirsuta* showed that ATPase activity was absent at the host–haustorial interface (Woods and Gay, 1983). Although no structure resembling a neckband was seen, the wall and neck area were cytochemically distinct from the other wall regions and marked domains in the interfacial membranes. There was a close association of the fungal and host plasma membranes with the fungal wall in this region, and the results suggested the presence of a neckband equivalent. Ultrastructural studies of the downy mildew fungus *Bremia lactucae* infecting *Lactuca sativa* also failed to detect a distinct neckband. However, two physiologically distinct domains of the host plasma membrane in infected cells were identified (Woods *et al.*, 1988b), since the invaginated region of the host plasma membrane around the haustoria lacked ATPase activity. The transition in membrane characteristics was abrupt, suggesting the presence of an extrinsic stabilizing structure equivalent to a neckband.

Overall, the results suggest that the EHMs of powdery mildews, downy mildews and rust fungi lack the ATPase activity characteristic of the host plant plasma membrane. As mentioned earlier, the EHM lacks the 10 nm particles that are present in the normal plant plasma membranes. It has been suggested that these may be a transport complex involving the ATPase (Spencer-Phillips and Gay, 1981), or they could be involved in wall microfibril formation (e.g. cellulose synthetase (Bushnell and Gay, 1978)). Ultrastructural evidence indicates the absence of fibrillar material in the extrahaustorial matrix (Bushnell and Gay, 1978), and negative results with probes for cellulose, such as gold-labelled cellobiohydrolase (Hajlaoui *et al.*, 1991), suggest that the EHM lacks cellulose synthase activity.

The neckbands, or neckband equivalents, are clearly important for the maintenance of the EHM as a domain distinct from the host plasma membrane. Neckbands of *E. pisi* contain chitin, β-glucans, lipidic substances, protein, sulphydryl and phenolic groups (Stumpf and Gay, 1990). Lateral diffusion of membrane components between the two membrane domains of the EHM and the uninvaginated plant plasma membrane is probably prevented by these neckbands, which have been suggested to act as 'domain delimiters', being the equivalent of the tight junctions found in polarized animal cells (Manners and Gay, 1983). However, the results of the experiments on the interface of *Puccinia poarum* (Woods and Gay, 1987) appear to indicate that a neckband is not absolutely necessary for this type of domain separation and maintenance. It was suggested that the gradual transition of ATPase activity observed could be achieved by a decrease in membrane fluidity, causing slower diffusion of membrane proteins of all or part of the invaginated membrane, such as has been reported in senescing or damaged biological membranes (e.g. Houslay and Stanley, 1982).

Role of the invaginated plasma membrane in transport of solutes

As discussed above (see Section 2), almost nothing is known about the detailed molecular mechanisms by which solutes are transferred across host–pathogen interfaces, and any specific role attributed to the host plasma membrane in these processes is largely circumstantial. The general physiological approaches to characterizing transport systems which may be involved in biotrophic interactions have been outlined by Hall *et al.* (1992), and involve either observing the kinetics of labelling of the fungus following the application of radiolabelled solutes such as sugars to intact tissues or the application of solutes to isolated haustoria. The main drawback of the first approach is that the exogenous labelled solutes must first be taken up by the host cells that normally supply the pathogen structures, and the possibility of apoplastic transport directly to the fungus cannot be eliminated The drawback of the second approach is that isolated HCs may be damaged.

Sucrose and glycerol were reported to be either the main translocates or metabolic inter-mediates of this process in pea mildew (Manners and Gay, 1982, 1983). However, more recent reports measuring the uptake of fluorescent potentiometric dyes, in the presence of sugars, by haustoria of *E. graminis* in epidermal strips of barley showed that there was a more rapid increase in fluorescence in haustorial mitochondria in the presence of glucose than in the presence of sucrose, suggesting that glucose was the preferred translocate in these infections and that it was transferred to the haustorium (Mendgen and Nass, 1988). This is consistent with experiments reporting the direct uptake of glucose but not sucrose by isolated haustoria of *E. graminis* (Hall *et al.*, 1992). Mannitol appears to be the major primary sink metabolite of powdery mildews, and it is probable that its rapid formation in the haustorium creates the necessary concentration gradient for simple or facilitated diffusion of the putative translocate, glucose, across the EHM. Whether this diffusion is aided by specific channel proteins inserted into the EHM remains to be seen.

In any model of nutrient transfer mechanisms, the neckbands need to be considered. Isolated HCs of *E. pisi* are impermeable to uranyl ions (Gay and Manners, 1987), and this and other data have shown that the seal of the EHM to the haustorial neckbands acts as a permeability barrier between the apoplast of the plant and the extrahaustorial matrix. Thus the haustoria of these fungi are localized within an apoplastic compartment which is separate from the general leaf apoplast. In powdery mildew infections, nutrients must therefore enter the symplasm of the epidermal cell before uptake by the haustorium. The sealed extrahaustorial matrix compartment means that import into the fungus can be tightly coupled to the release of nutrients from the host.

A hypothesis incorporating the above concepts of solute transport through the plasma mem-branes of epidermal cells infected by haustoria of *E. pisi* (Spencer-Phillips and Gay, 1981) has since been applied to a range of biotrophic plant–pathogen interactions (Manners and Gay, 1982, 1983; Smith and Smith, 1990). The epidermal cells in which the haustoria are confined contain only a few undeveloped chloroplasts, so that photoassimilates must enter the epidermis before entering the haustoria. In pea plants there are relatively few plasmodesmatal connections between mesophyll and epidermal cells (Bushnell and Gay, 1978), and this suggests that the high levels of ATPase activity present in the wall-lining region of the plasma membrane are required in order to scavenge photosynthate from the apoplast by active transport. The lack of ATPase activity at the EHM causes depolarization of this domain, and results in net efflux of solutes into the extrahaustorial matrix by either passive or facilitated diffusion.

Further evidence for the ATPase-deficiency coupled transport hypothesis was provided by the demonstration that the uptake of cationic cyanine potentiometric dyes by haustoria of *E. graminis* in isolated epidermal strips from barley coleoptiles was sensitive to depolarizing agents, suggesting that an ionic gradient exists between the host and the fungus (Bushnell *et al.*, 1987). Gay *et al.* (1987), using the same system, showed that the non-invaginated region of the host plasma membrane is responsible for the ionic gradient established during haustorial infection. Their results suggest that the non-invaginated region of the host plasma membrane provides the proton gradient for subsequent movement of solutes across the host–pathogen interface.

Assembly of the EHM

Overall, the observed structure and activities of the EHM formed in biotrophic interactions have shown that this membrane differs from the invaginated host plant plasma membrane in a number of respects. Some proteins or glycoproteins that are normally present in the plant plasma membrane are absent from the EHM (e.g. the glycoprotein recognized by mAb UB9, and particles observed by freeze-fracture) and, in the case of *E. pisi*, the EHM contains a specific glycoprotein component, so there must be a drastic change in host targeting of membrane components if the EHM is indeed of host origin. The formation of a functional HC, and in particular the EHM, must require some co-operation between the plant and fungal partners, even though it is the fungus which benefits from the interaction. However, the cellular mechanisms involved in the development and differentiation of the EHM are still a mystery. Since the EHM increases in size with haustorial development, new membrane material must be produced and incorporated either with new glycoproteins already formed or with novel modifications occurring after incorporation. It is possible that the plant host provides this membrane material, possibly under fungal direction, in a similar way to that in which bacteroids can influence the plant to change the protein composition of the peribacteroid membrane in the *Rhizobium*–legume symbiosis (Werner *et al.*, 1988). It is also conceivable that this new membrane material or specific glycoproteins (e.g. the glycoprotein recognized by the mAb UB11) could originate from the invading fungus with a mechanism for incorporation into the EHM similar to the biogenesis of Gram-negative bacterial outer membranes (Bayer *et al.*, 1979; Davis and Tai, 1980). Studies in the authors' laboratory are attempting to clarify these aspects for pea powdery mildew. This requires determination of the identity of novel proteins or glycoproteins inserted into the EHM or haustorial plasma membrane, as detected by the mAb probes. The preferred route to this, bearing in mind the limitations of carrying out conventional biochemical characterizations with the very small amounts of material involved, is through the corresponding gene sequences isolated by immunoscreening cDNA expression libraries.

There is clearly a polarization of activity of the epidermal cell during development of the HC and the EHM, and there is some evidence for alterations in host cellular organization. For example, there is often an extensive host endoplasmic reticulum (ER) network adjacent to haustoria; in pea downy mildew, smooth ER with flattened cisternae appears to be associated around the EHM (Hickey and Coffey, 1977) and in the cereal rust caused by *Puccinia coronata*, the EHM is connected to the host ER (Harder and Chong, 1984). Recent studies of the pea powdery mildew interaction have used an ER-specific probe (anti-HDEL mAb) in conjunction with confocal microscopy to localize the ER in infected epidermal cells (Leckie *et al.*, 1995). The anti-HDEL mAb recognizes a tetrapeptide at the C-terminus of proteins retained in the ER (Napier *et al.*, 1992). In uninfected pea epidermal cells, the ER was an open network evenly distributed throughout the cytoplasm, and it showed features typical of the cortical ER. However, in infected cells, the ER became concentrated in a dense network in the vicinity of the HC, particularly during the early stages of infection, and there was often intense ER staining very close to the EHM (Leckie *et al.*, 1995).

It has been suggested that the ER may be involved in the modification of the invaginated region, and that constituents of the normal plant plasma membrane may be incorporated into the EHM via association with the ER (Woods and Gay, 1987). Vesicles are also often profuse in the area around the EHM, although they may be involved in endocytosis or exocytosis (Manners and Gay, 1983). A future, more comprehensive understanding of these intracellular transport events is likely to involve studies of the underlying molecular controls such as the involvement of low-molecular-weight GTP-binding proteins, for example, the *rab* and *ras* gene families.

References

Al-Khesraji TO, Losel DM. (1981) The fine structure of haustoria, intracellular hyphae and intercellular hyphae of *Puccinia poarum*. *Physiol. Plant Path.* **19:** 301–311.

Allen FHE, Coffey MD, Heath MC. (1979) Plasmolysis of rusted flax: a fine structural study of the host–pathogen interface. *Can. J. Bot.* **57:** 1528–1533.

Ashcroft FM, Roper J. (1993) Transporters, channels and human disease. *Curr. Opin. Cell Biol.* **5:** 677–683.

Askerlund P, Larsson C, Widell S, Moller IM. (1987) NAD(P)H oxidase and peroxidase activities in purified plasma membranes from cauliflower inflorescences. *Physiol. Plant.* **71:** 9–19.

Atkinson MM, Baker CJ. (1989) Role of the plasmalemma H^+-ATPase in *Pseudomonas syringae*-induced K^+/H^+ exchange in suspension-cultured tobacco cells. *Plant Physiol.* **91:** 298–303.

Basse CW, Fath A, Boller T. (1993) High affinity binding of a glycopeptide elicitor to tomato cells and microsomal membranes and displacement by specific glycan suppressors. *J. Biol. Chem.* **268:** 14724–14731.

Bayer ME, Thurow H, Bayer MH. (1979) Penetration of the polysaccharide capsule of *E. coli* by bacteriophage K29. *Virology* **94:** 95–118.

Bent AF, Kunkel BN, Dahlbeck D, Brown KL, Schmidt R, Giraudat J, Leung J, Staskawicz BJ. (1994) *RPS2* of *Arabidopsis thaliana*: a leucine-rich repeat class of plant disease resistance genes. *Science:* **265:** 1856–1860.

Bracker CE. (1968) Ultrastructure of the haustorial apparatus of *Erysiphe graminis* and its relationship to the epidermal cell of barley. *Phytopathology* **58:** 12–30.

Bushnell WR, Gay JL. (1978) Accumulation of solutes in relation to the structure and function of haustoria in powdery mildews. In: *The Powdery Mildews* (Spencer DM, ed). London: Academic Press, pp. 183–235.

Bushnell WR, Mendgen K, Liu Z. (1987) Accumulation of potentiometric and other dyes in haustoria of *Erysiphe graminis* in living cells. *Physiol. Mol. Plant Path.* **31:** 237–250.

Callow JA. (1986) Models for host–pathogen interaction. In: *Genetics and Plant Pathogenesis* (Day PR, Jellis GJ, ed.). Oxford: Blackwell Science Ltd, pp. 274–286.

Callow JA, Green JR (ed.). (1992) *Perspectives in Plant Cell Recognition*. SEB Seminar Series 48. Cambridge: Cambridge University Press.

Callow JA, Mackie A, Roberts AM, Green JR. (1992) Evidence for molecular differentiation in powdery mildew haustoria through the use of monoclonal antibodies. *Symbiosis* **14:** 237–246.

Chai HB, Doke N. (1989) Superoxide anion generation — a response of potato leaves to infection with *Phytophthora infestans*. *Phytopathology* **77:** 645–649.

Chard JM, Gay JL. (1984) Characterisation of the parasitic interface between *Erysiphe pisi* and *Pisum sativum* using fluorescent probes. *Physiol. Plant Path.* **25:** 259–276.

Chauhan E, Cowan DS, Hall JL. (1991) Cytochemical localization of plasma membrane ATPase activity in plant cells. *Protoplasma* **165:** 27–36.

Cheong J-J, Hahn MG. (1991) A specific, high affinity binding site for the hepta-β-glucoside elicitor exists in soybean membranes. *Plant Cell* **3:** 137–147.

Clarke JSC, Spencer-Phillips PTN. (1990). Isolation of endophytic mycelia by enzymic maceration of *Peronospora*-infected leaves. *Mycol. Res.* **94:** 283–287.

Cosio EG, Frey T, Verduyn R, van Boom J, Ebel J. (1990) High affinity binding of a synthetic heptaglucoside and fungal glucan elicitors to soybean membranes. *FEBS Lett.* **271:** 223–226.

Croft KPC, Jutner F, Slusarenko AR. (1993) Volatile products of the lipoxygenase pathway evolved from *Phaseolus vulgaris* (L). leaves, inoculated with *Pseudomonas syringae* pv. *phaseolicola. Plant Physiol.* **101:** 13–24.

Davis BD, Tai PC. (1980) The mechanism of protein secretion across membranes. *Nature* **283:** 433–438.

Delmer DP. (1990) The role of the plasma membrane in cellulose synthesis. In: *The Plant Plasma Membrane* (Larsson C, Moller IM, ed.) Berlin: Springer-Verlag, pp. 256–268.

De Wit PJGM. (1995) Fungal avirulence genes and plant resistance genes: unravelling the molecular basis of gene-for-gene interactions. *Adv. Bot. Res.* **21:** 148–177.

Dixon RA, Lamb CJ. (1995) Function of the oxidative burst in hypersensitive disease resistance. *Proc. Natl Acad. Sci. USA* **92:** 4158–4163.

Ebel J, Cosio EG. (1994) Elicitors of plant defense responses. *Int. Rev. Cytol.* **148:** 1–36.

Epperlein MM, Noronha-Duta AA, Strange RN. (1986) Involvement of the hydroxyl radical in the abiotic elicitation of phytoalexin in legumes. *Physiol. Mol. Plant Pathol.* **28:** 67–77.

Fournier J, Pouenat M-L, Rickauer M, Rabinovitch-Chable H, Rigaud M, Esquerre-Tugaye M-T. (1993) Purification and characterisation of elicitor-induced lipoxygenase in tobacco cells. *Plant J.* **3:** 63–70.

Gabriel DW, Rolfe BG. (1990) Working models of specific recognition in plant–microbe interactions. *Annu. Rev. Phytopathol.* **28:** 365–391.

Gahrtz M, Stolz J, Sauer N. (1994) A phloem-specific sucrose-H^+ symporter from *Plantago major* L. supports the model of apoplastic phloem loading. *Plant J.* **6:** 697–706.

Gahrtz M, Schmelzer E, Stolz J, Sauer N. (1996) Expression of the *PmSUC1* sucrose carrier from *Plantago major* L. is induced during seed development. *Plant J.* **9:** 93–100.

Gay JL, Manners JM. (1987) Permeability of the host–haustorium interface in powdery mildews. *Physiol. Mol. Plant Path.* **30:** 389–399.

Gay JL, Salzberg A, Woods AM. (1987) Dynamic experimental evidence for the plasma membrane ATPase domain hypothesis of haustorial transport and for ionic coupling of the haustorium of *Erysiphe graminis* to the host cell (*Hordeum vulgare*). *New Phytol.* **107:** 541–548.

Gierlich A, Hermann H, Rohe M, Knogge W. (1993) Molecular characterization of a putative avirulence gene and its product from *Rhynchosporium secalis*. In: *Proceedings of the Sixth International Congress on Plant Pathology*, p. 185 (abstract).

Gil F, Gay JL. (1977) Ultrastructural and physiological properties of the host interfacial components of haustoria of *Erysiphe pisi in vitro* and *in vivo*. *Physiol. Plant Path.* **10:** 1–12.

Gilroy S, Trewavas A. (1990) Signal sensing and signal transduction across the plasma membrane. In: *The Plant Plasma Membrane* (Larsson C, Moller IM, ed.). Berlin: Springer-Verlag, pp. 203–232.

Green JR, Mackie AJ, Roberts AM, Callow JA. (1992) Molecular differentiation and development of the host–parasite interface in powdery mildew of pea. In: *Perspectives in Plant Cell Recognition* (Callow JA, Green JR, ed.). SEB Seminar Series 48. Cambridge: Cambridge University Press, pp. 193–212.

Hahn M, Mendgen K. (1992) Isolation by ConA binding of haustoria from different rust fungi and comparison of their surface qualities. *Protoplasma* **170:** 95–103.

Hajlaoui MR, Benhamou N, Belanger RR. (1991) Cytochemical aspects of fungal penetration, haustorium formation and interfacial material in rose leaves infected by *Sphaerotheca pannosa* var. *rosae*. *Physiol. Mol. Plant Path.* **39:** 341–355.

Hall JL, Aked J, Gregory AJ, Storr T. (1992) Carbon metabolism and transport in a biotrophic fungal association. In: *Carbon Partitioning Within and Between Organisms* (Pollock CJ, Farrar JF, Gordon AJ, ed.). Oxford: BIOS Scientific Publishers, pp. 181–198.

Hammond-Kosack KE, Jones JDG. (1995) Plant disease resistance genes — unravelling how they work. *Can. J. Bot.* **73** (Suppl. 1): S495–S505.

Harder DE, Chong J. (1984) Structure and physiology of haustoria. In: *The Cereal Rusts, Vol. I* (Bushnell WR, Roelfs AP, ed.). New York: Academic Press, pp. 431–476.

Harper JF, Manney L, DeWitt ND, Yoo MH, Sussman MR. (1990) The *Arabidopsis thaliana* plasma membrane H^+-ATPase multigene family. *J. Biol. Chem.* **265:** 13601–13608.

Harrison MJ, van Buuren ML. (1995) A phosphate transporter from the mycorrhizal fungus *Glomus versiforme*. *Nature* **378:** 626–629.

Heath MC. (1976) Ultrastructural and functional similarity of the haustorial neckband of rust fungi and the Casparian strip of vascular plants. *Can. J. Bot.* **54:** 2484–2489.

Hickey EL, Coffey MD. (1977) A fine-structural study of the pea downy mildew fungus *Peronospora pisi* in its host *Pisum sativum*. *Can. J. Bot.* **56:** 2845–2858.

Horn MA, Heinstein PF, Low PS. (1989) Receptor-mediated endocytosis in plant cells. *Plant Cell* **1:** 1003–1009.

Houslay MD, Stanley KK. (1982) *Dynamics of Biological Membranes. Influence on Synthesis, Structure and Function.* Chichester: J. Wiley & Sons.

Johal GS, Briggs SP. (1992) Reductase activity encoded by the *HM1* disease resistance gene in maize. *Science* **258:** 985–987.

Johansson F, Sommarin M, Larson C. (1993) Fusicoccin activates the plasma membrane H^+-ATPase by a mechanism involving the C-terminal inhibitory domain. *Plant Cell* **5:** 321–327.

Karnovsky ML. (1980) Active O_2 species and the function of phagocytic leucocytes. *Annu. Rev. Biochem.* **49:** 695–726.

Kato T, Shiraishi T, Toyoda K, Saitoh K, Saitoh Y, Tahara M, Yamada T, Oku H. (1993) Inhibition of ATPase activity in pea plasma membranes by fungal suppressors from *Mycosphaerella pinodes* and their peptide moieties. *Plant Cell. Physiol.* **34:** 439–445.

Kauss H. (1990) Role of the plasma membrane in host–pathogen interactions. In: *The Plant Plasma Membrane* (Larsson C, Moller IM, ed.). Berlin: Springer-Verlag, pp. 320–350.

Kauss H, Jeblick W, Conrath U. (1992) Protein kinase inhibitor K-252a and fusicoccin induce similar initial changes in ion transport of parsley suspension cells. *Physiol. Plant.* **85:** 483–488.

Keppler LD, Novacky A. (1989) Changes in cucumber cotyledon membrane lipid fatty acids during paraquat treatment and a bacteria-induced hypersensitive response. *Phytopathology* **79:** 705–708.

Knauf GM, Welter K, Muller M, Mendgen K. (1989) The haustorial host–parasite interface in rust infected bean leaves after high pressure freezing. *Physiol. Mol. Plant Path.* **34:** 519–530.

Kombrink E, Somssich IE. (1995) Defense responses of plants to pathogens. *Adv. Bot. Res.* **21:** 1–34.

Lamb CJ, Dixon RA. (1994) Molecular mechanisms underlying induction of plant defence gene transcription. *Biochem. Soc. Symp.* **60:** 241–248.

Langen G, Dammers W, Romme Y, Kogel KH. (1993) Functional analysis of Pgt elicitor binding sites in wheat plasma membranes. In: *Proceedings of the Sixth International Plant Pathology Congress*, p. 194 (abstract).

Lawrence GJ, Ellis JG, Finnegan EJ. (1994) Cloning a rust resistance gene in flax. In: *Advances in Molecular Genetics of Plant–Microbe Interactions* (Daniels MJ, ed.). Current Plant Science and Biotechnology in Agriculture, Vol. 3. Dordrecht: Kluwer Academic Publishers, pp. 303–306.

Leckie CP, Callow JA, Green JR. (1995) Reorganisation of the endoplasmic reticulum in pea leaf epidermal cells infected by the powdery mildew fungus *Erysiphe pisi*. *New Phytol.* **131:** 211–221.

Legrendre L, Rueter S, Heinstein PF, Low PS. (1993) Characterisation of the oligogalacturonide-induced oxidative burst in cultured soybean (*Glycine max*) cells. *Plant Physiol.* **102:** 233–240.

Low PS, Legrendre L, Heinstein PF, Horn MA. (1993) Comparison of elicitor and vitamin-receptor mediated endocytosis in cultured soybean cells. *J. Exp. Bot.* **44:** 269–274.

Maathuis FJM, Sanders D. (1992) Plant membrane transport. *Curr. Opin. Cell Biol.* **4:** 661–669.

Mackie AJ, Roberts AM, Callow JA, Green JR. (1991) Molecular differentiation in pea powdery-mildew haustoria. Identification of a 62-kDa N-linked glycoprotein unique to the haustorial plasma membrane. *Planta* **183:** 399–408.

Mackie AJ, Roberts AM, Green JR, Callow JA. (1993) Glycoproteins recognised by monoclonal antibodies UB7, UB8 and UB10 are expressed early in the development of pea powdery mildew haustoria. *Physiol. Mol. Plant Pathol.* **43:** 135–146.

Manners JM, Gay JL. (1982) Transport, translocation and metabolism of ^{14}C-photosynthates at the host–parasite interface of *Pisum sativum* and *Erysiphe pisi*. *New Phytol.* **91:** 221–244.

Manners JM, Gay JL. (1983) The host–parasite interface and nutrient transfer in biotrophic parasitism. In: *Biochemical Plant Pathology* (Callow JA, ed.). Chichester: J. Wiley & Sons Ltd, pp. 163–195.

Marré. (1979) Fusicoccin: a tool in plant physiology. *Annu. Rev. Plant Physiol.* **30:** 273–288.

Martin GR, Brommonschenkel SH, Chunwongse J, Frary A, Ganal MW, Spivey R, Wu T, Earle ED, Tanksley SD. (1993) Map-based cloning of a protein kinase gene conferring disease resistance in tomato. *Science* **262:** 1432–1436.

Mendgen K, Nass P. (1988) The activity of powdery mildew haustoria after feeding the host cells with different sugars, as measured with a potentiometric cyanine dye. *Planta* **174:** 283–288.

Mendgen K, Schneider A, Sterk M, Fink W. (1988) The differentiation of infection structures as a result of recognition events between some biotrophic parasites and their hosts. *J. Phytopathol.* **123:** 259–272.

Meyer C, Feyerbrand M, Weiler EW. (1989) Fusicoccin binding protein in *Arabidopsis thaliana* (L.). Heynb. Characterisation, solubilisation and photoaffinity labelling. *Plant Physiol.* **89**: 692–699.

Mindrinos M, Katagiri F, Yu GL, Ausubel FM. (1994) The *A. thaliana* disease resistance gene *RPS2* encodes a protein containing a nucleotide-binding site and lecuine-rich repeats. *Cell* **78**: 1089–1099.

Möller IM, Crane FL. (1990) Redox processes in the plasma membrane. In: *The Plant Plasma Membrane* (Larsson C, Motter IM, ed.). Berlin: Springer-Verlag, pp. 93–126.

Napier RM, Fowke LC, Hawes C, Lewis M, Pelham HRB. (1992) Immunological evidence that plants use both HDEL and KDEL for targeting proteins to the endoplasmic reticulum. *J. Cell Sci.* **102**: 261–271.

Novacky A. (1983) The effects of disease on the structure and activity of membranes. In: *Biochemical Plant Pathology* (Callow JA, ed.). Chichester: J. Wiley & Sons Ltd, pp. 347–366.

Nurnberger T, Nennstiel D, Jabs T, Sacks WR, Hahlbrock K, Scheel D. (1994) High affinity binding of a fungal oligopeptide elicitor to parsley plasma membranes triggers multiple defense responses. *Cell* **378**: 449–460.

Parker CW. (1987) Lipid mediators produced through the lipozygenase pathway. *Annu. Rev. Immunol.* **5**: 65–84.

Preisig CL, Kuć JA. (1987) Inhibition by salicylhydroxamic acid, BW755C, eicosatetranoic acid and disulfiram of hypersensitive resistance elicited by arachidonic acid or poly-L-lysine in potato tuber. *Plant Physiol.* **84**: 891–894.

Riesmeier JW, Willmitzer L, Frommer WB. (1992) Isolation and characterisation of a sucrose carrier cDNA from spinach by functional expression in yeast. *EMBO J.* **11**: 4705–4713.

Riesmeier JW, Hirner B, Frommer WB. (1993) Potato sucrose transporter expression in minor veins indicates a role in phloem loading. *Plant Cell* **5**: 1591–1598.

Roberts AM, Mackie AJ, Hathaway V, Callow JA, Green JR. (1993) Molecular differentiation in the extrahaustorial membrane of pea powdery mildew haustoria at early and late stages of development. *Physiol. Mol. Plant Pathol.* **43**: 147–160.

Sauer N, Stadler R. (1993) A sink-specific H⁺/monosaccharide co-transporter from *Nicotiana tabacum*: cloning and heterologous expression in baker's yeast. *Plant J.* **4**: 601–610.

Sauer N, Stoltz J. (1994) SUC1 and SUC2: two sucrose transporters from *Arabidopsis thaliana*; expression and characterization in baker's yeast and identification of the histidine tagged protein. *Plant J.* **6**: 67–77.

Serrano R. (1989) Structure and function of plasma membrane ATPase. *Annu. Rev. Plant Physiol.* **400**: 61–94.

Shibuya N, Kakku H, Kuchitsu K, Maliarik MJ. (1993) Identification of a novel high-affinity binding site for *N*-acetylchito-oligosaccharide elicitor in the membrane fraction from suspension cultured rice cells. *FEBS Lett.* **329**: 75–78.

Shiraishi T, Saitoh K, Kim HM, Kato T, Tahara M, Oku H, Yamada T. (1992) Two suppressors, supprescins A and B, secreted by a pea pathogen *Mycosphaerella pinodes*. *Plant Cell Physiol.* **33**: 663–667.

Smith CJ. (1996) Accumulation of phytoalexins: defence mechanisms and stimulus response system. *New Phytol.* **132**: 1–45.

Smith SE, Smith FA. (1990) Structure and function of the interfaces in biotrophic symbioses as they relate to nutrient transport. *New Phytol.* **114**: 1–38.

Spencer-Phillips PTN, Gay JL. (1981) Domains of ATPase in plasma membranes and transport through infected plant cells. *New Phytol.* **89**: 393–400.

Stadler R, Brandner J, Schulz A, Gahrtz M, Sauer N. (1995) Phloem loading by the PmSUC2 sucrose carrier from *Plantago major* L. occurs only in companion cells. *Plant Cell.* **7**: 1545–1554.

Staswick PE. (1992) Jasmonate, genes and fragrant signals. *Plant Physiol.* **99**: 804–807.

Stumpf MA, Gay JL. (1990) The composition of *Erysiphe pisi* haustorial complexes with special reference to the neckbands. *Physiol. Mol. Plant Path.* **37**: 125–143.

Tomiyama K, Okamoto H, Katou K. (1983) Effect of infection by *Phytophthora infestans* on the membrane potential of potato cells. *Physiol. Plant Pathol.* **22**: 233–243.

Toyoda K, Shiraishi T, Hino I, Kato T, Yamada T, Oku H. (1993) Regulation of signal transduction for defense response of pea by elicitor and suppressor from *Mycosphaerella pinodes*. In: *Proceedings of the Sixth International Congress on Plant Pathology*. p. 196 (abstract).

Vera-Estrella R, Blumwald E, Higgins VJ. (1992) Effect of specific elicitors of *Cladosporium fulvum* on tomato suspension cells — evidence for the involvement of active oxygen species. *Plant Physiol.* **99**: 1208–1215.

Vera-Estrella R, Barkia BJ, Higgins VJ, Blumwald E. (1994a) Plant defense response to fungal pathogens. I. Activation of host plasma membrane H⁺-ATPase by elicitor-induced enzyme dephosphorylation. *Plant Physiol.* **104**: 209–215.

Vera-Estrella R, Higgins VJ, Blumwald E. (1994b) Plant defense response to fungal pathogens. II. G-protein-mediated changes in host plasma membrane redox reactions. *Plant Physiol.* **106:** 97–102.

Werner D, Morschel E, Garbers C, Bassarab S, Mellor RB. (1988) Particle density and protein composition of the peribacteroid membrane from soybean root nodules is affected by mutation in the microsymbiont *Bradyrhizobium japonicum. Planta* **174:** 263–270.

Wevelsiep L, Kogel KH, Knogge W. (1991) Purification and characterization of peptides from *Rhynchosporium secalis* inducing necrosis in barley. *Physiol. Mol. Plant Pathol.* **39:** 471–482.

Wevelsiep L, Rupping E, Knogge W. (1993) Stimulation of barley plasmalemma H+-ATPase by phytotoxic peptides from the fungal pathogen *Rhynchosporium secalis. Plant Physiol.* **101:** 297–301.

Whitham S, Dinesh-Kumar SP, Choi D, Hehl R, Corr C, Baker B. (1994) The product of the tobacco mosaic virus resistance gene *N*: similarity to Toll and the interleukin-1 receptor. *Cell* **78:** 1101–1115.

Woods AM, Gay JL. (1983) Evidence for a neckband delimiting structural and physiological regions of the host plasma membrane associated with haustoria of *Albugo candida. Physiol. Mol. Plant Path.* **23:** 73–88.

Woods AM, Gay JL. (1987) The interface between haustoria of *Puccinia poarum* (monokaryon) and *Tussilago farfara. Physiol. Mol. Plant Path.* **30:** 167–185.

Woods AM, Fagg J, Mansfield JW. (1988a) Fungal development and irreversible membrane damage in cells of *Lactuca sativa* undergoing the hypersensitive reaction to the downy mildew fungus *Bremia lactucae. Physiol. Mol. Plant Pathol.* **32:** 483–497.

Woods AM, Didehvar F, Gay JL, Mansfield JM. (1988b) Modification of the host plasmalemma in the haustorial infections of *Lactuca sativa* by *Bremia lactucae. Physiol. Mol. Plant Path.* **33:** 299–310.

Index

ORDERING DETAILS

Main address for orders

BIOS Scientific Publishers Ltd
9 Newtec Place, Magdalen Road,
Oxford OX4 1RE, UK
Tel: +44 1865 726286
Fax: +44 1865 246823

Australia and New Zealand
DA Information Services
648 Whitehorse Road, Mitcham, Victoria 3132, Australia
Tel: (03) 9210 7777
Fax: (03) 9210 7788

India
Viva Books Private Ltd
4325/3 Ansari Road, Daryaganj, New Delhi 110 002, India
Tel: 11 3283121
Fax: 11 3267224

Singapore and South East Asia
(Brunei, Hong Kong, Indonesia, Korea, Malaysia, the Philippines,
Singapore, Taiwan, and Thailand)
Toppan Company (S) PTE Ltd
38 Liu Fang Road, Jurong, Singapore 2262
Tel: (265) 6666
Fax: (261) 7875

USA and Canada
BIOS Scientific Publishers
PO Box 605, Herndon, VA 20172-0605, USA
Tel: (703) 435 7064
Fax: (703) 689 0660

Payment can be made by cheque or credit card (Visa/Mastercard, quoting number and expiry date). Alternatively, a *pro forma* invoice can be sent.

Prepaid orders must include £2.50/US$5.00 to cover postage and packing
(two or more books sent post free)